A Manual of Practical Zoology Chordates

For Undergraduate and Postgraduate Students

Dr. P.S. Verma
M.Sc., Ph.D., FESI, FAZ
Reader (Retd.)
Department of Zoology
Meerut College
Meerut

S. CHAND
PUBLISHING

S Chand And Company Limited
(ISO 9001 Certified Company)

S Chand And Company Limited
(ISO 9001 Certified Company)
Head Office: Block B-1, House No. D-1, Ground Floor, Mohan Co-operative Industrial Estate, New Delhi – 110 044 | Phone: 011-66672000
Registered Office: A-27, 2nd Floor, Mohan Co-operative Industrial Estate, New Delhi – 110 044
Phone: 011-49731800

www.**schandpublishing.com**; e-mail: **info@schandpublishing.com**

Branches

Chennai	:	Ph: 23632120; chennai@schandpublishing.com
Guwahati	:	Ph: 2738811, 2735640; guwahati@schandpublishing.com
Hyderabad	:	Ph: 40186018; hyderabad@schandpublishing.com
Jalandhar	:	Ph: 4645630; jalandhar@schandpublishing.com
Kolkata	:	Ph: 23357458, 23353914; kolkata@schandpublishing.com
Lucknow	:	Ph: 4003633; lucknow@schandpublishing.com
Mumbai	:	Ph: 25000297; mumbai@schandpublishing.com
Patna	:	Ph: 2260011; patna@schandpublishing.com

First Edition 1972
Subsequent Editions and Reprints 1973, 74, 75, 76, 77, 78, 79, 81, 83, 84, 85, 86, 87, 88, 90, 91, 92, 94, 95 (Twice), 97, 98, 99, 2000, 2001, 2003, 2004, 2005, 2006, 2007, 2008, 2009
Eleventh Revised and Enlarged Edition 2010
Reprints 2011, 2012, 2013, 2014 (Twice), 2015, 2016, 2018 (Twice), 2019, 2020
Reprint 2021 (Four times)

ISBN: 978-81-219-0830-6 **Product Code:** H6CDP68ZOOL10ENAK10O

PRINTED IN INDIA

By Vikas Publishing House Private Limited, Plot 20/4, Site-IV, Industrial Area Sahibabad, Ghaziabad – 201 010 and Published by S Chand And Company Limited, A-27, 2nd Floor, Mohan Co-operative Industrial Estate, New Delhi – 110 044.

Dedicated

To

my respected father

Late Th. C.B. Singh Ji

who could not survive to see the

fruits of this labour.

Prem Singh

Preface to the Eleventh Edition

It is the pleasant duty of the author to thank the kind and generous readers for the warm and cordial reception accorded by them to the previous editions. The suggestions and encouraging opinions from the colleagues and students have prompted the author to bring out an early and revised New Edition. The bringing out of the present edition of this book has given the author an opportunity to revise and improve it. Alterations and additions have been made practically in all the chapters. The following new topics have been added in the respective chapters : (i) several new birds and mammals in the chapters of museum specimens of birds and mammals; (ii) new microscopic slides in the respective chapters of microscopic slides; (iii) several new embryological slides in the respective chapters of embryological slides; (iv) a new chapter of Skeleton of *Scoliodon* has been added afresh in this edition; (v) several new exercises have been included in the section of Experimental Physiology. By the inclusion of the above chapters the present edition will be more useful to the readers. Although maximum efforts have been made to eradicate the misprints and mistakes of the previous edition but even then the author cannot claim the removal of all the misprints due to the printers devil. The author is extremely thankful to his wife Smt. Preeti Singh for her kind co-operation during the revision of this book. The author further expresses his thanks to the generous readers who helped him by giving valuable suggestions for the improvement of the book.

Suggestions for the further improvement of the book will be thankfully acknowledged and incorporated in future editions.

Dr. P.S. Verma

"PREETI"
121, New Mohanpuri
Colony – Meerut, 250001

PREFACE TO THE FIRST EDITION

The book **"A Manual of Practical Zoology Chordates"** has been written to meet the needs of those studying Practical Zoology at both Undergraduate and Post-graduate levels in Indian Universities. The book is divided in seven parts. The first part includes the general directions for laboratory work. The second part deals with the dissections. In this part the procedure for many dissections along with their diagrams have been given. The third and fourth parts incorporate the methods of permanent mounting and some important mountings. The fifth part of elaborately deals with the study of museum. The study of museum includes the general characters, outline classification in tabular form, and a detailed one of various groups and classes of phylum Chordata with examples of each order or suborder, if any, together with the distinguishing characters and geographical distribution of museum specimens. In the sixth part, the study of permanent slides has been undertaken. The last, seventh part deals with the osteology of representatives of various classes, *e.g.*, Amphibia, Reptilia, Aves and Mammalia. The important characters and geographical distribution of museum specimens and permanent slides have been given pointwise along with their suitable illustrations in order to make it easily understandable to the students. A large number of types have been selected from each group and class to fulfil the needs of students at different Universities. Text is illustrated with 522 figures. The figures have been drawn with full clarity. Special care has been taken to co-ordinate the account with the figures. It is hoped that the present book will serve the purpose of those for whom it is written.

The author is thankful to Dr. A.P. Tyagi, D.A.V. College, Muzaffarnagar, Dr. D.P. Singh, R.B.S. College, Agra, Dr. R.G. Sharma, Dr. (Mrs.) R.K. Sharma, Dr. M.P. Singh, Agra College, Agra and Dr. O.P. Raghuvanshi, Janta College, Bakewar (Etawah) for valuable suggestions during the course of preparation of this book. Thanks are due to late Prof. Vimal Sharma, M.A. for preparing the illustrations and cover design. The author is also grateful to Shri T.N. Goel (S. Chand & Company Ltd., New Delhi) for expediting the publication of this book.

Suggestions for the improvement of the book shall be highly appreciated from teachers and students.

Department of Zoology,
Meerut College, Meerut.

Dr. Prem Singh Verma

CONTENTS

Chapter 1

GENERAL DIRECTIONS FOR LABORATORY WORK

STUDENTS' EQUIPMENTS

Each student, while coming to the laboratory for the practical work, is required to bring certain necessary equipments. However, it is not possible to list all what is required, but the following list is considered necessary.

1. Practical note book.
2. A manual of Practical Zoology for self guidance.
3. Drawing pencils (preferably H, B and HB).
4. Pencil sharpner (sand paper).
5. Pencil eraser.
6. Measuring scale.
7. Dissecting box containing the following instruments :
   (*i*) **Scalpels** two or three, sharp and of different sizes.
   (*ii*) **Scissors** two pairs, a fine pointed pair for fine dissection and other larger for general work.
   (*iii*) **Forceps** two, one fine pointed and other with blunt ends.
   (*iv*) **Dissecting needles** two, mounted in handles.
   (*v*) **Blow pipe** one.
   (*vi*) One edged safety razor or blade.
8. Brush one.
9. Dropper one.
10. Hand lens one.
11. A piece of clean cloth.
12. Pins.

Do not mind bringing all these things on every turn, you do not know when you need them.

DIRECTIONS TO THE STUDENTS

1. While coming to the laboratory for practical class check that you possess the **practical note book**, **pencil**, **pencil eraser**, **sharpner**, **scale**, **brush** and complete set of **dissecting instruments**.

2. Instruments should be sharp and according to the requirements.
3. Come prepared with the work you are supposed to do in the laboratory.
4. Keep your instruments, practical note book and seat well arranged and tidy.
5. Never encourage the habit of lending either to or from your class fellows. Bring all the requirements of the day.
6. Before starting the work, listen carefully the instructions given by your teacher.
7. Remove your all difficulties from your teacher. Do not consult your class fellows for any help.
8. Never rub your pencil on the floor or on the top of the table, always use the sharpner.
9. Maintain complete silence in the laboratory.
10. Clean and well arrange your seat before leaving the laboratory.

Dissections

1. Before starting the dissection listen to the instructions given by the teacher and follow them carefully during the dissection.
2. Draw a diagram of the dissection to be done.
3. Dissection should be done under water and should be left completely submerged in water.
4. Always keep the water clean in the dissecting dish during dissection by changing it whenever stained with blood, etc., so that visibility remains clear.
5. Put every waste of dissection in the petridish or in the sink slowly. Do not throw it carelessly on the top of the table or everywhere.
6. Dissect **along** and not **across** the blood vessels and nerves.
7. While doing dissection, the required parts should not only be exposed but these should be properly displayed by cleaning and black papering (fine strips of paper).
8. A well-labelled diagram of the dissection should be left along with dissection only if it is required or demanded.
9. Do the dissection patiently (not in haste) and confidently.
10. During the course of dissection, concentrate only on your own dissection with confidence.

Flag-labelling

Frequently, the students are asked to dissect some organs, blood vessels or nerves, etc., and flag-label them. In flag-labelling, following instructions are to be followed :

1. Very small pieces of paper (1.5 × 0.7 cm) are prepared and a needle is passed through each close to one end.
2. The names of organs, blood vessels or nerves are written (in pencil) on these flags.
3. Then each flag is inserted in the dissecting dish close to the organ, blood vessel or nerve bearing the name on the flag.

Mountings

While making the permanent slides (mountings), the following instructions should be followed strictly :

1. Never keep your mounting material for less or more time than the desired time in an alcoholic grade or stain.
2. Keep the material for slightly more time in 90% alcohol and absolute alcohol for complete dehydration.

3. Ensure the complete dehydration after absolute alcohol by putting the material in xylol or benzene. If it gives turbidity with xylol then dehydrate it again.
4. Do the dehydration in closed specimen tubes or in covered cavity blocks.
5. Always use a brush and never forceps for holding the mounting material.
6. Put the required amount, one or two drops of **Canada balsam** on the slides for mounting. Excess of Canada balsam makes the slides dirty.

Study of Museum Specimens

1. First study the characteristics of the specimen to be drawn from the book.
2. Try to find out those characteristics in the specimens.
3. Draw only that view (*e.g.*, dorsal, ventral or lateral) which exhibits maximum details.
4. Diagrams should be drawn from the original specimen although help of the book can be taken for comparing the details and labelling of various parts.
5. While drawing, special care should be taken in maintaining the proportion of length and width of the animal body.
6. Draw only line diagrams, shading should be avoided.
7. Always write down the classification of all the specimens.

Study of Slides

1. While studying microscopical preparations (permanent mounts) under the microscope, do not disturb the slide focussed by the teacher.
2. Draw from the slide. Compare and label it with your book.
3. Whenever there is a mount of an entire animal do not forget to write its classification.

Practical Note Book (Practical Record)

1. Practical note book should be neat and clean, and uptodate.
2. Put date on the left hand corner of the page of the note book and details of the work on the top in centre.
3. Diagrams should be correctly drawn and well-labelled.
4. Diagrams of all dissections, museum specimens and slides should be drawn and comments of all should be written.

Spotting

While doing spotting during the practical examination, be careful about the following points :

1. First of all, write the spot number.
2. Draw a rough but well-labelled diagram of the spot.
3. Write down the classification if it is entire animal, whether a museum specimen or entire animal mounted on the slide.
4. Give characteristic features of the phylum, subphylum, class, subclass, order, suborder, etc.
5. Only interesting and important features should be written in comments.

Chapter 2

MICROSCOPES

Because of the minute size of cells, they cannot be observed by human naked-eyes. The human eyes have limited distinguishing or resolving power and they cannot distinguish the objects smaller than 0.1 mm (100 μ). Moreover, the living cells are transparent in ordinary light and it becomes difficult to discriminate among various cellular components. To overcome these practical difficulties early cell biologists investigated various methods of staining and magnifying the cell and its components. The object of magnification of cells and their components was achieved by the **lenses**. The lenses could magnify the minute objects up to a limited extent, so two or more lenses were combined together to form an instrument, known as **microscope** (Gr., *mikros* = small; *skopein* = to see). Microscopes are important in the study of cells and microorganisms, because they possess the characteristics of magnification and resolving power.

Resolving Power of Microscope

Important in magnification is the ability of the microscope to resolve or to distinguish between the objects lying close together, thus, the smallest object '*d*' resolved by the light microscope may be determined by the following equation :

$$d = \frac{\lambda}{NA + na} \quad \ldots(i)$$

In this equation, λ is the wave-length of light used, *NA* is the numerical aperture of the objective lens and *na is* the numerical aperture of the condenser. Since for critical study, the numerical aperture of the condenser is chosen equal to that of the objective, the equation is often written :

$$d = \lambda / 2NA \quad \ldots(ii)$$

The numerical aperture is given by the equations :

$$NA = n \sin \alpha \quad \ldots(iii)$$

Here, *n is* the refractive index of the medium and α is the half angular aperture of the light beam entering the condenser. Now, because the power of resolution (resolving power) of the microscope is inversely related to its limit of resolution. The limit of resolution of a microscope can be defined as the minimum distance between two points in order that they can be discriminated as such.

$$\text{Limit of resolution} = \frac{0.61\lambda}{NA} \quad \ldots(iv)$$

Thus, higher the resolving power, the smaller is the limit of resolution.

Types of Microscopes

1. Simple dissecting microscope;
2. Compound microscope;
3. Binocular microscope;
4. Ultramicroscope;
5. Ultraviolet microscope;
6. Fluorescence microscope;
7. Dark-field microscope;
8. Phase-contrast microscope;
9. Interference microscope;
10. Polarizing microscope;
11. X-ray microscope;
12. Electron microscope.

1. Simple or Dissecting Microscope

The simple or dissecting microscope has a single lens system. It is composed of a single lens through which upturned image of the object is seen. In this microscope, in a sense, the magnifying lens is a simple microscope. Simple microscope has the following parts : (*i*) foot, (*ii*) stand, (*iii*) vertical limb, (*iv*) folded arm, (*v*) a simple convex lens, (*vi*) glass stage, (*vii*) reflecting mirror, (*viii*) clips, and (*ix*) adjustment screw. The **foot** is also called a supporting stand. It bears the weight of the microscope. The **vertical limb** on the stand is fitted with an adjustment screw for focussing. The simple convex lens is mounted on a vertical limb through a folded arm. The glass stage is a sort of platform. The slide with the specimen or the entire specimen is kept over the stage. The clips hold the slide firmly. The reflecting mirror is used for getting the light focussed on the object. The magnification of simple microscope ranges from 4 to 40 times, depending upon the magnification of the lens. The simple microscope is used to magnify small animals like, bed bug, ant, louse, etc., or to see large sections, or to perform dissections of small animals.

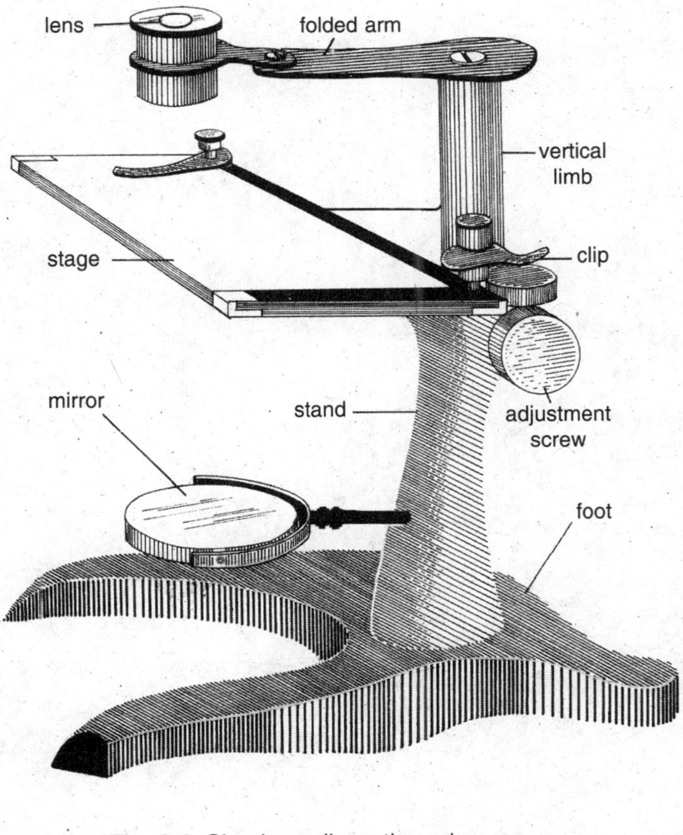

Fig. 2.1. Simple or dissecting microscope.

Operation of Simple Microscope

To see any object, clean and dry the stage, place the object on it and place the eye close to one side of the lens and adjust the mirror. Turn the focussing screw up and down as the need may be to get a sharp and distinct image of the object.

2. COMPOUND MICROSCOPE

The compound microscope of twentieth century is the microscope of much improved and modified kind. It consists of three types of lens systems which are as follows : (*i*) **Condenser lens system** occurs beneath the specimen and its function is the collection and focussing of light rays on the object or specimen, which is placed on the stage of the microscope. (*ii*) **Objective lens system** remains near and above the specimen. It produces and magnifies the image of the specimen. (*iii*) **Eye-piece lens system** or **ocular lens** system remains near the eyes of the observer and it magnifies and forms the image (secondary) of the (primary) image previously produced by the objective.

Parts of Compound Microscope

The compound microscope comprises of the following parts :

1. Foot. The foot gives support to the microscope and bears its weight. The foot comprises two parts, the tripod foot and horse-shoe-shaped foot.

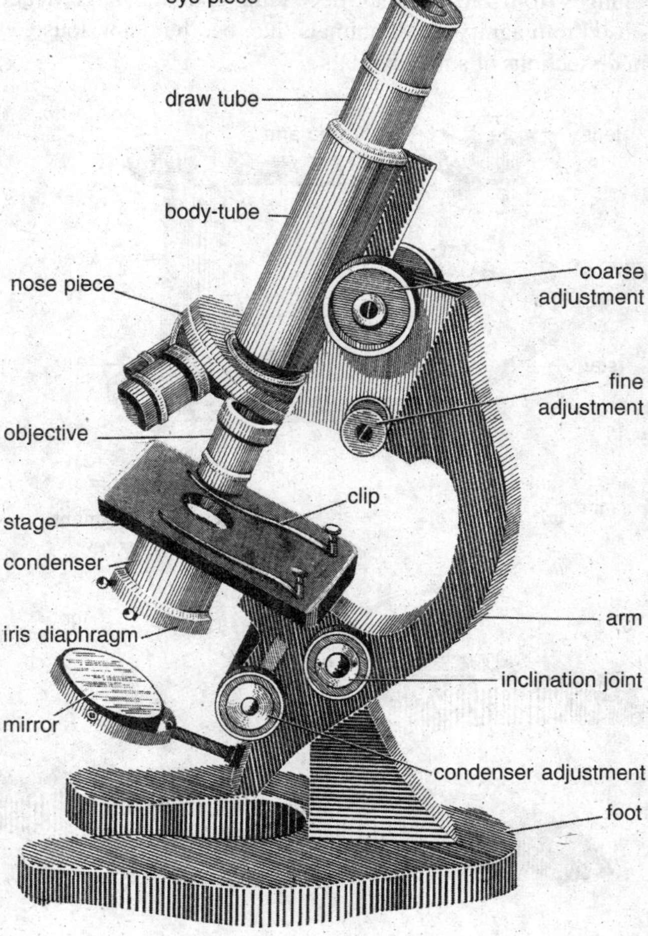

Fig. 2.2. Compound microscope.

2. Mirror. The mirror is fitted below the stage having concave and plane surfaces. It reflects light upward through the diaphragm and the hole in the stage.

3. Stage. The stage is a rectangular platform with the sides attached to the arm by means of a broad rigid fork or bracket. There is a hole in the stage, which admits light from the mirror through the condenser. There is a pair of clips fitted on the stage which keep the slide in a fixed position over the stage.

4. Body-tube. In majority of the microscopes, the body-tube is 160 mm long. It carries the objectives at the lower end and the eye-piece at the upper end. It also holds the eye-piece and objectives at the proper working distance from each other. It possesses a rack and pinion arrangement to bring the object in the proper view.

5. Nose-piece. In the nose-piece, low power and high power lenses are fitted.

6. Arm. It supports the body-tube and coarse adjustment.

7. Condenser. It is present in between the mirror and stage. It is used for focussing light and adjusting the intensity of illumination.

8. Coarse adjustment. The coarse adjustment is very important for the rapid and precise focussing of the object. It is attached to the body-tube. It moves the body-tube up and down approximately to the correct distance from the specimen. The coarse adjustment has a large head. By turning the head of coarse adjustment in the clock-wise direction, the body-tube moves downward, while by turning it in anti-clock-wise direction, the body-tube moves upwards.

9. Fine adjustment. The fine adjustment is very essential for the perfect image of the object. It is attached with the arm on either side in the form of horizontal small heads parallel and below to the coarse adjustment heads. It also moves the body-tube up and down like the coarse adjustment. A slight revolution of fine adjustment is required for the exact and sharp focussing of the object.

10. Eye-piece or ocular. The eye-piece or ocular lies at the top of the body-tube. The eye-piece may be replaced with another of lower or higher magnification. The eye-pieces are generally 5*x*, 6*x*, 8*x*, 10*x*, 12*x*, and 15*x*.

11. Objectives. The objectives are attached with the nose-piece. These contain lenses of different magnifications, *e.g.*, 1*x*, 10*x*, 40*x*, and oil immersion (100*x*).

Setting the Microscope

1. Take out the microscope from its case and place it gently on the working table, keeping the arm towards yourself and stage away. The base of microscope should by kept several centimetres away from the edge of the table.
2. Clean the body and stage of the microscope with neat cloth and the lenses with lens paper.
3. Rotate the nose-piece so as to make the low-power objective in line with the body-tube.
4. By using the coarse adjustment raise the body-tube about 2.5 cm above the stage.
5. Open the diaphragm of the condenser for passing the light on the stage.
6. Look through the eye-piece with your left eye. Hold the edge of the mirror and tilt and turn it towards the light source, adjust its position so that it reflects light upwards through the hole in the stage. Plane mirror should be used and avoid direct sun-light.
7. Do not use coarse adjustment when viewing through high power objective.
8. The slide should not be dirty or wet when it is to be observed under the microscope.
9. Microscope should be kept usually in the upright position.
10. When not in use the microscope should be kept in its case.

Focussing of Microscope

1. Place the prepared slide to be examined on the stage of the microscope, and try to keep the object in the centre of the hole of stage.
2. Lower the body-tube with the help of coarse adjustment until the low power objective comes close to the cover glass or until it cannot be moved down.
3. Look through the eye-piece and raise the objective with coarse adjustment until the object comes in view.
4. Make the focus sharp with the help of fine adjustment.
5. If the object is to be examined under high power, first focus it under low power and then change to high power objective by using the fine adjustment for sharp focussing. Do not try to focus directly under high power.
6. While focussing under high power never use coarse adjustment.

Care and Precautions in Handling the Microscope

Microscope is a delicate instrument. It must be handled with all care and precautions. Some of them are following :

1. Take out the microscope from its case only when it is to be used and replace it immediately afterwards.
2. When taking out the microscope from its case, hold its arm with one hand and support the base with the other. Do not take out the microscope carelessly with one hand.
3. Place the microscope on the working table gently atleast a few centimetres away from the edge of the table in order to prevent its accidental fall on the ground.
4. The arm of the microscope should always be towards you and the stage away from you.
5. Make it a point to check all the parts of the microscope before use. In case of damage or absence of any part, inform the teacher incharge.
6. Clean the parts of the microscope before use. Lens paper or muslin cloth is to be used for cleaning the objectives and eye-pieces.
7. Avoid tilting of the microscope, keep it in upright position.
8. Do not allow direct sun-rays to strike the mirror. Plane mirror is used for natural day light and concave mirror for artificial light.
9. Keep the stage always dry and clean.
10. Do not put wet and dirty slides on the stage. The material to be examined should be properly mounted and covered with cover-glass.
11. Clean the objective lenses with a mixture of xylol and absolute alcohol after use.
12. After use make the objective free and lower the body-tube. Transfer it in the case.
13. While observing through the microscope, always keep both the eyes open.
14. Do not use the microscope for long at a time. Rest in between your work, if the work is to be done throughout the day.

3. Binocular Microscope

The low power stereoscopic binocular microscope consists of a combination of two separate and converging microscopes, mounted equidistant from the vertical axis so as to include an angle of 15° to 16°. The body-tubes comprise matched objectives and eye-pieces focussing simultaneously on the same area of object, each eye viewing the field from different angle. Prisms are mounted between each objective and eye-piece, so that the image is straight. These unique features of stereoscopic binocular microscope facilitate the studies of several objects illuminated by reflected light. The various parts and layers of the object are revealed clearly. The natural straight image and the three-dimensional effect are useful in the studies, which renders binocular microscope invaluable and practically essential for the dissection of small animals and plant materials, and mounting under the microscope.

4. Ultramicroscope

The ultramicroscope is the improved type of compound microscope in which the object is illuminated by a strong beam of light which comes parallel to the surface of the stage or at the right angle to the direction of vision. This microscope is used in observing many smaller particles.

5. Ultraviolet Microscope

In ultraviolet microscope, beside the ordinary light, the invisible **ultraviolet rays**, which have short wave-lengths, are employed for the illumination of the object. Further, the glass lenses are replaced by the lenses of fused quartz, fluorite or lithium fluoride, so that ultraviolet rays of short wave-lengths can pass easily through them. In such microscopes, invisible ultraviolet rays are employed and the image cannot be observed directly but can be photographed.

Ultraviolet microscopes have proved very efficient in the qualitative determinations of different cellular components.

6. Fluorescence Microscope

The fluorescence microscope is based on the phenomenon of fluorescence. It is found that when certain chemical compounds are illuminated by ultraviolet rays they become excited at molecular level and start to emit light of visible range. The phenomenon is known as **fluorescence**. The source of illumination of fluorescence microscope is ultraviolet rays of higher wave-lengths (3500 to 4000Å). The different chemical compounds of the object to be seen, when fluoresced by these ultraviolet rays, they emit rays of different wave-lengths which can be detected and the chemical nature of the given object can be tested.

7. Dark-field Microscope

The dark field microscope is an ordinary microscope with a special condenser. By use of a dark stop in the condenser the light from the centre of the field is removed and the object is illuminated only by an oblique beam of light. The object is, therefore, seen by light reflected and scattered from interfaces in the object. With dark-field microscope the outline of the cell, nucleus, mitochondria, oil droplets, vacuoles are readily detected. Moreover, in cells undergoing division, the centrosomes, asters, spindles and chromosomes may be observed.

8. Phase-contrast Microscope

The phase-contrast microscope has the resolving power similar to ordinary light microscope and it enables to see structures in the living cell. It is based on the principle of phase-contrast. The principle of phase-contrast was discovered by **Zernike** (1955). The living cells are heterogeneous in their contents, because each of their cellular component has different thickness. Due to different thickness, each cellular component has different refractive indices. The difference of refractive indices of cellular components produces difference in their phases. The phase-contrast microscope exaggerates the small difference between the phases or refractive indices of different cellular components and enables to distinguish adjacent structures.

The phase-contrast microscope contains an **annular phase plate** in the objective and an **annular diaphragm** in the condenser. The annular diaphragm of the phase-contrast microscope permits light to pass through the condenser as a hollow cone, the remaining light being absorbed. This cone is focussed on the object. The **phase plate**, which is placed at the back focal plane of the objective lens, is a transparent disc containing a groove (or elevation) of such a size and shape as coincide with the direct image of the substage annular diaphragm which is formed at the back focal plane when no object is viewed. If an object is placed between the condenser and the objective, in addition to direct image a number of overlapping diffracted images of the diaphragm then appear at the back focal plane of the objective. The depth of groove (or the elevation) in the phase plate of the objective is so made that the two sets of rays forming the direct image and the diffracted image differ in optical path by a quarter wave-length of illuminating beam of light. Under these conditions, the phase difference which is not seen by the eye, is converted to intensity difference which we see. In the bright contrast phase optical system, the two sets of light rays are added to make a brighter image and in the dark contrast phase they partially cancel one another making a more contrasting darker image. The phase-contrast microscope helps in the observation of behaviour of chromosomes of living cells during mitosis and various other cellular components.

9. Interference Microscope

Interference microscope is designed on the principle of interference. It is the interference between two beams of light which gives the familiar colours observed in thin films of oil on the surface of road. If two beams of light from the same source are in phase, they supplement each other and produce a region of brightness. But, when both remain out of phase, they neutralise each other and the field appears dark. The causation of brightness and darkness by two rays due to their subsequent phases is known as **interference** or **interposition**.

10. Polarising Microscope

The source of light of polarised microscope is **polarised light**. The polarised light is produced when the light waves lie in one plane. For example, light will be polarised if it vibrates in planes parallel to the plane of the paper. Light of this kind is produced by a **calcite prism**, which splits the light into two beams or by selective absorption of one of the polarised component in a polaroid plastic film in which the molecules are oriented parallel to each other. The polarising microscope is like a light microscope, except that the condenser and the ocular are equipped with polarising optics each of which transmits only plane polarised light. The condenser is known as **polariser** and the ocular as **analyser**. The polariser and analyser remain adjusted in such a fashion that when the analyser is rotated 360 degrees, the visual field alternates between bright and dark at every 180 degree turn. The two positions of maximum light transmission are obtained when the analyser is set parallel to the polariser. The polarising microscope is used in analysing the ultrastructure (molecular and macromolecular organisation of the cell) in measuring thickness of objects and in measuring concentration of matter and water contents in the object.

11. X-ray Microscope

The X-ray microscope has been found to be a very efficient cytological tool to study the cell at molecular level. In X-ray microscope, the X-rays are used as source of illumination. The X-rays possess extremely short wave-lengths and great penetration power comparatively in thin living tissues. In X-ray microscope, the electromagnetic lenses or reflecting curved mirrors focus X-rays beam and the image is formed on the film. By X-ray microscope, the object can be studied by following methods :

1. Direct contact method of microradiography. In this method the thin sections of the object 1μ or less thickness are cut and are placed on the fine grain photographic emulsion. These are exposed to soft X-rays in the range of 1-10Å.

2. Projection method. In this method the object is placed in between the photographic plate and the source of X-ray beam. The image is formed on the photographic film.

3. Diffraction method. In the diffraction method a beam of X-rays is passed through the material and a photographic plate is placed beyond this to record the diffraction pattern. Due to interference between the different diffracted rays, a series of concentric spots or bands may appear on the photographic plate. By counting the distance between the spots from the centre, the units of object can be easily counted.

With the help of X-ray microscope, the molecular structure of DNA, haemoglobin and proteins are quantitatively measured.

12. Electron Microscope

The resolving power (or capacity to detect) of light microscope is limited to objects of approximately the same dimensions as the wave-length of the light used. Objects between 2,000Å and 3,000Å in diameter can just be detected, but may not be sharply resolved. Therefore, the biologists searched for some source of light which might have very short wavelength. They found it in the electrons. It has been observed that when a metal filament placed in a vacuum tube is heated, it emits electrons which can be accelerated by an electrical potential. These electrons can be controlled and focussed by electromagnetic lenses. In the condition of high vacuum, the stream of electrons follows a straight path and has properties similar to light. Like light, it has corpuscular and vibratory characters, but unlike that the electron beam has very short-wavelength (*i.e.*, the wave-length of electron beam = 0.05 Å and the wave-length of light rays = 5500 Å). For efficient working, the vacuum must be so great that the electrons may not collide with the atoms of oxygen or nitrogen in the air.

The electron microscope has been designed on the same principle by **Knoll** and **Ruska** of Germany in 1932; **Marton** of Belgium in 1934; and **Prebus** and **Miller** of Canada in 1934. Because the early electron microscopes have employed the transmitted electrons, so are known as **transmission microscopes**.

Transmission electron microscopes. In transmission type electron microscope, the metal filament or cathode emits the electron beam in the highly evacuated cathode tube. The electrons are collected and focussed in the plane of object by an electromagnetic condenser lens. After the electrons passed through the object, they are collected again by an electromagnetic coil, which acts as an objective lens. The electromagnetic objective lens produces a magnified image of the object, which is received by a third electromagnetic lens, which in its turn acts as a ocular or projection lens. The ocular lens magnifies the image formed by the objective lens. The final image can be visualised on fluorescent screen or can be recorded on photographic plate. The image results from differential scattering of electrons from the molecular constituents of the cell. Scattering is greater the denser the material, regardless of its chemical composition. A transmission type electron microscope has a resolving power 400 (theoretically 2000) times that of the light microscope. The electron microscopes are used in the study of ultrastructures of cell and its components. It has contributed to cell biology in the deep understanding of cell.

Scanning electron microscope. The scanning electron microscope is a comparatively recent development. In this microscope, a very fine pencil of electrons scans the surface of object, much as a screen is scanned in a television tube. Secondary electrons are emitted from the surface at the points where the beam is scanning. These are collected by the positively charged grid. The signal from the grid is transferred to a television tube which scans in synchrony with the beam in the microscope. An image of specimen is produced on the screen. The scanning microscope produced an image in much the same way as the image seen with the naked eye. The eye collects light which is reflected from solid objects somewhat like the grid in the scanning microscope. The image formed by this microscope, therefore, has a remarkable three-dimensional appearance.

Chapter 3

PREPARATION OF LABORATORY REAGENTS

FIXATIVES OR FIXING AGENTS

1. Formalin solution

It is prepared 0.5 to 0.1% or 5%.

1. Formalin 40% ...10 ml.
2. Distilled water ...90 ml.

This fixative is used for the fixation of animal material.

2. Alcohol

70%, 90% or absolute alcohols are used as fixatives chiefly for protozoans and other micro-organisms. A few drops of the alcohol are poured on the object.

3. Formalin acetic alcohol (F.A.A.)

1. Formalin 40% ...10 ml.
2. Glacial acetic acid ... 5 ml.
3. Ethyl alcohol 70% ...85 ml.

This fixative is generally used for fixing the plant material and nematodes.

4. Bouin's fluid (Aqueous)

1. Saturated aqueous solution of picric acid ...75 ml.
2. Formalin 40% ...25 ml.
3. Glacial acetic acid ... 5 ml.

The tissue fixed in aqueous Bouin's fluid must be thoroughly washed in water to remove the traces of picric acid so that it may not interfere in staining and the process of dehydration.

5. Bouin's Fluid (Alcoholic)

1. Saturated solution of picric acid in 70% alcohol ...75 ml.
2. Formalin 40% ...25 ml.
3. Glacial acetic acid ... 5 ml.

The tissue fixed in alcoholic Bouin's fluid should be thoroughly washed in 70% alcohol after fixation to remove the traces of picric acid. Bouin's fluid fixatives are used in microtomic preparations.

6. Chromo-acetic acid (Flemming solution)

1. Chromic acid ... 0.2 to 0.25 ml.
2. Acetic acid ... 0.1 ml.
3. Distilled water ... 100 ml.

7. Formalin-picric-trichloracetic fixative

1. Formalin 40% ... 15 ml.
2. Trichloracetic acid ...0.50 gm.
3. 2% picric acid solution in 50% alcohol ... 85 ml.

8. Carnoy's fluid

(*a*) Formula No. 1

1. Glacial acetic acid ... 25 ml.
2. Absolute alcohol ... 75 ml.

(*b*) Formula No. 2

1. Glacial acetic acid ... 10 ml.
2. Chloroform ... 30 ml.
3. Absolute alcohol ... 60 ml.

Carnoy's fluid is used for animal as well as plant material.

9. Zenker's fluid

1. Potassium dichromate ... 2.5 gm.
2. Mercuric chloride ... 5.0 gm.
3. Sodium sulphate ... 1.0 gm.
4. Glacial acetic acid ... 5.0 ml.
5. Water ... 100 ml.

The common practice of making a stock solution of first two ingredients and adding the acetic acid just before use has been found to be unnecessary when chemically pure acetic acid is used.

10. Chromo-aceto-osmic acid

(*a*) Weak formula

1. Chromic acid ... 0.25 ml.
2. Osmic acid ... 0.1 ml.
3. Glacial acetic acid ... 0.1 ml

This fluid is good for fixing the smaller animals.

(*b*) Strong formula

1. Chromic acid 1 % ... 15 ml.
2. Osmic acid 2% ... 4 ml.
3. Glacial acetic acid ... 1 ml.

The stock solution of the above constituents is prepared and mixed at the time of use. This fluid is good for fixing the large animals or objects.

11. Corrosive sublimate acetic fixative

1. 5% or saturated solution of mercuric chloride in water ... 95 ml.
2. Glacial acetic acid ... 5.0 ml.

12. Formol-sublimate fixative

1. Saturated solution of mercuric chloride in 10% formalin.

13. Henking fluid

1. Osmic acid ... 1 ml.
2. Formic acid ... 3 ml.

3. Glycerine ...16 ml.
4. Dahlia ...0.4 gm.
5. Distilled water ... 80 ml.

14. Mayer's albumen

1. White of egg (Albumen) ...50 ml.
2. Glycerine ...50 ml.
3. Sodium salicylate ...1.0 gm.

Shake them well together and filter into a clean bottle for use.

Mayer's albumen is used to fix the small organisms and microtome sections on the slide.

15. Normal saline

1. Sodium chloride (analytical) ...0.78 gm.
2. Distilled water ...100 ml.

Dissolve sodium chloride in water and mix it thoroughly before use.

GRADES OF ALCOHOL

The alcoholic grades are usually prepared from rectified spirit which contains approximately 95% alcohol or from 90% or absolute alcohol.

(i) **30% Alcohol**

Rectified spirit ...31.5 ml.
Distilled water ...68.5 ml.

OR

90% alcohol ...100 ml.
Distilled water ...206.22 ml.

(ii) **50% Alcohol**

Rectified spirit ...52.6 ml.
Distilled water ...47.4 ml.

OR

90% alcohol ...100 ml.
Distilled water ...84.71 ml.

(iii) **70% Alcohol**

Rectified spirit ...73.6 ml.
Distilled water ...26.4 ml.

OR

90% alcohol ...100 ml.
Distilled water ...31.05 ml.

(iv) **90% Alcohol**

Rectified spirit ...94.7 ml.
Distilled water ...5.3 ml.

Acid alcohol

1. Alcohol 70% ...100 ml.
2. Hydrochloric acid concentrated ...1 ml.

STAINS (Dyes)

1. Borax carmine

1. Borax ...4 gm.
2. Carmine ...3 gm.
3. Distilled water ...100 ml.

The solution of borax and carmine is boiled for about half an hour so that it changes its colour. Now add equal volume (100 ml) of 70% alcohol, allow it to stand and filter. After 2-3 days, it becomes mature enough to be used in the laboratory. It is a cytoplasmic stain.

2. Picro-indigo-carmine

1. Saturated solution of picric acid in 90% alcohol ...1 part
2. Saturated solution of indigo carmine in 70% alcohol ...2 parts

This stain is very conveniently used for the staining of chitinous parts such as mouth parts of insects. It is used after dehydration of material in 70% alcohol.

3. Aceto-carmine or Propionic carmine

1. Glacial acetic acid ...45 ml.
2. Distilled water ...55 ml.
3. Dry carmine stain ...0.5-1.0 gm.

Mix 45 ml of glacial acetic acid or pure propionic acid in 55 ml of distilled water and heat to boiling and then add 0.5-1.0 gm of carmine. Shake well, cool and filter.

This is one of the most commonly used stain for chromosomal studies.

4. Aceto-orcein

1. Glacial acetic acid ...45 ml.
2. Distilled water ...55 ml.
3. Orcein stain ...1 gm.

Mix the above three things together. It is an ideal stain for the study of chromosomes.

5. Haematoxylin

(I) Delafield's haematoxylin

1. Haematoxylin crystals ...4 gm.
2. Absolute alcohol ...25 ml.
3. Saturated solution of ammonia alum ...400 ml.
4. Glycerine ...100 ml.
5. Methyl alcohol ...100 ml.

Dissolve 4 gm of haematoxylin crystals in 25 ml of absolute alcohol and add to it 400 ml of saturated solution of ammonia alum. Leave it exposed to light and air in unstoppered bottle for about a week Filter and add 100 ml of glycerine and 100 ml of methyl alcohol. Allow the solution to stand (uncorked) until the colour is sufficiently dark, then filter and use.

(II) Ehrlich's acid haematoxylin

1. Haematoxylin crystals ...2 gm.
2. Absolute alcohol ...100 ml.
3. Glycerine ...100 ml.
4. Glacial acetic acid ...10 ml.
5. Distilled water ...100 ml.
6. Alum in excess

Dissolve the haematoxylin in absolute alcohol, add the acetic acid and then the glycerine and water. Let the mixture ripen in the light (for about 5 to 10 days) until it acquires a dark red colour. It will then keep with constant power for years, if kept in a well stoppered bottle. Haematoxylin is used as a nuclear stain.

6. Eosin

(I) Aqueous eosin

1. Eosin powder ...1.0 gm.
2. Distilled water ...100 ml.

(II) Alcoholic eosin

1. Eosin powder ...1.0 gm.
2. Alcohol 70% or 90% ...100 ml.

It is a cytoplasmic stain and used in double staining along with haematoxylin.

7. Safranin

1. Safranin ...1.0 gm.
2. Absolute alcohol ...100 ml.

Safranin is popular for its brilliancy and permanency in balsam. It is used to colour plant tissues.

8. Light green

1. Light green stain ...0.82 gm.
2. Alcohol 90% ...100 ml.

This is a very good plasma stain.

9. Fast green

1. Fast green stain ...1 gm.
2. Alcohol 90% ...100 ml.

Mix these together. It is a very good cytoplasmic stain.

10. Acid fuchsin

1. Acid fuchsin ...0.5 gm.
2. Distilled water ...100 ml.

This stain is mostly used for the staining of chitinous parts such as mouth parts and other sclerites of arthropods.

11. Leishman stain

1. Leishman stain ...0.15 gm.
2. Absolute alcohol ...100 ml.

Dissolve the stain in absolute alcohol and leave it for ripening for 48 hours. Leishman stain is very good for the staining of protozoans.

12. Mallory's triple stain

Sol. A. 1. Acid fuchsin ...1.0 gm.
2. Distilled water ...100 ml.

Sol. B. 1. Aniline blue ...0.5 gm.
2. Orange G ...2.0 gm.
3. Phosphomolybdic acid ...2.0 gm.
4. Distilled water ...100 ml.

Sol. C. 1. Phosphomolybdic acid ...1 gm.
2. Distilled water ...100 ml.

For Mallory's triple stain, three solutions are prepared. In solution A, sections are stained for 30 seconds and after it washed in distilled water for 2 minutes and then stained in solution B for 1–5 minutes and subsequently washed in distilled water for 2-5 minutes. After staining the sections in A and B solutions, sections are kept in solution C for 30 minutes and are washed again in distilled water for 1-2 minutes and then dehydrated and mounted in Canada balsam. Mallory's triple is a very good stain for staining the microtomy sections of animal tissues. It gives very good results.

13. Congo red vital stain

1. Congo red ...0.5 gm.
2. Distilled water ...100 ml.

Stir the congo red into the water until all dye is dissolved. Filter to remove any large particles.

14. Crystal violet stain

1. Crystal violet (Gentian violet) ...2.0 gm.
2. Ethyl alcohol 90% ... 20 ml.
3. Distilled water ...180 ml.

Dissolve the crystal violet in the alcohol and add the water. This is a simple substitute for Gram's stain.

15. Giema's stain

1. Giema's powder ..0.5 gm.
2. Glycerine ...33 ml.
3. Absolute methyl alcohol ...33 ml.

It takes 1 to 2 hours to dissolve the powder in the glycerine. Then add the acetone free absolute methyl alcohol. This stock solution is diluted 1:10 with distilled water before using.

16. Methylene blue

1. Methylene blue ...1.5 gm.
2. Ethyl alcohol 95% ...100 ml.

This stain is used for staining the small animals in life.

17. Methyl green

1. Methyl green ...8.5 gm.
2. Distilled water ...100 ml.
3. Glacial acetic acid ... 1 ml.

OTHER REAGENTS

1. Acid water

1. Concentrated HCl ... 1 drop
2. Distilled water ...100 ml.

4 Chapter

DISSECTIONS

INSTRUCTIONS FOR DISSECTIONS

The meaning of **"dissection"** is to cut open the animal in order to ascertain the structure of its various parts. The object of dissection is to separate the several parts from one another, so as to define their boundaries and display clearly their mutual relations. Dissection consists mainly in removing the connective tissue (mesenteries) which binds the several parts together.

The following precautions should be followed during the dissection :

1. For every dissection, some instructions are given by the teacher. Carefully follow them.
2. Have the theoretical knowledge of the dissection which you are performing.
3. Have a well-labelled diagram of the dissection to be done.
4. Keep your instruments clean and sharp.
5. Pin down the animal firmly in the dissecting dish. Never attempt to dissect a specimen that is not fixed so.
6. In fixing animal, stick the pins obliquely not vertically so that their heads do not come in the way or obscure the dissection.
7. Never cut away anything until you are quite certain what it is you are removing.
8. Slightly stretch the part which you are dissecting, *e.g.*, when dissecting the blood vessels or nerves of the neck, distend it by passing a small roll of cotton wool down the oesophagus.
9. In cleaning blood vessels or nerves always dissect **along them** and not **across them**, and avoid laying hold of them with the forceps.
10. The organs, blood vessels and nerves are connected with each other by delicate tissues. These should be removed away with great care in order to expose the required structures.
11. Dissections should be done in water. The water must completely immerse the dissection.
12. Always keep the water clean in the dissecting dish during dissection by changing it whenever stained with blood, so that visibility remains good.
13. Always black paper your dissection for good display.
14. Vertebrates are always dissected from the **ventral side** due to presence of nerve cord and vertebral column on the dorsal side.

Scoliodon (Dog-fish)

External Features

Take a preserved **Scoliodon** *and wash it with tap water to remove formalin smell. Lay down the fish in a dissecting dish. Examine the external features and draw the lateral view of the fish and note the following points :*

1. **Size.** It measures from below 1feet (30 cm) to 2 feet (60 cm) in length. Males are smaller than females.

2. **Shape.** The body is spindle-shaped and laterally compressed. Head is dorso-ventrally flattened.

3. **Colour.** Dorsal and lateral surfaces are dark grey in colour, while the ventral surface is yellowish white.

4. **Texture.** The entire surface is rough due to the presence of the backwardly directed placoid scales.

5. **Divisions of the body.** The body is divisible into **head**, **trunk** and **tail**.

(*i*) **Head** is dorso-ventrally flattened having a blunt snout. It bears the following structures :

(*a*) **Mouth** is large, crescentic opening situtated ventrally. The mouth is guarded by upper and lower jaws bearing one or two rows of backwardly directed conical teeth.

(*b*) **External nostrils** are obliquely placed small openings on either side of the head anterior to the mouth.

(*c*) **Eyes** are large and laterally placed on the head. Each eye has an upper and lower eyelids and a thin transparent nictitating membrane.

(*d*) **Gill-slits** are five on either side of the head behind the eyes.

(*ii*) **Trunk** is the largest part of the body extending from behind the gill-slits up to the cloacal aperture. Trunk bears the median unpaired fins and lateral paired fins.

(*a*) **Median unpaired fins** are as follows :

(*i*) First dorsal fin larger, (*ii*) second smaller dorsal fin, and (*iii*) ventral fin or anal fin.

(*b*) **Lateral paired fins** are as follows :

(*i*) A pair of larger pectoral fins just behind the gill-slits and (*ii*) a pair of pelvic fins. These are smaller in size and located on either lateral side of cloacal aperture.

(*iii*) **Tail** is that portion of the body which lies behind the cloacal opening. Tail is heterocercal (*i.e.*, two lobes of caudal fin of different size and tail extends in dorsal lobe) and bears a caudal fin.

Caudal fin extends along dorsal and ventral surfaces of the tail in the median line and forms dorsal and ventral lobes. Dorsal lobe is larger than the ventral lobe.

6. **Lateral line.** A faint line runs on either lateral side of the body extending from the head to the posterior end of the tail.

7. **Claspers.** The male possesses a pair of **claspers** connected on the inner edge of the pelvic fins.

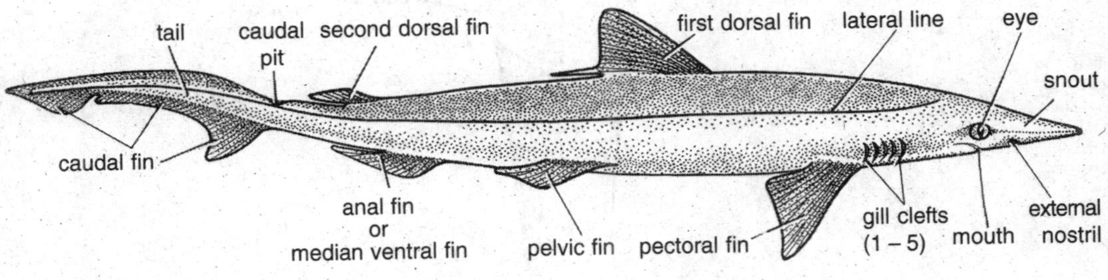

Fig. 4.1. External features of *Scoliodon*.

Scoliodon (Dog-fish)

General Anatomy

Procedure for dissection. *Take preserved* **Scoliodon** *and wash it with tap water. Lay down the fish in a dissecting dish* **ventro-dorsally**, *i.e., ventral surface upwards. Fix the specimen by pins. Pins are to be fixed in pectoral fins. The abdominal cavity is opened by a median-ventral incision in the body-wall extending from the anterior border of the pelvic fins to the anterior border of the pectoral fins. Cut transversely along the line of pectoral and pelvic fins and pin the flaps of the body-wall.*

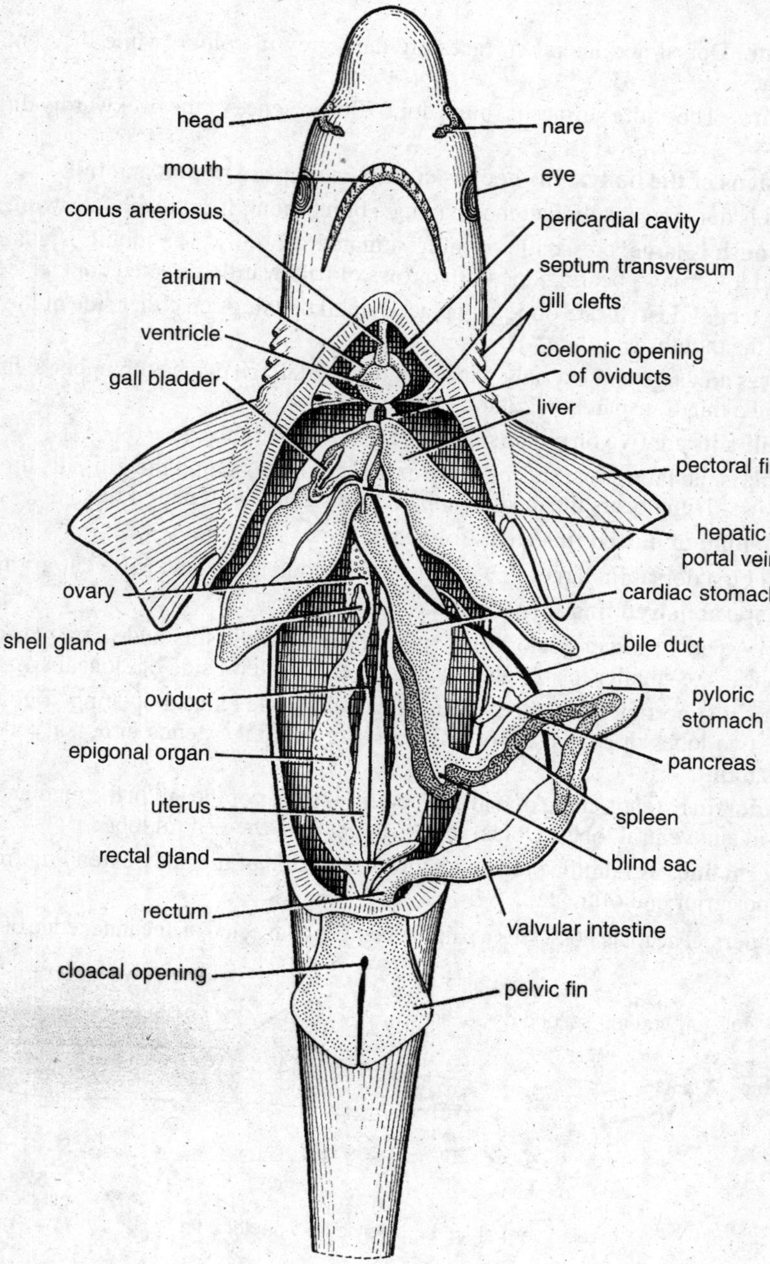

Fig. 4.2. Dissection of general anatomy of female *Scoliodon*.

Study the internal viscera and note the following organs :

1. **Digestive organs.** The following digestive organs are seen :
   - (*i*) **Oesophagus** is short opening anteriorly into pharynx and posteriorly into stomach.
   - (*ii*) **Stomach** is j-shaped, consists of long proximal limb extending almost to the posterior end of the abdominal cavity and is called **cardiac stomach**, while the short narrow distal limb is called **pyloric stomach**. At the junction of cardiac and pyloric stomach a blind sac is present.
   - (*iii*) **Intestine** is a wide tube about the diameter of cardiac stomach. It runs straight backwards in the abdominal cavity. It has a longitudinal scroll valve in its lumen, which is the folding of internal wall of intestine.
   - (*iv*) **Rectum** is formed by narrowing of the intestine. It opens into the cloaca.
   - (*v*) **Rectal gland** is a small diverticulum from the dorsal wall of the rectum.
   - (*vi*) **Liver** is an elongated gland, lies below the cardiac stomach. It consists of two lobes which extend back along the greater part of the abdominal cavity.
   - (*vii*) **Gall bladder** is V-shaped and thin-walled sac, embedded in the anterior portion of the right lobe of the liver.
   - (*viii*) **Pancreas** is a compact bilobed gland situated in the angle between the two limbs of stomach.
   - (*ix*) **Spleen** is a large brownish-red body attached to the loop of the pyloric stomach.
2. **Vascular organs.** The following vascular organs are seen :

   (*i*) Pericardial cavity, (*ii*) heart (atrium and ventricle), (*iii*) conus arteriosus, and (*iv*) ventral aorta.
3. **Urinogenital organs.** The following reproductive organs are seen :
   - (*i*) **Female urino-genital organs** include ovaries, shell glands, oviducts, uteri, epigonal organs, kidneys, ureters and cloaca.
   - (*ii*) **Male urino-genital organs** include testes, vasa efferentia, vasa deferentia, vesicula seminalis, sperm sacs, kidneys, ureters and cloaca.

After completing the dissection, draw a neat and well-labelled diagram of the dissection with the help of figure No. 4.2. (previous page).

Scoliodon (Dog-fish)

Heart and Afferent Branchial Arteries

HEART

Procedure for dissection. *Lay down the fish in a dissecting dish* **ventro-dorsally**, *i.e., ventral surface facing upwards. Fix the fish by pinning it in pectoral fins. Cut open the body-wall in a triangular shape in between the pectoral fins. Carefully remove the muscles and expose the pericardium which encloses the heart. Remove the pericardium carefully and, thus, heart will be exposed.*

Note down the following structures and draw a diagram :

(*i*) **Pericardial cavity** is a median triangular cavity in which heart is lodged.

(*ii*) **Heart** is dorso-ventrally bent, muscular tube, consisting of four chambers, *viz.*, **dorsal sinus venosus, atrium, ventricle** and **conus arteriosus**.

- (*a*) **Sinus venosus** is a triangular thin-walled chamber elongated transversely and lying along the base of pericardial cavity.
- (*b*) **Atrium (auricle)** is a large triangular sac, situated in front and dorsal to the sinus venosus.
- (*c*) **Ventricle** is the most prominent chamber and lies ventral to the atrium or auricle.
- (*d*) **Conus arteriosus** is a stout muscular tube extending from the ventricle to the anterior apex of the pericardial cavity. The conus arteriosus is continued forward through the wall of the pericardium as the **ventral aorta**.

AFFERENT BRANCHIAL ARTERIES

Procedure for dissection : *After exposing the heart carefully remove the body-wall and muscles from the front end and trace the conus arteriosus which continues forward as* **ventral aorta**. *From the ventral aorta arise five pairs of afferent branchial arteries.*

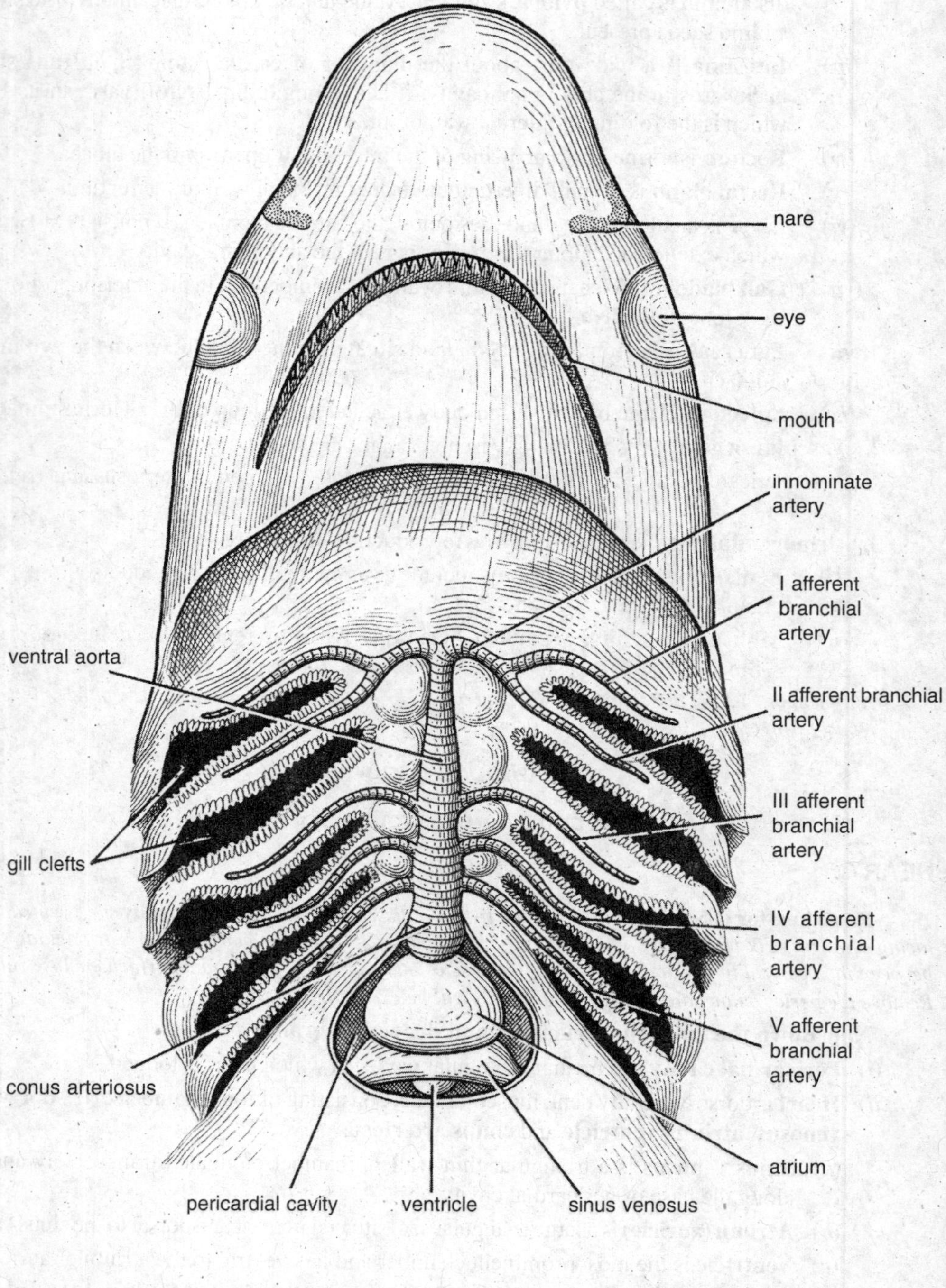

Fig. 4.3. Dissection of heart and afferent branchial arteries of *Scoliodon*.

Expose the following arteries :

(*i*) **Ventral aorta** extends along the ventral surface of the pharynx right up to the border of the hyoid arch, where it bifurcates into two branches the **innominate arteries**, each of which again divides into **first** and **second afferent branchial arteries**.

(*ii*) **First afferent branchial artery** runs along the posterior border of hyoid arch and supplies to all the gill-lamellae of the hyoidean demibranch.

(*iii*) **Second afferent branchial artery** follows a similar course along the first branchial arch supplying branches to both the anterior and posterior gill-lamellae.

(*iv*) **Third, fourth and fifth afferent branchial arteries** arise from the ventral aorta almost equidistant from one another and run along the outer borders of the second, third and fourth branchial arches. Fifth pair arises from the dorsal side of the ventral aorta as a common vessel that immediately bifurcates. Other afferent branchial arteries arise from lateral sides of the ventral aorta.

After completing the dissection, insert black paper beneath the ventral aorta and each afferent branchial artery for the display of the dissection.

Draw a neat and well-labelled diagram of the dissection with the help of figure No. 4.3. (previous page).

Scoliodon (Dog-fish)

Efferent Branchial Arteries

Procedure for dissection. *Lay down the fish in a dissecting dish* **ventro-dorsally**, *i.e., ventral surface facing upwards. Fix the fish by pinning it in pectoral pins. Insert the scissors through the left corner of the mouth and cut the branchial arches and continue cutting posteriorly to the level of pectoral girdle. Give a transverse cut near the pectoral fins and open it. Pin the floor of the pharynx on the left hand side exposing the dorsal surface of the buccopharynx. Carefully remove the mucous membrane lining the buccal cavity and the pharynx and expose the dorsal aorta. Cut open the gills and expose the efferent branchial arteries which are forming loops around each gill-cleft.*

Study the dissection and note down the following :

1. There are nine efferent branchial arteries on each side running along the anterior and posterior border of five gill-clefts as pre-trematic (anterior) and post-trematic (posterior).

2. I + II, III + IV, V + VI and VII + VIII efferent branchial arteries form four complete loops round the first four gill-clefts.

3. The single IX efferent branchial artery runs along the anterior side of V gill pouch and joins with the VIII efferent branchial artery.

4. The four loops are connected with one another by short longitudinal connectives which connect the posterior efferent of a gill-cleft with the anterior efferent of the next gill-cleft behind.

5. Each of the four efferent branchial loops is continued into an epibranchial artery.

6. The four pairs of epibranchial arteries unite to form median dorsal aorta.

From hyoidean (first efferent) arises three vessels (external carotid, afferent spiracular and hyoidean epibranchial) which supply blood to head region.

After completing the dissection, insert the black paper beneath the dorsal aorta, each epibranchial and efferent branchial arteries for the display of the dissection.

Draw a neat and well-labelled diagram of the dissection with the help of figure No. 4.4. (next page).

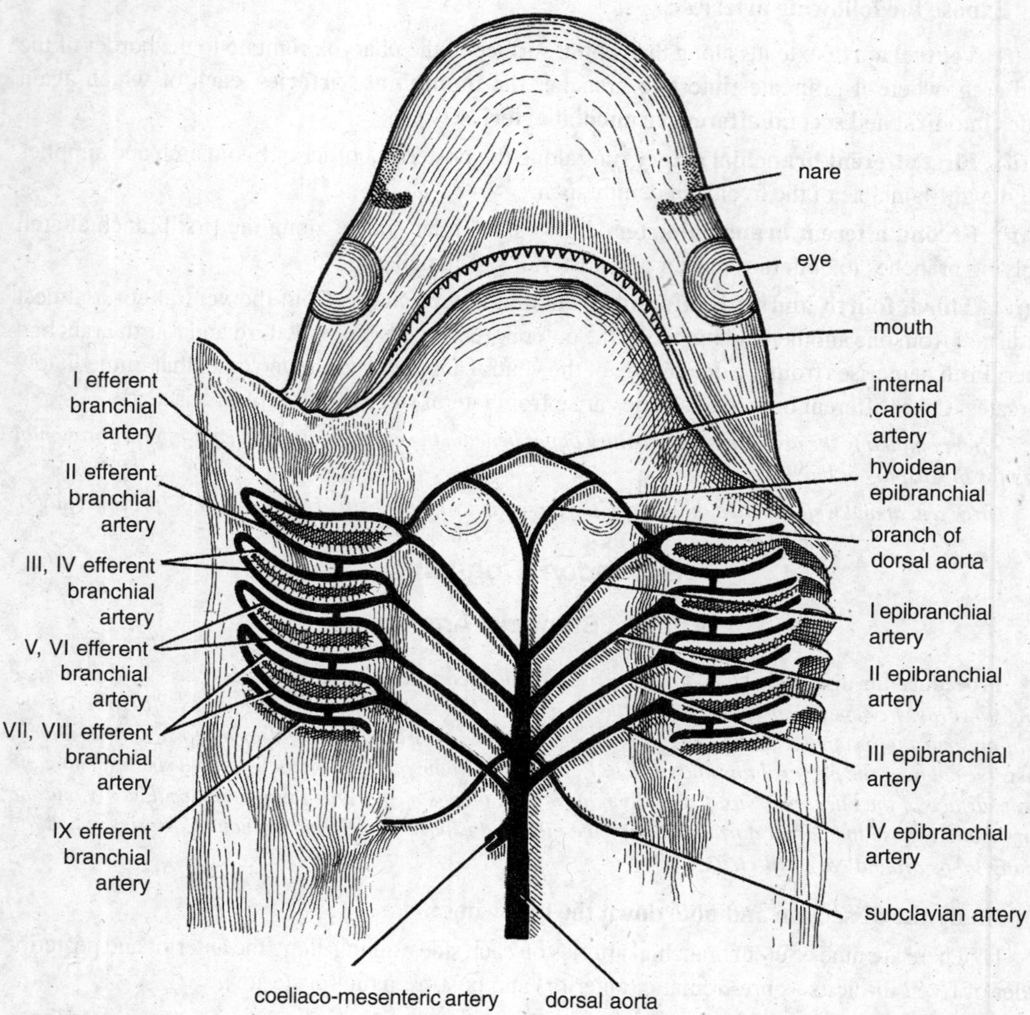

Fig. 4.4. Dissection of efferent branchial arteries of *Scoliodon.*

Scoliodon (Dog-fish)

Cranial Nerves

Procedure for dissection. *Take a preserved fish and make a longitudinal incision in the integument in the middle of the head from the pectoral fin region up to the snout. Carefully remove the flap of integument by giving the transverse incision at the level of pectoral fins. Remove the cranial roof in order to expose the brain. Expose the cranial nerves on the right side and trace them from their origin to their innervation. There are ten pairs of cranial nerves but generally the students are asked to expose* V, VII, IX *and* X *cranial nerves.*

Study and note down the following cranial nerves :

1. V or **Trigeminal nerve** is a large mixed nerve arising from antero-lateral side of medulla along with VII and VIII nerves. It has four branches :

(*i*) **Ophthalmicus superficialis** passes forward above the orbit.

(*ii*) **Maxillaris** has two branches :

(*a*) **Maxillaris superior** runs over the floor of orbit.

(*b*) **Maxillaris inferior** supplies skin of posterior part of upper jaw.

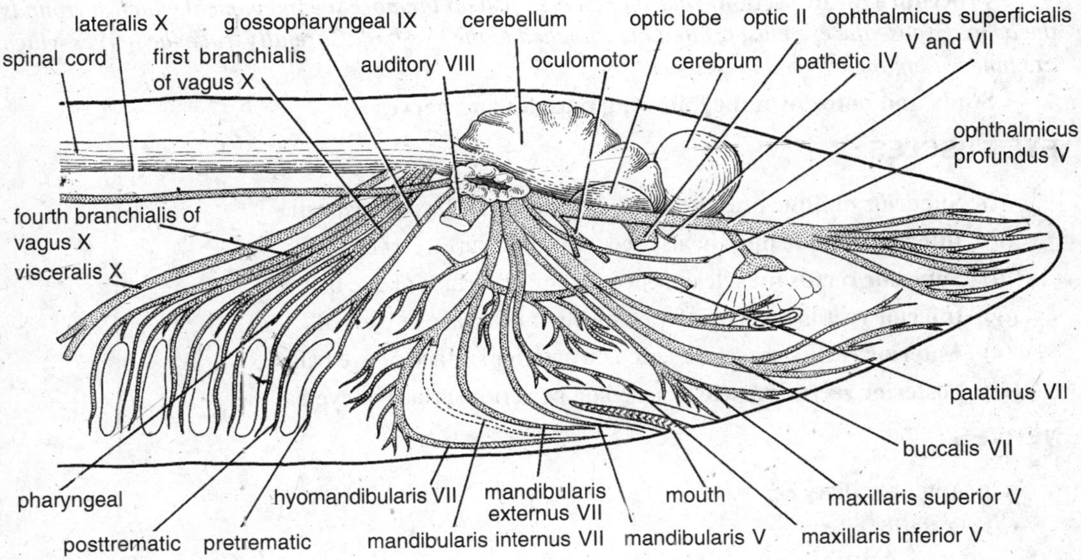

Fig. 4.5. Dissection of cranial nerves of *Scoliodon*.

(*iii*) **Mandibularis** runs along the posterior wall of orbit.

(*iv*) **Ophthalmicus profundus** passes forward along inner surface of eye-ball.

2. **VII** or **Facial nerve** arises from the side of medulla and has four branches :

   (*i*) **Ophthalmicus superficialis** runs along upper border of orbit along with V.

   (*ii*) **Ramus palatinus** runs over the floor of orbit.

   (*iii*) **Ramus buccalis** runs over floor of orbit along with maxillaris superior of V.

   (*iv*) **Ramus hyomandibularis** runs along posterior wall of orbit and has three branches:

   (*a*) **Mandibularis externus** to mandibular canal.

   (*b*) **Mandibularis internus** to buccal floor.

   (*c*) **Hyoidean** to muscles of throat.

3. **IX** or **Glossopharyngeal nerve** arises ventrolaterally from medulla and has three branches :

   (*i*) **Pre-trematic** in anterior border of I gill-pouch.

   (*ii*) **Post-trematic** in posterior border of I gill-pouch.

   (*iii*) **Pharyngeal**.

4. **X** or **Vagus nerve** arises by several roots from lateral side of medulla and has three branches :

   (*i*) **Branchialis** are four, supplying the remaining four gill pouches.

   (*ii*) **Viceralis** enter the body cavity.

   (*iii*) **Lateralis** runs along the entire length of body under lateral line canal.

After completing the dissection, insert black paper beneath the cranial nerves for the display of the dissection.

Draw a neat and well-labelled diagram of the dissection with the help of figure No.4.5.

Scoliodon (Dog-fish)

Eye Muscles and Nerves

Procedure for dissection. *Take a preserved fish and remove the integument which surrounds the orbit. Expose the eye muscles that are attached to the eye-ball. Carefully trace the nerves which are innervating the eye and muscles.*

Study and note down the following muscles and nerves :

EYE MUSCLES :

(*i*) **Superior oblique muscle** inserted on dorsal surface of eye ball.
(*ii*) **Inferior oblique muscle** inserted on ventral surface of eye ball.
(*iii*) **Superior rectus muscle** attached on dorsal surface of eye ball.
(*iv*) **Inferior rectus muscle** inserted on ventral surface of eye ball.
(*v*) **Anterior rectus muscle** inserted on anterior surface of eye ball.
(*vi*) **Posterior rectus muscle** attached on posterior surface of eye ball.

NERVES :

(*i*) Anterior ciliary nerve.
(*ii*) Posterior ciliary nerve.
(*iii*) V, VII, VIII, IX and X cranial nerves.

After completing the dissection, draw a neat and well-labelled diagram with the help of figure No. 4.6.

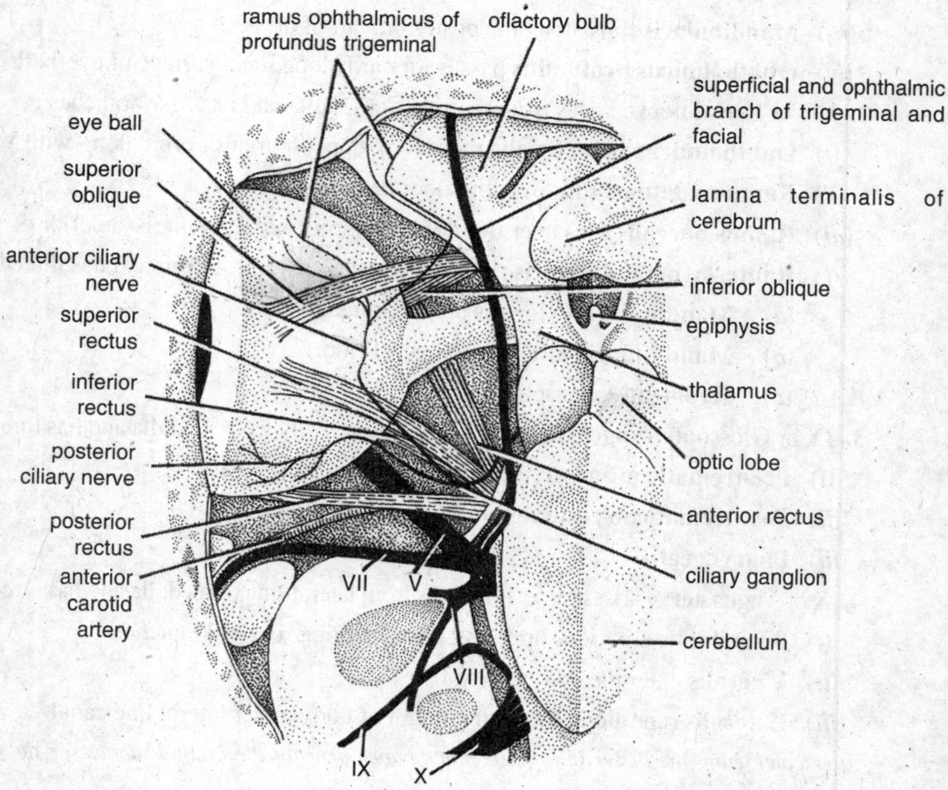

Fig. 4.6. Dissection of eye muscles and nerves of *Scoliodon*.

Scoliodon (Dog-fish)

Internal Ear

Procedure for dissection. *Take a preserved fish and remove the skin of the portion between the eye and the first gill-cleft up to the mid-dorsal line. Remove the muscles completely so as to expose the cartilage of auditory capsule that lodges the internal ear. The auditory capsule shows three ridges of semi-circular canals. Carefully remove the cartilage of the ridges with the help of forceps or needle without injurying or breaking the canals. After locating the canals remove the cartilage present between the canals so as to expose the other parts of the ear. When the ear is completely exposed try with the forcep whether it is completely free from the capsule or not ? If not then further scrap the cartilage and make it free. When it is free from auditory capsule take it out and keep it in water in a watch glass.*

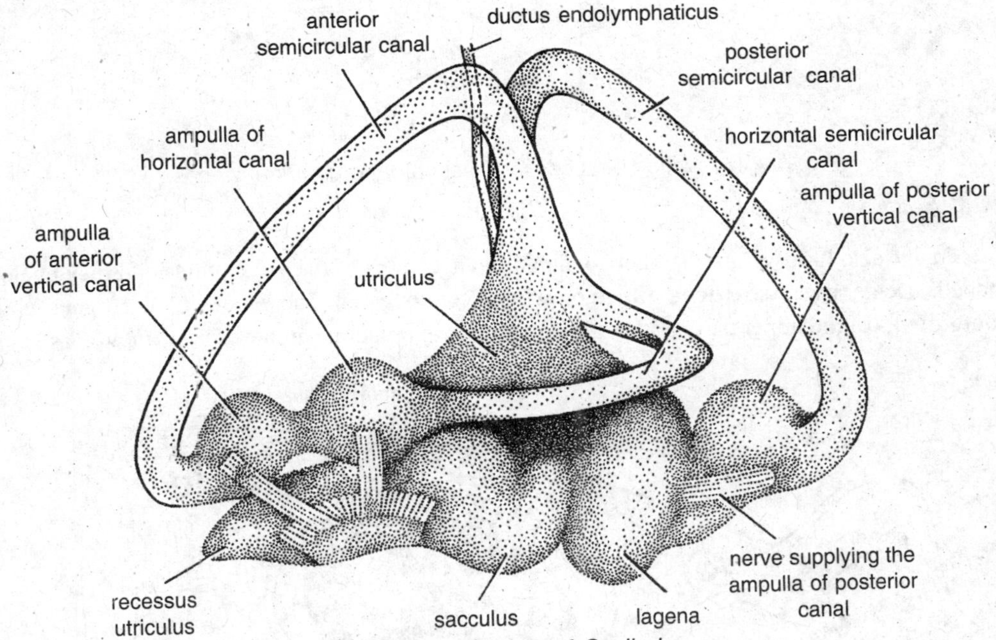

Fig. 4.7. Internal ear of *Scoliodon.*

The internal ear or membranous labyrinths main body is **vestibule** made of dorsal **utriculus** and ventral **sacculus** whose posterior outgrowth is **lagena cochlea**. Anterior outgrowth of utriculus is **recessus utriculi.** From vestibule arises three semicircular canals. Ductus endolymphaticus arises from the sacculus.

Study the internal ear and draw a neat and well-labelled diagram of it with the help of figure No. 4.7.

Accessory Respiratory Organs

Anabas

There are two air chambers, one on each side of the head. These are the dorsal extensions of the branchial cavities (gill-chambers). They contain concentrically arranged wavy bony plates **(labyrinthiform organs)** which are outgrowths of the upper part of the first branchial arch. The plates are covered with vascular mucous membrane. Each chamber communicates with the pharynx by the first gill-slit and with the gill-cavity by an opening situated between the hyoid and the first branchial arch. Air is drawn in by the mouth into the air chamber by the first gill-slit and it passes out by the branchial aperture.

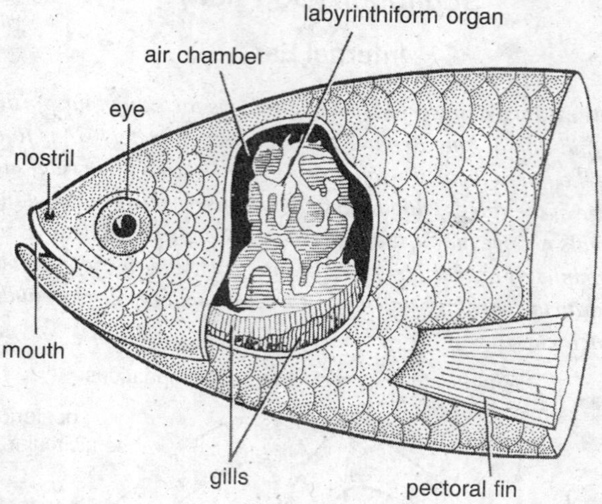

Fig. 4.8. Dissection of accessory respiratory organs of *Anabas.*

Clarias

Clarias also has two air chambers which are outgrowths of the gill cavities and they enclose branched vascular processes **(dendriform** or **arborescent organs).** There are two such organs which are outgrowths of the upper parts of the second and fourth branchial arches. They function as lungs.

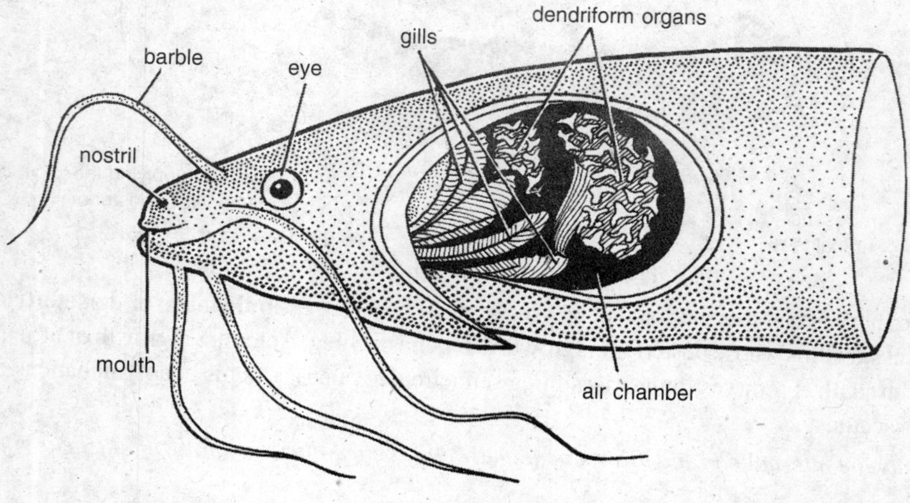

Fig. 4.9. Dissection of accessory respiratory organs of *Clarias.*

Heteropneustes or Saccobranchus

Heteropneustes or *Saccobranchus* has a pair of tubular **extrabranchial diverticulum** as outgrowths of the gill-chambers extending as far as the middle of the tail region. At the hind end of the extrabranchial diverticulum, there are folds forming a sort of air chamber. The air chamber communicates with the buccal cavity by a slit through which the air passes in and out.

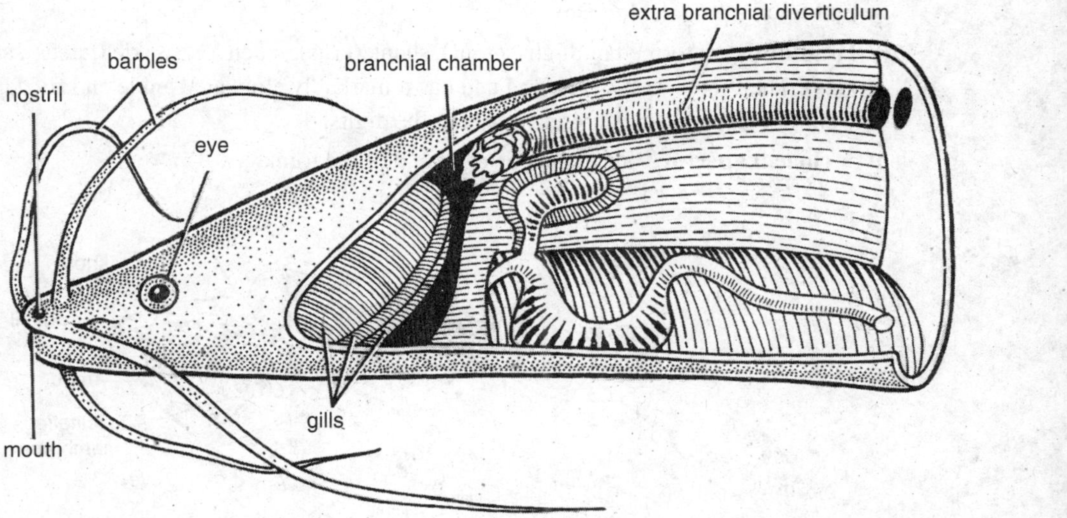

Fig. 4.10. Dissection of accessory respiratory organs of *Heteropneustes.*

Study the accessory respiratory organs and draw neat and well-labelled diagram with the help of figure No. 4.8, 4.9 and 4.10.

Rana tigrina (Frog)

External Features

Take a chloroformed **frog** *and put it in a dissecting dish.*

Study and note down the following :

Size. *Rana tigrina* (*Phyllobatracus tigrinus*) measures up to 7.5 – 9.0 cm in length.

Texture. Skin is smooth and slippery due to mucus secreted by mucous glands of skin.

Colour. The dorsal surface is green or olive coloured with dark patches, while ventral surface is yellowish or whitish.

Division of body. Body is divisible into **head** and **trunk**. Neck and tail are completely absent.

(*i*) **Head** is triangular, dorso-ventrally compressed and possesses a blunt snout. The following structures are noteworthy in the head region :

(*a*) **Nostrlis** or **external nares** are small openings situated at the anterior end of the snout.

(*b*) **Eyes** are large and seen protruding from the dorsolateral sides of the head. Each eye is guarded by two eyelids : an immovable upper eyelid and a semi-transparent movable lower eyelid. From the lower eyelid arises the transparent nictitating membrane, that covers and protects the eye during swimming.

(*c*) **Tympanum** or **ear-drum** is more or less circular area found posterior to each eye.

(*d*) **Mouth** is a wide transverse opening at the anterior end bounded by upper and lower jaws. In male frog, throat bears a pair of bluish patches of skin called **vocal sacs.** They intensify the croaking sound during breeding period.

(*ii*) **Trunk** forms the major portion of the body and contains forelimbs and hindlimbs.

(*a*) **Forelimbs** are shorter and formed of upper arm (brachium), forearm (antebrachium) and hand (manus). The hand consists of carpus (wrist), metacarpus and four digits. The digits are without web. Thumb is vestigial. In male frog, base of first inner finger bears a **nuptial pad** for clasping the female.

(*b*) **Hindlimbs** are formed of thigh (femur), shank (crus), much long ankle (tarsus), and foot (pes). The foot is well-developed and has remarkably elongated ankle and five digits. A well- developed web is present between the digits.

Vent or **cloacal aperture** lies at the posterior end of the trunk.

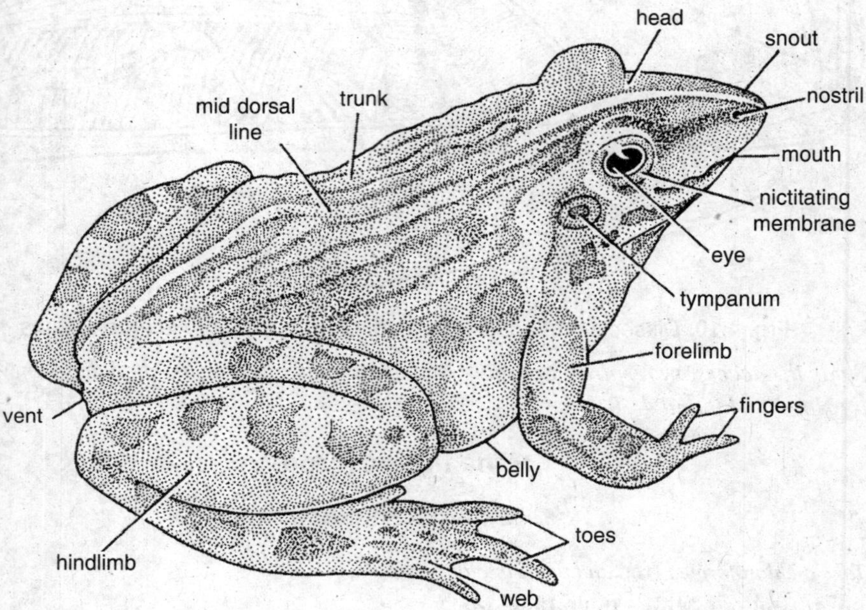

Fig. 4.11. External features of *Rana tigrina*.

Draw a neat and well-labelled diagram of the external features of frog with the help of figure No. 4.11.

Rana tigrina (Frog)

General Anatomy

Procedure for dissection. *Lay down the chloroformed* **frog** *in a dissecting dish* **ventro-dorsally**, *i.e., ventral surface facing upwards. Fix it in the dissecting dish by pinning it through the limbs. First give a median logitudinal incision in the skin from the cloacal region to the pectoral girdle and pin the skin on both the sides. Also cut the skin of the hindlegs up to thigh to expose the muscles. Make two lateral incisions one on either side of the mid-ventral line so as to leave a strip of body-wall about 6 mm in breadth in between the two incisions. Extend these incisions from the region of the pelvic girdle to the region of pectoral girdle. Give the lateral incisions at the two ends and stretch the body-wall on both the sides and pin them. See that the abdominal vein is closely applied beneath the strip of body-wall and anteriorly entering the liver leaving the muscles. Cut the strip of body-wall near the liver avoiding the puncture of abdominal vein. Carefully separate the abdominal vein from the strip of body-wall and remove the strip. Ligature the abdominal vein near its entrance in the liver and also near the pelvic region. Cut the abdominal vein between two ligatures. Thus, the abdominal cavity is exposed which contain the viscera.*

Study and note down the following organs :

(*i*) **Heart** is enclosed in pericardium, situated in the mid-ventral line in between two forelimbs.

(*ii*) **Liver** is a large reddish-brown bilobed organ lies behind and at the sides of the heart.

(*iii*) **Lungs** are two thin-walled spongy sacs, lying at the sides of the heart.

(*iv*) **Gall-bladder** is a small greenish lobe, lying betweea the two lobes of liver.

(*v*) **Stomach** lies in the left side dorsal to the liver and lungs.

(*vi*) **Intestine** is long convoluted tube following the stomach. Rectum is a short wide tube.

(*vii*) **Pancreas** lies in between the duodenum and stomach.

(*viii*) **Spleen** is a small rounded body attached with the large intestine by mesenteries.

(*ix*) **Urinary bladder** is a thin-walled sac opening into the cloaca which is a sac-like part.

(*x*) **Kidneys** are flat, elongated oval bodies of reddish colour, attached to the dorsal body-wall, close to the median line, one on each side of the vertebral column.

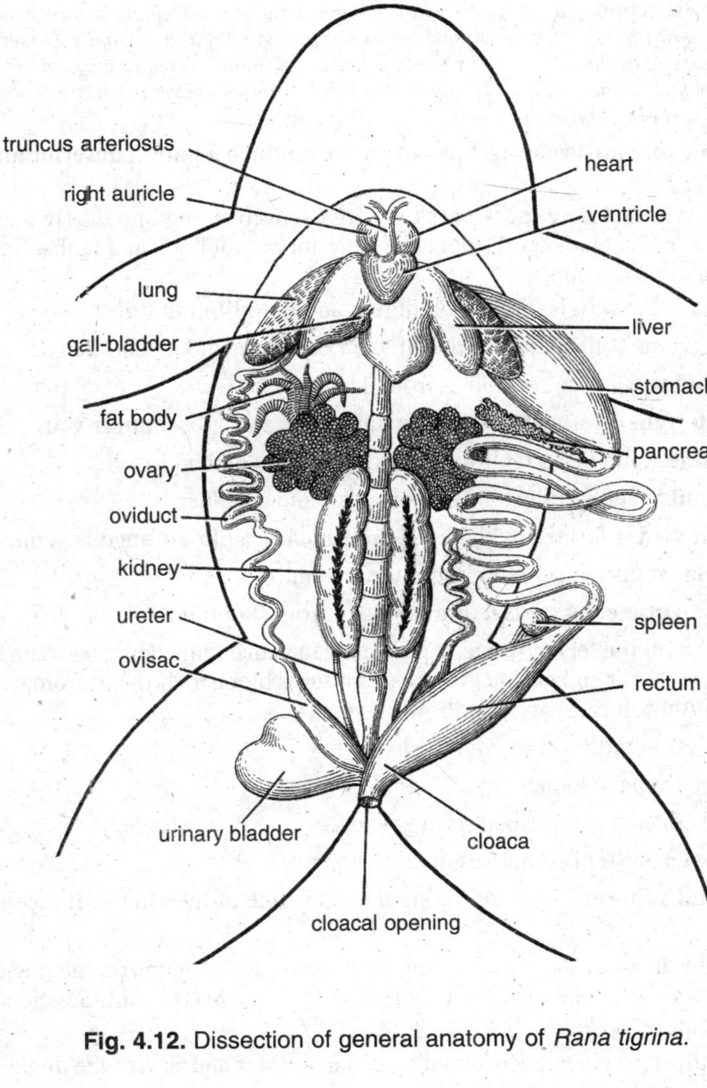

Fig. 4.12. Dissection of general anatomy of *Rana tigrina.*

Reproductive organs

(*i*) **Male reproductive organs** comprise a pair of testes.

Testes are a pair of ovoid bodies of pale yellow colour, attached to the dorsal wall of the body cavity.

(*ii*) **Female reproductive organs** include the ovaries and oviducts.

Ovaries are two large bodies of irregular shape, each consisting of a mass of spherical black and white eggs especially in breeding season.

Oviducts are two long much convoluted tubes lying at the sides of the body cavity.

Draw a neat and well-labelled diagram of the dissection with the help of figure No. 4.12.

Rana tigrina (Frog)

Venous System

Procedure for dissection. *Lay down the chloroformed* **frog** *in a dissecting dish* **ventro-dorsally**, *i.e., ventral surface facing upwards. Fix it in the dissecting dish by pinning through the limbs. Dissect open the body cavity as before and expose the heart. Carefully remove the pericardium. In cleaning a vein take hold the tissue surrounding it with the forcep and take special care not to injure or prick the vein. Always dissect along and not across a blood vessel and pin out the parts so as to stretch it slightly.*

Study and note the sinus venosus formed by the union of a pair of anterior and a posterior veins.

A. Precavals or **anterior venacava** are two large veins opening into the right and left side of the sinus venosus and collect the blood from head, anterior part of body and limbs. Each precaval is formed by the union of three veins :

(*a*) **External jugular vein** is formed by **lingual** and **mandibular** veins.

(*i*) **Lingual veins** collect the blood from the floor of the mouth and tongue.

(*ii*) **Mandibular veins** collect blood from the lower jaw.

(*b*) **Innominate vein** is formed by the **internal juglar** and **subscapular veins.**

(*i*) **Internal jugular veins** returning blood from orbit and brain.

(*ii*) **Subscapular veins** collect blood from arms and shoulder.

(*c*) **Subclavian vein** is formed by the **branchial** and **musculo-cutaneous** veins.

(*i*) **Branchial veins** collect blood from the forelimbs.

(*ii*) **Musculo-cutaneous veins** returning blood from skin snd muscles.

B. Post-caval or **posterior venacava** is a median vein commencing between kidneys and opening into the posterior end of sinus venosus. It returns blood to the heart from liver, kidneys, viscera and hindlimbs. It receives the following veins :

(*a*) **Right** and **left hepatic veins** from the liver.

(*b*) **Renal veins** from the kidneys.

(c) **Gonadial veins** returning blood from gonads,

Renal portal vein system comprises the following veins :

(*a*) **Renal portal vein** runs forwards along the outer side of the kidney. It receives following branches :

(*i*) **Femoral vein** is outer and sciatic vein is inner in the leg. Femoral on entering the abdominal cavity divides into a dorsal **renal portal** and ventral **pelvic vein**. Sciatic joins the renal portal vein.

(*ii*) **Dorso-lumbar vein** collects blood from the muscles and skin of the thigh. It unites with the renal portal vein while running on the outer border of kidney.

Hepatic portal system is formed by the following veins :

(*a*) **Anterior abdominal vein** is a median vein formed by the union of two pelvic veins. It receives blood from bladder and ventral abdominal wall. Anterior abdominal vein joins with the hepatic portal vein in the region of liver.

(*b*) **Hepatic portal vein** is a large vein that collects blood from the following veins :

(*i*) **Gastric vein** from stomach.

(*ii*) **Intestinal vein** from intestine.

(*iii*) **Splenic vein** from spleen.

(*iv*) **Pancreatic vein** from pancreas.

After completing the dissection, draw a neat and well-labelled diagram with the help of figure no. 4.13.

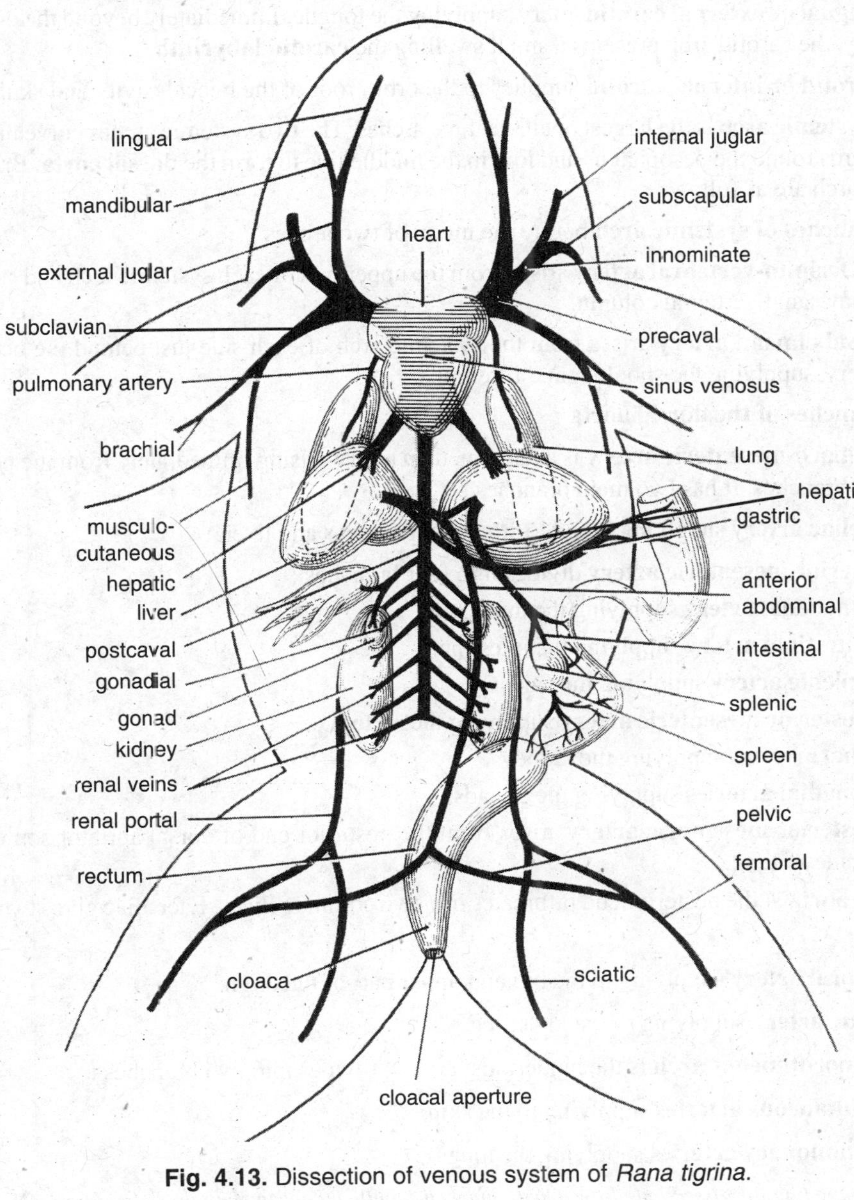

Fig. 4.13. Dissection of venous system of *Rana tigrina*.

Rana tigrina (Frog)

Arterial System

Procedure for dissection. *Lay down the chloroformed* ***frog*** *in a dissecting dish ventro-dorsally, i.e., ventral surface facing upwards. Fix it in the dissecting dish by pinning through the limbs. Dissect open the body cavity as before and expose the heart. Carefully remove the pericardium. Carefully clean the three aortic arches commencing at the truncus arteriosus and follow the several arteries to their distribution.*

Study and note the division of truncus arteriosus in front into right and left branches which again divides into three aortic arches—the common carotid arch, systematic arch and pulmocutaneous arch.

1. Common carotid arch is the most anterior of the three arches and divides into two arteries :

(*i*) **Lingual or external carotid** artery supplying the tongue. Immediately beyond the origin of lingual artery, the carotid arch presents a small swelling the **carotid labyrinth**.

(*ii*) **Carotid or internal carotid** supplies to the orbit, roof of the buccal cavity and skull.

2. Systemic arch is the largest of all the three arches. The two systemic arches curve upwards and downwards round the oesophagus and join in the middle line to form the **dorsal aorta**. Branches of systemic arch are as follows :

(*a*) **Branches of systemic arch** before the union of two arches :

(*i*) **Occipito-vertebral artery** arises from the upper part of each systemic arch and supplies to the head, jaws and vertebral column.

(*ii*) **Subclavian artery** arises from the systemic arch of each side just behind the occipito-vertebral artery, supplying the shoulder and forelimbs.

(*b*) **Branches of the dorsal aorta** :

I. Coeliaco-mesenteric artery is a large median artery arising immediately from the point of union of two branches. It has two main branches :

(*a*) **Coeliac artery** supplying blood to stomach, pancreas and liver.

(*b*) **Anterior mesenteric artery** divides into four branches :

(*i*) **Duodenal artery** supplying the duodenum.

(*ii*) **Intestinal artery** supplying the intestine.

(*iii*) **Splenic artery** supplying the spleen.

(*iv*) **Posterior mesenteric artery** supplying the rectum.

II. Renal arteries supplying the kidneys.

III. Gonadial arteries supplying the gonads.

IV. Posterior mesenteric artery arises from the posterior end of dorsal aorta or sometimes from anterior mesenteric.

Dorsal aorta at the posterior end bifurcates into two common iliacs. Each iliac gives two main branches :

(*a*) **Femoral artery** supplying to the hip and upper part of the thigh.

(*b*) **Sciatic artery** supplying to the lower leg.

2. Pulmocutaneous arch is the hindermost arch. It divides into two branches :

(*i*) **Cutaneous arteries** supplying to the skin.

(*ii*) **Pulmonary arteries** supplying the lungs.

After completing the dissection, draw a neat and well-labelled diagram with the help of figure No. 4.14. (next page).

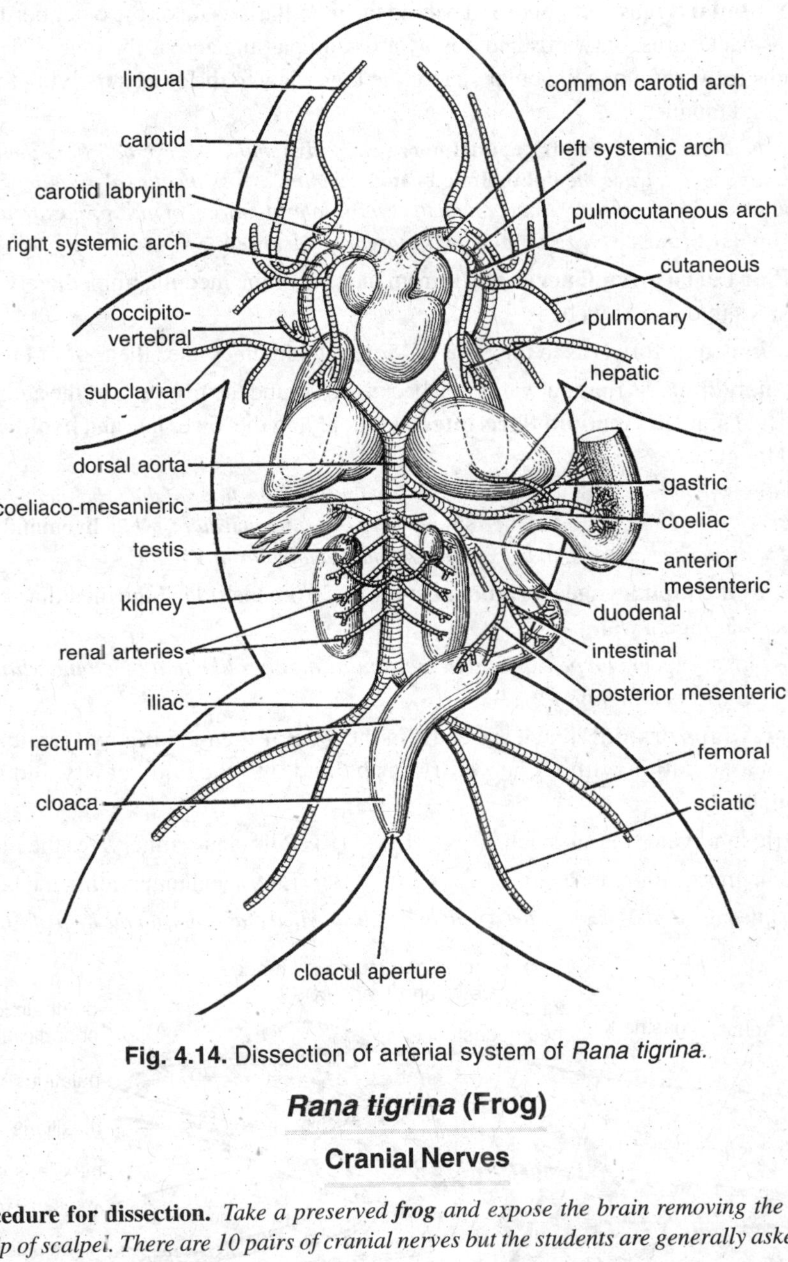

Fig. 4.14. Dissection of arterial system of *Rana tigrina.*

Rana tigrina (Frog)

Cranial Nerves

Procedure for dissection. *Take a preserved* ***frog*** *and expose the brain removing the roof of the skull with the help of scalpel. There are 10 pairs of cranial nerves but the students are generally asked to expose only* V, VII, IX, X *cranial nerves and their branches.*

Expose and note the following cranial nerves :

1. **V or Trigeminal cranial nerve** arises from the anterior lateral part of the medulla. It divides into three branches :

(*i*) Ophthalmicus superficialis runs forwards through the orbit lying close to its inner side between the skull wall and the eye. At the anterior end of the orbit it divides into two branches, which pass through the walls of the nasal capsule and supply the skin of the snout.

(*ii*) Maxillaris runs forwards and outwards of the orbit behind and below the eye-ball to the margin of the upper jaw and divides into two branches namely **maxillaris externus** and **maxillaris internus**.

(*iii*) **Mandibularis** runs parallel to and behind the maxillaris as far as the outer border of the eye-ball, then turns backwards, outwards and downwards and passing across the inner side of the upper jaw, it then runs forwards along the outer side of the lower jaw to the chin, supplying the muscles of the lower jaw and tongue.

To trace the branches of V *or* **trigeminal nerve** *carefully remove the eye-ball from the orbit and also remove the nasal bone and trace the* **ophthalmic branch** *running across the dorsal margin of the orbit. The* **maxillaris branch** *can be exposed by tracing it on the ventro-lateral-margin of the orbit. Carefully expose and trace the* **mandibular branch** *which runs along the margin of the lower jaw.*

2. VII or Facial cranial nerve arises from the side of the medulla immediately behind the V nerve and divides into two branches :

(*i*) **Palatinus** runs forwards in the floor of the orbit and innervates the roof of buccal cavity.

(*ii*) **Hyomandibularis** runs outwards and backwards round the front end of the auditory capsule. It divides into two branches **mandibularis internus** supplying the lower jaw and **hyoidean** supplying the hyoid and tongue.

The **palatinus** *found attached with the mucous membrane of the floor of the orbit and can be separated from that with the help of needle from posterior side where it enters medulla, while* **hyomandibular** *can be exposed by breaking the squamosal and tracing behind the mandibular of V nerve.*

3. IX or Glossopharyngeal cranial nerve arises from the side of the medulla and supplies to the tongue and pharyngeal wall.

The glossopharyngeal *can be exposed in the similar manner like that of hyomandibular because glossopharyngeal lies just behind it.*

4. X or Vagus cranial nerve arises from the side of the medulla by forming a ganglion at the origin in association with IX nerve. It runs behind into the body cavity and then divides into four branches :

(*i*) Gastric innervates the stomach.

(*ii*) Pulmonary innervates the lungs.

(*iii*) Cardiac innervating the heart.

(*iv*) Laryngeal innervating the larynx.

After completing the dissection, draw a neat and well-labelled diagram with the help of Figure No. 4.15.

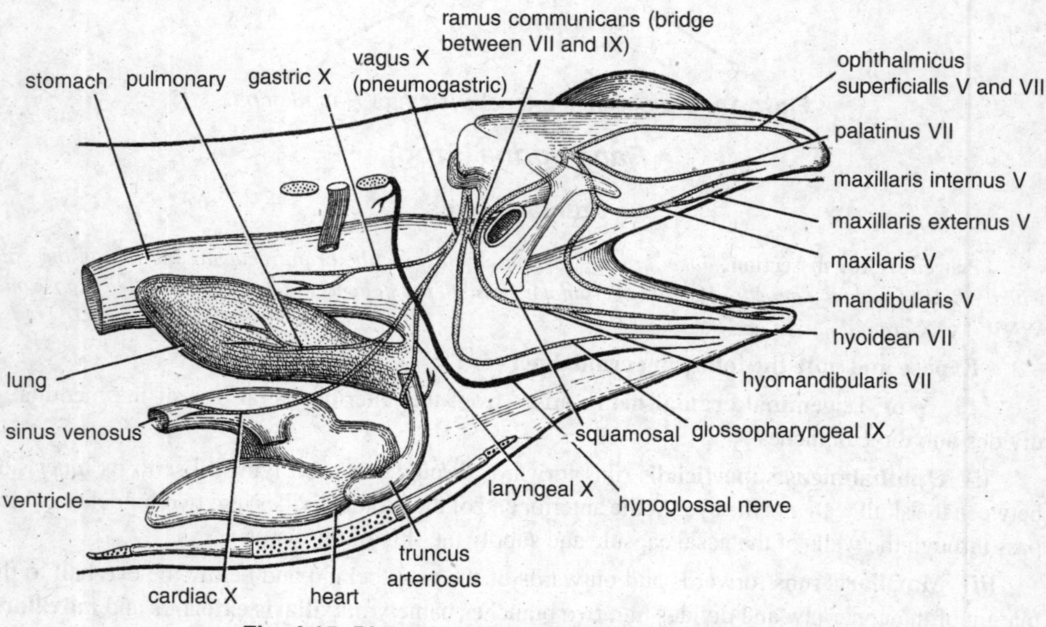

Fig. 4.15. Dissection of cranial nerves of *Rana tigrina*.

Rana tigrina (Frog)

Internal Ear

Procedure for dissection. *Take a preserved* **frog** *and remove the skin and muscles from the lateral sides of the head behind the eyes.Expose the auditory capsule and observe the buldging ridges of three semi-circular canals. Now break the cartilage of these ridges by forcep or hard needle and expose the three semi-circular canals. After exposing the canals proceed for the utriculus and sacculus. These can be exposed in the same manner. Test with the help of the needle whether the internal ear is free from the wall of the auditory capsule or not ? If not try to remove the cartilaginous portion in between the parts of the ear and the auditory capsule and detach it from the capsule. Put it in a watch glass in water.*

Study and note the following parts of the internal ear :

1. The **vestibule** is a membranous sac lying in the cavity of the auditory capsule. It is divided by a constriction into two main divisions :

(*i*) The **utriculus** is upper and larger oblong division.

(*ii*) The **sacculus** is the inferior and smaller oval division, from its posterior part arises two smaller saccular dilations, called lagena and pars basilaris. Above these two is the pars neglecta which arises from utriculus.

2. The **semicircular canals** are three tubular branches of the utriculus into which they open at both ends.

(*i*) **Anterior vertical semicircular canal** has its ampulla at its anterior end.

(*ii*) **Posterior vertical canal** has its ampulla at its outer end, while its inner end joins the posterior end of the anterior vertical canal.

(*iii*) **Horizontal canal** has its ampulla at its anterior end.

All these three are arranged at right angles with one another.

Draw a neat and well-labelled diagram of the internal ear with the help of figure No. 4.16.

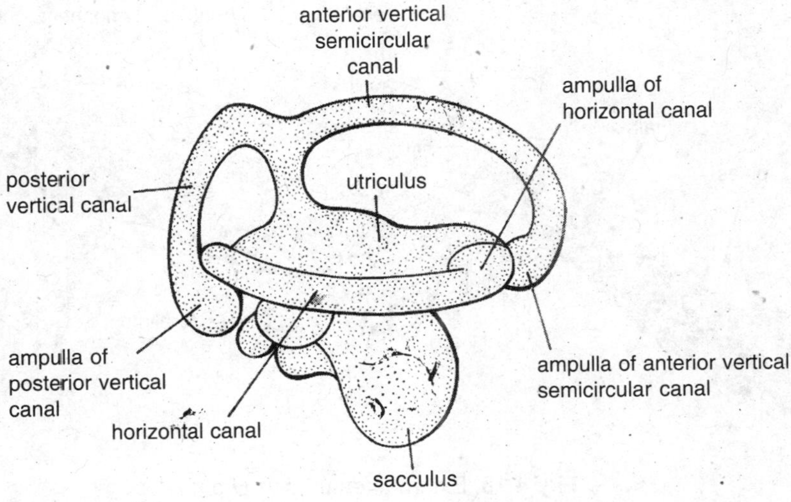

Fig. 4.16. Internal ear of frog.

Columella Auris

Procedure for dissection. *First of all locate the tympanum which lies behind the eyes of frog.For taking out the columella auris, cut the tympanum from its all sides by sharp scissor. After cutting it from sides remove* *the skin and muscles from its margins and postero-dorsal portion of the skull. Now break the prootic bone from its dorsal side by a scalpel or hard needle. As the prootic bone is broken a rod is seen inside its canal which is connected with the tympanum at its anterior end. Break the prootic bone further till the posterior end of rod (columella), by which it is attached to inner wall of the cranium is seen. As soon as the posterior end is seen, detach the rod slowly and carefully and keep it in the watch glass.*

Draw a neat and well-labelled diagram of columella auris *with the help of figure No. 4.17.*

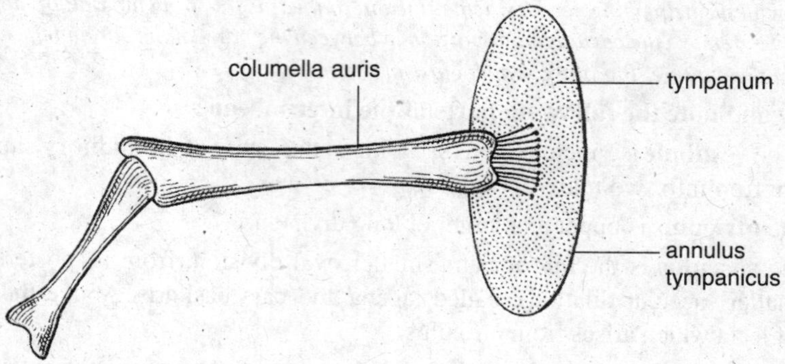

Fig. 4.17. Columella auris of frog.

Uromastix (Lizard)

External Features

Take a chloroformed **Uromastix** *(Sanda in Hindi) and put it in a dissecting dish.*

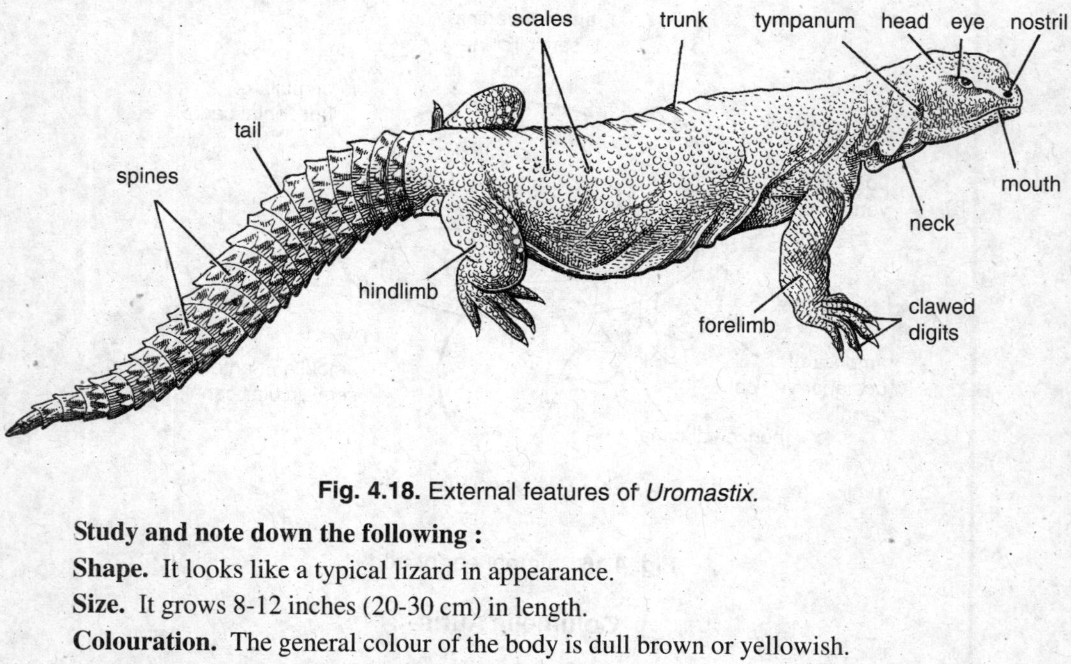

Fig. 4.18. External features of *Uromastix.*

Study and note down the following :

Shape. It looks like a typical lizard in appearance.

Size. It grows 8-12 inches (20-30 cm) in length.

Colouration. The general colour of the body is dull brown or yellowish.

Exoskeleton. The body is covered with epidermal scales.

Divisions of body. The body is divisible into **head**, **neck**, **trunk** and **tail**.

1. Head is small, triangular produced into a short snout bearing the following structures :

(*a*) **Nostrils.** A pair of nostrils lie in front of the eyes, which open into the buccal cavity.

(*b*) **Eyes.** A pair of eyes lie dorso-laterally almost in the middle of the head. Each eye has two movable eyelids and a nictitating membrane.

(*c*) **Tympanum.** Behind each eye lies a sunken pit. At the base of pit is present a tympanic membrane (tympanum).

(*d*) **Mouth.** It is a large anterior opening bounded by upper and lower lips covered with horny scales. Inside the upper and lower lips lie the respective jaws provided with cutting teeth.

2. Neck is short but has a fold of skin hanging on its ventral and lateral sides.

3. Trunk forms the major elongated part of the body. It bears two pairs of limbs fore-and hind-limbs.

(*a*) **Forelimbs** are paired. Each forelimb consists of upper arm, fore arm and hand. The hand bears five horny clawed digits.

(*b*) **Hindlimbs** are also paired. Each hindlimb consists of thigh, shank and foot. The foot bears five horny clawed digits. In males on the ventral surface of the thighs are present 12-18 pores of femoral glands arranged in a single row from cloaca to the knee joint. These pores are known as **preanal femoral pores** which are more conspicuous in males than in females. Femoral glands secrete horny secretions that forms temorary spines for grasping the female during mating.

(*c*) **Cloacal opening** is a median opening lying ventrally at the junction of trunk with the tail. The cloaca has a pair of **copulatory penes.**

4. Tail is elongated and thick and lies behind the cloaca. It bears about 27 rings of transversely arranged spines.

Draw a neat and well-labelled diagram of external features of ***Uromastix*** *with the help of figure No. 4.18. (previous page).*

Uromastix (Lizard)

Venous System

Procedure for dissection. *Take a chloroformed* ***Uromastix*** *and lay down it in the dissecting dish* ventro-dorsally, *i.e., ventral surface facing upwards. Fix it by pinning through its limbs. Give a median longitudinal incision into the skin from cloacal opening to the anterior end up to the snout. Give transverse cuts posterior to the forelimbs and posteriorly in between two hindlimbs and fix the flaps of skin by pins. Now make a median longitudinal incision in the body-wall (muscles) and cut the pectoral girdle and sternum. Carefully remove the body-wall without injuring the blood vessels and other organs. Remove the pericardium and expose the heart. Now expose the roots of* **veins** *entering the heart.*

Identify the following veins and expose them carefully.

1. Precaval veins collect blood from the anterior body region by the following branches :

(*i*) **Right external jugular** collects the blood from the upper and lower jaws. Left external jugular is atrophied.

(*ii*) **Internal jugular** collects blood from the head (brain and orbit). External and internal jugular veins unite at the level of tympanum to form the common jugular vein. It forms jugular sinus at the base of neck.

(*iii*) **Subclavian** collects blood from the forelimb and joins the jugular sinus at its posterior end.

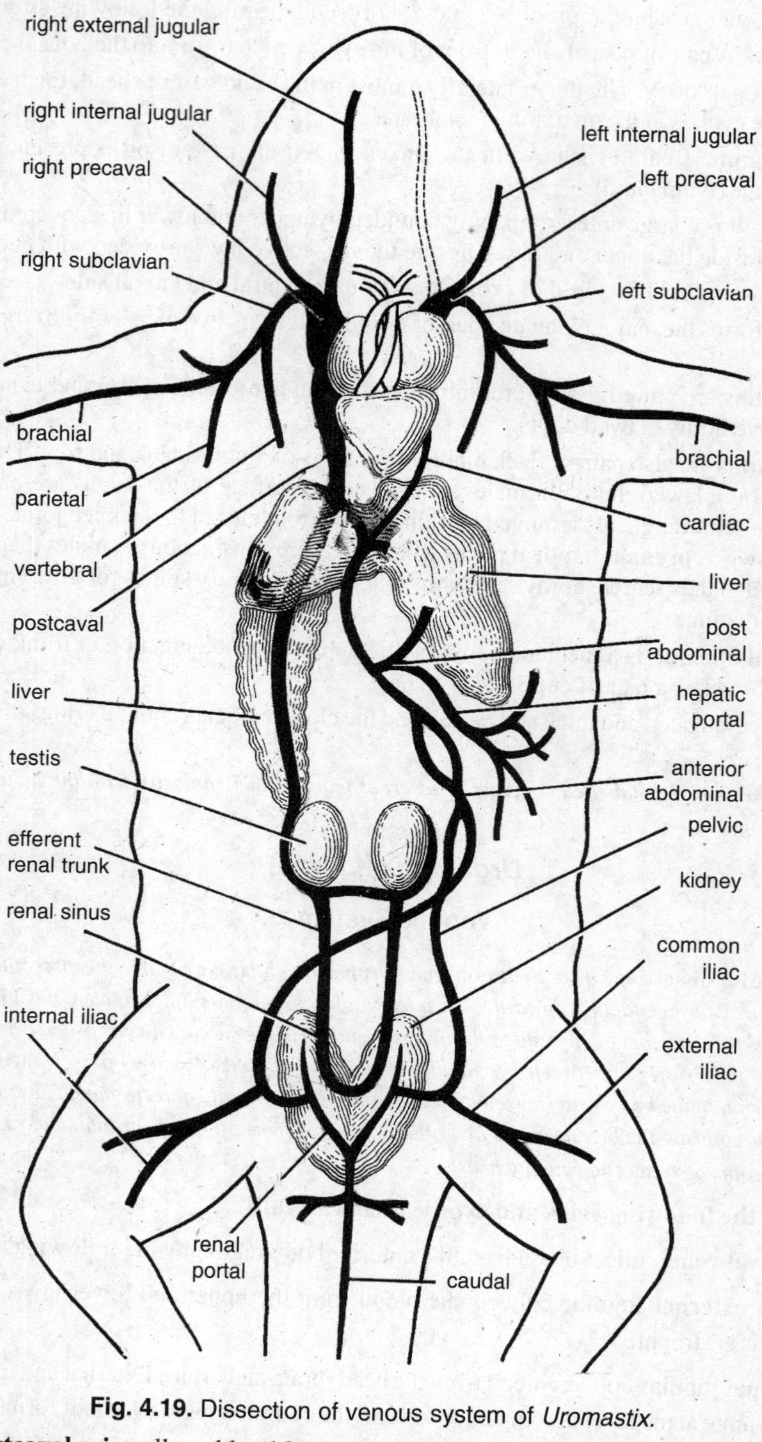

Fig. 4.19. Dissection of venous system of *Uromastix*.

2. Postcaval vein collects blood from the posterior region of body and pours it into the sinus venosus. The posterior veins which unite to form postcaval are as follows :

(i) **Caudal vein** collects blood from the tail. It runs up to kidney and divides into two afferent renal or renal portal veins which run forward through ventral surface of kidneys.

(*ii*) **Iliac veins** collect blood from the legs, the **external iliac** (femoral) from the outer side and **internal iliac** (sciatic) from the inner side. Both unite in the abdominal cavity to form **common iliac** on each side of body. Common iliac joins the renal portal vein.

(*iii*) **Pelvin vein** is the anterior continuation of common iliac vein.

(*iv*) **Anterior abdominal vein** is formed by the union of two pelvic veins. At anterior end near liver it bifurcates, one enters left liver lobe and the other joins the hepatic portal vein to form **postabdominal vein** which enters left liver lobe.

(*v*) **Renal sinus.** Blood from both kidneys collects into large central **renal sinus** which gives off two efferent **renal veins**. These run up to gonads and both unites to form **commissural vein** that receives blood from gonads. Right efferent vein runs forward through right liver lobe as **postcaval**.

(*vi*) **Hepatic portal vein** is formed by the union of gastric, branch of intestinal, splenic and pancreatic veins. It joins with one branch of portoabdominal vein to open into left liver lobe.

After completing the dissection, draw a neat and well-labelled diagram of the dissection with the help of figure No. 4.19 (previous page).

Uromastix (Lizard)

Arterial System

Procedure for dissection. *The procedure for dissecting the arterial system is more or less the same as that of the venous system described earlier. After exposing the heart locate the aortic arches, i.e., pulmonary, systemic and dorsal aorta. The blood is distributed to all parts of the body by the following arteries :*

1. Pulmonary arch gives rise to two right and left pulmonary arteries supplying the blood to the lungs.

2. **Systemic arches** are two, *i.e.*, left (arises from right side) and right systemic arches. Both the systemic arches in front of heart cross over to the opposite sides and at the point of cross both communicate through a **foramen of Panizae.** The systemic arches give rise to the following arteries :

(*i*) **Innominate artery** arises from the right systemic arch and divides into right and left common carotids. Each carotid artery divides into external and internal carotid arteries. These supply blood to neck and head region. Since carotid arteries arise from right systematic arch, hence, it is also called **carotico-systemic arch**.

(*ii*) **Ductus caroticus** connects the internal carotid artery with the systemic arch on both the sides.

(*iii*) **Common subclavian artery** arises from the right systemic arch and divides into right and left subclavian arteries and supplying blood to the forelimbs.

3. **Dorsal aorta** is formed by the union of right and left systemic arches. The dorsal aorta supplies the general viscera by several arteries which are as follows :

(*i*) **Posterior oesophageal artery** supplies blood to oesophagus. Anterior oesophageal artery arises from left systematic arch.

(*ii*) **Unpaired gastric arteries** supply blood to the stomach.

(*iii*) **Unpaired coeliac artery** supplies the blood to the stomach.

(*iv*) **Unpaired pancreato-hepatic artery** supplies blood to pancreas and liver.

(*v*) **Unpaired anterior mesenteric artery** supplies blood to the intestine.

(*vi*) **Unpaired posterior mesenteric artery** supplies blood to the caecum and rectum.

(*vii*) **Gonadial arteries** one or two pairs supply to gonads.

(*viii*) **Unpaired rectal** and **cloacal arteries** supply blood to rectum and cloaca respectively.

(*ix*) **Renal arteries** (3 pairs) supply blood to kidneys.

(*x*) **Common iliac arteries** (one pair) supply to legs.

(*xi*) **Caudal artery** supplies blood to tail.

After completing the dissection, draw a neat and well-labelled diagram of the dissection with the help of figure No. 4.20.

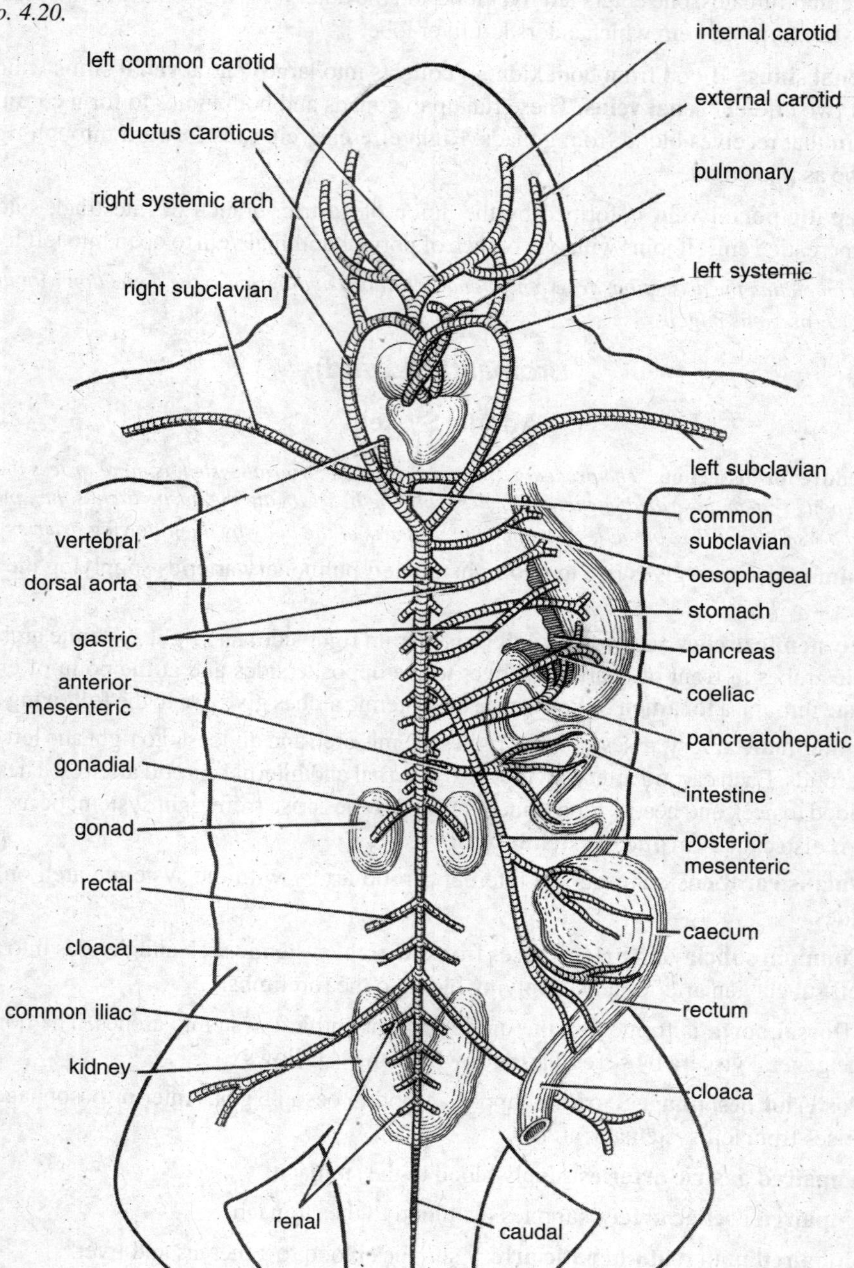

Fig. 4.20. Dissection of arterial system of *Uromastix.*

Columba livia (Pigeon)

External Features

Take a freshly chloroformed ***pigeon*** *and note that the whole body, except beak and feet, is covered with feathers.*

1. Size. It grows about 13 inches (32.5 cm) in length.

2. Colouration. Rock pigeon is slaty grey with glistening metallic green, purple and magenta sheen on the neck and upper breast. Two dark bars on wings and a band across the end of tail.

3. Division of the body. The body is divisible into **head**, **neck**, **trunk** and **tail**.

(*i*) **Head** bears numerous structures which are as follows :

(*a*) **Beak** is strong, horny, straight and pointed. At the base of which is present **cere** (naked swollen skin).

(*b*) **Nostrils** are obliquely situated at the base of beak overhung by the cere.

(*c*) **Eye.** Each rounded eye is surrounded by three eyelids, an upper and lower and a nictitating membrane.

(*d*) **External ear openings** lie posterior to the eyes and covered by feathers called **auricular**.

(*ii*) **Neck** is long and flexible and can be bent to form an S-shape.

(*iii*) **Trunk** is the major part of body and divided into a large **thorax** and a small **abdomen**. The thorax is strengthened ventrally by a large sternum from which projects mid-ventrally a **keel** or **carina**. Trunk also bears two pairs of limbs.

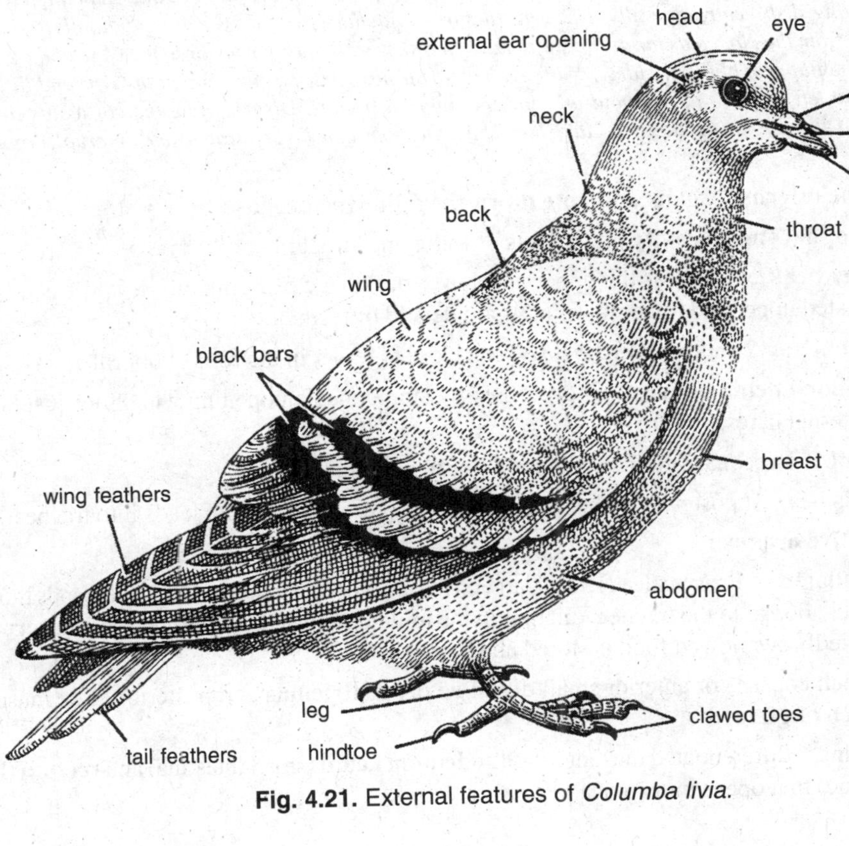

Fig. 4.21. External features of *Columba livia*.

(*a*) **Forelimbs** are modified into wings. Each wing has upper arm, fore arm and hand.

(*b*) **Hindlimbs** support the body for bipedal locomotion; the body is borne obliquely on them. The two feet are not covered by feathers but by horny scales and the digits end each in a horny claw.

(*iv*) **Tail** is short, a transverse cloacal aperture is located ventrally at its junction with the abdomen. On the dorsal surface of the tail is an **oil** or **uropygial gland** that secretes oil to preen feathers with the aid of beak.

4. Feathers are of three principal kinds :

(*a*) **Contour-feathers** which appear on the surface of the entire body. The strongest contour feathers are the **quill-feathers**, of which those attached with the wings are called **remiges**, and those attached with the tail are called **rectrices**. Remiges (23) attached to bones of hand are called **primaries** (11) and those attached to ulna are called **secondaries** or **cubitals** (12).

(*b*) **Filoplumes** or **hair feathers** are very thin hair-like degenerate feathers lying at the base of contour feathers specially the quill-feathers.

(*c*) **Down-feathers** or **plumes** are situated below the contour feathers. They have no hooks on the barbules and, thus, appear fluffy.

After studying the external features, draw a neat and well-labelled diagram with the help of figure No. 4.21. (previous page).

Columba livia (Pigeon)

General Anatomy

Procedure for dissection. *Take a freshly chloroformed* **pigeon** *and remove its feathers completely. Place it in the dissecting dish ventro-dorsally, i.e., ventral surface facing upwards. Cut longitudinally through its abdominal wall from the cloaca as far as the sternum, then transversely just posterior to the edge of the sternum and pin down the flaps on both the sides. With the help of strong scissors, cut through the ribs and lift up the sternum with the left hand so as to avoid injuring the internal organs. Disarticulate the coracoids from the sternum and cut the furcula at the two clavicles. Thus, sternum can be separated and internal organs are exposed.*

Study the internal organs and note down the following organs :

1. Buccopharyngeal cavity. It consists of following structures :

(*i*) **Horny beak** consists of the upper and lower beaks enclosing mouth. It is formed by the elongated upper and lower jaws. Jaws are devoid of teeth.

(*ii*) **Internal nares** are two narrow slits lying close together in the roof of buccal cavity.

(*iii*) **Common opening of eustachian tubes** is a single median opening which lies just behind the internal nares.

(*iv*) **Tongue** is large, narrow, triangular and sharply pointed.

(*v*) **Glottis** lies just posterior to the tongue at the floor of pharynx that leads into trachea.

2. Digestive organs :

(*i*) **Oesophagus.** Pharynx leads into a large, wide and very long oesophagus. It extends through the neck, dorsal to the trachea, dilating in the middle to form large bilobed sac, the **crop**. In it hurriedly swallowed food is stored and moistened.

(*ii*) **Stomach** consists of anterior small digestive **proventriculus** and posterior large muscular stomach or **gizzard**.

(*iii*) **Intestine** is differentiated into anterior **duodenum, ileum** (small intestine) and **rectum** (large intestine) that opens into cloaca.

(*iv*) **Associated glands** are liver, pancreas and spleen.

(*a*) **Liver** consists of two lobes, right larger than the left. Gall-bladder is absent.

(*b*) **Pancreas** is reddish in colour from which arises three pancreatic ducts.

(*c*) **Spleen** is a small oval red body attached to the right side of the proventriculus.

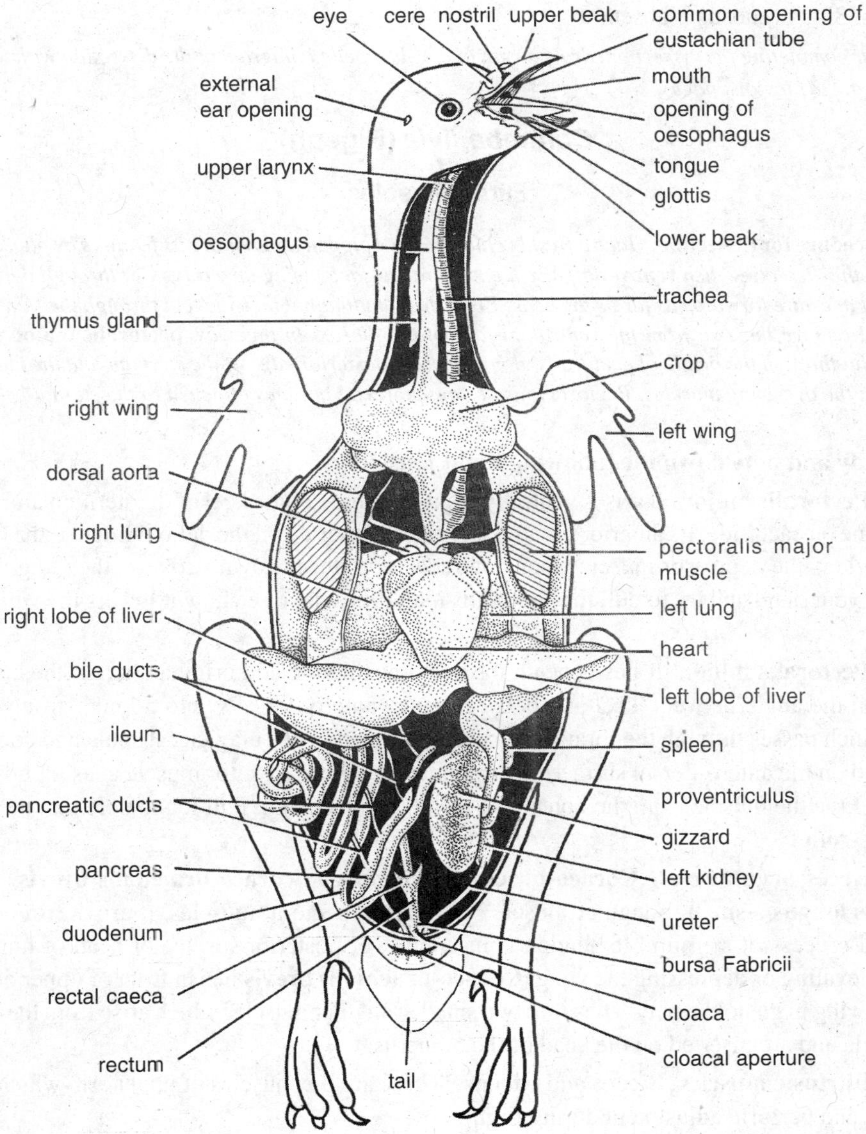

Fig. 4.22. Dissection of general anatomy of *Columba livia*.

3. Respiratory organs comprise the following organs :

(*i*) **Larynx** opens by glottis into pharynx. Larynx is voiceless chamber.

(*ii*) **Trachea** is long, flexible and lies ventral to the oesophagus. In thorax, it divides into two **bronchi** which pass to the two lungs. At the base of trachea lies the **syrinx** or **voice box**, that is concerned with sound production.

(*iii*) **Air-sacs** are membranous extensions of the main branchial branches. Nine air-sacs are found, *viz,* cervical air-sacs, interclavicular air-sac, anterior thoracic air-sacs, posterior thoracic air-sacs and abdominal air-sacs.

4. Circulatory organs :

Heart enclosed within pericardium, is four chambered consisting of two auricles and two ventricles. Sinus venosus is absent.

After completing the dissection, draw a neat and well-labelled diagram of the dissection with the help of figure No. 4.22 (previous page).

Columba livia (Pigeon)

Flight Muscles

Procedure for dissection. *Take a freshly chloroformed* **pigeon** *and remove its feathers by plucking them. Place it in the dissecting dish* **ventro-dorsally**, *i.e., ventral surface facing upwards. Cut through the skin from the cloacal aperture forwards as far as the beak. From this longitudinal incision, cut through the skin along the wings, and remove the skin from the ventral surface of the thorax. In this way,* **pectoralis major muscle** *is exposed. Cut through the origin of pectoralis along the keel, posterior edge of the sternum and the clavicle and free it from the overlying muscles. Pectoral muscles are powerful muscles attached to the sternal keel and the wings.*

Study and note down the following flight muscles :

1. Pectoralis major. It arises ventrally from the whole of the keel of the sternum and from the clavicle one on each side. Its anterior broad flat tendon is inserted on the deltoid ridge of the humerus. This muscle is the depressor muscle when contracts it produces down stroke of the wing. Hold the pigeon in your hand and try to pull the pectoralis major muscle to see how it lowers the wing **(down stroke)**.

2. Pectoralis minor. It lies beneath the pectoralis major, and originates from the keel of the sternum in the angle between keel and body. It tapers antero-dorsally into a long, strong and thin tendon which passes through the foramen triosseum formed by the clavicle, scapula and coracoid, to be inserted on the antero-dorsal surface of the humerus. It is an elevator muscle causing upstroke of the wing. Free the muscle from the bones like the pectoralis major. Pull it and note that it raises the wing **(upstroke)**.

3. Accessory muscles. Coraco-brachialis longus and **coraco-brachialis brevis**. **Coraco-brachialis longus** a small triangular muscle lies beneath pectoral muscles. It arises from coracoid and costal process of sternum. Its tendon is inserted over posterior surface of head of humerus. It helps in elevating or depressing the wing. **Coraco-brachialis brevis** lies in front of upper one. They rotate the wing in glenoid cavity. These are two small triangular muscles which arise from the coracoid and scapula and are inserted on the head of the humerus.

4. Intrinsic muscles. Biceps and **triceps.** These are the muscles of upper arm which operate the elbow and perform adjustment during flight.

Extensor carpi radialis and **extensor carpi ulnaris.** These are the muscles of the forearm, help to stretch or fold the wing.

Brachio-radialis. These muscles rotate medially the radius bone of the forearm.

Digital muscles of hand move the digits and feathers individually. **Policis muscles** attached to the first digit move the bastard wing.

5. Tensor muscles. Three muscles called **tensor longus**, **tensor brevis** and **tensor accessorius** keep the prepatagium stretched when the wing is extended in flight. A similar **tensor posterioris alae** keeps the postpatagium tensed during flight.

After completing the dissection, draw a neat and well-labelled diagram of the dissection with the help of the figure No. 4.23.

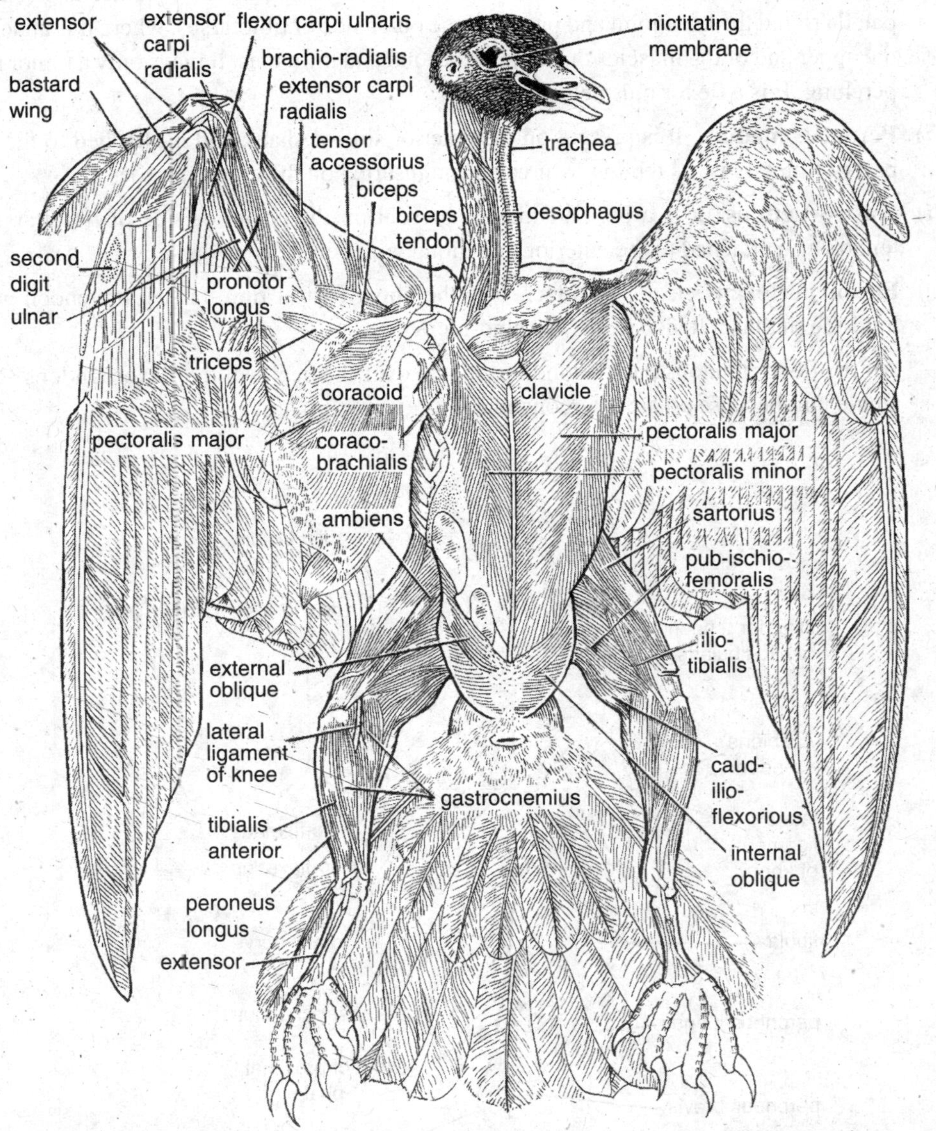

Fig. 4.23. Pigeon dissected from the ventral surface showing flight muscles.

Columba livia (Pigeon)

Perching Muscles

Procedure for dissection. *Take a freshly chloroformed* **pigeon** *and remove the feathers from the hind-legs. Remove the skin by giving a median longitudinal incision throughout the length of the leg and transverse incision at the base* of *the leg.*

Study and note down the following perching muscles :

There are following muscles which control the perching mechanism of the pigeon.

1. Flexor muscles. The calf muscles are chiefly responsible for bringing about the flexion of the toes during perching and also form an elaborate system of tendons attached to the phalanges. Following are the calf muscles :

(*i*) **Ambiens.** It is the main muscle that brings about the flexion of the toes. It arises from the anterior part of pubis and lies on the inner side of the thigh. Its tendon extends beneath the patella round the knee joint and passes to the outer side of tibiotarsus, where it is attached to the upper end of the muscles that flex the second and third toes. It plays only a minor role in perching. It is a flexor muscle.

(*ii*) **Peroneus medius.** It is present on the anterior side of shank and is attached to the upper part of tibiotarsus. Its tendon trifurcates and inserted on three anterior toes.

(*iii*) **Gastrocnemius.** It is a big calf muscle present on the posterior side of tibio-tarsus. Its tendon also attached to the anterior toes along with peroneus medius.

(*iv*) **Flexor perforans.** It is also attached to the upper part of tibio-tarsus. Its tendon goes to hind-toe (hallux). It also flexes the hind-toe.

All the flexor tendons pass behind the shank, over the heel. All the flexor tendons act as a single unit because all are interconnected.

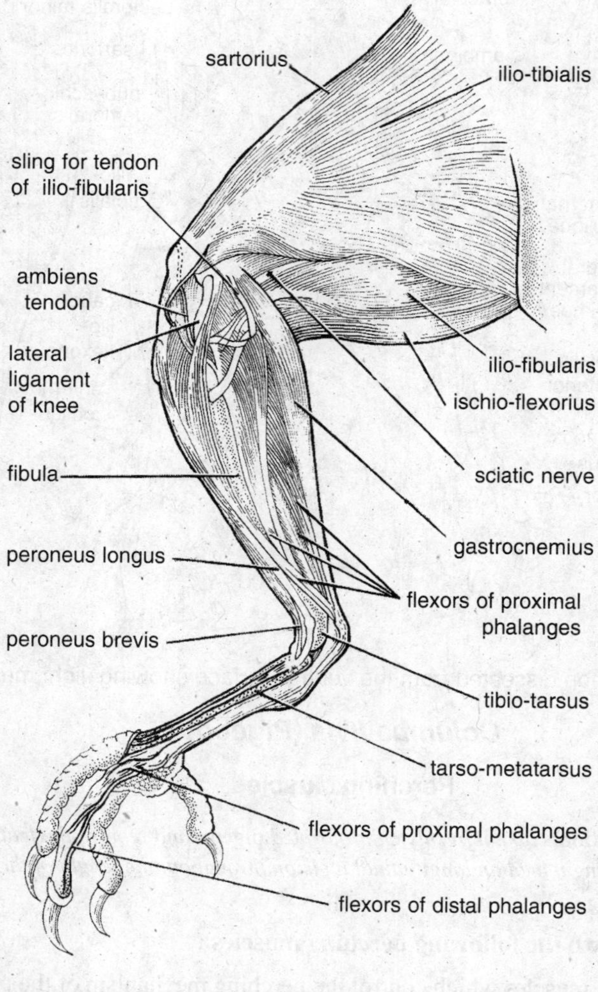

Fig. 4.24. Leg of pigeon dissected from the lateral side to show the perching muscles.

2. **Extensor muscles.** A number of extensor muscles are present on the front side of tibiotarsus. **Tibialis anterior muscles** tendon runs down in front of inter-tarsal joint and gives off three branches, each passes to each anterior toe. They are attached to upper surfaces of phalanges. Their contraction opens the toes.

After completing the dissection, draw a neat and well-labelled diagram of the dissection with the help of the figure No. 4.24 (previous page).

Columba livia (Pigeon)

Venous System

Procedure for dissection. *The procedure for dissection is more or less similar to that of the general anatomy. After exposing the heart, remove the pericardium and cut away the fat and connective tissue from the chief vessels at the base of heart. Heart is four chambered, consisting of two auricles and two ventricles. Expose the roots of* **veins** *entering the heart. Identify the following veins and expose them carefully :*

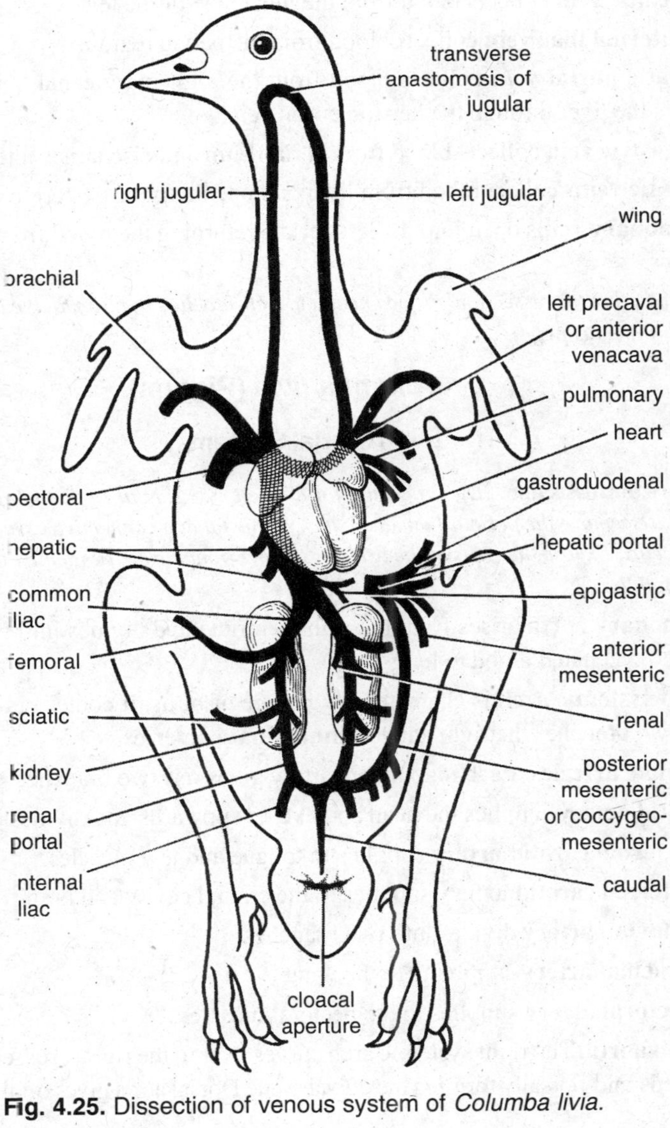

Fig. 4.25. Dissection of venous system of *Columba livia*.

1. **Anterior venacava** or **precaval** is formed by the following three chief veins :

(*a*) **Jugular vein** collects blood from the head and neck. The right and left jugular veins unite in the neck forming **transverse anastomosis of jugular**.

(*b*) **Brachial veins** collect blood from the wings.

(*c*) **Pectoral veins** collect blood from pectoral muscles.

2. **Posterior venacava** collects blood from the posterior part of the body and comprises the following veins :

(*i*) **Common iliac veins** are formed by the following veins :

(*a*) **Caudal vein** collects biood from the tail which divides into the following veins :

(*i*) **Renal portal veins** enter the kidneys and divide into renal veins.

(*ii*) **Posterior mesenteric** or **coccygeomesenteric vein** collects blood from the intestine.

(*b*) **Femoral vein** collects blood from the outer side of the leg.

(*c*) **Sciatic vein** collects blood from the inner side of the leg.

(*d*) **Internal iliac vein** collects blood from the pelvic region.

(*ii*) **Hepatic portal vein** collects blood from the alimentary canal, pancreas and spleen and enters the liver through the hepatic portal vein.

(*iii*) **Epigastric vein** collects blood from peritoneum and discharges into hepatic vein.

(*iv*) **Hepatic veins** collect blood from liver and open into the posterior venacava.

3. **Pulmonary veins** open into the left auricle returning the blood from the lungs. These form four short veins.

After completing the dissection, draw a neat and well-labelled diagram of the dissection with the help of figure No. 4.25 (previous page).

Columba livia (Pigeon)

Arterial System

Procedure for dissection. *The procedure for dissection is similar to that of the general anatomy. After exposing the heart, remove the pericardium and cut away the fat and connective tissue from the chief vessels at the base of the heart. Expose the roots of the arteries arising from the heart. Identify the following arteries and expose them carefully :*

1. **Pulmonary aorta** arises from the right ventricle and divides into two pulmonary arteries which carry deoxygenated blood to lungs.

2. **Right systemic arch** is characteristic feature of birds. It comes out from the left ventricle and gives off two branches the right and left **innominate arteries**.

Innominate arteries. Each innominate artery gives rise two branches.

(*i*) **Carotid artery** supplies the head and extends upwards and divides into two branches :

(*a*) **External carotid artery** supplies to tongue and jaw muscles.

(*b*) **Internal carotid artery** supplies to the cranial cavity and brain.

(*ii*) **Subclavian artery** divides into two branches.

(*a*) **Brachial artery** supplying to the wing.

(*b*) **Pectoral artery** supplies to the pectoral muscles.

3. **Dorsal aorta.** The right systemic arch curves over to the right, crosses the pulmonary artery and then proceeds mid-dorsally forming the dorsal aorta. Dorsal aorta gives off the following branches :

(i) **Coeliac artery** supplies to stomach and part of intestine.

(ii) **Anterior mesenteric artery** supplies to the intestine.

(iii) **Anterior renal arteries** supplying the anterior lobes of kidneys.

(iv) **Femoral arteries** supply to the outer sides of the legs.

(v) **Sciatic arteries** supply to the inner side of the legs. Each sciatic artery gives off two renal arteries supplying to the middle and posterior kidneys.

(vi) **Posterior mesenteric artery** supplies to the rectum and cloaca.

(vii) **Internal iliac arteries** supply to the pelvis.

(viii) **Caudal artery** supplies to the tail.

After completing the dissection, draw a neat and well-labelled diagran of the dissection with the help of figure No. 4.26.

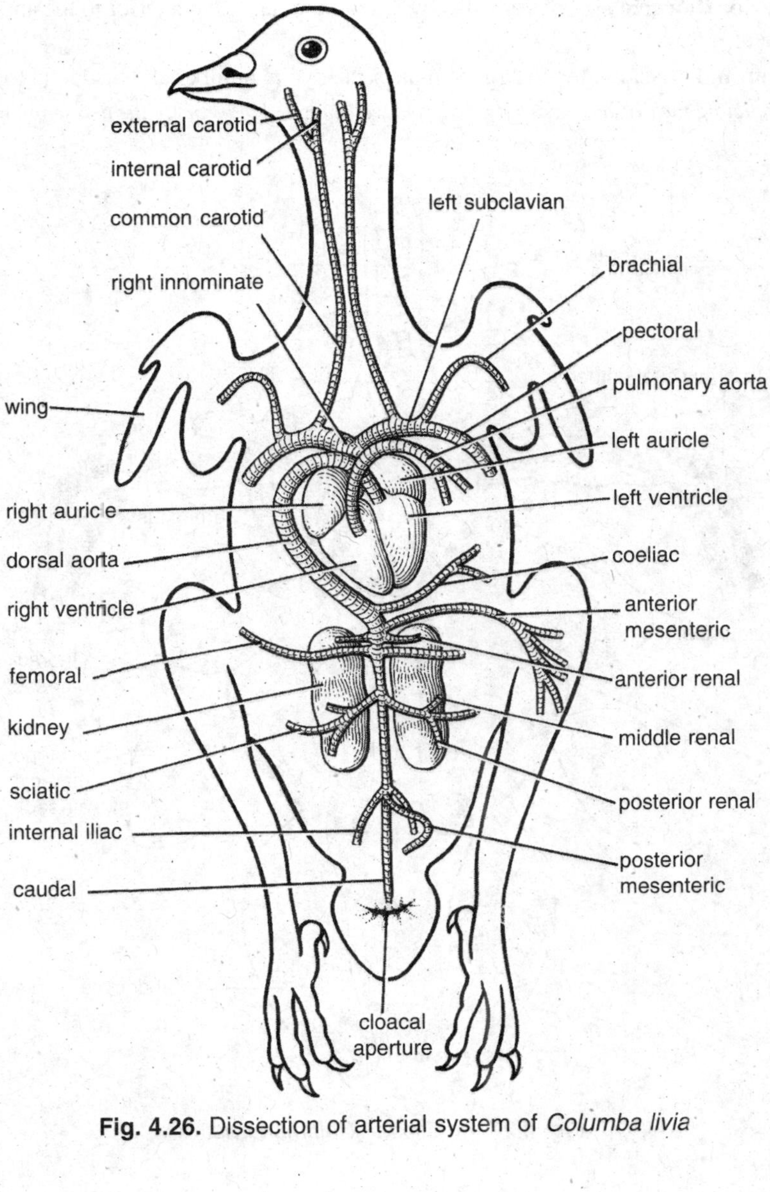

Fig. 4.26. Dissection of arterial system of *Columba livia*

Columba livia (Pigeon)

Air-sacs

Procedure for dissection. *Take a freshly chloroformed* **pigeon** *and remove its feathers completely. Put it in dissecting dish* **ventro-dorsally**, *i.e., ventral surface facing upwards. Make a mid-ventral incision in the neck and stretch the flaps of skin on both sides and pin down them. Open the trachea, insert a blow pipe in it by making a small cut and then inflate the air sacs, tie a ligature and cut longitudinally in the mid-ventral line through the abdominal wall and now locate the following air sacs :*

1. **Cervical air-sacs** are two in number, lying at the base of the neck.
2. **Interclavicular air-sac** is single and lies between two clavicles. It has two lateral branches called axillary air-sacs which communicate with the cavities of both the humerus.
3. **Anterior thoracic air-sacs** are two in number and cover the ventral surfaces of the anterior parts of lungs.
4. **Posterior thoracic air-sacs** are also two in number and lie posterior to the anterior thoracic air-sacs.
5. **Abdominal air-sacs** are the largest of air-sacs, two in number and lie dorsal to the intestine.

After completing the dissection, draw a neat and well-labelled diagram of the dissection with the help of figure No. 4.27.

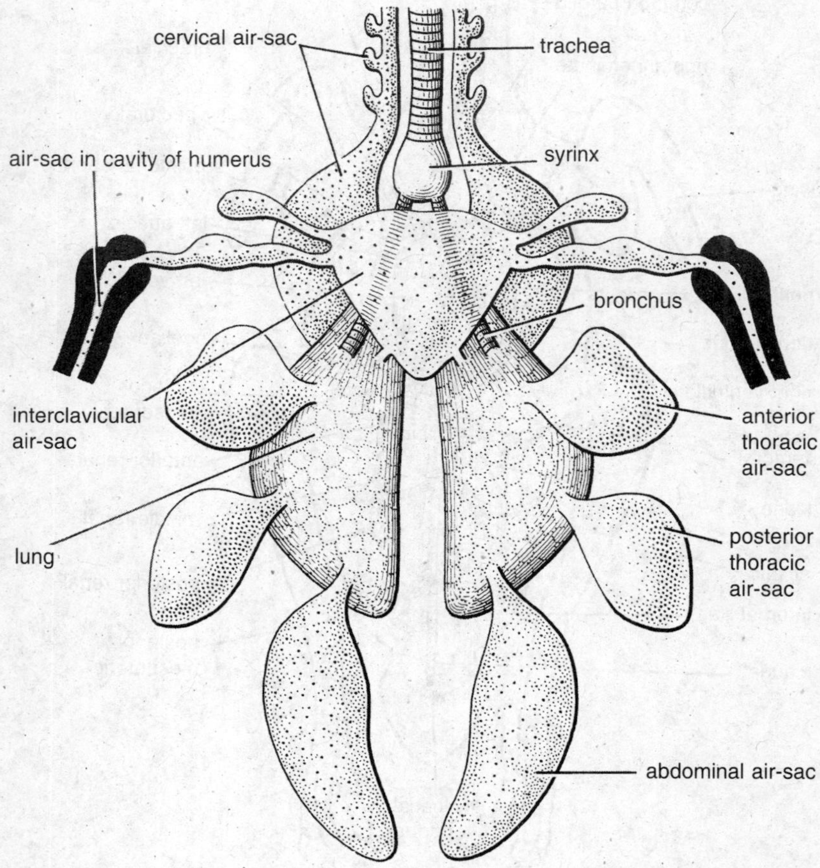

Fig. 4.27. Air sacs of *Columba livia.*

Columba livia (Pigeon)

Pecten

Procedure for dissection. *Take out the complete eye-ball of* **pigeon** *by cutting the eye-muscles. The muscles can be cut or removed with the help of a scalpel or a pair of scissors. Now locate the optic nerve entering the eye-ball and cut the eye-ball into two halves, i.e., one anterior half and one posterior half, at the point of entering the optic nerve, the* **pecten** *is seen hanging obliquely. It is a thin black structure. Take it out with the help* of *forceps and scissors.*

Pecten is composed of vertical vascular and pigmented plates arising from the axis. It projects into the cavity of the eye from its orbital surface ventral to blind spot at the site of the entrance of the optic nerve. It is folded longitudinally like a fan.

After taking out the pecten from the eye, study it and draw a well lebelled diagram of it with the half of figure No. 4.28.

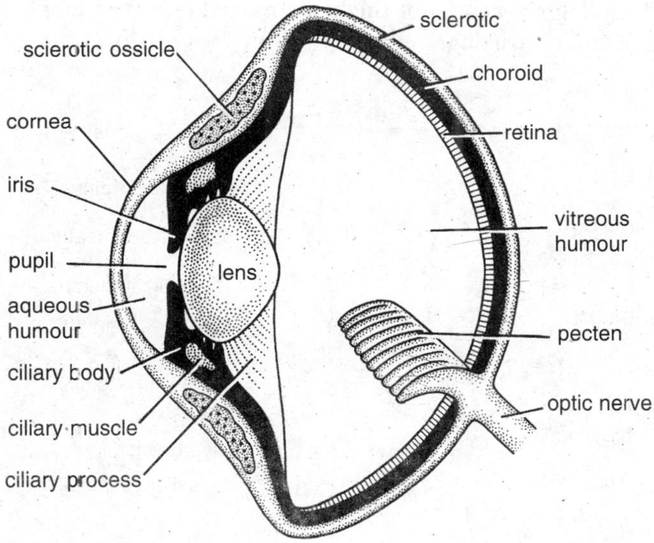

Fig. 4.28. Dissection of eye of *Columba livia* showing the pecten.

Rattus rattus (Rat)

External Features

Take a freshly chloroformed **rat** *and immerse it in the solution of some antiseptic. Lay down it in the dissecting dish.*

Examine the external features and note down the following :

Size. Rat measures about 25 cm including tail.

Shape and texture. The body of rat is slender with thick-set fine fur mixed with stiff hairs.

Colour. It is usually grayish or brownish in colour.

Division of body. Body is divisible into four distinct regions : **head**, **neck**, **trunk** and **tail**.

1. Head. The head is elongated, somewhat conical in shape and tapering anteriorly to a terminal blunt snout. The head bears the following structures :

(*i*) **Nostrils.** A pair of nostrils or nares is situated at the tip of the fleshy nostrils.

(*ii*) **Mouth.** It is somewhat ventral in position. It is bounded by an **upper** and a **lower** lip. Upper lip is also called hair lip because it is divided into two lateral halves by a median vertical cleft.

(*iii*) **Eyes.** There are two eyes, one on each side. These are dorso-lateral, situated about mid-way between the muzzle and the base of ears. Each eye is provided with movable upper and lower eyelids bearing eye lashes. Movable nictitating membrane is also present at the anterior corner of the eye.

(*iv*) **External ears or pinna.** There are two external ears or pinna lying at the postero-lateral sides of the head.

(*v*) **Vibrissae.** These are long hairs developed on the lateral sides of the upper lip and near eyes.

2. Neck. It is a short flexible region joining the head to the trunk.

3. Trunk. It is the largest part of the body. It is divided into the **thorax** and **abdomen**. Trunk bears two pairs of limbs, *viz.,* short **forelimbs** and long **hindlimbs.** It also bears certain apertures.

(*i*) **Forelimbs.** The forelimbs include the upper arm, forearm and the hand. The hand comprises the palm and the digits five in number. The first digit or thumb is greatly reduced.

(*ii*) **Hindlimbs.** The hindlimbs consist of **thigh**, **shank** and **foot**. The foot bears four clawed digits. First toe (hallux) is lacking.

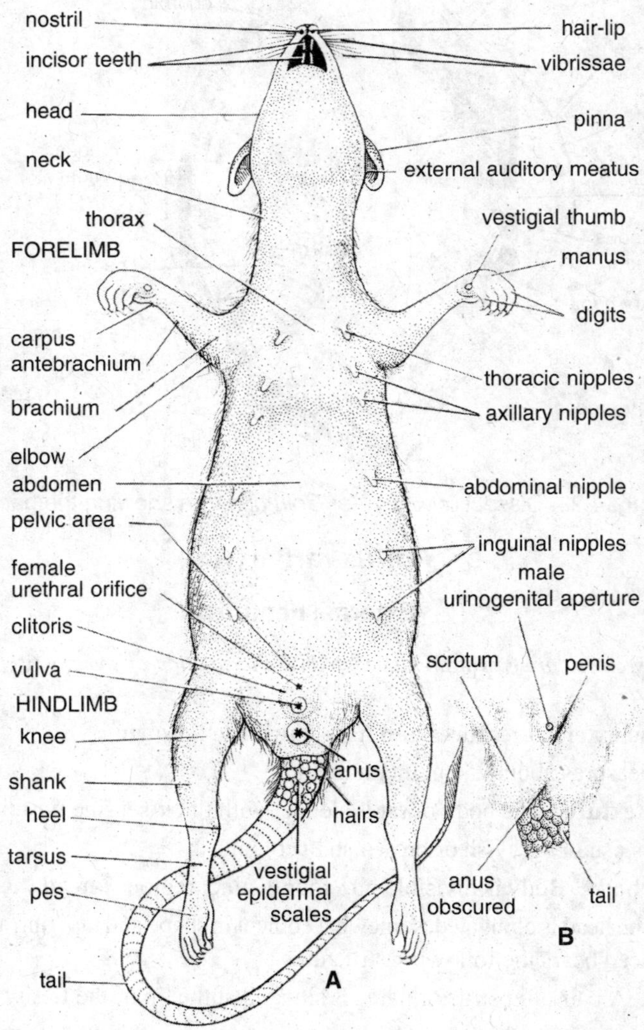

Fig. 4.29. External features of *Rattus rattus.* A—Female rat in ventral view; B—Pelvic region of male rat in ventral view.

(*iii*) **Nipples.** In female, one pair of thoracic nipples, two pairs of axillary nipples and one pair of abdominal nipples and two pairs of inguinal nipples are present on the ventral surface of the body.

(*iv*) **Urethral opening and vulval opening.** In female, urethral opening and vulval opening and anus lie close together in the pelvic region.

(*v*) **Penis and testes.** In male, the penis is cylindrical erectile organ in front of anus, which is enclosed in a loose sheath of skin, the **prepuce**. Penis bears opening of urinogenital canal. Two large testes are situated in the scrotal sacs.

(*vi*) **Anus.** The anus is small median aperture below the base of the tail.

4. Tail. The tail is elongated, cylindrical and tapering and often slightly longer than body. It is nearly devoid of hair and covered with overlapping vestigial scales. It acts as a balancing organ.

After studying the external features of rat, draw a neat and well-labelled diagram of the ventral view of rat with the help of figure No. 4.29 (previous page).

Rattus rattus (Rat)

General Anatomy

Procedure for dissection. *Take a freshly chloroformed* **rat** *and immerse it in the solution* of *some antiseptic. Lay down it in the dissecting dish* **ventro-dorsally**, *i.e., ventral surface facing upwards. Fix it in the dissecting dish by pinning it in the limbs. Make a longitudinal incision in the skin from the pubic symphysis to the lower jaw and transversely at the anterior as well as posterior ends. Separate the skin from the muscular body-wall and note that two are connected by loose subcutaneous connective tissue. Reflect the flaps of skin and pin them down.*

Make a median longitudinal incision in the abdominal wall from the pubic symphysis to the xiphisternum taking care not to injure the internal viscera. Cut parallel to the ribs on both the sides reflect the flaps of abdominal wall and fix them down.

Now make a slight incision in the diaphragm with the tip of scissors. Thus, cut through the diaphragm so as to separate it from ribs all round. Make two longitudinal cuts through the ribs from behind forwards until you reach the first rib. Keep the first right rib intact but cut through the left one carefully without damaging any of the blood vessels of the neck and then reflect the cut median strip of the ribs and sternum to the left. Cut away a considerable part of the remainder of a thoracic wall on either side.

Study the general anatomy and note down the following :

Neck region. In the neck region, following structures are seen :

(*i*) Lymph nodes.

(*ii*) Extra-orbital lacrimal glands.

(*iii*) Submaxillary glands.

Thoracic region. In the thoracic region, following structures are seen :

(*i*) Heart is surrounded by the pericardium.

(*ii*) Thymus glands.

(*iii*) Lungs.

Abdominal region. In the abdominal region, following structures are seen :

(*i*) Liver, (*ii*) Stomach, (*iii*) Coils of intestine, (*iv*) Caecum, (*v*) Colon, (*vi*) Urinary bladder, (*vii*) Pancreas, (*viii*) Fat bodies, (*ix*) Vesicular gland, (*x*) Prostate, (*xi*) Penis, etc.

After completing the dissection, make a neat and well-lebelled diagram of the dissection with the help of figure No. 4.30 (next page).

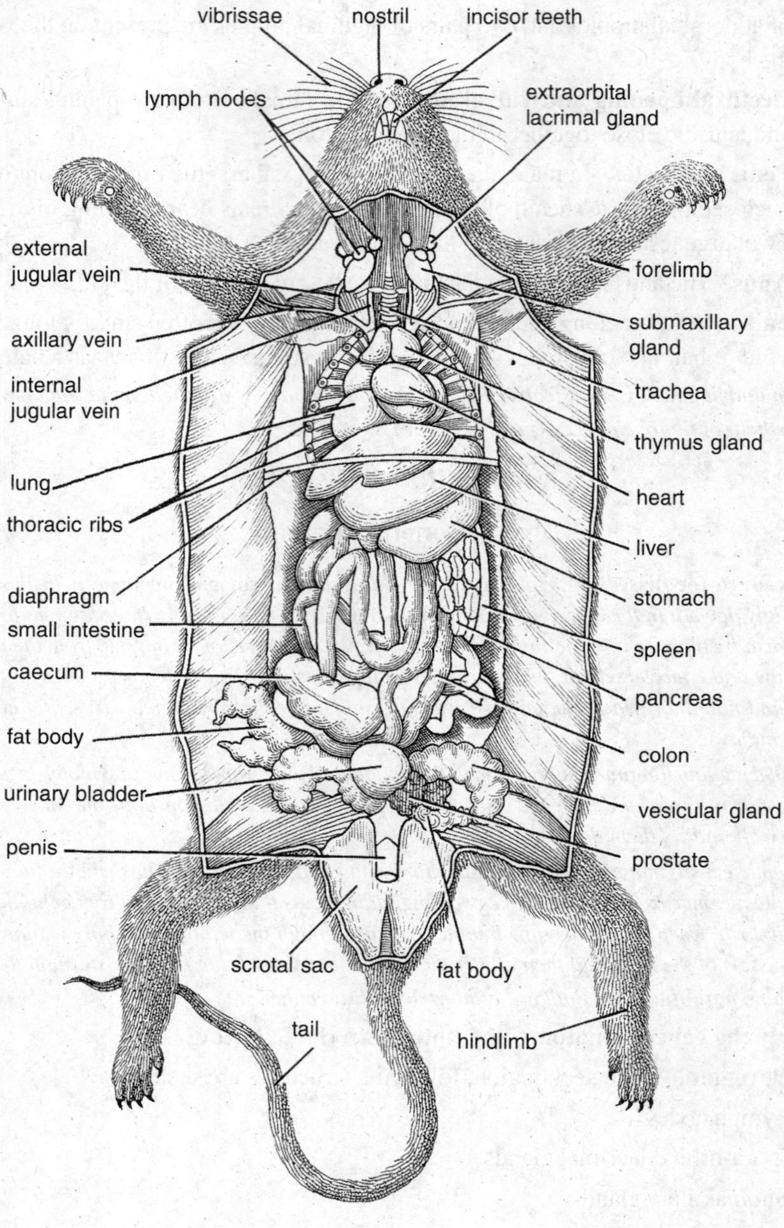

Fig. 4.30. Dissection of general anatomy of *Rattus rattus.*

Rattus rattus (Rat)

Venous System

Procedure for dissection. *The procedure for dissection of* **venous system** *is more or less the same as that of the general anatomy described earlier. After exposing the heart, remove the thymus gland and pericardium and cut away the fat and connective tissue from the main vessels at the base of the heart. Expose the roots of the veins entering the heart. Identify the following* **veins** *and expose them carefully :*

1. Anterior venacava is formed by the union of the following veins :

(*i*) **Internal jugular vein** collects blood from the brain, and occipital region and runs in the neck.

(*ii*) **External jugular vein** collects blood from the head, neck and shoulders.

(*iii*) **Subclavin veins** collect blood from the forelimbs and shoulders.

(*iv*) **Internal mammary vein** collects blood from the mammary glands.

(*v*) **Anterior intercostal vein** collects blood from anterior five to six intercostal spaces.

(*vi*) **Anterior phrenic vein** collects blood from diaphragm.

(*vii*) **Azygos vein** collects blood from the posterior 8 or 9 intercostal spaces of both the sides.

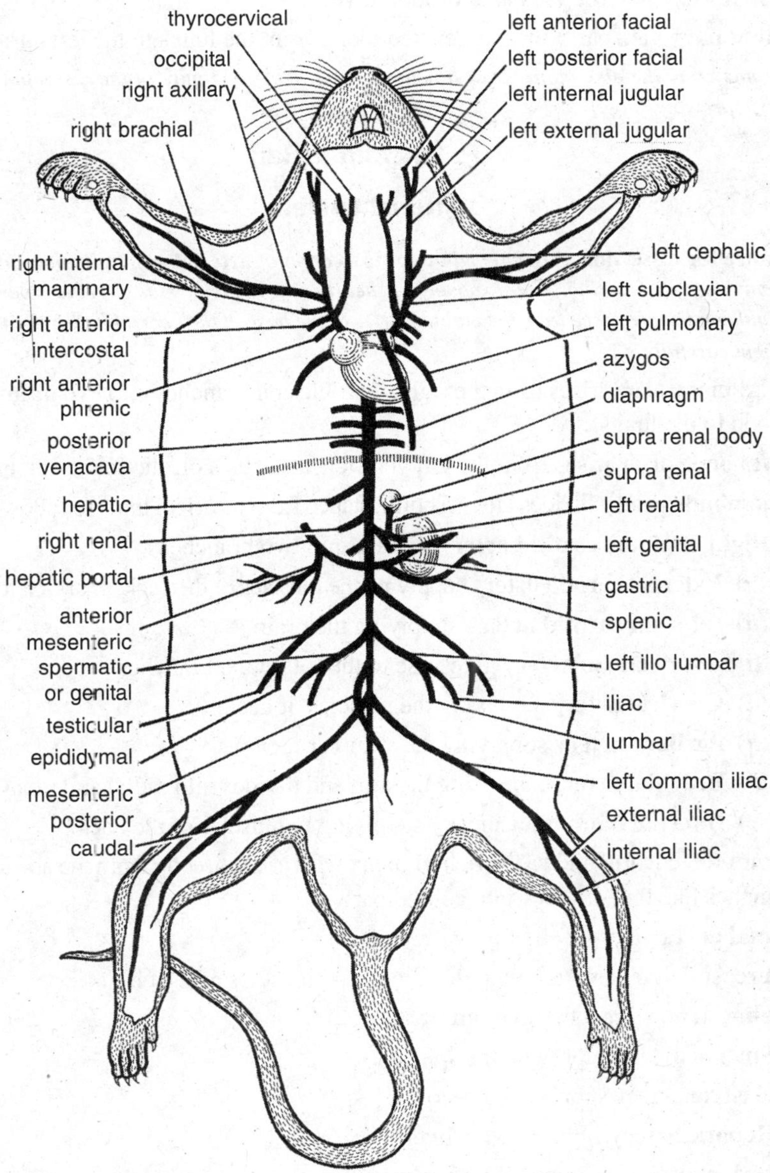

Fig. 4.31. Dissection of venous system of *Rattus rattus*.

2. **Posterior venacava** is formed by the union of the following veins :

(*i*) Two pairs of **hepatic veins** collect blood from liver.

(*ii*) A pair of **renal veins** collect blood from kidneys.

(*iii*) Paired **spermatic** or **genital veins** collect blood from gonads.

(*iv*) A pair of **ilio-lumbar veins** collects blood from the lumbar region.

(*v*) **Common iliac veins** collect blood from the legs and are formed by the union of **external iliac** and **internal iliac**.

(*vi*) **Caudal vein** collects blood from the tail.

3. **Hepatic portal vein** is formed by the union of **anterior** and **posterior mesenteric veins**, **gastric vein**, **splenic vein** and **duodenal vein,** etc.

4. **Pulmonary vein** carry the oxygenated blood from the lungs to the left auricle.

After completing the dissection, draw a neat and well-lebelled figure of the dissection with the help of figure No. 4.31 (previous page).

Rattus rattus (Rat)

Arterial System

Procedure for dissection. *The procedure for dissection of* **arterial system** *is the same as that for the general anatomy described earlier. After exposing the heart, remove the thymus gland and pericardium and cut away the fat and connective tissue from the main arteries at the base of the heart. Identify the following arteries and expose them carefully* :

1. **Pulmonary arch** arises from the right ventricle and branches into two **pulmonary arteries** to supply blood to the lungs.

2. **Left aortic arch** arises from the left ventricle and gives off the following branches :

(*i*) **Innominate artery** divides after a short distance from its origin into the following branches :

(*a*) **Right common carotid artery** divides into two branches :

(*i*) **External carotid artery** supplying the muscles of the neck, pharynx, tongue and jaws.

(*ii*) **Internal carotid artery** supplying the brain.

(*b*) **Right subclavian artery** gives rise to the following branches :

(*i*) **Cervical artery** supplying the muscles of neck.

(*ii*) **Axillary artery** supplying the scapular region.

(*iii*) **Brachial artery** supplying the arm and divides into radial and ulnar arteries.

(*iv*) **Internal mammary artery** supplying to muscles of the back.

Left common carotid and **left subclavian arteries** arise directly from the aortic arch and have the same branches like that of their right counterparts.

3. **Dorsal aorta** gives off the following branches :

(*i*) **Intercostal arteries** supplying the intercostal muscles of the chest.

(*ii*) **Coeliac artery** branches into three arteries :

(*a*) **Splenic artery** supplying the spleen.

(*b*) **Gastric artery** supplying the stomach.

(*c*) **Hepatic artery** supplying the liver.

(*iii*) **Anterior mesentric artery** supplying the intestine.

(*iv*) **Renal arteries** supply the kidneys.

(*v*) **Spermatic** or **genital arteries** supplying the gonads.

(*vi*) **Ilio-lumbar arteries** supplying the muscles of ilio-lumbar region.

(*vii*) **Posterior mesentric artery** supplying the colon and rectum.

(*viii*) **Common iliac artery** divides into two branches :

(*a*) **External iliac artery** supplying the outer part of thigh.

(*b*) **Internal iliac artery** supplying the muscles of inner part of thigh and shank.

(*ix*) **Caudal artery** supplies the tail.

After completing the dissection, draw a neat and well-labelled diagram of the dissection with the help of figure No. 4.32.

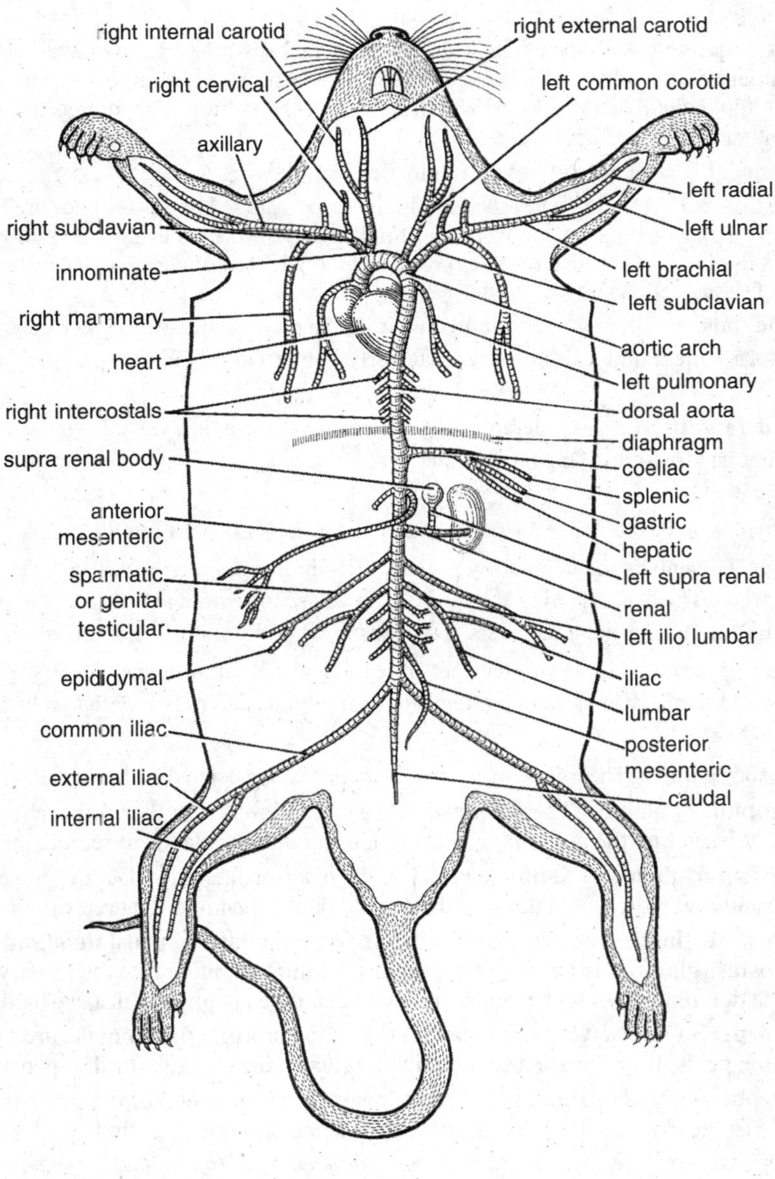

Fig. 4.32. Dissection of arterial system of *Rattus rattus.*

Rattus rattus (Rat)

Male Urinogenital Organs

Procedure for dissection. *Take a freshly chloroformed rat and immerse it in the solution of some antiseptic. Put it in the dissecting dish* **ventro-dorsally,** *i.e., ventral surface facing upwards. Fix it in the dissecting dish by pinning it in the limbs. Make a longitudinal incision in the skin from the pubic symphysis to the lower jaw and make transverse cuts at the ends of longitudinal cut. Separate the skin from the muscular body-wall and reflect the flaps of the skin and pin them down. Now make a median longitudinal incision in the abdominal wall from the pubic symphysis to the xiphisternum and reflect the flaps of the abdominal body-wall and fix them down. Pull the alimentary canal to one side and pin it down in the dissecting dish. Cut the bones of the sides of the pubic symphysis and remove cut pieces so as to expose the posterior part of the male urinogenital system.*

Study the male urinogenital organs and note down the following :

The urinogenital organs comprise the urinary and genital organs.

Urinary organs or Kidneys. The kidneys are dark brown and bean-shaped structures. The point of attachment of the kidney with the ureter is known as hilus. At the anterior end of each kindey is situated an adrenal gland. Cut the longitudinal section of the kidney and study its internal structure which shows outer cortex and inner medulla.

Genital organs. The genital organs comprise the following parts :

1. Testes. A pair of testes lies embedded in the scrotal sacs. The scrotal sacs are in the form of pouches lying in front of the anus between the hindlegs. Both the testes are exposed by cutting the scrotal sacs longitudinally. Each testis is an elongated and ovoid body attached to the posterior end of the sac by a band of tissue called gubernaculum.

2. Epididymis. It is an irregular and convoluted tube lying along the inner edge of the testis. Anteriorly it forms the caput epididymis, posteriorly cauda epididymis and in the middle corpus epididymis.

3. Vasa deferentia. Each vas deferens arises from the cauda epididymis. Both the vasa deferentia are elongated tubular structures lying on the inner side of the testes. They open into the urethra posterior to the urinary bladder.

4. Urethra. It is the common duct for sperms and urine and opens into the penis.

5. Penis. The penis is the copulatory organ which discharges the sperms into the vagina of the female genital organ. The penis is covered by loose skin whose loose fold at the tip end of penis is called **prepuce**. The urethra runs through the penis and opens at its lip by a slit-like aperture.

6. Spermatic cord. It is an elongated cord-like structure consists of connective tissue, blood vessels and nerves. The spermatic cord arises from the caput epididymis and runs into the body cavity through the inguinal canal.

7. Accessory glands. The following accessory glands are associated with the male genital system :

(*i*) **Ampullary glands.** The outer end of the vas deferens near the entrance into the urethra is enclosed into the ampulla which contains ampullary glands to secrete the mucus.

(*ii*) **Vesicular glands** or **seminal vesicles.** In front of the ampullae are present two large vesicular glands, which through their small ducts open into the urethra.

(*iii*) **Prostate gland.** The prostate gland comprises the **lateral prostate gland** and **ventral prostate gland**. The lateral gland lies just behind the junction of the urinary bladder and vasa deferentia, while the ventral prostate gland lies below the urinary bladder.

(*iv*) **Cowper's glands.** Cowper's glands are two in number, arise from the urethra at the base of the penis, they secrete a fluid which lubricates the passage for the sperms.

(*v*) **Preputial glands.** These glands arise near the terminal end of the penis from the inner fold of the skin, the prepuce. They secrete some odorous secretion.

After completing the dissection, draw a neat and well-labelled diagram of the dissection with the help of figure No. 4.33 (next page).

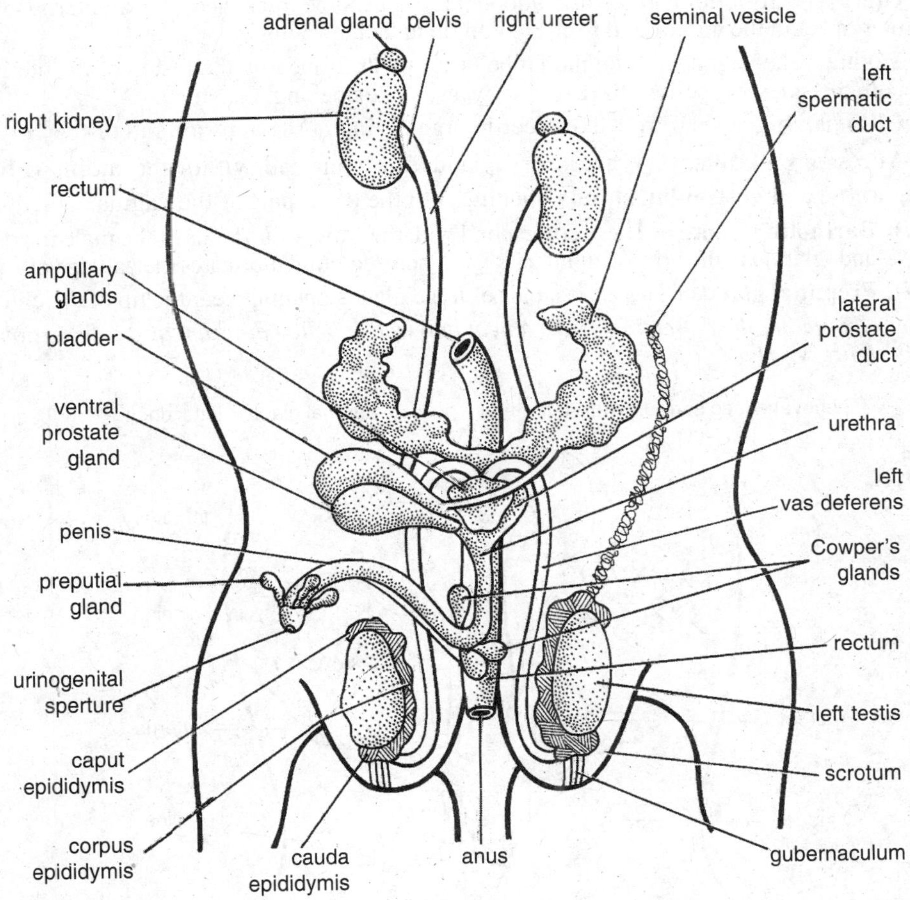

Fig. 4.33. Dissection of male urinogenital system of *Rattus rattus.*

Rattus rattus (Rat)

Female Urinogenital Organs

Procedure for dissection. *Take a freshly chlorformed* **rat** *and immerse it in the solution of some antiseptic. Fix it in the dissecting disc* **ventro-dorsally**, *i.e., ventral surface facing upwards by pinning it in the limbs. Make a longitudinal incision in the skin from the pubic symphysis to the lower jaw and make transverse cuts at the ends of longitudinal cut. Separate the skin from the muscular body-wall and reflect the flaps of the skin aud pin them down. Now make a longitudinal incision in the abdominal wall from the pubic symphysis to the xiphisternum and reflect the flaps of the abdominal body-wall and pin them down. Pull the alimentary canal to one side and pin it down in the dissecting dish. Cut the bones of the sides of the pubic symphysis and remove the cut pieces so as to expose the posterior part of the female urinogenital system.*

Study the female urinogenital organs and note down the following :

The urinogenital organs constitute the urinary and genital organs.

Urinary organs. The urinary organs comprise a pair of kidneys, ureters and urinary bladder which are like those of male.

Genital organs. The genital organs comprise the following parts :

1. **Ovaries.** The ovaries are small, multilobed and compact yellowish bodies of the size of pea situated on the outer side of the kidneys. These are suspended in the body cavity by a fold of peritoneum, the mesovarium.
2. **Fallopian tubes.** The fallopian tubes are narrow, coiled and convoluted tubes. Their anterior ends dilated into the oviducal funnel. They convey the ova to the uteri.

3. **Uteri.** The posterior ends of the fallopian tubes become thickened to form uteri. Both the uteri meet in the middle and form a common tube, the vagina.
4. **Vagina.** The vagina is a common tube or chamber formed by the fusion of the uteri from both the sides. It opens to the exterior by a slit-like opening, the vulva.
5. **Clitoris.** It is a small rod-like erectile organ lying at the anterior end of the vulva.
6. **Accessory glands.** The accessory glands are confined within the uterus **(uterine glands)** and **vestibular glands** opening into the outer part of the vagina.
   (*i*) **Bartholin's glands.** These correspond with the Cowper's glands of the male in position and relation to the urino-genital passage. Their secretion lubricates the passage of vagina.
   (*ii*) **Preputial glands.** These are a pair of large glands opening near the tip of the clitoris.

After completing the dissection, make a neat and well-labelled diagram of the dissection with the help of figure No. 4.34.

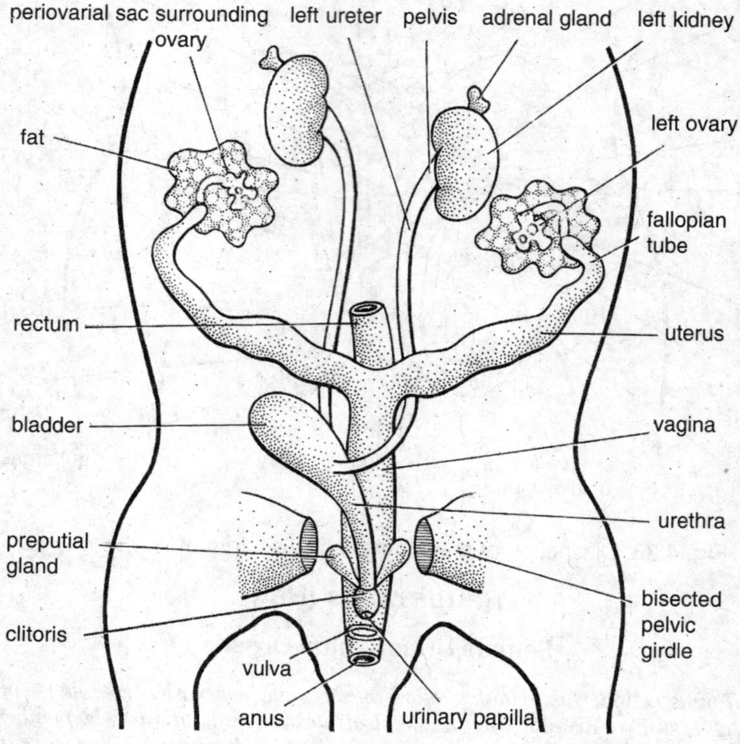

Fig. 4.34. Dissection of female urinogenital organs of *Rattus rattus.*

Rattus rattus (Rat)

Neck Nerves

Procedure for dissection. *Take a chloroformed* **rat** *and put it in a dissecting dish* **ventro-dorsally**, *i.e., ventral surface facing upwards. Fix it by pinning through the limbs. Remove the skin from the neck and thoracic region. Expose the hyoid and remove it. After removing the hyoid locate common carotid artery and note the nerves present on its sides. Expose the vagus nerve present on its outer side and then trace following nerves :*

(*i*) **Vagus verve (X)** leaves the skull by the foramen lacerum posterius, has a distinct ganglion near its origin. It runs posteriorly along the neck and the thorax into the abdomen and innervates the lungs, oesophagus, stomach, etc.

(*ii*) **Anterior laryngeal nerve** (Superior laryngeal) arises from the vagus, close to the vagus ganglion ending in the larynx and thymus.

(*iii*) **Recurrent laryngeal nerve** (Posterior laryngeal) arises from the vagus nerve just above the heart and runs alongside the trachea to end in the muscles of larynx.

(*iv*) **Cardiac depressor nerve** arises from the anterior laryngeal and runs backwards along the neck and innervates the heart.

(*v*) **Cervical sympathetic nerve** is running between the cardiac depressor and vagus nerves. It runs throughout the length of the body up to tail on the lateral sides of vertebral column. It forms **anterior cervical ganglion** anteriorly, **middle** and **posterior cervical ganglia** posteriorly, *i.e.*, 3 ganglia in neck region.

(*vi*) **Phrenic nerve** arises from the fourth cervical spinal nerve and innervates the diaphragm of its side.

After completing the dissection, insert the black paper below the neck nerves. Draw a neat and well-labelled diagram of the dissection with the help of figure No. 4.35.

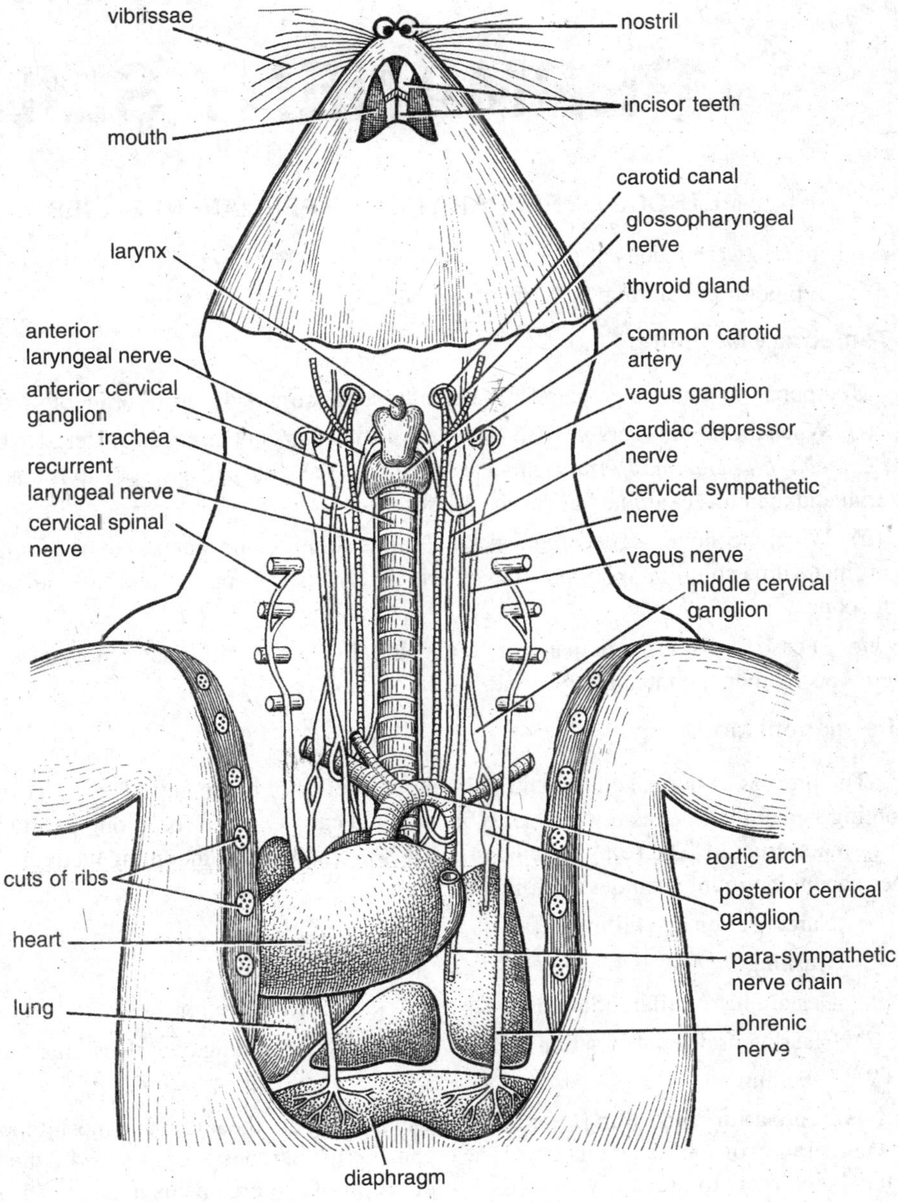

Fig. 4.35. Dissection of neek nerves of *Rattus rattus*.

Chapter 5

PREPARATION OF PERMANENT SLIDES

METHODS OF PREPARATION OF PERMANENT SLIDES

There are two methods of preparation of permanent slides, *viz.,*

1. Temporary mounting; 2. Permanent mounting.

1. Temporary Mounting

Temporary mounts are prepared in water, normal saline and glycerine, or Iugol's iodine solution.

(*i*) When the mount is prepared in water, a drop of water containing the material to be examined, *e.g., Amoeba, Euglena and Paramecium* etc., is kept on the slide and a cover slip is placed up on the material and then it is examined under the microscope.

(*ii*) When the mount is prepared in normal saline solution, the material to be examined is put in the normal saline solution upon the slide on which cover slip is placed and then studied under the microscope.

(*iii*) In case of glycerine preparation, the material is first of all washed in water and stained in aqueous eosin, then mounted in glycerine.

2. Permanent Mounting

The process of permanent mounting involves a series of treatments. The utility of permanent mounting is that the organism which is to be mounted can be stored for a long period without any change in its structure.The **Canada balsam** or **DPX** is employed as mounting medium. The process of permanent mounting includes the following sreps :

1. Narcotisation and killing ;
2. Fixing ;
3. Washing ;
4. Staining ;
5. Destaining or differentiation ;
6. Dehydration ;
7. Clearing or dealcoholisation ;
8. Mounting over the slide ;
9. Labelling.

1. Narcotisation and killing. Narcotisation and killing are the initial and most important stages in the preparation of permanent mounts. The organisms are narcotised or killed with the use of some chemicals in order to prevent the changes in the form of the organisms as they had in the living condition during their life.

The narcotics used are absolute alcohol, chloroform, menthol, ether, cocaine hydrochloride, acetone or by passing a current of CO_2 through the water containing the organism.

For killing the organisms, two methods are employed depending upon the purpose of study :

1. If the specimen is to be dissected, chloroform is generally used to kill the organism.

2. If the specimen is to be converted into permanent microscopical preparation where only small organisms or small pieces or larger animals are concerned—a reagent is employed which not only kills but penetrates and fixes the tissue so that they are slightly affected by subsequent treatment.

In some cases, smaller organisms are killed by heating. This method of killing also brings about the simultaneous fixation of the organisms. There are some of the chemicals which perform both the functions of killing and fixing, *e.g.*, formol sublimate mixture.

2. Fixing. Fixing is most essential for all types of microscopical preparations, *e.g.*, whole mounts and sections of parts of the body. The fixation of tissues and entire organisms is done by various types of fixatives. The important functions of the fixatives or fixing fluids are as follows :

1. The fixing fluids penetrate the tissues by inward diffusion so that the tissues can be preserved as nearly as possible in natural shape and check the further post-mortem changes.

2. The fixatives harden the tissues and set intra-cellular elements.

3. The fixing fluids render insoluble the various cell constituents and bring out differences in the refractive indices of different portions.

Some of the fixing fluids or fixatives or fixing agents are alcohol, 90% alcohol + glycerine, formalin, corrossive sublimate, acetic acid, picric acid and osmic acid, etc., used in various combinations and strengths. The time of fixation may vary in different cases.

3. Washing. Washing of the tissue or organism, after fixation, is an essential process to remove the excess of fixative. If the fixative is not removed or washed completely from the tissue it may interfere in the subsequent processes of dehydration and staining. Washing is done by different liquids depending upon the type of fixative used. The aqueous Bouin's fluid is washed in running tap-water, while alcoholic Bouin's fluid is washed with the 70% alcohol or alcohol of the same strength which is present in the fixative. Formalin and other fixatives containing large amount of water are mostly removed by washing in distilled water. Sublimate is washed in alcohol.

4. Staining. By staining, the parts or components of cell or of organism can be clearly differentiated. In staining, the tissues or organisms are coloured by particular dyes. The dyes which are commonly used for staining in the laboratory are : borax carmine, picro-indigo carmine, methylene blue, aniline blue, haematoxylin, eosin, orange G, Victoria blue, Sudan black, Congo red and alizarin, etc. The stains can be classified into the following types :

(*i*) **Nuclear stains** are used to stain the nucleus of the cell, *e.g.*, **haematoxylin**.

(*ii*) **Cytoplasmic stains** are used to stain the cytoplasm, *e.g.*, **eosin**.

(*iii*) **General stains** are those which stain the entire cell or organism, *e.g.*, **borax carmine, picro-indigo carmine**, **methylene blue** and **orange G**, etc.

The stains are prepared in aqueous (water), alcoholic and acid media.

There are two types of staining, *viz.*, single staining and double staining.

Single staining. When only one stain, such as **borax carmine** or **picro-indigo carmine**, is used, the process is called single staining.

Double staining. When two stains, one nuclear stain such as haematoxylin and another cytoplasmic stain such as eosin are used, the process is called double staining.

5. Destaining or differentiation. Sometimes the tissues or organisms are overstained and render difficulty during observations under the microscope. To avoid this difficulty, excess of stain is removed.

This process of removing excess of stain is called **destaining** or **differentiation**. Destaining is done by acid water or acid alcohol. **Acid water** is used with the **aqueous stains**, while **acid alcohol** is used with the **alcoholic stains**.

6. Dehydration. Dehydration means the removal of water from the tissues. Dehydration is an essential process due to the following reasons :

1. It prevents the tissues from decomposition by removing the water and by killing the germs harboured by tissues.

2. It is necessary because the mounting medium Canada Balsam is not miscible with water and the dehydration renders the mounting medium to penetrate the tissue.

Dehydration is done by processing the tissues in various grades of alcohol. The need of passing the tissues in different grades of alcohol is to avoid the shrinkage and distortion of tissues by abrupt use of strong alcohol. Therefore, dehydration is carried out gradually by passing the tissues from lower grade of alcohol to subsequent higher grade of alcohol such as 30, 50, 70, 90 and 100 per cent. The dehydration is done in corked specimen or staining tubes. The time in each grade of alcohol is to be adjusted according to the thickness of tissue.

7. Clearing or dealcoholisation. After dehydration, clearing or dealcoholisation of tissues is necessary due to the following reasons :

1. To remove the alcohol from the tissue or organism and to facilitate the penetration of the mounting medium, which is mostly Canada Balsam.

2. To make the tissue or organism more clear and transparent so that very minute details can be observed under the microscope.

The clearing agents which are commonly used in the laboratories are xylol, benzene, clove oil and cedar wood oil, etc. Sometimes the turbidity is observed with the use of clearing agent, which indicates the presence of moisture, *i.e.,* incomplete dehydration. In such cases, tissue or organism should be again dehydrated in 90% and 100% alcohol for complete dehydration and then transferred in clearing agent.

Clove and cedar wood oil are the best clearing agents used for clearing the soft tissues and organisms.

8. Mounting. It is the last step in the permanent mounting and is essential for the study and preservation of the material for longer time. The commonest mounting medium is **Canada Balsam**. The use of Canada Balsam in the permanent mounting is due to the fact that the refractive index of Canada Balsam is the same as that of the glass of slide and cover slip.

Procedure of mounting. (*i*) Clean the slide and cover-slip with a neat cloth.

(*ii*) Determine the centre of the slide by drawing a diagram of the slide and sketching its diagonals as in fig. 5.1 A.

(*iii*) Put a drop of mounting medium (Canada Balsam) in the centre of the slide with a glass rod.

(*iv*) Transfer the tissue or organism from the clearing agent (xylol or clove oil) on the slide by a brush or forcep.

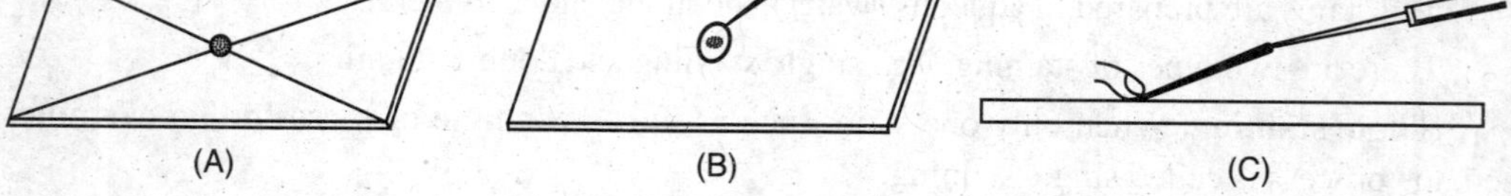

(A) (B) (C)

Fig. 5.1. Diagram showing the procedure of mounting.

(*v*) Take a cover slip and place its one edge on the slide touching the drop of Canada Balsam and hold its other edge obliquely either with needle or forcep.

(*vi*) Now gently lower down the needle or forcep till the cover slip completely touches the Canada Balsam and covers the material.

(*vii*) After mounting, the slide may be dried in hot chamber at 40°–50°C.

(*viii*) The excess of balsam which comes out around cover slip, should be washed with cotton soaked in xylol or in a mixture of xylol and absolute alcohol.

Precautions. 1. The slide, cover-slip, needle, brush, etc., should be cleaned.

2. The cover slip should be in the centre of slide.
3. The material should be in the centre of cover-slip.
4. There should be no air bubble in the Canada Balsam.
5. Excess of Canada Balsam should not come out beyond the cover-slip.

9. Labelling. After making the permanent mount, the slide should be properly labelled, *i.e.*, identification of material should be pasted on one edge of the slide and name of the student on the other edge of the slide.

Summary of Permanent Mounting

Single staining

Fix the material in fixative
↓
Wash in water to remove fixative
↓
Dehydrate in 30% alcohol (5 minutes)
↓
Dehydrate in 50% alcohol (5 minutes)
↓
Dehydrate in 70% alcohol (5 minutes)
↓
Stain in borax carmine
or
↓
Picro-indigo carmine
or eosin (3-4 minutes)
↓
Destain in acid alcohol (few seconds)
↓
Dehydrate in 90% alcohol (5 minutes)
↓
Absolute Alcohol (5-10 minutes)
↓
Xylol or clove oil (5 minutes)
↓
Mount in Canada Balsam or DPX

Double staining

Fix the material in fixative
↓
Wash in water to remove fixative
↓
Stain in haematoxylin (2-5 minutes)
↓
Wash in water
↓
Destain in acid water if it is overstained
↓
Wash in water
↓
Dehydrate in 30% alcohol (5 minutes)
↓
Dehydrate in 50% alcohol (5 minutes)
↓
Dehydrate in 70% alcohol (5 minutes)
↓
Stain in eosin (5-7 minutes)
↓
Dehydrate in 90% alcohol (5 minutes)
↓
Absolute alcohol (5 -10 minutes)
↓
Xylol or clove oil (5 minutes)
↓
Mount in Canada Balsam or DPX

Chapter 6

IMPORTANT MOUNTINGS

PROTOCHORDATES

1. Spicules of *Herdmania*

Procedure. Take a small piece of **test** of *Herdmania* and boil it in 10% KOH solution (potassium hydroxide solution), decant and wash thoroughly in water. Now dehydrate up to 70% alcohol and stain in **picro-indigo carmine** or **borax carmine** or proceed further without staining. After staining again dehydrate up to absolute alcohol and dealcoholise or clear in **xylol** or **clove oil** and finally mount in Canada balsam.

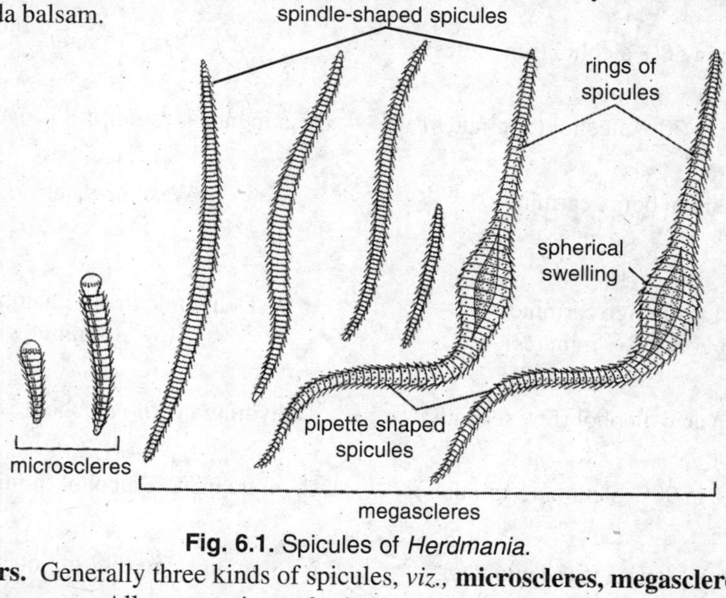

Fig. 6.1. Spicules of *Herdmania.*

Characters. Generally three kinds of spicules, *viz.,* **microscleres, megascleres** and **pipette-shaped** spicules are seen. All possess rings of spines.

2. Neural gland, Nerve ganglion and Dorsal tubercle (Neural complex) of *Herdmania*

Procedure. Take complete *Herdmania* and fix it by pins passing through the siphons and stretch the **inter-siphonal region** as far as possible. Expose the neural gland from above and proceed along the sides till the whole of the nerve ganglion is exposed and then trace the nerves at the two ends as far as possible. Snip off the mantle lying on either side of these organs with a small pair of forceps till the wall of the branchial-sac is reached. Remove the wall of the peripharyngeal zone and carefully expose the dorsal tubercle. The three organs, *viz.*, **neural gland**, **nerve ganglion** and **dorsal tubercle** now be dissected out keeping the neural duct intact. Wash in water, dehydrate up to 70% alcohol, stain lightly with **borax carmine**, again dehydrate up to absolute alcohol, clear in **xylol** or **clove oil** and finally mount on a cavity slide in **Canada balsam**.

For details of structure see figure No. 6.2.

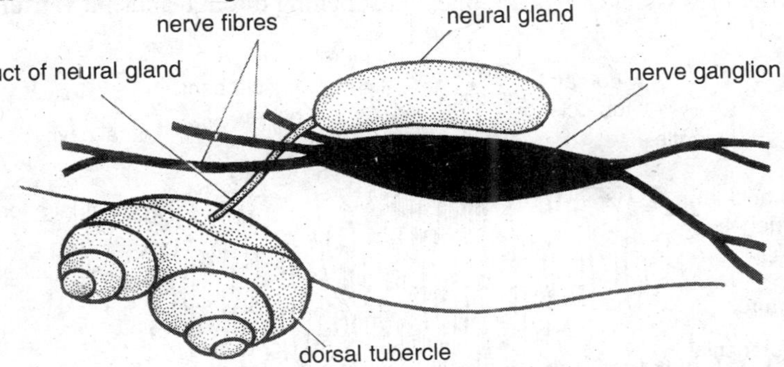

Fig. 6.2. Neural gland, nerve ganglion and dorsal tubercle (collectively called neural complex) of *Herdmania*.

3. *Salpa* (Whole mount)

Procedure. Take the entire animal and wash it in 70% alcohol (because the materials for mounting are usually preserved in 70% alcohol). After washing in 70% alcohol, stain it with **borax carmine** or **eosin** and then again dehydrate up to absolute alcohol, clear in **xylol** or **clove oil** and finally mount in **Canada balsam**.

Compare the structure with figure No. 6.3.

1. *Salpa* is a tunicate showing polymorphism exists in two phases : asexual or oozoid phase and sexual or blastozooid phase, both alternate.
2. Oozoid possesses branchial and atrial apertures at opposite ends, and dorsal nerve ganglion, subneural gland. In the pharynx are present dorsal lamina, endostyle, peripharyngeal bands and gill-slits.
3. Body wall is transparent having outer test, inner mantle and muscle bands (circular).
4. Budding stolon is ventral.

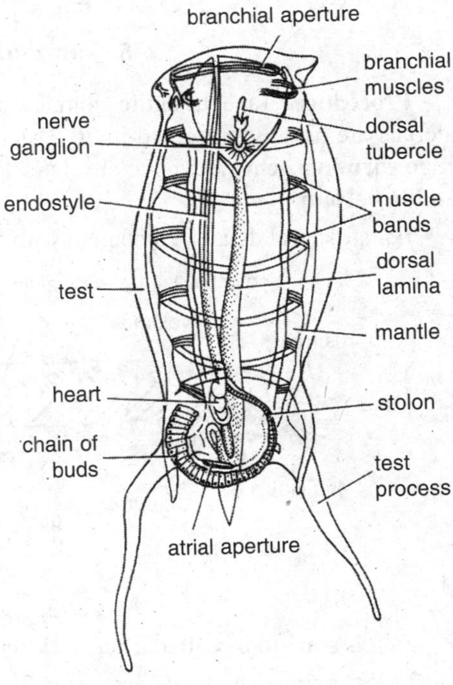

Fig. 6.3. *Salpa* (Asexual phase).

4. *Doliolum* (Whole mount)

Procedure. Take the entire animal, wash it thoroughly in 70% alcohol (because the mounting materials are usually preserved in 70% alcohol), stain it with **borax carmine** or **eosin**, dehydrate it up to absolute alcohol, clear in **xylol** or **clove oil** and finally mount in **Canada balsam**.

Compare the structure with figure No. 6.4.

1. *Doliolum* is a tunicate, present in two phases : solitary or gonoozoid phase and gregarian or oozoid phase, both alternate during reproduction.

2. Oozoid is barrel-shaped covered by transparent test and inner mantle and muscle bands.

3. Branchial and atrial openings at the opposite ends.

4. Pharynx possesses gill-slits, dorsal tubercle, dorsal lamina endostyle and peripharyngeal bands.

5. Dorsal nerve ganglion, subneural gland are also present.

6. Other structures visible are heart, intestine, opening through anus into atrium.

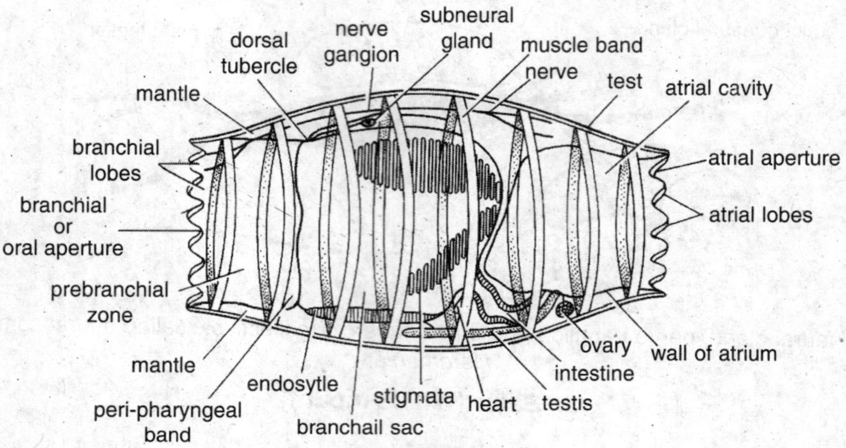

Fig. 6.4. *Doliolum* (Sexual).

5. *Amphioxus* (Whole mount)

Procedure. Take the entire animal, wash it thoroughly in 70% alcohol (because the mounting materials are usually preserved in 70% alcohol), stain it with **borax carmine** or **eosin** or **picro-indigo carmine**, dehydrate up to absolute alcohol, clear it in **clove oil** or **xylol** and finally mount in **Canada balsam**.

For structural details, compare it with figure No. 6.5.

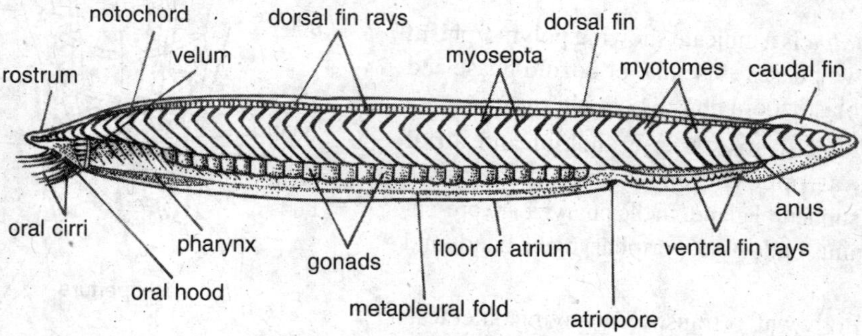

Fig. 6.5. *Amphioxus.*

1. It is commonly called lancelet (lance-shaped).

2. It is a simple chordate possessing paired pharyngeal gill-slits, dorsal nerve cord and ventral notochord.

3. Presence of dorsal, ventral and caudal fins.

4. Body wall has outer single-layered epidermis and dermis and beneath it are myotomes (muscles).

5. Anterior end bears oral cirri around mouth which opens into oral hood.

6. Pharynx possesses paired gill-slits supported by gill-bars, dorsal epipharygeai band, periphery pharyngeal bands and ventral endostyle.

7. Gill-slits open into atrium which opens outside through atriopore.
8. Anus lies ventral at the base of caudal fin.

6. Oral hood of *Amphioxus*

Procedure. Take a large specimen of *Amphioxus* and put it in water in a petridish. Dissect the anterior part of the body under a binocular microscope and cut out the oral hood with the help of a sharp blade. After taking out the oral hood, wash it thoroughly in water, dehydrate it up to 70% alcohol and then stain it with **borax carmine** or **eosin** again dehydrate up to absolute alcohol, clear in **xylol** or **clove oil** and finally mount in **Canada balsam**.

For details of structure, refer to figure No. 6.6.

1. It is the anterior most part of the animal.
2. Oral hood bears 10—11 pairs of oral cirri or buccal tentacles around a large opening that leads into vestibule or buccal cavity.
3. Oral cirri get support from skeletal rods.
4. Oral cirri bear sensory papillae.

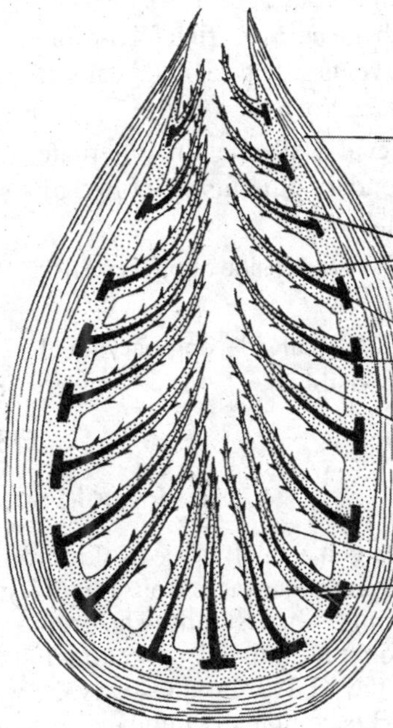

Fig. 6.6. Oral hood of *Amphioxus.*

7. Velum of *Amphioxus*

Procedure. Dissect the anterior end of *Amphioxus* under a binocular microscope. The velum is found at the base of vestibule. Take out the velum, wash it thoroughly in water, dehydrate up to 70% alcohol, stain in **borax carmine** or **eosin**, again dehydrate up to absolute alcohol, clear in **xylol** or **clove oil** and mount in **Canada balsam**.

For details of structure, refer to figure No. 6.7.

1. Velum is present at the base of vestibule or in between vestibule and pharynx.
2. It bears a central perforation called enterostome.
3. Enterostome is surrounded by about 12 velar tentacles which are directed backward forming a sieve over enterostome.
4. Velar tentacles bear sensory papillae and cilia.

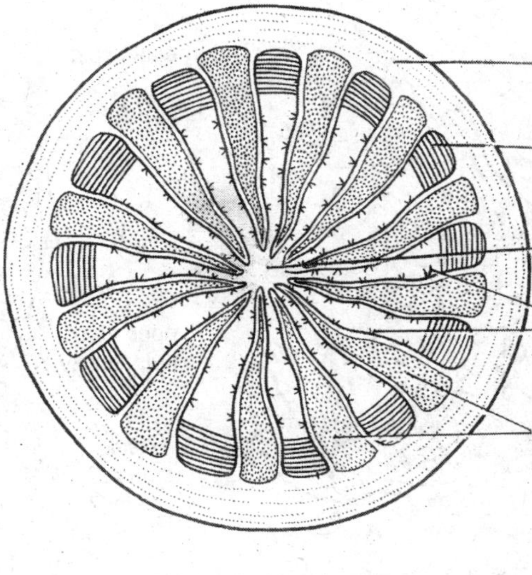

Fig. 6.7. Velum of *Amphioxus.*

8. Pharyngeal wall of *Amphioxus*

Remove the body wall and muscles from over the pharynx under a dissecting microscope. Then cut out a portion of pharynx and unfold it with the help of a fine forcep and needle and observe its structure. For details of structure, see figure No. 6.8.

1. Pharyngeal wall contains alternate primary and secondary gill-slits supported by gelatinous skeletal rods.

2. Along the ventral side of pharynx lies the endostyle

3. Along the dorsal side, epipharyngeal band may also be visible.

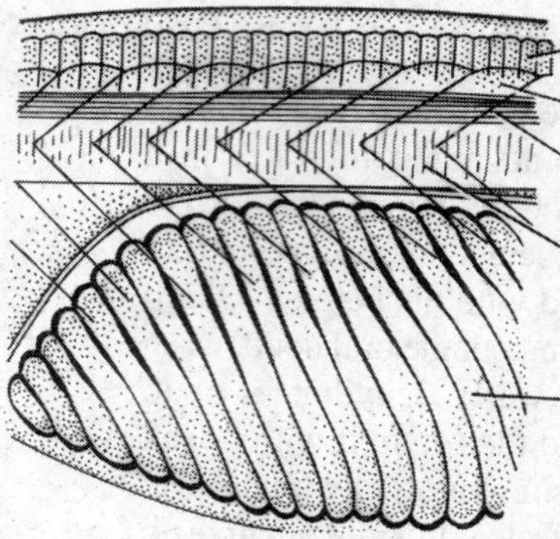

Fig. 6.8. Pharyngeal wall of *Amphioxus.*

FISHES

9. Placoid scales of *Scoliodon*

Procedure. Cut a small piece of skin from the dorsal surface of *Scoliodon* and remove all the muscles from its ventral surface. Boil it in 5% solution of KOH in a test tube until a brown residue settles at the bottom of the test tube. The brown residue is nothing but the placoid scales which have been detached from the skin during boiling. Let the residue be settle down on the bottom of the test tube and remove the KOH solution slowly. Now thoroughly wash the placoid scales in water 4 or 5 times so that KOH is completely removed from them.

Proceed for dehydration up to 70% alcohol, stain in **borax carmine** or **picro-indigo carmine**, again dehydrate up to absolute alcohol, clear in **xylol** or **clove oil** and finally mount in **Canada balsam**.

For details of structure, see figure No. 6.9.

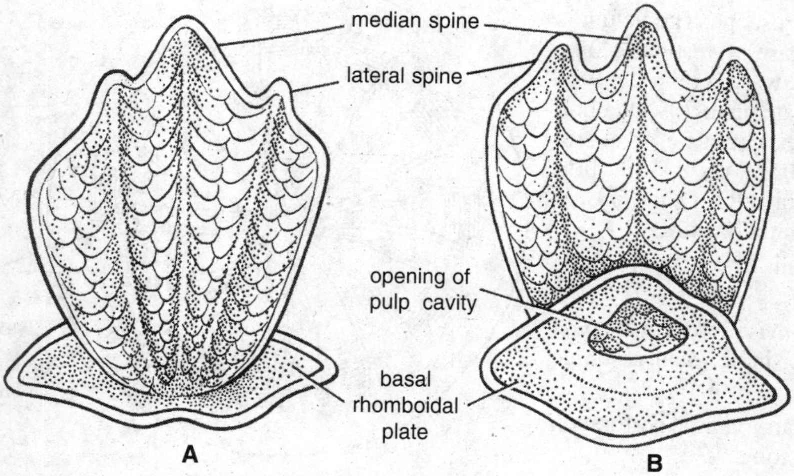

Fig. 6.9. Placoid scales of *Scoliodon*.A–Dorsal view; B–Ventral view.

1. Placoid scales are minute, transparent dermal denticles embedded in the skin in oblique rows.

2. Each placoid scale has a diamond-shaped basal plate which bears trident spines projecting beyond the skin.

10. Cycloid scales of *Labeo* (Rohu)

Take one or twc scales from the body of fish and stain in **picro-indigo carmine**. Dehydrate and mount in **Canada balsam**.

Cycloid scale (Fig. 6.10) bears concentric lines of growth around a central nucleus. These lines of growth show its age.

For details of structure, see figure No. 6.10.

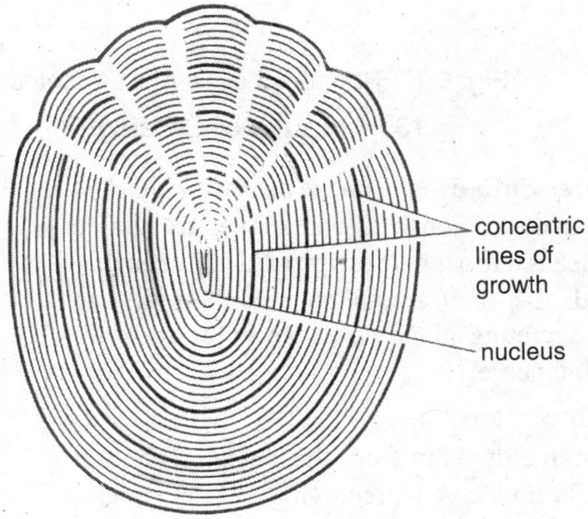

Fig. 6.10. Cycloid scale of *Labeo*.

11. Ctenoid scales of *Nandus*

Prepare its mount as before. Each etenoid scale (Fig. 6.11) has basal teeth or denticles and lines of growth around a central nucleus.

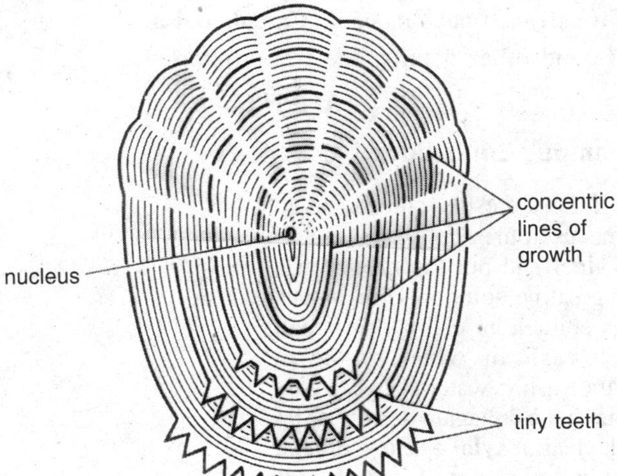

Fig. 6.11. Ctenoid scale of *Nandus*.

12. Rhomboid scales of *Lepidosteus*

Prepare its mount like placoid scales.

Scale (Fig. 6.12) is roughly rectangular in shape and bears lines of growth around a central nucleus.

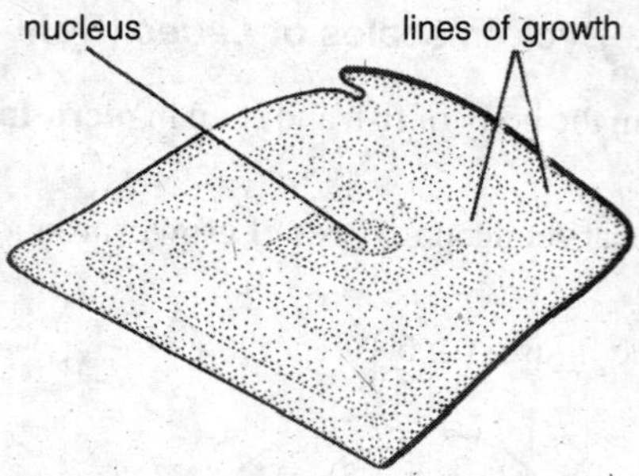

Fig. 6.12. Rhomboid scale of *Lepidosteus.*

13. Ampulla of Lorenzini

Ampullae of Lorenzini are found on the head of *Scoliodon.* There are several pores in the skin on the dorsal and ventral surface of the snout, each pore leads into a tubule of **ampulla of Lorenzini**. Ampulla is composed of 8 or 9 ampullary chambers having receptor and mucous secreting gland cells. Receptor cells are innervated by 7th cranial nerve.

These are thermoreceptors.

Procedure. Remove the skin from the dorsal or ventral surface of the snout of *Scoliodon.* After removing the skin, several tubules (creamish white) are seen. Take out a small bunch of ampullae with the help of forcep and put it on a slide and then tease the bunch with the help of needles and separate the ampulla from one another. Take two or three complete ampullae, wash them thoroughly in water, dehydrate up to 70% alcohol, stain in **borax carmine** or **eosin**, again dehydrate up to absolute alcohol, clear in **xylol** or **clove oil** and finally mount in **Canada balsam.**

For structure and other details, see figure No. 6.13.

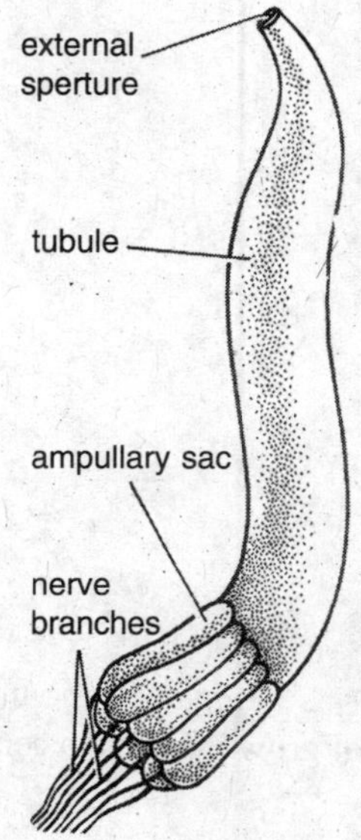

Fig. 6.13. Ampulla of Lorenzini.

AMPHIBIA

14. Squamous Epithelium

Procedure. Take a freshly chloroformed frog and scrap the mucus from its dorsal and ventral surfaces by the scalpel and put the mucus in a petridish containing saline solution. The mucus contains squamous epithelium cells. Take small piece of mucus, wash in water, stain in **haematoxylin**, again wash in water, dehydrate up to 70% alcohol, stain in **eosin,** again dehydrate up to absolute alcohol, clear in **xylol** or **clove oil** and finally mount in **Canada balsam**. (Living frog if submerged in water in a trough also discards small flakes of epidermal cells). These cells are thin, flat, polygonal and nucleated.

For structural details, see figure No. 6.14.

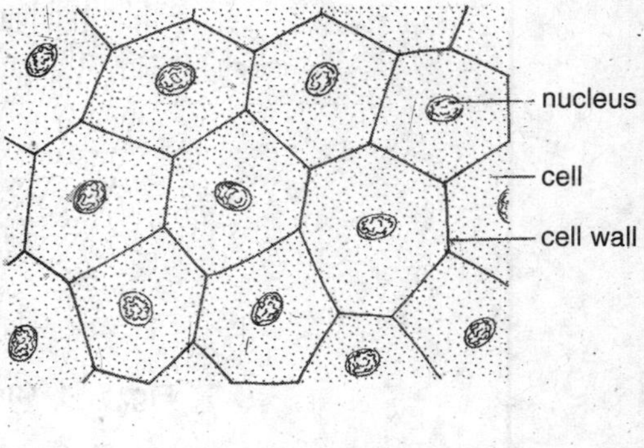

Fig. 6.14. Squamous epithelium.

15. Ciliated epithelium

Procedure. Take a freshly chloroformed frog and gently scrap the roof of the buccal cavity with the help of scalpel. Put the scrapping over a slide, stain in **borax carmine**, dehydrate, clear and mount in **Canada balsam**.

1. These cells are columnar in shape having nucleus and nucleolus.
2. Their free ends bear fine vibratile cilia. In between ciliated cells are present mucus secreting goblet cells.

For structural details, see figure No. 6.15.

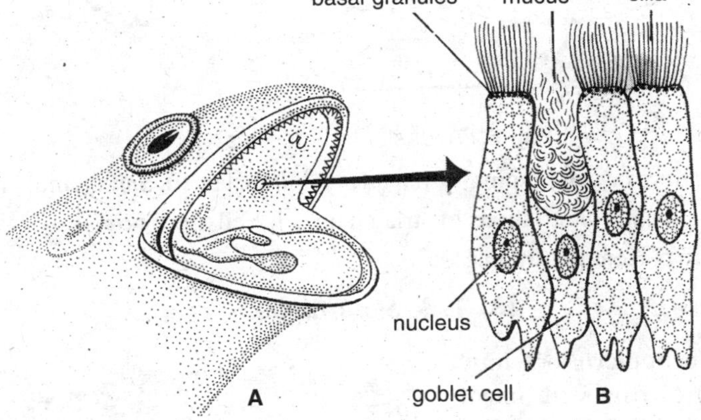

Fig. 6.15. Ciliated epithelium. A– Buccal cavity of frog; B– Ciliated epithelium.

16. Striated (Striped) muscles

Procedure. Take a few muscles from the thigh region of freshly chloroformed frog. Tease the muscles with the help of needles. Take a few fibres, fix them in **absolute alcohol** or **corrosive sublimate**, wash them in water, stain in **haematoxylin** again wash in water, dehydrate up to 70% alcohol, stain in **eosin**, again dehydrate up to absolute alcohol, clear in **xylol** or **clove oil** and finally mount in **Canada balsam**.

For details of structure see figure No. 6.16.

1. Striated muscle fibres or cells are covered by a delicate membrane called sarcolemma and are multinucleated.
2. These cells show alternate dark and light bands due to which these are called striated.
3. Myofibrils are present in sarcoplasm.

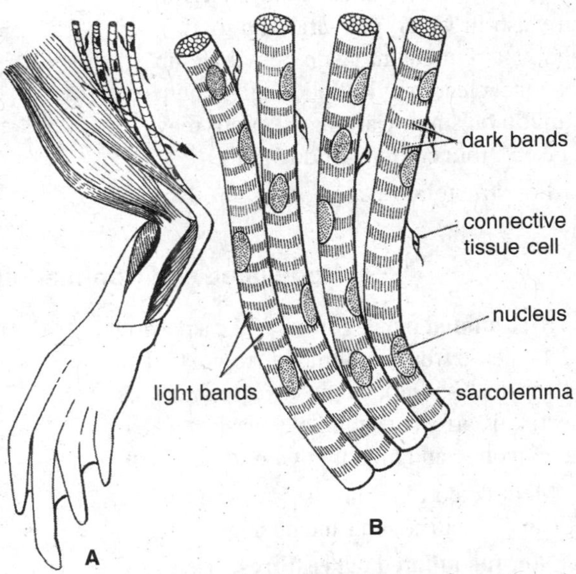

Fig. 6.16. Striated muscles.

17. Unstriated (Unstriped) muscles

Procedure. Take a small piece of urinary bladder of freshly chloroformed frog, wash it in water, fix in **absolute alcohol** or **corrosive sublimate**, remove the fixative, dehydrate up to 70% alcohol, stain in **borax carmine**, again dehydrate up to absolute alcohol, clear in **xylol** and mount in **Canada balsam**.

For structural details, see figure No. 6.17.

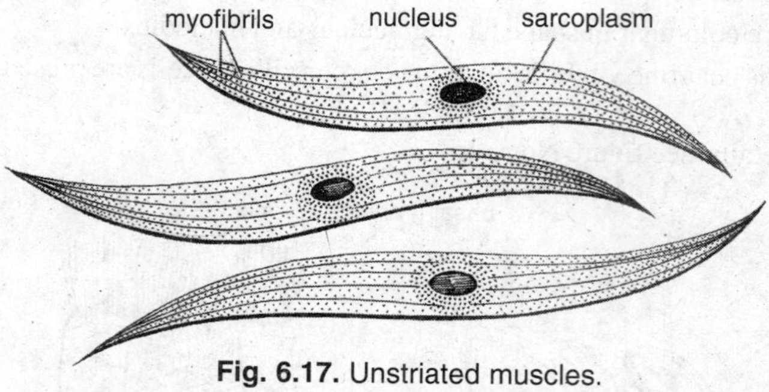

Fig. 6.17. Unstriated muscles.

1. Each muscle fibre or cell is spindle-shaped and uninucleated. Sarcolemma is absent.
2. Dark and light bands or striations of striated muscle cell are lacking.
3. Myofibrils are present in sarcoplasm.

18. T. S. of cartilage

Procedure. Take **humerus** or **femur** (or hyoid apparatus) of frog with intact cartilage on their heads. Wash the head in water and then cut very thin transverse sections of cartilage with a sharp razor or blade. Take one or two sections, wash thoroughly in water, stain in **haematoxylin**, again wash in water, dehydrate up to 70% alcohol, stain in **eosin**, again dehydrate up to absolute, clear in **xylol** and finally mount in **Canada balsam**. (Cartilage sections may also be cut from cranium of dog fish).

For structural details, see figure No. 6.18.

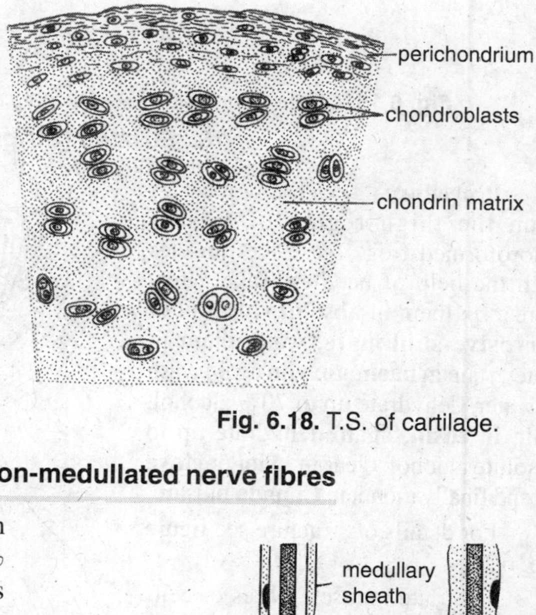

Fig. 6.18. T.S. of cartilage.

19. Medullated and Non-medullated nerve fibres

Medullated nerve fibres. Take a portion of sciatic nerve from hindlimb of frog in 0.78% NaCl (normal saline). Tease it to separate its fibres, fix in corrosive sublimate, wash in water, stain, dehydrate and mount in **Canada balsam**.

Medullated fibre has outermost medullar sheath, axon, neurilemma and nucleus.

Non-medullated nerve fibres. These are taken from sympathetic nerve. Proeeed as before and observe under microscope.

Non-medullated nerve fibre lacks medullary sheath, while other structures are the same.

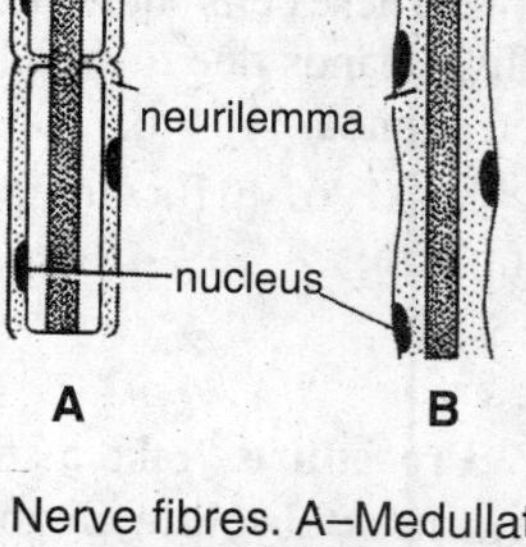

Fig. 6.19. Nerve fibres. A–Medullated; B–Non-medullated.

20. Blood of frog

Procedure. Take a drop of blood from a freshly-killed frog on a slide. Now hurriedly spread the blood over the slide by the edge of another slide in such a way that very thin film of blood is formed. Dry the film of blood, pour two or three drops of absolute alcohol and let it be dry again. After drying wash it in water, stain in **haematoxylin**, wash again in water, dehydrate up to 70% alcohol, stain in **eosin**, again dehydrate up to absolute alcohol, clear in **xylol** and finally mount in **Canada balsam**.

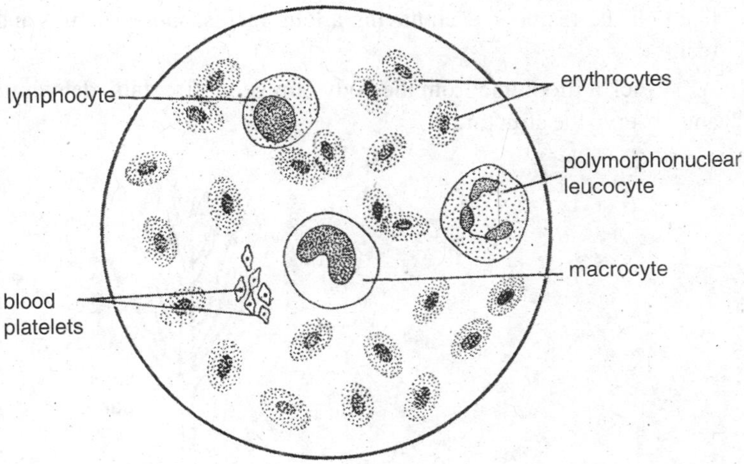

Fig. 6.20. Blood of frog.

For structural details, see figures No. 6.20.

Blood film shows oval nucleated erythrocytes, granular and agranular leucocytes and platelets.

AVES

21. Pecten of Pigeon

Procedure. Take out the complete eye-ball of pigeon by cutting the eye-muscles. The muscles can be cut or removed with the help of a scalpel or a scissor. Now locate the optic nerve entering the eye-ball and cut the eye-ball into two halves, *i.e.,* one anterior half and one posterior half. In the posterior half, at the point of entering the optic nerve, the **pecten** is seen hanging obliquely. It is a thin black structure. Take it out with the help of forcep and scissor and proceed for permanent mounting.

Wash thoroughly in water, dehydrate up to 70% alcohol, stain in **picro-indigo carmine**, again dehydrate up to absolute alcohol, clear in **xylol** or **clove oil** and mount finally in **Canada balsam**.

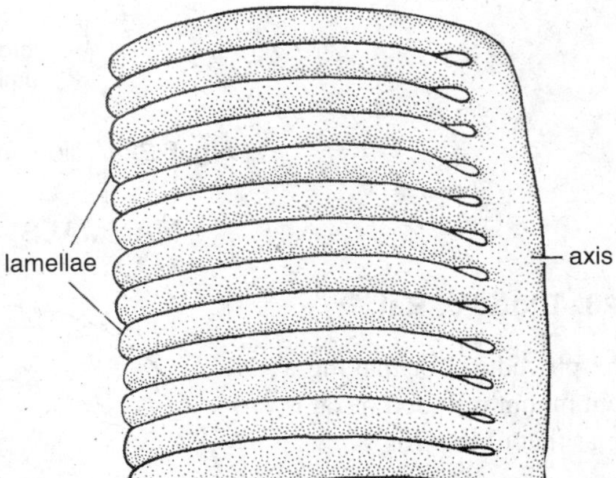

Fig. 6.21. Pecten of pigeon.

For structural details, see figure No. 6.21.

Pecten has an axis from which arise several parallel vertical lamellae.

22. Filoplume

Filoplumes are present over the body of pigeon at the bases of contour feathers. These are hair-like delicate feathers, each having a long rachis, short calamus and a few weak barbs at the tip of rachis.

Pick a filoplume from the body of pigeon, fix, stain, dehydrate and mount in **Canada balsam** and observe the structure.

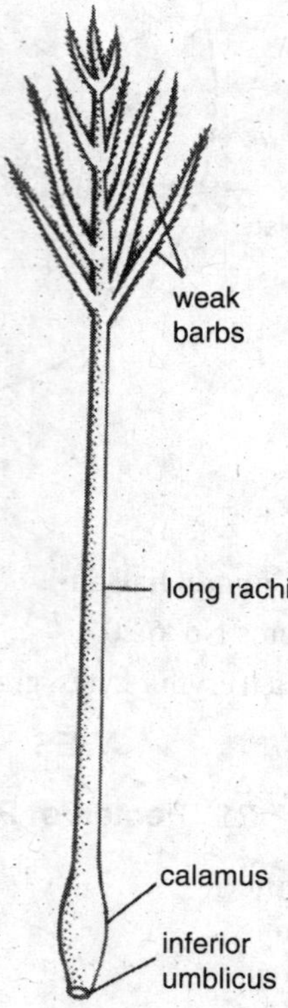

Fig. 6.22. A filoplume.

MAMMALS

23. Blood of Rat

The procedure for making a permanent mount of blood of rat is the same as that of the frog.

For structural details, see figure No. 6.23.

Blood film of rat shows enucleated, circular, red blood corpuscles, nucleated granular and non-granular white blood corpuscles.

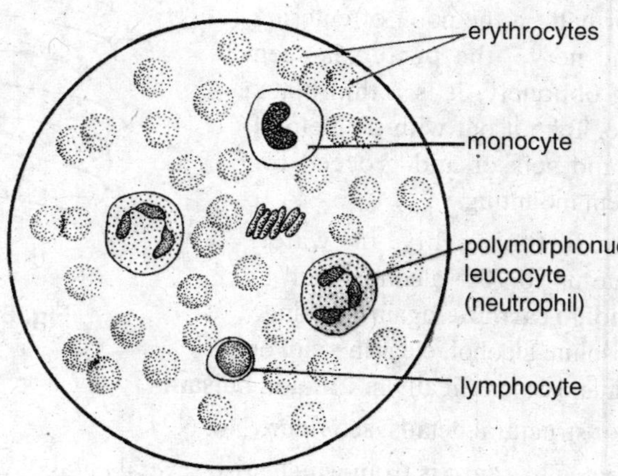

Fig. 6.23. Blood of rat.

7

Chapter

STUDY OF MUSEUM SPECIMENS

MUSEOLOGY

There are numerous chordate animals. Amongst chordate animals, only a few are studied as type specimens. The study of morphology (external form and structure) of animals from the preserved animals in the museum or laboratory constitutes the study of museum. The smaller animals are preserved wet in 5–10 per cent formalin solution or in 90 per cent alcohol, while the larger animals are stuffed and kept dry in almost all museums of zoology departments. The collection, preservation and maintenance of museum is called **museology**. The various types of specimens kept in a museum fall into following categories :

1. Wet preservations (Formalin preserved specimens in jars);
2. Dry specimens (Dry preservation);
3. Stuffed animals (Taxidermy);
4. Mounted specimens (Dry preservation of skeleton, etc.)

Before studying the museum specimens, it is very essential to know about collection and preservation of various specimens.

Collection

Animals are collected in various ways such as the following :

1. Catching by hand. Small terrestrial and freshwater animals are caught with hands. It is essential to wear hand gloves while catching small animals.

2. Netting. Most of the aquatic animals such as frogs, etc., are collected with the help of fine-meshed or coarse-meshed **nets**. Freshwater fishes and marine fishes are collected with the help of different kinds of nets.

3. Digging and draining. Burrowing animals such as *Amphioxus* are collected by digging and draining or sieving sand.

4. Trapping. The small rodents and carnivores are collected by employing different types of traps. The traps may be wooden traps, steel traps and wonder traps. For instance, common house trap (**chuhedan**) is used for trapping the house mouse. Some kind of bait is kept inside the trap to attract the animal. In wonder traps for catching black rats, pellets (bait) made by mixing besan (gram flour) with mustard oil are used. In house traps for catching house mouse, small pieces of bread or chapati are kept.

5. Shooting. Animals such as snakes, birds, fast running rodents and carnivore mammals are collected by shooting with small gun. After shooting, the animals are collected in canvas cloth bags.

Preservation

After collection, the animals are brought to laboratory and are preserved either by **wet method** or **dry method**.

1. Wet preservation. The preservation of animals in formalin or alcohol is called wet preservation.

(i) **Requirements.** The requirements for wet preservation are : formalin, glycerine, 90% alcohol, copper sulphate, chalk, thread, rectangular glass plates, glass jars of different sizes (100 ml, 250 ml, 1000 ml, 5000 ml and 10000 ml), plastic sheet jars of different sizes, wooden boards and printed labels.

(ii) **Procedure for wet preservation.** Stock solution of 5% formalin (100 ml of distilled water + 5 ml of formalin) is prepared. 5 ml of glycerine is added. The formalin solution is used as a fixative as well as a preservative.

For preserving specimens, the living animals are first of all anaesthesised with chloroform in a jar. For example, a living frog is anaesthesised with chloroform, washed and then 5% solution of formalin is injected into its thigh muscles and in thoracic region in order to preserve its internal organs. Then a small slit is made in the abdomen and finally the frog is mounted on a glass plate with ventral surface facing yourself. The frog is tied at forelimbs and hindlimbs. Then the mounted frog is put in a glass jar or plastic jar and filled with 5% solution of formalin. The mouth of jar is covered with a glass plate and sealed with wax to prevent evaporation of formalin solution. Finally, a label depicting date, place of collection and the classification is pasted on the jar.

2. Dry preservation. Usually birds and small rodents are preserved by dry preservation. Dry preservation is done by two methods :

(i) **Case-skin mount.** It is good for small rodents such as rats. For example, pick up a black rat, anaesthesise, make a mid-ventral incision on the thorax and abdomen and also a mid-ventral incision along both the limbs. Carefully remove the skin, treat it with borax solution and then spread over a wooden board. Stretch the skin, limbs and tail with dorsal surface facing you. Let the skin dry. Finally, a label depicting the date and classification of dried specimen is pasted on the wooden board.

(ii) **Taxidermy.** In taxidermy, three-dimensional view is maintained by stuffing the animal with cotton or saw dust mixed with chalk and copper sulphate. For example, take a living pigeon and anaesthesise it with chloroform. After anaesthesising the pigeon, keep it in a dissecting tray. Make a vertical incision along the mid-ventral abdominal axis and carefully remove the internal organs. Remove muscles and bones of the fore- and hindlimbs. Fill up the cavity by packing it with cotton mixed with chalk and copper sulphate. Let the pigeon dry. Classify and label the specimen. The label must depict the date of collection, place of collection and name of specimen. Let the specimen become completely dry and then keep it in the museum.

Dried articulated and disarticulated skeletons such as those of fish, frog, *Varanus,* snake, fowl and rabbit, etc., are mounted on wooden board with classification and labelling.

Collection and Preservation in Chordate Groups

Different types of chordate animals are collected for the observation and studies in the laboratries. The process of collection and preservation of different chordate animals groups is described below :

1. Urochordates. *Herdmania, Ascidia, Pyrosoma,* etc.

(*i*) **Collection.** *Herdmania* (sessile tunicate) is collected by careful digging along the substratum. *Pyrosoma* (pelagic tunicate) is collected by netting.

(*ii*) **Preservation.** Urochordates are preserved by wet method. Wash the specimens in water and then transfer in glass jars containing 5% formalin solution. Seal the jar carefully with wax, classify, write date of collection and place of collection.

2. Cephalochordates. *Amphioxus* (= *Branchiostoma*)

(*i*) **Collection.** *Amphioxus* is found buried in the sand. These are located by digging. Collect them by sieving the sand in coarse mesh sieve.

(*ii*) **Preservation.** By wet method.

3. Fishes. *Scoliodon, Labeo rohita Wallago attu,* etc.

(*i*) **Collection.** Both marine and freshwater fishes are collected by a variety of coarse mesh nets specially designed for catching fishes. Transfer the collected fishes in large containers.

(*ii*) **Preservation.** By wet method.

4. Reptiles. *Hemidactylus* (wall-lizard), Snakes and Turtles, etc.

(*i*) **Collection.** Wall-lizards are collected by sudden pressing them with stick or towel and catching them with large forceps. Snakes are collected by expert snake catchers by hooked stick. Turtles are collected by netting.

(*ii*) **Preservation.** Wet preservation method is employed for the preservation of reptiles.

5. Birds. Pigeon, Parrot, Koel, Sparrow, etc.

(*i*) **Collection.** Birds are collected by trapping and shooting with shotgun and collected in canvas bags.

(*ii*) **Preservation.** Dry preservation by taxidermy as described earlier.

6. Mammals. There are several mammals belonging to different groups. Important specimens are small rodents such as squirrel, rats and carnivores such as mongoose, *Lutra* (udbilao), weasels and shrews.

(*i*) **Collection.** Mammals are collected by steel traps, wonder traps (for black rat) and mole traps.

(*ii*) **Preservation.** Dry mounting by case-skin mount or taxidermy is employed as described earlier. Label the specimen and keep it in the museum.

Arrangement and Care of Specimens in Museum

1. Classify all the specimens.
2. Keep different specimens of different phyla or groups in different shelves according to classification.
3. All the specimens of museum should be regularly examined whether wet or dry.
4. In case of wet perservation, if the formalin is lost or dried, it should be immediately replaced.
5. Always use post-mortem gloves while collecting and preserving the specimens.
6. Always keep exhaust fan on while working in the museum.

Instructions to Draw Museum Specimens

1. While going to zoology practical laboratory, you should have a zoology practical note book, a book of practical zoology, HB pencil, pencil sharpner, pencil eraser (rubber) and a piece of soft cloth.

2. For studying the museum, the students are required to see the actual preserved animals either in laboratory or museum.

3. While drawing the pencil diagrams of specimens, special care should be taken to give a very correct proportion about the dimension (length and breadth) of the specimen.

4. Only line diagrams should be drawn. Shading should be avoided as far as possible.

5. Important features of specimen should be exhibited in a diagram.

6. Each diagram should be properly labelled with the help of practical book. Labelling should be horizontal.

7. Classification should be written on the right hand top of the diagram.

8. Write both zoological and common name of each specimen below the diagram.

9. Finally write the important characters of specimens regarding their external features, habits and habitat and geographical distribution.

10. Usually draw only two diagrams on one page belonging to the same class.

In the following pages, the museum specimen of **Phylum Chordata** have been described. The classification of classes into subclasses, orders and suborders, if any, and their general characters have been given. The classification of representatives of each class along with their general characters (comments), habits, habitat and their geographical distribution have also been described in detail.

8

Chapter

PHYLUM–CHORDATA

(Gr., *chorda,* a string, or L., *chordatus,* a chord, *i.e.*, notochord)

General Characters

This is the largest phylum of deuterostome. This includes a majority of living and extinct animals like tunicates, lancelets, lampreys, fishes, amphibians, reptiles, birds and mammals.

1. Bilaterally symmetrical and metamerically segmented.
2. Triploblastic, *i.e.*, presence of three germ layers.
3. Dorsal tubular **nerve cord** is present.
4. **Notochord** is always essentially present at some stage of life. It is completely or partly replaced by vertebral column in majority of animals.
5. Paired pharyngeal **gill-slits** are present on either side of the pharynx at some stage of life.
6. **Coelomate**, *i.e.*, a well-developed coelom is always present.
7. Hepatic portal system is present.
8. A post-anal tail is present at some stage of life.

Thus, chordates possess three important common characters which separate them from non-chordates. These three fundamental common characters are as follows :

1. Notochord (Gr., *noton,* back; *chorde,* cord). It is a delicate longitudinal rod of connective tissue located in between nerve cord and digestive tract. It gives support to the body. It persists throughout life in protochordates and cyclostomes, but in fishes, amphibians, reptiles, birds and mammals it is replaced by vertebral column in the adult.

2. Hollow dorsal nerve cord. Nerve cord in chordates is always hollow and dorsal to the notochord. It extends throughout the length of the body and arises from the ectoderm.

3. Pharyngeal gill-clefts. Paired gill-clefts or slits are always present on the lateral sides of pharynx at some stage of life in all chordates. In lower chordates they persist throughout life, but in higher chordates they persist only in embryonic stages and close off before hatching or birth.

Phylum Chordata has been divided into following groups, sub-phyla and classes, etc.

Outline Classification of Phylum Chordata
Phylum - Chordata

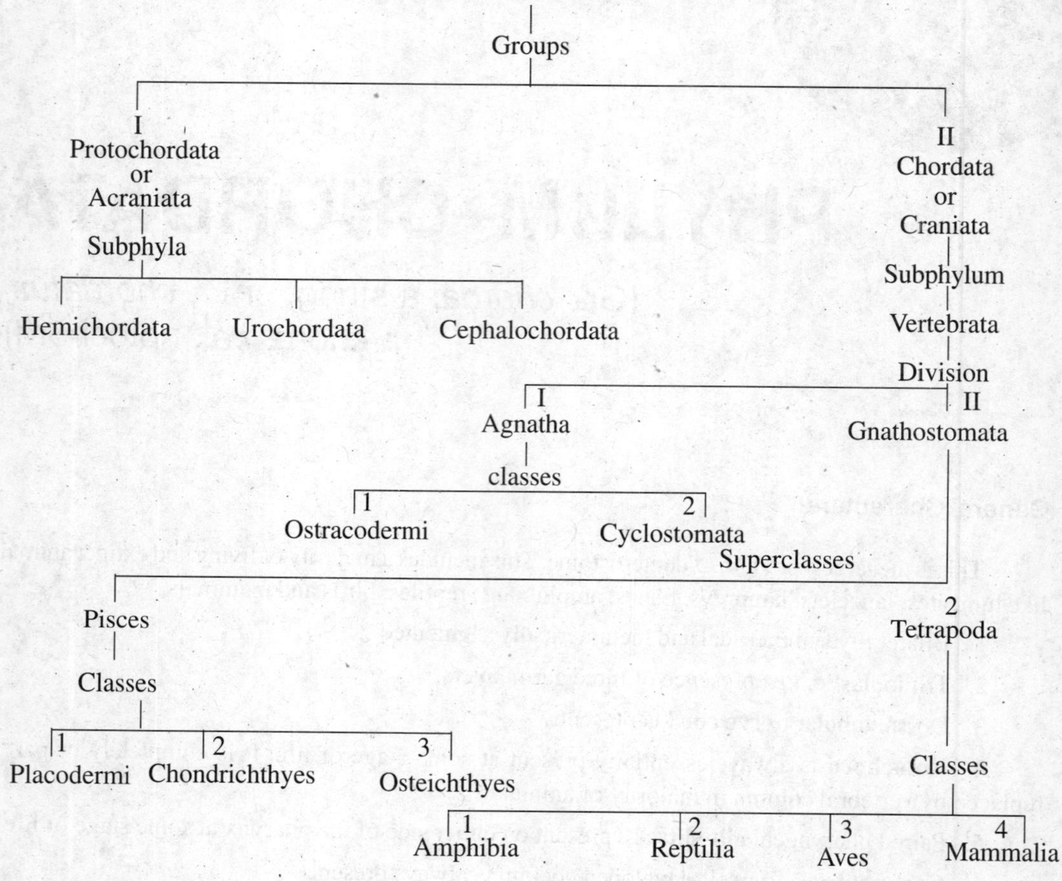

GROUP I. PROTOCHORDATA OR ACRANIATA

Exclusively marine. Mostly small-sized. Cranium or brain box, vertebral column and paired appendages absent. Nerve cord, notochord, gill-slits and myotomes present.

Subphylum 1. Hemichordata

Exclusively marine. Body divisible into proboscis, collar and trunk. Notochord occurs only in the anterior end of body. Numerous paired pharyngeal gill-slits present.

Class I. Enteropneusta. Body divisible into proboscis, collar and trunk. Numerous gill-slits and gonads present. Two rows of hepatic caecae present. **Examples :** *Balanoglossus, Saccoglossus, Ptychodera.*

Class II. Pterobranchia. Proboscis bears ciliated tentacles. Only one pair of gill-slits or none. Gonads few in number.

Order 1. Cephalodiscida. Two gill-slits, single pair of gonads and 4-6 tentaculated arms peresent. **Example :** *Cephalodiscus.*

Order 2. Rhabdopleurida. Gill-slits absent. Single gonad present. Single pair of tentaculated arms. **Example :** *Rhabdopleura.*

Subphylum 2. Urochordata

Exclusively marine. Body covered by **tunic** or **test**. Notochord and nerve cord present in larval stages and absent in adults. Numerous gill-slits present.

Class I. Larvacea. Free-swimming pelagic forms. Neotenic forms retain the larval form throughout adult-life. Test temporary. Single pair of gill-slits present. **Examples:** *Oikopleura, Appendicularia.*

Class II. Ascidiacea. Fixed or free-swimming marine forms. Test permanent and well-developed. Branchial sac perforated by numerous gill-slits.

Order 1. Ascidiae simplices. Fixed and solitary forms. Tail well-developed. Exhibits retrogressive metamorphosis. **Examples :** *Herdmania, Ciona, Ascidia.*

Order 2. Ascidiae compositae. Fixed and colonial forms. Larva develops within the atrial cavity. **Example :** *Botryllus.*

Order 3. Ascidiae luciae. Free-swimming pelagic colonial forms. Test transparent and gelatinous. Colony phosphorescent. **Example :** *Pyrosoma.*

Class III. Thaliacea. Free-swimming pelagic forms. Solitary or colonial. Test permanent and transparent. Muscles of body wall in the form of circular bands. Branchial sac has either two or many gill-slits.

Order 1. Cyclomyaria (Doliolida). Free-swimming pelagic forms. Branchial and atrial apertures at opposite ends. Muscle bands complete surrounding body. **Example :** *Doliolum.*

Order 2. Hemimyaria. Free-swimming pelagic forms. Pharynx communicate freely with the atrial cavity. Muscle bands always incomplete ventrally. **Example :** *Salpa.*

Subphylum 3. Cephalochordata

Exclusively marine, solitary forms. Notochord and nerve cord extend the entire length of body and persistent. Pharynx perforated by numerous gill-slits.

Class Leptocardii. Example : *Branchiostoma* (= *Amphioxus*)

GROUP II. CHORDATA OR CRANIATA

This group is characterised by distinct head, cranium, visceral arches, vertebral column and brain. It includes a single subphylum **vertebrata** in which notochord breaks up into a series of vertebrae.

Subphylum Vertebrata

Brain enclosed in cranium and notochord is replaced partly or wholly by vertebral column.

Division I. Agnatha

No true jaws or appendages.

Class 1. Ostracodermi. Extinct fishes. Body wall covered by large bony plates or scales. **Example :** *Cephalaspis.*

Class 2. Cyclostomata. Skin naked without scales. Mouth suctorial. Gill-slits 5-16 pairs. Skeleton cartilaginous. **Example :** *Petromyzon, Myxine, Bdellostoma.*

Division II. Gnathostomata

With true jaws and paired appendages.

Superclass 1. Pisces. Paired fins, gills and skin with scales. **Example :** Fishes.

Class 1. Placodermi. Extinct fishes with primitive jaws.

Class 2. Chondrichthyes (Elasmobranchi). Cartilaginous fishes. Skin with placoid scales. Gill-slits exposed (not covered with operculum). **Examples :** Sharks, rays and skates.

Class 3. Osteichthyes (Teleostomi). Bony fishes. Skin covered with cycloid or ctenoid scales. Gill-slits covered with operculum. **Examples :** Perch, Catfish, ganoids and lung fishes.

Superclass 2. Tetrapoda. Paired limbs, lungs, cornified skin and bony skeleton. It includes animals from Amphibia to Mammalia.

Class 1. Amphibia. Skin moist, soft, glandular and without external scales. **Examples :** Frogs, toads, salamanders.

Class 2. Reptilia. Skin dry, with external scales or scutes. Cold blooded. **Examples :** Lizards, snakes, turtles and crocodiles.

Class 3. Aves. Skin covered with feathers. Forelimbs form wings. Warm blooded. **Examples :** Birds.

Class 4. Mammalia. Skin covered with hair. Females have mammary glands to suckle young. Warm blooded. **Examples :** Duck-bill platypus, bats, moles, shrews, dogs, cats, horse, elephant, lion, cow, monkey, buffalo and man.

Chapter 9

GROUP–PROTOCHORDATA (Acraniata)

General Characters

1. Exclusively marine and small forms.
2. Cranium or brain-box, jaws, vertebral column and paired appendages are absent.
3. Dorsal tubular nerve cord, notochord, gill-slits and myotomes are usually present.
4. Sexes may be separate or united.
5. This group includes the lowly-organised members of Phylum Chordata.

GROUP—PROTOCHORDATA

Subphyla

- I Hemichordata
- II Urochordata or Tunicata
- III Cephalochordata **Ex.** *Amphioxus*

Subphylum—Hemichordata

Classes

- I Enteropneusta
  Ex. 1. *Balanoglossus*
  2. *Saccoglossus*
  3. *Ptychodera*
- II Pterobranchia
  Orders
  - 1 Cephalodiscida **Ex.** *Cephalodiscus*
  - 2 Rhabdopleurida **Ex.** *Rhabdopleura*

Subphylum—Urochordata

Classes

- I Larvacea — **Ex.** *Oikopleura*
- II Ascidiacea
- III Thaliacea

Class—Ascidiacea

Orders

- 1 Ascidiae simplices — **Ex.** 1. *Herdmania* 2. *Ciona* 3. *Ascidia* 4. *Clavelina* 5. *Molgula*
- 2 Ascidiae compositae — **Ex.** *Botryllus*
- 3 Ascidiae luciae — **Ex.** *Pyrosoma*

Class—Thaliacea

Orders

- 1 Cyclomyaria — **Ex.** *Doliolum*
- 2 Hemimyaria — **Ex.** *Salpa*

SUBPHYLUM I—HEMICHORDATA

1. Exclusively marine and soft-bodied forms.
2. Body is divisible into **proboscis**, **collar** and **trunk**.
3. Notochord occurs only in the anterior end of the body. Recently it has been called "**buccal diverticulum**" due to its doubtful nature.
4. Numerous paired pharyngeal gill-slits are present.
5. Nervous tissues lie embedded in the epidermis and occur both on the dorsal and ventral surfaces.
6. Coelom is usually divided into three distinct portions corresponding to the three regions of the body.
7. Blood vascular system is simple.
8. Sexes are separate and the development may be direct or indirect.

Class I—ENTEROPNEUSTA

1. Solitary and burrowing worm-like marine forms commonly known as **acorn** or **tongue worms**.

2. Body consists of the usual divisions, *viz.*, **proboscis** separated by the narrow stalk from the ring-shaped **collar**, which is succeeded by an elongated **trunk**.
3. Epidermis is ciliated and glandular.
4. Numerous pharyngeal gill-slits and gonads are present.
5. Alimentary canal straight with a terminal anus.
6. Two rows of hepatic caeca are present in the middle of the trunk.

Examples : 1. *Balanoglossus,* 2. *Saccoglossus,* 3. *Ptychodera.*

Class II—PTEROBRANCHIA

1. Sedentary, solitary or colonial and marine forms.
2. Proboscis bears ciliated tentacles to produce ciliary feeding currents of water.
3. One pair of pharyngeal gill-slits or none.
4. Alimentary canal is U-shaped with dorsal anus situated near the mouth.
5. Gonads are few in number.
6. Reproduction by budding.

Order 1. Cephalodiscida

1. Solitary or living in a common gelatinous house.
2. Two pharyngeal gill-slits are present.
3. Single pair of gonads is present.
4. 4–6 tentaculated arms are present.

Example : *Cephalodiscus.*

Order 2. Rhabdopleurida

1. Colonial forms.
2. Gill-slits are absent.
3. Single gonad is present.
4. Single pair of tentaculated arms.

Example : *Rhabdopleura.*

SUBPHYLUM II—UROCHORDATA

1. Exclusively marine commonly known as **sea squirts**.
2. Solitary or colonial.
3. Fixed or free-swimming.
4. Body is covered by a cuticular **tunic** or **test** in adult stage.
5. Notochord present in the larval stages and absent in the adults.
6. Dorsal tubular nerve cord is present in the larval forms, while degenerates in the form of small ganglion in adults.
7. Numerous pharyngeal gill-slits are present.
8. Heart is ventral, simple and tubular.
9. Sexes are united, *i.e.,* hermaphrodite.

Class I—LARVACEA

1. Free-swimming pelagic forms.
2. Neotenic forms which retain the larval form throughout adult-life.
3. Test is temporary.

4. Posterior part of the body takes the form of a large locomotory appendage, the tail.
5. Single pair of pharyngeal gill-slits is present.
6. Anus opens ventrally on the surface of the body.
7. Sexes are united, *i.e.*, hermaphrodite.
8. No metamorphosis.

Examples : 1. *Oikopleura,* 2. *Appendicularia.*

Class II—ASCIDIACEA

1. Fixed or free-swimming marine forms.
2. Simple or compound, solitary or colonial.
3. Adults are never provided with a locomotary appendage or tail and have no trace of notochord.
4. Test is permanent and well-developed.
5. Branchial-sac is large and well-developed with its walls perforated by numerous pharyngeal gill-slits.
6. Reproduction is both asexual and sexual.

Order 1. Ascidiae simplices

1. Fixed and solitary forms.
2. Test is well-developed.
3. Reproduction by gemmation is very rare.
4. Usually exhibit retrogressive metamorphosis.

Examples : 1. *Herdmania,* 2. *Ciona,* 3. *Ascidia.*

Order 2. Ascidiae compositae

1. Fixed and colonial forms.
2. Test is commonly called "synascidiae".
3. Reproduction by gemmation.
4. Larva develops within the atrial cavity and subsequently hatches out as tailed-larva.

Example : *Botryllus.*

Order 3. Ascidiae luciae

1. Free-swimming pelagic colonial forms.
2. Test transparent and gelatinous.
3. Colony consists of a hollow cylinder closed at one end.
4. Zooids of the colony are enclosed in a common test in such a manner that the branchial apertures open on the outer surface and atrial apertures on the inner surface.
5. Colony is **phosphorescent** due to the presence of symbiotic bacteria.

Example : *Pyrosoma.*

Class III—THALIACEA

1. Free-swimming pelagic forms.
2. Solitary or colonial.
3. Test is permanent and transparent, may be slightly or well-developed.
4. Musculature of the body wall is in the form of circular bands.
5. Branchial-sac has either two large or many small pharyngeal gill-slits.
6. Tail and notochord absent in adults.
7. Life-history exhibits an alternation of generations.

Order I. Cyclomyaria (Doliolida)

1. Free-swimming pelagic forms.
2. Body is cask-shaped with branchial and atrial apertures are at opposite ends.
3. Test is moderately well-developed never much thickened.
4. Muscle bands are always complete surrounding the body.
5. Stigmata are usually present in the posterior part of the body.
6. Hermaphroditic. Life history exhibits an alternation of generations.
7. Sexual generation is always polymorphic.
8. Larva possesses notochord and tail.

Example : *Doliolum.*

Order 2. Hemimyaria

1. Free-swimming pelagic forms.
2. Body is more or less fusiform with the branchial and atrial apertures nearly terminal and opposite.
3. Test is well-developed and transparent.
4. Muscle bands are always incomplete ventrally.
5. Pharynx communicates freely with the atrial cavity.
6. Life history exhibits an alternation of generations.
7. No tailed-larva.

Example : *Salpa.*

SUBPHYLUM III—CEPHALOCHORDATA

1. Exclusively marine and solitary forms.
2. Notochord and nerve cord extend the entire length of the body.
3. Notochord persists as such.
4. Limbs or paired fins are absent. Median fins are present.
5. Mouth is ventral and anterior, while anus is ventral and posterior.
6. Pharynx is large, perforated by numerous pharyngeal gill-slits and surrounded by atrium.
7. Exoskeleton, head, brain, auditory organs, and jaws are absent.
8. Sexes are separate.
9. Gonads are metamerically arranged on the body-wall.

Class. Leptocardii

Example : *Amphioxus* (= *Branchiostoma*)

MUSEUM SPECIMENS OF PROTOCHORDATA

1. *Balanoglossus*

Classification :

Phylum	...	**Chordata**	→	Dorsal tubular nerve cord, notochord, paired gill-slits and coelom are present.
Group	...	**Protochordata**	→	Marine small forms. Cranium, jaws and paired appendages are absent.
Subphylum	...	**Hemichordata**	→	Solitary, soft-boiled, coelomate with numerous gill-slits.
Class	...	**Enteropneusta**	→	Alimentary canal straight, two rows of caeca.
Genus	...	*Balanoglossus*		

Habit and habitat. *Balanoglossus* is a burrowing and exclusively marine animal. It is found in shallow waters between tide marks along the coasts of warm and temperate oceans.

Comments :

1. *Balanoglossus* (Fig. 9.1) is commonly known as **acorn worm**.

2. Body is soft and cylindrical having ciliated epidermis reaching a length of 10-50 cm.

3. The body is divisible into short conical **proboscis**, **collar** and a long **trunk**.

4. **Proboscis** has thick muscular walls and its cavity–proboscis coelom opens to the exterior by a probosics pore.

5. **Collar** is short, muscular cylinder-like, enclosing a pair of coelomic cavities (collar coelom) opening by a pair of collar pores on the dorsal surface.

6. **Trunk** is superficially ringed (devoid of segmentation). It is divisible into anterior branchio-genital region, a middle hepatic region and a posterior abdominal region.

7. In the branchiogenital region are present a pair of genital wings formed by internal gonads and a branchial groove having numerous paired gill-slits arranged into two rows.

8. In the hepatic region are present double rows of hepatic caeca.

9. Alimentary canal is straight and anus is present on the posterior end of the body.

10. Sexes are separate and fertilisation is external.

11. Development includes a free-swimming pelagic larva, the **tornaria**.

Geographical distribution. *Balanoglossus* is cosmopolitan or world-wide.

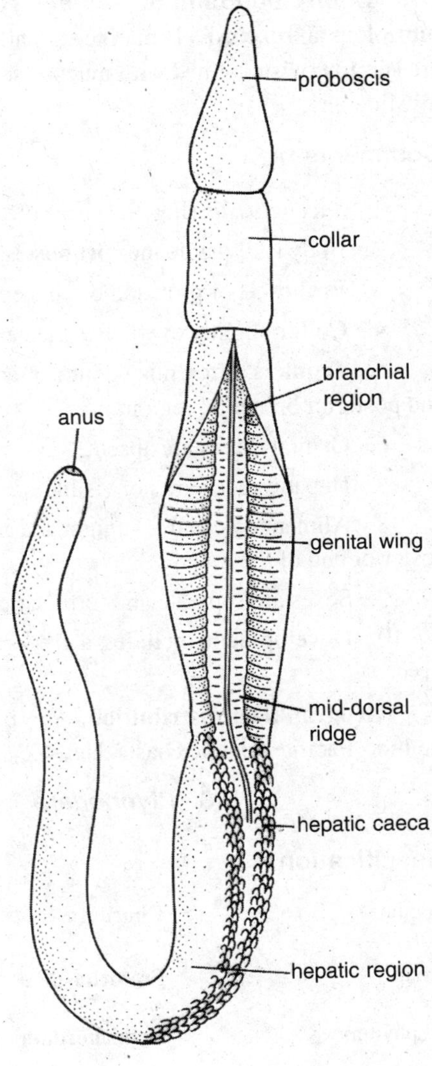

Fig. 9.1. *Balanoglossus.*

2. *Saccoglossus*

Classification :

Phylum	...	**Chordata**	→	Dorsal tubular nerve cord, notochord and paired gill-slits are present.
Group	...	**Protochordata**	→	Marine small forms. Cranium, jaws and paired appendages are absent.
Subphylum	...	**Hemichordata**	→	Solitary or colonial, soft-bodied, coelomate with numerous gill-slits.
Class	...	**Enteropneusta**	→	Alimentary canal straight, two rows of hepatic caeca.
Genus	...	*Saccoglossus*		

Habit and Habitat. *Saccoglossus* is a burrowing, tubicolous and exclusively marine animal. The burrows in sand are spirally twisted, lined with mucous secretions and lie near low tide line.

Comments :

1. *Saccoglossus* (Fig. 9.2) is a soft, cylindrical animal.
2. Body is divisible into **proboscis**, **collar** and **trunk**.
3. **Proboscis** or protosome is exceptionally elongated.
4. **Collar** slightly overhangs the beginning of the trunk.
5. **Trunk** is differentiated into anterior branchial region and posterior branchial region.
6. Genital ridges are absent.
7. Hepatic caeca are not visible.
8. Alimentary canal is straight and anus is present on the posterior end of the body.
9. Sexes are separate and fertilisation is external.
10. Development includes a free-swimming **tornaria** larva.

Geographical distribution. *Saccoglossus* is found in the Indo-Pacific Atlantic coasts, New Zealand and Australia.

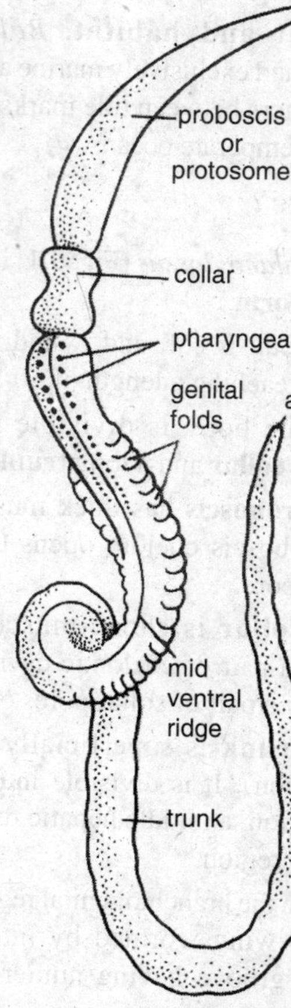

Fig. 9.2. *Saccoglossus.*

3. *Ptychodera*

Classification :

Phylum	...	**Chordata**	→	Dorsal tubular nerve cord, notochord and paired gill-slits are present.
Group	...	**Protochordata**	→	Marine small forms, Cranium, jaws and paired appendages are absent.
Subphylum	...	**Hemichordata**	→	Solitary, soft-bodied, coelomate with numerous gill-slits.
Class	...	**Enteropneusta**	→	Alimentary canal straight, two rows of hepatic caeca.
Genus	...	*Ptychodera*		

Habit and Habitat. *Ptychodera* is a burrowing and exclusively marine animal.

Comments:

1. *Ptychodera* (Fig. 9.3) is a soft, cylindrical animal.
2. The general colour of the animal is yellowish.
3. Its body is elongated and divisible into **proboscis, collar** and **trunk**.
4. **Proboscis** and **collar** are short, while **trunk** is quite elongated.
5. Gonads are situated in the genital wings.
6. **Hepatic** caeca are distinct.
7. **Sexes** are separate.
8. It exhibits bioluminescence.

Geographical distribution. *Ptychodera* is found in the seas of tropical and subtropical regions.

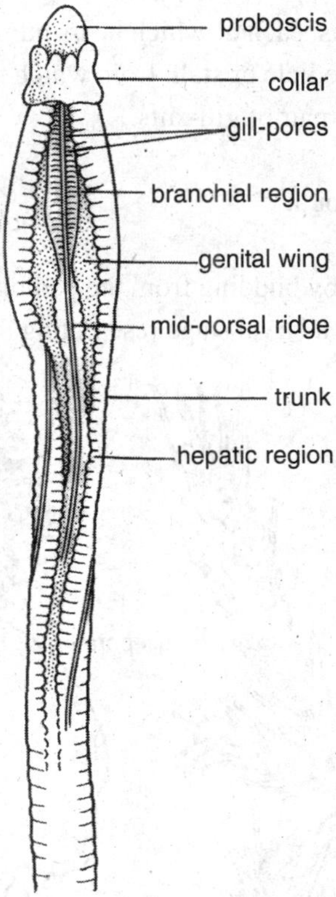

Fig. 9.3. *Ptychodera.*

4. *Cephalodiscus*

Classification :

Phylum	...	**Chordata**	→	Dorsal tubular nerve cord, notochord and paired gill-slits are present.
Group	...	**Protochordata**	→	Marine small forms, Cranium, jaws and paired appendages are absent.
Subphylum	...	**Hemichordata**	→	Solitary or colonial soft-bodied, coelomate with numerous gill-slits.
Class	...	**Pterobranchia**	→	Proboscis with ciliated tentacle, alimentary canal U-Shaped
Order	...	**Cephalodiscida**	→	Colonies with gill-slits and tentaculated arms are present.
Genus	...	*Cephalodiscus*		

Habit and Habitat. *Cephalodiscus* individuals or zooids occur together in large aggregations attached to objects on the sea bottom at a depth of 50 to 60 metres. It is a marine colonial animal.

Comments :

1. *Cephalodiscus* (Fig. 9.4) is a colonial and sedentary animal.
2. The colony consists of minute zooids, each living in a chitinous tube.
3. The chitinous tubes of a colony are held together by a gelatinous mass.
4. The zooids (individuals) have no organic connection with one another.

5. Each zooid is differentiated into **proboscis**, **collar** and **trunk**.
6. **Proboscis** is a shield-shaped lobe overhanging the mouth.
7. **Collar** bears 8–16 arms, each of which bears numerous pinnately arranged tentacles.
8. **Trunk** bears a small pedicle or stalk from which new zooids arise by budding.
9. **Pharynx** has only one pair of gill-slits.
10. Intestine is U-shaped.
11. Anus is close to the mouth.
12. Sexes are separate.
13. Asexual reproduction by budding from the pedicle.
14. Development includes a larval stage resembling the **cyphonaustes**.

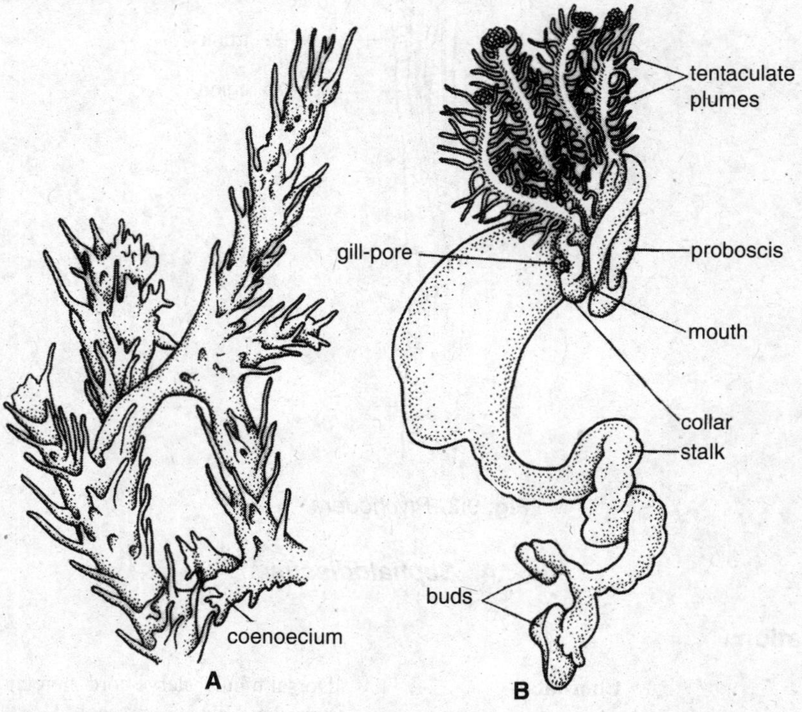

Fig. 9.4. *Cephalodiscus.* A—A part of colony; B—An individual zooid.

Geographical distribution. *Cephalodiscus* is found at various depths in the Indian Ocean. Antarctic and also off the coasts of Japan and Korea.

5. *Rhabdopleura*

Classification :

Phylum	...	**Chordata**	→	Dorsal tubular nerve cord, notochord and paired gill-slits are present.
Group	...	**Protochordata**	→	Marine small forms. Cranium, jaws and paired appendages are absent.
Subphylum	...	**Hemichordata**	→	Solitary or colonial, soft-bodied, coelomate with numerous gill-slits.
Class	...	**Pterobranchia**	→	Proboscis with ciliated tentacles, alimentary canal U-shaped.
Order	...	**Rhabdopleurida**	→	Colonies without gill-slits and single pair of tentaculated arms.
Genus	...	*Rhabdopleura*		

Habit and Habitat. *Rhabdopleura* is a marine and colonial form. The colonies of *Rhabdopleura* are found attached to stones, corals and sessile animals at about 5–6 metres deep in the sea water.

Comments :

1. *Rhabdopleura* (Fig. 9.5) is a true colonial form.
2. The colony has a thin horizontal branching tube called **coenoecium** from which short erect tubes arise, each containing a zooid.
3. Each zooid is minute, divisible into **proboscis**, **collar** and a **trunk**.
4. **Proboscis** is disc-shaped and overhangs the mouth.
5. **Collar** bears a pair of hollow arms, each with a double row of slender tentacles.
6. **Trunk** has a stout pedicle or stalk.
7. Gill-slits are absent.
8. Intestine is U-shaped.
9. Anus is close to the mouth.

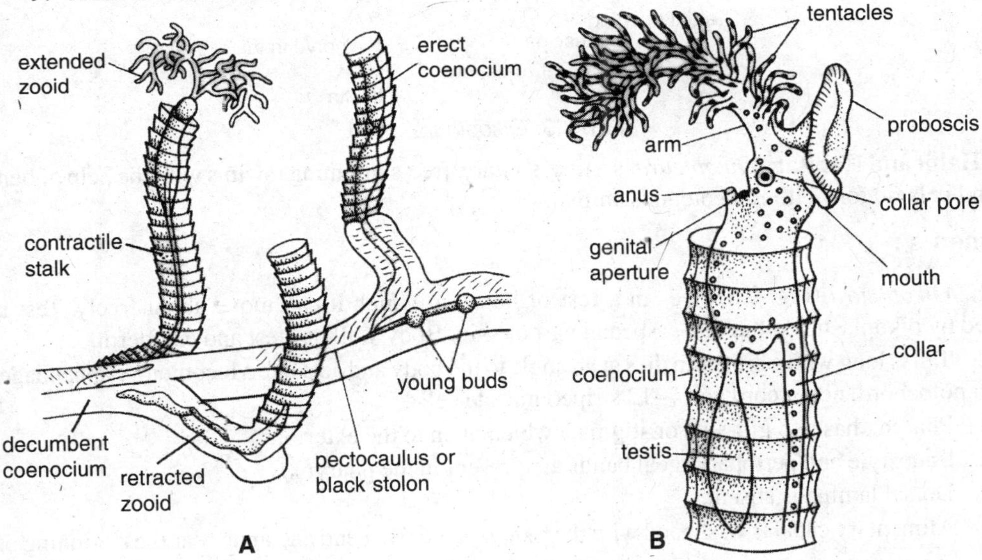

Fig. 9.5. *Rhabdopleura.* A—A portion of a colony; B—An individual zooid.

10. Sexes are separate, each sex has a single gonad with a gonopore on the right side.
11. Asexual reproduction by budding.
12. Development includes a free-swimming larval stage.

Geographical distribution. *Rhabdopleura* is found at moderate depths in Norway, North Atlantic, France, Celebes and South Australia.

6. *Oikopleura*

Classification :

Phylum	...	**Chordata**	→	Dorsal tubular nerve cord, notochord and paired gill-slits are present.
Group	...	**Protochordata**	→	Marine small forms, Cranium, jaws and paired appendages are absent.
Subphylum	...	**Urochordata**	→	Solitary or colonial, fixed or free-swimming with cuticular test containing tunicin.
Class	...	**Larvacea**	→	Neotenic pelagic forms with temporary test.
Genus	...	*Oikopleura*		

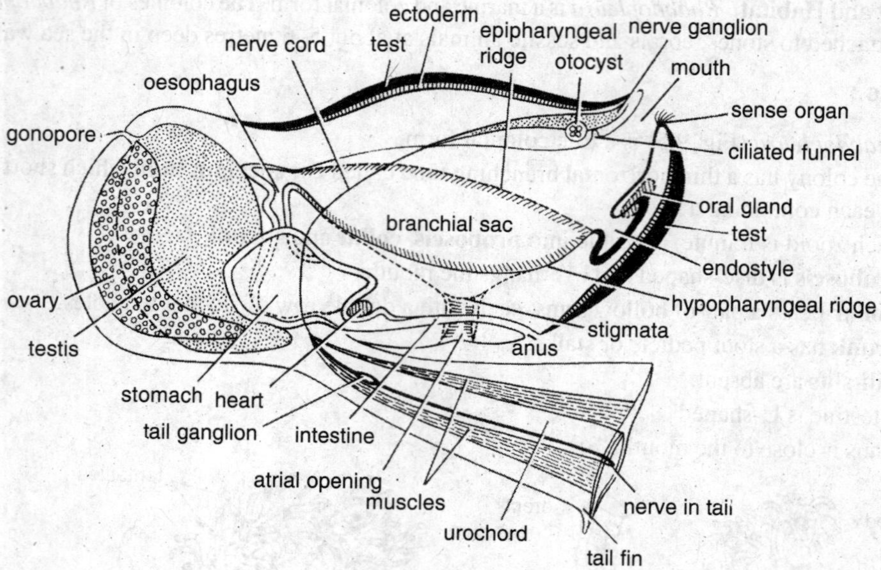

Fig. 9.6. *Oikopleura.*

Habit and Habitat. *Oikopleura* is a tiny, solitary, free-swimming (swims with the help of bent tail) and pelagic form. It is a neotenous animal.

Comments :

1. *Oikopleura* (Fig. 9.6) lives in a **test** or house in which it can move about freely. Test is secreted by oikoplastic epithelium—special part of skin. Body wall has test and ectoderm.
2. Tail is long with a broad fin, lies at an angle to the body and acts like a locomotory appendage. It has a notochord, nerve cord and 7–12 striped muscle cells.
3. Pharynx has two gill-slits or stigmata which open to the exterior.
4. Endostyle and peripharyngeal bands are present in the pharynx.
5. Dorsal lamina is absent.
6. Alimentary canal is U-shaped with the mouth at the front end and anus near the beginning of tail.
7. Dorsal nerve cord extends throughout the length of the body including the tail.
8. Nerve ganglion and otocyst present.
9. The house is provided with a filter apparatus by which the food particles are captured. Food includes microorganisms.
10. Hermaphrodite. Testis and ovary lie in close approximation. Gonopore at the postero-dorsal surface of the body.

Geographical distribution. *Oikopleura* is widely distributed in open sea.

Identification. Body is of peculiar shape with a persistent tail having nerve, notochord and muscle bands.

7. *Herdmania*

Classification :

Phylum	…	**Chordata**	→	Dorsal tubular nerve cord, notochord and paired gill-slits are present.
Group	…	**Protochordata**	→	Marine and small forms, Cranium, jaws and paired appendages are absent.

Subphylum	...	**Urochordata** →	Solitary or colonial, fixed or free-swimming with cuticular test containing tunicin.
Class	...	**Ascidiacea** →	Well-developed and permanent test. Branchial sac large perforated by numerous gill-slits.
Order	...	**Ascidiae simplices** →	Solitary forms, exhibit retrogressive metamorphosis.
Genus	...	*Herdmania*	
Species	...	*pallida*	

Habit and Habitat. *Herdmania pallida* is a solitary, sedentary marine form. They sometime live as commensal with gastropod shells like *Xanchus* species (Shankh and Conch).

Comments :

1. *Herdmania* (Fig. 9.7) is commonly called **sea squirt**.
2. Animal is attached at its postero-ventral end by a foot, which is large, rough, leathery with attached foreign objects.
3. Body is roughly oblong in outline, about 9.5 to 12 cm long and 5.2 to 6.9 cm in breadth, and enclosed in a soft leathery test or tunic formed of tunicin and protein.
4. The colour of the body of a fresh specimen is pink.
5. Free end (antero-dorsal) of the body is provided with **branchial** and the **atrial** apertures located over branchial and atrial siphons respectively.
6. Mouth opens by branchial aperture, while anus by atrial aperture.
7. Pharynx is sac-like perforated by numerous stigmata.
8. Alimentary canal is U-shaped. Food comprises chiefly microscopic plants and animals, *viz.,* diatoms, algae and infusorians.
9. Respiration by branchial-sac. Test also acts as an accessory respiratory organ.
10. Blood vascular system is open type.
11. Excretory organ is neural gland situated above the nerve ganglion.
12. Sexes are united or hermaphroditic and protogynous. Gonads ony a single pair.
13. Fertilisation is external. Development includes a free-swimming tadpole larva having all the chordate characters (dorsal nerve cord, and notochord).
14. Metamorphosis is **retrogressive** in which nerve cord, notochord, tail with tail fins degenerate. Thus, it looses chordate characters.

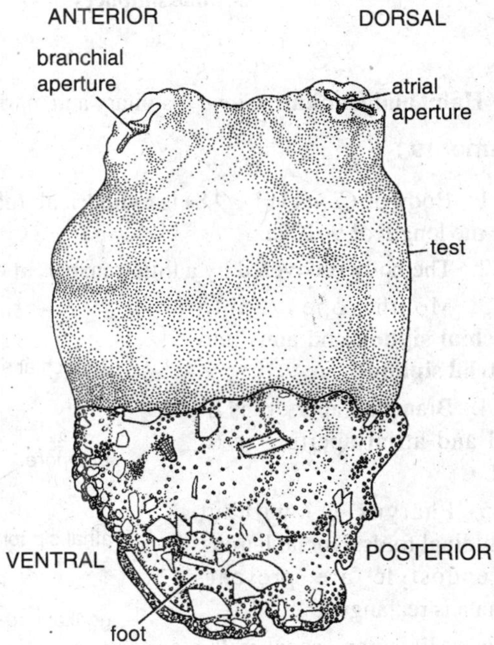

Fig. 9.7. *Herdmania pallida.*

Geographical distribution. *Herdmania pallida* is found in Indian, Pacific and the Atlantic Oceans and also from Malya and West Indies.

8. *Ciona*

Classification :

Phylum	...	**Chordata**	→	Dorsal tubular nerve cord, notochord and paired gill-slits are present.
Group	...	**Protochordata**	→	Marine and small forms. Cranium, jaws and paired appendages are absent.
Subphylum	...	**Urochordata**	→	Solitary or colonial, fixed or free-swimming with cuticular test containing tunicin.
Class	...	**Ascidiacea**	→	Well-developed and permanent test. Branchial sac large perforated by numerous gill-slits.
Order	...	**Ascidiae simplices**	→	Solitary forms, exhibit retrogressive metamorphosis.
Genus	...	*Ciona*		

Habit and Habitat. *Ciona* is solitary and marine form, found attached to rocks, etc, (sedentary).

Comments :

1. Body of *Ciona* (Fig. 9.8) is cylindrical, tubular, transparent and longer than broad. It may reach the length of 9–12 cm.

2. The body is covered by a thin transparent outer **test** and inner **mantle**.

3. Mouth opens into **branchial siphon** and anus into the **atrial siphon**.

4. **Branchial aperture is 8-**lobed and atrial aperture is 6-lobed.

5. Pharynx without longitudinal folds. Dorsal lamina and endostyle are present. Stigmata is rectangular.

6. Alimentary canal is U-shaped and consists of pharynx, oesophagus, stomach and intestine. It lies in the pericardial cavity.

7. Heart is surrounded by pericardium.

8. Longitudinal muscle bands are clearly seen.

9. Dorsal tubercle is horse-shoe-shaped.

10. Hermaphroditic and protogynous. Gonads lie in the loop of intestine and open by genital pore through genital duct.

11. Fertilisation is external. Eggs develop into an oikopleura type larva that settle and undergo metamorphosis to form adults.

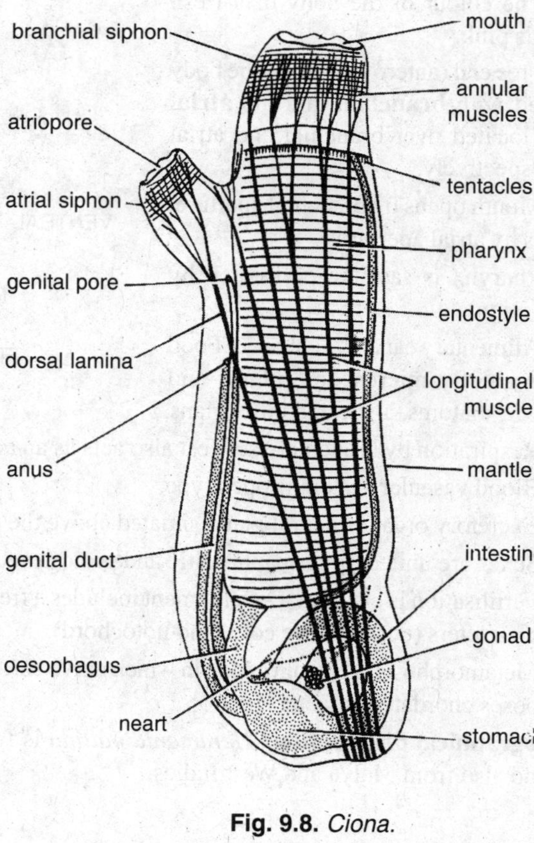

Fig. 9.8. *Ciona.*

12. Blood of *Ciona* contains a green pigment **haemovanadin** that contains **vanadium** produced by vanadocytes. It has no oxygen carrying capacity.

Geographical distribution. *Ciona* is commonly found in Western regions in shallow waters.

9. *Ascidia*

Classification :

Phylum	...	**Chordata**	→	Dorsal tubular nerve cord, notochord and paired gill-slits are present.
Group	...	**Protochordata**	→	Marine and small forms. Cranium, jaws and paired appendages are absent.
Subphylum	...	**Urochordata**	→	Solitary or colonial, fixed or free-swimming with cuticular test containing tunicin.
Class	...	**Ascidiacea**	→	Well-developed and permanent test. Branchial sac large perforated by numerous gill-slits.
Order	...	**Ascidiae simplices**	→	Solitary forms, exhibit retrogressive metamorphosis.
Genus	...	*Ascidia* (Sea squirt)		

Habit and Habitat. *Ascidia* is solitary, occurs in large association attached to the rocky shores or wharf piles in shallow water, while some occur deep up to 5500 metres.

Comments :

1. *Ascidia* (Fig. 9.9) is commonly called **sea squirt**.

2. The shape of the body is like a short cylinder with a broad base by which it is attached to the rocks.

3. Outer covering of the body is wrinkled, brown and transluscent and is known as **test** beneath which lies the mantle.

4. The free terminal end bears a large 8-lobed rounded aperture, the **branchial aperture** and on one side just behind the branchial aperture lies the six lobed **atrial aperture.**

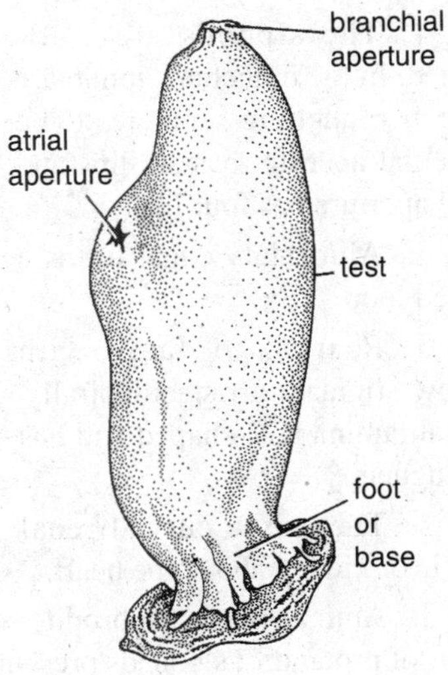

Fig. 9.9. *Ascidia.*

5. Branchial aperture leads into the branchial chamber or pharynx which is perforated by a number of stigmata. Pharynx contains dorsal lamina and endostyle.

6. Alimentary canal consists of mouth, pharynx, stomach, intestine, rectum and anus.

7. Heart is a simple muscular tube, situated near the stomach in the pericardium.

8. Nervous system comprises the single nerve ganglion.

9. Neural gland lies on the ventral side of the nerve ganglion.

10. Sexes are united. Ovary and testes are situated close together on the left side of the body.

11. Reproduction is sexual. Exhibits retrogressive metamorphosis. Tailed-larva occurs in development which shows all the chordate characters.

Geographical distribution. *Ascidia* is distributed in cold temperate regions especially in Europe.

10. *Molgula*

Classification :

Phylum	...	**Chordata**	→	Dorsal tubular nerve cord, notochord and paired gill-slits are present.
Group	...	**Protochordata**	→	Marine and small forms. Cranium, jaws and paired appendages are absent.
Subphylum	...	**Urochordata**	→	Solitary or colonial, fixed or free-swimming with cuticular test containing tunicin.
Class	...	**Ascidiacea**	→	Well-developed and permanent test. Branchial sac perforated by numerous gill-slits.
Order	...	**Ascidiae simplices**	→	Solitary forms, exhibit retrogressive metamorphosis.
Genus	...	*Molgula*		

Habit and Habitat. *Molgula* is a solitary, sedentary marine form. It is found attached to wharfs, piers, stones and ship's bottom in shallow coastal waters of Atlantic ocean.

Comments :

1. *Molgula* (Fig. 9.10) is oval or spherical in outline. It is about 1 to 3 cm in diameter.
2. Foot is absent.
3. Test is thin and transparent. It is often encrusted with shells and sand.
4. The siphons are well-formed. In the branchial siphon are sixteen branching tentacles. The branchial aperture has six lips and atrial aperture has four lips.
5. Alimentary canal forms a curved loop.
6. Pharynx is large with narrow stigmata arranged spirally. Dorsal lamina is C-shaped and has no lauguets.
7. There is a curved renal vesicle or kidney above the heart.
8. Animal is hermaphrodite. A pair of protandrous gonads present.

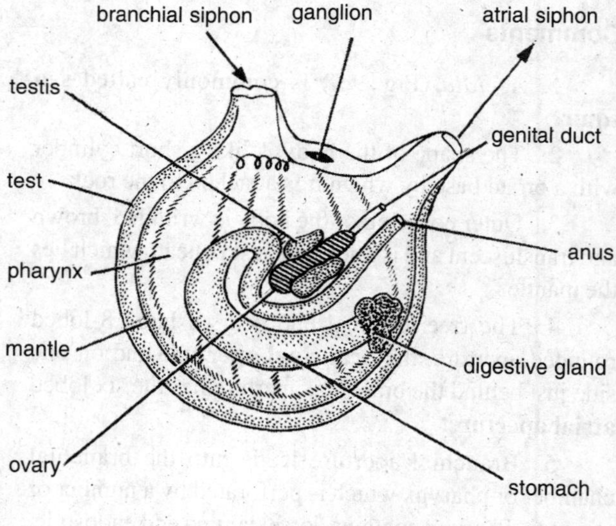

Fig. 9.10. *Molgula.*

9. *Molgula* is usually oviparous, the tadpole larva has no ocelli and in some species no tail. Some species of *Molgula* are viviparous giving birth to larva which may or may not have a tail in different species.

Geographical distribution. *Molgula* is found abundantly in the Atlantic and less widely in Indian ocean being fixed to wharps, piers and rocks, etc.

11. *Botryllus*

Classification :

Phylum	...	**Chordata**	→	Dorsal tubular nerve cord, notochord and paired gill-slits are present.
Group	...	**Protochordata**	→	Marine and small forms. Cranium, jaws and paired appendages are absent.
Subphylum	...	**Urochordata**	→	Solitary or colonial, fixed or free-swimming with cuticular test containing tunicin.
Class	...	**Ascidiacea**	→	Well-developed and permanent test. Branchial sac-large perforated by numerous gill-slits.
Order	...	**Ascidiae composite**	→	Colonial forms. Larva develops in thin atrial cavity.
Genus	...	*Botryllus*		

Habit and Habitat. *Botryllus* is sedentary, compound or colonial ascidian. It is specially abundant in harbours attached to rocks, wharfs, piers and ship's bottom. *Botryllus* exhibits colour changes.

Comments :

1. The zooids of a colony are embedded in a common gelatinous test.
2. The zooids or individuals form groups arranged like rosette or in star-shaped fashion in a colony.
3. Body of a zooid is short and compact and does not exhibit division into different regions.
4. Each zooid has a separate oral or branchial aperture directed outwards. Atrial apertures of the zooids of a colony open into a central common cloaca, which opens outside by a common cloacal opening.
5. Dorsal lamina is without languets.
6. Gill-slits are rectangular.
7. Intestine is U-shaped.
8. Neural gland is dorsal to nerve ganglion.
9. Two hermaphrodite gonads are present.
10. Asexual reproduction by budding occurs inside the atrial cavity. Zooids are formed by budding and are embedded in a common gelatinous mass.

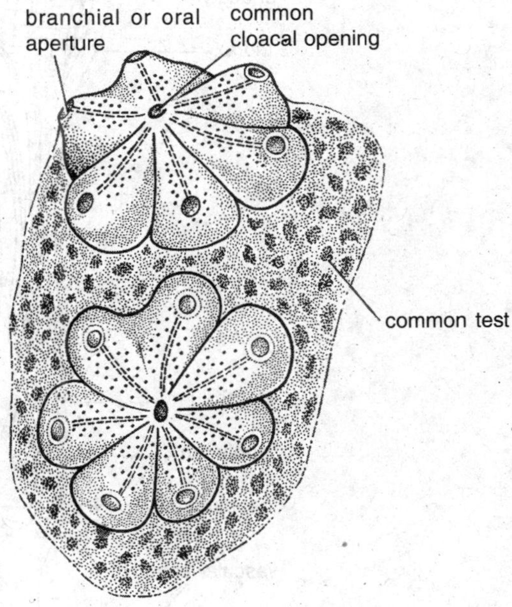

Fig. 9.11. *Botryllus.*

Geographical distribution. *Botryllus* is distributed in the Atlantic and Mediterranean seas.

12. *Clavellina*

Classification :

Phylum	...	**Chordata**	→	Dorsal tubular nerve cord, notochord and paired gill-slits are present.

Group	...	**Protochordata**	→	Marine and small forms. Cranium, jaws and paired appendages are absent.
Subphylum	...	**Urochordata**	→	Solitary or colonial, fixed or free-swimming with cuticular test containing tunicin.
Class	...	**Ascidiacea**	→	Well-developed and permanent test. Branchial sac-large perforated by numerous gill-slits.
Order	...	**Ascidiae composite**	→	Colonial forms. Larva develops within atrial cavity.
Genus	...	*Clavellina*		

Habit and Habitat. *Clavellina* is a colonial ascidian forming large clusters and found on rocks in shallow seas.

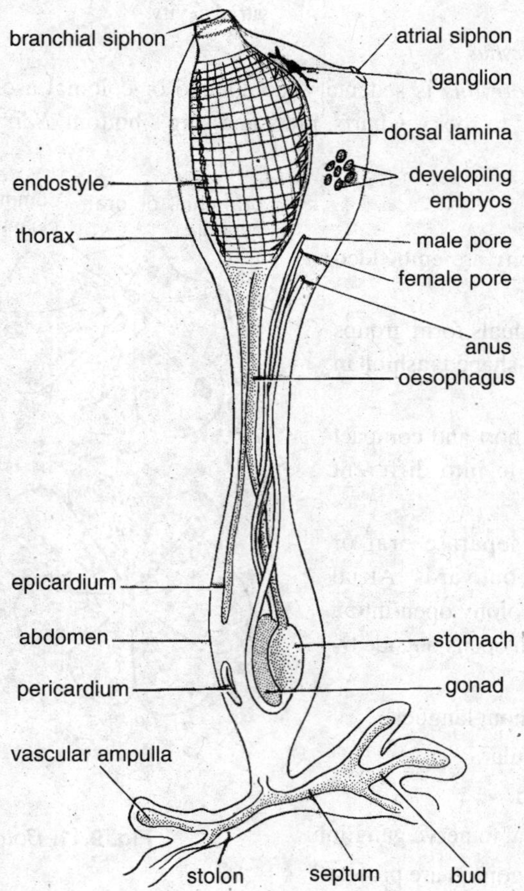

Fig. 9.12. *Clavellina (single zooid).*

Comments :

1. The colony of *Clavellina* (Fig. 9.12) consists of several independent individuals called **zooids** attached loosely by basal part of the test.

2. The basal tubular extensions of the test and visceral region form branching and horizontally creeping stolon which is fixed to a rock. Zooids project from the stolon.

3. The zooids are free from each other, but all zooids are connected by a common blood system.
4. Each zooid is about 2.5 cm long having a transparent test.
5. A zooid has an upper **thorax** and a lower **abdomen**.
6. Each zooid has two short siphons branchial and atrial. Pharynx with dorsal lamina with languets and a dorsal tubercle.
7. The abdomen is narrow and long, it contains the rest of the alimentary canal, heart and gonad.
8. A narrow tubular outgrowth of the visceral region extends from the pharynx through the abdomen, it is called **epicardium** and forms the stolon.
9. Reproduction is both sexual and asexual.
10. The fertilised eggs develop in the atrial cavity, the tadpole larvae escape for a short free-swimming existence. Thus, *Clavellina* is **viviparous**.
11. The larva gets fixed and metamorphoses into an **oozoid**. The oozoid forms a stolon; new zooids arise asexually from the terminal branches of the stolon to form a colony.
12. In autumn, the zooids of a colony disappear, leaving the stolon into which the substance of zooids is passed as food-laden **trophocytes**, which survive the winter, new zooids arise from the stolon next spring.
13. *Clavellina* is of interest because it links the simple ascidians to compound ones.

Geographical distribution. *Clavellina* is found in shallow waters of European seas.

13. *Pyrosoma*

Classification :

Phylum	...	**Chordata**	→	Dorsal tubular nerve cord, notochord and paired gill-slits are present.
Group	...	**Protochordata**	→	Marine and small forms. Cranium, jaws and paired appendages are absent.
Subphylum	...	**Urochordata**	→	Solitary or colonial, fixed or free-swimming with cuticular test containing tunicin.
Class	...	**Ascidiacea**	→	Well-developed and permanent test. Branchial sac-large perforated by numerous gill-slits.
Order	...	**Ascidiae luciae**	→	Colonial forms, phosphorescent.
Genus	...	*Pyrosoma*		

Habit and Habitat. *Pyrosoma* is a free-swimming, pelagic colonial form.

Comments :

1. Colony of *Pyrosoma* (Fig. 9.13) consists of numerous individuals (blastozooids) forming a transluscent gelatinous tube like a hollow cylinder of 25 cm to 1 metre in length. Individuals are embedded in a common test having test processes and terminal opening.
2. One end of the hollow cylinder is open and truncate, while the other end is closed and rounded.
3. The oral apertures of the zooids are open on the outer surface of the cylinder, while the atrial apertures of the zooids open into the **common cloaca**, the hollow of the cylinder.
4. Common cloaca opens to the outside by single aperture which is guarded by a velum or diaphragm.
5. Pharynx is divided into anterior short and posterior branchial sac having numerous elongated gill-slits, endostyle and dorsal lamina.

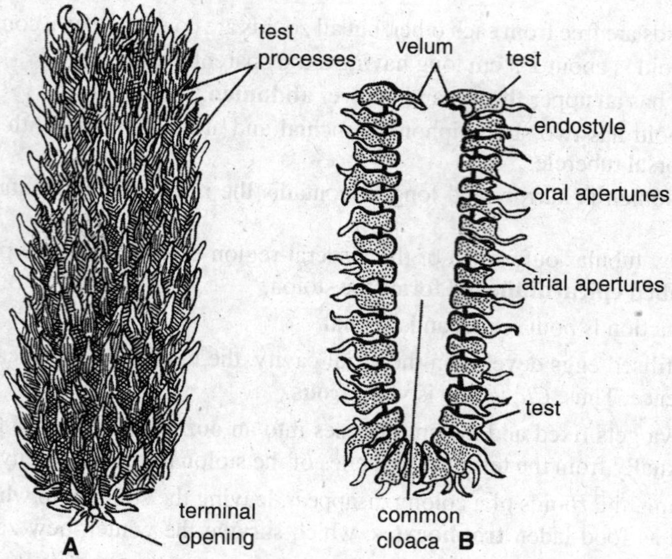

Fig. 9.13. *Pyrosoma.* A—Colony ; B — Saggital Section of colony.

6. Hermaphrodite. Reproduction sexual and asexual.
7. Development direct without a tailed-larva.
8. Phosphorescence (emission of light) is a characteristic feature of the colony. The light is produced by the symbiotic bacteria living in cells in various parts of the body.
9. Colony moves by jet propulsion by rhythmic contraction of muscle bands.

Geographical distribution. *Pyrosoma* is found in warmer seas of tropical and subtropical regions up to a depth of 500 meters, few are abyssal.

14. *Doliolum*

Classification :

Phylum	...	**Chordata**	→	Dorsal tubular nerve cord, notochord and paired gill-slits are present.
Group	...	**Protochordata**	→	Marine and small forms. Cranium, jaws and paired appendages are absent.
Subphylum	...	**Urochordata**	→	Solitary or colonial, fixed or free-swimming with cuticular test containing tunicin.
Class	...	**Thaliacea**	→	Free-swimming pelagic, solitary or colonial. Muscles in the form of circular bands. Exhibit an alternation of generations.
Order	...	**Cyclomyaria**	→	Muscle bands are always complete surrounding the body.
Genus	...	*Doliolum*		

Habit and Habitat. *Doliolum* is a free-swimming, pelagic form inhabiting the warm marine water.

Comments :

GENERAL ORGANISATION

1. *Doliolum* (Fig. 9.14) is polymorphic having two phases : solitary phase or gonozooid and asexual form or oozooid.

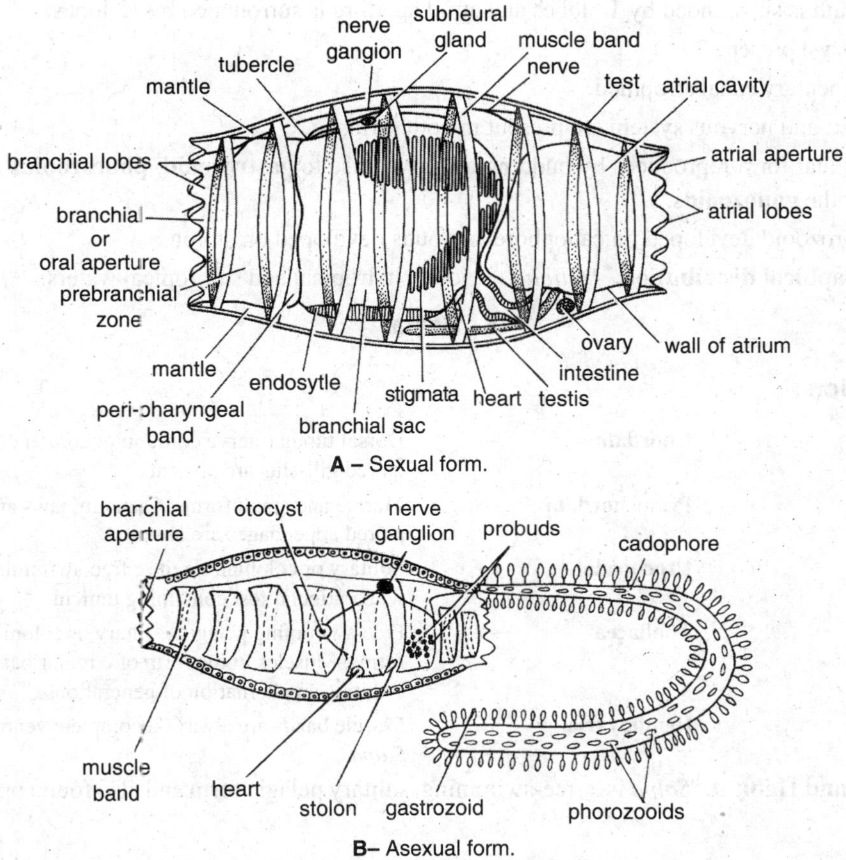

Fig. 9.14. *Doliolum* A—Sexual form; B—Asexual form.

Sexual form or Gonozoid :

2. Body is barrel-shaped measuring 1–1.5 cm with mouth and anus at opposite ends.

3. The test or tunic is thin, transparent and contains no cells. Muscle bands are circular and eight in number.

4. The mouth or branchial aperture and atrial aperture are surrounded by 12 and 10 lobes respectively.

5. Alimentary canal includes large mouth that leads into spacious pharynx, oesophagus, stomach and intestine opening into atrial cavity through anus.

6. Pharynx is perforated by dorsal and ventral rows of stigmata and also bears ventral endostyle and peripharyngeal ciliated bands. Dorsal lamina and tentacles are absent.

7. Dorsal ganglion, subneural glands and dorsal tubercle are situated near the anterior end.

8. Cadophore and otocyst absent.

9. Hermaphrodite. Gonads unpaired. Fertilised ovum gives rise to a tailed-larva which metamorphoses into an asexual zooid, the **oozooid**.

10. Life cycle exhibits an alternation of sexually mature adult, the **gonozooid** with the asexually reproducing phase, the **oozoid**.

Asexual form or Oozooid :

1. It possesses nine circular muscular bands.

2. Mouth is surrounded by 10 lobes and atrial aperture is surrounded by 12 lobes.
3. Otocyst present.
4. Alimentary canal atrophied.
5. Heart and nervous system are present in both forms.
6. Asexual form reproduces by budding and gives rise to **gastrozooid**, **phorozooids** and the sexual forms the **gonozooids**.
7. Phorozooid developed on cadophore and buds developed on stolon.

Geographical distribution. *Doliolum* is found in tropical and subtropical waters.

15. *Salpa*

Classification :

Phylum	...	**Chordata**	→	Dorsal tubular nerve cord, notochord and paired gill-slits are present.
Group	...	**Protochordata**	→	Marine and small forms. Cranium, jaws and paired appendages are absent.
Subphylum	...	**Urochordata**	→	Solitary or colonial, fixed or free-swimming with cuticular test containing tunicin.
Class	...	**Thaliacea**	→	Free-swimming pelagic, solitary or colonial forms. Muscles in the form of circular bands. Exhibits an alternation of generations.
Order	...	**Hemimyaria**	→	Muscle bands are always incomplete ventrally.
Genus	...			*Salpa*

Habit and Habitat. *Salpa* is a free-swimming, solitary pelagic form and also found up to 200 metres deep.

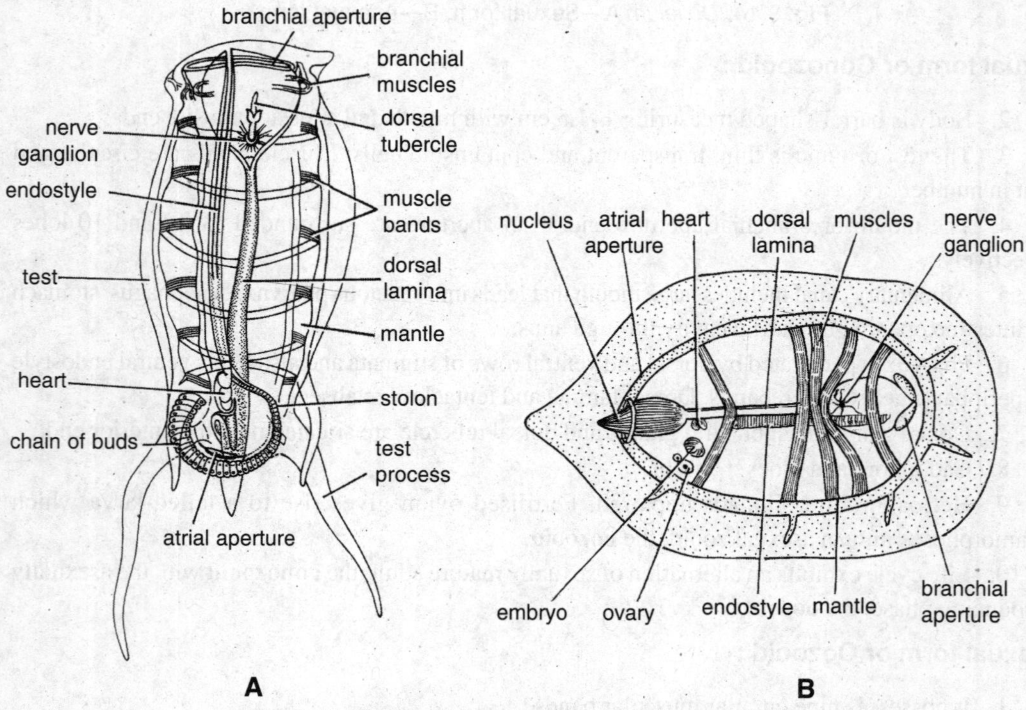

Fig. 9.15. Salpa. A-Asexual phase; B-Sexual phase.

Comments :

General Organisation :

Asexual form or Oozooid :

1. Body is barrel-shaped measuring 1–8 cm in length with mouth and atrial apertures at topposite ends.
2. Test is transparent and produces into two atrial processes posteriorly. Mantle lies beneath the test.
3. Pharynx has ventral endostyle, peripharyngeal bands, dorsal lamina and a pair of gills.
4. A single tentacle is present on the dorsal surface.
5. Oesophagus, stomach, intestine and heart form a compact mass at the postero-ventral side called the **nucleus**.
6. Neural complex is present as usual.
7. From the ventral body wall is given off an outgrowth known as **stolon**. Stolon gives rise to numerous buds which arise as a chain of zooids.
8. The chain of zooids develop sex organs and break off from the parent colony.
9. Animal moves by jet propulsion with the help of muscle bands which contract rhythmically.

Sexual phase or Blastozooid :

1. Hermaphroditic, protogynous. Gonads lying in the nucleus, gonoducts opening separately into the atrium.
2. Ovary and paired testis mature at different times.
3. Only one egg is produced at a time which develops in the atrial cavity.
4. No tailed-larva in the development.
5. Embryo develops into an asexual form and separates from the parent. Thus, alternation of generations take place in the life history of *Salpa*.

Geographical distribution. *Salpa* is generally found in all tropical seas.

16. *Amphioxus*

Classification :

Phylum	...	**Chordata**	→	Dorsal tubular nerve cord, notochord and paired gill-slits are present.
Group	...	**Protochordata**	→	Marine and small forms. Cranium, jaws and paired appendages are absent.
Subphylum	...	**Cephalochordata**	→	Solitary forms. Notochord and nerve cord extend the entire length of the body.
Class	...	**Leptocardii**	→	Numerous paired pharyngeal gill-slits.
Genus	...	*Amphioxus* (*Branchiostoma*)		
Species	...	*lanceolatum*		

Habit and Habitat. *Amphioxus* is a fish-like burrowing marine animal. It burrows in the sand by its posterior end, anterior end protrudes out of the sand for feeding. It is a ciliary feeder.

Comments :

1. *Amphioxus* (Fig. 9.16) is commonly called **lancelet**.
2. Body is elongated, measuring 5 cm in length, laterally compressed and pointed at both ends. The anterior end projects forwards as the rostrum.

3. The fins are low and continuous with each other, a dorsal, a ventral and a caudal fin. There are two **metapleural folds**, extending up to atriopore.

4. The **myotomes** are arranged on both the sides separated by V-shaped connective tissue partitions, the **myosepta** or **myocommata**.

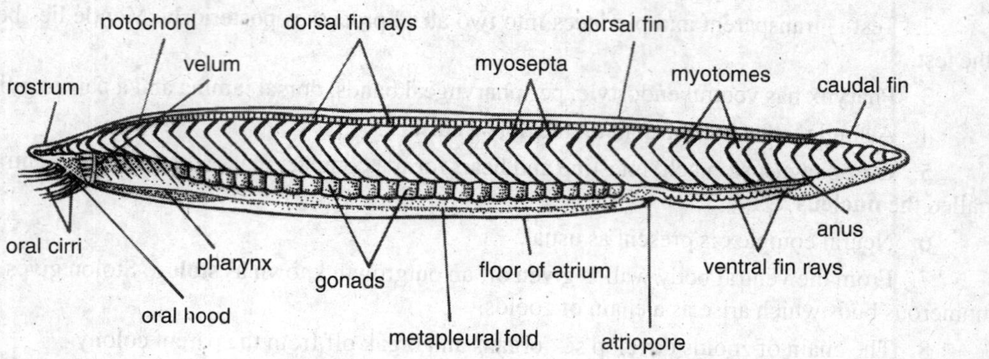

Fig. 9.16. *Amphioxus.*

5. The **atriopore** is median and ventral and lies at the junction of metapleural folds and ventral fin.

6. Alimentary canal is straight. The mouth is ventral to rostrum and is guarded by the **oral hood** bearing numerous **oral cirri**. The anus lies on the left side, a short distance in front of the posterior end.

7. Notochord extends throughout the length of the body and nerve cord lies above the notochord whose anterior end is swollen forming the unspecialised brain.

8. The gonads consist of 26 pairs and are metamerically arranged on both the sides of the pharynx, beneath the myotomes.

9. The sexes are separate but externally not distinguishable (absence of sexual dimorphism).

Geographical distribution. *Amphioxus lanceolatum* is worldwide in distribution being found on the shores of all continents and many oceanic islands.

Special features. *Amphioxus* shows primitive, specialised and degenerate characters.

Primitive characters are notochord extending into snout, segmented myotomes, absence of paired fins and jaws, straight intestine, ciliary feeding and tubular heart.

Specialised characters are eleborate oral hood and velum and several pharyngeal gill-slits.

Degenerate features are unspecialised, reduced brain and sense organs.

Chapter 10

CLASS—CYCLOSTOMATA

GENERAL CHARACTERS

1. Body is long, rounded and eel-like.
2. Skin is soft, smooth and without exoskeleton.
3. Nostril is single and median.
4. Mouth is suctorial devoid of functional jaws.
5. Paired fins or lateral appendages are absent.
6. Notochord is persistent.
7. Skeleton is cartilaginous.
8. Heart is two chambered and aortic arches are many.
9. Single gonad without duct.
10. Development is direct or indirect.

CLASS—CYCLOSTOMATA

Orders

1	2
Petromyzontia	Myxinoidea
Ex. *Petromyzon*	**Ex. 1.** *Myxine*
	2. *Bdellostoma*

Order 1. Petromyzontia

1. Dorsal fin well-developed.
2. The naso-hypophysial sac terminates posteriorly in a blind sac, *i.e.*, it does communicate with the mouth.
3. Mouth suctorial with rasping tongue.
4. Seven pairs of gill-slits. The gills open into a respiratory tube below the oesophagus.
5. Branchial basket complete.
6. Development is indirect.

Example : *Petromyzon.*

Order 2. Myxinoidea

1. Dorsal fin absent.
2. The naso-hypophysial sac opens posteriorly in the mouth.
3. Mouth is terminal and surrounded by 3 or 4 pairs of small tentacles.
4. Gill-slits 6–14 pairs.
5. Branchial-basket is reduced.
6. Development is direct.

Example : 1. *Myxine,* **2.** *Bdellostoma.*

MUSEUM SPECIMENS OF CYCLOSTOMATA

1. *Petromyzon (= Lampetra)*

Classification :

Phylum	...	**Chordata**	→	Dorsal tubular nerve cord, notochord and paired gill-slits are present.
Group	...	**Vertebrata**	→	Notochord is replaced by vertebral column;two pairs of appendages; circulatory system closed; hepatic portal system present; blood red containing RBCs.
Subphylum	...	**Agnatha**	→	Jaws and appendages are absent.
Class	...	**Cyclostomata**	→	Mouth suctorial devoid of functional jaws.
Order	...	**Petromyzontia**	→	Well-developed dorsal fin; complete branchial-basket; naso-hypophysial sac blind; seven pairs of gill-slits.
Genus	...	*Petromyzon*		
Species	...	*marinus*		

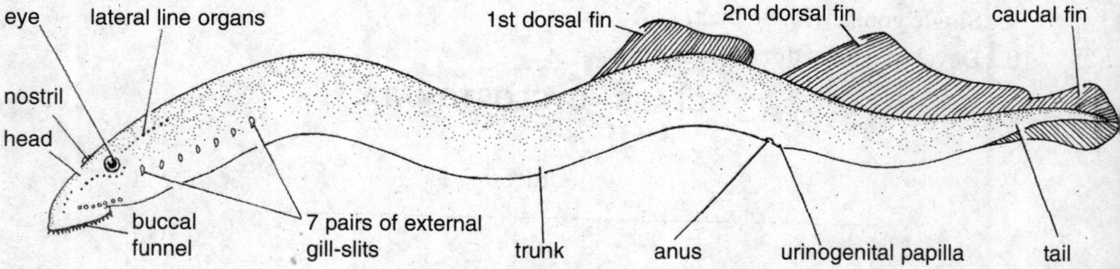

Fig. 10.1. *Petromyzon marinus.*

Habit and habitat. *Petromyzon* is found in freshwater and salt waters. They lead an ectoparasitic life on other fishes, being attached by buccal funnel and secrete an anticoagulant for sucking blood.

Comments:

1. *Petromyzon* (Fig. 10.1) is commonly known as **lamprey** or **lamper eel**.
2. Body is eel-like measuring about 1 metre and differentiated into head, trunk and tail.
3. The surface of the body is smooth and slimy and generally heavily pigmented.
4. Two dorsal fins and one caudal fin, all are supported by cartilaginous rays and are confluent.
5. Mouth is circular and surrounded by a large, ventral suctorial funnel with numerous horny teeth. Tongue is piston-like and toothed.
6. The paired eyes are relatively large and functional. Two small median eyes—the pineal and parietal eyes.

7. **Nostril** is single and dorsal.

8. Seven pairs of **external gill-apertures** and well-developed branchial-basket.

9. Jaws and paired fins are absent.

10. Sexes are separate in adults and there is only single large gonad. Female with large anal fin and male with urinogenital (copulatory) papilla.

11. Fertilisation is external. Development includes **ammocoete** larva. It is regarded as connecting link between cyclostomes and *Amphioxus*.

Economic value. Lampreys have no food value, destroy fishes by sucking their blood and cause infection also. Larval lampreys are used as bait for fishing.

Special features. The nearest allies of lampreys are ancient ostracoderms (Silurian and Devonian periods). They have no fossil record.

Geographical distribution. *Petromyzon* has an almost worldwide distribution, being present in both salt and freshwaters of North America, Europe, West Africa, Japan, Chili, Australia, New Zealand and Tasmania.

2. *Myxine*

Classification :

Phylum	...	**Chordata**	→	Dorsal tubular nerve cord, notochord and paired gill-slits are present.
Group	...	**Vertebrata**	→	Notochord is replaced by vertebral column; two pairs of appendages; circulatory system closed; hepatic portal system present; blood red containing RBCs.
Subphylum	...	**Agnatha**	→	Jaws and appendages are absent.
Class	...	**Cyclostomata**	→	Mouth suctorial devoid of functional jaws.
Order	...	**Myxinoidea**	→	Dorsal fin absent; branchial-basket reduced; naso-hypophysial sac opens in the mouth; mouth terminal surrounded by 3 or 4 pairs of small tentacles; 6–14 pairs of gill-slits.
Genus	...	*Myxine*		
Species	...	*glutinosa*		

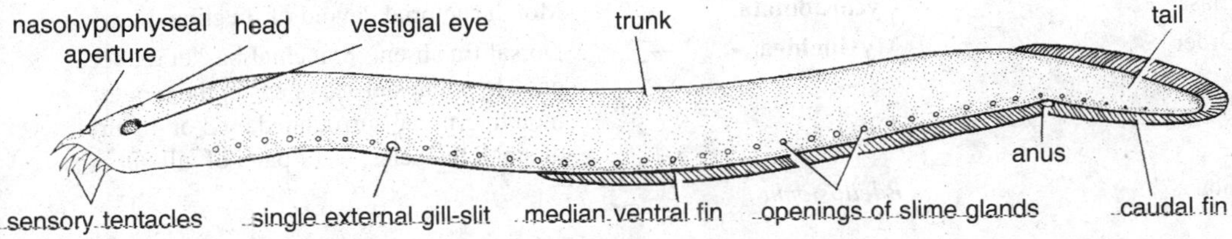

Fig. 10.2. *Myxine glutinosa.*

Habit and Habitat. *Myxine* is found buried in the muddy sea bottom. They are nocturnal feeders, parasitic or quasi-parasitic because they are sometimes found within the bodies of their prey, which are fishes of various types to eat their viscera and muscles. During the daytime, they live buried in the bottom mud at depths of over 650 metres.

Comments :

1. *Myxine* (Fig. 10.2) is commonly known as **hagfish**.

2. Body is eel-like, measuring about 60 cm in length and differentiated into head, trunk and tail.

3. The surface of the body is soft and smooth without scales. Large mucous glands are present opening by **mucous pores**.

4. The mouth is terminal and surrounded by soft lips and four pairs of short tentacles supported by skeletal rods.

5. Buccal funnel and jaws are absent. Branchial-basket is also reduced.

6. Nostril is single, lies very close to the mouth and opens terminally.

7. Single pineal eye is visible on the top of the head. Paired eyes are vestigial or degenerated due to bottom dwelling habit.

8. Six to fourteen pairs of gills which do not open separately to the outside but open by a single ventral **external gill-opening**.

9. Single median fin runs from about the middle of the ventral surface extending around the tail region.

10. Hermaphroditic and protandrous with single **ovotestis**, anterior part being ovary and posterior testis. Eggs are enclosed in horny shell having hooks for attaching with the weeds.

11. Development is direct.

12. Hagfishes are injurious to fish industry.

Identification. *Myxine* is eel-like and has 4 pairs of tentacles, single ventral gill-opening and no buccal funnel.

Geographical distribution. *Myxine* has a wide distribution along sea coasts of both Atlantic and Pacific Oceans, occurring in the waters of Northern Europe, North Atlantic, America, Chile and Japan, etc.

3. *Bdellostoma*

Classification :

Phylum	...	**Chordata**	→	Dorsal tubular nerve cord, notochord and paired gill-slits are present.
Group	...	**Vertebrata**	→	Notochord is replaced by vertebral column; two pairs of appendages; circulatory system closed; hepatic portal system present; blood red containing RBCs.
Subphylum	...	**Agnatha**	→	Jaws and paired appendages are absent.
Class	...	**Cyclostomata**	→	Mouth suctorial devoid of functional jaws.
Order	...	**Myxinoidea**	→	Dorsal fin absent; branchial-basket reduced, naso-hypophysial sac opens in the mouth; mouth terminal surrounded by 3 or 4 pairs of small tentacles, 6–14 pairs of gill-slits.
Genus	...	*Bdellostoma*		

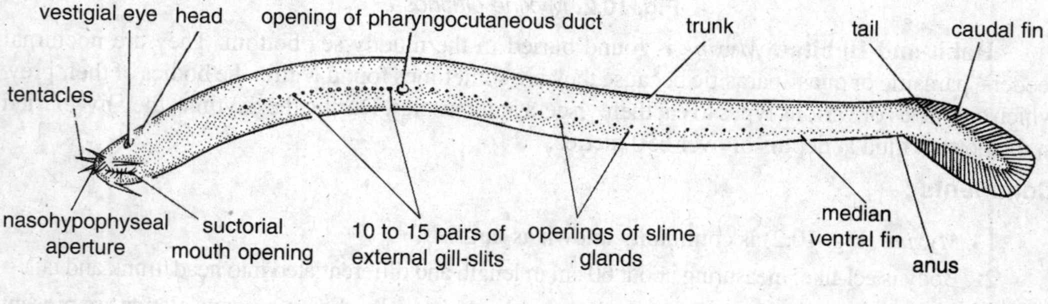

Fig. 10.3. *Bdellostoma*.

Habit and Habitat. *Bdellostoma* is found buried in the bottom mud of sea during day; otherwise it is nocturnal feeder and ectoparasitic and highly adapted for sucking.

Comments :

1. *Bdellostoma* (Fig. 10.3) is commonly known as **hagfish**.
2. The long eel-like body has a soft integument without scales. Large slime or mucous glands are present opening by mucous pores. Body is divided into head, trunk and tail.
3. The mouth is terminal surrounded by soft lips.
4. Four pairs of short tentacles, supported by skeletal rods, are present on the lateral sides of the mouth.
5. Buccal funnel and jaws are absent.
6. The single nostril lies very close to the mouth and opens terminally.
7. Single pineal eye is present on the top of the head. Paired eyes are vestigial or degenerated due to the bottom dwelling habit.
8. The **gill-openings** are 6–14 in number which open independently by pores.
9. The median fin is confined to the caudal region.
10. Blood is isosomatic with sea water.
11. Adult with persistent pronephros kidney but mesonephros is functional kidney.
12. Hermaphroditic. Single **ovotestis**, the anterior part being ovary and the posterior testis.
13. Egg is yolky, cleavage partial forming embryo placed on a mass of yolk.

Geographical distribution. *Bdellostoma* occurs off the Pacific coasts of both North and South America, South Africa and New Zealand.

4. Ammocoete Larva

Classification :

Phylum	...	**Chordata**	→	Dorsal tubular nerve cord, notochord and paired gill-slits are present.
Group	...	**Vertebrata**	→	Notochord is replaced by vertebral column; two pairs of appendages; circulatory system closed; hepatic portal system present; blood red containing RBCs.
Subphylum	...	**Agnatha**	→	Jaws and paired appendages are absent.
Class	...	**Cyclostomata**	→	Mouth suctorial devoid of functional jaws.
Order	...	**Petromyzontia**	→	Well-developed dorsal fin; complete branchial basket; naso-hypophysial-sac blind; seven pairs of gill-slits.
Type	...	Ammocoete larva of *Petromyzon*		

Comments :

1. Ammocoete larva (Fig. 10.4) hatches out of the egg of *Petromyzon.*
2. It looks like a minute transluscent *Amphioxus*, about 10 mm long, and lies bured in sand.
3. Eel-like body is divisible into head, trunk and tail.
4. Head contains a median **nostril**, an **eye**, an **auditory vesicle** and **brain**.
5. The median fin is continuous along the greater part of the body, forming a continuous **dorsal** and a **caudal** fin around the tail.

6. The **buccal cavity** is surrounded by upper and lower lips and provided with a number of buccal tentacles or oral cirri but devoid of teeth and tongue.

7. At the posterior end of the buccal cavity, there is **velum** which is followed by the pharynx.

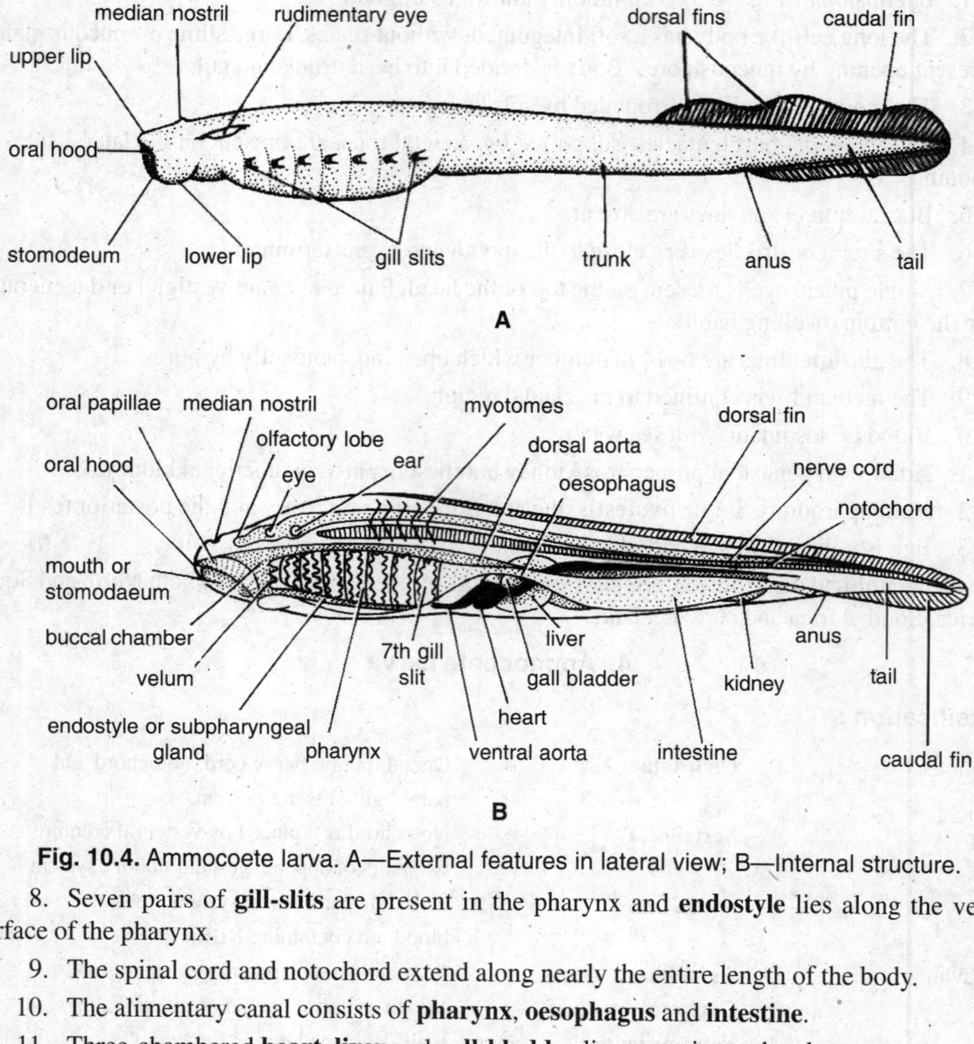

Fig. 10.4. Ammocoete larva. A—External features in lateral view; B—Internal structure.

8. Seven pairs of **gill-slits** are present in the pharynx and **endostyle** lies along the ventral surface of the pharynx.

9. The spinal cord and notochord extend along nearly the entire length of the body.

10. The alimentary canal consists of **pharynx**, **oesophagus** and **intestine**.

11. Three-chambered **heart**, **liver** and **gall-bladder** lies posterior to the pharynx.

12. Myotomes are segmentally arranged.

13. The larva lies in mud and feeds on small organisms brought by the currents of water produced by the muscular action of the branchial apparatus.

14. The larval stage lasts for a long period of about 3 or 4 years and during this period it grows and becomes adult.

15. Ammocoete larva shows intermediate characters between Cephalochordata and Cyclostomata.

Geographical distribution. Ammocoete larva has an almost world wide distribution, being present in both salt and fresh waters of North America, Europe, West Africa, Japan, Chile, Australia, New Zealand and Tasmania.

11

Chapter

SUPERCLASS–PISCES (= FISHES)

General Characters

1. Aquatic, either freshwater or marine; cold blooded vertebrates.
2. The skin is covered with scales, dermal denticles or bony plates.
3. Body usually streamlined but some are elongated snake-like and few are dorso-ventrally compressed.
4. Presence of paired and unpaired fins supported by soft or spiny rays. Dorsal, caudal and anal fins are unpaired, while the pectorals and pelvics (ventrals) are paired.
5. Tail is muscular and used for propulsion.
6. Nostrils are paired but do not open into pharynx except lung fishes.
7. Presence of barbles (excellent tactile organs) in some Teleostomi.
8. The endoskeleton is cartilaginous or bony.
9. Organs of respiration are gills but accessory respiratory organs may also be present.
10. Gills are borne by true gill-arches. They open outside by gill-slits which are covered by operculum in Teleostomi and uncovered in elasmobranchs.
11. The gill-slits are never more than seven pairs. Usually five pairs are present in elasmobranchs and 4 pairs in Teleostomi.
12. A swim-bladder or air-bladder is usually present, but secondarily lost in some species.
13. Lateral line system is well-developed.
14. Heart is venous and two chambered, *i.e.*, one auricle and one ventricle.
15. Ten pairs of cranial nerves are present.
16. Only internal ear is present.
17. The kidney is mesonephros.
18. The gonads possess true gonoducts.
19. Sexes are separate and development is indirect.
20. Amnion and allantois are absent in the embryonic stages. Pisces are anamniotes.

Outline Classification of Superclass Pisces (= Fishes)

Superclass Pisces

Classes

- I. Chondrichthyes (= Elasmobranchi)
  - Subclasses
    - A. Selachi
      - Order
        - I. Squaliformes or Pleurotremata
          Ex. 1. *Scoliodon*, 2. *Stegostoma*, 3. *Zygaena*, 4. *Scyllium*, 5. *Heptanchus*, 6. *Mustelus*, 7. *Rhineodon*, 8. *Chiloscylium*, etc.
        - II. Batoida or Hypotremata
          Ex. 1. *Rhinobatus*, 2. *Trygon*, 3. *Myliobatis*, 4. *Pristis*, 5. *Raia or Raja*, 6. *Torpedo*, 7. *Urobatus*, 8. *Aetobatus*, etc.
    - B. Brachydonti or Fossil and recent fishes
      - Order
        - Eubrachydonti
          Extinct fishes
        - Holocephali
          Ex. *Hydrologus* (= *Chimaera*)
- II. Osteichthyes
  - Subclasses
    - A. Actinopterygii
    - B. Sacropterygii or Choanichthyes

Subclass A. Actinopterygii

Superorders

- I. Chondrostei
  - Orders
    - 1. Polypteriformes
      Ex. *Polypterus*
    - 2. Acipenseriformes
      Ex. *Acipenser*, *Polyodon*
- II. Holostei
  - Orders
    - 1. Amiiformes
      Ex. *Amia*
    - 2. Lepidosteformes
      Ex. *Lepidosteus*
- III. Teleostei
  It includes 12 Orders

Superorder III. Teleostei

Classes

1. Clupeiformes
   Ex. *Clupea, Salmo, Tarpon, Essox, Hilsa*
2. Scopeliformes
   Ex. *Harpodon*
3. Cypriniformes or Ostariophysi
   Ex. *Cyprinus, Labeo, Catla, Clarias, Heteropneustes, Wallago, Mystus, Notopterus*
4. Anguilliformes
   Ex. *Anguilla, Muraena*
5. Beloniformes or Synentognathi
   Ex. *Belone* or *Xenentodon, Hemirhamphus, Exocoetus, Cypselurus*
6. Syngnathiformes or Solenichthyes
   Ex. *Hippocampus, Syngnathus Fistularia*
7. Ophiocephaliformes or Channiformes
   Ex. *Ophiocephalus* (= *Channa*)
8. Symbranchiformes
   Ex. *Amphipnous, Symbranchus*
9. Mastacembeliformes
   Ex. *Mastacembelus, Micrognathus*
10. Perciformes or Percomorphi
    Ex. *Anabus, Perca, Lates*
11. Scorpaeniformes
    Ex. *Pterois*
12. Pleuronectiformes
    Ex. *Synaptura, Pleuronectes, Solea*
13. Echeneiformes or Discocephali
    Ex. *Echeneis* or *Remora*
14. Tetradontiformes or Plectognathi
    Ex. *Ostracion, Tetradon, Diodon*
15. Lophiformes or Pediculati
    Ex. *Lophius, Antennarius*

Subclass B. Sarcopterygii (= Choanichthyes)

Orders

1. Crossopterygii
   Ex. *Latimeria* (Extinct)
2. Dipnoi
   Ex. *Neoceratodus, Protopterus, Lepidosiren*

CLASS I. CHONDRICHTHYES (= ELASMOBRANCHII)

1. Skin is covered with placoid scales or naked.
2. Endoskeleton cartilaginous, often calcified.
3. Gill-slits 5–7 on each side. First gill-slit a spiracle.
4. Operculum or gill-cover is absent.
5. Air-bladder and lungs are absent.

6. Mouth and 2 nostrils.
7. Cloaca is present.
8. Male usually possesses claspers.
9. Ova with much yolk. Cleavage meroblastic.
10. Mostly marine.

Subclass A. SELACHII

1. Paired fins without a median axis.
2. Base of pectoral fin constricted.
3. Pectoral radials jointed.
4. Vertebral centra usually differentiated.
5. Males with claspers.

Sharks. Spiracle behind each eye. Multiple gill-slits on either side.

Order 1. Squaliformes or Pleurotremata

1. 5–7 pairs of gill-slits on lateral sides.
2. Anterior margin of pectoral fins free. Dorsal fins two and one caudal.
3. Hyomandibular bearing branchial rays.
4. Right and left halves of pectoral girdle dorsally separated.

Examples : 1. *Scoliodon* (Dog fish), 2. *Stegostoma* (Tiger shark), 3. *Zygaena* or *Sphyrna* (Hammer-headed shark), 4. *Scyllium,* 5. *Heptanchus* (Seven-gilled shark), 6. *Chiloscylium.*

Order 2. Batoida or Hypotremata

1. 5 pairs of gill-slits are ventral.
2. Anterior margin of pectoral fin fused with sides of the body or head.
3. No branchial rays on hyomandibular.
4. Two halves of pectoral girdle united either to each other or to the vertebral column. Body depressed.

Examples : 1. *Rhinobatus* (Skate), 2. *Trygon* (Sting-ray), 3. *Myliobatus* (Eagle-ray), 4. *Pristis* (Saw-fish), 5. *Raja* or *Raia,* 6. *Torpedo.*

Subclass B. BRACHYODONTI

Fossil and recent fishes

Order 1. Eubrachydonti

Example : Extinct fishes.

Order 2. Holocephalii

1. Palato-quadrate fused to the neurocranium.
2. Four gill-slits on each side covered by fleshy operculum.
3. An inter-orbital septum is present dorsal to the brain case.
4. Cloaca, spiracle and scales are absent.
5. Two halves of the pelvic girdle not fused.
6. Males with claspers.
7. Tail whip-like.

Example : *Chimaera.*

CLASS II. OSTEICHTHYES

1. Skeleton chiefly bony (cartilaginous in *Sturgeon*).
2. Skin with dermal bony scales (cycloid, ctenoid or ganoid).

Subclass A. ACTINOPTERYGII

Ray finned fishes. Fins with dermal fin rays.

1. Radials of paired fins not arranged biserially.
2. Internal nares are absent except in stargazer.
3. Squamosal bone absent.

Superorder I. Chondrostei

1. Primitive ray-finned fishes or cartilaginous ganoids.
2. Heterocercal tail fin.

Order 1. Polypteriformes

1. Skin covered with typical rhombic ganoid scales.
2. Dorsal fin 8 or more, each preceded by a spine. Caudal fin arrow-shaped.
3. Pectoral fins with a small prominent base.
4. Air-bladder bilobed opening into the intestine ventrally.

Example : *Polypterus.*

Order 2. Acipenseriformes

1. Body covered with 8 rows of bony ganoid scutes or body naked with few vestigial scales.
2. Snout elongated, or paddle-shaped.
3. Endocranium cartilaginous.
4. Caudal fin is heterocercal.

Examples : *Acipenser, Polyodon.*

Superorder II. Holostei

1. These fishes are intermediate between Chondrostei and Teleostei.
2. Ray-finned fishes.
3. Scales ganoid.

Order 1. Amiiformes

1. Caudal fin short heterocercal. Snout normal.
2. Premaxillary not protractile, firmly articulated with the cranium.
3. Vertebral centra non-opisthocoelus.
4. Pectoral radials are attached to the scapula-coracoid cartilage.
5. Scales thin, overlapping and cycloid.

Example : *Amia* (bow fin)

Order 2. Lepidosteiformes

1. Body elongated. Skin covered with rhombic ganoid scales.
2. Nasal openings at the end of much elongated snout.
3. Caudal fin abbreviate heterocercal.
4. Vertebrate completely ossified, opisthocoelus.
5. Air-bladder cellular.

Example : *Lepidosteus* (Gar-pike).

Superorder III. Teleostei

1. Modern ray-finned fishes.
2. Skeleton light.
3. Ganoid scales reduced or absent.
4. Hydrostatic swim-bladder.

Order 1. Clupeiformes

1. Scales are well-developed and cycloid.
2. Caudal fin homocercal.
3. Dorsal and anal fins are without spines. Pelvic fin abdominal.
4. Vertebral centra completely ossified.
5. Weberian apparatus absent.
6. Air-bladder connected with pharynx.

Examples : *Clupeia, Salmo, Tarpon, Esox, Sardinops.*

Order 2. Scopeliformes

1. Dorsal and anal fins without spines.
2. An adipose fin is present.
3. Mouth large and provided with numerous small teeth.
4. Luminous organs are usually present.

Example : *Harpodon* (Bombay-duck).

Order 3. Cypriniformes or Ostariophysi

1. A peculiar Weberian apparatus, connecting the ear with the air-bladder, is present.
2. Air-bladder is connected with the pharynx by a duct.
3. Fins either without spines or dorsal, anal or pectoral have a spine each.
4. Ventral fins are abdominal.

Examples : *Cyprinus* (Carp), *Labeo rohita* (Rohu), *Catla, Botia, Carassius* (Gold fish), *Clarius* (Magur), *Heteropneustes* or *Saccobranchus* (Singhi), *Wallago* (Lachi), *Mystus* (Tengra), *Electrophorus* (Electric eel).

Order 4. Anguilliformes

1. Body slender, elongated eel-like.
2. Skin naked or with minute scales embedded in it.
3. Dorsal, caudal and anal fins are long and narrow and usually confluent.
4. Pelvic fins, if present, abdominal.
5. Fins devoid of spines.
6. Air-bladder connected with the intestine.

Examples : *Anguilla, Muraena.*

Order 5. Beloniformes or Synentognathi

1. Body elongated covered with cycloid scales.
2. Fins without spines.
3. Ventral fins abdominal.
4. Some of them are capable of jumping into the air and glide with the help of enlarged pectoral fins, located high on body.

Examples : *Belone* or *Xenentodon, Hemirhamphus* (Half beak), *Exocoetus* (Flying-fish), *Cypselurus,* etc.

Order 6. Syngnathiformes or Solenichthyes

1. Body covered with protective layer of scales or bony rings.
2. Mouth is terminal, at tubiform snout.
3. Fin rays of dorsal, anal and pectoral fins not branched.
4. Physoclistic, Swin-bladder closed.

Examples : *Hippocampus* (Sea-horse), *Fistularia* (Flute-fish), *Syngnathus* (Pipe-fish).

Order 7. Ophiocephaliformes or Channiformes

1. Body covered with cycloid scales.
2. Head depressed, covered with large plate-like scales.
3. Fins without spines.
4. Accessory respiratory organs are present.
5. Metapterygoid articulating with the sphenotic or frontal in front of the hyomandibular.
6. Air-bladder very long, prolonged into caudal region without duct.
7. Physoclistic.

Example : *Ophiocephalus* or *Channa*.

Order 8. Symbranchiformes

1. Body eel-like or snake-like devoid of scales.
2. Air-bladder is absent.
3. Fins are without spines.
4. Dorsal, caudal and anal fins are continuous.
5. Gill-slits single, transverse and ventral.

Examples : *Amphipnous cuchia, Symbranchus bengalensis.*

Order 9. Mastacembeliformes

1. Freshwater eel-shaped fishes.
2. Dorsal, caudal and anal fins are confluent. Sometimes a small caudal fin is separate.
3. Some free spines present before dorsal fin.
4. Anal fin with three spines.
5. Ventral fins absent.
6. Pectoral fins are present.
7. Buccal cavity enlarged for air-breathing.
8. Anterior nostril tubular opening in tentacle on each side of the fleshy rostral appendage.
9. Physoclistic.

Examples : *Mastacembelus, Micrognathus.*

Order 10. Perciformes or Percomorphi

1. Fins usually with spines.
2. Two dorsal fins. Pectoral fin placed high on body.
3. Ventral fins thoracic with not more than 6 rays.
4. Weberian apparatus absent.
5. Physoclistic. Air bladder without duct.

Examples: *Anabas* (Climbing-perch), *Perca* (Yellow perch), *Lates* (Bhetki).

Order 11. Scorpaeniformes

1. Enlarged head and pectoral fins.
2. Projecting spines from gill-covering.

Examples : *Pterois* (Scorpion fish).

Order 12. Pleuronectiformes

1. Body flat adapted for bottom living.
2. Both the eyes are on the same side.
3. Head asymmetrical.

4. Fins usually without spines.
5. Air-bladder is absent in adults. Dorsal and anal fins fringing the body.

Examples : *Synaptura, Pleuronectes, Solea.*

Order 13. Echeneiformes

1. First dorsal fin is modified into an adhesive disc for attachment overhead.
2. Scales cycloid.
3. No spines in second dorsal and anal fins.
4. Air-bladder is absent.

Examples : *Echeneis* or *Remora.*

Order 14. Tetrodontiformes (= Plectognathi)

1. Variable body shape, some globose.
2. Scales are modified into spines.
3. Scutes or bony plates cover the body.
4. Gill-slits are restricted.
5. Air-bladder present or absent.
6. Ventral fins thoracic or subthoracic.
7. Strong jaws with sharp teeth.

Examples : *Ostracion, Tetrodon, Diodon, Fieraster, Molamola.*

Order 15. Lophiiformes or Pediculati

1. First ray of the spinous dorsal fin is placed on the head and transformed into **illicium** (lure and bait).
2. Orbitosphenoid, basisphenoid and opisthotic bones are absent.
3. Air-bladder is absent.
4. Mouth wide.

Examples : *Lophius, Antennarius.*

Subclass B. SARCOPTERYGII (= CHOANICHTHYES)

1. Paired fins limb-like or lobed with a fleshy bony central axis covered with scales.
2. Two dorsal fins. Caudal fin heterocercal with an epichordal lobe.
3. Olfactory sacs usually connected to mouth cavity by internal nostrils or choanae.

Order 1. Crossopterygii

1. Lobed finned fishes.
2. Paired fins with scale covered lobe containing radials.
3. Teeth conical and well-developed.
4. Squamosal bone present.

Examples : *Latimeria* (Extinct), *Coelacanthus* (Extinct).

Order 2. Dipnoi

1. Lung fishes. Air-bladder paired lung-like.
2. Cycloid scales covering the skin.
3. Paired fins are lobate. Tail diphycercal.
4. Single gill-slit on either side present.
5. Internal nares are present.
6. Operculum, air-bladder and cloaca are present.

Examples : *Neoceratodus, Protopterus, Lepidosiren.*

MUSEUM SPECIMENS OF FISHES
CLASS: CHONDRICHTHYES (CARTILAGINOUS FISHES)

1. *Scoliodon*

Classification :

Phylum	...	**Chordata**	→	Dorsal tubular nerve cord, notochord and paired gill-slits are present.
Group	...	**Craniata**	→	Defined head with cranium and brain.
Subphylum	...	**Vertebrata**	→	Notochord is replaced by vertebral column; two pairs of appendages; circulatory system closed; hepatic portal system present; blood red containing RBCs.
Division	...	**Gnathostomata**	→	Jaws and paired appendages are present.
Superclass	...	**Pisces**	→	Presence of paired or unpaired fins supported by soft or spiny rays and gill-slits.
Class	...	**Chondrichthyes (= Elasmobranchi)**	→	Placoid scales in most of the species, endoskeleton cartilaginous, gill-slits 5–7 on each side, operculum absent, air-bladder absent, male with claspers, cloaca present, marine.
Subclass	...	**Selachii**	→	Sharks and rays. Paired fins without a median axis, pectoral radials jointed, Gills in separate clefts.
Order	...	**Squaliformes or Pleurotremata**	→	Gill-slits lateral, pectoral fins, small.
Genus	...	*Scoliodon*		

Habit and habitat. *Scoliodon* is marine but some live in estuaries and also ascend in rivers. They are active swimmers, predaceous and voracious feeders.

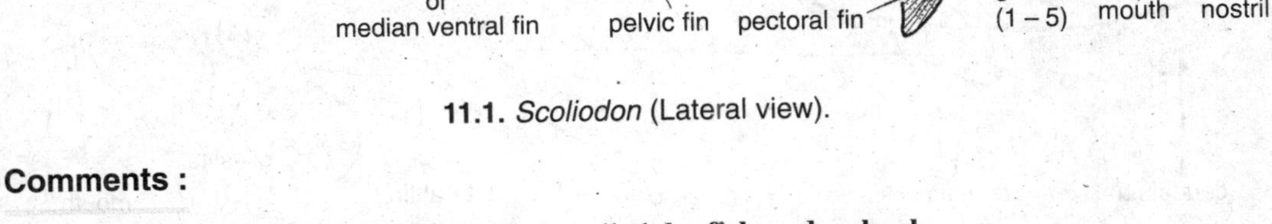

11.1. *Scoliodon* (Lateral view).

Comments :

1. *Scoliodon* (Fig. 11.1) is commonly called **dogfish** or **dogshark**.

2. The body is long, laterally compressed, spindle-shaped and it measures 30 to 60 cm in length and its body is divisible into **head**, **trunk** and **tail**.

3. The colour of the body is dark grey on the dorsal and lateral surfaces, while it is pale on the ventral surface.

4. The head is strongly compressed dorso-ventrally and produced in front into a wedge-shaped snout.

5. The trunk is the largest part of the body extending from behind the gill-slits up to the cloacal aperture. Tail is heterocercal with caudal fin.

6. Head bears a pair of eyes, wide ventral crescentic **mouth**, a pair of external **nostrils** and five pairs of lateral **gill-clefts**.

7. Trunk bears two dorsals, a caudal and a ventral or anal fins.

8. Paired fins are pectorals and pelvics. In between pelvic fins lies the cloacal aperture.

9. In male, each pelvic fin has a copulatory organ called **clasper** connected with its inner edge.

10. A faint line runs on either side of the body from the head to the posterior end of the tail called **lateral line**.

11. Shark has a food value in many countries. Its dried skin is called shagreen used for polishing furnitures and metals, etc. It is used in laboratories for dissections.

Geographical distribution. *Scoliodon* has a wide geographical distribution; it has been recorded from Zanzibar to Sri Lanka and Sri Lanka to the Malay Archipelago in the Indian Ocean from the Bay of Bengal, the East Indies and the Philippine Islands, from Mexico to Panama in the Eastern Pacific, from Labrador to Brazil in Atlantic, from Cuba in West Indies and also off the Eastern coast of South America.

2. *Chiloscyllium*

Classification :

Phylum	...	**Chordata**	Characters same as that of *Scoliodon.*
Group	...	**Craniata**	
Subphylum	...	**Vertebrata**	
Division	...	**Gnathostomata**	
Superclass	...	**Pisces**	
Class	...	**Chondrichthyes (= Elasmobranchi)**	
Subclass	...	**Selachii**	
Order	...	**Squaliformes or Pleurotremata**	
Genus	...	*Chiloscyllium*	

Habit and habitat. *Chiloscyllium* is a marine fish found up to the depth of 400 fathoms.

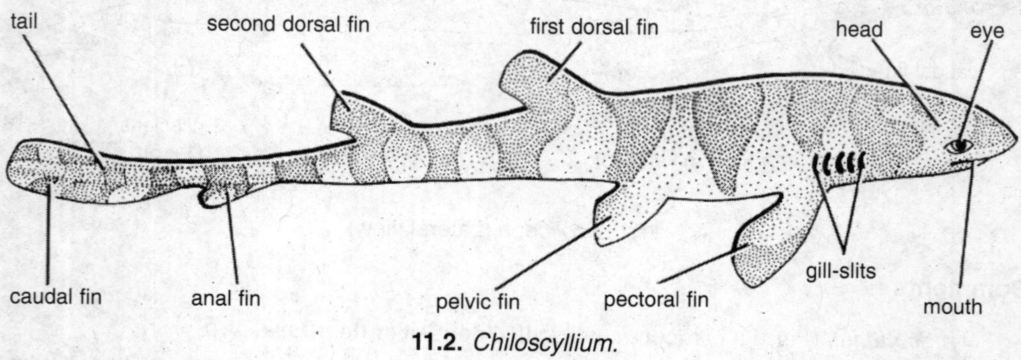

11.2. *Chiloscyllium.*

Comments :

1. *Chiloscyllium* (Fig. 11.2) is commonly called true dogfish.

2. The body is elongated, laterally compressed and distinguished into **head, trunk** and **tail**. Its body has alternate grey and white bands.

3. Head is produced into snout and possesses a pair of eyes, spiracles behind the eyes, ventral mouth and nasal openings.

4. Trunk contains median fins-2, dorsal fins, a ventral anal fin and caudal fin around the tail and paired fins as pectorals behind gill-slits and pelvics on lateral sides of cloaca.

5. Tail is straight and not bent upwards.

6. It exhibits sexual dimorphism. Male possesses the **claspers** for copulation attached with the inner ends of pelvics.

7. Fertilisation is internal. Oviparous and lays eggs in egg cases (mermaids purse). Egg cases are large, quadrate with long tendrils for attachment.

8. Its liver oil is rich in vitamin A.

Geographical distribution. *Chiloscyllium* is widely distributed from the cape of Good Hope to Indian Ocean and coasts of Australia, China and Japan. Lower carboniferous to recent.

3. *Stegostoma*

Classification :

Phylum	...	**Chordata**	Characters same as that of *Scoliodon*.
Group	...	**Craniata**	
Subphylum	...	**Vertebrata**	
Division	...	**Gnathostomata**	
Superclass	...	**Pisces**	
Class	...	**Chondrichthyes (= Elasmobranchi)**	
Subclass	...	**Selachii**	
Order	...	**Pleurotremata**	
Genus	...	*Stegostoma*	
Species	...	*tigrinum*	

Habit and habitat. *Stegostoma* is an active swimmer and predaceous, feeds on crabs, fishes and turtles, etc.

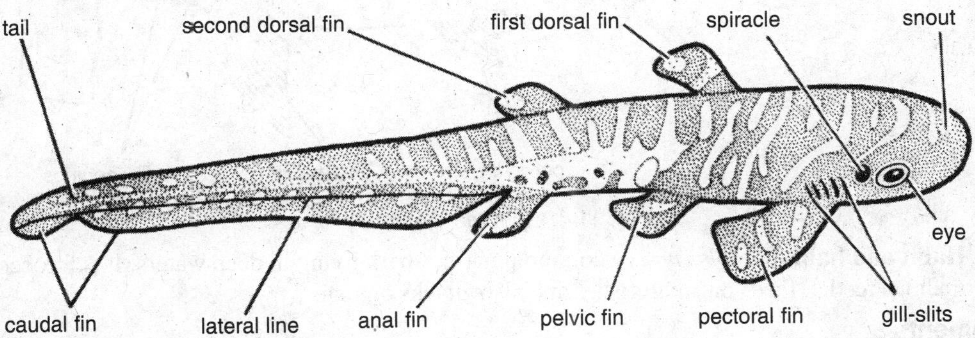

11.3. *Stegostoma.*

Comments :

1. *Stegostoma tigrinum* (Fig. 11.3) is commonly called **tiger-shark** or **zebra-shark** due to colour pattern of the body, showing dark bands on a yellow ground.

2. Body is elongated reaching a length of 5–7 metres.

3. Body is divisible into **head**, **trunk** and **tail**.

4. Head is broad and rounded in front, and bears a ventral mouth, a pair of eyes, without nictating membranes, a pair of spiracles behind the eyes, and five pairs of gill-slits on the lateral sides.

5. Trunk bears two dorsal fins, a caudal fin and a median fin. Paired fins are also present.
6. Tail is heterocercal.
7. Viviparous.

Geographical distribution. *Stegostoma tigrinum* is found in abundance in Indian Oceans. Lower carboniferous to recent.

4. *Sphyrna*

Classification :

Phylum	...	**Chordata**	Characters same as that of *Scoliodon.*
Group	...	**Craniata**	
Subphylum	...	**Vertebrata**	
Division	...	**Gnathostomata**	
Superclass	...	**Pisces**	
Class	...	**Chondrichthyes (= Elasmobranchi)**	
Subclass	...	**Selachii**	
Order	...	**Pleurotremata**	
Genus	...	*Sphyrna*	
Species	...	*zygaena*	

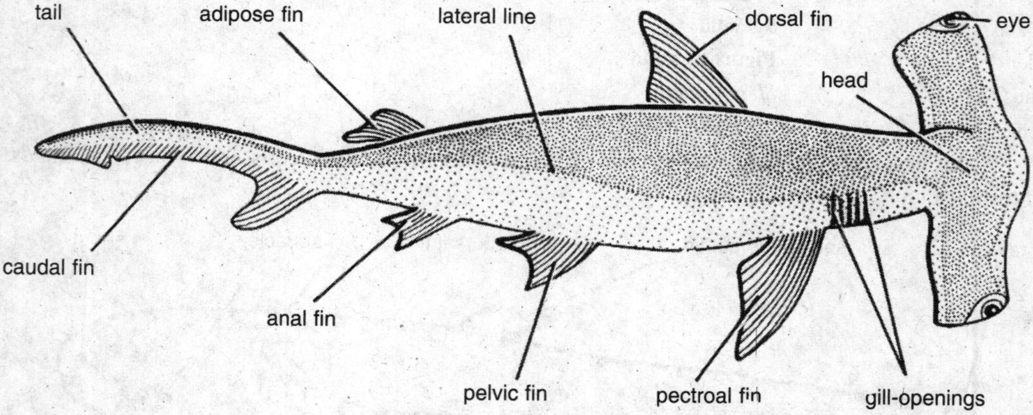

Fig. 11.4. *Sphyrna zygaena.*

Habit and habitat. *Sphyrna* is a common marine shark living in deep waters. It is feroceous and voracious feeder, feeds on small fishes and also attacks on man.

Comments :

1. *Sphyrna* (Fig. 11. 4.) is commonly called **hammer-headed shark**.
2. Body is elongated, measuring about 4–5 metres in length and divided into head, trunk and tail. Colour of the body is greyish dorsally and yellowish on ventral side.
3. Head is flattened in front and expanded sideways into two conspicuous lateral lobes, hence, resembling a hammer. Eyes lie at the tips of lateral lobes and provided with 3 eye-lids and nictitating membrane. Mouth is crescentic and ventral, nostrils ventral at the base of lateral lobes and five pairs of lateral gill-slits. Spiracles are absent.

4. Trunk bears median as well as paired fins. Median fins are two dorsals, caudal and an anal and paired fins are pectorals and pelvics.

5. Vertebrae are asterospondylus.

6. Viviparous, produces about 40 youngs.

7. *Sphyrna* is caught for its oil, although the yield is low.

Geographical distribution. *Sphyrna* is distributed in tropical and subtropical seas and coastal warm waters of Pacific ocean. Lower carboniferous to recent.

5. *Scyllium*

Classification :

Phylum	...	**Chordata**	Characters same as that of *Scoliodon*.
Group	...	**Craniata**	
Subphylum	...	**Vertebrata**	
Division	...	**Gnathostomata**	
Superclass	...	**Pisces**	
Class	...	**Chondrichthyes (=Elasmobranchi)**	
Subclass	...	**Selachii**	
Order	...	**Pleurotremata**	
Genus	...	*Scyllium*	

Habit and habitat. *Scyllium* is a marine true dog-fish.

Comments :

1. Body is elongated, laterally compressed and of moderate size.

2. The body is distinguished into head, trunk and tail.

3. Head is provided with a pair of eyes without nictitating membranes, spiracles behind the eyes, ventral mouth and nasal openings and 5 pairs of lateral gill-slits.

4. Trunk is provided with median and paired fins. Median fins are two dorsals, a caudal and anal behind the cloaca.

5. Tail bent slightly upward.

6. Sexes are separate. Male possesses **claspers**.

7. Fertilisation is internal.

8. Oviparous. Egg-cases are large and quadrate with long tendrils for attachment.

Geographical distribution. *Scyllium* is widely distributed in temperate and tropical seas. *S. canicula* and *S. catulus* are common on the British coasts.

6. *Rhinobatus*

Classification :

Phylum	...	**Chordata**	Characters same as that of *Scoliodon*.
Group	...	**Craniata**	
Subphylum	...	**Vertebrata**	
Division	...	**Gnathostomata**	
Superclass	...	**Pisces**	

Class	...	**Chondrichthyes (=Elasmobranchi)**	→	Placoid scales in most species, endoskeleton cartilaginous, gill-slits 5–7 on each side, operculum and air-bladder absent, male with claspers, cloaca present, marine.
Subclass	...	**Selachii**	→	Paired fins without a median axis, pectoral radials jointed, base of pectoral fin constricted.
Order	...	**Hypotremata**	→	Gill-slits ventral, anterior margin of pectoral fin fused with sides of body or head, no branchial rays on hyomandibular.
Genus	...	*Rhinobatus*		

Habit and habitat. *Rhinobatus* is a marine, bottom dwelling fish.

Comments :

1. *Rhinobatus* (Fig. 11.5) is commonly known as **guitar fish**.

2. Body is dorso-ventrally flattened and sub-rhomboidal in shape. Body is divided into head, trunk and tail.

3. Head is depressed wedge-shaped and produced into an elongated triangular snout.

4. Mouth is ventral in position. Teeth very obtuse.

5. Spiracles are present behind the eyes on the dorsal side.

6. Five pairs of gill-slits are present on the ventral side.

7. Pectoral fins are expanded and fused with the head and body. Paired pelvic fins are close to pectoral fins. Dorsal fins near tail region.

8. Tail is strong and possesses caudal fin.

9. Electric organs are absent.

10. Viviparous.

11. It is used mainly as food.

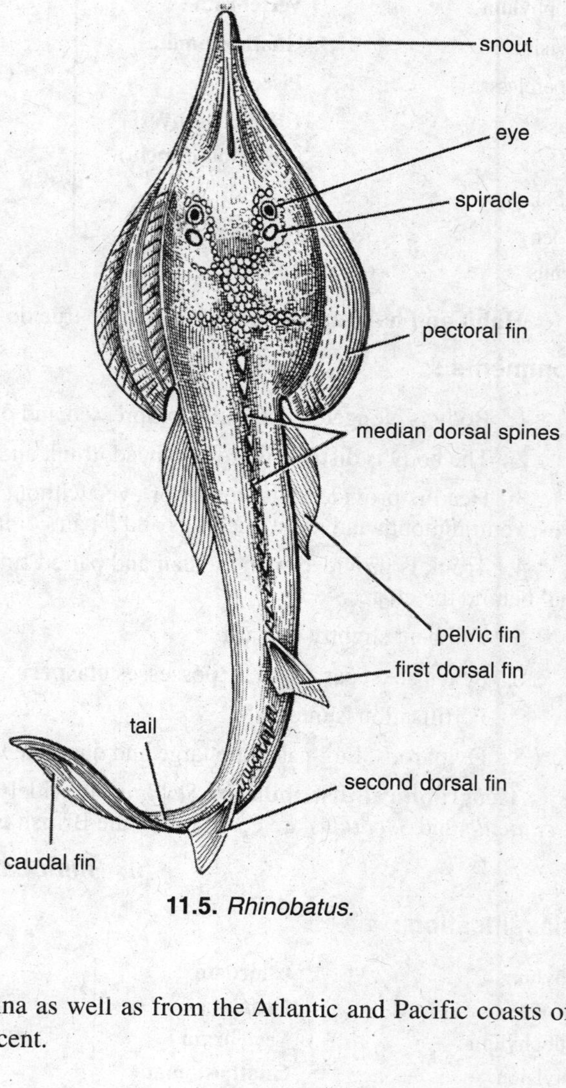

11.5. *Rhinobatus.*

Geographical distribution. *Rhinobatus* is cosmopolitan in distribution but commonly found in Mediterranean, Red Sea, the West coast of Africa, Indian Ocean, Australia, China as well as from the Atlantic and Pacific coasts of America and Gelapagos. Upper Jurassic to recent.

7. *Trygon* (Sting-ray)

Classification :

Phylum	...	**Chordata**	Characters same as that of *Rhinobatus*.
Group	...	**Craniata**	
Subphylum	...	**Vertebrata**	
Division	...	**Gnathostomata**	
Superclass	...	**Pisces**	
Class	...	**Chondrichthyes (= Elasmobranchi)**	
Subclass	...	**Selachii**	
Order	...	**Hypotremata**	
Genus	...	*Trygon* Or *Dasyatis*	

Habit and habitat. *Trygon* lies quietly on the sea bottom and occasionally swims in search of prey or in self defence. It is **carnivorous** and feeds on small fishes, crustaceans and molluscs. It also has protective colouration to protect itself from enemies.

Comments :

1. *Trygon* or *Dasyatis* (Fig. 11.6) is commonly known as **sting-ray** or **whip-tailed-ray** because of the presence of stings over the tail.
2. Mouth is ventral in position and a rectangular naso-frontal flap is present in front the mouth.

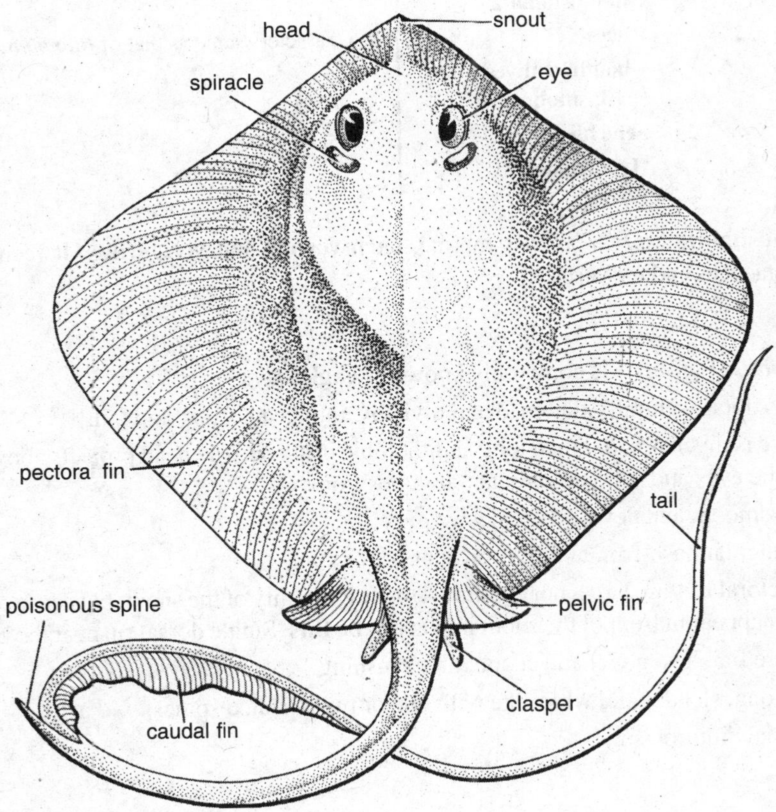

11.6. *Trygon.*

3. Head and body dorso-ventrally compressed. Pectoral fins are confluent with the sides of the head and body. Paired pelvic fins are small.

4. Spiracles are present behind the eyes of the dorsaı side.

5. Five pairs of gill-slits are present on the ventral side.

6. Tail is long, slender and whip-like terminating in a small caudal fin and armed with sharp serrated poisonous spines or stings.

7. Claspers are present near the pelvic fin in the male.

8. Viviparous.

9. Carnivorous. Food cheifly comprises crustaceans, molluscs and small fishes.

10. It can inflict wounds on its victims by means of the sting on the tail.

11. The flesh of these fishes is edible and liver yields a fair amount of oil.

Geographical distribution. *Trygon* or *Dasyatis* is found in tropical parts of Atlantic and Pacific Oceans including the Pacific coasts of North and South America. It has also been reported from India, Japan, China and Australia. Upper jurassic to recent.

8. *Myliobatis* (Eagle-ray)

Classification :

Phylum	...	**Chordata**	Characters same as that of *Rhinobatus*.
Group	...	**Craniata**	
Subphylum	...	**Vertebrata**	
Division	...	**Gnathostomata**	
Superclass	...	**Pisces**	
Class	...	**Chondrichthyes (= Elasmobranchi)**	
Subclass	...	**Selachii**	
Order	...	**Hypotremata**	
Genus	...	*Myliobatis*	

Habit and habitat. *Myliobatis* is marine, **carnivorous** and **viviparous**. It feeds on small organisms, crustaceans and molluscs.

Comments :

1. *Myliobatis* (Fig. 11.7) is commonly known as **eagle-ray**.

2. Body is dorso-ventrally flattened and divisible into head, trunk and tail.

3. The disc is much broader than long and rhombic in shape. The head projects above the level of the disc and the eyes and spiracles are lateral in position.

4. Skin is smooth and devoid of scales.

5. A rectangular naso-frontal fold is present.

6. The pectoral fins are large, not continued to the extremity of the snout, but cease on the sides of the head and reappear in front of the snout into **cephalic fins**. Single dorsal fin near the base of tail.

7. Teeth are flat, hexagonal and adapted for crushing.

8. Tail is long, slender and whip-like with one or two serrated spines.

9. No sexual dimorphism.

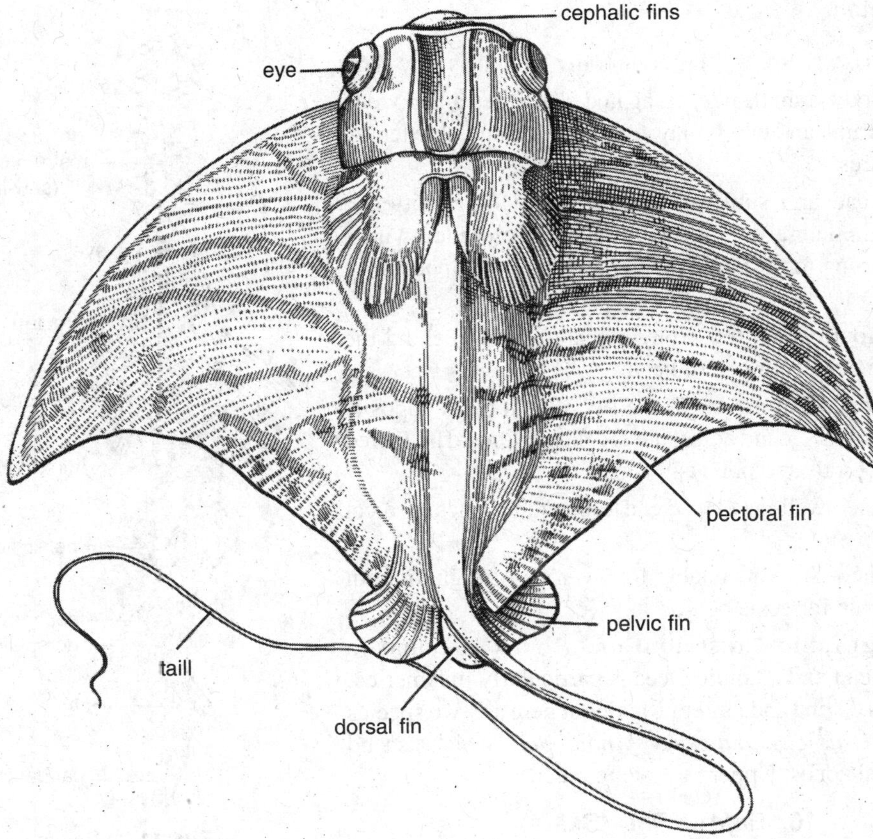

Fig. 11.7. *Myliobatis.*

Geographical distribution. *Myliobatis* is found in tropical and subtropical seas, in Mediterranean sea and eastern and southern coasts of England. *Myliobatis nichofii* occurs in the mouth of Ganges and Chilka lake and also along the east and west coasts. Upper jurassic to recent.

9. *Pristis* (Saw-fish)

Classification :

Phylum	...	**Chordata**	Characters same as that of *Rhinobatus*.
Group	...	**Craniata**	
Subphylum	...	**Vertebrata**	
Division	...	**Gnathostomata**	
Superclass	...	**Pisces**	
Class	...	**Chondrichthyes (Elasmobranchi)**	
Subclass	...	**Selachii**	
Order	...	**Hypotremata**	
Genus	...	*Pristis*	

Habit and habitat. *Pristis* is found in warm marine waters. It often ascends in rivers. It is **predaceous**, feeds on fishes and other marine animals by slashing them with its saw-like minute teeth.

Comments :

1. *Pristis* (Fig. 11.8) is commonly known as **saw-fish**.

2. Body is shark-like, elongated, depressed and divided into head, trunk and tail. It may attain the considerable length of 3–6 metres.

3. Head and skull prolonged into a long flattened **rostrum**, the lateral margins of which are provided with a series of strong tooth-like denticles giving it the appearance of a saw.

4. Mouth is ventral. Spiracles are present behind the eyes on the dorsal surface of head.

5. Dorsal fins are large, the first dorsal fin is opposite to the pelvic fin. Second dorsal pin is called **adipose fin**. Paired fins are pectorals and pelvics.

6. Tail is well-developed and terminating in heterocercal caudal fin.

7. The fish is invaluable for liver oil of high vitamin value and skin for scale boards.

Geographical distribution. *Pristis* is found in Mediterranean and Atlantic Oceans particularly in America, West Indies, China and Gulf of Mexico. There are two species in India, *Pristis cuspidata and P. microdon,* which ascend beyond tidal zones. Upper jurassic to recent.

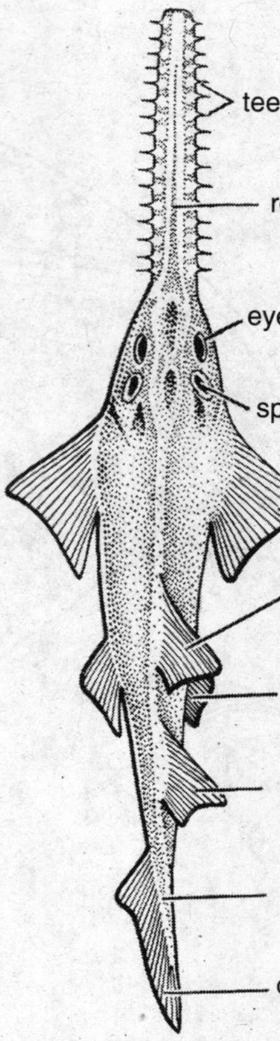

Fig. 11.8. *Pristis.*

10. *Raja* or *Raia* (Skate)

Classification :

Phylum	...	**Chordata**	Characters same as that of *Rhinobatus.*
Group	...	**Craniata**	
Subphylum	...	**Vertebrata**	
Division	...	**Gnathostomata**	
Superclass	...	**Pisces**	
Class	...	**Chondrichthyes (= Elasmobranchi)**	
Subclass	...	**Selachii**	
Order	...	**Hypotremata**	
Genus	...	*Raja* or *Raia*	

Habit and habitat. *Raja* is a marine, sluggish, bottom dwelling fish. It is **carnivorous** feeding on small fishes and crustaceans. It preys by dropping its body over the animal.

Comments :

1. *Raja* or *Raia* (Fig. 11.9) is commonly known as **skate** and measures 2–3 metres in width.

2. Body is dorso-ventrally depressed, rhomboidal and differentiated into disc and tail. Disc includes head and trunk.

3. Large pectoral fins are confluent with the head and trunk. Pelvic fins are small.

4. The mouth is on the ventral side of head and provided with numerous sharp, rasping teeth with which they tear up their prey.

5. Spiracles are present behind the eyes.

6. Five pairs of gill-slits are ventral in position.

7. Tail is slender, sharply marked off from the trunk, and provided with two small dorsal fins. Caudal fin is very small.

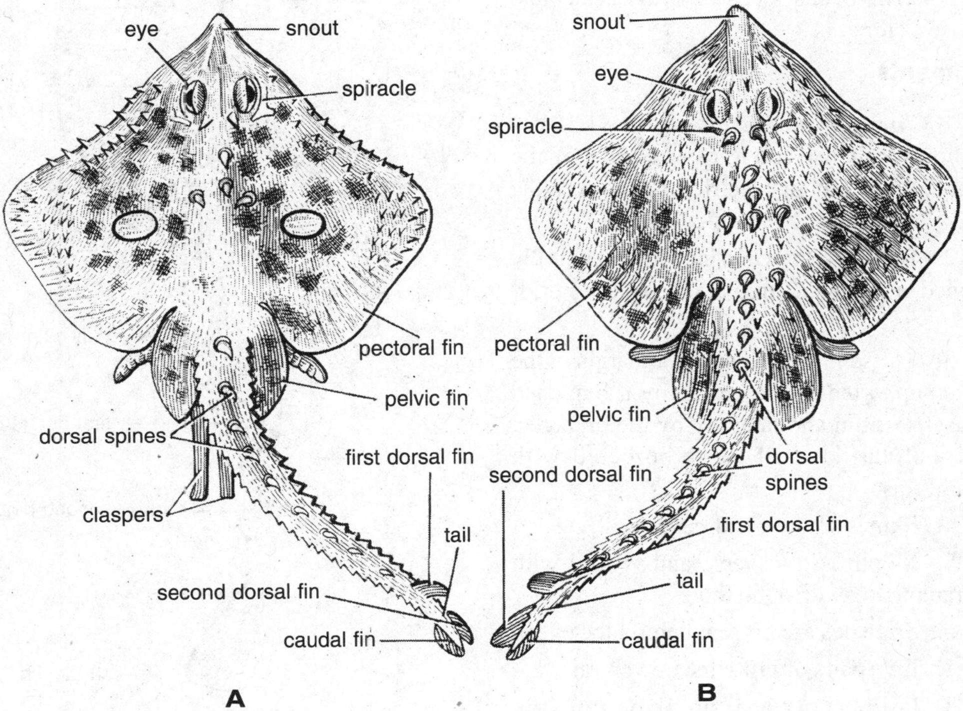

Fig. 11.9. *Raja* or *Raia*. A—Male; B—Female.

8. Caudal electric organs (modified caudal muscles) are present on each side of the terminal portion of the tail.

9. Larger or smaller spines are generally present on disc and tail.

10. Sexual dimorphism is well-marked. Male is provided with a pair of **claspers** present near the pelvic fin.

11. Oviparous. Egg-cases four horned without tendrils.

Geographical distribution. *Raja* or *Raia* is found in temperate seas but is more abundant in northern hemisphere, Arctic and Antarctic regions. Upper jurassic to recent.

11. *Torpedo* or *Astrape* (Electric-ray)

Classification :

Phylum	...	**Chordata**	Characters same as that of *Rhinobatus*.
Group	...	**Craniata**	
Subphylum	...	**Vertebrata**	
Division	...	**Gnathostomata**	
Superclass		**Pisces**	
Class		**Chondrichthyes (= Elasmobranchi)**	
Subclass	...	**Selachii**	
Order	...	**Hypotremata**	
Genus	...	*Torpedo* or *Astrape*	

Habit and habitat. *Torpedo* or *Astrape* is a marine fish, found over sandy or muddy bottom at a depth of about 40 to 50 fathoms. Fish is **carnivorous** feeds on crustaceans and molluscs, etc.

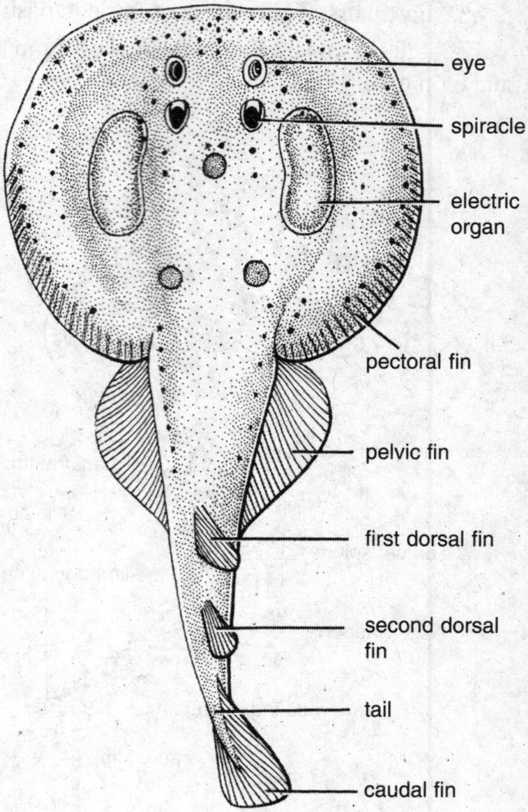

11.10. *Torpedo* or *Astrape*.

Comments :

1. *Torpedo* or *Astrape* (Fig. 11.10) is commonly called **electric-ray** because of the presence of a pair of electric organs over the head.
2. Body is regionated into dorso-ventrally flattened anterior subcircular disc and posterior tail.
3. The semicircular anterior margin of the disc is supported in the centre by a branched prenasal rostrum and laterally by the branched preorbital cartilages. Disc is bordered with pectoral fins.
4. Skin is smooth without scales.
5. Mouth is transverse and ventral with quadrangular naso-frontal lobe.
6. Spiracles are present behind the eyes.
7. Five pairs of gill-slits are ventral.
8. Pelvic fins are small behind the pectroals.
9. Electric organs consist of muscle fibres arranged in blocks and serve as batteries. These fishes are capable of giving a heavy electric shock.
10. Tail is relatively short with two dorsal fins and a caudal fin.
11. Viviparous and produces live young ones.

Geographical distribution. *Torpedo* is found in Mediterranean, Red Sea, Atlantic and Pacific Oceans, East Indies, China, Tasmania, Australia, Japan, South Africa, North and South America. *T. marmorata* is found in Indian Ocean. Upper Jurassic to Recent.

12. *Chimaera* (Rat-fish)

Classification :

Phylum	...	**Chordata**	}	Characters same as that of *Scoliodon*.
Group	...	**Craniata**		
Subphylum	...	**Vertebrata**		
Division	...	**Gnathostomata**		
Superclass	...	**Pisces**		
Class	...	**Chondrichthyes** (**= Elasmobranchi**)	→	Endoskeleton cartilaginous. Notochord rudimentary. Spiral valve in intestine.
Subclass	...	**Brachyodonti**	→	Crushing type dentition.
Order	...	**Holocephali**	→	Spiracles absent. Operculum present. Skull autostylic with 2 dorsal fins and an anal fin.
Genus	...	*Chimaera*		
Species	...	*monstrosa*		

Habit and habitat. *Chimaera* (Rat-fish) is a marine fish found at a depth of 350 to 2200 metres and feeds on fishes, invertebrates and seaweeds.

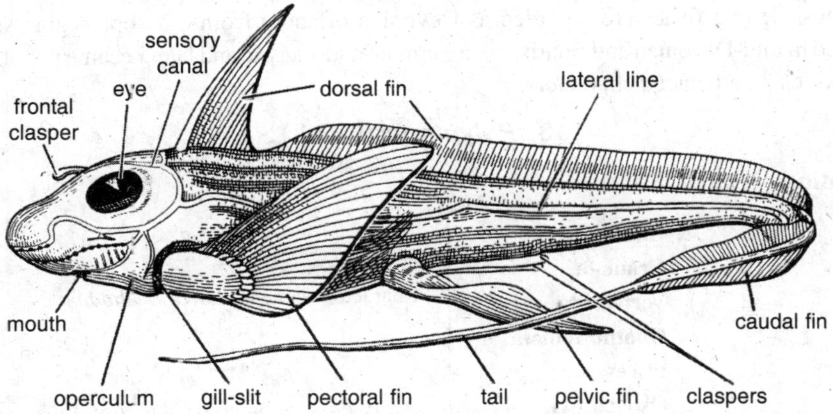

Fig. 11.11. *Chimaera monstrosa.*

Comments :

1. *Chimaera* (Fig. 11.11) is commonly known as **rat-fish** or **king of herrings**. Body is divisible into head, trunk and tail.
2. Body is elongated and shark-like, measuring up to 1 metre in length. Head is large and compressed with a blunt conical snout.
3. Skin is naked with a characteristic open groove lodging the lateral line system.
4. Mouth and nostrils are ventral in position.
5. Gill-slits covered with non-bony operculum. Spiracles are absent.
6. Pectoral and pelvic fins are large. Ventral fin small.
7. Two large dorsal fins, anterior over the pectoral fins with a stout spine in front and posterior one is continuous and non-erectile.
8. Caudal fin consists of nearly equal-sized dorsal and ventral lobes.
9. Tail is long, tapering whip-like and diphycercal.
10. Sexual dimorphism exists. In male, head possesses a **frontal clasper** armed with a pad of recurved denticles, a pair of claspers present behind the pelvic fins and a pair of claspers in front of pectoral fins.
11. Eggs enclosed in egg shell. Young rat-fish is with placoid scales.

Geographical distribution. *Chimaera* occurs off the coasts of Europe from Norway to Portugal including the Mediterranean and also in the neighbourhood of Azores, Cape of Good Hope and coasts of Japan and North America. Upper Devonian to Recent.

Chimaera shows selachian and certain specialised characters. **Selachian characters** are cartilaginous endoskeleton, limb girdles, dermal denticles, absence of air bladder, presence of spiral valve in the intestine, conus arteriosus and clapsers in males. **Specialised characters** are autostylic skull, fused palatoquadrate with cranium, dental plates for crushing, and five claspers in male. Gill-slits covered with operculum, suppressed spiracles and absence of cloaca (Osteichthyes characters). Thus, *Chimaera* is an intermediate fish between sharks and bony fishes.

CLASS : OSTEICHTHYES

(BONY FISHES)

Bony fishes have bony endoskeleton, scales, plates, air bladder and advanced brain. They are found in freshwater or brackish water, salt water, warm water or cold water.

Lung fishes have short jaws, teeth form crushing plates, internal nares present, reduced skeletal system, tail diphycercal and one or two lungs (air bladders). Their three genera are distributed discontinuously. Lung fishes are regarded as Devonian offshoot from crossopterygian stem. They first appeared in mid-Devonian and flourished in Permian and Triassic and later became rare. Lungfishes show primitive and advanced characters.

13. *Polypterus* (Bichir)

Classification :

Phylum	...	**Chordata**	Characters same as that of *Scoliodon.*
Group	...	**Craniata**	
Subphylum	...	**Vertebrata**	
Division	...	**Gnathostomata**	
Superclass	...	**Pisces**	
Class	...	**Osteichthyes** →	Endoskeleton bony; skull usually hyostylic; single external gill-slit on each side of head; operculum and air-bladder present; cloaca and claspers absent, scales dermal.
Subclass	...	**Actinopterygii** →	Ancient fishes. Dermal fin rays numerous. Nostril not connected with mouth cavity.
Superorder	...	**Chondristei** →	Primitive or ray-finned fishes or cartilaginous ganoides; heterocercal tail fin.
Order	...	**Polypteriformes** →	Skin covered with rhombic ganoid scales; dorsal fin with many peculiar finlets; pectoral fins separate; air-bladder lobed opening into the intestine ventrally.
Genus	...	*Polypterus* (Bichir)	

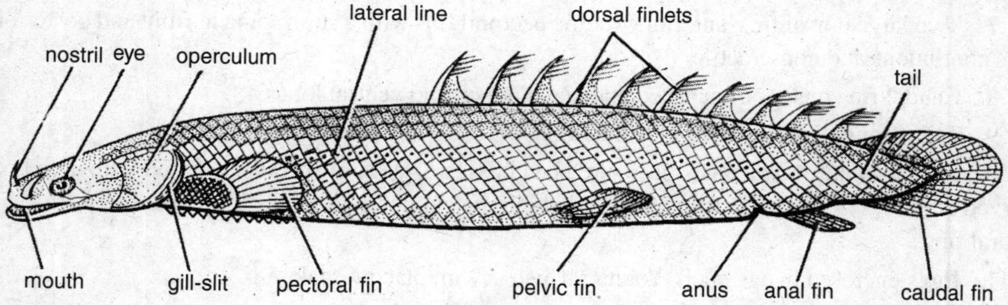

11.12. *Polypterus.*

Habit and habitat. *Polypterus* is marine, active in night and feeds on small fishes, amphibians and crustaceans.

Comments :

1. *Polypterus* (Fig. 11.12) is one of the most primitive living bony fish.
2. Body is elongated about 1 metre long and covered with rhombic ganoid scales.
3. Head is covered with a horny shield, bearing tubular nostrils and anteriorly placed eyes.
4. Spiracle is present and there are four gill-arches.
5. **Dorsal fin** consists of number of (5–18) finlets, each with a spine.
6. **Pectoral fins** have a scale-covered fleshy-lobe at the base.
7. **Pelvic** and **anal fins** are very small.
8. **Caudal fin** is diphycercal.

9. Air-bladder is bilobed and cellular serving for respiration.

10. It does not readily live out of water, rarely longer than three to four hours, and then only when covered with damp grass or weeds.

11. Larva has plume-like external gills attached to hyoid arches.

Polypterus show certain **primitive characters** such as non-overlapping scales, spiracles, certain skull bones, spiral valve in intestine and paired fins resemble to broad-based crossopterygian fins and certain **advanced features** like dorso-lateral nostrils, pyloric caecum in gut and urinogenital system like that of bony fishes. Tail is not heterocercal and air bladder connected with pharynx acted as respiratory organ.

Geographical distribution. *Polypterus* is found in the rivers of tropical Africa, the Nile, Nigeria and Congo. Paleozoic and recent, but none known species.

14. *Acipenser* (Sturgeon)

Classification :

Phylum	...	**Chordata**	Characters same as that of *Scoliodon*.
Group	...	**Craniata**	
Subphylum	...	**Vertebrata**	
Division	...	**Gnathostomata**	
Superclass	...	**Pisces**	
Class	...	**Osteichthyes**	Characters same as that of *Polypterus*.
Subclass	...	**Actinopterygii**	
Superorder	...	**Chondrostei**	
Order	...	**Acipenseriformes** →	Body covered with bony scutes or naked; snout elongated; endocranium cartilaginous; caudal fin heterocercal. Notochord unsegmented; vertebrae acentrous.
Genus	...	*Acipenser*	

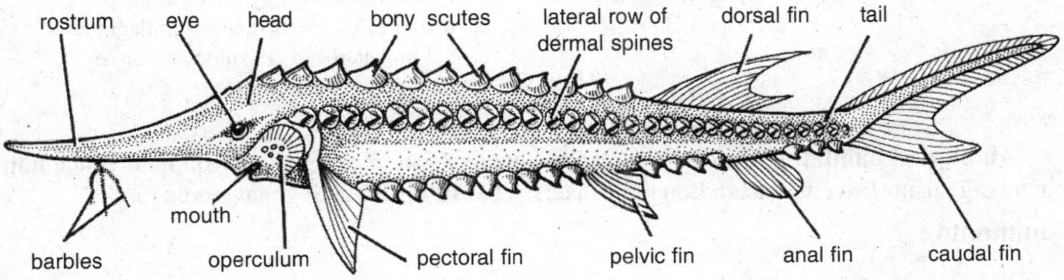

Fig. 11.13. *Acipenser.*

Habit and habitat. *Acipenser* is marine, bottom dweller fish. It stirs the bottom with its snout for searching small invertebrates, worms, molluscs, small fishes, etc. It is **anadromous**.

Comments :

1. *Acipenser* (Fig. 11.13) is commonly known as **sturgeon**.

2. Body is elongated about 2–4 metres in length and covered wtih five rows of bony scutes, one dorsal, two laterals and two ventrals.

3. Head is produced ahead as **rostrum** with a transverse row of four preoral barbles used for searching food.

4. The mouth is small, ventral and protrusible. Jaws without teeth.

5. Presence of functional open spiracles.
6. Operculum is represented by an opercular bone.
7. The air-bladder is smooth, oval and non-sacculated having a lining of ciliated epithelium.
8. Skeleton cartilaginous.
9. Dorsal fin single and posteriorly located. Anal fin opposite to dorsal fin and caudal fin bilobed (heterocercal). Paired fins are also present.
10. A single fish lays 2–3 million eggs in a breeding season. European sturgeon's eggs are black and are used to prepare a delicacy called **Caviar**. Their air-bladders are used to prepare **isinglass** which is used for cleaning wines and beers.

Geographical distribution. *Acipenser* is found in the Black Sea, Sea of Azor, Caspian and their tributary rivers and also on the coasts of Northern America, Europe and China. Lower Jurassic to recent.

Sturgeons resemble with elasmobranchs having open spiracles, heterocercal tail, mostly cartilaginous skeleton, intestinal spiral valve and conus arteriosus. They show affinity with ganoid fishes in having rhomboid scales and loss of bones.

15. *Polyodon* (Paddle-fish)

Classification :

Phylum	...	**Chordata**	Characters same as that of *Scoliodon.*
Group	...	**Craniata**	
Subphylum	...	**Vertebrata**	
Division	...	**Gnathostomata**	
Superclass	...	**Pisces**	
Class	...	**Osteichthyes**	Characters same as that of *Polypterus.*
Subclass	...	**Actinopterygii**	
Superorder	...	**Chondrostei**	
Order	...	**Acipenseriformes** →	Body covered with bony scutes or naked; snout elongated; endocranium cartilaginous; caudal fin heterocercal. Notochord unsegmented; vertebrae acentrous.
Genus	...	*Polyodon*	

Habit and habitat. *Polyodon* is a marine sluggish fish, feeding chiefly on mud containing minute organisms (invertebrates). Long gill-rakers act as filters for taking planktonic food.

Comments :

1. *Polyodon* (Fig. 11.14) is commonly known as **paddle-fish** or **spoon-bill**.
2. Body is elongated and fusiform measuring up to one and a half to two metres in length.

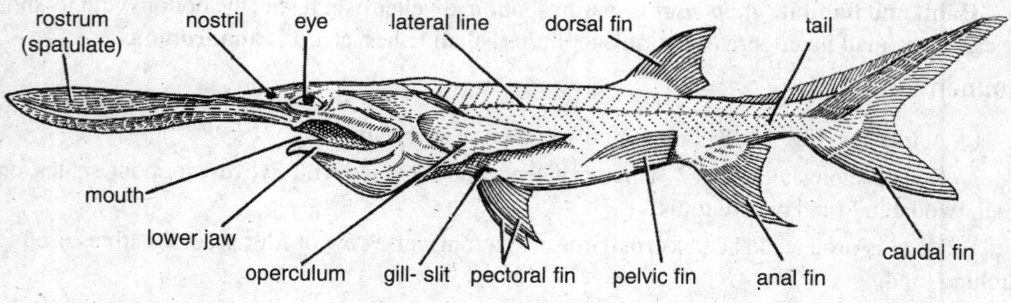

Fig. 11.14. *Polyodon.*

3. Skin without scales.

4. Head is produced in froni as rostrum that is exceptionally long and spatulate with a rigid axis and thinner and more flexible margins.

5. Mouth is wide and ventral. Teeth minute.

6. Head bears small eyes and open spiracles.

7. A small suboperculum is present in addition to a small-rayed operculum extending posteriorly.

8. Tail is heterocercal. Caudal fin bilobed.

9. Paired fins small. Dorsal fin opposite to anal fin.

Geographical distribution. *Polyodon* is found in the rivers of Southern states of North America, Mississippi, Ohio and Missouri and their tributary rivers and streams. Lower Jurassic to recent.

Polyodon also shows some primitive and advanced characters.

16. *Amia* (Bowfin)

Classification :

Phylum	...	**Chordata**	}	Characters same as that of *Scoliodon*.
Group	...	**Craniata**		
Subphylum	...	**Vertebrata**		
Division	...	**Gnathostomata**		
Superclass	...	**Pisces**	}	Characters same as that of *Polypterus*.
Class	...	**Osteichthyes**		
Subclass	...	**Actinopterygii**		
Superorder	...	**Holostei**	→	Bony ganoids.
Order	...	**Amiformes**	→	Tail heterocercal; premaxillary firmly articulated with cranium, centra non-opisthocoelus; pectoral radials are attached to the scapulo-coracoid cartilage.
Genus	...	*Amia*		
Species	...	*calva*		

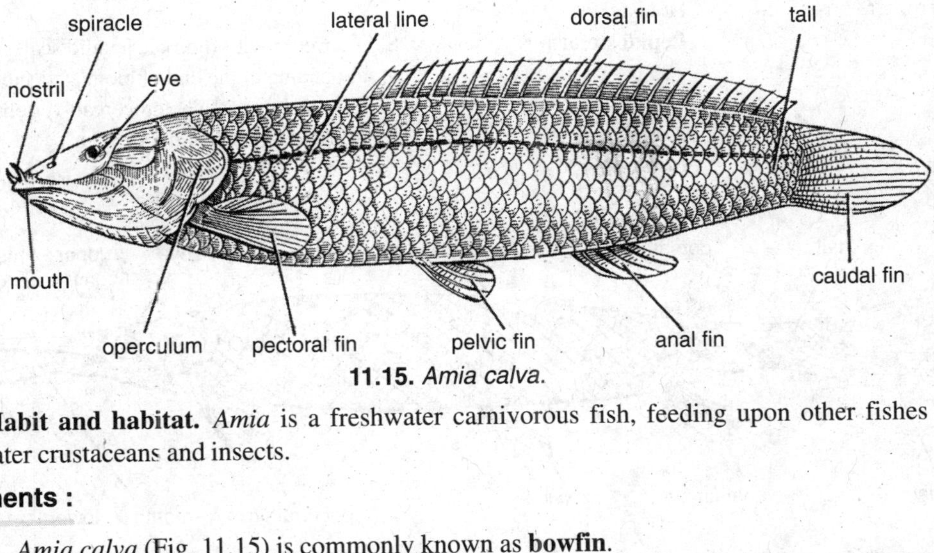

11.15. *Amia calva.*

Habit and habitat. *Amia* is a freshwater carnivorous fish, feeding upon other fishes and freshwater crustaceans and insects.

Comments :

1. *Amia calva* (Fig. 11.15) is commonly known as **bowfin**.

2. Body is fusiform and somewhat compressed, measuring about 60 cm in length. Body is divided into head, trunk and tail.

3. Scales are uniformly thin, cycloid and imbricate.

4. Head bears subterminal mouth, a pair of nasal barbels and dorsal eyes.

5. Moderately large conical teeth are present on the premaxillae, palatines, and dentaries and smaller teeth on the vomers, pterygoids, splenials and parasphenoids.

6. Single dorsal fin is long, low and continuous, anal fin is short and caudal fin is nearly homocercal with a rounded hinder margin.

7. Air-bladder is often used for respiration, the fish coming to the surface and gulping the air.

8. Pyloric caeca is absent.

9. Sexually dimorphic. The male is smaller than the female and is distinguished by the presence of a round black spot encircled by a margin of orange colour at the base of caudal fin. Two peculiar comb-like structures are present over the head.

10. Breeding season in May and June. Male builds a nest in water weeds in which the female lays her eggs. Male guards the youngs until they can move.

Geographical distribution. *Amia calva* is found in the rivers and lakes of central and southern North America. Upper Jurassic and lower Tertiary.

Amia calva is the only one living species. It shows some primitive features like dorsal cellular and bilobed air-bladder, vestigial spiral valve, jugal plate large, well developed branchiostegal rays, and some advanced features such as thin cycloid scales without ganoin, homocercal tail and vertebrae solid opisthocoelous.

17. *Lepidosteus* (Garpike)

Classification :

Phylum	...	**Chordata**	Characters same as those of *Scoliodon.*
Group	...	**Craniata**	
Subphylum	...	**Vertebrata**	
Division	...	**Gnathostomata**	
Superclass	...	**Pisces**	
Class	...	**Osteichthyes**	Characters same as that of *Amia.*
Subclass	...	**Actinopterygii**	
Superorder	...	**Holostei**	
Order	...	**Lepidosteformes** →	Skin covered with rhombic ganoid scales; nasal opening at the end of elongated snout; caudal fin abbreviate heterocercal; vertebrae opisthocoelus; air-bladder cellular.
Genus	...	*Lepidosteus*	

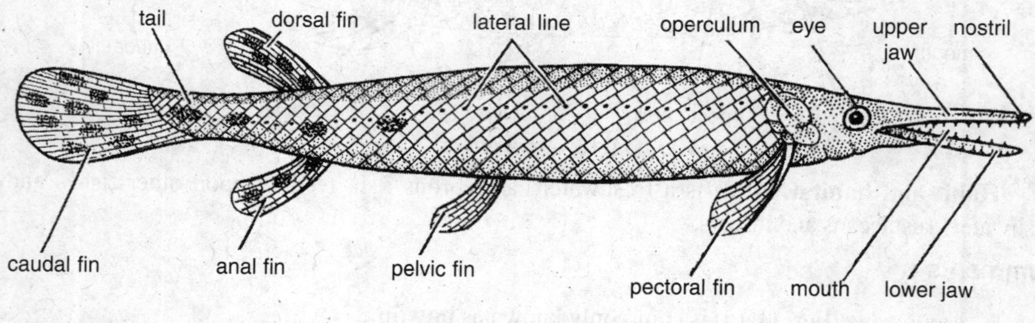

Fig. 11.16. *Lepidosteus.*

Habit and habitat. *Lepidosteus* is a freshwater fish. It is voracious fish, preying upon small fishes. Heavily armoured predaceous gars usually inhabits in shallow weedy waters.

Comments :

1. *Lepidosteus* (Fig. 11.16) is commonly known as **garpike**.
2. Body is elongated and cylindrical and divided into head, trunk and tail.
3. Scales are rhombic, thick, ganoin-coated and articulated by peg and socket joints.
4. Dorsal and anal fins are short and far behind the pelvic fins. Caudal fin is semi-heterocercal.
5. Head is produced into snout. Both the upper and lower jaws are elongated, forming a long tapering beak. Teeth are numerous, slender and of unequal size. The nostrils are placed at the anterior end of the beak. Eyes are small.
6. Vertebrae are opisthocoelous and are like reptiles, a unique feature. Vomer paired.
7. The air-bladder is used as a lung and air is gulped through the mouth. Spiracles closed.
8. Pyloric caeca is branched.
9. Breeding season is in the month of May. Eggs are extremely sticky and adhere to the rocks and stones on which they are deposited.
10. Gar's skin is used for the manufacture of hand bags and arts.

Geographical distribution. *Lepidosteus* inhabits the freshwaters of North America. Upper Cretaceous to Recent.

Cellular air-bladder, ganoin coated scales, absence of jugal plate and closed spiracles are its primitive features. Its advanced characters are branched pyloric caeca, reduced maxilla, tooth-bearing infraorbitals and opisthocoelous vertebrae.

18. *Salmo* (Trout)

Classification :

Phylum	...	**Chordata**	}	Characters same as that of *Scoliodon*.
Group	...	**Craniata**	}	
Subphylum	...	**Vertebrata**	}	
Division	...	**Gnathostomata**	}	
Superclass	...	**Pisces**	}	
Class	...	**Osteichthyes**	}	Characters same as that of *Polypterus*.
Subclass	...	**Actinopterygii**	}	
Superorder	...	**Teleostei**	→	Modern ray-finned fishes; hydrostatic swim-bladder.
Order	...	**Clupeiformes**	→	Fins without fin rays. Air-bladder communicates with pharynx. Maxillae included in gap of mouth.
Genus	...	*Salmo*		
Species	...	*sola*		

Habit and habitat. *Salmo* is found in sea water. They are migratory in habits, young migrate to sea and mature in 2 to 8 years and then return to freshwater to spawn once and die.

Comments :

1. *Salmo sola* (Fig. 11.17) is commonly known as **salmon** (trout) and weighs 10–15 kgs.
2. Body is elongated and covered with overlapping scales.
3. Head is large with conical snout and eyes and nostrils.
4. Mouth is large with powerful dentition.
5. First dorsal fin large and in middle, second dorsal fin small and above anal fin. Anal fin with 8–12 anal rays.

6. Pelvic fin abdominal and pectoral fin close to operculum.
7. Fins without spiny rays.
8. Tail homocercal.
9. Air-bladder opens into pharynx.
10. Salmon is prized for food (fresh or canned).

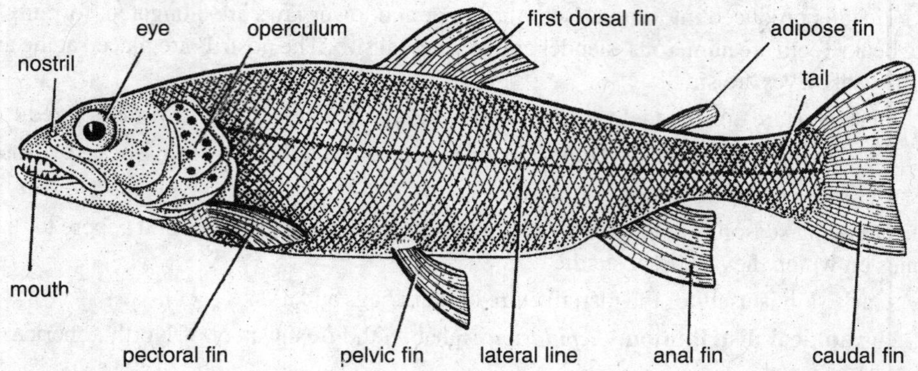

11.17. *Salmo.*

Geographical distribution. *Salmo* is found in the seas and freshwaters of Europe, Asia and North America extending southwards to North West Africa, Asia Minor, Northern Persia. It has been introduced into Kashmir and Nilgiris. Jurassic to Recent.

19. *Esox* (Pike)

Classification :

Phylum	…	**Chordata**	}	Characters same as that of *Scoliodon.*
Group	…	**Craniata**		
Subphylum	…	**Vertebrata**		
Division	…	**Gnathostomata**		
Superclass	…	**Pisces**		
Class	…	**Osteichthyes**	}	Characters same as that of *Polypterus.*
Subclass	…	**Actinopterygii**		
Superorder	…	**Teleostei**	→	Modern ray-finned fishes; hydrostatic fin rays.
Order	…	**Clupeiformes**	→	Fins with soft rays. Pelvic fin abdominal, Air bladder communicates with pharynx.
Genus	…	*Esox*		

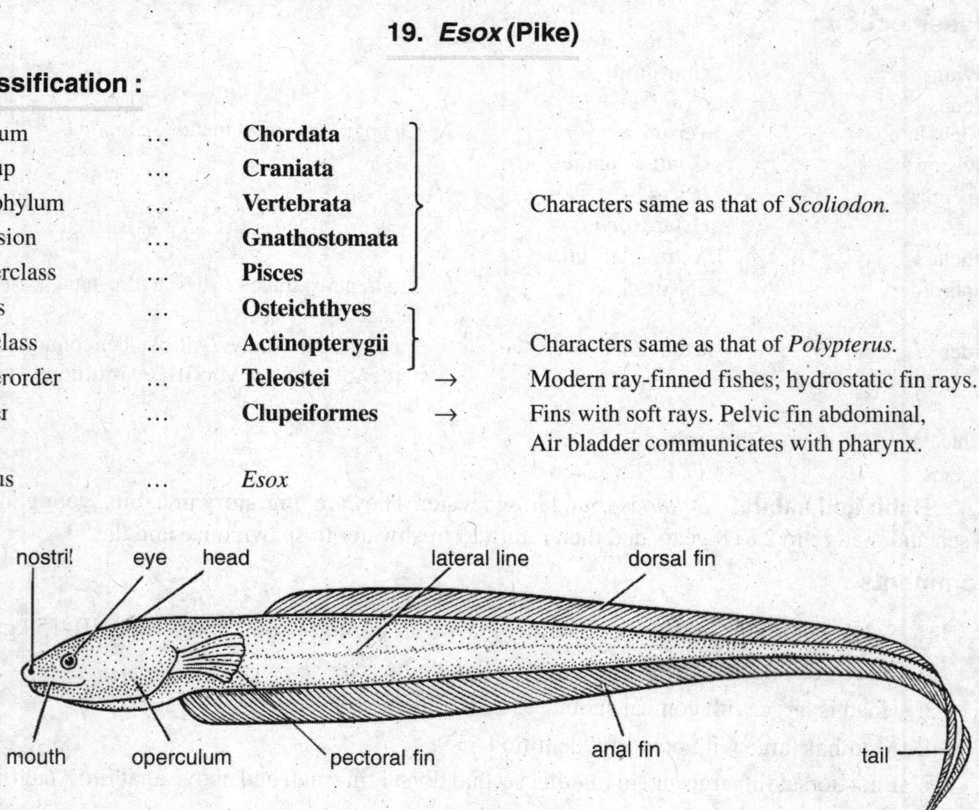

11.18. *Esox.*

Habit and habitat. *Esox* is a marine, carnivorous and predaceous fish.

Comments :

1. *Esox* (Fig. 11.18) is commonly called **pike**.
2. Body is elongated, measuring up to two and a half metres in length and weighs about 45 kgs.
3. Head is produced into snout. Mouth is large and margin of the upper jaw is constituted by premaxillaries and toothless maxillaries.
4. Supra-occipital is in contact with frontals. Opercular bone is well-developed.
5. Adipose fin absent. Dorsal and anal fins are confluent with body continuing up to head. Caudal fin present.
6. Pectoral fins are inserted very low and ventrals (pelvic) provided with 6–11 rays.
7. Pyloric caeca is absent.
8. These fishes have slender body with large mouth and conspicuous teeth.

Geographical distribution. *Esox* is found in Europe, Northern Asia, and North America. Jurassic to Recent. Its fossils have been found in Europe belonging to Oligocene period.

20. *Hilsa*

Classification :

Phylum	...	**Chordata**	Characters same as that of *Scoliodon*.
Group	...	**Craniata**	
Subphylum	...	**Vertebrata**	
Division	...	**Gnathostomata**	
Superclass	...	**Pisces**	
Class	...	**Osteichthyes**	Characters same as that of *Polyplerus*.
Subclass	...	**Actinopterygii**	
Superorder	...	**Teleostei**	Characters same as that of *Salmo*.
Order	...	**Clupeiformes**	
Genus	...	*Hilsa*	
Species	...	*ilisha*	

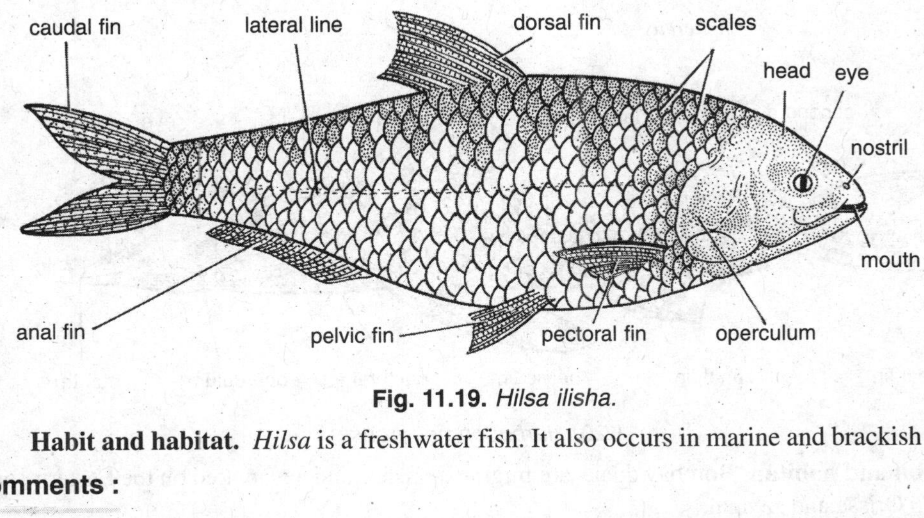

Fig. 11.19. *Hilsa ilisha.*

Habit and habitat. *Hilsa* is a freshwater fish. It also occurs in marine and brackish waters.

Comments :

1. *Hilsa ilisha* (Fig. 11.19) is commonly called hilsa. It measures about 31–38 cm in length.
2. Body is laterally compressed and covered by large silvery scales. Vertical faint dark bars across the back and upper part of the sides.

3. Pectoral, pelvic, dorsal, anal and caudal fins are present. Caudal fin is deeply forked and partly covered by scales.
4. Head bears eyes with thick adipose eyelids. Barbels absent.
5. Belly possesses saw-like edge.
6. Air-bladder is connected with ear.
7. This fish is used for food because it forms the delicious food for those who eat fish.
8. *Hilsa ilisha* swarm up in large rivers of India to spawn at the time of local monsoon. Mature fishes are available from May to October in the Ganges when they start leaving for sea.
9. **Fin formula.** D. 19, P. 15–16, V. 8, A. 21–22 and C. 19.

Geographical distribution. *Hilsa* is found in temperate and tropical seas and fresh waters of Ethiopian and Oriental regions including China and Japan.

21. *Harpodon*

Classification :

Phylum	...	**Chordata**		Characters same as that of *Scoliodon*.
Group	...	**Craniata**		
Subphylum	...	**Vertebrata**		
Division	...	**Gnathostomata**		
Superclass	...	**Pisces**		
Class	...	**Osteichthyes**		Characters same as that of *Polypteius*.
Subclass	...	**Actinopterygii**		
Superorder	...	**Teleostei**	→	Modern ray-finned fishes; hydrostatic swim-bladder.
Order	...	**Scopeliformes**	→	Dorsal and anal fins without spines; an adipose fin is present; mouth large provided with numerous small teeth; luminous organs are usually present.
Genus	...	*Harpodon*		
Species	...	*nehereus*		

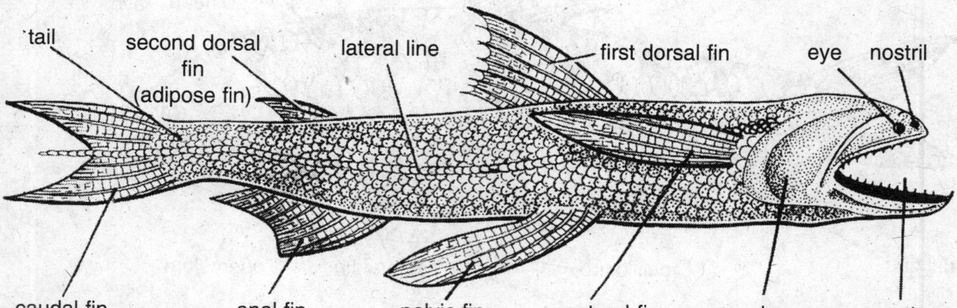

11.20. *Harpodon nehereus.*

Habit and habitat. Bombay ducks are migratory fishes and are spotted on the Eastern coasts as north in Orissa and Bengal.

Comments :

1. *Harpodon nehereus* (Fig. 11.20) is commonly known as **Bombay-duck (Nehari)**.
2. Body is elongated, abdomen nearly clindrical and is silvery white in colour. Scales deciduous.

3. Snout is short, bluntly rounded and eyes are small over the anterior part of head.

4. Mouth is very wide and armed with numerous small teeth. Jaws subequal, lower jaw longer.

5. Dorsal fin lies in the middle of the body. Dorsal and anal fins are without spines. Small adipose (second dorsal) fin is present. Caudal fin is trilobed. (D. 12–14 and A. 13–15). Pectoral fins very long.

6. Gill-opening is wide.

7. Lateral line slightly keeled.

8. It is commercially important as food fish. Its fishery is very important on the Bombay coasts from Ratnagiri to Broach (Saurashtra).

Geographical distribution. *Harpodon nehereus* is abundantly found along the Bombay coast.

22. *Notopterus* (Chitala)

Classification :

Phylum	...	**Chordata**		Characters same as that of *Scoliodon.*
Group	...	**Craniata**		
Subphylum	...	**Vertebrata**		
Division	...	**Gnathostomata**		
Superclass	...	**Pisces**		
Class	...	**Osteichthyes**		Characters same as that of *Polypterus.*
Subclass	...	**Actinopterygii**		
Superorder	...	**Teleostei**	→	Modern ray-finned fishes; hydrostatic swim-bladder.
Order	...	**Ostariophysi (Cypriniformes = Ostariophysi)**	→	Anterior vertebrae fused. Weberian ossicles present between air-bladder and ear.
Genus	...	*Notopterus*		
Species	...	*chitala*		

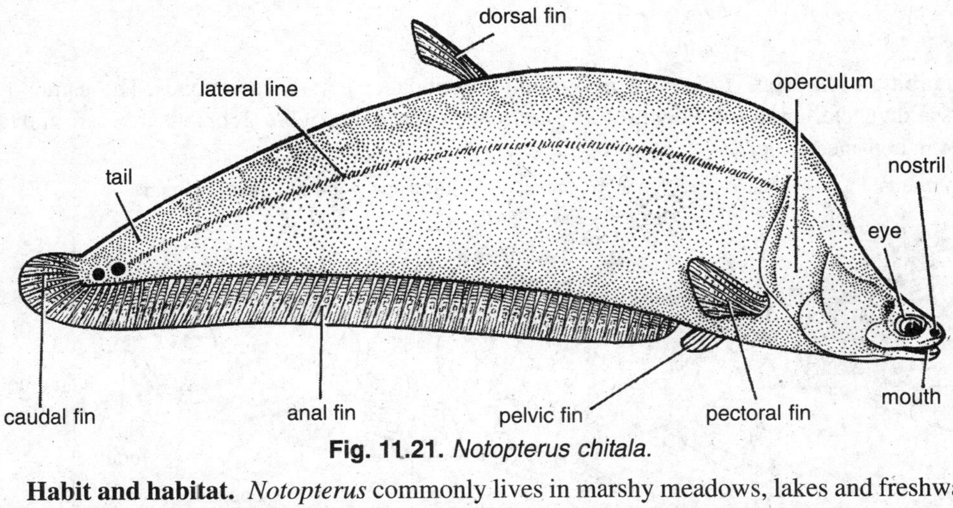

Fig. 11.21. *Notopterus chitala.*

Habit and habitat. *Notopterus* commonly lives in marshy meadows, lakes and freshwater. It is bottom feeder, **carnivorous** feeding on small organisms, crustaceans and insects.

Comments :

1. *Notopterus chitala* (Fig. 11.21) is commonly known as **chitala**.

2. Body is strongly compressed and covered with minute scales. It measures up to a metre in length.

3. The colour is coppery brown or greyish along the back with 15 or 16 silvery transverse bars.
4. Head is small and bears large mouth, white eyes and nostrils. Snout is prominent.
5. Dorsal fin is small. Anal fin is much elongated and confluent with the caudal fin. Rays of anal and caudal fins vary from 85–100. Pectoral fin is prominent and pelvic fin very small.
6. Musciferous channels on the head are well-developed.
7. The air-bladder is very large with several divisions.
8. Operculum is scaly. Lateral line scales 120–180.
9. It is commercially important as food fish, its flesh is said to be uncommonly rich and well-flavoured.
10. **Fin formula:** D. 9, P. 16, V. 6, A. 110–118, C. 12.

Geographical distribution. *Notopterus* lives in marshes and lakes of freshwater and brackish water of West Africa, India, Myanmar and Malaya. *N. chitala* is found exclusively in freshwaters of India. Eocene to Recent.

23. *Labeo rohita* (Rohu)

Classification :

Phylum	...	**Chordata**		Characters same as those of *Scoliodon.*
Group	...	**Craniata**		
Subphylum	...	**Vertebrata**		
Division	...	**Gnathostomata**		
Superclass	...	**Pisces**		Characters same as those of *Polypterus.*
Class	...	**Osteichthyes**		
Subclass	...	**Actinopterygii**		
Superorder	...	**Teleostei**	→	Moderately finned fishes; hydrostatic swim-bladder.
Order	...	**Cypriniformes (= Ostariophysi)**	→	Characters same as that of *Notopterus.*
Genus	...	*Labeo*		
Species	...	*rohita*		

Habit and habitat. *Labeo rohita* is found in abundance in rivers and ponds. These are bottom feeders and vegetarian. It is bred with *Catla catla* and *Cirrhina mrigala.* It breeds in rivers and tanks and not in confined waters.

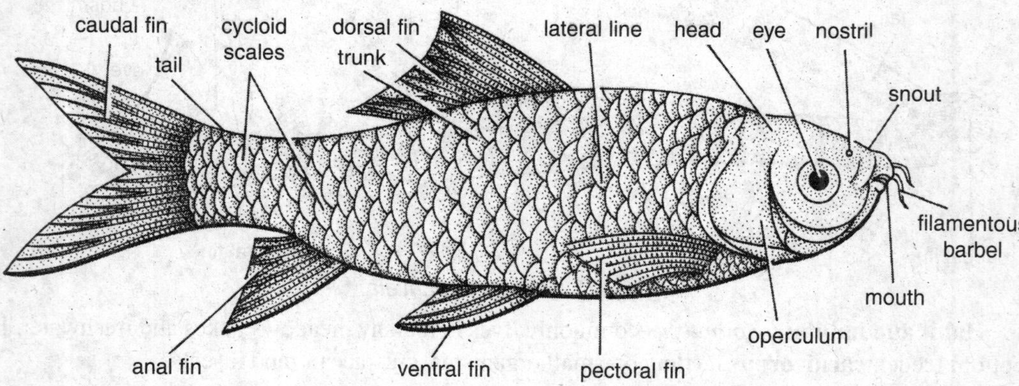

Fig. 11.22. *Labeo rohita.*

Comments :

1. *Labeo rohita* (Fig. 11.22) is the most famous carp and commonly known as **rohu** in Hindi.

2. Body is elongated with moderately rounded abdomen, measuring up to 1 metre in length and weighs about 4 kg.

3. Colour brownish grey to black above and silvery white below.

4. Scales (cycloid) are large and orange to reddish in colour in the centre.

5. Head is prominent with blunt snout. Mouth is transverse and semi-oval with fringed thick lips. A pair of filamentous barbels arise from the upper lip.

6. Lateral line is distinct.

7. Paired fins are present. Dorsal fin single just above the pelvic fins, caudal fin is forked.

8. Air-bladder is usually large and divided into an anterior and posterior part.

9. It is economically important due to its food value. It is relished very much in food.

10. **Fin formula:** D. 16, P. 17, V. 9, A. 7, C. 19. Scales on lateral line are 40–41.

Geographical distribution. *Labeo rohita* is found all over Northern and Central India. It is now found in Godawari and Krishna rivers. Eocene to Recent.

24. *Catla*

Classification :

Phylum	...	**Chordata**	}	Characters same as that of *Scoliodon*.
Group	...	**Craniata**		
Subphylum	...	**Vertebrata**		
Division	...	**Gnathostomata**		
Superclass	...	**Pisces**		
Class	...	**Osteichthyes**	}	Characters same as that of *Polypterus*.
Subclass	...	**Actinopterygii**		
Superorder	...	**Teleostei**	→	Characters same as that *Labeo*.
Order	...	**Cypriniformes**	→	Characters same as that of *Notopterus*.
Genus	...	*Catla*		
Species	...	*catla*		

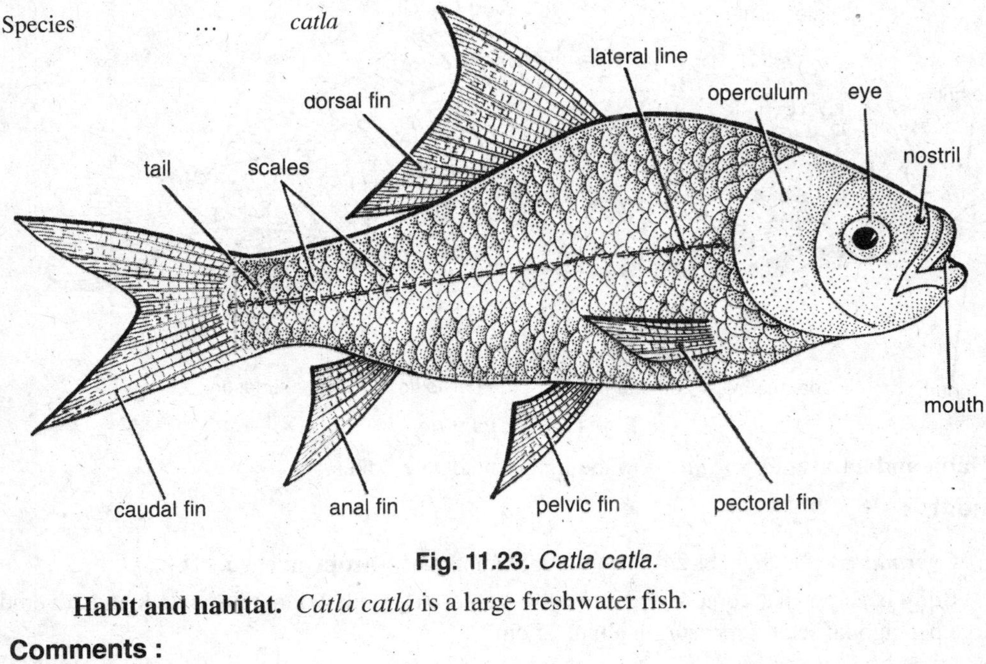

Fig. 11.23. *Catla catla.*

Habit and habitat. *Catla catla* is a large freshwater fish.

Comments :

1. *Catla* (Fig. 11.23) has only one species. *Catla* is the largest Indian carp and commonly known as **Katla** in Hindi.

2. Body is short, deep with rounded abdomen, measuring more than a metre in length. Dorsal profile of the body is more convex.

3. Colour blackish grey above and silvery on the sides.

4. Scales are pink or coppery in the centre on the dorsal side and whitish below.

5. Head is broadly large having large rounded eyes, mouth is wide with prominent lower lip. Upper lip absent. Barbels absent.

6. Dark black paired and median fins present. Dorsal fin is quite large and caudal fin is bilobed.

7. Air-bladder is usually large and divided into an anterior and posterior part.

8. It is also important as a food fish.

9. **Fin formula:** D. 11, P. 15, V. 9, C. 19. Lateral line scales 46.

Geographical distribution. *Catla catla* is found throughout India, although rather scarce to south of the Krishna river.

25. *Cyprinus* (European carp)

Classification :

Phylum	...	**Chordata**	Characters same as that of *Scoliodon.*
Group	...	**Craniata**	
Subphylum	...	**Vertebrata**	
Division	...	**Gnathostomata**	
Superclass	...	**Pisces**	
Class	...	**Osteichthyes**	Characters same as that of *Polypterus.*
Subclass	...	**Actinopterygii**	
Superorder	...	**Teleostei**	→ Characters same as that of *Labeo.*
Order	...	**Cypriniformes**	→ Characters same as that of *Notopterus.*
Genus	...	*Cyprinus*	
Species	...	*carpio*	

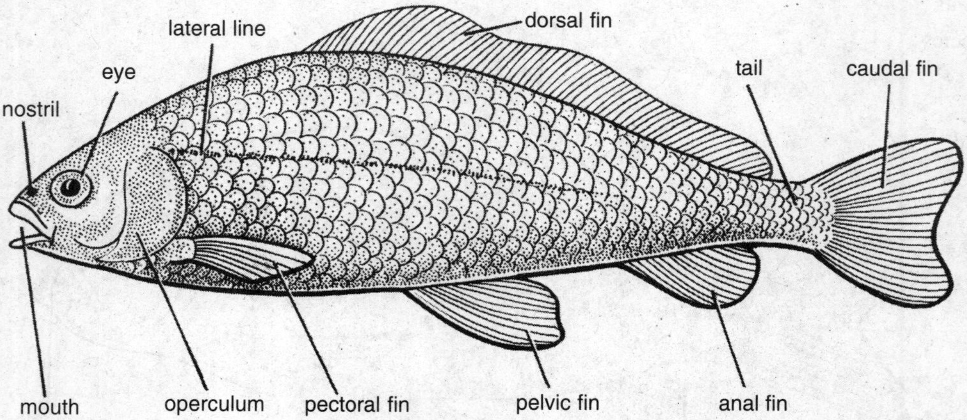

Fig. 11.24. *Cyprinus.*

Habit and habitat. *Cyprinus* is found in the freshwater lakes.

Comments :

1. *Cyprinus carpio* (Fig. 11.24) is commonly known as **European carp**.

2. Body is somewhat compressed and robust anteriorly and abdomen rounded, and covered with large pentagonal scales, measuring about 75 cm.

3. Head is large with wide mouth having fleshy lips. Upper jaw projected anteriorly. Barbels two pairs.

4. Dorsal fin is quite large reaching up to the tail.

5. Paired and median fins present. Dorsal fin very long, caudal fin pointed. Ventral fin is abdominal.

6. Air-bladder is connected with the ear by means of Weberian ossicles and communicates with the gut.

Geographical distribution. *Cyprinus* is not a native of India, it has been introduced into the lakes of Nilgiri mountain ranges from China and South East Asia. It was first introduced into Sri Lanka and later in South India.

26. *Cirrhina mrigala* (Mrigal)

Classification :

Phylum	...	**Chordata**	}	Characters same as that of *Scoliodon.*
Group	...	**Craniata**		
Subphylum	...	**Vertebrata**		
Division	...	**Gnathostomata**		
Superclass	...	**Pisces**		
Class	...	**Osteichthyes**	}	Characters same as that of *Polypterus.*
Subclass	...	**Actinopterygii**		
Superorder	...	**Teleostei**	→	Characters same as that of *Labeo.*
Order	...	**Cypriniformes**	→	Characters same as that of *Notopterus.*
Genus	...	*Cirrhina*		
Species	...	*mrigala*		

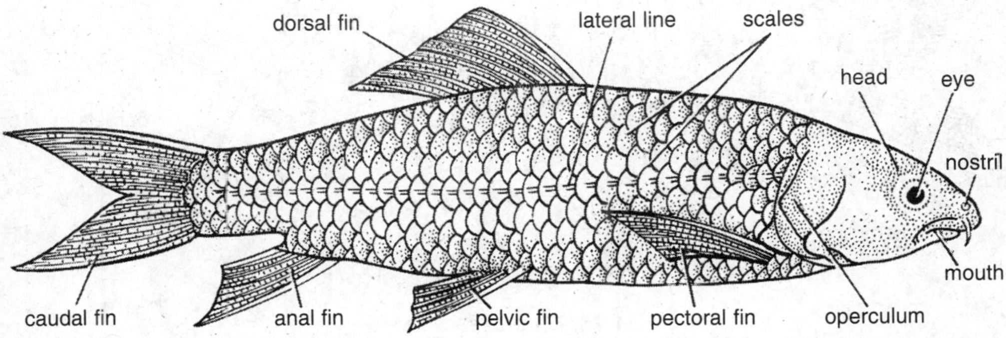

Fig. 11.25. *Cirrhina mrigala.*

Habit and habitat. *Cirrhina mrigala* is found in freshwater of north India, Pakistan and Myanmar.

Comments :

1. *Cirrhina mrigala* (Fig. 11.25) commonly called **mrigal** is a freshwater carp of India.
2. The length of the body reaches up to 66 cm and weighs up to 1.4 to 2.8 kg.
3. The colour of the body is silvery, dark-grey along the back, sometimes with coppery tinge.
4. The body is covered by large hexagonal cycloid scales. The scales are absent on the head.
5. Head large with eyes and mouth is wider with thinner lips. Barbels small, one pair.
6. Pectoral, pelvic and anal fins are orange in colour tipped with black.
7. Caudal fin is deeply forked.
8. It gives best sport on rod and line and is used as food.
9. **Fin formula:** D. 11, P. 16, V. 9, A. 8, C. 19. Lateral line scales 36–37.

Geographical distribution. *Cirrhina mrigala* is found throughout India, Pakistan and Myanmar.

27. *Barbus sarana* (= *Puntius*)

Classification :

Phylum	...	**Chordata**	Characters same as those of *Scoliodon.*
Group	...	**Craniata**	
Subphylum	...	**Vertebrata**	
Division	...	**Gnathostomata**	
Superclass	...	**Pisces**	
Class	...	**Osteichthyes**	Characters same as those of *Polypterus.*
Subclass	...	**Actinopterygii**	
Superorder	...	**Teleostei**	→ Characters same as that of *Labeo.*
Order	...	**Cypriniformes (= Ostariophysi)**	→ Characters same as that of *Notopterus.*
Genus	...	*Barbus* (= *Puntius*)	
Species	...	*sarana*	

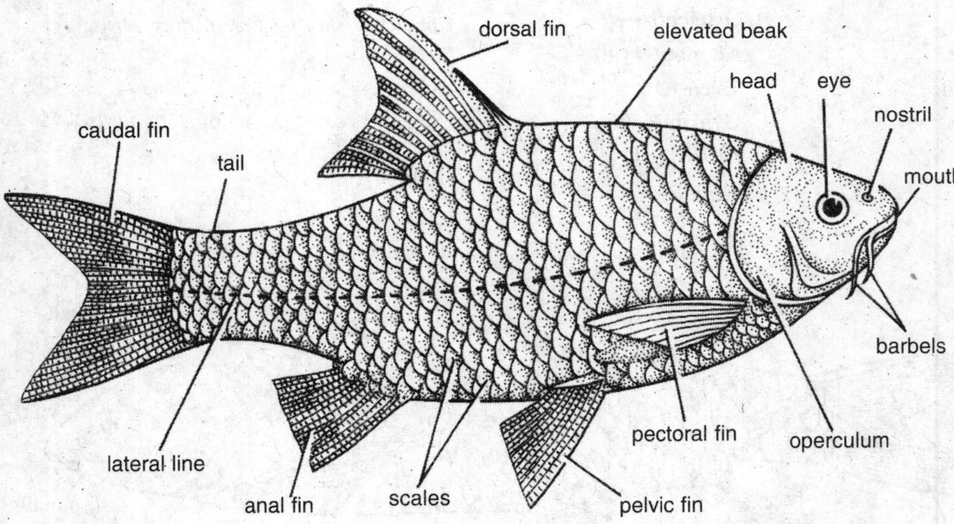

11.26. *Barbus sarana.*

Habit and habitat. *Barbus* is a common freshwater fish of ponds, lakes and rivers.

Comments :

1. *Barbus sarana* (Fig. 11.26) is commonly called **punti**, minnow or carp.
2. The colour of the fish is silvery-white or with dark black along dorsal side which gets fainter on sides.
3. The body is covered by large silvery scales. The scales are absent on the head.
4. Head is with mouth, large eyes and small barbels (2 pairs).
5. Dorsal, pectoral, pelvic, anal and caudal fins are present. Caudal fin forked.
6. The operculum is stout with golden colour.
7. Air-bladder is divided into two lobes by a constriction and enclosed in a bony capsule.
8. *Barbus* is used as a food. It ranks very high as food fish.
9. **Fin formula:** D. 11-12, P. 15–16, V. 8–9, A. 8, C. 19. Lateral line scales 33–34.

Geographical distribution. *Barbus* is distributed in India and Myanmar. Eocene to Recent.

28. *Arius*

Classification :

Phylum	...	**Chordata**		Characters same as that of *Scoliodon.*
Group	...	**Craniata**		
Subphylum	...	**Vertebrata**		
Division	...	**Gnathostomata**		
Superclass	...	**Pisces**		Characters same as that of *Polypterus.*
Class	...	**Osteichthyes**		
Subclass	...	**Actinopterygii**		
Superorder	...	**Teleostei**	→	Characters same as that of *Labeo*.
Order	...	**Cypriniformes**	→	A peculiar Weberian apparatus connecting ear with the air-bladder; air-bladder is connected with the alimentary canal by a duct; fins either without spines or dorsal, anal or pectoral have a spine each; ventral fins are abdominal.
Division	...	**Siluri**	→	Body naked, not covered by scales; maxillary bone reduced supporting the barbles; second, third, fourth and sometimes the fifth vertebrae are fused.
Genus	...	*Arius*		

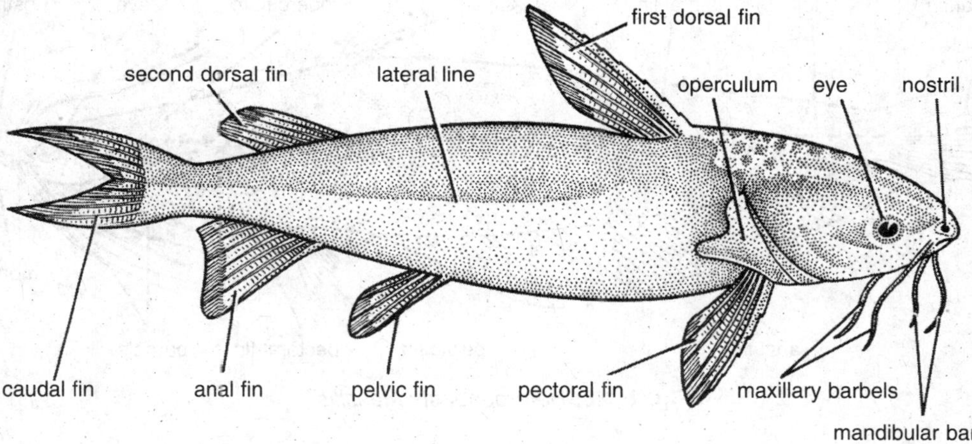

Fig. 11.27. *Arius.*

Habit and habitat. *Arius* is a marine fish.

Comments :

1. Body of *Arius* (Fig. 11.27) is elongated and scaleless.
2. Head is osseus and covered with very thin skin.
3. Head with eyes with free orbital margins, mouth is wide, upper jaw generally longer. Barbles six (two maxillary and four mandibular).
4. First dorsal fin bears a serrated spine. Second dorsal fin is adipose. Caudal fin is forked.
5. Pectoral fin with a strong serrated spine.
6. Gill-openings are wide, covered with operculum.
7. Air-bladder is long and longitudinally divided and connected with the gut.
8. It is used as food of inferior quality.

Geographical distribution. *Arius* is found in the seas of tropical countries, *e.g.*, Asia and America.

29. *Heteropneustes* (*Saccobranchus*)

Classification :

Phylum	...	**Chordata**	Characters same as that of *Scoliodon.*
Group	...	**Craniata**	
Subphylum	...	**Vertebrata**	
Division	...	**Gnathostomata**	
Superclass	...	**Pisces**	
Class	...	**Osteichthyes**	Characters same as that of *Polypterus.*
Subclass	...	**Actinopterygii**	
Superorder	...	**Teleostei**	→ Characters same as that of *Labeo*.
Order	...	**Cypriniformes**	Chanacters same as those of *Arius*
Division	...	**Siluri**	
Genus	...	*Heteropneustes*	
Species	...	*fossilis*	

Habit and habitat. *Heteropneustes* is found in freshwater of India, Sri Lanka, Myanmar and China.

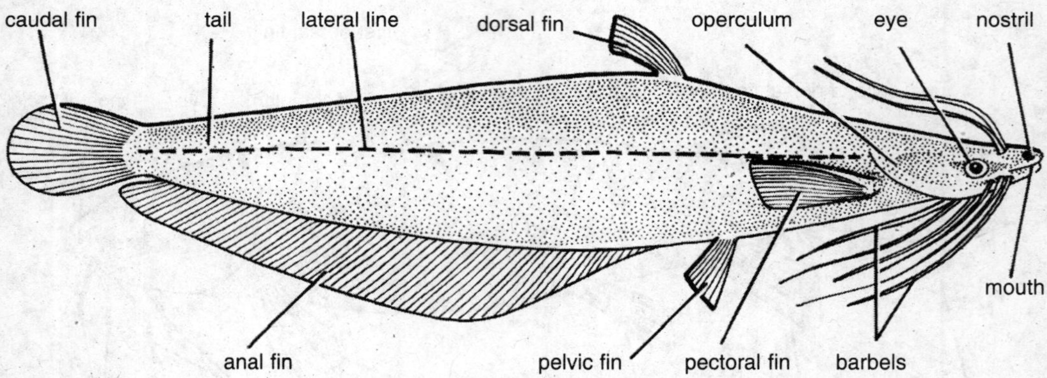

Fig. 11.28. *Heteropneustes fossilis.*

Comments :

1. *Heteropneustes foosilis* (Fig. 11.28) is commonly called **singhi** in Hindi.
2. Body is scaleless, elongated and laterally compressed, measuring about 30 cm in length.
3. Head is flattened and covered with thin skin. Head is provided with eyes, flat snout with mouth and 4 pairs of barbles – one pair maxillary, one pair nasal and 2 pairs mandibular. Eyes with free circular margins.
4. Dorsal fin is short without spine, ventral fin situated at the level of the dorsal fin, pectoral fins are strong with poison spine, and anal fin is elongated, reaches up to the rounded caudal fin and separated from it by a notch.
5. Gill-opening is wide, its membranes not being confluent with the skin of isthumus and separated by deep notch.
6. Accessory breathing organs are present.
7. Air-bladder is also present.

8. **Fin formula:** D. 6, P. 1/7, V. 6, A. 62—66, C. 19.

Geographical distribution. *Heteropneustes* (*Saccobranchus*) is found in freshwaters of India and Myanmar.

30. *Clarias* (Mangur)

Classification :

Phylum	...	**Chordata**	Characters same as that of *Scoliodon.* (Phylum to Superclass)
Group	...	**Craniata**	
Subphylum	...	**Vertebrata**	
Division	...	**Gnathostomata**	
Superclass	...	**Pisces**	
Class	...	**Osteichthyes**	Characters same as that of *Polypterus* (Class and Subclass)
Subclass	...	**Actinopterygii**	
Superorder	...	**Teleostei**	→ Characters same as that of *Labeo.*
Order	...	**Cypriniformes (= Ostariophysi)**	→ Characters same as that of *Arius.*
Genus	...	*Clarias*	
Species	...	*batrachus*	

Habit and habitat. *Clarias batrachus* is found in fresh and brackish waters. It feeds on clams, crustaceans and insect larvae and acts as scavanger.

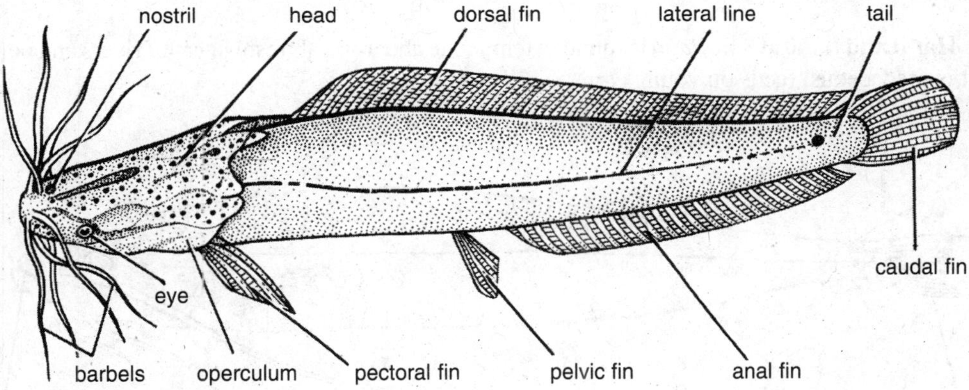

Fig. 11.29. *Clarias batrachus.*

Comments :

1. *Clarias batrachus* (Fig. 11.29) is called catfish or magur.
2. Body is elongated, scaleless and measuring up to 45 cm in length.
3. The general colour of the body is uniform brown or greyish-black.
4. Head depressed and covered with osseus plates dorsally and laterally.
5. Barbles are four pairs—one pair maxillary, one pair nasal and 2 pairs mandibular.
6. Upper jaw is longer.
7. Dorsal fin is long and without spines, extending up to the caudal fin. Anal fin also long. No adipose fin. Caudal fin is more or less rounded. Pectoral fins are provided with spines.
8. Tail is diphycercal.
9. Accessory respiratory organs are branched tree-like in the branchial chamber. It can live out of water for a long time with the help of its accessory respiratory organs.

10. The air-bladder is connected with internal ear by Weberian ossicles. It is physostomous.
11. It is highly nourishing and esteemed as food.
12. It is also used in laboratories for experimental purposes.
13. **Fin formula:** D. 65–70, P. 1/8, V. 6, A. 47, C. 17.

Geographical distribution. *Clarias* is found in Africa and South and West Asia. *Clarias batrachus* occurs in fresh and brackish waters throughout India. Eocene to Recent.

31. *Wallago* (Lachi)

Classification :

Phylum	...	**Chordata**		Characters same as that of *Scoliodon.*
Group	...	**Craniata**		
Subphylum	...	**Vertebrata**		
Division	...	**Gnathostomata**		
Superclass	...	**Pisces**		
Class	...	**Osteichthyes**		Characters same as that of *Polypterus.*
Subclass	...	**Actinopterygii**		
Superorder	...	**Teleostei**	→	Characters same as that of *Labeo.*
Order	...	**Cypriniformes (= Ostariophysi)**	→	Characters same as that of *Arius.*
Genus	...	*Wallago*		
Species	...	*attu*		

Habit and habitat. *Wallago* is found in temperate and tropical freshwaters of rivers and ponds, etc. It is predaceous, feeds on young carps.

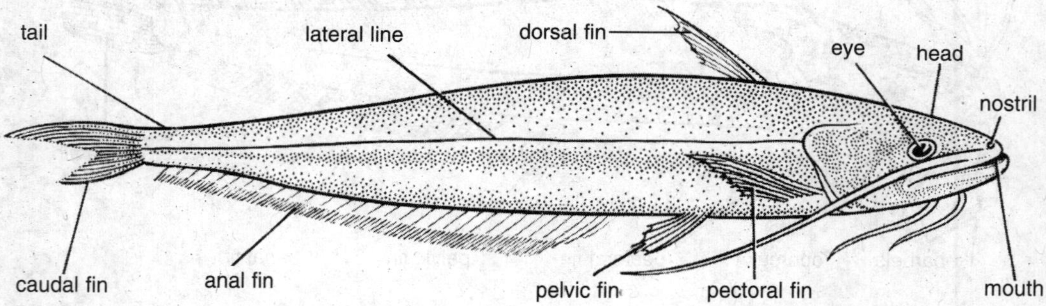

Fig. 11.30. *Wallago attu.*

Comments :

1. *Wallago* (Fig. 11.30) is commonly called **lachi** or **catfish**.
2. Body colour is generally white with grey-yellowish along the back.
3. Body is elongated and laterally compressed with a straight back. It may reach in length up to 2 metres but the specimens generally available are 60–90 cm in length.
4. Head is very large, trunk small and tail long and tapering.
5. Mouth is wide extending below and even behind the eyes. Jaws are provided with teeth.
6. Barbels long and 4, two maxillary and two mandibular.
7. Eyes with free eyelids.
8. Dorsal fin is short and spineless. Anal fin is very long. Caudal fin deeply forked with longer upper lobe.

9. Weverian ossicles are present. Gill membranes free.
10. **Fin formula:** D. 5, P. 1/14, V. 10, A. 86, C. 17.

Geographical distribution. *Wallago attu* is found throughout India. Eocene to Recent.

32. *Mystus*

Classification :

Phylum	...	**Chordata**	}	Characters same as that of *Scoliodon*.
Group	...	**Craniata**		
Subphylum	...	**Vertebrata**		
Division	...	**Gnathostomata**		
Superclass	...	**Pisces**		
Class	...	**Osteichthyes**	}	Characters same as that of *Polypterus*.
Subclass	...	**Actinopterygii**		
Superorder	...	**Teleostei**	→	Chatacters same as that of *Labeo*.
Order	...	**Cypriniformes (= Ostariophysi)**	→	Chatacters same as that of *Arius*.
Genus	...	*Mystus*		
Species	...	*seenghala*		

Habit and habitat. *Mystus* is a freshwater fish found in rivers and ponds. It is predaceous, feeds on small carps, and prawns, etc.

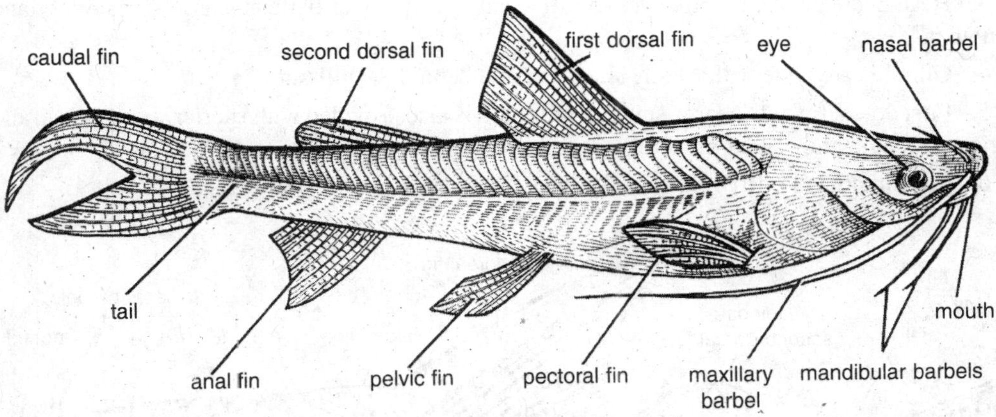

Fig. 11.31. *Mystus seenghala.*

Comments :

1. The body colour of *Mystus seenghala* (Fig. 11.31) is brownish above with silvery white on sides.
2. Body is elongated, measuring up to one metre in length. It reaches a length of 120 cm.
3. Head with distinctly long snout, mouth is transverse and terminal, upper jaw is generally longer and barbels 8–two nasal, two maxillary and four mandibular.
4. Adipose dorsal fin is well-developed having a circular black spot. Pectoral fin with a stronger spine than that of dorsal. Caudal fin is deeply forked with a longer upper lobe.
5. Employed as food by the poor natives.
6. **Fin formula:** D. 1/7/0; P. 1/9; V. 6; A. 11–12; C. 19–21.

Geographical distribution. *Mystus seenghala* is widely distributed in India being found in Jamuna, Ganga and the Deccan rivers as well as Assam rivers. It is also found in Yunnan and Myanmar.

33. *Rita*

Classification :

Phylum	...	**Chordata**	Characters same as that of *Scoliodon*.
Group	...	**Craniata**	
Subphylum	...	**Vertebrata**	
Division	...	**Gnathostomata**	
Superclass	...	**Pisces**	
Class	...	**Osteichthyes**	Characters same as that of *Polypterus*.
Subclass	...	**Actinopterygii**	
Superorder	...	**Teleostei**	→ Characters same as that *Labeo*.
Order	...	**Cypriniformes (= Ostariophysi)**	→ Characters same as that *Arius*
Genus	...	*Rita*	
Species	...	*rita*	

Habit and habitat. *Rita* also inhabits freshwaters.

Comments :

1. Body of *Rita* (Fig. 11. 32) is scaleless and elongated, measuring from 45 cm to 1 metre in length.
2. Body colour brownish olive green and lighter along sides and below.
3. Head bears a wide mouth, eyes, nostrils and three pairs of barbels—nasals, maxillary and mandibular.
4. Gill-slits are wide with a bony shield present behind the gill-slits.
5. First dorsal fin with a large spine, pectoral fin is also provided with shorter spine, pelvic and anal fins are small, and caudal fin is forked.
6. **Fin formula:** D. 1/6/0; P. 1/10; V. 8; A. 13; C. 19.

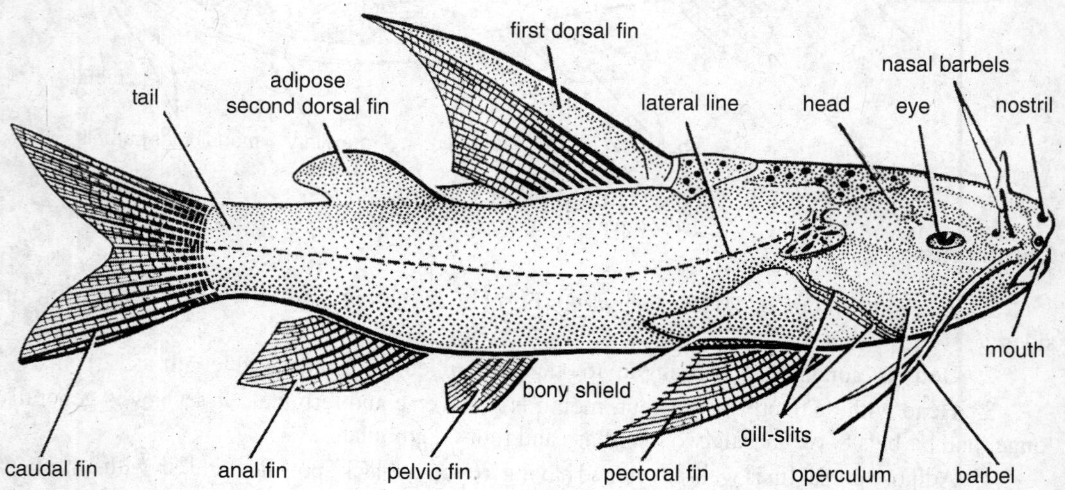

Fig. 11.32. *Rita rita.*

Geographical distribution. *Rita rita* is found in freshwaters throughout northern India and Myanmar.

34. *Anguilla* (Eel)

Classification :

Phylum	...	**Chordata**	Characters same as that of *Scoliodon*.
Group	...	**Craniata**	
Subphylum	...	**Vertebrata**	
Division	...	**Gnathostomata**	
Superclass	...	**Pisces**	
Class	...	**Osteichthyes**	Characters same as that of *Polypterus*.
Subclass	...	**Actinopterygii**	
Superorder	...	**Teleostei**	→ Characters same as that of *Labeo*.
Order	...	**Anguilliformes (= Apodes)**	→ Body elongated eel-like; skin naked or with minute scales embedded in it; dorsal, caudal and anal fins are long narrow and usually confluent behind; pelvic fins, if present, abdominal fins devoid of spines; air-bladder connected with the intestine. Gill-openings small.
Genus	...	*Anguilla*	

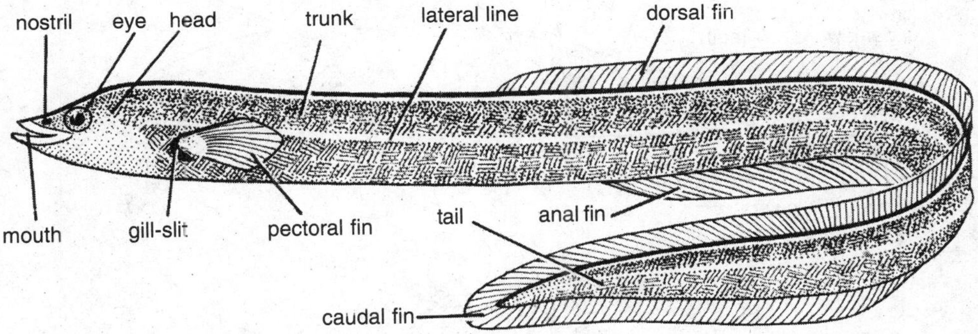

Fig. 11.33. *Anguilla.*

Habit and habitat. *Anguilla* is a freshwater catadromous fish. It can live for hours out of water. Adults live in ponds, rivers, estuaries and damp grass, etc.

Comments :

1. *Anguilla* (Fig. 11.33) is commonly known as **eel**.
2. Body is elongated, cylindrical and snake-like. It measures up to one metre in length.
3. The colour is brown on the back and yellowish below.
4. Small more or less linear rudimentary scales are embedded in the skin, arranged in small groups which are placed obliquely at right angles to one another forming a curious pattern.
5. Dorsal, caudal and anal fins are confluent, pectoral fin small and pelvic fin is absent. Fins are supported by rays, spines being absent.
6. Gill-opening is narrow and minute.
7. Air bladder has ductus pneumaticus.
8. Oviducts absent.
9. Its eggs hatch into pelagic leptocephalic larvae.

Geographical distribution. *Anguilla* is found in Gangetic estuaries, the Bay of Bengal, the Indo-Pacific region of the seas of China and beyond. Cretaceous to Recent.

35. *Belone* (*Xenentodon*)

Classification :

Phylum	...	**Chordata**	Characters same as that of *Scoliodon.*
Group	...	**Craniata**	
Subphylum	...	**Vertebrata**	
Division	...	**Gnathostomata**	
Superclass	...	**Pisces**	
Class	...	**Osteichthyes**	Characters same as that of *Polypterus.*
Subclass	...	**Actinopterygii**	
Super order	...	**Teleostei**	→ Characters same as that *Labeo.*
Order	...	**Beloniformies**	→ Body elongated covered with cycloid scales; fins without spines; ventral fins abdominal; some of them are capable of jumping into the air and glide with the help of enlarged pectoral fins.
Genus	...	*Belone* (*Xenentodon*)	
Species	...	*cancila.*	

Habit and habitat. *Belone* is a freshwater fish of India.

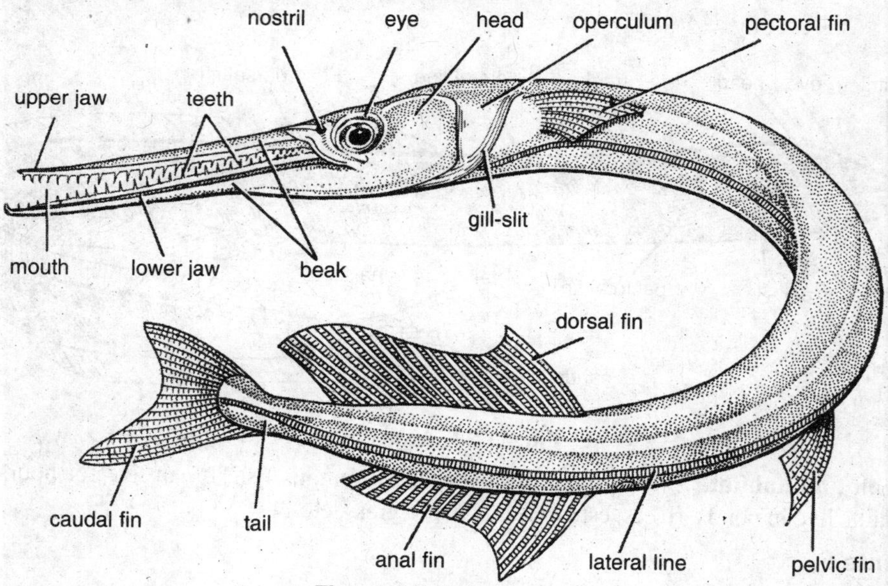

Fig. 11.34. *Belone cancila.*

Comments :

1. Body of *Belone* (Fig. 11.34) is elongated, sub-cylindrical, compressed and covered with small irregularly arranged scales. It grows up to 50 cm in length. Body is divided into head, trunk and tail. A silvery streak with black margin is present from eye to tail over the body.

2. Head is provided with lateral eyes, nostrils and wide elongated mouth Both the jaws are produced into a long slender beak having a row of large sharp teeth.

3. Gill-slit is wide, one on each side covered by operculum.

4. Pectoral and pelvic fins are small and widely apart from each other. Dorsal fin opposite to the anal fin. Caudal fin is forked.

5. Air-bladder is generally present.

6. It is esteemed as food.

Geographical distribution. *Belone cancila* is a freshwater fish found in India, Myanmar and Sri Lanka.

36. *Hemirhamphus*

Classification :

Phylum	...	**Chordata**		Characters same as that of *Scoliodon.*
Group	...	**Craniata**		
Subphylum	...	**Vertebrata**		
Division	...	**Gnathostomata**		
Superclass	...	**Pisces**		
Class	...	**Osteichthyes**		Characters same as that of *Polypterus.*
Subclass	...	**Actinopterygii**		
Superorder	...	**Teleostei**	→	Characters same as that of *Labeo.*
Order	...	**Beloniformies**	→	Characters same as that of *Belone.*
Genus	...	*Hemirhamphus*		

Habit and habitat. *Hemirhamphus* are mostly marine and some of them ascend in freshwater rivers. It is herbivorous, feeding on green algae.

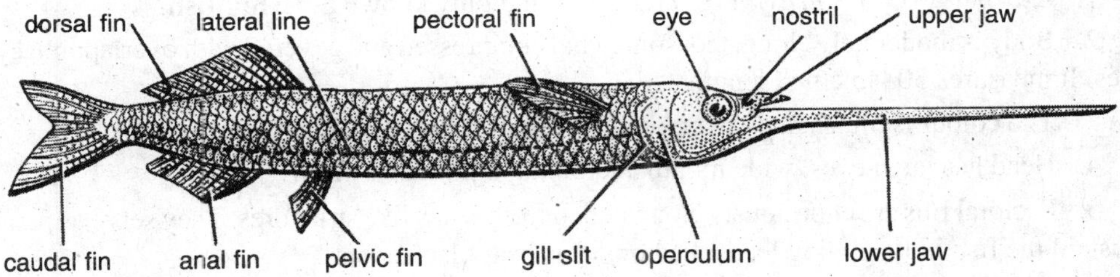

Fig. 11.35. *Hemirhamphus.*

Comments :

1. Body of *Hemirhamphus* (Fig. 11.35) is elongated and sub-cylindrical covered with cycloid scales, measuring up to 30 cm in length. Body is divided into head, trunk and tail.
2. Colour is greenish blue above with two bands along each side, one silvery and other black.
3. Upper jaw is short, more or less triangular in shape, while lower jaw is elongated forming the beak.
4. Head is provided with lateral eyes, nostrils and mouth.
5. Gill-opening is wide covered by operculum.
6. Tail is homocercal and caudal fin forked.
7. Dorsal and anal fins are opposite to each other.
8. It is used as food fish.
9. **Fin formula:** D. 14–15; P. 10; V. 6; A. 15; C. 19–20. Lateral line 50–54 scales.

Geographical distribution. *Hemirhamphus* is found in seas from India to China and abundant on the south-east coast of India.

37. *Exocoetus*

Classification :

Phylum	...	**Chordata**	Characters same as that of *Scoliodon*.
Group	...	**Craniata**	
Subphylum	...	**Vertebrata**	
Division	...	**Gnathostomata**	
Superclass	...	**Pisces**	
Class	...	**Osteichthyes**	Characters same as that of *Polypterus*.
Subclass	...	**Actinopterygii**	
Superorder	...	**Teleostei**	→ Characters same as that of *Labeo*.
Order	...	**Beloniformies** (= **Synentognathi**)	→ Flying fish, dorsal fin above anal, pectorals high on body.
Genus	...	*Exocoetus*	

Habit and habitat. *Exocoetus* is a marine pelagic fish living in shoals and feeding on prawns, young fishes and their eggs. Oviparous.

Comments :

1. *Exocoetus* (*Cypselurus*) (Fig. 11.36) is commonly known as **flying-fish**.
2. Body is moderately elongated, somewhat compressed and covered with overlapping cycloid scales. It measures 30–45 cm in length.
3. The colour is bluish and silvery below. Pectoral fins have black spots.
4. Head has large eyes, wide mouth and both the jaws bear teeth.
5. Pectoral fins are enormously elongated to form wing-like structures. They serve as parachute to sustain the fish in its gliding leaps. Pectoral fins bear black spots.
6. Dorsal and anal fins are short and supported by fin rays.
7. Tail is **hypobatic**, *i.e.*, ventral lobe of the tail fin is large.
8. Air-bladder is present.
9. It is considered as good edible fish and provide fishery in certain seasons.

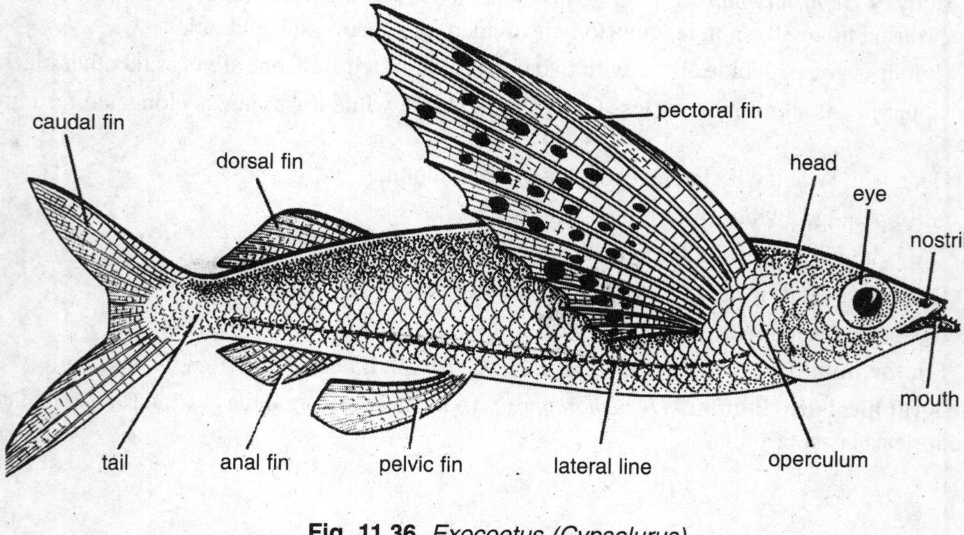

Fig. 11.36. *Exocoetus (Cypselurus).*

Geographical distribution. *Exocoetus* is found in tropical and warm parts of Atlantic and Indian Oceans. *E. pecilopterus* is found from the Indian Ocean to the China seas.

38. *Hippocampus* (Sea-horse)

Classification :

Phylum	...	**Chordata**	Characters same as that of *Scoliodon*.
Group	...	**Craniata**	
Subphylum	...	**Vertebrata**	
Division	...	**Gnathostomata**	
Superclass	...	**Pisces**	
Class	...	**Osteichthyes**	Characters same as that of *Polypterus*.
Subclass	...	**Actinopterygii**	
Superorder	...	**Teleostei**	→ Characters same as that of *Labeo*.
Order	...	**Syngnathiformes** (= **Solenichthyes**)	→ Body covered with protective layer of scales or bony rings; mouth is terminal, snout tubiform; fin-rays of dorsal, anal and pectoral fins not branched; physoclistic.
Genus	...	*Hippocampus*	

Habit and habitat. *Hippocampus* swims in vertical position swaying its tail and gyrating its trunk. Its prehensile tail encoil the surrounding weed. It feeds on minute organisms.

Comments :

1. *Hippocampus* (Fig. 11.37) is commonly known as **sea-horse**, resembling its head and neck with the horse.

2. Body is more or less elongated having an exoskeleton of ring-like bony plates. Size 5–17 cm.

3. Head is produced anteriorly into a tubular snout bearing terminal mouth. Eyes lie dorsally on the head.

4. Trunk is compressed, somewhat elevated with 10–12 exoskeletal rings. Some species have trailing filaments over the body.

5. Dorsal fin is single and small, pectoral fins are small on either side of head. Pelvic and caudal fins are absent. Females have a small anal fin.

6. Operculum is fused with the body wall ventrally and posteriorly, leaving a small upwardly directed branchial aperture.

7. Male possesses a **brood pouch** on the abdomen. In the brood pouch, eggs are retained until they hatch as young ones showing parental care.

8. It exhibits **mimicry** by having numerous prominent spines or cutaneous trailing filaments on the body, resembling the sea-weeds.

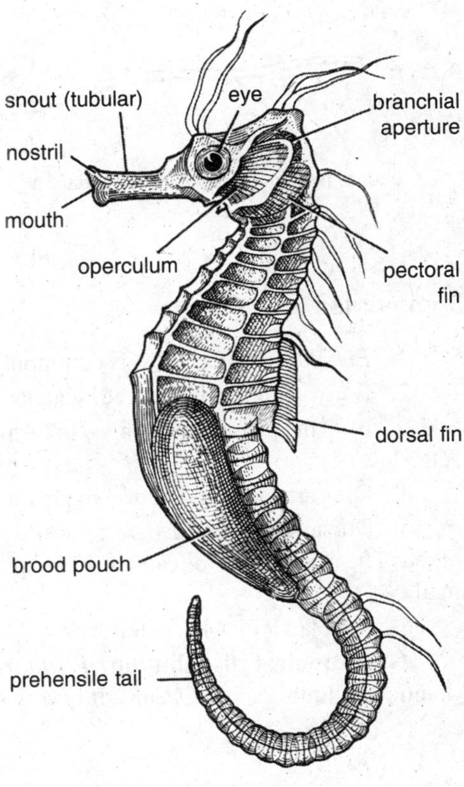

Fig. 11.37. *Hippocampus.*

Geographical distribution. *Hippocampus* is found in tropical and temperate seas, particularly in Indian and Atlantic seas, Japan, Malaya, Archipelago and China. Cretaceous to Recent.

39. *Fistularia* (Flute-fish)

Classification :

Phylum	...	**Chordata**	Characters same as that of *Scoliodon.*
Group	...	**Craniata**	
Subphylum	...	**Vertebrata**	
Division	...	**Gnathostomata**	
Superclass	...	**Pisces**	
Class	...	**Osteichthyes**	Characters same as that of *Polypterus.*
Subclass	...	**Actinopterygii**	
Order	...	**Syngnathiformes**	→ Characters same as that of *Hippocampus.*
Superorder	...	**Teleostei**	→ Characters same as that of *Labeo.*
		(= Solenichthyes)	
Genus	...	*Fistularia*	

Habit and Habitat. *Fistularia* is a marine fish.

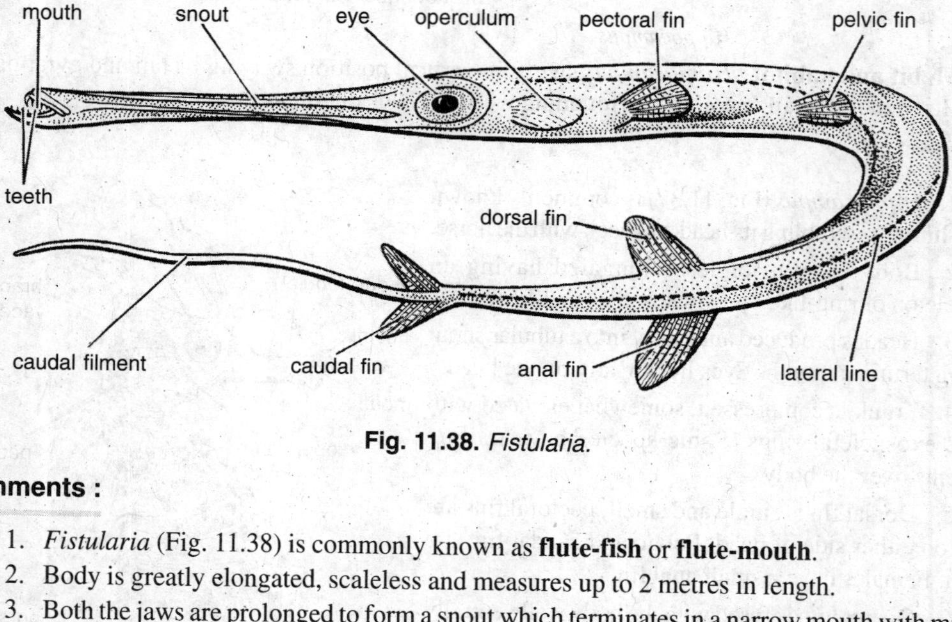

Fig. 11.38. *Fistularia.*

Comments :

1. *Fistularia* (Fig. 11.38) is commonly known as **flute-fish** or **flute-mouth.**
2. Body is greatly elongated, scaleless and measures up to 2 metres in length.
3. Both the jaws are prolonged to form a snout which terminates in a narrow mouth with minute teeth.
4. Eyes are large and lateral in position.
5. Dorsal and anal fins are opposite to each other. Anal fin with six soft rays. Caudal fin is forked. The middle rays of caudal fin are produced into a long caudal filament. Pelvic fins are very small.
6. Air-bladder is physocleistous.

Geographical distribution. *Fistularia* is found in the tropical and subtropical parts of the Atlantic and Indo-Pacific Oceans. It is very common at Chennai Coast.

40. *Syngnathus* (Pipe-fish)

Classification :

Phylum	...	**Chordata**	Characters same as that of *Scoliodon*.
Group	...	**Craniata**	
Subphylum	...	**Vertebrata**	
Division	...	**Gnathostomata**	
Superclass	...	**Pisces**	
Class	...	**Osteichthyes**	Characters same as that of *Polypterus*
Subclass	...	**Actinopterygii**	
Superorder	...	**Teleostei** →	Characters same as that of *Labeo*.
Order	...	**Syngnathiformes** →	Characters same as that of *Hippocampus*.
		(= Solenichthyes) →	Mouth at the end of tubular snout. Brood pouch in male; fins very small. Pipe-fish and sea-horses.
Genus	...	*Syngnathus*	

Habit and habitat. *Syngnathus* is a marine fish found in brackish waters. Oviparous. Fish swims in vertical position.

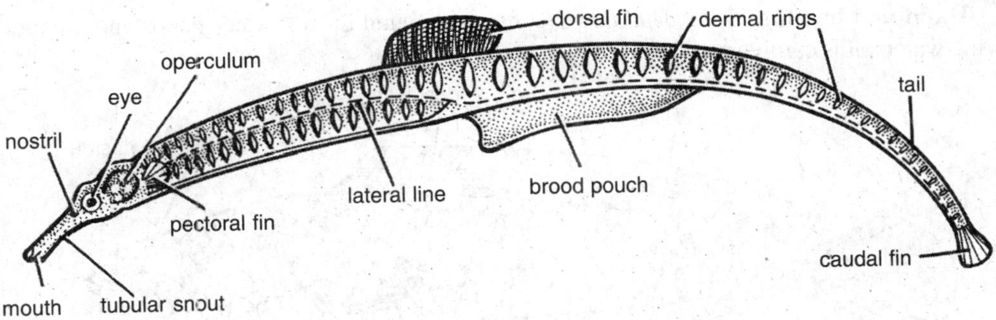

Fig. 11.39. *Syngnathus.*

Comments :

1. *Syngnathus* (Fig. 11.39) is commonly known as **pipe-fish**.
2. Body is more or less elongated having an exoskeleton of enlarged plates.
3. Head is produced into small and tubular snout.
4. The mouth is small, terminal, toothless and suctorial, lies at the extremity of the snout. Eyes are lateral in position.
5. Gill-slit is reduced to a very small opening near the upper posterior angle of the operculum.
6. Pectoral fins are very small. Single dorsal fin having 16–20 rays. Ventral fins are absent.
7. Tail is long and not prehensile, with a poorly developed rounded caudal fin.
8. Male possesses a **brood pouch** on the ventral side of the abdomen. In the brood pouch eggs are retained until they hatch as young ones, thus, exhibiting parental care.

Geographical distribution. *Syngnathus* is found mainly in the Oceans of India and China. It also inhabits Atlantic, Pacific and Indian Oceans.

41. *Ophiocephalus* (*Channa*)

Classification :

Phylum	...	**Chordata**	Characters same as that of *Scoliodon.*
Group	...	**Craniata**	
Subphylum	...	**Vertebrata**	
Division	...	**Gnathostomata**	
Superclass	...	**Pisces**	
Class	...	**Osteichthyes**	Characters same as that of *Polypterus.*
Subclass	...	**Actinopterygii**	
Superorder	...	**Teleostei** →	Characters same as that of *Labeo.*
Order	...	**Ophiocephaliformes** →	Body covered with cycloid scales; head depressed, covered with large scales; fins without spines; accessory respiratory organs are present; metapterygoid articulating with the sphenotic or frontal in front of the hyomandibular; air-bladder very long prolonged into caudal region. Teeth present on jaws and palate.
Genus	...	*Ophiocephalus* (*Channa*)	

Habit and habitat. *Ophiocephalus* is commonly found in freshwater ponds and are rare in flowing waters. It is carnivorous.

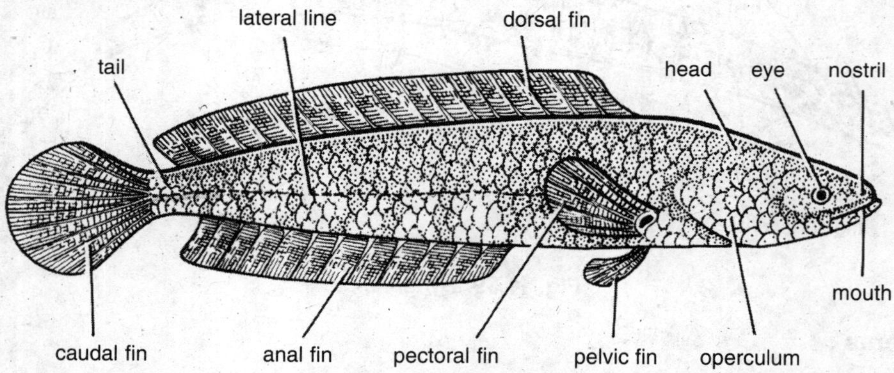

11.40. *Ophiocephalus (Channa).*

Comments :

1. *Ophiocephalus* (*Channa*) (Fig. 11.40) is commonly called live-fish and popularly known as **snake-headed fish**.
2. Body is elongated, anteriorly cylindrical and posteriorly compressed and covered with cycloid scales. It may grow to the length of 45 cm.
3. Body colour is variable–greenish black, side yellowish and abdomen stripped.
4. Head depressed and covered with large cycloid scales.
5. Dorsal and anal fins are single, long and without spines. Pectoral fins near pelvic fins. Caudal fin is rounded and fan-shaped.
6. Gill-opening is wide and lateral in position and covered with operculum.
7. Accessory breathing organs are present. On the roof of its pharynx, the fish has a pair of cavities which have folded linings richly supplied with blood vessels for taking in air.
8. Accessory respiratory organs enable these fishes to survive out of water for a few hours or migrate from one pool to another.

9. Air-bladder is much elongated.
10. **Fin formula:** D. 41–43; P. 16–18; V. 6; A. 24–25; and lateral line 58–60.

Geographical distribution. *Ophiocephalus* (*Channa*) is found in the freshwaters of India, Myanmar, Sri Lanka, Malaya, Thailand, China, Formosa, Islands of East Archipelago, Tropical Africa, etc.

42. *Amphipnous*

Classification :

Phylum	...	**Chordata**	Characters same as that of *Scoliodon*.
Group	...	**Craniata**	
Subphylum	...	**Vertebrata**	
Division	...	**Gnathostomata**	
Superclass	...	**Pisces**	
Class	...	**Osteichthyes**	Characters same as that of *Polypterus*.
Subclass	...	**Actinopterygii**	
Superorder	...	**Teleostei** →	Characters same as that of *Labeo*.
Order	...	**Symbranchiformes** →	Body eel-like or snake-like devoid of scales; air-bladder absent; fins are without spines; dorsal, caudal and anal fins are continuous; gill-slit single, transverse and ventral.
Genus	...	*Amphipnous*	
Species	...	*cuchia*	

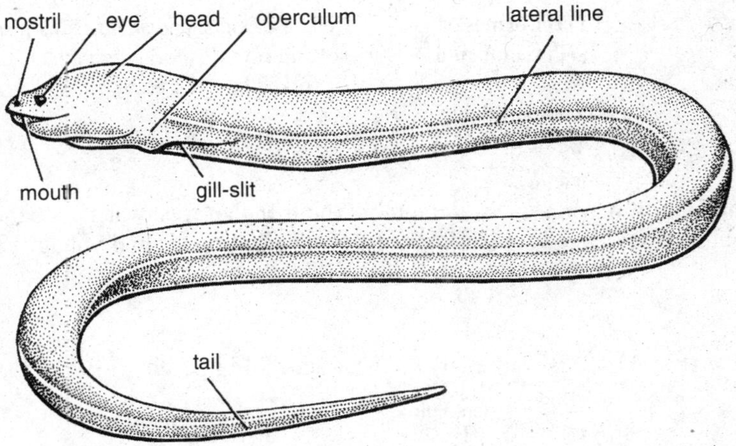

Fig. 11.41. *Amphipnous cuchia.*

Habit and habitat. *Amphipnous* is a fresh and brackish water amphibious fish. It frequently engulfs air from outside the water and often found in grassy banks of ponds.

Comments :

1. *Amphipnous cuchia* (Fig. 11.41) is commonly known as **blind-serpent**.
2. Body is elongated, eel-like or snake-like and slimy, and grows up to 60 cm or 1 metre in length. Body is divided into head, trunk and tail.
3. Skin appears almost naked because the minute scales are buried deep in it.
4. Head bears the nostrils, eyes and small mouth.

5. Gill-openings are covered by operculum and united into one slit on the ventral surface. The gills are rudimentary and the fish has evidently lost its power of aquatic respiration.

6. The air-breathing organs consist of a pair of lung-like sacs growing out of the pharynx above the gills and connected with gill-cavity.

7. Paired and median fins absent. Caudal fin is rudimentary and a membranous flap present around the tail.

8. It spends most of its time in grass on the banks of ponds and rivers.

9. When in water it constantly rises to the surface for the purpose of respiration.

Geographical distribution. *Amphipnous cuchia* is found in the rivers and ponds of Bihar and Bengal and also found in Myanmar. *Amphipnous* is also found in Africa, Indo-Australian Archipelago and South America.

43. *Anabas*

Classification :

Phylum	...	**Chordata**	}	Characters same as that of *Scoliodon.*
Group	...	**Craniata**		
Subphylum	...	**Vertebrata**		
Division	...	**Gnathostomata**		
Superclass	...	**Pisces**		
Class	...	**Osteichthyes**	}	Characters same as that of *Polypterus.*
Subclass	...	**Actinopterygii**		
Superorder	...	**Teleostei**	→	Characters same as that *Labeo.*
Order	...	**Perciformes (=Percomorphi)**	→	Perches. Dorsal and anal fins with spiny and soft fin rays. Pectoral arch attached to skull by forked post-temporal.
Family	...	**Anabantidae**		
Genus	...	*Anabas*		
Species	...	*scandens*		

Habit and habitat. *Anabas* is a common South Indian freshwater fish and can live out of water for a long period. It feeds on shrimps, ostracods, gastropods and young fishes. Male shows parental care.

Comments :

1. *Anabas* (Fig. 11.42) is commonly known as **climbing perch**.

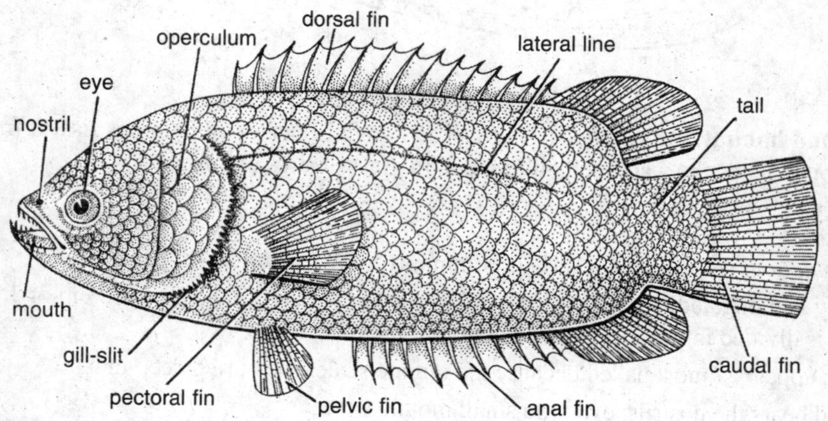

Fig. 11.42. *Anabas scandens.*

2. The body is laterally compressed and covered with ctenoid scales. It attains the length of 23–26 cm.

3. The colour of the body is greyish black along the dorsal side, becoming lighter along the sides and abdomen.

4. Dorsal (two) and anal fins are long and spinous. The caudal fin is fan-shaped. Pectoral and pelvic fins are smaller.

5. The operculum bears backwardly directed spines.

6. The accessory respiratory superbranchial organs are well-developed. These are formed of thin and folded bony laminae covered with mucous membrane.

7. The fish can live for a long time out of water. When it remains out of water it respires by the accessory respiratory organs.

8. It is highly esteemed food fish at several places.

9. This fish can walk on land with the help of fin spines and opercular spines.

10. **Fin formula** of *A. testudineus*: D. 17/9–10; P. 15; V. 1/5; A. 10–11; C. 17. Lateral line scales 29.

Geographical distribution. *Anabas scandens* is found in estuaries and freshwaters of India, Sri Lanka, Myanmar, Malaya Archipelago, Africa and Philippines.

44. *Trichiurus*

Classification :

Phylum	...	**Chordata**	Characters same as that of *Scoliodon*.
Group	...	**Craniata**	
Subphylum	...	**Vertebrata**	
Division	...	**Gnathostomata**	
Superclass	...	**Pisces**	
Class	...	**Osteichthyes**	Characters same as that of *Polypterus*.
Subclass	...	**Actinopterygii**	
Superorder	...	**Teleostei**	→ Characters same as that of *Labeo*.
Order	...	**Perciformes (= Percomorphi)**	→ Characters same as that of *Anabas*.
Genus	...	*Trichiurus*	

Habit and habitat. *Trichiurus* is a marine fish, it enters estuaries.

Comments :

1. *Trichiurus* (Fig. 11.43) is commonly known as **ribbon-fish**.

2. Body is scaleless, elongated, very thin, ribbon-shaped and flattened from side to side. It grows up to 30 cm or more in length.

3. A single long dorsal fin extending the whole length of the back. Caudal fin and ventral fins absent.

4. Pectoral fins are inserted low down the sides.

5. Tail is tapering.

6. Lateral line is seen clearly.

7. Pseudobranchiae are present.

8. These fishes are mostly sun dried in the open for the market.

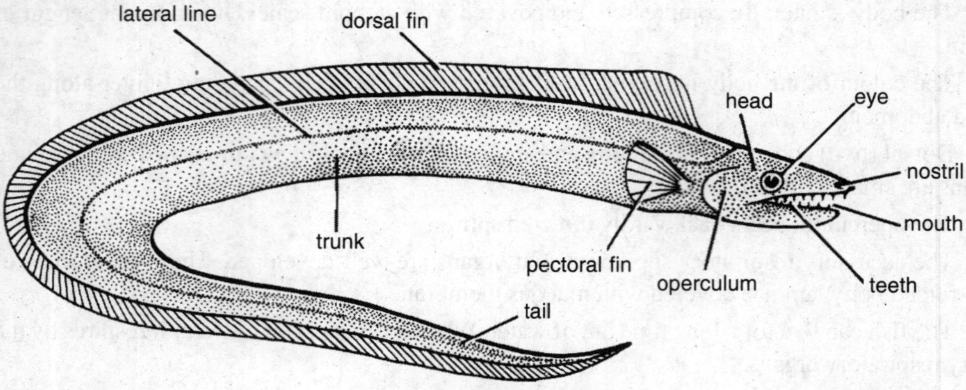

Fig. 11.43. *Trichiurus.*

Geographical distribution. *Trichiurus* is found in subtropical seas. *T. mutius* and *T. savala* are more common in Kerala, Konkan, Maharashtra (Mumbai) and Orissa.

45. *Cybium* (*Scomberomorus*)

Classification :

Phylum	...	**Chordata**		Characters same as that of *Scoliodon.*
Group	...	**Craniata**		
Subphylum	...	**Vertebrata**		
Division	...	**Gnathostomata**		
Superclass	...	**Pisces**		
Class	...	**Osteichthyes**		Characters same as that of *Polypterus*
Subclass	...	**Actinopterygii**		
Superorder	...	**Teleostei**	→	Characters same as that of *Labeo.*
Order	...	**Perciformes (= Percomorphi)**	→	Characters same as that of *Anabas*.
Genus	...	*Cybium* (*Scomberomorus*)		

Habit and habitat. *Cybium* is a marine, fast swimming fish. It is carnivorous, predatory on smaller fishes such as sardines and mackerels.

Comments :

1. *Cybium* (*Scomberomorus*) (Fig. 11.44) is commonly known as **seer-fish**.

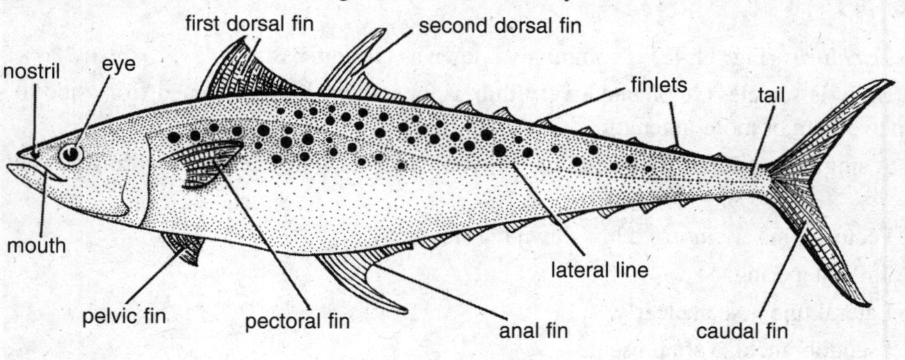

Fig. 11.44. *Cybium.*

2. Body is very much elongated and beautifully coloured. Length up to 150 cm.
3. Colour blue along back, abdomen silvery white.

4. First dorsal fin is small and spinous. Second dorsal and anal fins are longer and broken up into finlets. Caudal fin with two large and equal lobes.

5. Pectoral fins are inserted high up the sides.

6. Pseudobranchiae are present.

7. It is greatly esteemed as food fish throughout India. The flesh is reddish with high fat contents.

8. Seer fishes are caught all along the coast on long line baited with pieces of sardines or prawns.

Geographical distribution. *Cybium* is found in Arabian Sea and Bay of Bengal.

46. *Trichogaster*

Classification :

Phylum	...	**Chordata**	Characters same as that of *Scoliodon.*
Group	...	**Craniata**	
Subphylum	...	**Vertebrata**	
Division	...	**Gnathostomata**	
Superclass	...	**Pisces**	
Class	...	**Osteichthyes**	Characters same as that of *Polypterus.*
Subclass	...	**Actinopterygii**	
Superorder	...	**Teleostei**	→ Characters same as that of *Labeo.*
Order	...	**Perciformes (= Percomorphi)**	→ Characters same as that of *Anabas.*
Genus	...	*Trichogaster* (= *Colisa*)	

Habit and habitat. *Trichogaster* is a freshwater fish. It is mostly used for decorative purposes in the aquaria.

Comments :

1. Body of *Trichogaster* (Fig. 11.45) is laterally compressed and divided into head, trunk and tail. It measures 2.5 – 3.5 cm in length.

2. The colour of the body is greenish or bluish above and dirty white on the ventral side and bluish spots on gill cover.

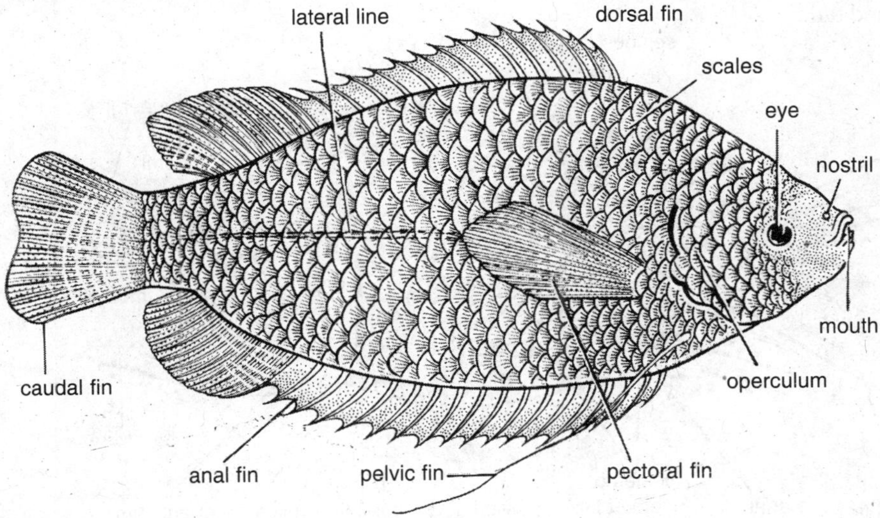

Fig. 11.45. *Trichogaster.*

3. Fourteen or more orange bands descend obliquely downwards from the back of the body. These bands alternate with blue bands.

4. Dorsal, caudal and anal fins are spotted orange. Anal fin is edged with deep red. Caudal fin is truncated.

5. Pelvic fin comprises a single filiform long ray.

6. The eyes are red in colour.

7. Lateral-line is interrupted.

8. **Fin formula:** D. 16–17/10–11; P. 10; V. 1; A. 16–17/15-16; C. 15. Lateral line scales 29–30.

Geographical distribution. *Trichogaster* is distributed in the freshwaters of India, Thailand and Cambodia.

47. *Pterois (Scorpion-fish)*

Classification :

Phylum	...	**Chordata**	Characters same as that of *Scoliodon.*
Group	...	**Craniata**	
Subphylum	...	**Vertebrata**	
Division	...	**Gnathostomata**	
Superclass	...	**Pisces**	
Class	...	**Osteichthyes**	Characters same as that of *Polypterus.*
Subclass	...	**Actinopterygii**	
Superorder	...	**Teleostei** →	Characters same as that *Labeo.*
Order	...	**Scorpaeniformes** →	Sub-orbital extends across the cheek to the pre-operculum.
Genus	...	*Pterois*	

Habit and habitat. *Pterois* is a marine, carnivorous and viviparous fish. It adopts the surrounding colour.

Comments :

1. *Pterois* (Fig. 11.46) is commonly known as **scorpion-fish** or **mail-cheeked** fish.

2. Body is oblong and more or less compressed and covered with scales. Body is divided into head, trunk and tail.

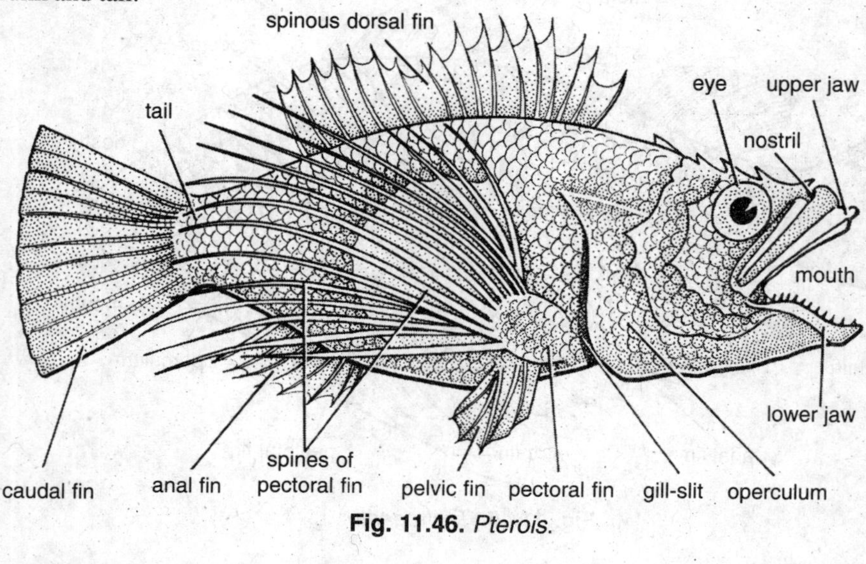

Fig. 11.46. *Pterois.*

3. Head usually bears spines and covered with bony cuirass (membranous processes), wide mouth, large eyes and nostrils.
4. Dorsal fin is strong and spinous having poison glands. Pectoral fin bears very large spines.
5. Gills are 3½ or 4, pseudobranchiae.
6. Air-bladder is physoclistic.
7. Spiracles are absent.
8. **Fin formula :** D. 8–18, A. 2–5, P. 15–25, V. 2–8.

Geographial distribution. *Pterois* is found in the warmer parts of Pacific and Indian Oceans. Cretaceous to Recent.

48. *Synaptura* (Flat-fish)

Classification :

Phylum	...	**Chordata**	}	Characters same as that of *Scoliodon*.
Group	...	**Craniata**		
Subphylum	...	**Vertebrata**		
Division	...	**Gnathostomata**		
Superclass	...	**Pisces**		
Class	...	**Osteichthyes**	}	Characters same as that of *Polypterus*.
Subclass	...	**Actinopterygii**		
Superorder	...	**Teleostei**	→	Characters same as that of *Labeo*.
Order	...	**Pleuronectiformes**	→	Flat-fish. Body flat adapted for bottom living; both the eyes are on the upper side; skull asymmetrical; fins usually without spines; air-bladder is absent in adults.
Genus	...	*Synaptura*		

Habit and habitat. *Synaptura* is a bottom dwelling marine fish and feeds on molluscs. Flatfish of Indian waters shows banded colouration.

Comments :

1. *Synaptura* (Fig. 11.47) is commonly known as **flat-fish**.
2. Body is thin, laterally compressed and flat.
3. Scales covering the body are usually imbricate, cycloid or ctenoid.

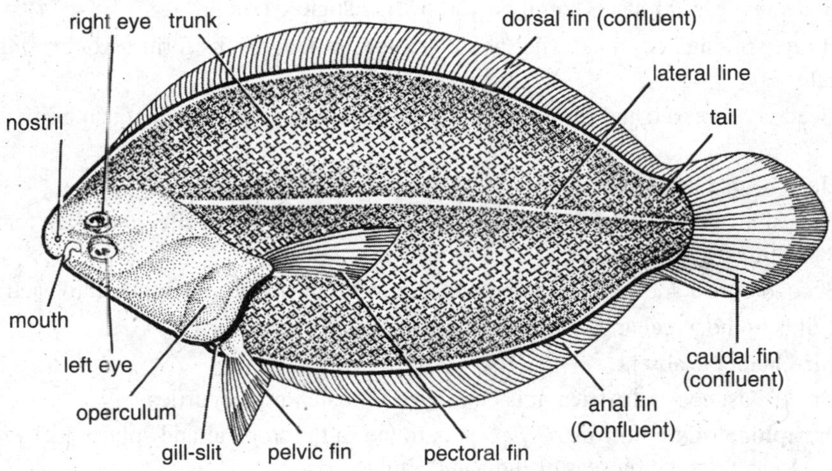

Fig. 11.47. *Synaptura.*

4. The side of the body bearing the eyes turned upwards being coloured, while the lower side is white.

5. Head is asymmetrical (posterior cranium is normal and anterior carnium is twisted, hence, both eye orbits lie on one side in adults). Both the eyes are situated on the upper (left) side. Nostrils also on left side.

6. Mouth is more or less protractile. Jaws bear sharp chisel-like teeth and flat crushing teeth in the pharynx.

7. Dorsal and anal fins are long without spines and confluent with the caudal fin. Caudal fin is fan-shaped. Pectoral and pelvic fins well-developed.

8. Gills four, a slit behind the fourth, pseudobranchiae are present.

9. It is economically important as food-fish.

Geographical distribution. *Synaptura* is found in Atlantic, Pacific and Indian Oceans, *i.e.*, South America, India, Malayasia and Southern hemisphere.

49. *Echeneis* (Sucker-fish)

Classification :

Phylum	...	**Chordata**		Characters same as that of *Scoliodon.*
Group	...	**Craniata**		
Subphylum	...	**Vertebrata**		
Division	...	**Gnathostomata**		
Superclass	...	**Pisces**		
Class	...	**Osteichthyes**		Characters same as that of *Polypterus.*
Subclass	...	**Actinopterygii**		
Superorder	...	**Teleostei**	→	Characters same as that of *Labeo.*
Order	...	**Echeneiformes (= Discocephali)**	→	First dorsal fin is modified into an adhesive-disc for attachment; other fins normal: air-bladder is absent.
Genus	...	*Echeneis*		

Habit and habitat. *Echeneis* is a marine fish feeding on small fishes. It attaches itself by means of its adhesive organ to boats, sharks, turtles, cetaceans and other large swift swimming animals.

Comments :

1. *Echeneis* (Fig. 11.48) is commonly known as **sucker-fish**.

2. Body is elongated, measuring about 1 metre in length, fusiform and covered with small cycloid scales.

3. Head is depressed and furnished above with an adhesive organ that is modified first dorsal fin.

4. Head bears lateral eyes and wide and deep mouth.

5. Adhesive disc is flat, oval and transversely furrowed and is an effective organ for attachment. It bears 21–28 transverse lamellae.

6. Second dorsal and anal fins are elongated, without spines and opposed to each other.

7. Tail is homocercal and caudal fin bilobed.

8. Air-bladder is absent.

9. On the east coast of Africa, it is employed for catching the turtles.

Geographical distribution. *Echeneis* is found in the tropical and subtropical seas (Atlantic and Pacific Oceans) including seas of India and China.

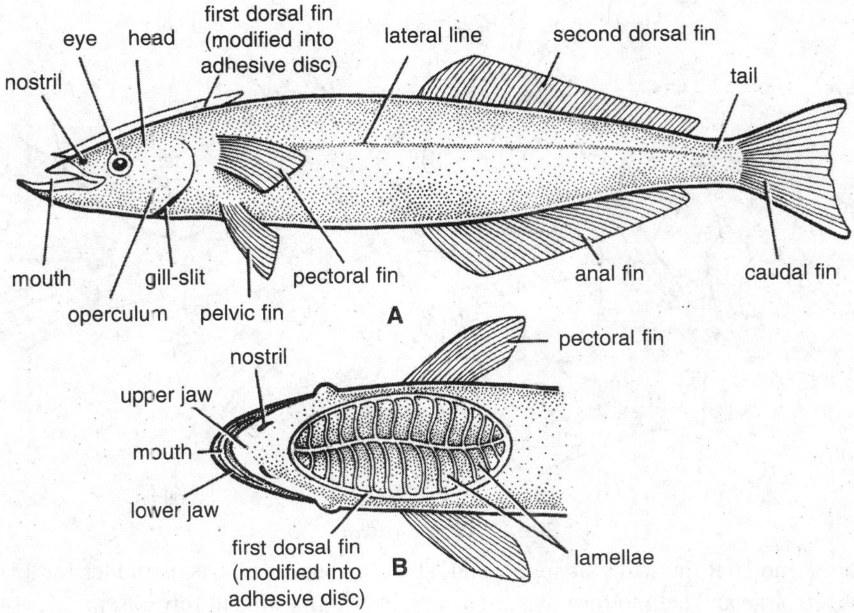

Fig. 11.48. *Echeneis.* A—Lateral view; B—Dorsal view of head.

50. *Ostracion*

Classification :

Phylum	...	**Chordata**		Characters same as that of *Scoliodon.*
Group	...	**Craniata**		
Subphylum	...	**Vertebrata**		
Division	...	**Gnathostomata**		
Superclass	...	**Pisces**		
Class	...	**Osteichthyes**		Characters same as that of *Polypterus.*
Subclass	...	**Actinopterygii**		
Superorder	...	**Teleostei**	→	Characters same as that of *Labeo.*
Order	...	**Tetrodontiformes (= Plectognathi)**	→	Scales are modified into spines; scutes or bony-plates cover the body; gill-slits are small; Body globose, jaws short; teeth strong incisors.
Genus	...	*Ostracion*		

Habit and habitat. *Ostracion* is found at the bottom of shallow sea water.

Comments :

1. *Ostracion* (Fig. 11.49) is commonly known as **coffer-fish** or **trunk-fish**.
2. The body is short, roughly triangular and variedly coloured. Body is divided into head, trunk and tail.
3. Body is covered in a carapace formed of large juxtaposed hexagonal bony plates. Each bony plate has blue spot in the centre. Carapace is closed behind the anal fin.
4. The contour of the body is three ridged, the median ridge is elevated into a sharp dorsal spine which is triangular in shape.
5. A compressed supra-orbital spine above the orbit is directed upwards or a little backwards.

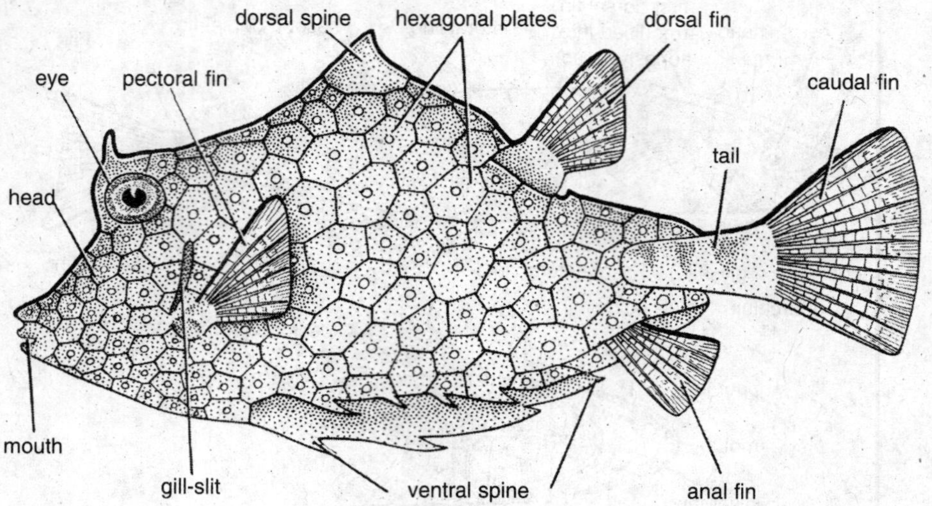

Fig. 11.49. *Ostracion.*

6. Dorsal and anal fins without spine. Caudal fin is round and acts as rudder for swimming. Pectoral fins are large and help to form water current. Pelvic fins are entirely absent.

7. Ventral side has backwardly directed spines.

8. Gills are four in number, gill-slit is situated near the pectoral fin.

9. Exhibits warning colours and acts as danger signal to predaceous fishes. Some trunk fishes discharge a toxic substance **ostracitoxin** that kills other fish in confined water. It is also toxic to trunk fish itself.

Geographical distribution. *Ostracion* is widely distributed in Red sea, Indian seas to the Malaya Archipelago, African seas, Atlantic and Pacific Oceans.

51. *Tetrodon*

Classification :

Phylum	...	**Chordata**	Characters same as that of *Scoliodon*.
Group	...	**Craniata**	
Subphylum	...	**Vertebrata**	
Division	...	**Gnathostomata**	
Superclass	...	**Pisces**	
Class	...	**Osteichthyes**	Characters same as that of *Polypterus*.
Subclass	...	**Actinopterygii**	
Superorder	...	**Teleostei**	Characters same as that of *Labeo*.
Order	...	**Tetrodontiformes** → **(= Plectognathi)**	Scales are modified into spines; scutes or bony-plates cover the body; gill-slits are restricted; body globose, jaws strong; teeth sharp.
Genus	...	*Tetradon*	

Habit and habitat. *Tetradon* is a marine fish.

Comments :

1. *Tetrodon* (Fig. 11.50) is commonly known as **globe-fish** or **puffer.** Body is divided into head, trunk and tail.

2. The body is spherical and covered with subimbricate bony laminae. Head and anterior part of body is large and scaleless. Head bears eyes, nostrils and mouth.

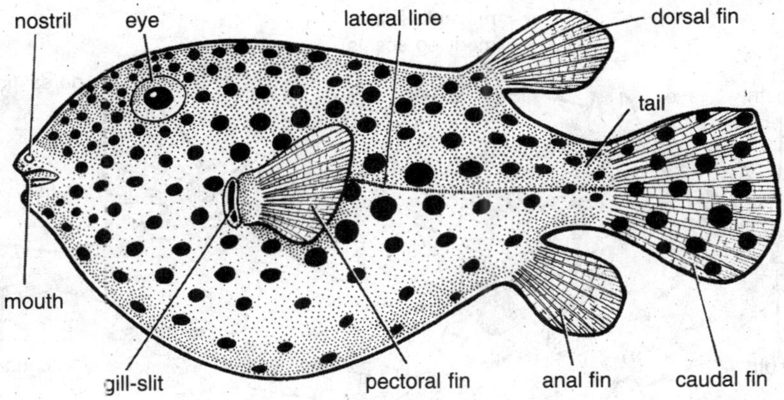

Fig. 11.50. *Tetrodon.*

3. The colour is light brown with dark brown spots throughout the body.

4. Dorsal fin opposite to anal fin and have few fin-rays. Pectoral fins present and pelvic fins are entirely absent.

5. The teeth in jaws are fused to form beak but with a median suture.

6. Gills are three. Gill-slit or gill-opening is situated near the pectoral fin.

7. The air-bladder is horse-shoe-shaped.

8. The oesophagus is dilatable. The air is engulfed and, thus, the fish can float on the surface of water.

9. The flesh of this fish is poisonous.

Geographical distribution. *Tetrodon* is found in all tropical and subtropical seas (Atlantic and Indian seas).

52. *Diodon*

Classification :

Phylum	...	**Chordata**		Characters same as that of *Scoliodon.* (Phylum to Division)
Group	...	**Craniata**		
Subphylum	...	**Vertebrata**		
Division	...	**Gnathostomata**		
Superclass	...	**Pisces**		Characters same as that of *Polypterus.* (Superclass to Subclass)
Class	...	**Osteichthyes**		
Subclass	...	**Actinopterygii**		
Superorder	...	**Teleostei**	→	Characters same as that *Labeo.*
Order	...	**Tetrodontiformes (= Plectognathi)**	→	Scales are modified into spines; scutes or bony-plates cover the body; gill-slits are restricted; body globose, jaws strong; teeth sharp.
Genus	...	*Diodon*		

Habit and habitat. *Diodon* is also a marine fish. It is carnivorous, feeds upon molluscs and corals, etc.

Comments :

1. *Diodon* (Fig. 11.51) is commonly known as **porcupine-fish**.

2. Body is nearly globular, covered with stiff and movable dermal spines which are organs of defence.

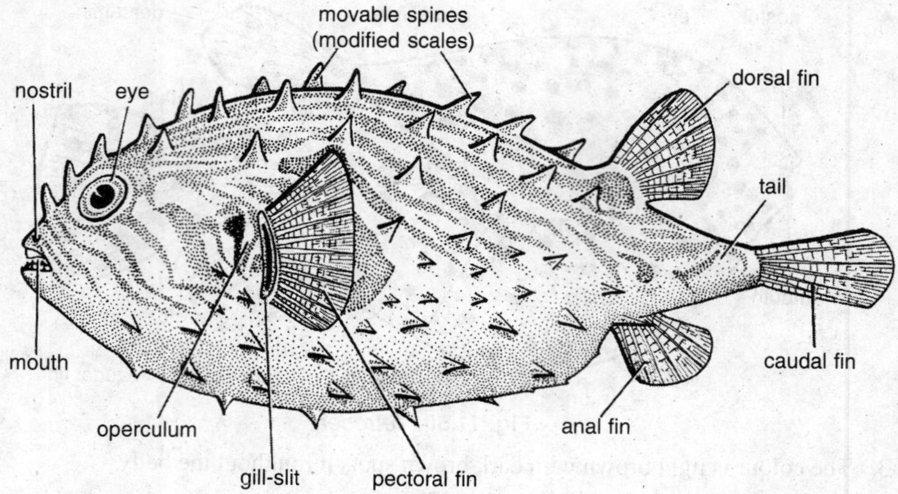

Fig. 11.51. *Diodon.*

3. Head bears eyes, nostrils and small mouth opening and jaws without median suture.

4. Dorsal and anal fins are placed opposite to each other. Caudal fin is rounded and slightly tilted upwards. Pectoral fin near operculum and pelvic fins are absent.

5. Gills are three in number covered by operculum. Gill-slit is situated near the pectoral fin.

6. Air-bladder is present.

7. A thin-walled inflatable gastric diverticulum is present which allows the whole body to be puffed into a globular shape and the spines become defensively erected.

8. The flesh of this fish is regarded as poisonous.

Geographical distribution. *Diodon* is found in the tropical seas.

53. *Lophius*

Classification :

Phylum	...	**Chordata**		Characters same as that of *Scoliodon.*
Group	...	**Craniata**		
Subphylum	...	**Vertebrata**		
Division	...	**Gnathostomata**		
Superclass	...	**Pisces**		
Class	...	**Osteichthyes**		Characters same as that of *Polypterus.*
Subclass	...	**Actinopterygii**		
Superorder	...	**Teleostei**	→	Characters same as that of *Labeo.*
Order	...	**Lophiiformes (= Pediculate)**	→	First ray of the spinous dorsal fin is placed on the head and transformed into illicium (line and bait); orbitosphenoid, basisphenoid and opisthotic bones are absent.
Genus	...	*Lophius*		
Species	...	*piscatorius*		

Habit and habitat. *Lophius* is a marine shore-fish living up to 60 fathoms deep. Deep sea anglers live up to a depth of 6000 feet. It captures fish, marine invertebrates, etc.

Comments :

1. *Lophius* (Fig. 11.52) is commonly known as **fishing-frog** or **angler-fish** or **sea-devil**.

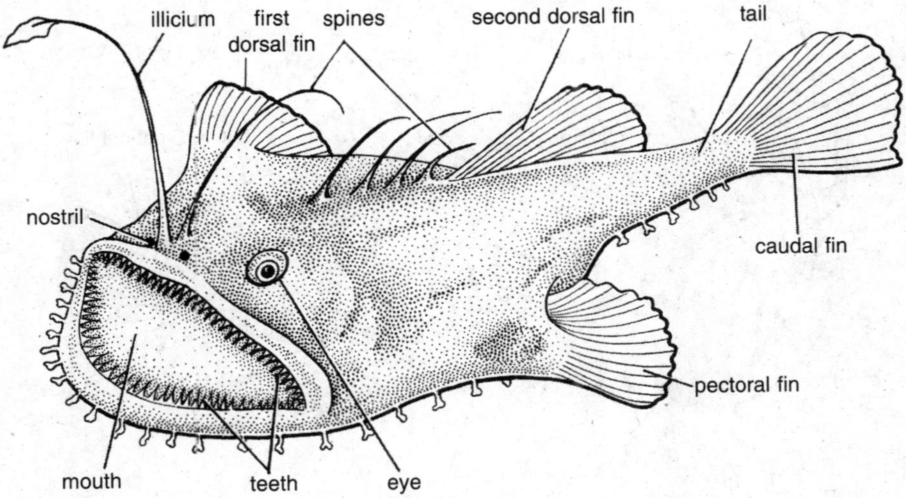

Fig. 11.52. *Lophius piscatorius.*

2. Body is elongated, depressed and naked and grows to a length of over 1 metre and 60 cm. Anterior part of body is large and scaleless.

3. Mouth is extremely large and terminal with very strong recurved teeth.

4. Eyes are lateral in position and nostrils small.

5. Dorsal fin is spinous. First ray of the dorsal fin, inserted on the snout, is very long, movable and ends in a flap modified into **illicium** which acts as **bait** or **lure** to attract small worms and other fishes. Rest of the first dorsal fin and second dorsal fin is supported by fin rays.

6. Pectoral fin is scarcely geniculated.

7. Gill-opening lies in the lower axil of pectoral fin. Pseudobranchiae are present.

8. Male is small in size and attached to the body of the female as ectoparasite.

Geographical distribution. *Lophius piscatorius* is widely distributed in Atlantic, Indian and Pacific Oceans. It also occurs on the coasts of Europe and North America.

54. *Antennarius*

Classification :

Phylum	...	**Chordata**	Characters same as that of *Scoliodon.*
Group	...	**Craniata**	
Subphylum	...	**Vertebrata**	
Division	...	**Gnathostomata**	
Superclass	...	**Pisces**	
Class	...	**Osteichthyes**	Characters same as that of *Polypterus.*
Subclass	...	**Actinopterygii**	
Superorder	...	**Teleostei**	→ Characters same as that of *Labeo.*
Order	...	**Lophiiformes (= Pediculate)**	→ First ray of the spinous dorsal fin is placed on the head and transformed into illicium (line and bait); orbitosphenoid, basisphenoid and opisthotic bones are absent.
Genus	...	*Antennarius*	

Habit and habitat. *Antennarius* is a marine fish. It lives mostly in the coral grooves where it lies in wait for prey, well concealed by its colouration.

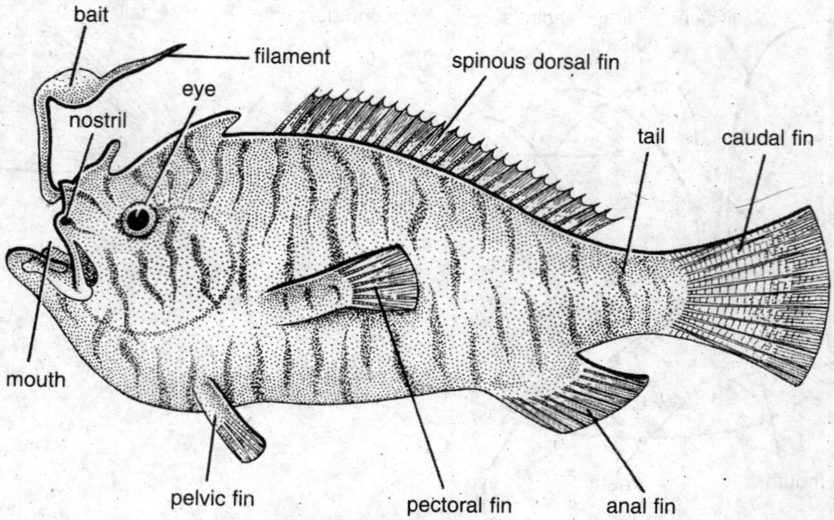

Fig. 11.53. *Antennarius.*

Comments :

1. *Antennarius* (Fig. 11.53) is commonly known as **angler-fish**.
2. Body is elongated, compressed and yellowish in colour with brown spots and lines all over it. Body is divided into head, trunk and tail.
3. Head is large, elevated and compressed, bearing mouth that is large nearly vertical and turned upwards, eyes are lateral in position and nostrils terminal.
4. Dorsal fin is spinous. The first three dorsal fin rays are isolated, the anterior of which ends in a fleshy knob, the **bait** situated on the snout, while the remaining two lie posteriorly.
5. Bait is phosphorescent and acts as lure to attract the worms and other preys.
6. Pectoral fin forming an elbow-like angle. Pectoral and pelvic fins are pediculate and help in hopping on moist mud.
7. Gill-opening is a small hole and lateral in position. Pseudobranchiae are absent.
8. **Sexual dimorphism** is exhibited. The male is much smaller than the female and often remains attached as a physical parasite to the body of the female.

Geographical distribution. *Antennarius* is found distributed from the seas of India to Malaya Archipelago.

55. *Neoceratodus*

Classification :

Phylum	...	**Chordata**		Characters same as that of *Scoliodon.*
Group	...	**Craniata**		
Subphylum	...	**Vertebrata**		
Division	...	**Gnathostomata**		
Superclass	...	**Pisces**		
Class	...	**Osteichthyes**	→	Bony fishes, Gill-slits covered by operculum.
Subclass	...	**Choanichthyes**	→	Nostrils communicate with the mouth cavity. Paired fins with larger median lobe.
Order	...	**Dipnoi**	→	Lung fish. Body long and slender. Air-bladder lung-like.
Genus	...	*Neoceratodus*		

Habit and habitat. *Neoceratodus* lives in quiet pools which become stagnant during dry season. Fish ascends to engulf fresh air into air-bladder. It feeds on worms and small crustaceans and molluscs.

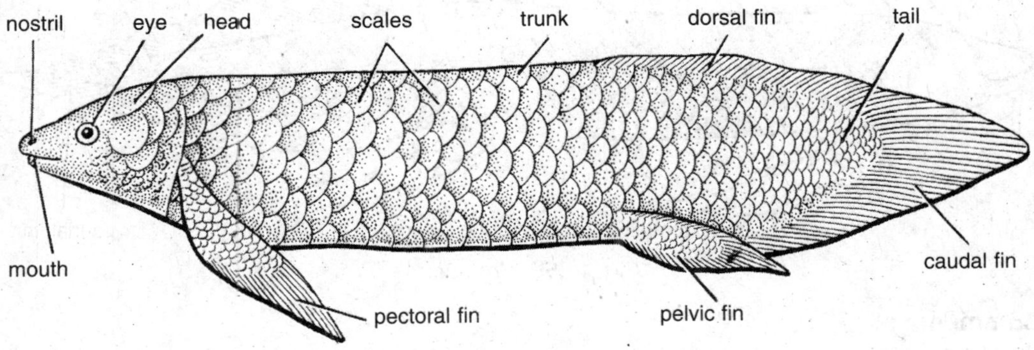

Fig. 11.54. *Neoceratodus.*

Comments :

1. *Neoceratodus* (Fig. 11.54) is commonly known as **Australian lung-fish** or **Burnett salmon**.
2. Body is elongated measuring about one and a half to two metres in length. Body divisible into head, trunk and tail.
3. Scales are cycloid which are large, thin and covered with spines.
4. Paired fins are rather leaf-like or paddle-like. Caudal fin is symmetrical and pointed and continuous with dorsal and anal fins.
5. Dental plates are oval crescentric or triangular terminating in a smooth or feebly denticulated biting margins.
6. Lower jaw with a small toothless dentary on each side.
7. Branchial arches are five in number and bisegmented. Hyobranchial cleft is open and associated with pseudobranch.
8. The first four branchial arches carry holobranchs.
9. Air-bladder is single, opening into oesophagus by a long duct.
10. The lateral line system is somewhat degenerated.

Geographical distribution. *Neoceratodus* is found in Burnett and Mary rivers in Queensland, Australia. Triassic to Cretaceous and Recent.

56. *Protopterus* (African Lung-fish)

Classification :

Phylum	...	**Chordata**	Characters same as that of *Scoliodon*.
Group	...	**Craniata**	
Subphylum	...	**Vertebrata**	
Division	...	**Gnathostomata**	
Superclass	...	**Pisces**	
Class	...	**Osteichthyes**	Characters same as that of *Neoceratodus*.
Subclass	...	**Choanichthyes**	
Order	...	**Dipnoi**	
Genus	...	*Protopterus*	

Habit and habitat. *Protopterus* lives in burrows in muddy water. In dry season, it undergoes summer sleep (aestivation) in nests (vertical burrows) in mud lined with mucus. It is **carnivorous** and voracious feeder.

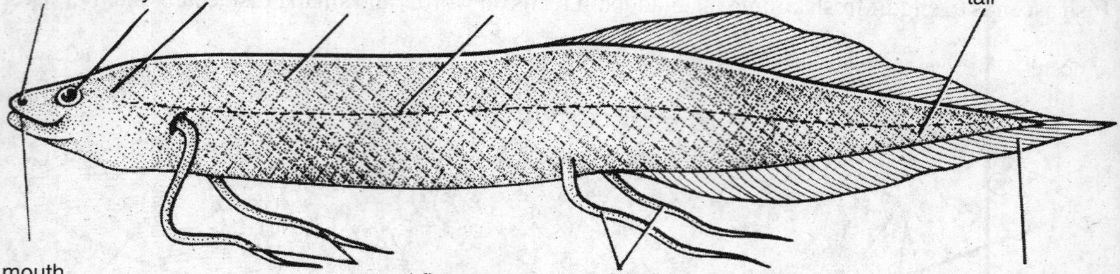

Fig. 11.55. *Protopterus.*

Comments :

1. *Protopterus* (Fig. 11.55) is commonly known as **Nile lung-fish**.
2. Body is elongated, cylindrical and more or less eel-like and grows to a length of 2 metres. Body is divisible into head, trunk and tail.
3. Head bears mouth, nostrils and small eyes.
4. Body covered with small cycloid scales completely enclosed in the skin.
5. The paired fins are long, thin filamentous and without fin-rays.
6. Dorsal, caudal and anal fins are united.
7. Tail is principal organ of locomotion.
8. Six branchial arches and five gill-slits are covered by operculum.
9. Air-bladder (lung) is double running throughout the length of the body. It is said to breathe by its lungs as well as by its gills.
10. Larval gills (4 pairs) are usually retained as vestiges throughout life in some species.
11. Lateral line is well-developed.

Geographical distribution. *Protopterus* has a wide distribution over the middle portion of great African continent ranging from the river Senegal and the White Nile on the north to the Congo basin, Lake Tanganijika and Zambesi on the south. Devonian to Recent.

57. *Lepidosiren* (American Lung-fish)

Classification :

Phylum	...	**Chordata**	Characters same as that of *Sceliodon.*
Group	...	**Craniata**	
Subphylum	...	**Vertebrata**	
Division	...	**Gnathostomata**	
Superclass	...	**Pisces**	
Class	...	**Osteichthyes**	Characters same as that of *Neoceratodus.*
Subclass	...	**Choanichthyes**	
Order	...	**Dipnoi**	
Genus	...	*Lepidosiren*	

Habit and habitat. *Lepidosiren* inhabits swampy places and make burrows in muddy water lined with mucus. It undergoes aestivation in such burrows. The entrance of burrow is plugged with a mud stopper having small round holes for air. Its chief food are snails and algae.

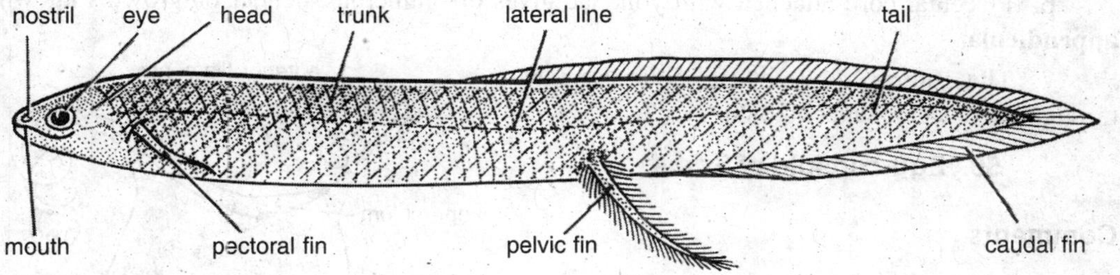

Fig. 11.56. *Lepidosiren.*

Comments :

1. *Lepidosiren* (Fig. 11.56) is commonly known as **South American lung-fish**.
2. Body is elongated (3 feet in length), cylindrical and more or less eel-like and divisible into head, trunk and tail.
3. The skin covering the body contains the small cycloid scales.
4. Paired fins are thin and filamentous. Dorsal, anal and caudal fins are united. During breeding season, vascular filaments develop on pelvic fins of male. These are supposed to be respiratory.
5. Eyes are small, reduced.
6. Five branchial arches and four gill-slits are covered by operculum.
7. Larval four gills disappear soon after the larval metamorphosis.
8. Air-bladder (lung) is double.
9. Lateral line is present.

Geographical distribution. *Lepidosiren* occurs along the course of main Amazon river and Paraguay basins, and plains of South America.

58. Shark Embryo with Yolk-sac

Comments :

1. *Scoliodon* is ovoviviparous. In *Scoliodon*, the embryo (3–7) develop inside the uterus in separate chambers and give birth to living youngs.
2. The fully-developed embryo more or less resembles the adult in external features.
3. Each chamber with in uterus is filled with fluid secreted by uterine lining.
4. In young stage, gut of the embryo is connected by a tubular yolk stalk with a yolk sac placenta.
5. In later stage, when blood vessels develop yolk stalk connection with embryo disappears and embryo is attached to uterine wall by a placental cord.

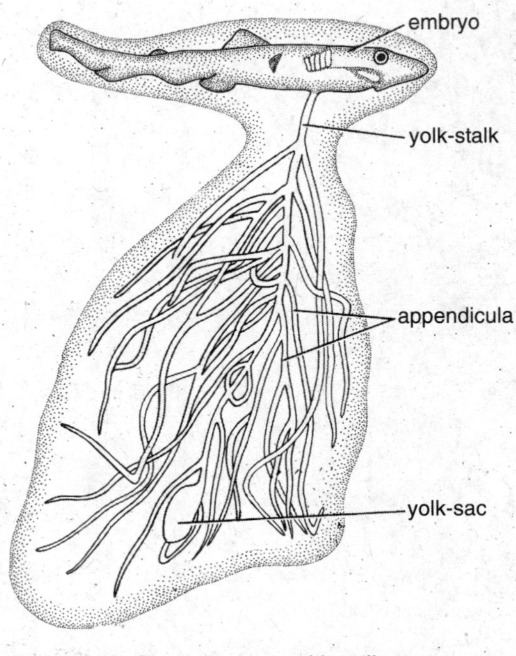

Fig. 11.57. Shark embryo with yolk sac.

6. Placental cord attached with yolk-sac gives off numerous slender outgrowths known as **appendicula**.

7. The embryo now gets its nourishment through the placental appendicula.

59. Egg-capsule of Shark

Comments :

1. Egg-capsule of shark (Fig. 11.58) is also known as **"Mermaid's purse"**.

2. It is more or less quadrangular and bag-like in appearance.

3. It is black in colour and composed of keratin.

4. It contains eggs and developing embryos.

5. All the four corners of the egg capsule bear tendril-like filaments which are attached with the sea weeds, etc.

6. Its anterior end has a wide opening, the **operculum** for entrance and exit of eggs and embryos.

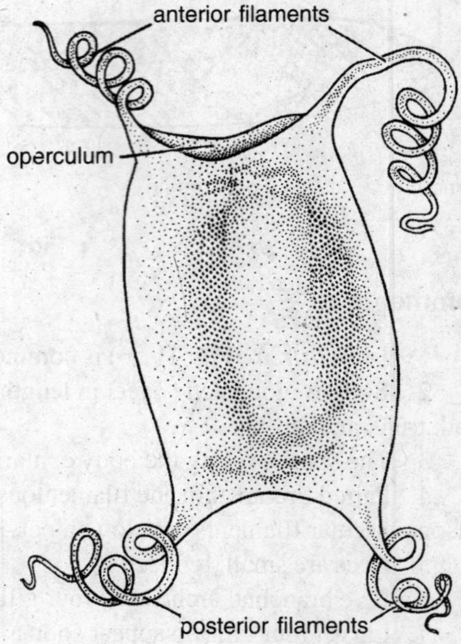

Fig. 11.58. Egg capsule of shark.

12

Chapter

CLASS–AMPHIBIA

Amphibians are amphibious–can live in water or damp places and on land. None is marine. They occur in tropical and temperate regions. One frog inhabits Arctic circle also. Toads live in dry habitat. *Cryptobranchus* (hell bender), *Necturus* (mud puppy) and *Amphiuma* (Congo eel) are entirely aquatic in habit.

General Characters

1. Cold blooded vertebrates.
2. Skin smooth or rough, rich in glands which keep it moist.
3. Scales, if present, are embedded in the skin.
4. Skull articulates with the atlas by two occipital condyles.
5. Auditory columellar apparatus fitting into fenestra ovalis.
6. Vertebrae are acentrous, pseudocentrous and notocentrous.
7. Limbs tetrapodus pentadactyle type. Some limbless.
8. Red blood corpuscles (R. B. Cs) are biconvex, oval and nucleated.
9. Heart 3-chambered (two atria and one ventricle) and it has a conus provided with valves.
10. Aortic arches are symmetrical.
11. Gills are present at least during some early larval stages of development.
12. Kidneys are provided with persistent nephrostome.
13. Lateral sense organs are present during early stages of development.
14. Cranial nerves are 10 pairs.
15. Eggs with gelatinous covering, usually laid in water.
16. The larvae pass through an aquatic stage before metamorphosing into adult.

CLASS–AMPHIBIA

Subclasses

A. Stegocephalia

Orders

1. Labyrinthodontia
2. Phyllospondyli
3. Lepospondyli

B. Lissamphibia

SUBCLASS B. LISSAMPHIBIA

Orders

I
Apoda
or
Gymnophiona

Ex. 1. *Ichthyophis*
2. *Uraeotyphlus*
3. *Siphonops*

II
Urodela
or
Caudata

III
Anura
or
Salientia

ORDER II URODELA

Suborder

1
Cryptobranchoidea
Ex. 1. *Cryptobranchus*
2. *Hynobius*

2
Ambystomoidea
Ex. 1. *Ambystoma*
2. Axolotl larva of *Ambystoma*

3
Salamandroidea
Ex. 1. *Salamandra*
2. *Amphiuma*
3. *Triton*

4
Proteida
Ex. 1. *Proteus*
2. *Necturus*

5
Meantes
Ex. 1. *Siren*

ORDER III ANURA

Suborder

1
Amphicoela
Ex. 1. *Ascaphus*
2. *Liopelma*

2
Opisthocoela
Ex. 1. *Alytes*
2. *Pipa*
3. *Bombinator*

3
Anomocoela
Ex. 1. *Pelobates*

4
Procoela
Ex. 1. *Bufo*
2. *Hyla*

5
Diplasiocoela
Ex. 1. *Rhacophorus*
2. *Rana*

Note: The classification is adopted from **G. Kingsley Noble** (1954) from "The Biology of Amphibia".

Subclass A. Stegocephalia

1. Extinct with pentadactyle limbs
2. Skin with scales and bony plates.

Order 1. Labyrinthodontia

1. Primitive tetrapods having large teeth-like crossopterygian ancestors.

Example : *Eryops*

Order 2. Phyllospondyli

1. Small, salamander-like.
2. Notochord and nerve cord enclosed in a common cavity.

Example : *Branchiosaurus.*

Order 3. Lepospondyli

1. Probable ancestors of modern caecilians.
2. Small, eel-like salamanders.
3. Vertebrae in a single piece.

Example : *Diplocaulus*

Subclass B. Lissamphibia

1. Modern amphibia
2. Exoskeleton absent.
3. Teeth small.
4. Respiration by skin, gills and lungs.

Order I. Apoda (Gymnophiona)

1. Primitive burrowing forms.
2. Body elongated eel-like
3. Limbs, girdles and tail are absent.
4. The skin is provided with numerous transverse grooves of wrinkles.
5. Minute scales are found embedded in the grooves of the skin in some forms.
6. Copulatory organs are present in males.

Examples : *1. Ichthyophis, 2. Uraeotyphlus, 3. Siphonops.*

Order II. Urodela (Caudata)

1. Body with distinct head, trunk and tail.
2. Limbs are of equal size. Gills permanent or absent in adults.
3. Skin devoid of scales.
4. Teeth are present in both the jaws.
5. Usually oviparous.
6. Larvae are aquatic like adults or resemble adults in form.

Suborder 1. Cryptobranchoidea

1. Permanently aquatic. Body is depressed.
2. Skin is soft and flabby.
3. Eye-lids are absent.
4. Gills are absent in adults.
5. Fertilisation is external.

Examples : *1. Cryptobranchus, 2. Hynobius, 3. Megalobatrachus.*

Suborder 2. Ambystomoidea

1. Adults terrestrial. Teeth present in transverse rows across the posterior margins of vomer.
2. Eye-lids are present.
3. Fertilisation is internal.
4. Exhibits the phenomena of **neoteny** or **paedogensis**, *i.e.*, larva becomes sexually mature and reproduces.

Examples : *1. Ambystoma, 2.* Axolotl larva of *Ambystoma.*

Suborder 3. Salamandroidea

1. Eye-lids present or absent.
2. Usually either gills or lungs are present but in some forms respiration by skin. Adults with only lungs.
3. Fertilisation is internal.
4. Vertebrae are amphicoelous.
5. Usually viviparous.

Examples :*1. Salamandra* (Salamander), *2. Amphiuma* (Congo eel), *3. Triton.*

Suborder 4. Proteida

1. Body is depressed.
2. Permanent larval forms wih 3 pairs of gills and 2 pairs of gill-slits.
3. Tail provided with fin.
4. External gills are always present. Lungs present.
5. Eye-lids are absent.
6. Permanently aquatic.

Examples : *1. Proteus* (Olms), *2. Necturus* (Mud puppy).

Suborder 5. Meantes (=Sirenoidea)

1. Body slender and eel-type.
2. Permanent larval forms.
3. Hindlimbs are absent. Forelegs small.
4. External gills are well-developed.
5. Eye-lids are absent.
6. Jaws with horny covering.
7. Fertilisation is external.
8. Completely aquatic.

Examples : *Siren* (Mud eel), *Pseudobronchus.*

Order III. Anura (Salientia)

1. Body short and broad.
2. Head and trunk fused.
3. Neck and tail absent.
4. Forelimbs are short, while hindlimbs are long.
5. Gills absent in aduts. Eye-lids and tympannum present. Respiration by lungs.
6. Fertilisation is usually external.

Suborder 1. Amphicoela

1. Vertebrae are amphicoelous.
2. Two tail wagging or relict muscles.
3. Fertilisation is internal.

Examples : *1. Ascaphus, 2. Liopelma.*

Suborder 2. Opisthocoela

1. Vertebrae are opisthocoelous.
2. Free ribs either in larva or adult.

   Examples : *1. Alytes, 2, Pipa, 3. Bombinator.*

Suborder 3. Anomocoela

1. Vertebrae procoelous or amphicoelous
2. Sacral vertebrae fused with urostyle.
3. No free ribs.
4. Skin is rough.

   Example : *Pelobates.*

Suborder 4. Procoela

1. Vertebrae are procoelous. No ribs.
2. Urostyle articulates with the last vertebra by two condyles.

   Examples : *1. Bufo, 2. Hyla.*

Suborder 5. Diplasiocoela

1. First seven vertebrae are procoelus.
2. Eighth vertebra is amphicoelous.
3. Sacral vertebra acoelous and bears two condyles posteriorly for the attachment with the urostyle.
4. Pectoral girdle fused with sternum.
5. Ribs absent.

   Examples : 1. *Rana*, 2. *Rhacophorus.*

MUSEUM SPECIMENS OF AMPHIBIA

1. *Ichthyophis* (Caecilian)

Classificiation :

Phylum	...	**Chordata**	→	Dorsal tubular nerve cord, notochord and paired gill-slits are present.
Group	....	**Craniata**	→	Definite head. Cranium with brain present.
Subphylum	..	**Vertebrata**	→	Notochord is replaced by vertebral column; circulatory system closed; hepatic portal system present; blood red containing R.B.Cs.
Division	....	**Gnathostomata**	→	Jaws and paired appendages are present.
Superclass	..	**Tetrapoda**	→	Paired limbs, lungs, cornified skin and bony skeleton.
Class	...	**Amphibia**	→	Cold blooded; aquatic and terrestrial; skin moist, soft glandular and devoid of external scales; two occipital condyles; heart three-chambered.
Order	....	**Apoda or Gymnophiona**	→	Burrowing; vermiform; without limbs and girdles ; skin with numerous grooves and wrinkles.
Genus	...	*Ichthyophis*		
Species	...	*glutinosa*		

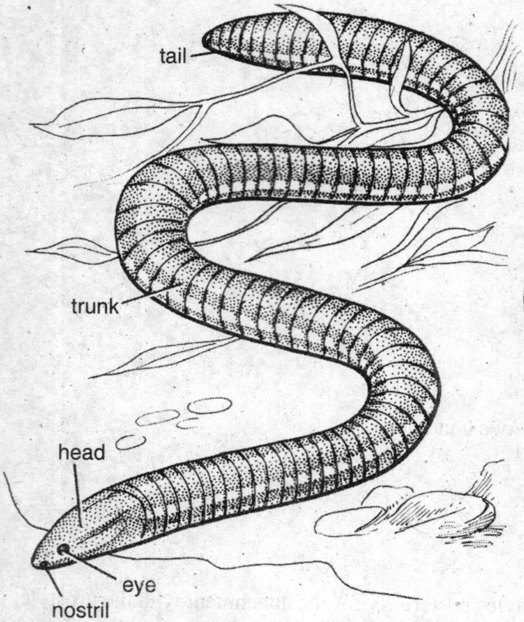

Fig. 12.1. *Ichthyophis* (Male).

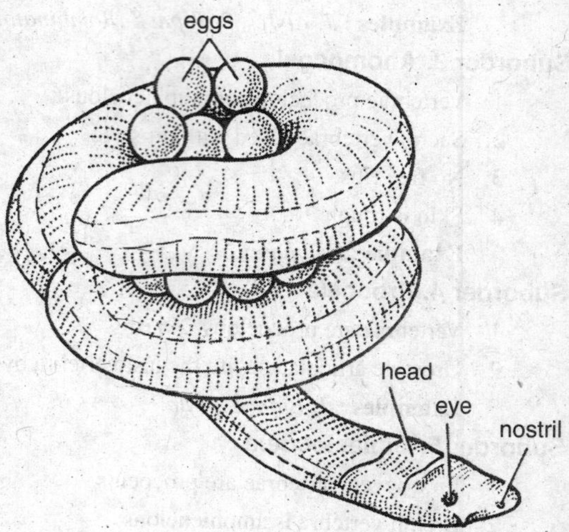

Fig. 12.2. *Ichthyophis.* (Female coiling around the eggs).

Habits and habitat. *Ichthyophis* lives in burrows in moist ground (fossorial). Animal is blind and adapted for burrowing life.

Comments :

1. *Ichthyophis* (Fig. 12.1) is a burrowing, elongated and eel-like and measures up to 30 cm in length. Body is divided into head, trunk and tail.
2. The colour of the body is dark brown or bluish–black with a yellow band along the side.
3. Skin is provided with numerous transverse grooves or wrinkles in which minute scales are embedded. Squirt glands present in the skin discharge an irritating fluid.
4. Limbs and limb-girdles are entirely absent.
5. Tail is short or vestigial.
6. Eyes are minute, devoid of lids buried deep in the head, are non-functional. A protrusible tentacle (sensory organ) present between the nostril and the eye.
7. Tympanic membrane and columella are absent.
8. Vertebrae amphicoelous.
9. Laryngotracheal chamber is prolonged into a distinct trachea. Right lung elongated and left lung is reduced.
10. Sexes separate, males possess large eversible copulatory organ provided with hooks. Fertilisation internal. Eggs are meroblastic.
11. Parental care is very well developed. Female coils herself around the gelatinous egg mass to protect it from ground burrowing animals.
12. Larva possesses three pairs of very long finely branched external gills.

Geographical distribution. *Ichthyophis* extends from the foot hills of Himalayas to Sri Lanka, the Malaya Archipelago and Thailand. It is the only representative of Gymnophiona living today in tropical countries.

2. *Uraeotyphlus*

Classification : Same as that of *Ichthyophis.*

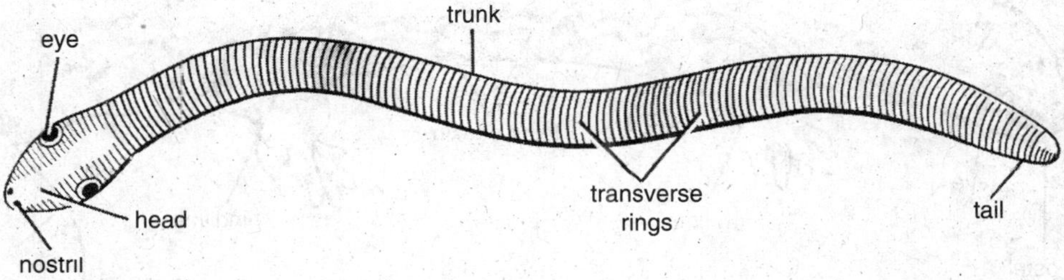

Fig. 12.3. *Uraeotyphlus.*

Habit and habitat. *Uraeotyphlus* also burrows in moist ground and feeds on small worms.

Comments :

1. *Uraeotyphlus* (Fig. 12.3) is a burrowing, elongated, cylindrical and vermiform animal.

2. Body is dark green or brown or black dorsally and white ventrally. Body is divided into head, trunk and tail.

3. Skin is transversely grooved containing the embedded cycloid scales.

4. Head is small, conical and bears lidless eyes covered by the skin, nostrils and mouth. A sensory tentacle is also present between eye and nostril.

5. Limbs and limb-girdles are entirely absent.

6. Tail is very short rather vestigial.

7. Cloaca is terminal.

8. Food chiefly comprises worms.

Geographical distribution. *Uraeotyphlus* is found in India, Malabar hills, Malaya, South Africa and South America.

3. *Cryptobranchus* (Hell-bender)

Classification :

Classification up to Amphibia is same as that of *Ichthyophis.*

Order	...	**Urodela (Caudata)**	→ Body scaleless with distinct head, trunk and tail; limbs two pairs and of equal size; with or without external gills and gill-slits in adults. oviparous; larvae aquatic like adults.
Suborder	....	**Cryptobranchoidea**	→ Body depressed; skin soft and flabby; eye-lids absent; gills absent in adults; fertilisation is external.
Genus	....	*Cryptobranchus*	

Habit and habitat. *Cryptobranchus* is found in running water with rocks and slabs. It is predaceous, feeding upon worms, crustaceans and small fishes.

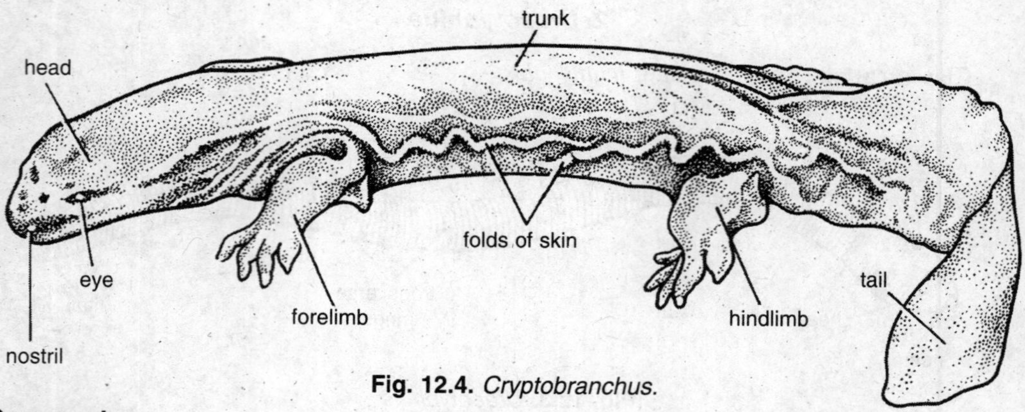

Fig. 12.4. *Cryptobranchus.*

Comments :

1. *Cryptobranchus* (Fig. 12.4) is commonly known as **hell-bender**. It is a giant, fully aquatic salamander. It is semi-larval (incompletely metamorphosed) in adult stage.
2. Body is large reaching a length of 60 cm or more and divided into head, trunk and tail.
3. Head is broad, bears a pair of small lidless eyes and nostrils.
4. Gills are absent in the adult but the first gill-cleft or spiracle remains open and acts as an outlet for water taken into the mouth during buccal respiration.
5. Vomerine teeth parallel to jaws. Angular and prearticular of lower jaw separate.
6. Two pairs of limbs are present.
7. Lateral line organs are well-developed.
8. Sexes are separate. Fertilisation external.
9. Eggs are laid in gelatinous sacs pulled out into two long strings and are fertilised when the males deposit sperm masses near the egg-laying females.

Geographical distribution. *Cryptobranchus* is widely distributed over Eastern United States from the Great lakes to Georgia and Louisiana, Japan, Eastern Asia. Miocene to Recent.

4. *Ambystoma* (Tiger Salamander)

Classification :

Classification up to order is the same as that of *Cryptobranchus.*

Suborder	....	**Ambystomoidea**	→ Teeth in transverse rows acrosss the posterior margins of vomer, eye-lids present; fertilisation internal, exhibit neoteny.
Genus	....	*Ambystoma*	
Species	....	*tigrinum*	

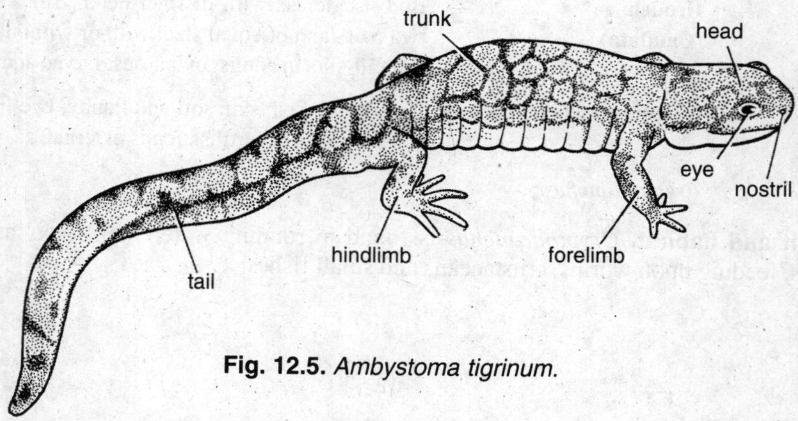

Fig. 12.5. *Ambystoma tigrinum.*

Habit and habitat. *Ambystoma* is a terrestrial animal.

Comments :

1. *Ambystoma tigrinum* (Fig. 12.5) is commonly known as **tiger-salamander**.

2. Body is elongated, lizard-like having transverse grooves , about 12 cm in length. Body is divided into head, trunk and tail.

3. The colour of the body is black with yellow patches. Skin is poisonous.

4. Head is depressed having large mouth, eyes with eye-lids, nostrils and a pair of paratoid glands (poisonous glands).

5. Limbs are well developed. Forelimbs bear 4 fingers and hindlimbs have 5 toes.

6. Tail fin and external gills are absent in the adult stage.

7. Vertebrae are amphicoelous. Prevomers devoid of posterior processes and teeth across near margins. Palatines devoid of teeth.

8. Sexes are separate. Fertilisation is internal.

9. Its larva is known as **Axolotl**.

Geographical distribution. *Ambystoma* is found in North America from Southern Alaska to Mexico. Neotenic forms occur in mountain region especially in Mexico. Upper Cretaceous to Recent.

5. Axolotl Larva

Classification : Same as that of *Ambystoma*.

Comments :

1. Axolotl (Fig. 12.6) is the larva of *Ambystoma tigrinum*. It was previously supposed as an adult form called *Siredon*.

2. It possesses three pairs of external gills and four pairs of gill-clefts. Tail with caudal fin.

3. Head has eyes, nostrils and mouth. Forelimbs and hindlimbs present.

4. Axolotl larva is a perennial form and develops gonads (sexually mature) and, thus, exhibits the phenomena of **neoteny** (failure to metamorphose because of environmental conditions).

5. When environmental conditions such as low temperature, abundance of nutrition and oxygen supply and lack of iodine in water inhibit the action of thyroid gland, the axolotl larva becomes sexually mature and reproduces like adult *Ambystoma*.

6. If the environmental conditions are changed, it is capable to metamorphose.

7. **Metamorphosis** in Axolotl can also be induced by injecting thyroid extract or by transferring to water with a higher iodine content.

8. During metamorphosis the gills and tail fin are lost.

Geographical Distribution. Axolotl larva is found in the mountain region lakes of Mexico.

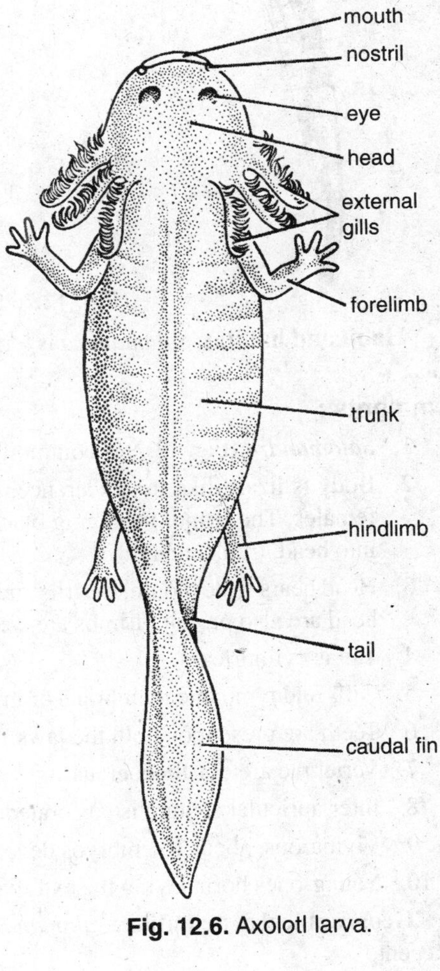

Fig. 12.6. Axolotl larva.

6. *Salamandra* (European Fire Salamander)

Classification : Same as that of *Cryptobranchus* up to order.

Suborder .... **Salamandroidea** → Eye-lids present or absent; either gills or lungs present; fertilisation is internal; viviparous.

Genus .... Salamandra

Species ... *maculosa*

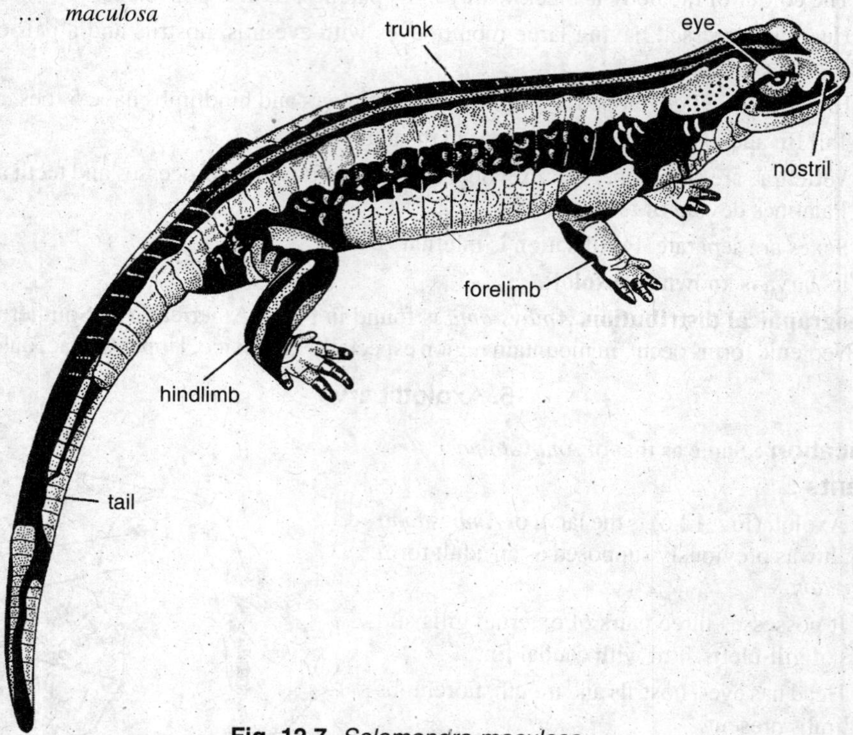

Fig. 12.7. *Salamandra maculosa.*

Habit and habitat. *Salamandra* is terrestrial and found under stones, logs and crevices of old walls.

Comments :

1. *Salamandra* (Fig. 12.7) is commonly known as **European fire-salamander**.
2. Body is lizard-like in appearance, measuring 12–15 cm in length. Males are shorter than females. The colour is shining black with yellow spots on back and limbs. Body divisible into head, trunk and tail.
3. Head bears wide mouth, nostrils and eyes with eye-lids. A pair of paratoid glands behind the head are also present. Limbs are well developed and strong.
4. Tail is cylindrical.
5. Gills and tympanic membrane or middle ear is absent. Lungs are present.
6. Teeth are present on both the jaws. Vomerine teeth are also present.
7. Vertebrae are opisthocoelous.
8. Inter-auricular septum is perforated.
9. Viviparous, about 15 embryos develop in the oviducts.
10. Young-ones born having the external gills which are shed before hatching.

Geographical distribution. *Salamandra* is found in Europe, Asia, and North America. Eocene to Recent.

7. *Amphiuma* (Congo Eel)

Classification : Same as that of Cryptobranchus up to order.

Suborder ..... **Salamandroidea** → Characters same as that of *Salamandra.*
Genus ..... *Amphiuma*

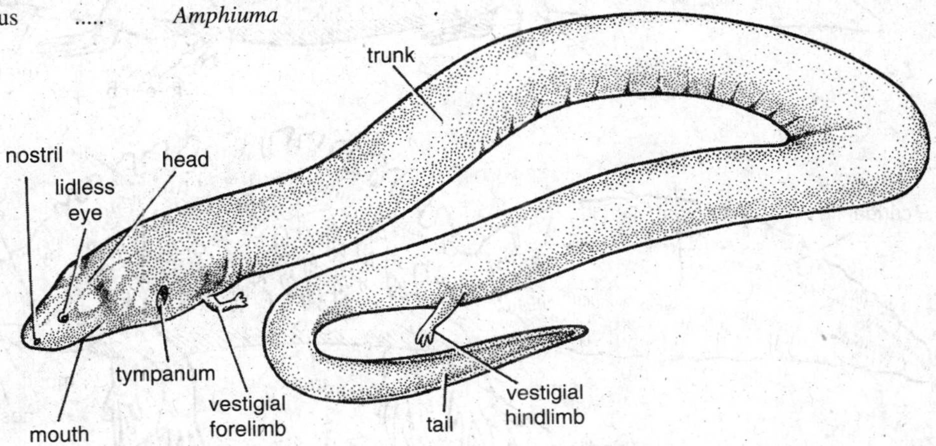

Fig. 12.8. *Amphiuma.*

Habit and habitat. *Amphiuma* is found in the swampy muddy water and rice fields and feeds on molluscs, crayfishes and small fishes.

Comments :

1. *Amphiuma* (Fig. 12.8) is commonly known as **Congo-eel**.
2. Body is very long and cylindrical, measuring about 90 cm in length, and divided into head, trunk and tail.
3. The general colour of the body is black. Head bears mouth, nostrils and eyes.
4. Eyes are without eye-lids and functional.
5. Fore- and hindlimbs are tiny and almost vestigial, which are of no value in locomotion.
6. A small pair of open gill-slits, lungs with long trachea supported by cartilages are present.
7. Sternum is absent.
8. It is a semi-larval form. *Amphiuma* lays its eggs in water and the female coils around them for protection (parental care).
9. *Amphiuma* is partly metamorphosed animal with small eyes. It has lidless eyes, four branchial arches, amphicoelous vertebrae and parallely arranged maxillary and vomerine teeth. Erythrocytes 90 microns in size (**Ponder**, 1924).

Geographical distribution. *Amphiuma* is found in South Eastern U.S.A. (Missouri and Virginia). Eocene to Recent.

8. *Triton* (European Crested Newt)

Classification :

Same as that of *Cryptobranchus* up to order.

Suborder ... **Salamandroidea** → Characters same as that of *Salamandra.*
Genus .... *Triton*

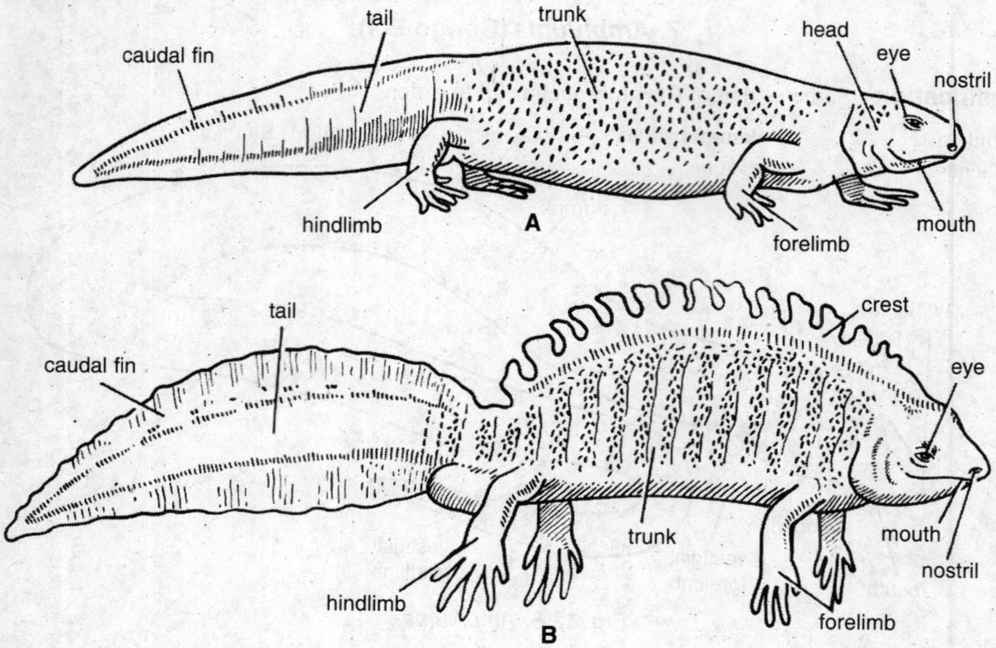

Fig. 12.9. *Triton.* A—Female; B—Male.

Habit and habitat. *Triton* is terrestrial. *Triturus* is also terrestrial.

Comments :

1. *Triton* (Fig. 12.9) is commonly known as **European-salamander** or **European crested newt**.
2. Body is elongated and cylindrical and divided into head, trunk and tail.
3. Skin is soft, scaleless and covered with slime.
4. Head bears wide mouth, extrernal nostrils and small eyes with upper and lower lids. Tympanum or ear-drum is absent.
5. Fore-and hindlimbs are well-developed. Forelimbs bear four fingers and hindlimbs bear five toes.
6. Tail is provided with a median caudal fin devoid of fin rays.
7. In adults the gills are absent and respiration cutaneous and by lungs.
8. Lateral line organs are found in the larva. Larva also possesses three pairs of external gills which are lost in adults.
9. Alimentary canal, excretory ducts and the gonadial ducts all open into the cloaca.
10. Vertebrae opisthocelous. Girdle and sternum primitive.
11. It exhibits marked **sexual dimorphism**. Male possesses a dorsal median **crest** over the back. It becomes brilliantly coloured in breeding season.

Geographical distribution. *Triton* is distributed in Europe and U.S.A. *Triturus* is also found in U.S.A from California to Southern Alaska and Europe. Eocene to Recent.

9. *Proteus* (European Blind Cave Salamander)

Classification : Same as that of *Cryptobranchus* up to order.

Suborder	...	**Proteida**	→ Body depressed; permanent larval forms, tail with fin; external gills always present; eye-lids absent; permanently aquatic.
Genus	...	*Proteus*	

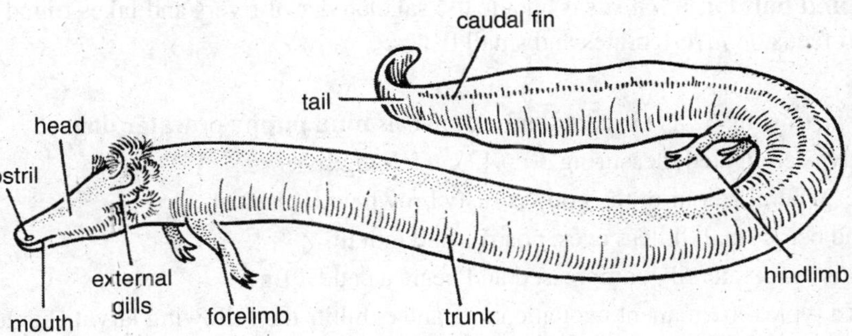

Fig. 12.10. *Proteus.*

Habit and habitat. *Proteus* lives in complete darkness in underground streams and feeds on small crustaceans.

Comments :

1. *Proteus* (Fig. 12.10) is commonly known as **olm** or **European blind cave-salamander**.
2. Body is elongated eel-type reaches up to the length of 45 cm. Body is divided into head, trunk and tail.
3. Due to the absence of pigment in the skin, the colour of body is pale whitish.
4. Head is provided with mouth, nostrils and eyes. Eyes are lost owing to the life in complete darkness, Minute eye rudiments sunk deep into the head and pigmentless skin.
5. Limbs are small, larval in character in the possession of only three fingers in forelimbs and two toes in hindlimbs.
6. Three pairs of branching external-gills and two pairs of gill-clefts are present in the adult. Lungs are also present.
7. Tail is compressed and has a caudal fin.
8. Skull cartilaginous. Teeth are present in both the jaws.
9. If the larvae of *Proteus* are reared in red light, fairly well-developed eyes appear and the skin becomes very heavily pigmented, showing that the genetic basis for these characters has not been lost in this species. In ordinary mixed light the eyes do not develop.
10. *Proteus* is a permanent neotenous larva having 3 pairs of external gills, tail with caudal fin, weak limbs and teeth in both the jaws.

Geographical distribution. *Proteus* is found in the subterranean and mountain caves of Dalmatia and Yugoslavia. Eocene to recent.

10. *Necturus* (Mud Puppy)

Classification : Same as that of *Cryptobranchus* up to order.

Suborder ... **Proteida** → Characters same as that of *Proteus*.

Genus ... *Necturus*

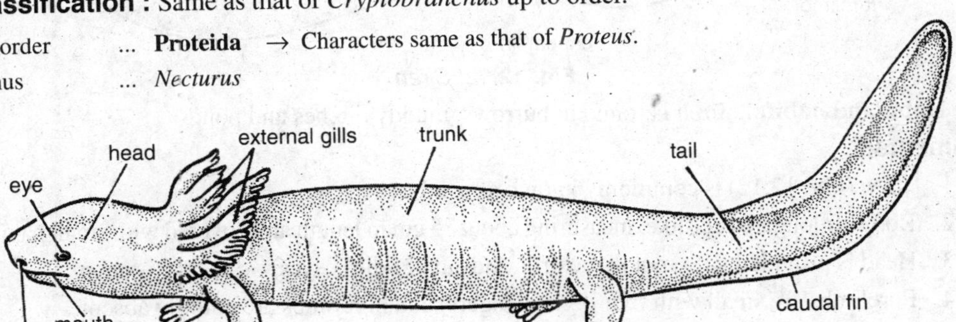

Fig. 12.11. *Necturus.*

Habit and habitat. *Necturus* is an aquatic salamander of rivers and lakes found crawling on the bottom. It feeds on invertebrates and small fishes.

Comments :

1. *Necturus* (Fig. 12.11) is commonly known as **mud puppy** or **water dog**.
2. Body is elongated measuring 30 – 42 cm in length.
3. The general colour of the body is rusty brown with blackish spots.
4. Head bears small lidless eyes, nostrils and mouth.
5. Tail is long, laterally compressed and bears a caudal fin.
6. It is a typical permanent neotenic larva and exhibits the following larval characters :
   (*i*) Three pairs of external-gills and two pairs of gill-clefts are present in the adult.
   (*ii*) Skull largely cartilaginous.
   (*iii*) Musculature larval.
   (*iv*) Larval circulatory system.
   (*v*) Lateral line organs are present and Jacobson's organ absent.
7. Lungs are present.
8. Attempts have been made to force *Necturus* to undergo metamorphosis but no success has been achieved.

Geographical distribution. *Necturus* is found in rivers of North America and Yugoslavia. Cretaceous to Recent.

11. *Siren* (Mud Eel)

Classification : Same as that of *Cryptobranchus* up to order.

Suborder ... **Meantes** or **Sirenidea** → Body slender eel-type; permanent larval forms; hindlimbs absent; external-gills well-developed; eyelids absent; jaws with horny covering; fertilisation external ; completely aquatic.

Genus ... *Siren*

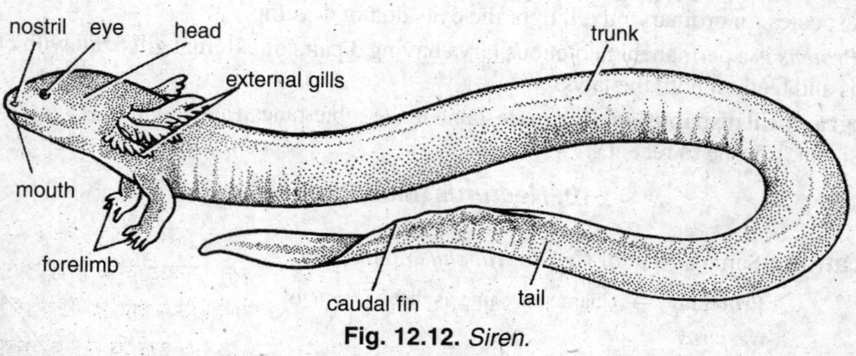

Fig. 12.12. *Siren.*

Habit and habitat. *Siren* is found in burrows, muddy ditches and ponds.

Comments :

1. *Siren* (Fig. 12.12) is commonly known as **mud-eel**.
2. Body is elongated eel-like, measuring about 75 cm in length and covered with small papillae.
3. Head is provided with mouth, nostrils and small lidless eyes.
4. Forelimbs are small with four digits (fingers) and hindlimbs are entirely absent.
5. Tail has a small caudal fin.
6. Three pairs of external gills and one pair gill-slits are present.
7. Teeth are absent in both the jaws. Jaws are provided with horny sheaths.

8. Fertilisation is external
9. It is a permanent larval form but exhibits few adult characters also.

Geographical distribution. *Siren* is found in muddy ponds and lakes of North America. (Virginia, Indiana, Florida and Texas). Pleistocene to Recent.

12. *Ascaphus*

Classification : Same as that of *Cryptobranchus* up to class

Order .... **Anura or Salientia** → Body short and broad; neck and tail absent; forelimbs short, hindlimbs long; gills absent; respiration by lungs; fertilisation usually external.

Suborder ... **Amphicoela** → Vertebrae amphicoelous ; two tail wagging muscles; fertilisation internal.

Genus ... *Ascaphus*

Habit and habitat. *Ascaphus* is amphibious in habit.

Comments :

1. Body is well developed and divided into head and trunk.
2. Head bears wide mouth, nostrils and eyes with eyelids.
3. Limbs are well-developed. Forelimbs bear four fingers and hindlimbs bear five toes.
4. Tail is lost but the two tail-wagging muscles are present.
5. Male possesses an eversible cloaca.
6. Vertebrae are amphicoelous.
7. Fertilisation is internal.

Geographical distribution. *Ascaphus* is found in North America.

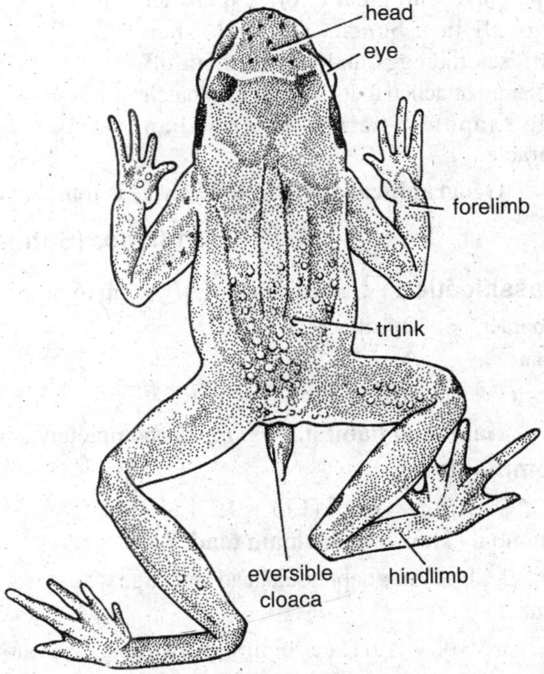

Fig. 12.13. *Ascaphus.*

13. *Alytes* (Mid-wife Toad)

Classification: Same as that of *Ascaphus.*

Order ... **Anura** → Body short and broad; head and trunk fused; neck and tail absent ; forelimbs short, hindlimbs long ; gills absent; respiration by lungs ; fertilisation usually external.

Suborder ... **Ophisthocoela** → Vertebrae opisthocoelous; free ribs either in larva or adult

Genus ... *Alytes*

Habit and habitat. *Alytes* is an amphibious toad.

Comments :

1. *Alytes* (Fig. 12.14) is commonly known as **mid-wife toad**.
2. Body measures 5 to 7.5 cm in length and divided into head and trunk.
3. Body surface bears papillae dorsally as well as ventrally.
4. Head bears wide mouth, nostrils and eyes with movable eyelids and large tympanum.
5. Vocal-sacs are absent in males.

6. Forelimbs with four fingers and hindlimbs with five toes.
7. Vertebrae are opisthocoelous and adults have ribs.
8. Upper jaw toothed.
9. Fertilisation is external.
10. Male *Alytes* exhibits an extraordinary **parental care** of eggs. The male is described as assiduously massaging the cloaca of the female with its hand. After this treatment the female apparently with great effort, expels her eggs all in a bunch. The male then fertilises the eggs and carries them off with him attached to its legs till they hatch into tadpoles. Larva has a median spiracle.

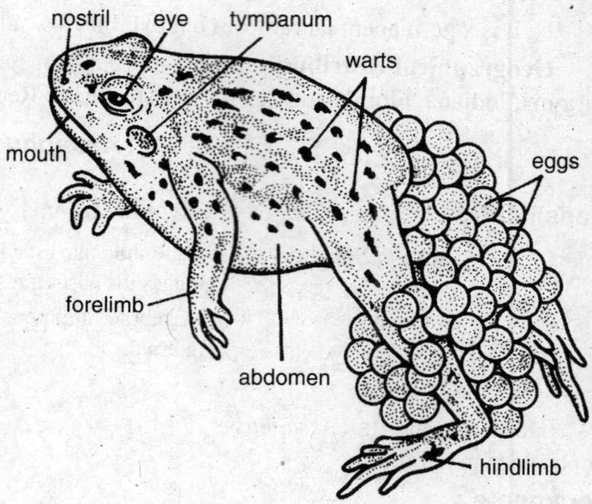

Fig. 12.14. *Alytes.*

Geographical distribution. *Alytes* is found in France and Italy. Miocene.

14. *Pipa* (Surinam Toad)

Classification : Same as that of *Alytes* up to order.

Suborder ... **Opisthocoela**
Genus ... *Pipa*
Species ... *pipa*

Habit and habitat. *Pipa pipa* is completely aquatic.

Comments :

1. *Pipa pipa* (Fig. 12.15) is commonly known as **Surinam toad**.
2. Body has depressed head and large trunk.
3. Body surface bears papillae throughout. Each papilla bears a horny spike at the tip and a poison gland at its base.
4. Head is very short, broad and depressed with a pair of small tentacles in front of eyes. Eyes are small with immovable eyelids and tympanum is not distinct.
5. Forelimb fingers are slender ending in star-shaped tips bearing sensory papillae and hindlimbs are exceedingly large and bear broadly webbed-toe.
6. Tongue is absent.
7. Eustachian tubes open into the pharynx by a median opening.
8. Jaws devoid of teeth and covered by horny sheath.
9. Vertebrae are opisthocoelous.

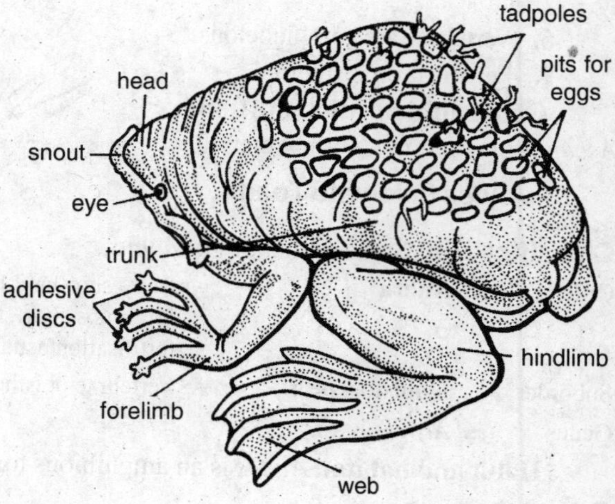

Fig. 12.15. *Pipa pipa (Female).*

10. In the female, the skin of the back is soft and puffy and forms pockets or small pits with lids in which the eggs undergo their complete development. Thus, female exhibits parental care.

Geographical distribution. *Pipa pipa* is found in Guiana and Brazil (South America). Miocene.

15. *Bombinator* (Fire-bellied Toad)

Classification: Same as that of *Alytes*.

Habit and habitat. *Bombinator* lives in standing water having hydrophytic vegetation.

Comments :

1. *Bombinator* (Fig. 12.16) is commonly called **fire-bellied toad**.
2. The body of the animal measures about 5 cm and divided into head and trunk.
3. The head bears wide mouth, nostrils and eyes with eyelids.
4. The dorsal surface of the body is dark and has papillae, but the ventral surface and limbs have black, red and orange patches which look like fire.
5. Tympanic membrane and vocal sacs are absent.
6. Fore-and hindlimbs are well-developed. Hindlimbs bear webbed toes and are enlarged for leaping.

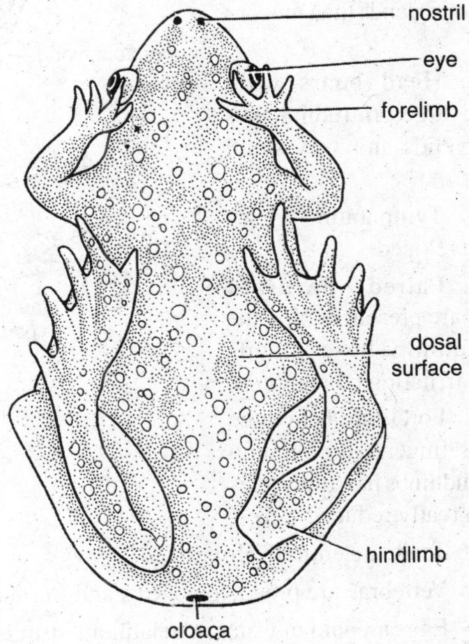

12.16. *Bombinator.*

7. Eustachian tubes are small and tongue is present.
8. Vertebrae opisthocoelous. Adults have ribs.
9. Fertilisation is external. Tadpole larva is found.
10. When it is disturbed it lies on its back assuming a warning attitude and showing its fiery patches as warning colours, then it feigns death to escape its enemies.

Geographical distribution. *Bombinator* is found in Europe, Eurasia and China.

16. *Bufo* (True Toad)

Classification: Same as that of *Alytes* up to order.

Suborder ... **Proceola** → Vertebrae procoelous; sacral vertebra fused with urostyle; skin is rough.

Genus ... *Bufo*

Species ... *melanostictus*

Habit and habitat: *Bufo melanostictus* is terrestrial, nocturnal and remains hidden under stones and logs or in burrows. Breeding occurs in water. Carnivorous, feeding on worms, insects and snails. etc.

Comments :

1. *Bufo melanostictus* (Fig. 12. 17) is commonly called **true-toad**.

2. Skin is rough, dry and warty on the dorsal surface of the body and has more poisonous glands than mucous glands.

3. Head bears wide mouth, large protruding eyes with eyelids and very small nostrils.

4. Tympanum is very well-developed.

5. Paired **paratoid glands** are present behind the tympanum. These glands secrete irritating poison fluid.

6. Forelimbs bear three webless fingers and a thumb pad. Hindlimbs have three toes with a greatly reduced web.

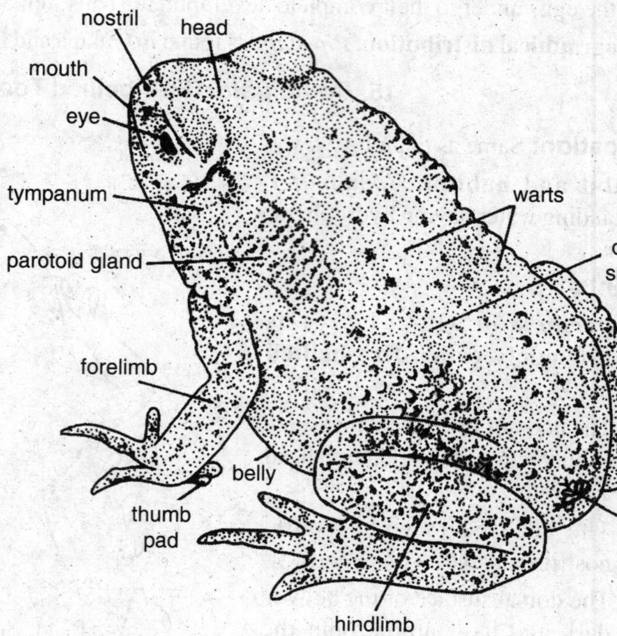

Fig. 12.17. *Bufo melanostictus.*

7. Teeth are entirely absent.

8. Vertebrae are procoelous. Ribs and sternum are absent.

9. Eggs are laid in water in gelatinous strings.

10. Paratoid glands secrete bufotalus and bufoqus toxins which cause nausea, disturbances in respiration and muscles and also effect in heart functioning.

Geographical distribution. *Bufo* is found all over the world except Australia and Medagascar. *Bufo melanostictus* is found in Himalayas up to 3000 metres.

17. *Hyla* (Tree Frog)

Classification: Same as that of *Alytes* up to order.

Suborder .... **Procoela** → Characters same as that of *Bufo*.

Genus ... *Hyla*

Habit and habitat. *Hyla* is arboreal, living on trees and rocks.

Comments:

1. *Hyla* (Fig. 12.18) is commonly called **tree-frog** due to its arboreal habits.

2. Body measures 2.5 to 7.5 cm in length and divided into head and trunk.

3. Skin is smooth on the dorsal surface but bears papillae on the ventral surface.

4. Head bears eyes, nostrils and tympanum.

5. Fingers and toes have **adhesive-pads** or **discs** for sticking on the smooth surfaces. Web is poorly developed.

6. Vocal sacs are greatly expanded and produce loud voice.

7. Teeth are present only in the upper jaw and absent in the lower jaw.

8. Vertebrae are procoelous.

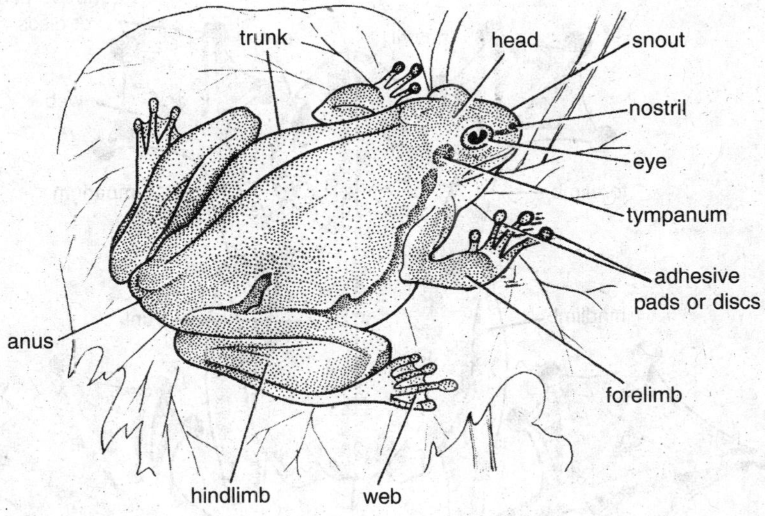

Fig. 12.18. *Hyla.*

9. Belly skin contains hygroscopic glands which help the animal to stick on leaf and twigs, etc.
10. They exhibit **mimicry** and are remarkably protectively coloured.
11. Fertilisation is external.
12. *Hyla arborea, H. versicolor* and *H. regita,* etc., are all tree frogs and adapted for amphibious and arboreal life.
13. In *Hyla goeldii,* the eggs are carried on the back of the female. *H. faber* also shows parental care.

Geographical distribution. *Hyla* is cosmopolitan in damp forests in India, China, Africa, U.S.A and Canada. Miocene.

18. *Rhacophorus* (Flying Frog)

Classification : Same as that of *Alytes* up to order.

Suborder ... **Displasiocoela** → First seven vertebrae procoelous, eighth vertebra amphicoelous, sacral vertebrae acoelous and bears two condyles posteriorly for the attachment with the urostyle; pectoral girdle fused with sternum; ribs absent.

Genus ... *Rhacophorus*

Habit and habitat. *Rhacophorus* also lives on trees or under stones and has power of rapidly changing colour.

Comments :

1. *Rhacophorus* (Fig. 12.19) is commonly known as **flying-frog**. Female is larger than male.
2. Body is slender and divided into head and trunk.
3. Head is broad and nearly conical bearing eyes with eye lids, nostrils and tympanum.
4. Limbs are long and thin with large feet. Web develops between the digits of all the four feet. Digits of all the four feet bear rounded **adhesive pads** or **discs** at tips.
5. It lives on trees. Usually large webbed feet are used as palms in gliding from branch to branch or tree to tree and can leap 6 – 10 metres in the air.
6. It also exhibits **parental care** by depositing eggs in the nests near standing pond water.

Geographical distribution. *Rhacophorus maculatus* and *Rhacophorus malabaricus* are found in India, while *R. pardalis* is found in the Malaya Archipelago. Miocene.

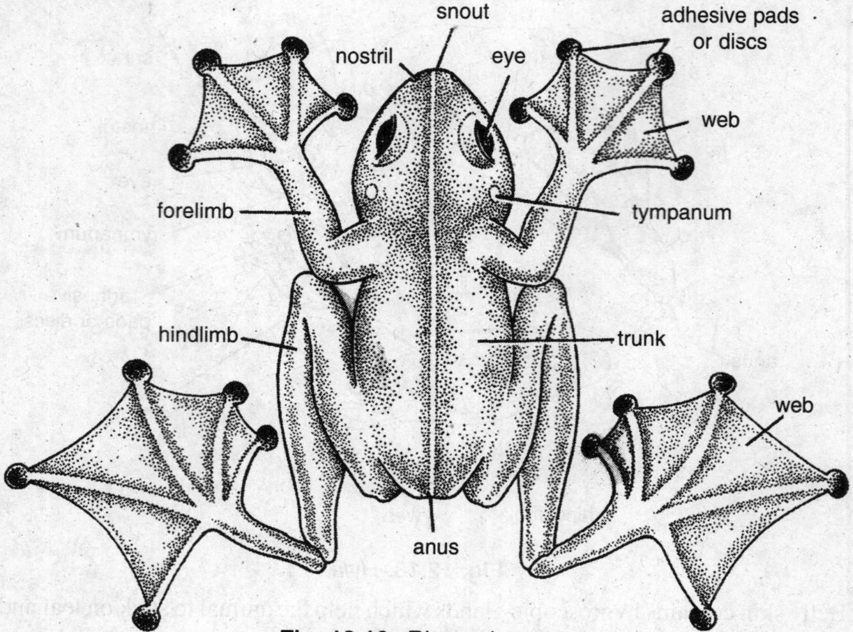

Fig. 12.19. *Rhacophorus.*

19. *Rana* (Indian Bull-Frog)

Classification: Same as that of *Alytes* up to order.

Suborder	...	**Diplasiocoela**	→ First seven vertebrae procoelus; peetoral girdle fused with sternum; ribs absent.
Genus	...	*Rana*	
Species	...	*tigrina*	

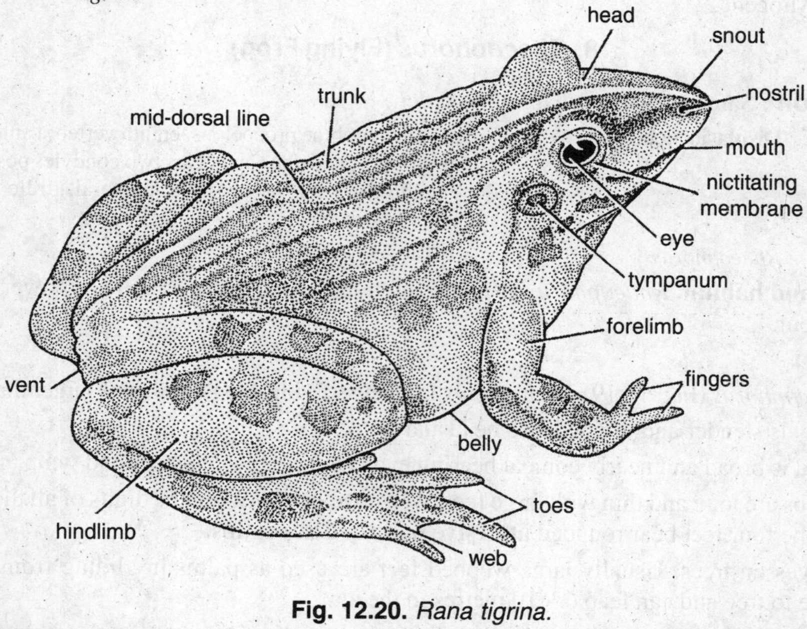

Fig. 12.20. *Rana tigrina.*

Habit and habitat. *Rana* is found in moist places, freshwater ponds and streams. Carnivorous, feeding on small worms, snails, slugs and aquatic insects.

Comments :

1. *Rana tigrina* (Fig. 12.20) is commonly called **Indian bull-frog.**

2. Body is green with black patches along with yellow midrib on the dorsal surface and pale yellow on the ventral surface and divided into head and trunk.

3. Head is flat and more or less triangular and bears nostrils, eyes with eyelids and tympanum behind the eyes.

4. Limbs pentadactyle. Forelimbs are smaller with four fingers and hindlimbs are larger and bear five toes which are webbed.

5. Lower jaw without teeth, while upper jaw teethed.

6. Vomerine teeth are present in two groups one on each side of the internal nostrils.

7. Tongue is large, muscular and forked, attached in front and free behind.

8. Vertebrae are procoelous.

Geographical distribution. *Rana* is world-wide in distribution.

13

Chapter

CLASS–REPTILIA

Reptilia (*L. reptum* = to creep) includes turtles and tortoises belonging to **order Chelonia**, lizards and snakes of **order Squamata**, alligators and crocodiles of **order Crocodilia** and tuatara of **order Rhynchocephalia**. These flourished in Mesozoic era. Reptiles are well-adapted on land. Reptiles show advance features over amphibians such as dry scaly skin, limbs adapted for rapid locomotion, separation of oxygenated and nonoxygenated blood in the heart, complete ossification of skull and eggs development on land with embryonic membranes. Adaptive radiation is found in reptiles, *i.e.*, they adapted to all kinds of habitat.

General Characters

1. Reptiles are cold blooded, terrestrial or aquatic animals.
2. Skin is dry and skin glands are absent.
3. Body is covered with horny epidermal scales or scutes, shields or plates.
4. Limbs tetrapodus, pentadactyle type with horny claws.
5. Skull with single median occipital condyle.
6. Mandible consists of many pieces and articulates with the cranium through the quadrate bone.
7. Vertebrae are gastrocentrous.
8. Ribs form a true sternum.
9. Respiration by lungs.
10. Heart is divided into two auricles and incompletely divided ventricle.
11. Right and left aortic arches are complete and functional.
12. Red blood corpuscles (R.B.Cs) are nucleated.
13. Kidneys without nephrostomes. Each kidney is provided with separate ureter.
14. Twelve pairs of cranial nerves.
15. Lateral sense organs are absent.
16. Typical cloaca is present.
17. Fertilisation is internal. Eggs are laid on land.
18. Embryonic membrances (amnion and allantois) are present.

CLASS— REPTILIA

Subclasses

- I Anapsida
- III Parapsida (Extinct)
- III Diapsida
- IV Synapsida (Extinct)

Subclass— ANAPSIDA

Orders

- 1 Cotylosauria (Extinct)
- 2 Chelonia
  Ex. 1. *Chelone*
  2. *Chrysemis*
  3. *Kachuga*
  4. *Trionyx*
  5. *Testudo*

Subclass—DIAPSIDA

Superorders

- I Lepidosauria
  - Orders
    - 1 Rhynchocephalia
      Ex. *Sphenodon*
    - 2 Squamata
      - Suborders
        - 1 Lacertilia
          Ex. Lizards
        - 2 Ophidia
          Ex. Snakes
- II Archosauria
  - Orders
    - Crocodilia
      Ex. 1. *Crocodilus*
      2. *Alligator*
      3. *Gavialis*

Subclass I. Anapsida

1. Solid roof of the skull.
2. No temporal vacuities or fossae behind eyes.
3. Quadrate is connected to the otic bones.
4. Limbs are generally strong.
5. Body is enclosed in the carapace and plastron.

Order 1. Cotylosauria

1. Primitive reptiles.
2. Resemble labyrinthodont amphibians.

Example : *Seymouria.*

Order 2. Chelonia

1. Terrestrial or aquatic.
2. Tetrapoda, pendactyle with walking limbs or paddles.
3. Trunk is enclosed in a bony shell composed of dorsal carapace and ventral plastron, which may or may not be covered with horny shields.
4. Tail is always present.
5. Jaws without teeth but with horny sheaths.
6. Quadrate bone is fixed.
7. Thoracic vertebrae and ribs are usually fused with the bony shell.
8. Two sacral vertebrae.
9. Ribs with capitular portion only.
10. Humerus with entepicondylar foramen.
11. Pubes and ischia forming symphyses.
12. Copulatory organ is unpaired.
13. Cloacal opening is longitudinal.
14. Oviparous.

Suborder. Thecaphora

1. Retractile neck bends in vertical direction.
2. Pelvis not fused with plastron.

Examples : *1. Chelone, 2. Chrysemis, 3. Kachuga, 4. Trionyx, 5. Testudo, 6. Terrapene, 7. Chelonia, 8. Coretta.*

Subclass II. Parapsida

Single temporal vacuity present behind the eye and bounded below by post-frontal and supra-temporal bones. **Ex.** *Icthyosaurus. (Extinct forms).*

Subclass III. Diapsida

1. Two temporal vacuities on each side behind the eyes.
2. Post-orbital and squamosal usually meet forming a bar between the temporal vacuities.
3. Antero-orbital opening or depression absent.

Superorder I. Lepidosauria

1. Two temporal vacuities are present behind the eyes.
2. Anterior orbital vacuities are absent.
3. Post-temporal fenestrae are usually present.
4. Humerus with two foramina.

Order 1. Rhynchocephalia

1. Two vacuities are present.
2. Teeth are acrodont.
3. An epipterygoid is present.
4. Quadrate bone is fixed or immovable.
5. Parietal foramen is present.
6. Vertebrae are amphicoelus.

7. External copulatory organs are entirely absent.

Example : *Sphenodon.*

Order 2. Squamata

1. Body is covered by horny epidermal scales or shields.
2. Single (supra or upper) temporal vacuity in lizards and no temporal vacuity in snakes.
3. Teeth are pleurodont.
4. Quadrate is movable.
5. Quadratojugal is absent.
6. Vertebrae are usually procoelous.
7. Limbs present (lizards) or absent (snakes).
8. Cloacal opening is transverse.
9. Male possesses a pair of eversible copulatory organs.

Suborder 1. Lacertilia or Sauria

1. Terrestrial, arboreal or burrowing forms.
2. Single (supra or upper) temporal vacuity is present.
3. Limbs pentadactyle and are usually present.
4. The rami of mandibles are fused in front.
5. Sternum is present.
6. Eyelids are movable.
7. Tympanum present.

Examples : *1. Hemidactylus, 2. Calotes, 3.Chaemeleon, 4. Draco, 5. Phrynosoma, 6. Varanus, 7. Crotaphytus, 8. Mabuiya, 9.Gecko, 10. Ophisaurus, 11. Rhineura, 12. Iguana, 13. Heloderma, 14, Molloch.*

Suborder 2. Ophidia or Serpentes

1. Terrestrial or aquatic, arboreal or burrowing.
2. Temporal vacuities are entirely absent.
3. Limbs are absent.
4. The rami of mandibles are united by a ligament.
5. Sternum is absent.
6. Eyelids are immovable.
7. Tympanum, sternum and urinary bladder are absent.
8. Tongue is bifid and protrusible.
9. Zygosphene and zygantra are present in the vertebra.
10. Poisonous and non-poisonous snakes.

Examples : *1. Typhlops* (Typhlopidae), *2. Python* (Boidae) *3. Eryx johni,* (Boidae) *4. Lycodon, 5. Bungarus* (Elapidae), *6. Naja naja* (Elapidae) *7. Enhydrina, 8.Hydrophis* (Hydrophiidae) *9. Viper* (Viperidae), *10. Ancistrodon* (Crotalidae), *11. Natrix* (Colubridae), *12. Zaemenis, 13. Dendrophis* (Colubridae), *14. Crotalus, 15. Elaps* (Elapidae) 16. *Uropeltis* (Uropeltidae).

Superorder II. Archosauria

1. Skull with closed upper temporal vacuity, but possesses both the temporal arches.
2. Anterior orbital vacuities are usually present.
3. Teeth are thecodont.
4. Humerus with foramina.
5. Vertebrae are amphicoelus or procoelous.

Order. Crocodilia

1. Freshwater predatory forms.
2. Body is covered with an exoskeleton of thick horny epidermal shields.
3. Tail is long and laterally compressed.
4. Maxillae, palatines and pterygoids are united along the middle to form a **secondary palate**.
5. Teeth is thecodont.
6. Quadrate is immovable.
7. Both supra-and infra-temporal arches are present.
8. Vertebrae are procoelous. Cervical vertebrae with two headed ribs.
9. Sternum is present.
10. Thoracic ribs possess uncinate process. Abdominal ribs are present.
11. Prepubis present. Pubis does not share in the formation of acetabulum.
12. Heart is four chambered that is ventricle is completely divided into two by a septum.
13. Lungs are spongy.
14. Cloacal opening is longitudinal.
15. Males possess a median unpaired penis.

Example : *1. Crocodilus, 2. Alligator, 3. Gavialis.*

Subclass IV. Synapsida

1. One temporal vacuity is present on each side behind eye bounded above by the post-orbital and below by the quadrato-jugal. Extinct mammal-like reptiles.

Example : *Dimetrodon, Varanosaurus.*

MUSUEM SPECIMENS OF REPTILIA

1. Chelone (= *Chelonia*) (Green Turtle)

Classification :

Phylum	.. **Chordata**	→ Dorsal tubular nerve cord, notochord and paired gill-slits are present.
Group	... **Craniata**	→ Definite head, cranium with brain.
Supphylum	... **Vertebrata**	→ Notochord is replaced by vertebral column; two pairs of appendages; circulatory system closed; hepatic portal system present; blood red containing R.B.Cs.
Division	... **Gnathostomata**	→ Jaws and paired appendages are present.
Superclass	... **Tetrapoda**	→ Paired limbs, lungs; cornified skin and bony skeleton.
Class	... **Reptilia**	→ Cold blooded, terrestrial or aquatic vertebrates; single occipital condyle; vertebrae gastrocentrous; respiration by lungs; heart with two auricles and incompletely divided ventricle ; right and left aortic arches complete and functional; cranial nerves 12 pairs and embryo with amnion and allantois.
Subclass	... **Anapsida**	→ Primitive reptiles. Skull having solid roof without temporal fossae; body enclosed in the carapace and plastron.
Order	... **Chelonia**	→ Jaws without teeth but with horny sheaths; thoracic vertebrae and ribs usually fused with the bony shell. Turtles, tortoises and terrapins.
Genus	... *Chelone*	
Species	... *mydas*	

Habit and habitat. *Chelone* is a marine turtle and comes shore only to lay eggs. Herbivorous, feeding mostly on marine algae.

Comments :

1. *Chelone mydas* (Fig. 13.1) is a marine form commonly called **green turtle**. It measures from 85 – 110 cm in length. Body is divided into head, trunk and tail.

2. The colour is olive or brown above with yellowish spots and pale yellow below.

3. Body is enclosed in a rigid shell. Carapace flat covered with smooth bony shields. Plastron joined with carapace by ligament.

4. Dorsal shields are juxtaposed, fitting closely into each other. Four pairs of costal shields.

5. Head is covered with one pair of prefrontal shields, the others are small. Head bears eyes having eyelids and nictitating membrane.

6. Head short, tail and limbs are retractile inside the carapace.

7. Fore-and hindlimbs form wing-like paddles. Only first digit of forelimb is clawed, while hindlimbs are clawed.

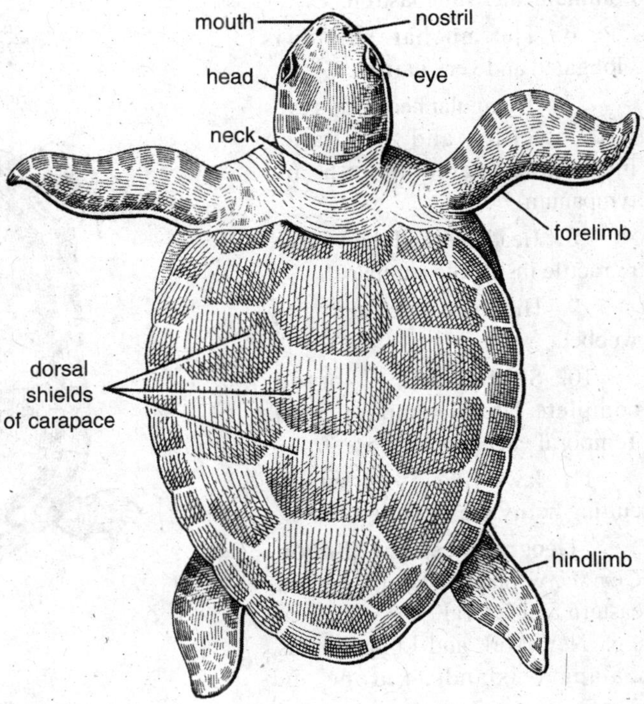

Fig. 13.1. *Chelone mydas.*

8. Jaws with denticulate edges.

Geographical distribution. *Chelone mydas* is found in the Atlantic, Indian and Pacific Oceans. But very commonly found in the Islands of Ascension, West Indies and in the Bay of Bengal. Upper Cretaceous to Recent.

2. *Chrysemys* (Painted Terrapin)

Classification : Same as that of *Chelone.*

Genus ... *Chrysemys*

Species ... *picta*

Habit and habitat. *Chrysemys* is a freshwater terrapin adapted for amphibious life. It feeds upon worms, insects and other animals.

Comments :

1. *Chrysemys picta* (Fig. 13.2) is commonly called **painted-terrapin**. Body is divided into head, neck, trunk and tail.

2. It measures about 15–20 cm in length.

3. The colour is dark olive-brown or blackish above with broad yellow bands across the anterior ends of the nuchal and costal shields.

4. Body is enclosed in a smooth depressed bony shell without a keel. The carapace is flat and plastron is immovable with a strongly developed bridge. Coastal and neural carapaces are convex and dome-shaped.

5. Marginal plates communicate with plastron.

6. The nuchal shield is elongated and very narrow.

7. Triangular head bears eyes with eyelids and nictitating membrane, nares, mouth and tympanum.

8. Head, feet and short tail are retractile inside the carapace.

9. Hindlimbs toes are partly webbed.

10. Skull with a broad, complete lateral temporal arch. Temporal vacuities are absent.

11. Jaws without teeth but with cutting horny sheaths.

Geographical distribution. *Chrysemys picta* is found in India, eastern and central North America, *e.g.*, New York and Long Islands, Galapago islands, Europe and Africa. Eocene to Recent.

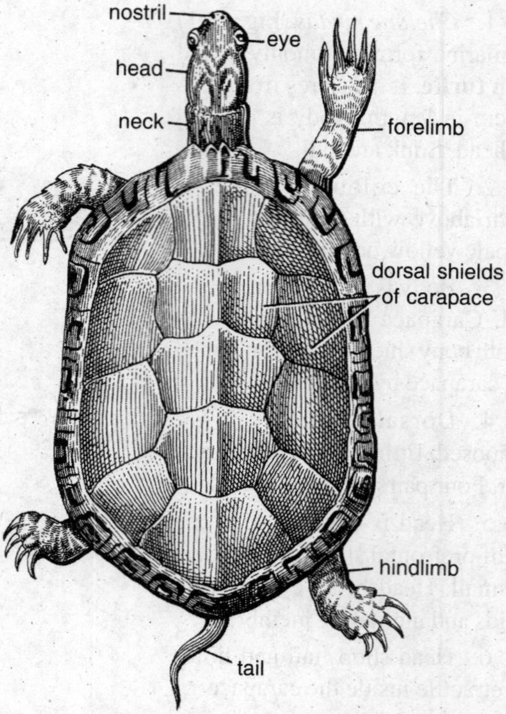

Fig. 13.2. *Chrysemis.*

3. *Kachuga* (Roofed Terrapin)

Classification : Same as that of *Chelone.*

Genus ... *Kachuga*
Species ... *tectum*

Habit and habitat. *Kachuga* is found in rivers especially in bathing places and areas where dead bodies remains are disposed. They feed both on plants and animals.

Comments :

1. *Kachuga tectum* (Fig. 13.3) is commonly known as **roofed-terrapin** (tortoise) and lives in warm deep waters.

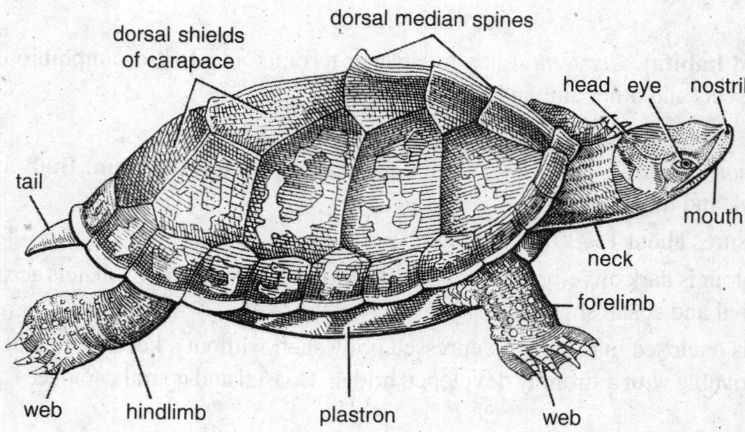

Fig. 13.3. *Kachuga tectum.*

2. Body is enclosed in a bony shell comprising the dorsal **carapace** and ventral **plastron**. Body is divided into head, neck, trunk and tail.

3. The first three central plates of the carapace are produced into backwardly directed **dorsal median spines**.

4. Head is provided with eyes having eyelids and nictitating membrane, nares, tympanum and mouth.

5. Jaws without teeth but with horny sheath.

6. Head, limbs and short stumpy tail are retractile inside the carapace.

7. Fore-and hindlimbs bear five partly webbed clawed digits.

Geographical distribution. *Kachuga tectum* is found in India in Indus and Ganges rivers. Eocene to Recent.

4. *Trionyx* (Soft River Terrapin)

Classification : Same as that of *Chelone.*

Genus ... *Trionyx*

Species ... *gangeticus*

Habit and habitat. *Trionyx gangeticus* is a freshwater tortoise. Carnivorous, feeds mainly on fish, frogs and molluscs.

Comments :

1. *Trionyx gangeticus* (Fig. 13.4) is commonly known a **tortoise** or **soft-river-terrapin**.

2. Body is enclosed in a bony shell comprising the dorsal carapace and ventral plastron. Skin is smooth and leathery.

3. Shell of this species is more than 60 cm in length.

4. Carapace has two nuchal shields and not jointed with the plastron.

5. The colour is olive green above and yellowish below. Head with black streaks.

6. Head is large, triangular having a proboscis and provided with mouth, eyes with eyelids, nares and external ear openings. Nostrils located at the tip of fleshy proboscis.

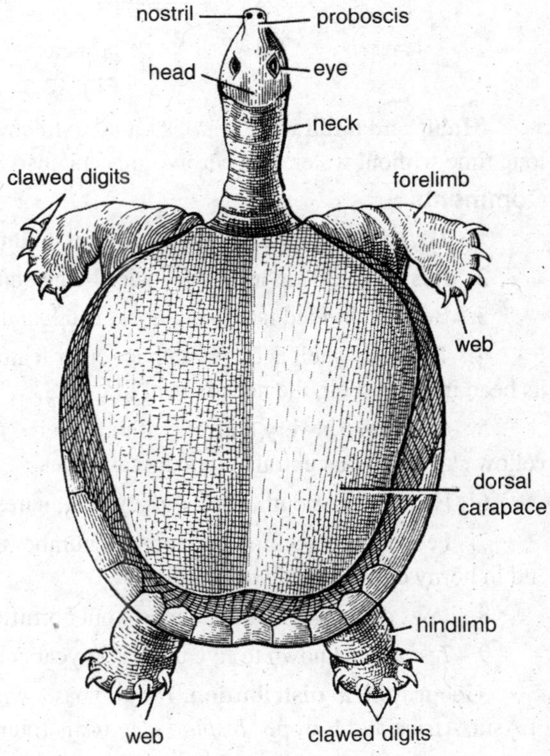

Fig. 13.4. *Trionyx gangeticus.*

7. Limbs are flipper-like, webbed and each limb bears five clawed digits.

8. Jaws are devoid of teeth but provided with horny sheaths.

9. Head, neck and limbs retractile into bony shell.

Geographical distribution. *Trionyx* has the greatest number of species and widest distribution. It is distributed in the rivers of India, Asia, Africa, Malayasia Archipelago and North America. *Trionyx gangeticus* is found in the rivers of North India. Cretaceous to Recent.

5. *Testudo* (Giant Turtle)

Classification : Same as that of *Chelone*.

Genus ... *Testudo*

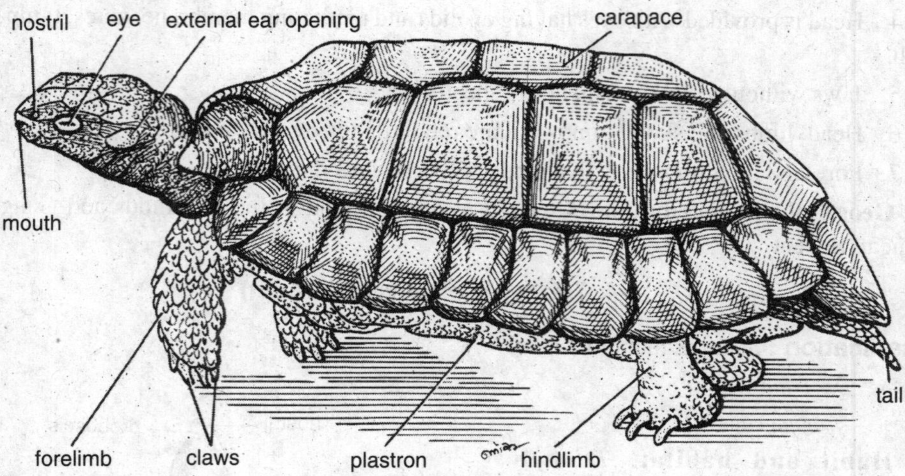

Fig. 13.5. *Testudo.*

Habit and habitat. *Testudo* is found in freshwater or salt water or terrestrial and can live for a long time without water. It is herbivorous and also feeds on worms and insects.

Comments :

1. *Testudo* (Fig. 13.5) is commonly called **land-tortoise** or **giant turtle**.
2. It is about 30 cm long, the males being smaller than females.
3. Body is divided into head, neck, trunk and tail.
4. The bony shell is ovoid with an outer leatnery skin or cornified scutes. When withdrawing its head into the shell, the neck is bent like an S.
5. **Carapace** is very convex with no nuchal plate, **plastron** is concave, on the shell are large yellow stars both on central and marginal scutes.
6. Head bears mouth, eyes with eyelids, nares and external ear openings behind eyes.
7. Feet are for walking on land, the number of phalanges reduced to two in each digit; digits end in horny claws and are not webbed.
8. Jaws devoid of teeth but have stout cornified sheath.
9. *Testudo* is known to live up to 100 years or more.

Geographical distribution. *Testudo* has a wide distribution in tropical and temperate regions of Asia, Africa and Europe. *Testudo elegans* is found in India and Sri Lanka in dry grassy-places, it goes to water in hot weather and hibernates in cold season. Jurassic to Recent.

6. Sphenodon (Tuatara)

Classification :

Phylum	... **Chordata**	→	Dorsal tubular nerve cord, notochord and paired gill-slits are present.
Group	... **Craniata**	→	Definite head, cranium with brain.
Subphylum	... **Vertebrata**	→	Notochord is replaced by vertebral column; two pairs of appendages; circulatory system closed; hepatic portal system present; blood red containing R.B.Cs.
Division	.... **Gnathostomata**	→	Jaws and paired appendages are present.

Class	... **Reptilia**	→ Cold blooded, terrestrial or aquatic vertebrates; single occipital condyle; vertebrae gastrocentrous; respiration by lungs; heart with two auricles and incompletely divided ventricle; right and left aortic arches complete and functional, cranial nerves 12 pairs and embryo with amnion and allantois.
Subclass	... **Diapsida**	→ Two temporal vacuities on each side; post-orbital and squamosal usually meet between temporal vacuities.
Superorder	... **Lepidosauria**	→ Two temporal vacuities present; anterior orbital vacuities absent; post-temporal fenestrae usually present; humerus with two foramina.
Order	... **Rhynchocephalia**	→ Two vacuities; teeth acrodont; an epipteryoid present; quadrate fused; parietal foramen present; vertebrae amphicoelous.
Genus	... *Sphenodon*	
Species	... *punctatum*	

Habit and habitat. *Sphenodon* leads a semiaquatic life and lives in burrows sharing with various kinds of petrels. It is sluggish, nocturnal and carnivorous feeding on insects, molluscs and other small invertebrates.

Comments :

1. *Sphenodon punctatum* or *Hatteria* (Fig. 13.6) is commonly known as **tuatara**.
2. The animal grows to a length of 75 cm, having a general appearances like a stout lizard. Body is divided into head, neck, trunk and tail.
3. The colour of the skin is dark olive green with small white or yellowish spots on the sides.
4. A series of slightly erectile yellowish spines form a **dorsal crest** in the mid-dorsal line from the head to the tail.
5. The belly is covered with numerous scales arranged in transverse rows.
6. Head bears a mouth, large photosensitive eyes with eyelids, nostrils and external ear openings behind the eyes.
7. Two pairs of pentadactyle limbs with clawed digits are present, adapted for walking. Humerus has an entepicondylar foramen and also an ectepicondylar foramen for radial nerve. Carpus has the primitive number of ten bones.
8. Tail is thick and slightly compressed laterally and capable of regeneration.
9. External copulatory organs are absent.

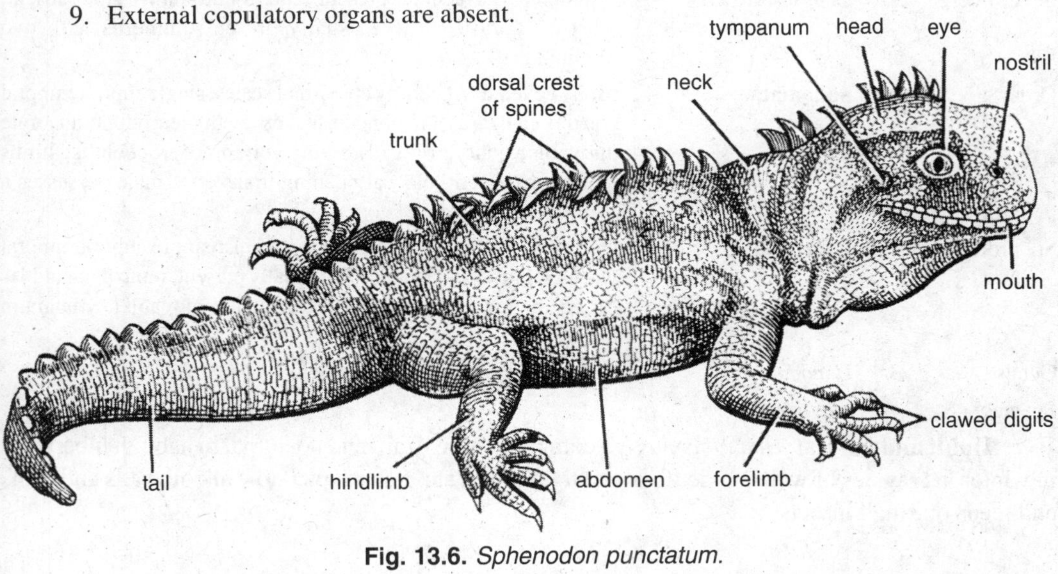

Fig. 13.6. *Sphenodon punctatum.*

10. Skull with two fossae, quadrate fixed, prefrontals are separate and teeth are **acrodont**. The temporal region is bridged by three **temporal arches**.

11. Vertebrae are **amphicoelus**.

12. Sternal and abdominal ribs are present.

13. Hard white shelled eggs are laid in holes in the ground. Eggs hatch in 13 months.

Sphenodon is a living fossil and survived from Permian to Recent. It is approaching towards extinction and, hence, protected by law. It has certain primitive features like two temporal fossae, vertebrae amphicoelous, pineal eye, ribs with uncinate processes, vomerine teeth in youngs, upper jaw with horny beak and absence of copulatory organ in male. It is near to the diapsid ancestor reptiles.

Geographical distribution. *Sphenodon punctatum* was widely distributed on the main islands of New Zealand, but it is now on the verge of extinction and is found only on some islands of New Zealand. Permian to Eocene and Recent.

7. *Hemidactylus* (Common House Lizard)

Classification :

Phylum	...	**Chordata**	→	Dorsal tubular nerve cord, notochord, and paired gill-slits are present.
Group	...	**Craniata**	→	Definite head, cranium with brain.
Subphylum	...	**Vertebrata**	→	Notochord is replaced by vertebral column; two pairs of appendages; circulatory system closed; hepatic portal system present; blood red containing R.B.Cs.
Division	...	**Gnathostomata**	→	Jaws and paired appendages are present.
Class	...	**Reptilia**	→	Cold blooded, terrestrial or aquatic vertebrates; single occipital condyle; vertebrae gastrocentrous; respiration by lungs; heart with two auricles and incompletely divided ventricle; right and left aortic arches complete and functional; cranial nerves 12 pairs and embryo with amnion and allantois.
Subclass	....	**Diapsida**	→	Two temporal vacuities on each side; post-orbital and squamosal usually meet between temporal vacuities.
Superorder	...	**Lepidosauria**	→	Two temporal vacuities present ; anterior orbital vacuities absent; post-temporal fenestrae usually present; humerus with two foramina.
Order	...	**Squamata**	→	Body covered with horny epidermal scales; single supra-temporal vacuity in lizards and none in snakes; teeth pleurodont; quadrate movable; quadratojugal absent; vertebrae procoelous; limbs present or absent; cloacal opening transverse; male possesses a pair of eversible copulatory organs.
Suborder	...	**Lacertilia**	→	Terrestrial, arboreal or burrowing forms; single supra-temporal vacuity; limbs pentadactyle and usually present; rami of mandibles fused in front; sternum present; eyelids movable; tympanum present.
Genus	...	*Hemidactylus*		
Species	...	*flaviviridis*		

Habit and habitat. *Hemidactylus* is a common house lizard and nocturnal in habit. It hibernates in winter in crevices of walls, under logs and light tubes, etc. It is adapted to walk on walls and roofs and feeds on small insects.

Comments :

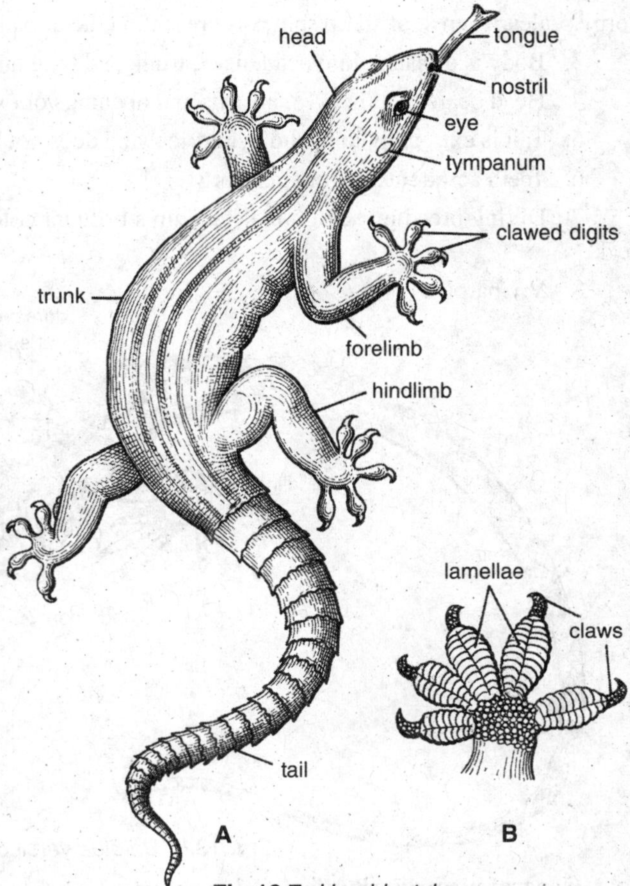

Fig 13.7. *Hemidactylus.*

A—Dorsal view; B—Ventral view of foot.

1. *Hemidactylus* (Fig. 13.7) is commonly known as **wall-lizard** or **common house lizard**. It is about 25 cm in length and pale green in colour.
2. Body is covered by scaly integument. Scales are minute and smooth. Body is divided into head, neck, trunk and tail.
3. Head is broad and flat and bears eyes with fused eyelids, nostrils, ear openings and mouth.
4. Tongue is short, sticky, slightly notched and protrusible.
5. Limbs are pentadactylous. Digits are clawed and dilated with two rows of ridged lamellae on the lower surface for the movement on ceilings and smooth surfaces.
6. Tail is moderately long and can break off at an unossified part of the caudal vertebrae. This phenomenon is known as **autotomy**. The tail possesses a great power of regeneration.
7. Only one supra-temporal arch is present in the skull. Quadrate bone is movable.
8. Vertebrae are amphicoelous.
9. Parietal organ is present.
10. Oviparous. It lays hard-shelled round eggs.
11. Male has several femoral pores on the underside of thighs.

Geographical distribution. *Hemidactylus* is found in India, Sri Lanka, China, Europe, Africa and U.S.A., and also common to all other warmer countries.

Another genus *Gecko* (Wall lizard) is especially found in Asian and European countries. It is like *Hemidactylus* in characters.

8. *Calotes* (Garden Lizard)

Classification : Same as that of *Hemidactylus.*

Genus ... *Calotes*

Species ... *versicolor*

Habit and Habitat. *Calotes versicolor* is arboreal in habit, found on trees in gardens. It is oviparous and feeds on insects. Remarkable for change of colour. The normal colour is olive-green but in courtship or threat the body becomes yellow and neck and sides of the head become red.

Comments :

1. *Calotes versicolor* (Fig. 13.8) is commonly known as **garden-lizard** or **girgit** and found in open fields and hedges.

2. Body is long, slender measuring 35 cm in length and covered with imbricate epidermal horny scales. A crest of sharp spines is present on the dorsal surface of the neck and back.

3. Body is divisible into head, neck, trunk and long tail.

4. Head bears mouth, eyes, nostrils and openings of external ear.

5. Tail is extremely long and cylindrical and does not break.

6. Teeth acrodont and tongue short.

7. During breeding season male acquires brilliant colours and gular pouch also develops in neck.

8. Viviparous.

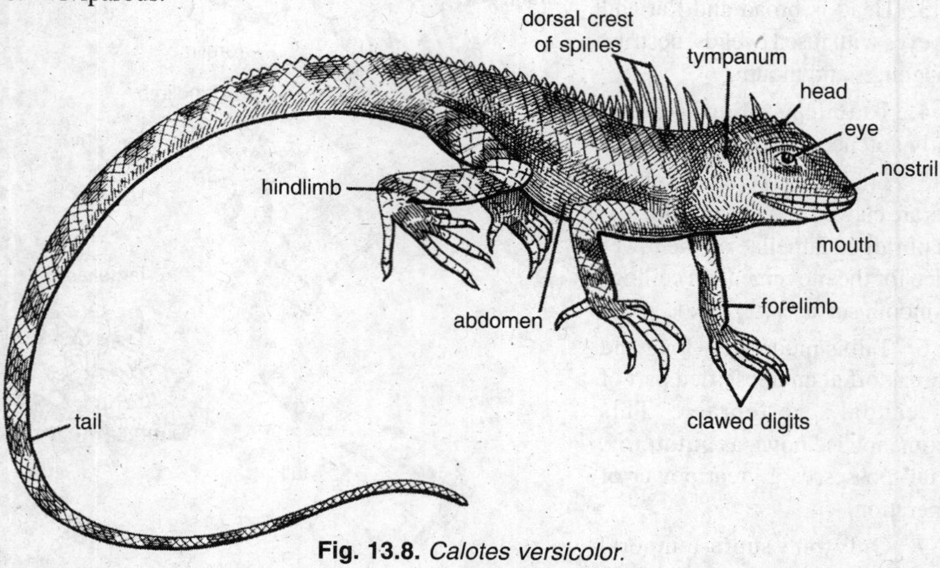

Fig. 13.8. *Calotes versicolor.*

Geographical distribution. *Calotes versicolor* is found in India, Malaya islands, Afghanistan and Southern China.

9. *Crotaphytus* (Collared Lizard)

Classification : Same as that of *Hemidactylus.*

Genus ... *Crotaphytus*

Species ... *collaris*

Habit and habitat. *Crotaphytus collaris* is a terrestrial fast running lizard and can jump like a frog. It is nocturnal, carnivorous, capable of changing colour and oviparous.

Comments :

1. *Crotaphytus collaris* (Fig. 13.9) is commonly known as **collared** or **inflated lizard** inhabiting rocks and open ground.

2. Body is lizard-like divided into head, neck, trunk and tail.

3. Head bears wide mouth, eyes with movable eyelids, nares, and external ear openings.

4. Forelimbs are smaller than hindlimbs and both are clawed.

5. Tongue is small and teeth are pleurodont.

6. Vertebrae are procoelous.

7. Tail is long and pointed. It exhibits **autotomy.** It can break off its tail at the time of danger.

8. Sexual dimorphism is present. Male is green having yellow spots, while female is slaty-grey. Male possesses a **double black collar** and female has a **slaty-grey narrow collar**.

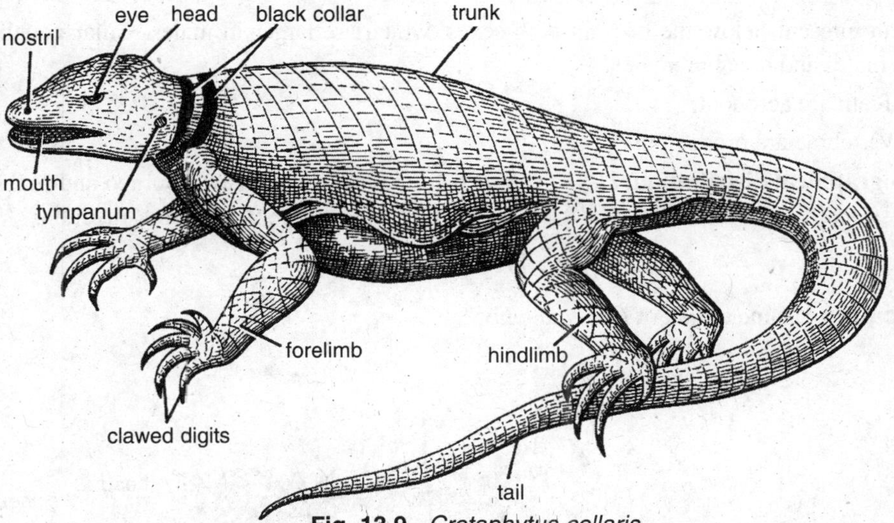

Fig. 13.9. *Crotaphytus collaris.*

Geographical distribution. *Crotaphytus* is found in Southern and Central America, Northern Mexico and West Indies.

10. *Draco* (Flying Dragon)

Classification : Same as that of *Hemidactylus.*

Genus ... *Draco*

Species ... *volans*

Habit and habitat. *Draco* is arboreal living on trees and feeds upon small insects.

Comments :

1. *Draco* (Fig. 13.10) is commonly known as **flying-dragon** or **flying lizard**.

2. Body is dorso-ventrally compressed and measuring about 25 cm in length of which 12 cm is taken by tail. Body is divided into head, neck, trunk and long, pointed tail.

3. The sides of the body, between the fore and hindlimbs, extend as a pair of large wing-like membranes, the **patagia** , which are supported by five or six much elongated posterior ribs. These are used as parachutes during flight. These wing-like membranes can be folded up like a fan.

4. Head bears wide mouth, eyes with eyelids, nostrils and openings of external ear.

5. On the throat are three pointed hooks, a short one on either side and a long one in the middle.

6. Sexual dimorphism exists. Male has a small nuchal crest. Distensible gular

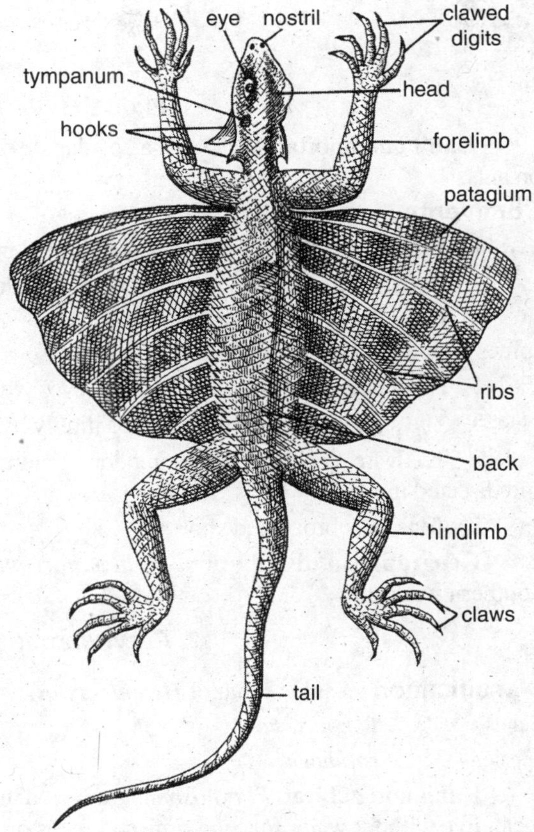

Fig. 13.10. *Draco volans.*

pouches are present below the neck in both sexes, which are larger in males. Gular pouches are orange in males and blue in females.

7. Teeth are acrodont.
8. Vertebrae are procoelous.

Geographical distribution. *Draco volans* is found in Sumatra, Java, Borneo and Malayasia. *Draco dussumieri* occurs in Chennai. *Draco* is also found in Europe, Africa and Australia.

11. *Moloch*

Classification : Same as that of *Hemidactylus.*

Genus ... *Moloch*

Species ... *horridus*

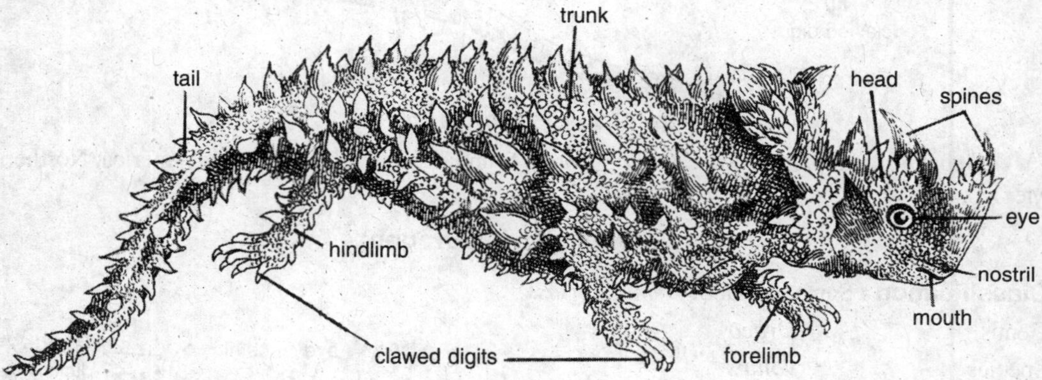

Fig. 13.11. *Moloch horridus.*

Habit and habitat. *Moloch* is also adapted for dry sandy habitat. It is insectivorous and feeds on ants.

Comments :

1. Body of *Molloch* (Fig. 13.11) is divided into head, neck, trunk and tail.
2. Body is much depressed and covered with small scales and large scaly spines.
3. Head bears two horn-like spines, trunk with thicker spines and tail and limbs also bear spines which are smaller in size.
4. Head bears small mouth, nostrils, eyes with eyelids and openings of external ears.
5. The skin is extremely rough and highly hygroscopic that absorbs moisture.
6. Teeth are pleurodont and homodont. Lateral teeth of the upper jaw are implanted horizontally and directed inwards.
7. Tongue is broad and short.

Geographical distribution. *Moloch horridus* is found in the sandy districts of Western and Southern Australia.

12. *Phrynosoma* (Horned Toad)

Classification : Same as that of *Hemidactylus.*

Genus .... *Phrynosoma*

Species .... *cornutum*

Habit and habitat. *Phrynosoma* is adapted in desert habitat and lives on dry and sandy places. It can live without water for long time and feeds on ants and small insects. During night it buries in the sand but head remains exposed.

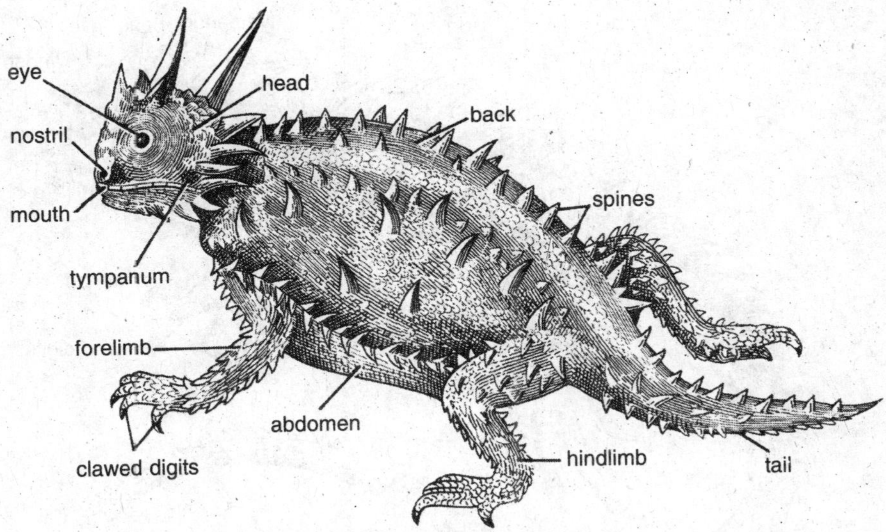

Fig. 13.12. *Phrynosoma cornutum.*

Comments :

1. *Phrynosoma* (Fig. 13.12) is commonly known as **horned-toad**.
2. Body is much flattened and broad and covered with larger and smaller strongly keeled scales. Head scales are large horn-like. Body is divided into head, neck, trunk and short tail, measuring about 12 cm.
3. The under-parts are covered with small very regular keeled scales.
4. The colour of the upper part is a mixture of yellow, grey, brown and black.
5. Head has five spikes on each side, one post-orbital, three temporal and one occipital. Head bears mouth, eyes, nostrils guarded by valves and ear openings.
6. The sides of the lower jaw project in the shape of prominent ledges and are protected by a series of small spines.
7. Both the sexes have a long row of pores on the under surface of the thighs.
8. Tongue is short, fleshy and protrusible
9. Teeth are pleurodont or homodont.
10. Vertebrae are procoelous.
11. Viviparous.
12. *Phrynosoma* when disturbed squirts a tiny stream of blood from the eyes extending up to 60 cm that protects it from its enemies.

Geographical distribution. *Phrynosoma* is found in Western half of the United States and Central America. *Phrynosoma cornutum* ranges from Illinois through Kansas and Texas to Northern Mexico.

13. *Iguana*

Classification : Same as that of *Hemidactylus.*

Genus ... *Iguana*

Species ... *tuberculata*

Habit and habitat. *Iguana tuberculata* is adapted for fast running and climbing. It lives on trees and feeds upon leaves, fruits, some insects and some invertebrates.

Comments :

1. *Iguana tuberculata* (Fig. 13.13) is arboreal living on trees and reaches the length of one and a half metre to two metres.

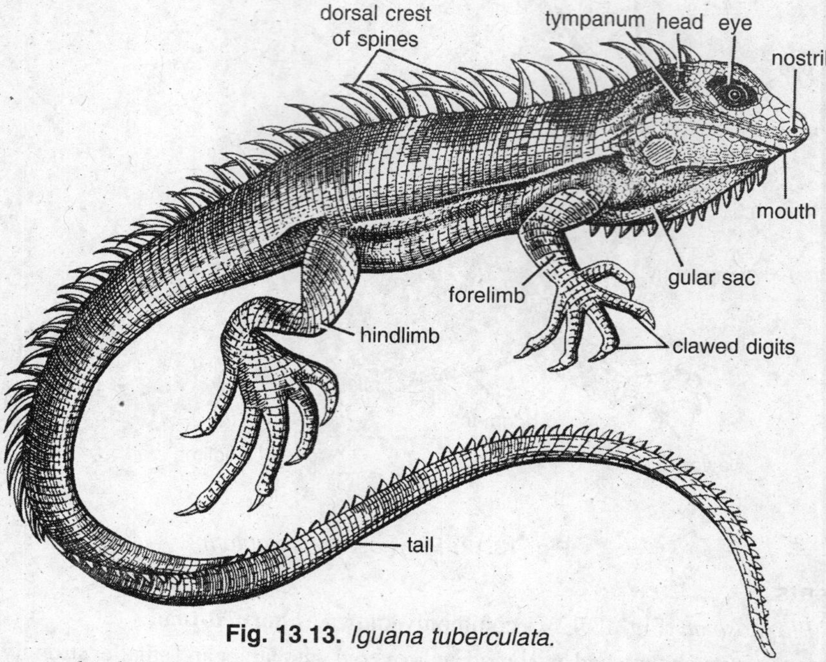

Fig. 13.13. *Iguana tuberculata.*

2. The general colour of upper-parts is a mixture of green and black and under-parts are pale greenish or whitish.

3. The body and tail are laterally compressed and covered with very small scales.

4. Head bears wide mouth, small eyes with eyelids, nostrils and openings of external ear. Midline spines present over jaw flap of skin.

5. The neck and back carry lance-like spines along mid-dorsal line extending upon tail.

6. Digits of fore- and hindlimbs are clawed.

7. Tongue is thick and non-portractile.

8. Teeth are pleurodont. The lateral teeth are finely serrated.

9. Both the sexes have long rows of femoral pores.

10. The delicate flesh of *Iguana* is much esteemed as food.

Geographical distribution. *Iguana tuberculata* is found in South and Central America and West Indies (Tropical countries.)

14. *Heloderma* (Gila Monster)

Classification : Same as that of *Hemidactylus.*

Genus ... *Heloderma*

Species .... *suspectum*

Habit and habitat. *Heloderma* inhabits dry localities, lives in concealment during the day and comes out in the evening. It is the only **poisonous lizard**. It is carnivorous, feeding on worms, centipedes, frogs, and the eggs of large lizards. Oviparous.

Comments :

1. *Heloderma suspectum* (Fig. 13.14) is commonly called **gila-monster** or **beaded-lizard** or **Hella monster**.

2. Body is stout, depressed and reaches about 60 cm in length and divided into head, neck, trunk and tail.

3. The skin of the upper surface is granular with many irregular tubercles of blackish brown and yellow or orange colour patches.

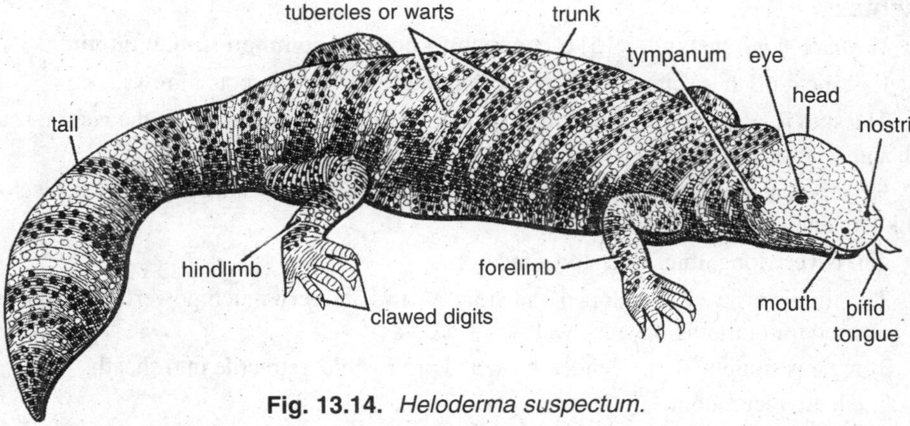

Fig. 13.14. *Heloderma suspectum.*

4. Head is short, stout with a wide mouth, eyes, nostrils and external ear openings.
5. Tail is short, thick with a fat reserve.
6. The limbs are short but strong and well-developed, digits clawed.
7. The tongue is smooth, protrusible and bifid in front.
8. Teeth are pleurodont, fang-like, curved with slightly swollen bases. Teeth have groove on their anterior surfaces through which labial glands secrete the poison open near the bases of the teeth of the lower jaw. Poison is fatal to small animals but rarely to humans.
9. Vertebrae are procoelous.

Geographical distribution. *Heloderma suspectum* is found in deserts in New Mexico and Arizona.

15. *Varanus* (Common Indian Monitor)

Classification : Same as that of *Hemidactylus.*

Genus ... *Varanus*

Species ... *monitor*

Habit and Habitat. *Varanus* lives on dry places beneath stones and rocks and burrowing in habit. It remains active in night. It lays eggs in a nest in the ground. Carnivorous, feeding on reptiles, mammals and deadbodies, etc.

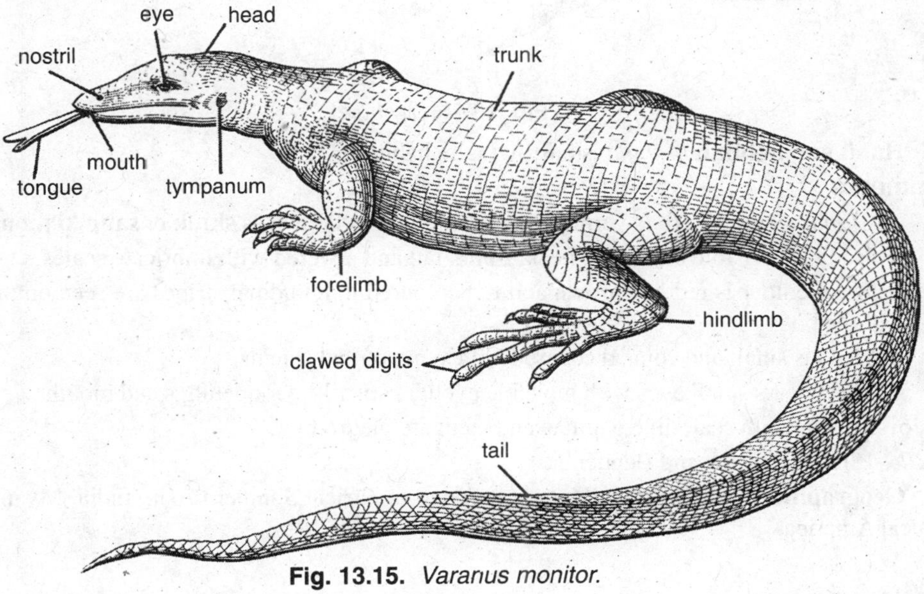

Fig. 13.15. *Varanus monitor.*

Comments :

1. *Varanus monitor* (Fig. 13.15) is commonly known as **common Indian monitor**.
2. It is about 60 to 90 cm and of brownish colour with darker spots in rows.
3. The skin is covered with very small scales and tubercles above, while the ventral scales are squarish and arranged in transverse rows.
4. Head is flattened and bears wide mouth, fixed eyes, nostrils and ear openings. Neck is long and trunk large and stout.
5. Tail is very long, thick and stores fat.
6. The limbs are well-developed and stout. Claws are very much powerful and as such the animal was used for climbing on steep walls.
7. Tongue is smooth, long, slender, bifid and protrusible, retractile into sheath.
8. Teeth are pleurodont.
9. Vertebrae are procoelous,

Geographical distribution. *Varanus* is found in Africa, Asia and Australia. *Varanus monitor* is found in India.

16. *Mabuia* (Skink)

Classification : Same as that of *Hemidactylus.*

Genus ... *Mabuia*

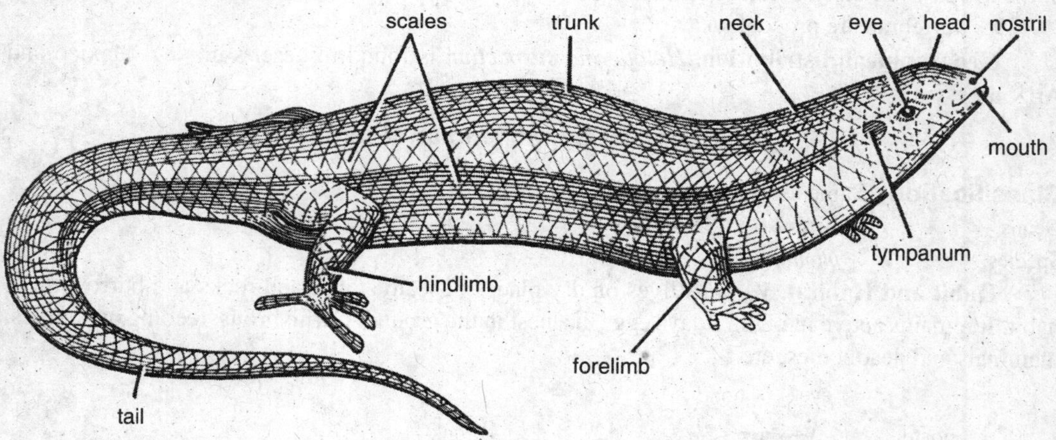

Fig. 13.16. *Mabuia.*

Habit and habitat. *Mabuia* is burrowing in habit.

Comments :

1. *Mabuia* (Fig. 13.16) is vermiform and commonly known as **skink** or **sanp ki mousi**.
2. Body is divisible into head, neck, trunk, tail and covered with imbricate scales.
3. Body colour is reddish brown and red coloured longitudinal stripes are seen on the dorsal surface of the body.
4. Head is small and conical covered with symmetrical shields.
5. Head bears small eyes with movable eyelids, nostrils, ear openings and mouth.
6. Tongue with scale-like papillae and teeth are pleurodont.
7. Limbs are weak and slender.

Geographical distribution. *Mabuia* is found in Africa, Southern Asia, India, Myanmar and Tropical America.

17. *Chamaeleon*

Classification : Same as that of *Hemidactylus*.

Genus ... *Chamaeleon*

Species ... *vulgaris*

Habit and habitat. *Chamaeleon* is an arboreal lizard which lives on trees and capable of changing colour of the skin according to the surroundings. It feeds on insects. Viviparous.

Comments :

1. *Chamaeleon's* (Fig. 13.17) body is laterally compressed and covered with minute tubercles or granules (scales).

2. Body is divided into head, neck, trunk and long tail.

3. Head has wide mouth, large eyes, nostrils, ear openings and backwardly directed hood (helmet) formed by squamosal and occipital bones. Proatlas is present between skull and atlas vertebra.

4. Eyes are large but the eyelids are united into one fold with a small central opening for pupil and adapted for binocular vision. The right and left eye can be moved separately from each other.

5. Tongue is club-shaped, extremely protrusible and covered with a sticky secretion and can be shot out for catching insects.

6. Limbs are relatively large and very slender. The digits of each limb form groups of three and two digits (syndactylous) which are used in grasping twigs.

7. Tail is long and **prehensile**.

8. Teeth are **acrodont**.

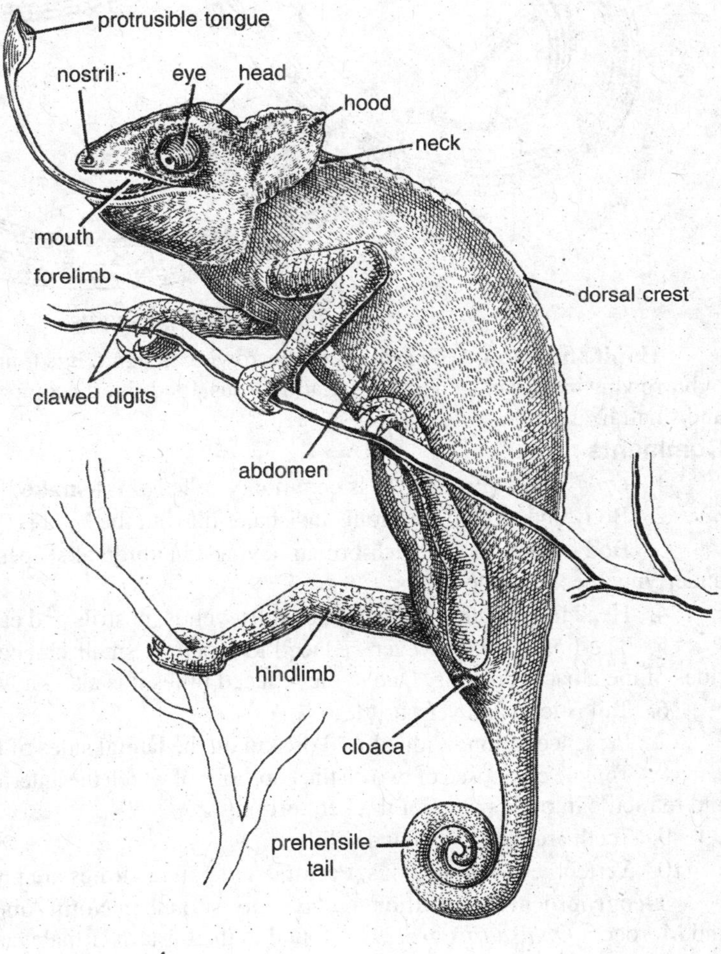

Fig. 13.17. *Chamaeleon*.

9. Vertebrae are procoelous.

10. Clavicles and interclavicles are absent, costal sternum is present and quadrate immovable.

11. Jacobson's organs are absent.

12. Lungs have air sacs.

Geographical distribution. *Chamaleons* are commonly found in Africa, Madagascar, South Arabia, Europe, South India and Sri Lanka. *Chamaeleon calcaratus* is found in the Southern India and Sri Lanka but it is not very common.

18. *Ophisaurus* (Glass Snake)

Classification : Same as that of *Hemidactylus*.

Genus .... *Ophisaurus*

Species ... *gracilis*

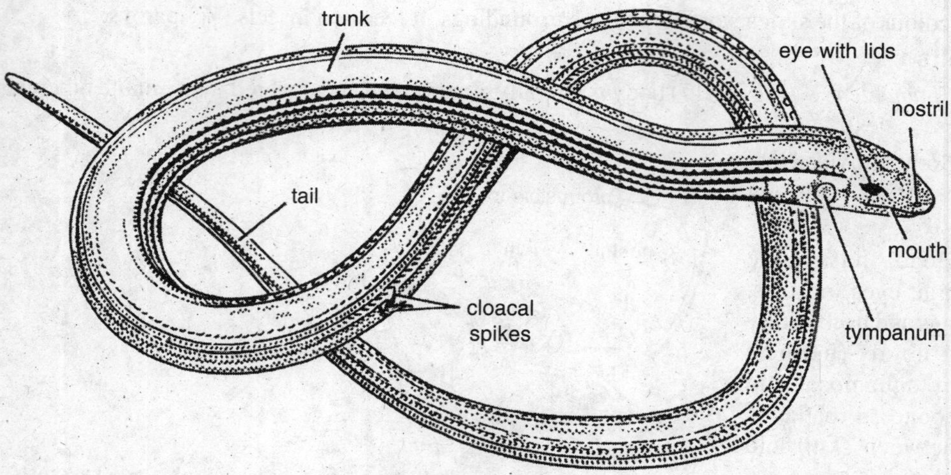

Fig. 13.18. *Ophisaurus.*

Habit and habitat. *Ophisaurus* is burrowing in habit. It is found inhabiting in bushy localities or burrowing in the sand. Oviparous. Carnivorous, food chiefly comprises snails, mice, insects, worms and small lizards.

Comments :

1. *Ophisaurus* (Fig. 13.18) is commonly called **glass-snake**.
2. Body and tail are very long and snake-like but the head is that of typical lizard.
3. Body colour is greenish-brown having longitudinal stripes and covered with bony plates underlying the imbricate scales.
4. Head has mouth, eyes with movable eyelids, nostrils and ear openings.
5. The limbs are, however, reduced to a pair of small **cloacal-spikes**, half concealed at the sides of the cloacal opening. Due to the reduced limbs it is also known as **"limbless lizard"**.
6. Tail is long and very brittle.
7. Presence of longitudinal fold of skin on the lateral sides of the body.
8. Tongue composed of two distinct portions of which the anterior is thin, emarginated, extensile and retractile into the posterior thicker portion.
9. Teeth are pleurodont and solid.
10. Vertebrae are procoelous. Pectoral and pelvic girdles are present.

Geographical distribution. *Ophisaurus* is found in North America, South Russia, Asia Minor and Morocoo. *Ophisaurus gracilis* is found in the Eastern Himalayas and Myanmar.

19. *Rhineura* (Worm Lizard)

Classification : Same as that of *Hemidactylus.*

Genus ... *Rhineura*

Habit and habitat. *Rhineura* is burrowing animal.

Comments :

1. *Rhinuera* (Fig. 13.19) is commonly called **worm-lizard**.
2. Body is worm-like resembling the earthworms.
3. The skin is covered with minute scales which are fused and form transverse numerous rings.

4. The general colour of the body is pale.

5. Head bears scales which are not fused and contains subcutaneous and rudimentary eyes due to burrowing mode of life. Eyelids are fused over the eyes, nostrils, tympanum and mouth present.

6. Limbs are absent.

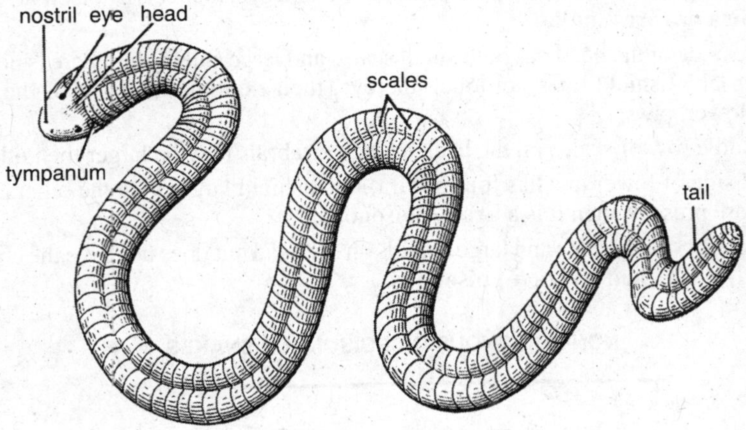

Fig. 13.19. *Rhineura.*

7. Tongue is long and bifid anteriorly.
8. Pectoral and pelvic girdles are rudimentary or absent.
9. Occipital condyle double, *i.e.*, skull dicondylic.
10. Vertebrae are procoelous.

Geographical distribution. *Rhineura* is found in Florida.

DISTINGUISHING FEATURES OF POISONOUS AND NON-POISONOUS SNAKES

The majority of snakes are non-poisonous and harmless animals. Some of the common poisonous snakes of India are cobra, king-cobra, krait, pit-viper, Indian viper, rattle-snake, russel's snake, coral-snake, and sea-snakes. Non-poisonous snakes are python, trinket snakes and racer snakes. In India snakes are worshipped on Nagpanchmi day. Poisonous snake's venom contains two types of toxins: **neurotoxins** and **haemotoxins**.

Neurotoxins act on motor nerves and cause paralysis. Venom of cobra develops convulsion and paralysis.

Haemotoxins develop haemmorrhage, destruction of tissue cells and red cells, blood vessels resulting in haemorrhage. In both cases patient feels difficulty in respiration, haemorrhage, dimness in vision, pulse rate increases, weakness, convulsions may be associated with vomiting. To counteract the effect of haemotoxins, **benadryl** acts as antidote. Medicines (antivenom) are injected intravenously. The identification of poisonous and non-poisonous snakes is very simple. The following key will serve to distinguish **poisonous snakes** from **non-poisonous snakes :**

1. If it is a **marine snake** with the tail laterally compressed it is poisonous.

2. In a **terrestrial snake** the tail is rounded or cylindrical and not compressed. **Examine its ventral scales.**

(*i*) If all the ventral scales are small or the ventral scales are somewhat broad, then it is a non-poisonous snake.

(*ii*) If the ventral scales are large transverse plates extending fully across the ventral side of belly, the snake may be poisonous or non-poisonous. Now examine the dorsal scales of the head.

(*a*) If all the dorsal scales of the head are small, then it may be a viper (poisonous).

(*i*) If there is a loreal-pit between the nostril and the eye, then it is **pit-viper** (poisonous).

(*ii*) If the **sub-caudals** are double and there is a **loreal-pit,** then it is a **russel's viper**. (poisonous).

(*b*) If dorsal side of the head has both small scales and large shields, the snake may or may not be poisonous, to ascertain look at side of the head.

3. If the **third supra-labiel shield** touches the nostril and eye, then it is a poisonous snake, may be **cobra**, **king-cobra** or **coral snake**.

4. If the upper side of the head has both small scales and large shields but there is no loreal-pit, and the third supra-labial shield does not touch the eye, then examine the back of the snake and ventral side of the lower jaw.

(*i*) The middle row of scales on the back called vertebrals may be larger then others.

(*ii*) Ventral side of lower jaw has fourth infra-labial shield larger than the others. If both (*i*) and (*ii*) characters are present, then it is a **krait** (poisonous).

5. If the snake has small scales and large shields on the head but does not have the characters of cobra, krait or coral-snake, then it is **non-poisonous**.

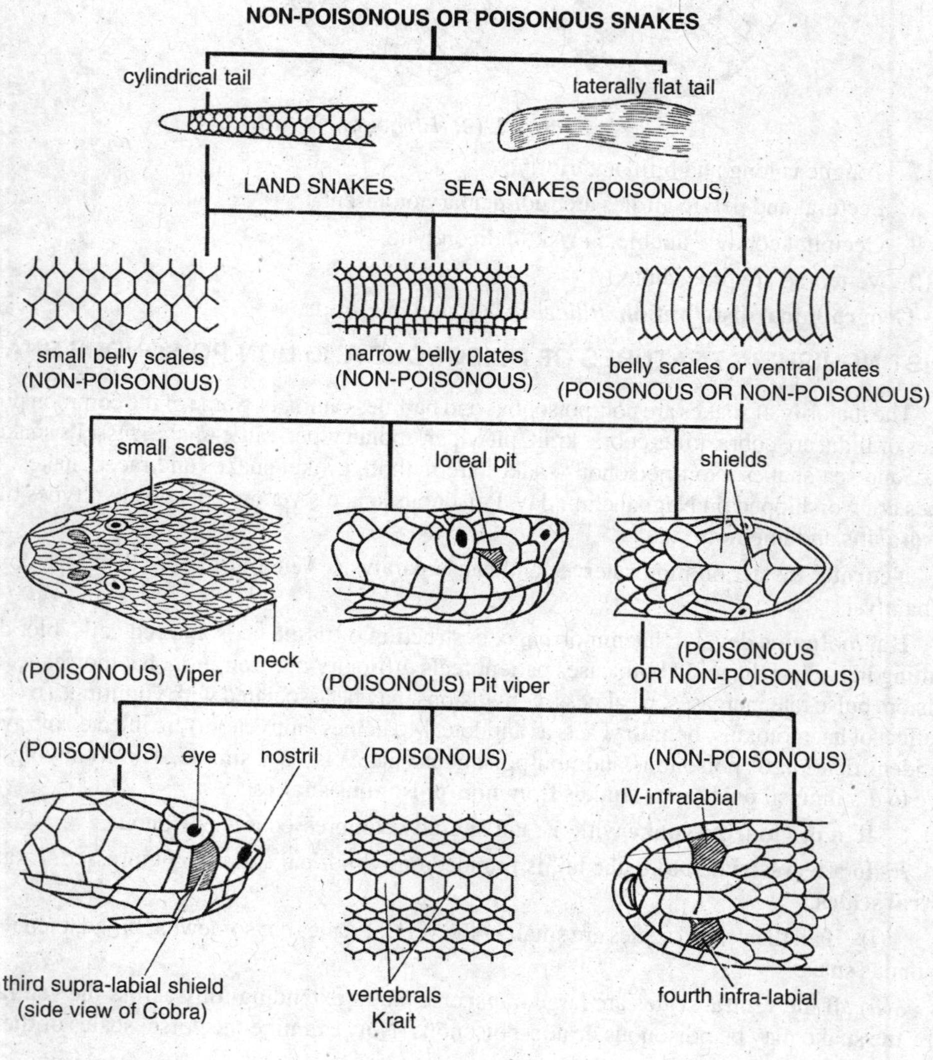

Fig. 13.20. Key to identification of poisonous and non-poisonous snakes.

Key for Identification of Poisonous and Non-poisonous Snakes

Snakes

- I–Tail flat and laterally compressed.
  - Sea snakes → **Poisonous**
- II – Tail cylindrical, rounded and not compressed.
  - Terrestrial or land Snakes → **Poisonous** or **Non–poisonous**
    - A. Small scales on the belly and on the back. → **Non–poisonous**
    - B. Belly scales not broad enough to extend right across it. → **Non-Poisonous**
    - C. Ventrals broad covering the entire width of the belly.
      - Small scales on the head. Pit absent. → **Viper** → **Poisonous**
      - Small scales or shields on the head and a loreal-pit between the nostril and eye. → **Pit-viper** → **Poisonous**
      - Shields on the head.
        - Ventrals not enlarged. Third supra-labial shield (upper-lip shield) touches the eye and nose shield. → **Poisonous**
          - Neck with hood and markings. → **Cobra** → **Poisonous**
          - Neck without hood, coral spots on the belly. → **Coral snakes** → **Poisonous**
        - Vertebrals (*i.e.*, scales on the middle of the back) are enlarged, subcaudals undivided and bands and half rings across the back. Fourth infra-labial is largest. → **Krait** → **Poisonous**
        - None of the characters of cobra, coral snake or krait. → **Non-poisonous**

20. *TYPHLOPS* (BLIND SNAKE)

Classification :

Phylum	...	**Chordata**	→ Dorsal tubular nerve cord, notochord and paired gill-slits are present.
Group	...	**Craniata**	→ Definite head with cranium and brain.
Subphylum	...	**Vertebrata**	→ Notochord is replaced by vertebral column; circulatory system closed; hepatic portal system present.
Division	...	**Gnathostomata**	→ Jaws and paired appendages are present.
Superclass	...	**Tetrapoda**	→ Paired limbs, lungs, cornified skin and bony skeleton.
Class	...	**Reptilia**	→ Cold blooded; single occipital condyle; vertebrae gastrocentrous; heart with two auricles and incompletely divided ventricle; crainal nerves 12 pairs and embryo with amnion and allantois.
Subclass	...	**Diapsida**	→ Two temporal vacuities on each side in skull ; post-orbital and squamosal usually meet between temporal vacuities.
Superorder	...	**Lepidosauria**	→ Two temporal vacuities present ; anterior orbital vacuities absent ; post-temporal fenestrae usually present ; humerus with two foramina.
Order	...	**Squamata**	→ Body covered with horny epidermal scales; single supra-temporal vacuity in lizards and none in snakes; teeth pleurodont; quadrate movable; quadrato-jugal absent; vertebrae procoelous; limbs present or absent; cloacal opening transverse; male possesses a pair of eversible copulatory organs.
Suborder	...	**Ophidia**	→ Terrestrial or aquatic, arboreal or burrowing; temporal vacuities entirely absent; limbs absent; rami of mandibles united by a ligament; sternum absent; eyelids immovable; tympanum absent; tongue bifid and protrusible, zygosphene and zygantra present in the vertebra.
Family	...	**Typhlopidae**	
Genus	...	*Typhlops*	
Species	...	*braminus*	

Habit and habitat. *Typhlops* is burrowing in habit, oviparous, feeding on worms, insects and their larvae. It is found in rotting vegetation in semitropical and tropical regions.

Comments :

1. *Typhlops braminus* (Fig. 13.21) is commonly called **blind-snake**.
2. The body is cylindrical and covered with uniform imbricate scales. It is about 170 – 180 mm in length.
3. Body colour is brown with a shining chocolate hue above and lighter below all over the body.
4. The head is not distinct from the body and bears mouth, nostrils and eyes covered with scales. Eyelids immovable.
5. The tail is blunt with small point.
6. Few teeth are present in the upper jaw only. Lower jaw toothless.
7. Presence of large rostral, nasal, ocular and preocular shields on the head.
8. Nasal shields are completely divided into an anterior lower and a posterior upper position.
9. Sebaceous glands are present on the body.
10. Some vestiges of pelvic bones are present inside the body. Limbs , ear openings, sternum and urinary bladder absent.
11. Parietal bone is paired. Ectopterygoid and squamosal absent and pterygoid separate from quadrate.
12. Mandibles united anteriorly by ligament.

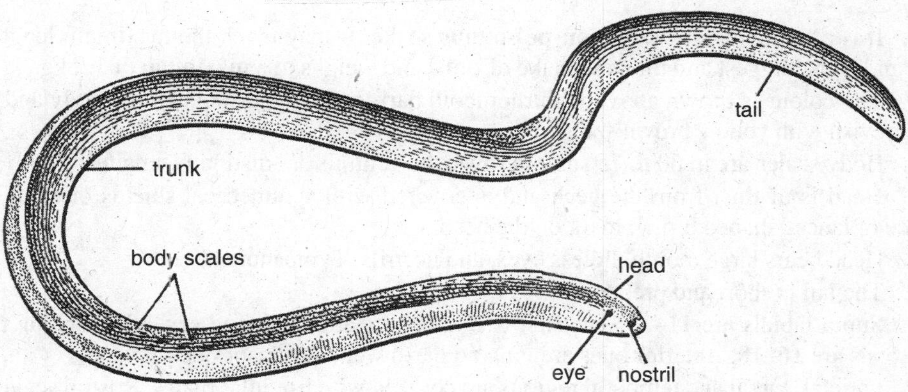

Fig. 13.21. *Typhlops braminus.*

Geographical distribution. *Typhlops braminus* is widely distributed in Asia, Malaya Islands, the islands in the Indian Ocean, in Southern Africa, Australia, America and throughout India.

21. *Python* (Ajgar)

Classification : Same as that of *Typhlops*.

Family	...	**Boidae**
Genus	...	*Python*
Species	...	*molurus*

Habit and habitat. *Python* is terrestrial found on dry, rocky or sandy places and on trees. Pythons are also aquatic found in lakes, ponds and swampy places full of vegetation. They feed upon reptiles, birds and mammals like sheep, goat, deer, cows, etc. They are oviparous.

Comments :

1. *Python molurus* (Fig. 13.22) is commonly called **Indian python** or **Ajgar**.

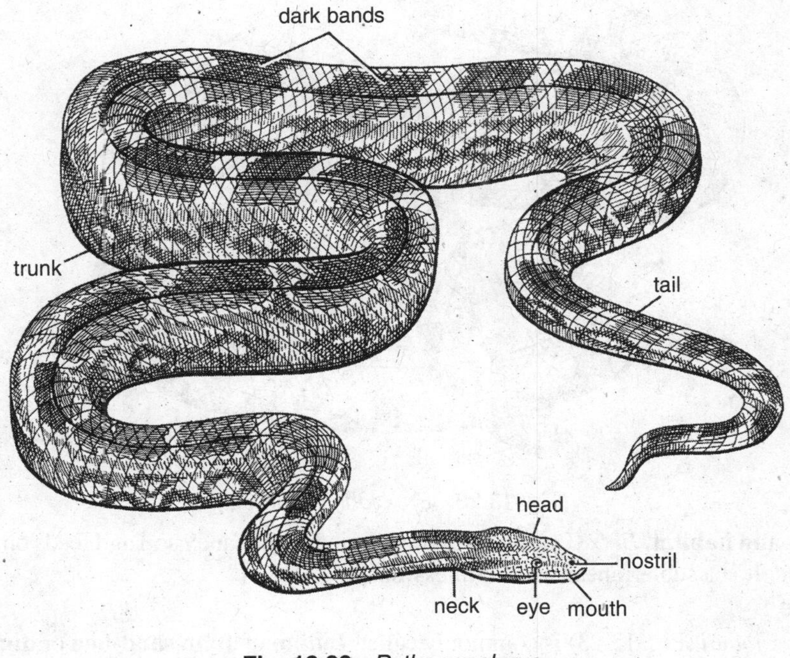

Fig. 13.22. *Python molurus.*

2. It is a large, massive and **non-poisonous** snake. It may reach the maximum length of 10 metres and is the biggest and thickest snake of India and weighs to a maximum of 125kg.

3. The colour is brown above with rhomboid dark-grey edged spots on the body and ventral side is greyish with yellow brown spots.

4. Body scales are in 60 to 75 smooth rows, the ventrals are distinctly smaller.

5. Head is distinct from the neck and is covered with symmetrical shields or small scales. Presence of lancet-shaped brown mark on the head.

6. Head bears large mouth, lidless eyes, and nostrils. Tympanum absent.

7. The tail is short and prehensile.

8. Supra-labials are 11–13. The first two have pits and sixth and seventh touching the eye. Infra-labials are 16–18. Anterior ones are long nd narrow and posterior ones have pits.

9. Parietal, loreal and temporal regions are covered with irregular plates. Rostral scales on the head bear deep pits which are probably thermoreceptive in function.

10. Pterygoids, palatines, maxillae, premaxillae and mandibles possess teeth. Maxillae, palatines and pterygoids are movable.

11. Rudiments of appendages in the form of two claw spurs are present near the cloaca.

Geographical distribution. *Python molurus* is distributed in India and also in Indo-China. *Pythons* are cosmopolitan in distribution except New Zealand.

22. *Eryx* (Indian Sand-Boa)

Classification : Same as that of *Typhlops.*

Family	...	**Boidae**
Genus	...	*Eryx*
Species	...	*johnii*

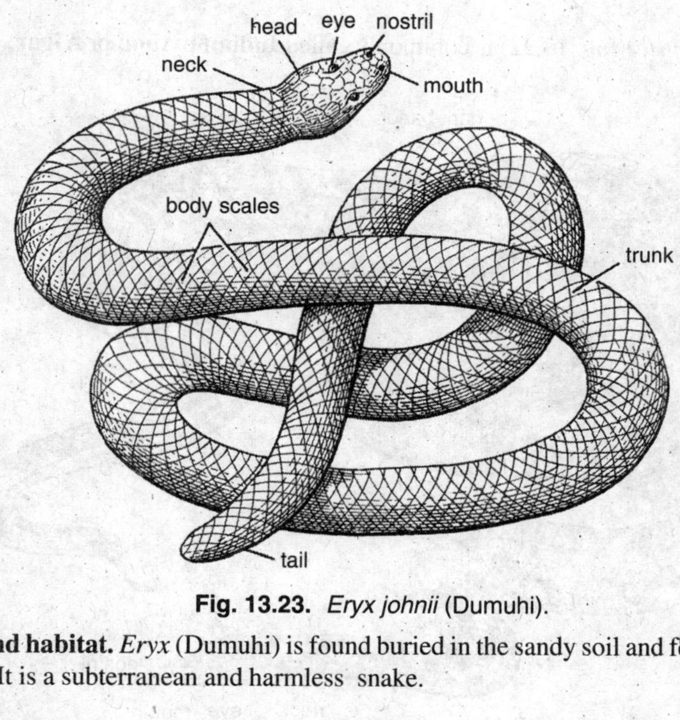

Fig. 13.23. *Eryx johnii* (Dumuhi).

Habit and habitat. *Eryx* (Dumuhi) is found buried in the sandy soil and feeds on frogs, lizards and mice, etc. It is a subterranean and harmless snake.

Comments :

1. *Eryx johnii* (Fig. 13.23) is commonly called **Indian** or **John sand-boa** or **dumuhi** (double -mouthed snake).

2. Body is elongated, measuring about one metre in length and chocolate in colour.

3. Head is not distinguished from the neck and bears very small eyes with a vertical pupil and slit-like nostrils. Tympanum is absent.

4. Tail is blunt and looks like the head end.

5. Only three scales on the head are enlarged and remaining are small. Body scales are 40 – 55 in rows often keeled particularly on the tail.

6. The ventral scales do not run across the body.

Geographical distribution. *Eryx johnii* is distributed in sandy parts of India, Sri Lanka and Africa.

23. *Natrix* (Water-snake)

Classification : Same as that of *Typhlops*.

Family ... **Colubridae**
Series ... **Aglypha**
Genus ... *Natrix*
Species ... *piscator*

Habit and habitat. *Natrix piscator* is found in freshwater ponds and paddy fields in rainy season. It is diurnal, oviparous and feeds on fishes, frogs and tadpoles.

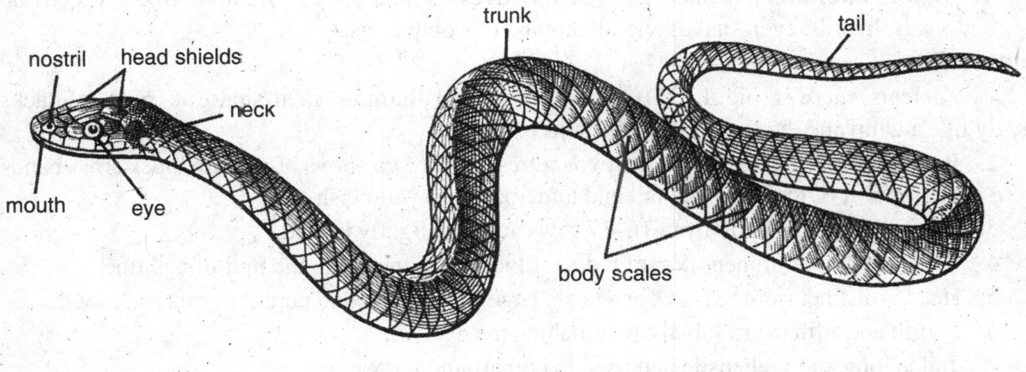

Fig. 13.24. *Natrix piscator.*

Comments :

1. *Natrix piscator* (Fig. 13.24) is commonly called **water-snake** or **pani-sanp**. It is non-poisonous but attacks like poisonous snakes.

2. Body is stout having a flat head, neck, trunk and tail.

3. The colour is olive with black spots arranged in the chess-board manner.

4. Head bears rounded eyes surrounded by white circles, nostrils and wide mouth. Tympanum is absent.

5. Scales are keeled and form 19 rows. Ventrals are plate-like and subcaudals are paired. Head is covered by shields.

6. Supra-labials are 9. Fourth and fifth are touching the eye.

7. Jaws contain teeth. Maxillary teeth 22–28 enlarged posteriorly. Fangs are absent.

8. Facial bones movable, squamosal is loosely attached to skull.

Geographical distribution. *Natrix* is distributed widely in United States, Mexico and Central America, North Africa and India. *N.piscator* is found all over India.

24. *Zamenis = Ptyas* (Dhaman)

Classification : Same as that of *Typhlops*.

Family ... **Colubridae**
Series ... **Agylpha**
Genus ... *Zamenis (Ptyas)*
Species ... *mucosus*

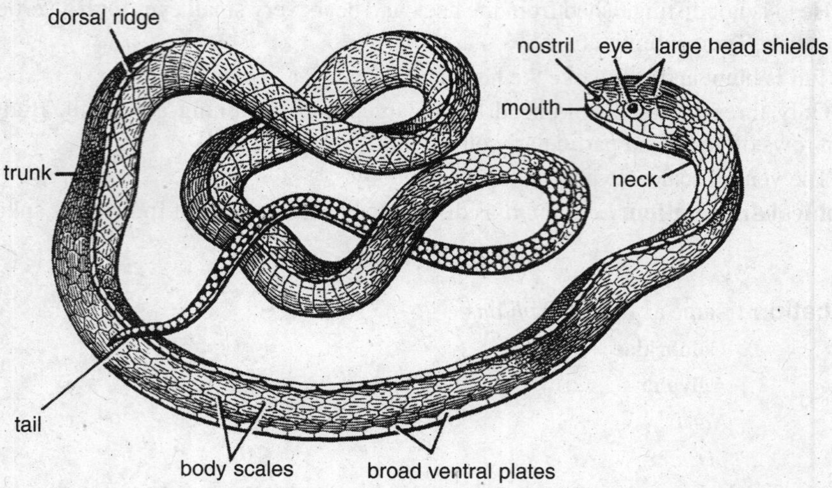

Fig. 13.25. *Zamenis mucosus.*

Habit and habitat. *Zamenis (Ptyas)* mainly lives on land but can climb on trees. It feeds on frogs and toads, lizards, even snakes and mammals. It is oviparous.

Comments :

1. *Zamenis mucosus* (Fig. 13.25) is commonly called **dhaman** or **rat-snake of India**. It attacks fiercely like a whip and, hence, called **rope snake**.
2. It grows to a length of more than two metres. It is brown above often with black cross bands on the posterior part of the body and tail and underparts are yellowish.
3. The scales on the body from 16–17 rows and are slightly keeled.
4. Presence of a prominent **dorsal ridge** of the back bone along the mid-dorsal line.
5. Head is distinct from the neck and bears large eyes with round pupil, nostrils and mouth.
6. Fourth and fifth supra-labials are touching the eye.
7. Tail is long and prehensile and used to coil around the victim.
8. Teeth are present but fangs are absent (non-poisonous).
9. It emits foul odour and secretes black secretion from anal glands.

Geographical distribution. *Zamenis* is very widely distributed. It is found in France, Italy, North-West Africa, America and India. *Z. mucosus* is found all over India in plains and also at some altitudes.

25. Dendrophis (Tree-snake)

Classification : Same as that of *Typhlops*.

Family	...	**Colubridae**
Series	...	**Aglypha**
Genus	...	*Dendrophis*

Habit and habitat. *Dendrophis* is arboreal in habit and lives on trees. Diurnal and feeds on frogs. Oviparous.

Comments :

1. *Dendrophis* (Fig. 13.26) is commonly called **tree-snake** and it is non-poisonous.
2. Body is elongated and slender, may reach the length of two metres, having yellow stripes.
3. Head is large covered by keeled shields arranged in 13–15 rows.
4. Head bears mouth, nostrils and eyes covered by a transparent membrane that is fused with eyes. Tympanic membrane is absent.
5. The scales of the vertebral row are enlarged.

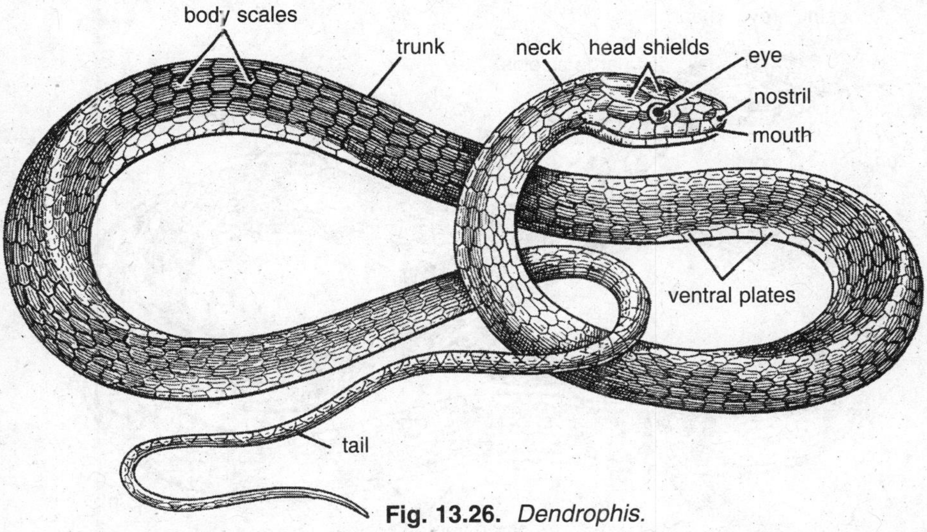

Fig. 13.26. *Dendrophis.*

6. Plate-like ventrals have a pair of suture-like lateral keels and notch on each side which are helpful in climbing.

7. Limbs and limb girdles are absent.

8. Teeth present over jaws but fangs absent.

Geographical distribution. *Dendrophis* is found in South East Asia, Australia and South America.

26. *Lycodon* (Common Wolf-snake)

Classification : Same as that of *Typhlops.*

Family ... **Colubridae**

Series ... **Aglypha**

Genus ... *Lycodon*

Habit and habitat. *Lycodon* is a garden snake and climbs on trees. On land it shows active movement and enters in human dwellings. It is nocturual in habit and feeds on frogs, lizards, skinks, small mammals and bird's eggs. Oviparous.

Comments :

1. *Lycodon* is commonly called **common wolf-snake**.
2. Body is elongated, measuring from 70 cm to 76 cm.
3. The dorsal surface is brown or purple with grey spots which form a reticulated pattern of cross-bars expanding laterally, while ventral surface is yellowish.
4. Upper lip is white with brown spots.
5. Head bears mouth, eyes with fused eyelids and nostrils. Tympanum is absent.
6. Head contains nine supra-labials, third, fourth and fifth are touching the eye. Loreal is in contact with inter-nasals.
7. Hypophyses are absent in the posterior part of the vertebral column.

Geographical distribution. *Lycodon* is found all over India.

27. *Bungarus* (Krait)

Classification : Same as that of *Typhlops.*

Family ... **Elapidae**

Genus ... *Bungarus*

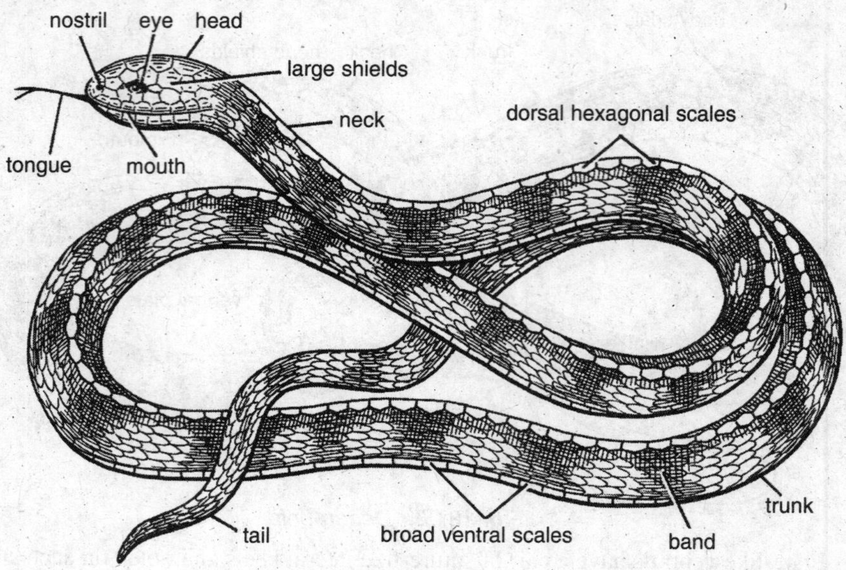

Fig. 13.27. *Bungarus.*

Habit and habitat. *Bungarus* is a very common snake, lives in cracks in walls and in heaps of bricks, etc. and feeds on rats, lizards and smaller snakes. It is nocturnal and oviparous. Female exhibits parental care.

Comments :

1. *Bungarus* (Fig. 13.27) is commonly called **krait**.
2. Body is elongated and slender, measuring about 1 metre in length. Body is divided into head, neck, trunk and tail.
3. The colour of the body is steel-blue with narrow cross bars or white specks dorsally and the underparts are uniform white.
4. Head with normal shields and is not differentiated from neck. Loreals are absent. Two post-oculars, one pre-ocular and seven supra-labials are present. Third and fourth supra-labials are touching the eye.
5. Head bears mouth, eyes of moderate size and nostrils. Fangs small.
6. The body scales are smooth and form 13–17 rows. Ventrals are 194–234 and 42–52 caudals.
7. Large hexagonal scales are along the mid-dorsal line of the body. Ventral scales beyond the anal region are in a single row.
8. *Bungarus* is **poisonous** and its venom is **neurotoxic**.

The krait "*Bungarus*" seems to cause more deaths in India than any other snake, since it is very common and often creeps into houses.

Geographical distribution. *Bungarus* is found in South Eastern Asia, India and Malayasia.

28. *Naja* (Indian Cobra)

Classification : Same as that of *Typhlops*.

Family	...	**Elapidae**
Genus	...	*Naja*
Species	....	*naja*

Habit and habitat. *Naja* is shy residing under stones, mud walls, thick vegetation and in holes. It is diurnal, oviparous and feeds on frogs, lizards, rats and other snakes. It hibernates in winters.

Comments :

1. *Naja naja* (Fig. 13.28) is commonly called **Indian cobra** or **nag**.

2. Body is elongated measuring one and a half to two metres in length and divided into head, neck, trunk and tail.

3. The colour of the body is brown or blackish.

4. Body is covered with smooth oblique scales without pits and are arranged in 15 – 25 rows. The subcaudals form only two rows.

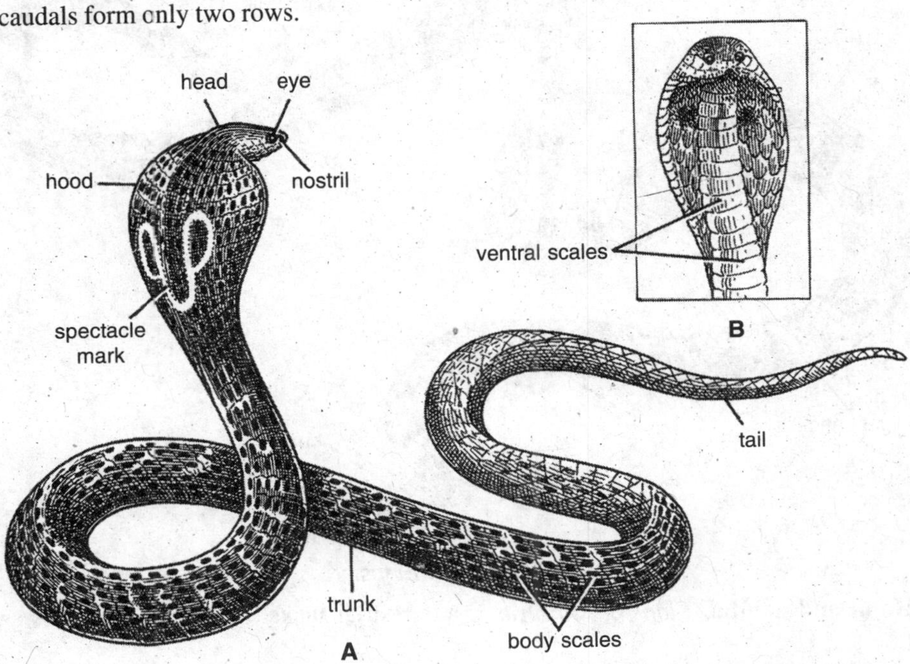

Fig. 13.28. *Naja naja*.A — Hood expanded; B — Ventral view of hood.

5. Head is not differentiated from the neck. Head bears mouth, very small eyes with round pupils, nostrils between nasals and internasal.

6. Neck is dilatable and the cervical ribs are elongated. The expansion of the neck and cervical ribs form the **hood**. The upper surface of the hood bears a **binocellate mark** to which people call a mark of **spectacle** or the figure of ten.

7. The lower surface of the hood bears two dark round spots running to four scales surrounded by white lateral borders.

8. Loreal is absent. Frontal shield is truncated. Three post-ocular scales are present.

9. Third supra-labial is large and touches the eye and the nasal shields.

10. Poison fangs are followed by 1–3 small teeth.

11. Tail is cylindrical and tapering posteriorly.

12. *Cobra* is **deadly poisonous** and its venom is **neurotoxic** and fatal. When it bites there is light pain and swelling, irritation and death due to respiratory failure.

Geographical distribution. *Naja* is widely distributed in China, Philippines, India, Malaya islands, Australia, Tasmania and Africa. *Naja naja* is found all over India.

Three types of cobras are found in India :

1. **Monocellate.** A single oval mark surrounded by elipses. Found in Bengal.
2. **Binocellate.** Spectacle-like mark connected by U. Found in Maharashtra.
3. **Noncellate.** No mark on hood. Found in Rajasthan, Madhya Pradesh and Gujarat.

29. *Elaps* = *Micrurus* (Coral Snake)

Classification : Same as that of *Typhlops.*

Family ... **Elapidae**
Genus ... *Elaps (Micrurus)*
Species ... *corallinus*

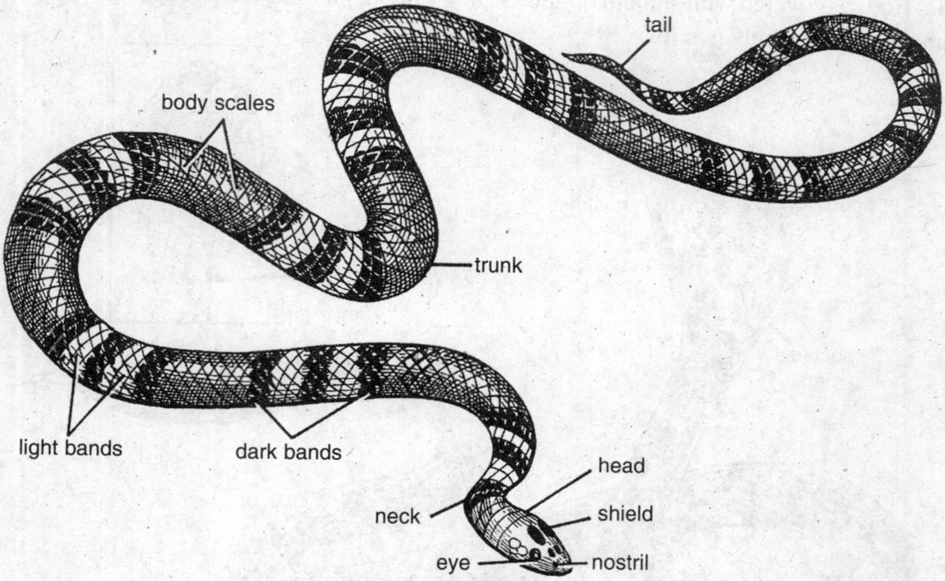

Fig. 13.29. *Elaps.*

Habit and habitat. *Elaps* or *Micrurus* is a terrestrial snake, feeding upon lizards and other snakes.

Comments :

1. *Elaps* or *Micrurus* (Fig. 13.29) is commonly called **coral-snake**.
2. Body is elongated, cylindrical and measuring about one to one and a half metre in length and divided into indistinct head, neck, trunk and tail.
3. The general colour of the body is extremely pretty black and red rings alternate with one another.
4. The scales of the body are smooth and form 15 rows.
5. The head is small and not distinct from the neck and bears mouth, nostrils and small eyes with vertical pupils.
6. The gape of the mouth is so limited that these snakes although possessing strong poison fangs are practically harmless to man.
7. Maxillaries carry no teeth behind the poison fangs.
8. The tail is short and round.
9. **Poisonous**, poison is **neurotoxic**.

Geographical distribution. *Elaps* is an entirely American genus. It is distributed in Tropical South America and the Lesser Antilles.

30. *Viper* (Russell's Viper)

Classification : Same as that of *Typhlops*.

Family ... **Viperidae**
Genus ... *Viper*
Species ... *russelli*

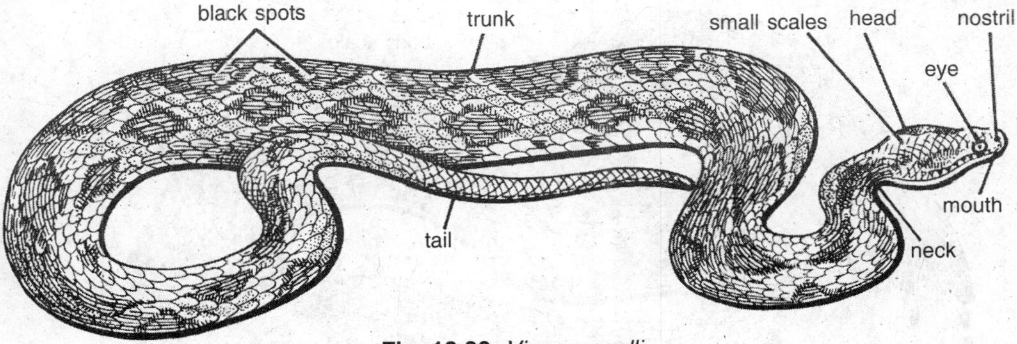

Fig. 13.30. *Viper russelli.*

Habit and habitat. *Viper* is found in rocky and bushy regions where its colour matches with the surroundings. It feeds on small mammals like mice and rats and lizards and birds. Viviparous.

Comments :

1. *Viper russelli* (Fig. 13.30) is commonly called **Russell's viper** or **daboia** or **pitless viper**.
2. Body is elongated, divided into head, neck, trunk and tail and measures up to one and a half metre in length.
3. The general colour is pale brown above with three longitudinal series of black spots and underparts are yellowish white.
4. About 30 rows of keeled scales are found on the body.
5. Head is triangular and covered with very small imbricate scales on the upper surface.
6. Head bears mouth, nostrils and eyes without eyelids and with golden iris and elliptical pupil.
7. Supra-labials are 10–12. Fourth supra-labial is the largest and it does not touch the eye. Ventral plates large enough across the belly.
8. Paired erectile fangs, one on each maxillary bone. These remain folded back when not in use.
9. No pit between eye and nostril.
10. Very **poisonous**. Venom is **haemotoxic**. It hisses loudly before striking and also swells its body.

Geographical distribution. *Viper russelli* is distributed in India, Sri Lanka, Myanmar and Thailand.

31. *Ancistrodon* = *Agkistrodon* (Himalayan Pit-viper)

Classification : Same as that of *Typhlops*.

Family ... **Viperidae**
Genus .... *Ancistrodon*
Species ... *himalayans*

Habit and habitat. *Ancistrodon himalayans* is a quite timid snake, remains hidden under stones, etc. It is viviparous.

Comments :

1. *Ancistrodon himalayans* (Fig. 13.31) is commonly called **Himalayan pit-viper**.
2. Body is not much elongated, measuring about 70 cm in length. Body is divided into head, neck, trunk and tail.
3. The general colour is bluish-brown above with dark-brown or black spots appearing like cross bars and the ventral side is brown with black and white spots.
4. Head is triangular, bearing grey mouth with black spots, eyes with golden iris and vertical pupil and nostrils.

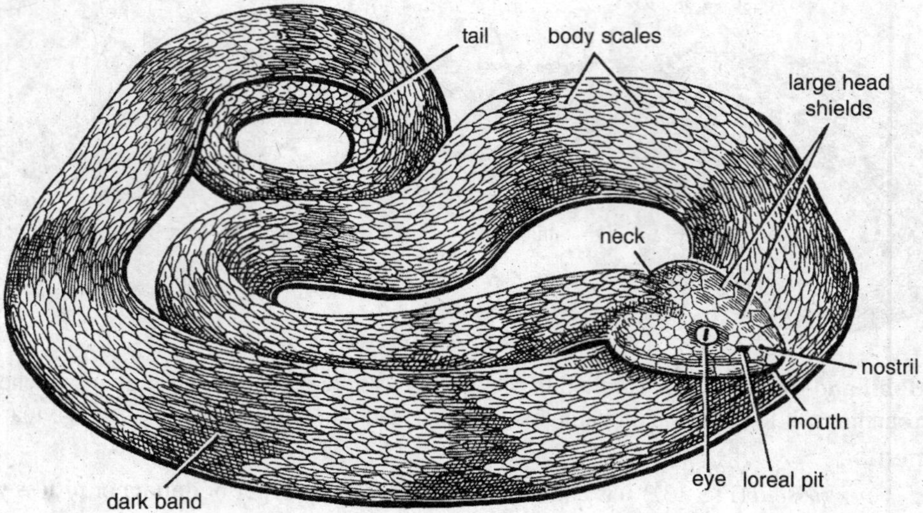

Fig. 13.31. *Ancistrodon* (Pit-viper).

5. The scales covering the body have apical sensory pits and head is covered with large shields.
6. Supra-labials are 5–7, third and fourth are enlarged but not touching the eye.
7. The upper lip is slightly raised in the rostral region. Post-oculars are two, the last greatly enlarged and preoculars are three.
8. A **loreal-pit** is present in between the eye and the nostril on each side.
9. Tail is cylindrical without rattles.
10. **Poisonous** having erectile sheathed fangs. Its poison is **haemotoxic**.

Geographical distribution. *Ancistrodon himalayans* is distributed in hilly regions of Eastern parts of India, Sri Lanka and Java. It is very common in Kashmir and is found even at an altitude of 4000 metres.

32. *Hydrophis* (Sea-snake)

Classification : Same as that of *Typhlops.*

Family ... **Hydrophiidae**

Genus ... *Hydrophis*

Habit and habitat. *Hydrophis* is a marine snake, feeding on fishes. Oviparous.

Comments :

1. *Hydrophis* (Fig. 13.32) is commonly called **sea-snake**.
2. The body is long and laterally compressed posteriorly, about 2 metres in length and covered by small scales.
3. Body is divided into head, neck, trunk and tail.
4. Head and neck are very slender. Indistinct head bears mouth, eyes and nostrils.
5. The general colour is dark olive-green above with yellowish cross bars and white below.
6. Head is covered with shields. Loreal-shield is absent. One pre-ocular, two post-oculars and 7– 8 supra-labials of which third and fourth are touching the eye. The ventral scales are small.
7. Tail is **laterally compressed** and acts like paddle in swimming.
8. Presence of 14–18 maxillary teeth behind the poison fangs.
9. *Hydrophis* is **deadly poisonous** and its venom is **neurotoxic**.

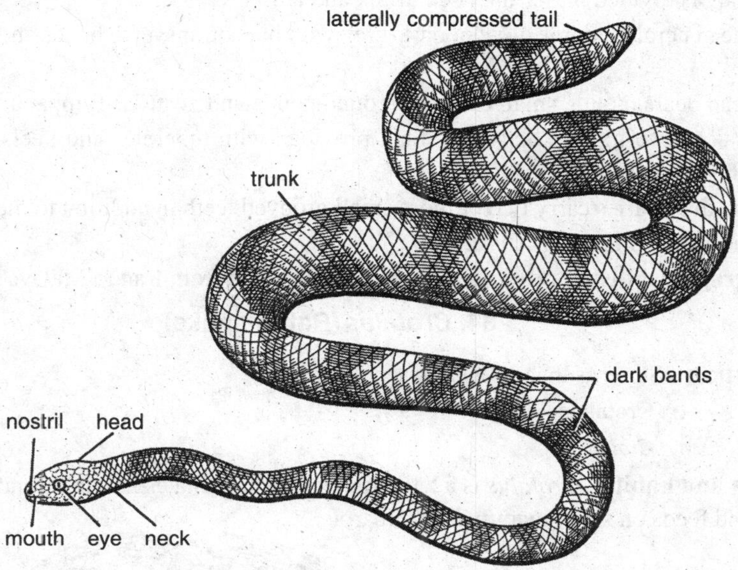

Fig. 13.32. *Hydrophis.*

Geographical distribution. *Hydrophis* occurs in the Bay of Bengal, Malaya Archipelago, Pacific Coast from Mexico to South America.

33. *Enhydrina*

Classification : Same as that of *Typhlops.*

Family ... **Hydrophiidae**
Genus ... *Enhydrina*
Species ... *bengalensis*

Habit and habitat. *Enhydrina* is found in sea. It feeds mainly on fishes. Viviparous.

Comments :

1. The body of *Enbydrina* (Fig. 13.33) is elongated with laterally compressed tail.

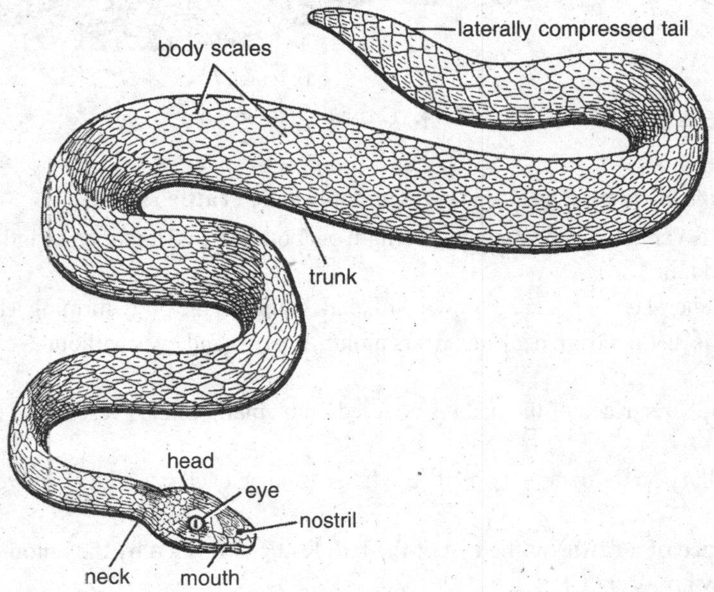

Fig. 13.33. *Enhydrina.*

2. Body is divided into head, neck, trunk and tail.

3. The general colour is olive or dark-grey with black transverse bands and the underparts are white.

4. Head bears mouth, small eyes with round pupils and nostrils. Tympanum is absent.

5. Scales covering the body are small provided with tubercles and keels. Ventrals are very small. Head is covered with shields.

6. The maxillaries carry two or more small grooved teeth in addition to the poison fangs.

7. It is a very **poisonous** snake.

Geographical distribution. *Enhydrina* is distributed from Iran to Malaya Islands.

34. *Crotalus* (Rattle-snake)

Classification : Same as that of *Typhlops.*

Family ... **Crotalinae**

Genus ... *Crotalus*

Habit and habitat. *Crotalus* is a terrestrial snake, hibernates in winter under the ground. It is nocturnal and feeds on small mammals. Viviparous.

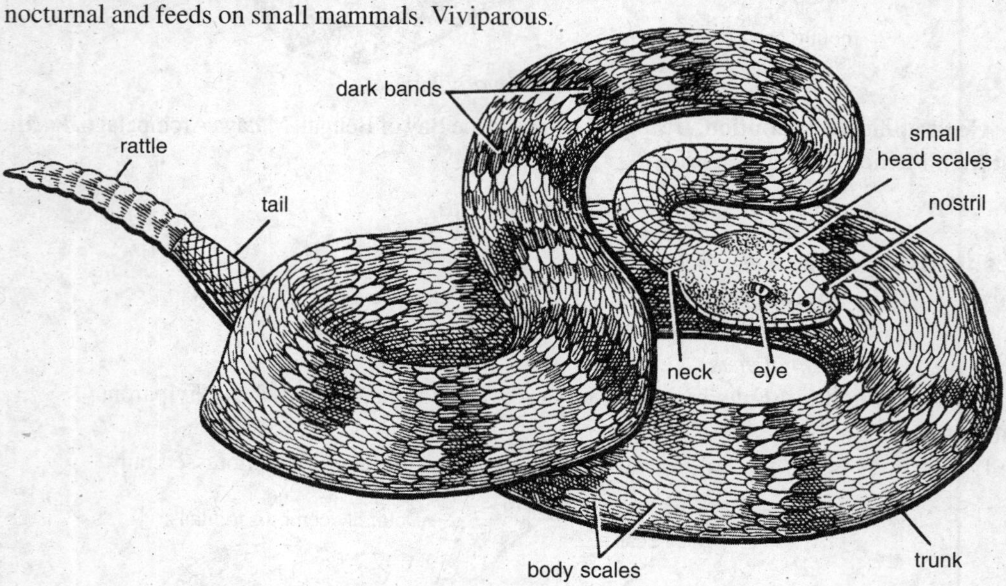

Fig. 13.34. *Crotalus* (Rattle-snake).

Comments :

1. *Crotalus durissus* (Fig. 13.34) is commonly called **rattle-snake**.

2. Body is very elongated reaching a length of about two and half metres and divided into head, neck, trunk and tail.

3. The general colour is grey-brown with dark bands on the body forming a handsome pattern.

4. Head is distinct from neck and bears mouth, nostrils and eyes without eyelids. Tympanum is absent.

5. The upper surface of the head is covered with small scales. Presence of two pairs of internasals.

6. Maxillary bones contain two erectile fangs (one on each maxillary) and remain folded back when not in use.

7. Presence of a **rattle** at the end of the tail. Rattle is formed by the union of 10 – 12 hollow segments of dry horny skin.

8. Rattle causes a noise which is useful to the snake as an instrument of warning of any approaching enemy.

9. **Deadly poisonous**.

Geographical distribution. *Crotalus* is distributed in Mexico and North America. *Crotalus horridus* is the common rattle snake of U.S.A. *C. confluentus* is found in Western America and *C. durissus* the common species in South Eastern North America.

35. *Crocodylus* (Mugger)

Classification :

Phylum	.... **Chordata**	→	Dorsal tubular nerve cord, notochord, and paired gill-slits are present.
Group	.... **Craniata**	→	Definite head, cranium with brain.
Subphylum	.... **Vertebrata**	→	Notochord is replaced by vertebral column, two pairs of appendages; circulatory system closed; hepatic portal system present; blood red containing R.B.Cs.
Division	.... **Gnathostomata**	→	Jaws and paired appendages are present.
Suplerclass	... **Tetrapoda**	→	Paired limbs; cornified skin; lungs and bony skeleton.
Class	.... **Reptilia**	→	Cold blooded, terrestrial or aquatic vertebrates; single occipital condyle; vertebrae gastrocentrous; respiration by lungs; heart with two auricles and incompletely divided ventricle; right and left aortic arches complete and functional; cranial nerves 12 pairs and embryo with amnion and allantois.
Subclass	.... **Diapsida**	→	Two temporal vacuities on each side; post-orbital and squamosal usually meet between temporal vacuities.
Superorder	.... **Archosauria**	→	Skull with closed upper temporal vacuity but possesses both temporal arches; anterior orbital vacuities are usually present; teeth thecodont; humerus with foramina; vertebrae amphicoelous or procoelous.
Order	... **Crocodilia**	→	Freshwater, predatory forms; body covered **with an exoskeleton of** thick horny epidermal scales; tail long laterally compressed; maxillae, palatines and pterygoids united along the middle to form a secondary palate; thecodont teeth; quadrate immovable; both supra- and infra-temporal arches present; vertebrae procoelous; sternum present; heart four chambered; lungs spongy; cloacal opening longitudinal and male with a median unpaired penis.
Genus	.... *Crocodylus*		
Species	.... *porosus*		

Habit and habitat. *Crocodylus* is found in lakes, rivers and marshes. It is predator, feeds on mammals and oviparous. It lays eggs in tunnels (10–13 metres long) below the level of water made by the animal.

Comments :

1. *Crocodylus porosus* (Fig. 13.35) is commonly called **mugger** or **marsh-crocodile**.

2. Body is stout, 4–5 metres long and covered with an exoskeleton of horny thick epidermal scales over dermal bony plates.

3. The general colour of the upper parts is dark olive-brown with black spots or bands.

4. The head is triangular and rough at the top, snout is broad, stout and provided with large wide mouth, nostrils at the tip of snount, eyes with eyelids and ear openings protected by a movable flap of skin.

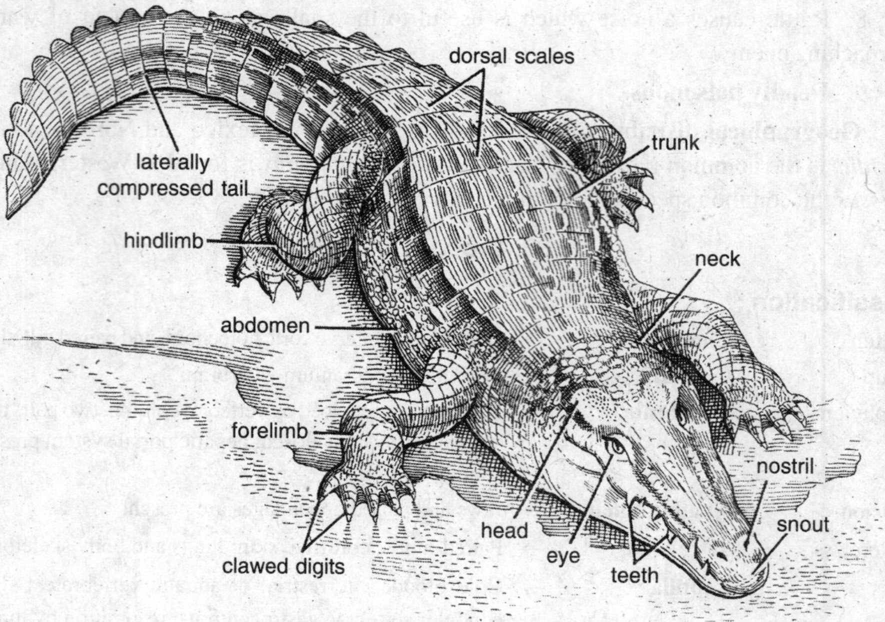

Fig. 13.35. *Crocodylus porosus.*

5. The tail is long, laterally compressed and powerful.

6. Limbs are pentadactyle with five fingers and four toes. The fingers are webbed at the base and the outer toes are broadly webbed.

7. Mandibular symphysis is short, extending only to the level of fourth or fifth tooth.

8. Upper and lower jaws, each containing 19 teeth on either side. The fourth mandibular tooth fits into a notch on the outer side of upper jaw. The fifth upper tooth is the largest. Dental formula is 16 – 19/14 – 15.

9. Vertebrae are procoelous.

10. The nuchal scutes are six in number which are packed closely together forming a square.

11. The dorsal shield is composed of four, sometimes of six rows of larger scuters of which the central pair is the broadest.

12. Heart is four chambered. Urinary bladder absent.

Geographical distribution. *Crocodylus porosus* has a wide distribution and found in India, Sri Lanka, Myanmar, Malacca, Malaya Islands and Beluchistan. Triassic to Recent.

36. *Alligator*

Classification : Same as that of *Crocodylus.*

Genus ... *Alligator*

Species .... *mississipiensis*

Habit and habitat. *Alligator* lives in shallow water and on slight disturbance enters the sand. Oviparous, eggs laid in nest.

Comments :

1. Body of *Alligator* (Fig. 13.36) measures two and a half to three and a half metres in length and is covered with an exoskeleton of horny epidermal scutes (exoskeleton) over bony scutes. The dorsal shield is formed by 6 or 8 longitudinal series of keeled bony scutes.

2. The general colour is greenish-black or dark-brown above and yellowish below.

3. Head is broad with short, broad and rounded snout having nostrils at the tip. Head bears eyes with movable eyelids and ear openings provided with movable flaps of skin.

4. Limbs pentadatyle with five fingers and four toes with webs.
5. The tail is long and laterally compressed.
6. Mandibular symphysis is short extending only to the level of fourth of fifth tooth. Fourth mandibular tooth fits into a pit in the upper jaw.
7. On either side 17–20 teeth in the upper jaw and 18–20 teeth in the lower jaw.
8. The nasal bones divide the nasal groove by a median bony septum.
9. Heart is four-chambered and urinary bladder absent.

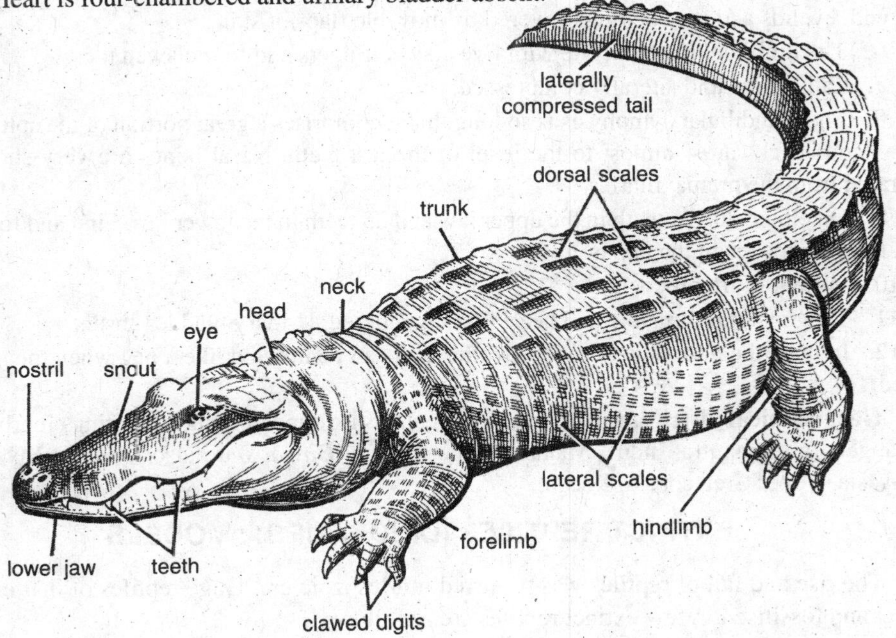

Fig. 13.36. *Alligator mississipiensis.*

Geographical distribution. *Alligator mississipiensis* is found in the rivers of Southern States of North America and *A. sinensis* occurs in the rivers of China.

37. *Gavialis* (Ghariyal)

Classification : Same as that of *Crocodylus.*

Genus .... *Gavialis*

Species ... *gangeticus*

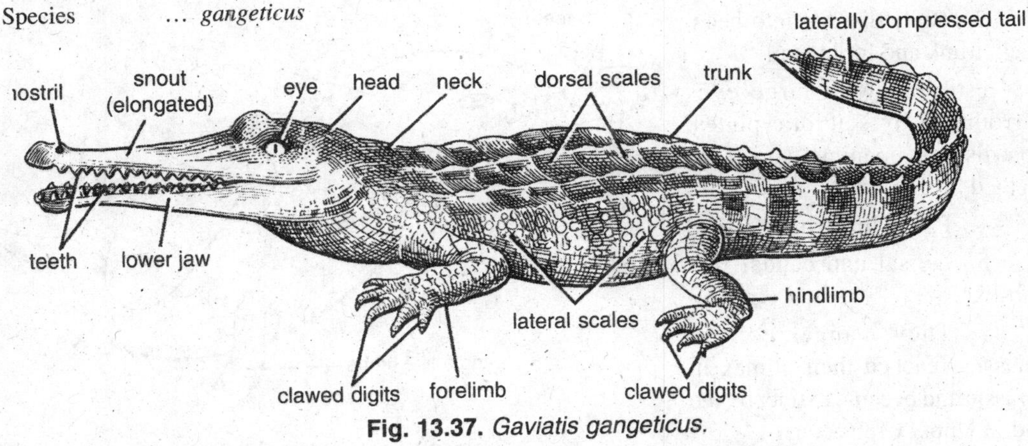

Fig. 13.37. *Gaviatis gangeticus.*

Habit and habitat. *Gavialis* lives in river Ganges and feeds on fishes.

Comments :

1. *Gavialis gangeticus* (Fig. 13.37) is commonly called **ghariyal** or **naka**.
2. Body length is about six and a half metre and divisible into head, neck, trunk and tail.
3. The general colour is dark olive-brown above with dark markings.
4. Body is covered with an exoskeleton of epidermal horny scutes over bony scutes.
5. Head is large produced into a long and narrow snout bearing nostrils. Head bears mouth, eyes with eyelids and ear openings covered by movable flaps of skin.
6. The limbs are pentadactyle with five clawed fingers and four clawed toes.
7. Tail is long and laterally compressed.
8. The mandibular symphysis is so long that it comprises a great portion of the splenial bones and extends backwards almost to the level of the last teeth. Nasal bones are very short and are separated from the premaxillaries.
9. On either side 28 teeth in the upper jaw and 25 teeth in the lower jaw. First and fourth lower teeth fit into grooves in the upper jaw.
10. Vertebrae are procoelus.
11. Tongue not protrusible and heart four chambered. Urinary bladder absent.
12. In male the nose is very much swollen and can be inflated like a bag when the nostrils are closed.

Geographical distribution. *Gavialis gangeticus* is found in India inhabiting chiefly the basin of Ganges, Brahmaputra, Indus, Mahanadi and Arakan. *Gavialis* are also found in Myanmar and Malayasia. Upper Cretacous to Recent.

EXTINCT REPTILES (DINOSAURS) : MODELS

The rise and fall of reptiles was occurred in Mesozoic era. Huge reptiles of that era became extinct and fossilised. A few extinct reptiles are as follows :

1. *Ichthyosaurus*

Classification : Same as that of *Crocodylus.*

Subclass ... **Parapsida** → Temporal openings bounded by supra-temporal and post-frontal.
Genus ... *Ichthyosaurus*

Comments :

1. *Ichthyosaurus* (Fig. 13.38) is a fish-like extinct reptile.
2. Body divided into head, neck, trunk and forked tail.
3. Head with large eyes surrounded by sclerotic plates, nostrils, ear openings and long, teethed beak.
4. Paired limbs paddle-like.
5. Dorsal and caudal fins fish-like.

They appeared in Triassic, reached their climax in Jurassic and became extinct by the end of Upper Cretaceous.

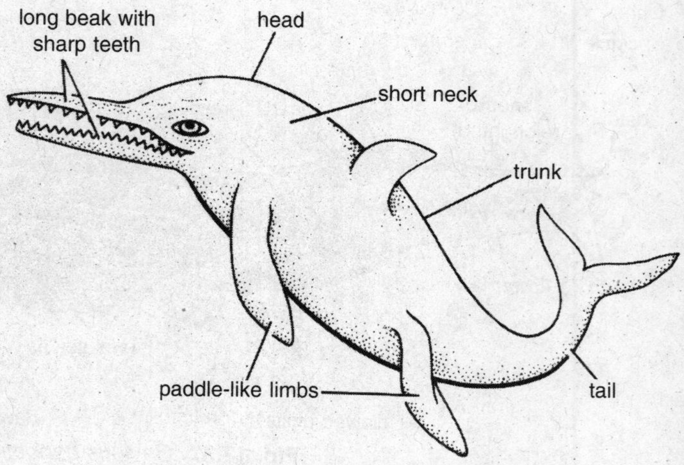

Fig. 13.38. *Ichthyosaurus.*

2. *Dimetrodon*

Classification : Same as that of *Crocodylus*.

Subclass ... **Synapsida** → Temporal opening behind eye bounded by post-orbital and squamosal.

Genus .... *Dimetrodon* (= *Pelycosaur*)

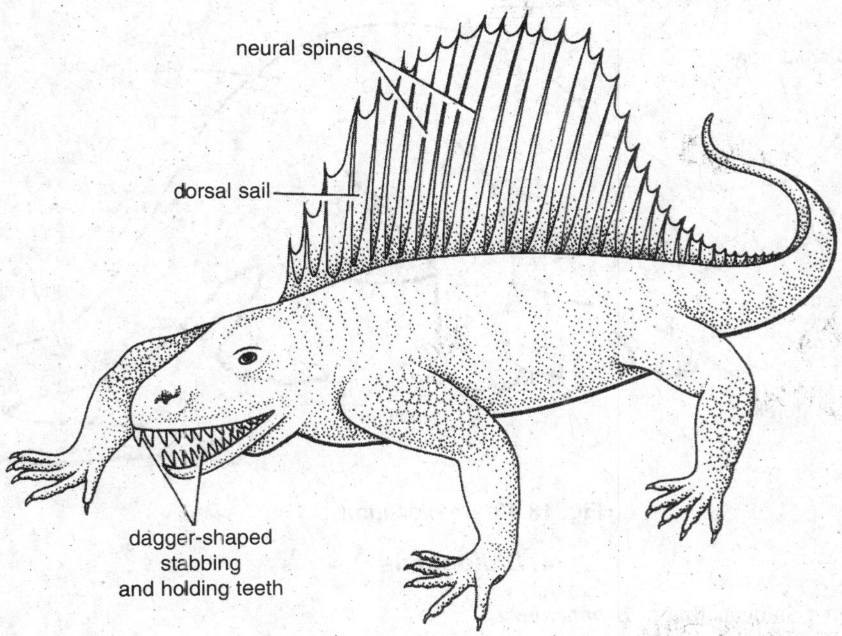

Fig. 13.39. *Dimetrodon.*

Comments :

1. *Dimetrodon* (Fig. 13.39) is extinct mammal-like reptile having single lateral temporal opening on either side in the skull.

2. Body more than 10 feet long and divisible into small head, neck, large trunk and comparatively small tail.

3. Jaws were provided with dagger-shaped teeth.

4. Limbs well developed

5. Dorsal sail over trunk formed by elongation of neural spines covered with skin.

6. Their extinction took place in Jurassic period. They might have given rise to ancestors of mammals.

3. *Brontosaurus*

Classification : Same as that of *Crocodylus*.

Subclass ... **Diapsida** → 2 temporal openings behind eyes separated by post-orbital and squamosal.

Order ... **Saurischia** → Ischium and pubis diverge.

Genus .... *Brontosaurus*

Comments :

1. *Brotosaurus* (Fig. 13.40) is commonly called **giant dinosaur** whose body was very massive.

2. Body divided into small head, elongated neck, massive trunk and long pointed tail.

3. Their body size was about 20–25 metre long and weighs about 50 tons.

4. Hindlimbs large, thick and stout than forelimbs.
5. Pelvis triradiate, Backbone massive with hollowed lateral vertebrae.
6. Brain small is size.
7. They lived in swamps in shallows water.

They faced extinction due to small brain and huge massive body.

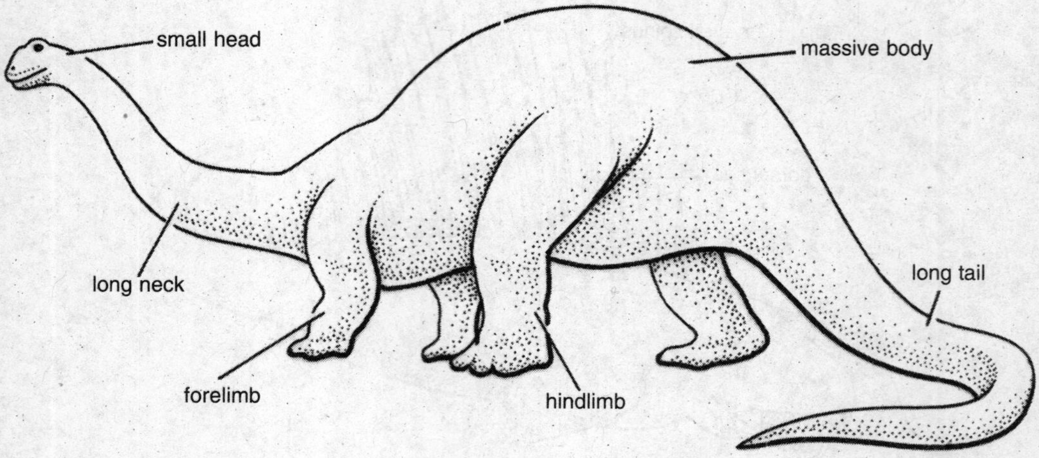

Fig. 13.40. *Brontosaurus.*

4. *Diplodocus*

Classification : Same as that of *Brontosaurus.*

Comments :

1. This was the longest dinosaur of Jurassic period measuring 25 metres long and 50 tons in weight.

2. Body was divisible into very small head, very much long and tapering neck, massive trunk d long massive tail.

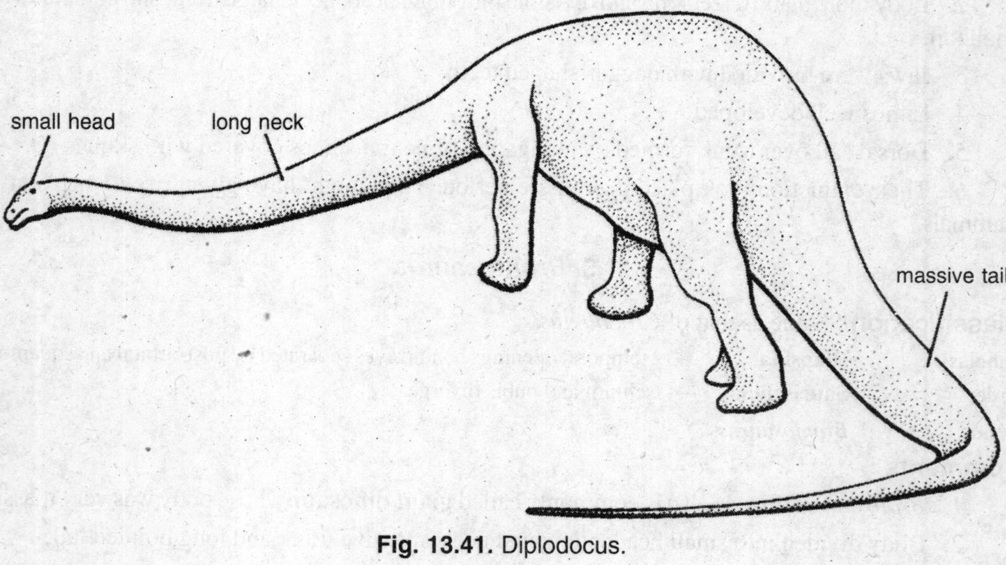

Fig. 13.41. Diplodocus.

3. Jaws small and brain exceptionally small (6 inches long).
4. Limbs pillar-like and hip region huge.
5. Quadrupedal, herbivorous and swamp dwellers.
6. They always needed constantly lush green pastures to support their generation due to bulky body.

5. *Tyrannosaurus*

Classification : Same as that of *Brontosaurus*.

Comments :

1. This was commonly called **tyrant dinosaur**, lived in plains of Western North America.
2. Body divided into large head, small neck, massive trunk and tail.
3. Body was about 15 metres long and 6–7 metres in height
4. Forelimbs very small used to hold prey, while hindlegs huge adapted for running. Feet were like those of mammoth bird having three powerful claws.
5. Jaws large with 3–6 inches long dagger-like teeth.

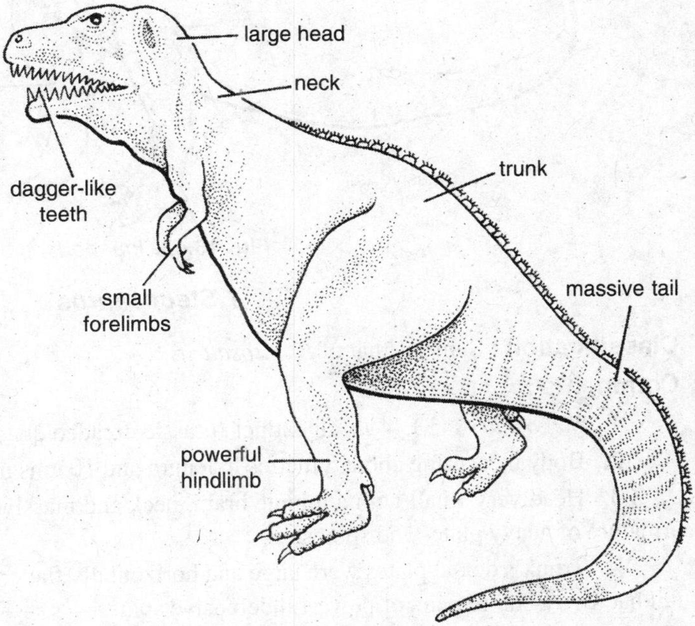

Fig. 13.42. *Tyrannosaurus.*

6. *Iguanodon*

Classification : Same as that of *Crocodylus*.

Order ... Ornithoschia

Genus .... *Iguanodon*

Comments :

1. *Iguanodon* (Fig. 13.43) is the most primitive bird-like dinosaur with typical tetraradiate pelvis.
2. Fairly large measuring approximately 5 metres in length.
3. Body divided into head, neck, trunk and tail.
4. Head and neck small, while trunk very heavy.
5. Hindlegs massive and forelegs not much reduced.
6. Peculiar feature of *Iguanodon* was its sharp dagger-like defensive thumb.

They were prevalent in Upper Cretaceous. Descendents of primitive thecodont stock.

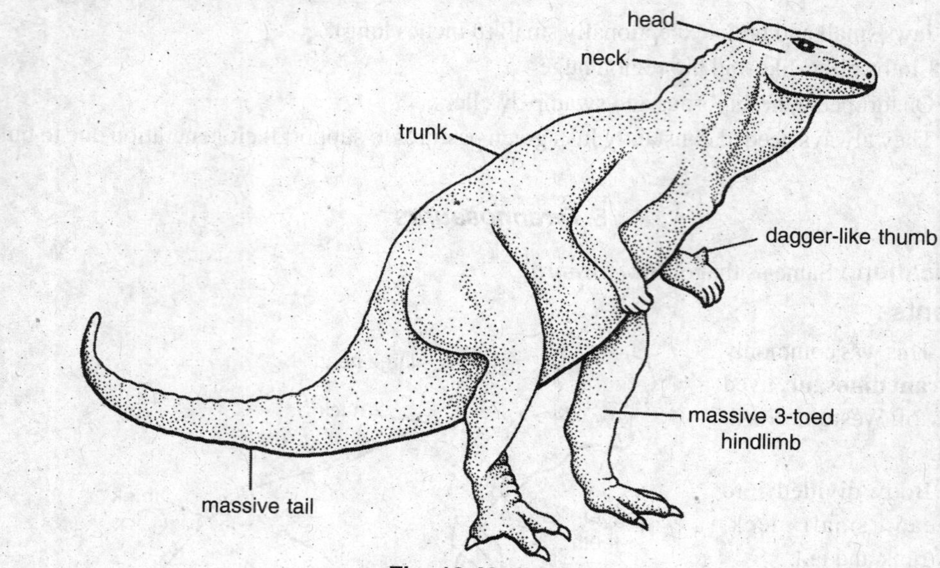

Fig. 13.43. *Iguanodon.*

7. *Stegosaurus*

Classification : Same as that of *Brontosaurus.*

Comments :

1. *Stegosaurus* (13.44) is an extinct Jurassic dreaded quadupedal dinosaur.
2. Body measuring about 8 metres in length and 10 tons in weight.
3. Head very small with a minute brain, neck and massive trunk and tail were provided with armature of heavy plates and spines.
4. Trunk's dorsal plates were large and horizontally flattened forming roof. Powerful massive tail had two or more pairs of powerful defensive spines.
5. Spinal cord at the base of tail was tremendously enlarged.

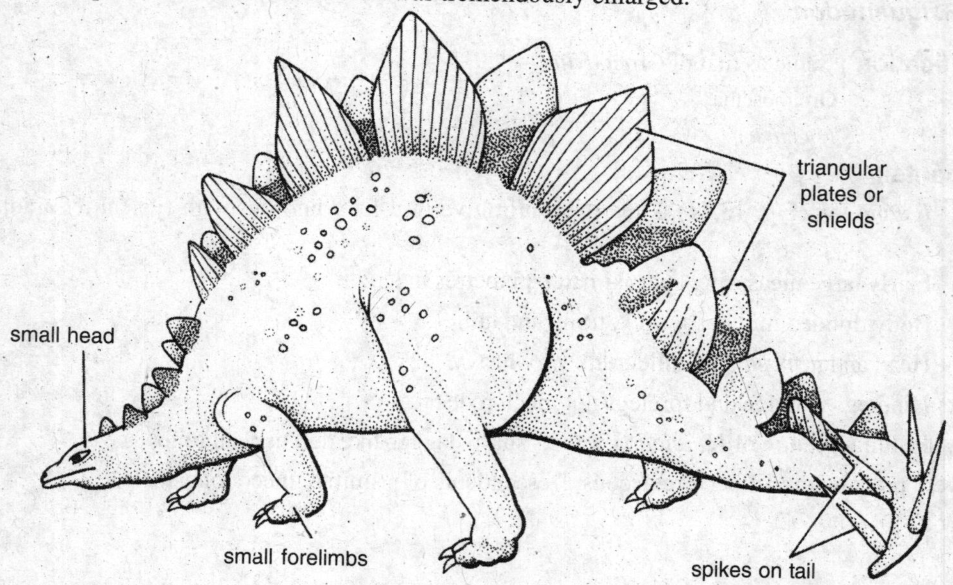

Fig. 13.44. *Stegosaurus.*

8. *Rhamphorhynchus*

Classification : Same as that of *Crocodylus*.

Order ... **Pterosuria** → Flying reptiles. Forelimbs with wing membranes.

Genus ... *Rhamphorhynchus*

Comments :

1. *Rhamphorhynchus* (Fig. 13.45) is commonly called **flying reptile** or **flying dragon**.
2. Primitive pterosaur had one metre wingspan.
3. Head with eyes, nostrils, ear openings and beak provided with teeth.
4. In forelimbs thumb and three fingers were small and clawed, while fifth finger was very much elongated and attached with body skin forming patagium used as wings during flight.

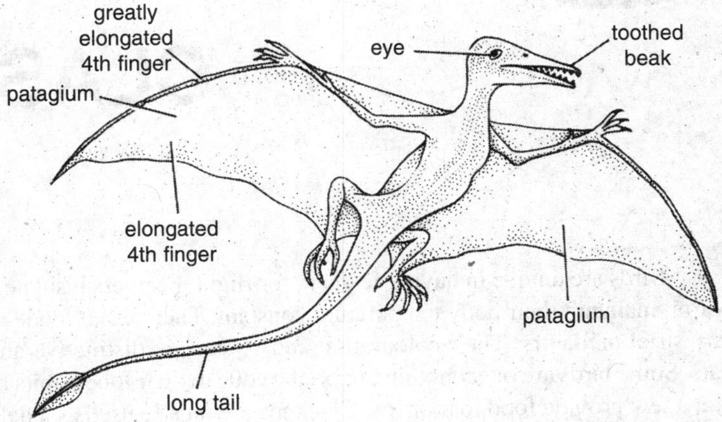

Fig. 13.45. *Rhamphorhynchus.*

5. Legs slender with five clawed toes, Tail very much elongated.
6. Their fossils were obtained from marine rocks.

9. *Pteranodon*

Classification : Same as that of *Rhamphorhynchus*.

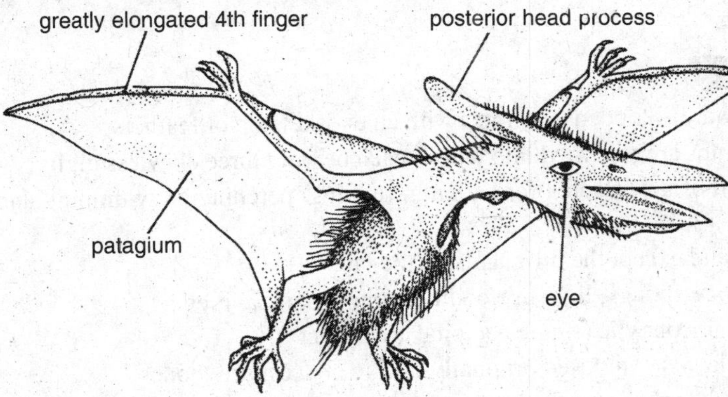

Fig. 13.46. *Pteranodon.*

Comments:

1. *Pteranodon* is bird-like giant extinct reptile having wing span about 27 feet.
2. Body divided into huge head, neck and trunk.
3. Head contained large eyes with sclerotic plates, nostrils, ear openings and toothless spear-like beak, Head also bear a posterior bony process.
4. On the back was present a long bony backwardly directed crest.
5. Fourth fingers of forelimbs elongated having patagia for soaring in the air.
6. Brain large with cerebellum very much like that of birds.
7. They appeared in Jurassic period and became extinct in late Cretaceous, thus, survived throughout Cretaceous.

Chapter 14

CLASS–AVES (BIRDS)

Birds are unique in having feathers for flight. Feathers insulate their bodies and keep the body warm, maintain their body temperature constant. Their aerial mode of life protects them from their terrestrial predators. Their colouration and voices are distinctive and pleasant to human eyes and ears. Some birds are of economic importance due to their food habits and certain birds are game birds and a few provide food to humans. Chickens are mostly used as a delicious food.

Birds are advanced over the lower vertebrates in certain characters, such as feathers covering the body, four-chambered heart, complete separation of arterial and venous blood, warm blooded, high metabolic rate due to flying habit, highly developed voice, sight and hearing.

Birds are distributed in all continents, Arctic and Antarctic regions, at all altitudes and in various habitats, like open ocean (*e.g.*, albatross), marshy places, Savanna, grassy fields, trees, etc. Certain birds have migratory habits and travel-thousand kilometers for breeding and nesting and avoid harsh weather climate, Feathers of birds are used in sports material, for example, in making shuttle cocks for badminton.

General Characters

1. Birds are warm-blooded verterates with an exoskeleton of feathers.
2. The forelimbs are modified into wings, each bearing three clawless digits.
3. The hindlimbs are adapted for walking, running, perching or swimming and bearing four clawed toes.
4. No skin gland except the oil-glands on the tail.
5. The bones are spongy, light in weight containing air cavities.
6. Skull is monocondylic (single occipital condyle).
7. The lower jaw is composed originally of five or six fused bones.
8. Modern birds have no teeth. A horny beak is present.
9. The cervical and free thoracic vertebrae are usually heterocoelous.
10. The sacral vertebrae are fused with the lumbar, fewer posterior thoracic and a few anterior caudals to form a **synsacrum**.
11. The posterior caudal vertebrae are usually fused to form a **pygostyle**.
12. Vertebral ribs are double-headed, with uncinate process.
13. The sternum is broad, usually with a longitudinal ventral keel for the attachment of flight muscles.
14. The coracoid is usually pillar-like; the scapula is sabre-shaped. The clavicles and interclavicle unite to form a V-shaped furcula.

15. Distal carpals and metacarpals of forelimbs are united to form a **carpo-metacarpus**.

16. Proximal tarsals are fused with the tibia to form a **tibio-tarsus** and distals with the metatarsals to form the **tarso-metatarsus**. In the leg, the ankle joint is inter-tarsal.

17. The gullet is dilated into a crop. Stomach is divided into a glandular proventriculus and a muscular gizzard. Cloaca is three-chambered.

18. Lungs are spongy and non-distensible. Air-sacs are present which communicate with air-cavities in the bones.

19. The voice is produced by the syrinx situated near the junction of the trachea with the bronchii. Lyrinx is without vocal cords (functionless).

20. The heart is four-chambered. Only right aortic arch is present in the adult. R.B.Cs. are oval, nucleated and biconvex.

21. Renal portal system is vestigial.

22. Kidneys are metanephric, three-lobed and ureters open into the cloaca.

23. The cerebral hemispheres and cerebellum are large and the optic lobes are displaced laterally.

24. Sexes are separate and sexual dimorphism is well-marked.

25. The left ovary alone is present. The right ovary and oviduct is more or less completely atrophied.

26. Fertilisation is internal. Oviparous. Large ovum containing much yolk invested with albumen.

27. The embryo has an amnion, an allantois and a large yolk-sac.

28. Parental care is highly developed.

CLASS—AVES

Subclasses

- I Archaeornithes
  Ex. *Archaeopteryx* (Extinct)
- II Neornithes

Subclass—NEORNITHES

Superorders

- I Odonotognathae
  Ex. 1. *Hesperornis*
  2. *Ichthyornis* } Extinct
- II Palaeognathae
- III Sphenisciformes
  Ex. *Aptenodytes*
- IV Neognathae

Superorder II—PALAEOGNATHAE

Orders

- 1 Struthioniformes
  Ex. *Struthio camelus* (Ostrich)
- 2 Casuariiformes
  Ex. 1. *Dromaeus*
  2. *Casuarius*
- 3 Aepyornithiformes
  Ex. *Aepyornis* (Extinct)
- 4 Dinornithiformes
  Ex. *Dinornis* (Extinct)
- 5 Apterygiformes
  Ex. *Apteryx* (Kiwi)
- 6 Rheiformes
  Ex. *Rhea*
- 7 Tinamiformes
  Ex. *Tinamus*

Superorder III — NEOGNATHAE OR CARINATAE

Orders

1. Gaviiformes
   Ex. *Gavia*
2. Podicipediformes
   Ex. 1. *Podiceps*
   2. *Podilymbus*
3. Procellariiformes
   Ex. 1. *Diomedea*
   2. *Puffinus*
4. Pelecaniformes
   Ex. 1. *Pelecanus*
5. Ciconiiformes
   Ex. 1. *Ardaea* (Great blue heron)
   2. *Phoenicopterus* (Flamingo)
6. Anseriformes
   Ex. Ducks, Geese, Swans
7. Falconiformes
   Ex. Vultures, Kites, Hawks, Eagles and Falcons
8. Galliformes
   Ex. Grouse, Quails, Phesants and Turkeys
9. Gruiformes
   Ex. Cranes, Rails and Coots
10. Diatrymiformes
    Ex. 1. *Diatryma* (Extinct)
11. Charadriiformes
    Ex. Shore birds, Gulls and Auks
12. Columbiformes
    Ex. Pigeons and Doves
13. Cuculiformes
    Ex. Cuckoos
14. Psittaciformes
    Ex. Parrots
15. Strigiformes
    Ex. Owls
16. Caprimulgiformes
    Ex. Night-hawks
17. Apodiformes
    Ex. Swifts and Humming birds
18. Coliiformes
    Ex. Collies (Mouse-birds)
19. Trogoniformes
    Ex. Trogons
20. Coraciiformes
    Ex. King-fishers
21. Piciformes
    Ex. Woodpeckers
22. Passeriformes
    Ex. Perching birds

Note: The classification is adopted from **Storer** and **Usinger** (1957).

Subclass I — ARCHAEORNITHES

1. Extinct Jurassic birds of Mesozoic age.
2. Tail long with more than 13 caudal vertebrae bearing rectrices arranged in two lateral rows.
3. Pygostyle was absent.
4. Vertebrae were amphicoelous.
5. Three fingers and metacarpals were free, each with a claw.
6. Teeth embedded in sockets were present in both jaws.

Examples : *1. Archaeopteryx,* *2. Archaeornis.* } Extinct.

Subclass II — NEORNITHES

1. Living as well as extinct birds.
2. Tail is usually short, ends in a pygostyle.
3. The rectrices are arranged in a semicircle around the pygostyle.
4. Teeth are absent except in few forms.
5. The metacarpals are fused with the distal carpals to form a carpometacarpus.
6. Except in one case, not more than two digits of the hand bear claws.
7. Claws are absent in the forelimbs.
8. The sternum is well-developed and usually provided with a keel or carina.
9. Vertebrae are heterocoelous in living forms.
10. Caudal vertebrae are 13 or less.

Superorder I — ODONTOGNATHAE

1. Extinct Cretaceous birds.
2. Teeth present in furrows on jaws.
3. Clavicles not fused. Sternum without keel.
4. Humerus vestigial.

Examples : *1. Hesperornis;* *2. Ichthyornis* (Extinct)

Superorder II — PALAEOGNATHAE

1. Mostly flightless, walking or running birds.
2. Wings are reduced or absent.
3. Teeth are absent.
4. Tail vertebrae are free.
5. Coracoid and scapula are small and fused at an obtuse angle.
6. Sternum devoid of keel or carina.

Order 1. Struthioniformes

1. Large flightless terrestrial birds.
2. Head, neck and legs sparsely feathered.
3. Feathers without aftershaft.
4. Head is comparatively small and the neck is very long and flexible.
5. The beak is short and broad.
6. Only two toes (3rd and 4th) on each foot.
7. Sternum without keel.
8. Pubic symphysis is present.
9. No pygostyle.

Example : *Struthio camelus* (Ostrich).

Order 2. Causuariformes

1. Large flightless terrestrial birds.
2. Head is sparsely feathered, while neck and body densely feathered.
3. Feathers with aftershaft nearly equal to shaft.
4. Wings small or rudimentary.
5. Three toes with typical claws on each foot.
6. Sternum is without keel.

Examples : *1. Dromaeus* (Emus), *2. Casuarius* (Cassowaries).

Order 3. Aepyornithiformes

1. Flightless, terrestrial **extinct** elephant birds.
2. Wings vestigial or tiny.
3. Legs powerful with 4 toes.
4. Sternum short, broad and without keel.
5. Eggs were surprisingly large.

Example : *Aepyornis* (Elephant bird) (Extinct).

Order 4. Dinornithiformes

1. Giant flightless, terrestrial, **extinct** birds.
2. Feathers with large aftershaft.
3. Wings almost absent.
4. The beaks were short.
5. Legs massive bearing 3 or 4 toes.
6. Sternum reduced and without keel.
7. Pectoral girdle completely absent.

Example : *Dinornis* (Moas) Extinct.

Order 5. Apterygiformes

1. Flightless terrestrial birds.
2. Feathers fluppy hair-like with large aftershafts.
3. Wings are rudimentary or degenerate. Humerus vestigial.
4. The beak is long and slender with nostrils at the tip of the maxilla.
5. The eyes are small.
6. The neck and legs are comparatively short.
7. Legs bear 4 toes.
8. Sternum is without keel.
9. Eggs are relatively largest of all the living birds.
10. Nocturnal, feeding on worms and insects.

Example : *Apteryx* (Kiwi)

Order 6. Rheiformes

1. Flightless terrestrial and swift running birds.
2. Head, neck and thighs are feathered.
3. Feathers without aftershaft.
4. Wings are better developed.
5. Legs bear 3 toes with heavy and typical claws.
6. Sternum is without keel.

Example : *Rhea americana* (American Ostrich).

Order 7. Tinamiformes

1. Small terrestrial, essentially cursorial with poor power of flight.
2. Wings are short, rounded and developed for flight.
3. Sternum is keeled.
4. Tail feathers and pygostyle are reduced.
5. Eggs are remarkable with highly glossy shells.
6. Vegetarian.

Examples : *1. Tinamus, 2. Rhyncotus.*

Superorder III — IMPENNAE

Order 1. Sphenisciformes

1. Flightless aquatic birds.
2. Feathers small scale-like covering the entire body.
3. Forelimbs (wings) are paddle-like adapted for swimming.
4. Feet are webbed.
5. Metatarsus incompletely fused, 4 toes all directed forward.
6. Bones are much compressed.
7. Thick layer of fat present beneath the skin.

Example : *Aptenodytes* (Penguin).

Superorder IV — NEOGNATHAE (CARINATAE)

1. Modern flying birds.
2. Skull is neognathus, *i.e.*, with short vomer and palatines extending posteriorly.
3. Teeth are entirely absent.
4. Feathers having interlocking mechanism.
5. Wings are well-developed.
6. Sternum with a well-developed keel.
7. Tail vertebrae are 5 or 6.
8. Pygostyle is present.
9. Forelimbs with carpals and metacarpals joined and fingers included in the wings.

Order 1. Gaviiformes

1. Legs are short and at the posterior end of the body.
2. Tail comprises 18–20 short stiff-feathers.
3. Toes are fully webbed.
4. Patella is reduced.
5. Powerful in flight.

Example : *Gavia immer* (Common loon).

Order 2. Podicipediformes

1. Freshwater divers.
2. Legs are placed far back on the body.
3. Feet are lobed.
4. Tail comprises a tuft of down feathers.
5. Patella is large.
6. Tarsus is compressed.

Examples : *1. Podiceps,* 2. Grebes, *3. Podilymbus.*

Order 3. Procellariiformes

1. Nostrils are tubular.
2. Horny sheath of bill consisting of several plates.
3. Skull with large nasal glands.
4. Hind toe is vestigial or none.
5. Feathers compact and oily in texture.
6. Wings are long and narrow.
7. Strictly marine, nest usually on islands.

Examples : *1. Diomedea* (Albatrosses), *2.Puffinus* (Petrels).

Order 4. Peleconiformes

1. Aquatic fish-eating and colonial nesters.
2. Nostrils are vestigial or absent.
3. Presence of a gular pouch on the throat.
4. Four-toed webbed feet.
5. Beaks are long with wide gapes for catching and swallowing the struggling prey (fish).

Examples : *Pelecanus* (Pelecans), Cormorants, Boobies and Gannets.

Order 5. Ciconiiformes

1. Long-necked and long-legged wading birds.
2. Feathers are decorative.
3. Beaks abruptly decurved at middle in flamingos.
4. Toes are without web.
5. Young born naked.
6. Food comprises fish and other aquatic animals.

Examples : 1. *Ardea* (Great blue heron), *2.Phoenicopterus* (Flamingo), Storks and Ibies.

Order 6. Anseriformes

1. Beak or bill is broad, covered with soft cornified epidermis.
2. Margins of the beak containing many transverse horny ridges (lamellae).
3. Tongue is fleshy.
4. Legs are short and feet are webbed.
5. Tail is usually short, consisting of many feathers.

Examples : Ducks, Geese, Swans, *Anas, Anser, Cycymas,* etc.

Order 7. Falconiformes

1. Beak is short, stout and curved at the tip.
2. Mandibles are sharp-edged.
3. Feet usually with sharp curved claws adapted for grasping and holding the prey.
4. Diurnal and strong fliers.

Examples : Vultures, Kites, Hawks and Falcons, *Gyps, Oragys, Milvus, Atharis.*

Order 8. Galliformes

1. Terrestrial game birds with short and powerful flight.
2. Beak is short.
3. Feathers with aftershaft.
4. Feet usually massive adapted for scratching and running.

5. Graminivorous.

Examples : *1. Gallus* (Jungle fowl), *2. Pavo cristatus* (Peacock) *3. Coturnix* (Quail), *4. Erancolinus* (Grey patridge), *5. Phasianus* (Pheasants), *6.* Turkeys, *7.* Grouse.

Order 9. Gruiformes

1. Small, medium and large-sized birds.
2. Flightless or weak or strong fliers.
3. Feathers with aftershaft.
4. Legs long and beaks heavy.

Examples : *1. Grus* (Crane), *2. Fulica atrica* (Common coot), *3. Antegona antegona* (Sarus crane), *4.* Rails.

Order 10. Diatrymiformes

1. Large flightless **extinct** birds.
2. Wings atrophied.
3. Massive head, beak and neck.
4. Foot bears 4 toes.

Example : *Diatryma* (Extinct).

Order 11. Charadriiformes

1. Shore dwelling aquatic birds with dense and firm plumage.
2. Hindlimbs are slender and enormously elongated.
3. Toes usually webbed.
4. Beaks mud-probing.
5. Feathers are dense and firm.

Examples : *1.Lobivanellus indicus* (Red-wattled lapwing), *2. Hydrophasianus chirugus* (Pheasant-tailed jacana), *3. Tringa glariola* (Sand-piper), *4. Capella* (Snipe), *5. Larus* (Gull), etc.

Order 12. Columbiformes

1. Skin thick and soft.
2. Beak usually short and slender.
3. Tarsus usually shorter than toes.
4. Crop large producing "pigeon milk" to feed young ones.
5. Graminivorous and frugivorous.

Examples : *1. Columba livia* (Blue-rock pigeon), *2. Goura cristata* (Crowned-pigeon), *3. Streptopelia risoria* (Ring-dove), *4. Streptopelia chinensis* (Spotted-dove), etc.

Order 13. Cuculiformes

1. Toes 2 in front and 2 behind (zygodactylous).
2. Feet not adapted for grasping.
3. Tail long and beak moderate.
4. Many cuckoos are parasitic, the female laying eggs in nest of other bird (crow).

Examples : *1. Cuculus canorus* (Cuckoo), *2. Eudynamys scolopaceous* (Koel).

Order 14. Psittaciformes

1. Feathers are green, blue, yellow or red.
2. Beak stout, narrow, sharp-edges and hooked at the tip.
3. Upper jaw is movably articulated with the skull.

4. Feet zygodactylus, *i.e.,* two toes in front and two toes behind.
5. Outer toe is not reversible.

Examples : *1. Psittacula eupatra* (Large Indian Parrakeet), *2. Psittacula krameri* (Green parrots).

Order 15. Strigiformes

1. Head large and rounded.
2. Eyes large and directed forwards. Each in a disc of radial feathers.
3. Retina contains principally the rods.
4. Ear opening large, often with flap-like cover.
5. Beak short, feet with sharp claws adapted for grasping.
6. Nocturnal and predators.

Examples : *1. Bubo bubo* (Great-horned-owl), *2. Ketupa zeylonensis* (Brown-fish owl), 3. *Tyto alba*.

Order 16. Caprimulgiformes

1. Beak is small and delicate.
2. Mouth is wide and margined with long-bristle-like sensory feathers.
3. Feathers are soft and lax.
4. Legs and feet small and weak, not adapted for grasping.
5. Nocturnal and insectivorous.

Examples : *Caprimulgus* — Goat suckers and Nightjars.

Order 17. Apodiformes

1. Usually small birds.
2. Legs are very short and feet very small.
3. Wings are pointed.
4. Beak is small and weak or slender with long tubular tongue.

Examples : *1.* Swifts, *2.* Humming birds.

Order 18. Coliiformes

1. Small passerine birds.
2. First and fourth toes are reversible.
3. Tail is very long.
4. Insectivorous and frugivorous.

Example : *Colius* (Mouse bird-colies)

Order 19. Trogoniformes

1. Beak short and stout with bristles at the base.
2. Feet are small and weak.
3. Feathers are brilliant in colour, often green.

Example : Trogons *(Trogon).*

Order 20. Coraciiformes

1. Beak is strong.
2. Third and fourth toes are fused at the base.

Examples : King-fishers, Hornbills.

Order. 21. Piciformes

1. Tail-feathers are stiff with pointed tips.
2. Beaks are stout and owl-like.

3. Protrussible tongue is roughened or with barbs at the tip.
4. Two toes in front and one behind, not reversible.

Examples: *1.Dryobates* (Yellow-front pied woodpecker), *Brachypternus bengalensis* (Golden-beak woodpecker), Fakamars, Puffbirds, Barbets, Honey guide, Toucans.

Order. 22. Passeriformes

1. Toes three in front and one behind, adapted for perching.
2. Beaks are adapted for cutting.

Examples: *1. Passer domesticus* (Common house-sparrow), *2. Corvus splendens* (Common house-crow), *3. Acridotherus tristis* (Common myna), *4. Saxicoloides fulicata* (Indian robin), *5. Oriolus oriolus* (Golden oriole), *6.* Bulbul, *7.* Swallows, *8.* Horned larks, *9.* Gous, *10.* Finches, *11.* Starlings, *12.* Pipits, *13.* Black birds, 14. Wood Warblers, *15.* Tenagers, *16.* Wax wings, *17.* Kinglets.

MUSEUM SPECIMENS OF AVES

1. *Phoenicopterus* (Flamingo)

Classification:

Phylum	...	**Chordata**	→	Dorsal tubular nerve cord, notochord, and paired gill-slits are present.
Group	...	**Craniata**	→	Definite head and cranium with brain.
Subphylum	...	**Vertebrata**	→	Notochord is replaced by vertebral column; two pairs of appendages; circulatory system closed; hepatic portal system present; blood red containing R.B.Cs.
Division	....	**Gnathostomata**	→	Jaws and paired appendages are present.
Superclass	...	**Tetrapoda**	→	Paired limbs, lungs, cornified skin and bony skeleton.
Class	...	**Aves**	→	Warm blooded with an exoskeleton of feathers; forelimbs modified into wings; hindlimbs for walking, perching or swimming; skull monocondylic, modern birds without teeth, a horny beak present; heart four chambered; oviparous; embryo with amnion, allantois and yolk-sac.
Subclass	...	**Neornithes**	→	Tail short ends in a pygostyle; rectrices are arranged in a semicircle around the pygostyle; teeth absent; metacarpals fused with distal carpals to form carpometacarpus; claws absent in forelimbs; sternum well developed with a keel; vertebrae heterocoelous.
Superorder	...	**Neognathae**	→	Skull neognathus; feathers having interlocking mechanism; wings well-developed; tail vertebrae are 5 or 6; forelimbs with metacarpals joined with fingers included in the wings.
Order	...	**Ciconiiformes**	→	Long-necked and long-legged birds; feathers decorative; beaks abruptly decurved at middle in flamingo; toes without web; young born naked; food fish and other aquatic animals.
Genus	...	*Phoenicopterus* (Flamingo)		

Habit and habitat. Flamingos live in flocks at jheels, brackish lagoons, on tidal mudflats, ponds and lakes. They wade in shallow water and feed on worms, small molluscs, crustaceans, insect larvae and seeds, etc. They fly rapidly.

Comments:

1. *Phoenicopterus* (Fig. 14.1) is commonly called **Flamingo** and **Boghans** or **Charaj baggo** or **Raj Hans** in Hindi.

2. It is a pale rosy-white bird stands one and half metres high. Wing coverts are brilliant scarlet with black wing-border.

3. It has elongated bare pink legs, long sinuous neck and heavy pink bill turned down at an angle (broken) from about half of its length.

4. The toes are webbed like ducks.

5. Head bears eyes, nostrils at the base of beak and decurved beak.

6. The fleshy-tongue works like a plunger sieving out the water through the comb-like fringes or lamellae along the edge of the mandibles, leaving the minute food particles behind.

7. The method of procuring food is to wade into shallow water and feed with their long necks bent down and head completely immersed.

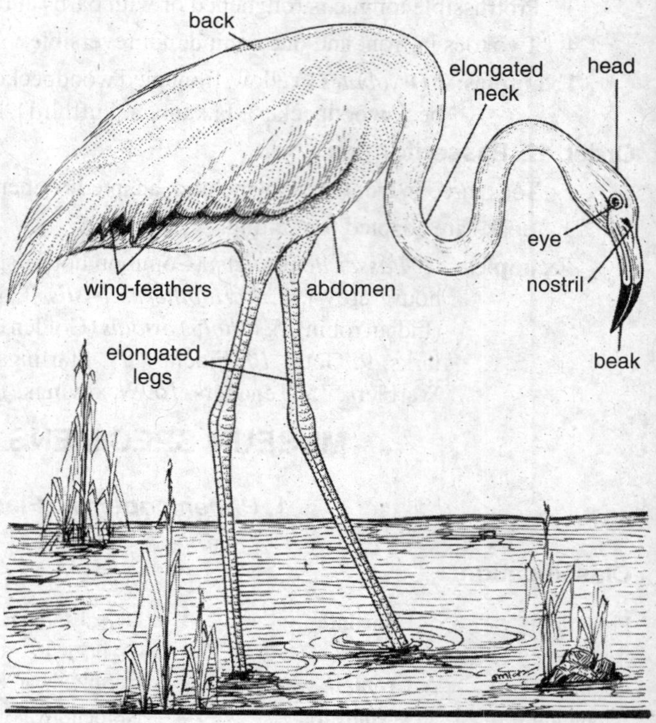

Fig. 14.1. *Phoenicopterus* (Flamingo).

8. Flamingos fly with fairly rapid wing beats in V-formation like geese. The slender neck is stretched in front, while the long red legs trail behind.

9. The nest is cone-shaped mound of scraped-up and plastered semi-liquid mud which becomes hard and sun-baked and has an average height of about 30 cm. A flat pancake-like depression is tamped on the top in which the eggs-2 or only 1-are laid. The incubating flamingo sits on this with its legs folded under and not standing astride the mound.

10. The only known breeding place of Flamingo is the Great Rann of Kutch whose vast numbers collect between October and March when water conditions are favourable.

Geographical distributuion. Flamingos are resident more or less throughout the Indian Union, Pakistan, Bangladesh. Sri Lanka, Afghanistan, France, East Africa, Siberia, etc. Eocene to Recent.

2. *Milvus* (Kite)

Classification: Same as that of Flamingo.

Order ... **Falconiformes** → Beak short and curved at the tip; mandibles sharp-edged; feet with sharp curved claws; diurnal and strong fliers.

Genus ... *Milvus*

Species ... *migrans*

Habit and habitat. *Milvus* (Kite) lives in the neighbourhood of human habitations and also in forests, etc. It feeds on offal and garbage, worms, winged termites, lizards, mice and rats, snakes and young birds, etc.

Comments :

1. *Milvus migrans* (Fig. 14.2) is commonly known as **Pariah-kite** and in Hindi it is called **Cheel**.

2. It is quite large in size grows up to 60 cm in length, and brown in colour.

3. Tail is forked, a distinguishing character from all other similar birds.

4. Head bears eyes, nostrils, ear openings and large, stout and decurved pointed beak adapted for tearing the prey.

5. Feet are feathered and toes are sharply clawed, 2 in front and 2 behind.

6. No sexual dimorphism. Both the sexes are alike.

7. Nesting season from September to April.

Geographical distribution. *Milvus* is distributed throughout India, Pakistan, Sri Lanka and Myanmar. Eocene to Recent.

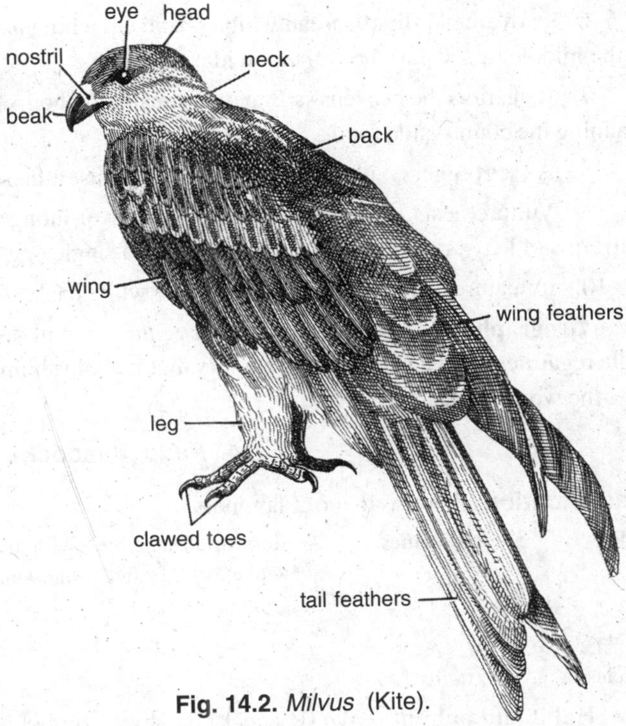

Fig. 14.2. *Milvus* (Kite).

3. *Gyps* (Vulture)

Classification: Same as that of Flamingo.

Order ... **Falconiformes** → Characters same as those of *Milvus.*

Genus ... *Gyps*

Species ... *bengalensis*

Habit and habitat. *Gyps* (Vulture) is found all over India except humid forests. It lives singly or in gatherings over trees. It feeds on variety of animals.

Comments :

1. *Gyps bengalensis* (Fig. 14.3) is commonly called **White-backed** or **Bengal vulture**. It is called **Gidh** in Hindi.

2. It is a heavy dirty blackish-brown rather repulsive-looking creature.

3. Head and neck are naked and scrawny. Head bears eyes, bill stout hooked at the tip and nostrils at soft naked cere.

4. Feet are provided with sharp claws and are usually adapted for grasping.

5. Mandibles are sharp-edged. Beak used for tearing muscles from dead animals.

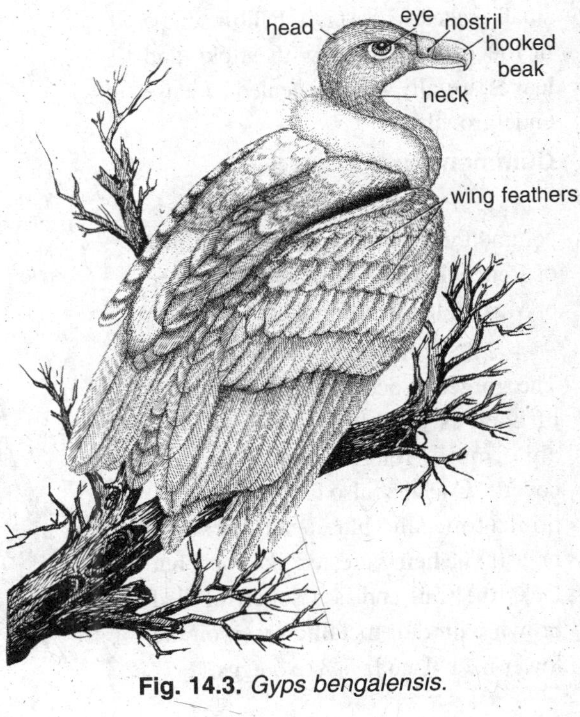

Fig. 14.3. *Gyps bengalensis.*

6. In overhead flight abroad whitish band stretching across the underside of the wings broken in the middle by the dark body, help in identification.

7. It quarters the heavens sailing majestically for hours on and on out spread motionless wings, scanning the countryside food.

8. As scavengers, vultures are of the greatest usefulness to man.

9. Vulture nests in large trees standing near or along the sides of roads building enormous platforms of leafy sticks and twigs and lays only a single egg.

10. Immature *Gyps* are brown without the white back.

Geographical distribution. *Gyps bengalensis* (Vulture) is found every where in peninsular India regardless of the nature of the country, but it avoids humid evergreen forest. *Gyps* are found all over the world. Eocene to Recent.

4. *Pavo* (Peacock)

Classification: Same as that of Flamingo.

Order ... **Galliformes** → Terrestrial game-birds with short and powerful flights; beak short; feathers with aftershaft; feet usually massive adapted for scratching and running; graminivorous.

Genus ... *Pavo*

Species ... *cristatus*

Habit and habitat. *Pavo* (Peacock) inhabits crop fields, dense scrubs, forest having streams and rivers. Peacocks are usually met within parties or droves of mixed sexes, chiefly in deciduous plains and foot hills forest. It feeds grains, tubers, vegetable shoots and also insects, lizards and small snakes. Its nest is a shallow scrape in the ground lined with sticks and leaves, usually well concealed in dense undergrowth.

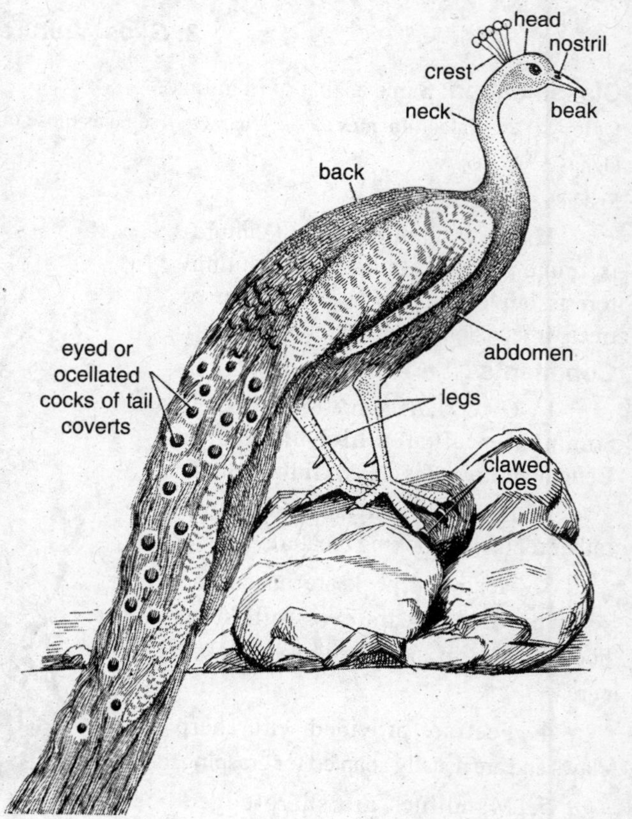

Fig. 14.4. *Pavo cristatus.*

Comments :

1. *Pavo cristatus* (Fig. 14.4) is commonly known as **Common peafowl** or **Peacock**. In Hindi it is called **Mor** or **Mayur**. It is a **national bird of India**.

2. Sexual dimorphism present. The gorgeous occellated or **eyed** trail of **cock** is actually not his tail but abnormally lengthened upper tail coverts. **Cock** has also fan-shaped crest, brilliant metallic blue head, neck and breast. The **hen** is crested like cock, but lacks the trail and is a sober mottled brown with dull metallic green on the lower neck. Hen lays 3 to 5 eggs.

3. Head bears eyes, nares at the base of short bill and ear-openings.

4. The feet are adapted for scratching and running.

5. At night they roost up in lofty trees, and at early dawn the jungle resounds with loud ugly screaming *may-awe* calls of the cocks.

6. The dance of male peacock with its georgeous tail-coverts spread like a fan is very famous. The peacock dances specially on cloudy and rainy days to attract hen.

Geographical distribution. *Pavo cristatus* is found in various parts of India. Eocene to Recent.

5. *Columba livia* (Pigeon)

Classification : Same as that of Flamingo.

Order ... **Columbiformes** → Beak usually short and slender; crop large producing pigeon milk.
Genus ... *Columba*
Species ... *livia*
Variety ... *domestica*

Habit and habitat. *Columba livia* is a very common and familiar bird found in the semi-domesticated condition, living as a commensal of man, warehouses, factory sheds, mosques, railway stations, goods yards and abandoned empty places. Its food are cereals, pulses and ground nuts, etc. Its nest is sketchy pad of twigs and straw. It breeds throughout the year.

Comments :

1. *Columba livia* (Fig. 14.5) is commonly known as **Blue-rock-pigeon** and in Hindi it is called **Kabutar**.

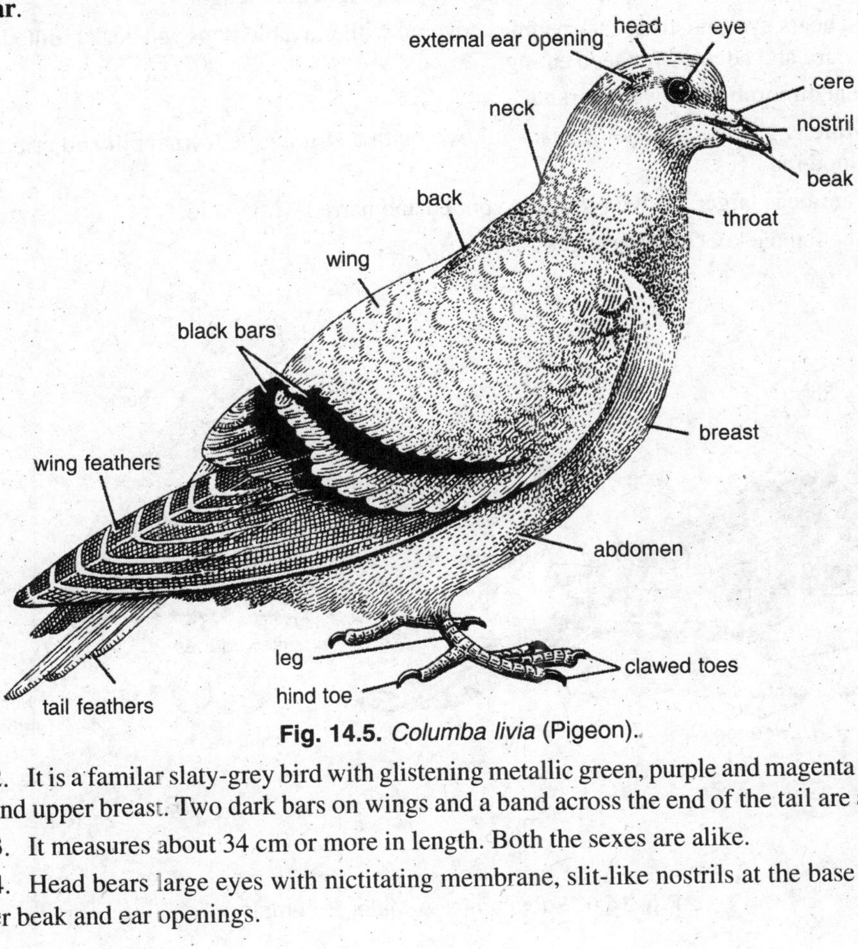

Fig. 14.5. *Columba livia* (Pigeon).

2. It is a familar slaty-grey bird with glistening metallic green, purple and magenta sheen on the neck and upper breast. Two dark bars on wings and a band across the end of the tail are also present.

3. It measures about 34 cm or more in length. Both the sexes are alike.

4. Head bears large eyes with nictitating membrane, slit-like nostrils at the base of short and slender beak and ear openings.

5. Forelimbs are modified into wings having skeleton and feathers and hindlimbs are covered with epidermal scales and adapted for bipedal locomotion.

6. Pigeons are domesticated for amusement. In olden times they were used as messengers for carrying informations. It is an example of Darwin's Theory of Artificial Selection of Evolution as a number of varieties have been produced by man.

Geographical distribution. *Columba livia* is distributed throughout India, Pakistan, Sri Lanka and Myanmar. Pigeons are also found in U.S.A. Eocene to Recent.

6. *Eudynamys* (Koel)

Classification : Same as that of Flamingo.

Order ... **Cuculiformes** → Toes two in front and two behind, out hind toe reversible, feet not adapted for grasping; tail long and moderate.

Genus ... *Eudynamys*

Species ... *scolopacea*

Habit and habitat. *Eudynamys* (Koel) lives on trees of gardens and grooves. Its (male) voice kuoo-kuoo is frequently heard during mango season. During winter its voice is not heard and presumed to have migrated. Its food consists of banyan and peepal figs, berries, insects and caterpillars.

Comments :

1. *Eudynamys* (Fig. 14.6 A, B) is commonly known as **Koel** or **Kokila.**
2. Body is slender with a longer tail, measuring up to 42 cm in length.
3. Head bears eyes, nostrils, ear openings covered with movable flaps and beak pointed at tip, curved downward and adapted for seed eating.
4. Sexual dimorphism is well marked.
   (*i*) **Male** is glistening metallic black all over with a striking yellowish bill and crimson or blood red eyes.
   (*ii*) **Female** is larger, brown profusely spotted and barred with white.
   (*iii*) The female koel has no song.

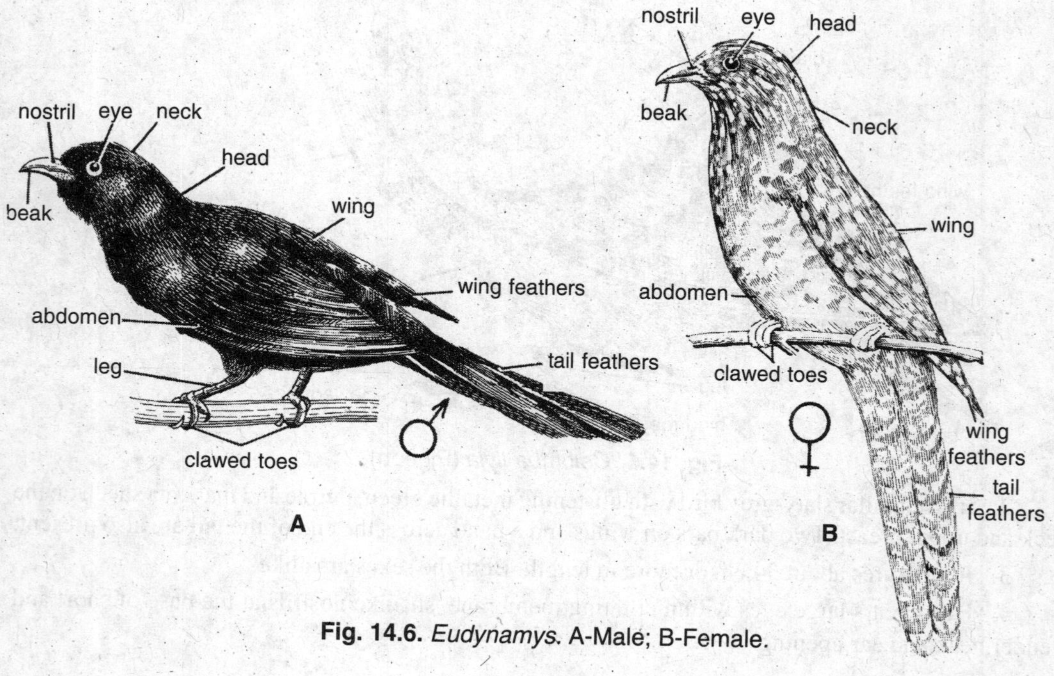

Fig. 14.6. *Eudynamys.* A-Male; B-Female.

5. It builds no nest of its own but deposits its eggs in crows nest leaving them to be hatched, and the young to be reared by the foster parents.

Geographical distribution. Koel is distributed throughout India, Pakistan, Sri Lanka, Myanmar. It is also found in China, Philippines and Australia.

7. *Psittacula* (Parrot)

Classification : Same as that of Flamingo.

Order ... **Psittaciformes** → Beak stout, narrow sharp-edged and hooked on the tip; upper jaw movably articulated with skull; feet zygodactylus, *i.e.,* two toes in front and two toes behind; outer toe not reversible.

Genus ... *Psittacula*

Species ... *krameri*

Habit and habitat. *Psittacula* (Parrot) lives in hollow tree trunks or one excavated by the bird itself but crevices and holes in buildings are also used for the purpose. It often bands itself into large flocks and is highly destructive at all times to crops and orchard fruits, gnawing and wasting far more than it actually eats. Food chiefly comprises the fruits and ripening grains.

Comments :

1. *Psittacula* (Fig. 14.7) is commonly known as **Rose-ringed Parakeet** or **parrot** and in Hindi it is called **Hiraman Tota**.

2. Body is slender with a long pointed tail feathers and measures about 25 cm in length.

3. It is grass green in colour, the red bill is short, stout and deeply-hooked, and black and rose-pink collar is present in the male and absent in the female. Female is entirely green.

4. Head bears eyes, nostrils at the base of decurved horny beak and ear openings.

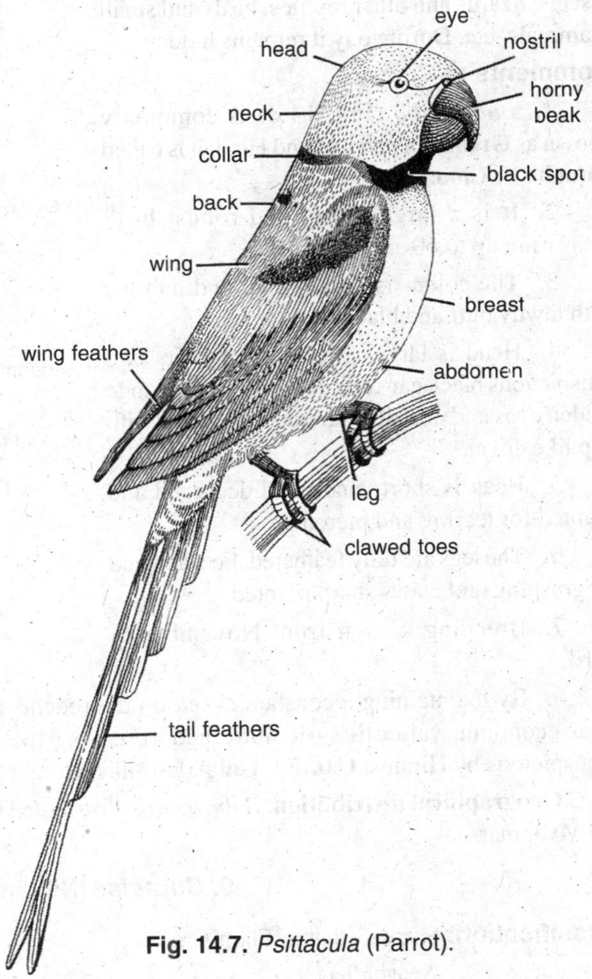

Fig. 14.7. *Psittacula* (Parrot).

5. Feet adapted for grasping and climbing. Foot zygodactylous–I and IV digits are directed backwards and II and III forward.

6. Nesting takes place from December to April.

7. It is a popular cage bird and can be taught to repeat a few words and sentences rather indistinctly.

Geographical distribution. *Psittacula krameri* is distributed throughout India, Pakistan, Sri Lanka, Myanmar and U.S.A. Eocene to Recent.

8. *Bubo* (Horned Owl)

Classification : Same as that of Flamingo.

Order ... **Strigiformes** → Head large and rounded; eyes large and directed forwards, each in a disc of feathers; retina contains principally the rods; ear-opening large often with flap-like cover; beak short; feet with sharp-claws ; nocturnal and predators.

Genus ... *Bubo*

Species ... *bubo*

Habit and habitat. *Bubo bubo* is nocturnal, residing in woody places or on shady rocky projections in a ravine or river. It feeds on large insects, lizards and other reptiles, birds and small mammals, etc. During day it remains hidden.

Comments :

1. *Bubo bubo* (Fig. 14.8) is commonly known as **Great-horned owl** and Hindi it is called **Ghughu** or **Uloo.**

2. It is a large heavy and robust bird, measuring up to 60 cm in length.

3. The colour is dark brown and mottled with tawny buff and black.

4. Head is large, rounded and bears two conspicuous black ear-tufts or horns, large orange golden eyes and nostrils. Ear openings often with flap like cover.

5. Beak is short, sharp and decurved and adapted for tearing and piercing.

6. The legs are fully feathered. Feet adapted for grasping and claws sharp pointed.

7. Breeding season from November to April.

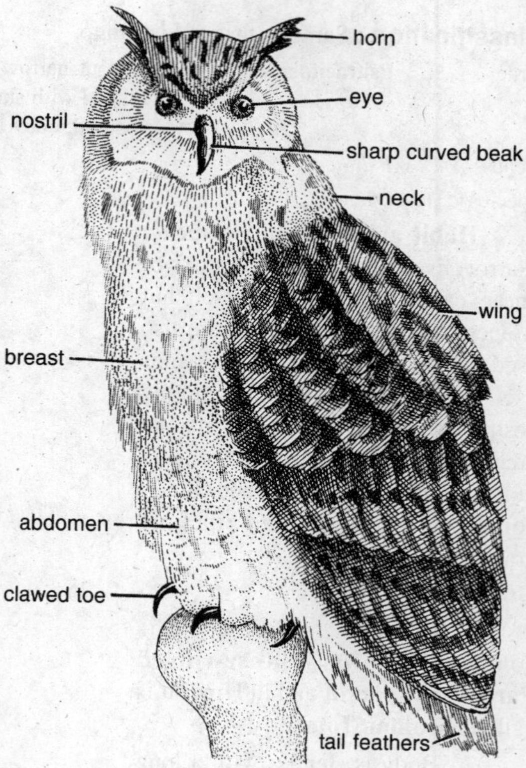

Fig. 14.8. *Bubo bubo* (Owl).

8. By maintaining a constant check on the rodents and other destructive vermin they are of great economic value to agriculture and deserving the strictest protection. Owl is considered unauspicious by Hindus. **Camillo Golgi** described Golgi apparatus from nerve net of owl.

Geographical distribution. *Bubo bubo* is distributed worldwide and throughout India, Pakistan and Myanmar.

9. *Coracias* (Nilkanth)

Classification: Same as that of Flamingo.

Superorder ... **Neognathae**

Order ... **Coraciformes** → Beak strong; third and fourth toes fused at the base.

Genus ... *Coracias*

Species ... *benghalesis*

Habit and habitat. *Coracias* (Neelkanth) affects open cultivated country and light deciduous forest. From a look out on a telegraph wire or other point of vantage it pounces upon some large insect, frog or lizard on the ground returning with it either to the same perch or flying leisurely across to another nearby. Here the quarry is battered to death and swallowed. Highly beneficial to agriculture since it destroys vast quantities of injurious insects.

Comments:

1. *Coracias benghalensis* (Fig. 14.9) is commonly known as **Indian roller** or **Blue Jay** and in Hindi it is called **Nilkanth** or **Sabzak**.

2. It is a Oxford and Cambridge blue bird about the size of a pigeon.

3. Head is big with heavy and black bill.

4. The breast is rufous-brown and pale-blue abdomen and under the tail.

5. The dark and pale blue portions of the wings show up as brilliant bands in flights.

6. Food chiefly comprises crickets, grasshoppers, beetles, and other insects but occassionally lizards, mice and frogs are also eaten.

Fig. 14.9. *Coracias* (Nilkanth).

7. The nest is a collection of straw, feathers and rubbish in natural tree hollows.

8. It is essentially an inhabitant of open cultivated country and avoids dense forests.

9. It is usually seen perched on an exposed tree stump or telegraph wires.

10. The bird does great service to agriculture by destruction of some injurious pests.

Geographical distribution. *Coracias benghalensis* (Nilkanth) is distributed throughout India, Pakistan, Sri Lanka and Myanmar.

10. *Dinopium* (Woodpecker)

Classification: Same as that of Flamingo.

Order ... **Piciformes** → Tail feathers stiff with pointed tips; beak stout and owl-like; protrusible tongue rough or with barbs at the tip; two toes in front and two or one behind, not reversible.

Genus ... *Dinopium*

Species .... *benghalense*

Habit and habitat. *Dinopium* (Woodpecker) inhabits the grooves, gardens and light scrub in villages. It ascends to tree trunks tapping on the bark and pear into cracks for ants, grubs, wood-boring beetles and other insects injurious to trees.

Comments:

1. *Dinopium benghalense* (Fig. 14.10) is commonly known as **Golden-backed woodpecker** and in Hindi it is called **Katphora**.

2. It is a small bird, about 22 cm in length.

3. Upper plumage is distinctive golden-yellow and black, while lower is buffy-white streaked with black.

4. The crown and occipital crest is crimson in male only partly so in the female.

5. Head bears large eyes, nostrils, ear openings and long, stout pointed beak adapted for wood chiselling and piercing.

6. The tongue is long, protrusible and barb-tipped, for extracting insects, etc.

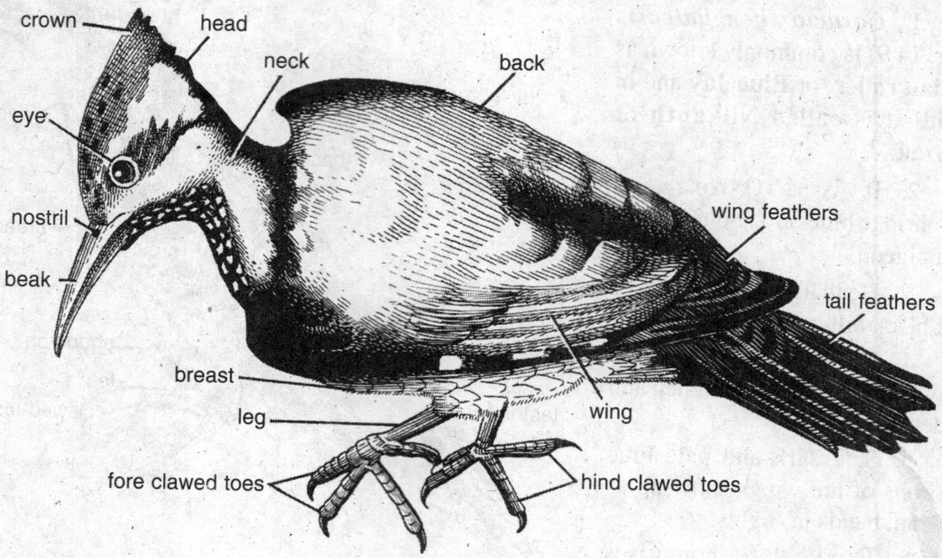

Fig. 14.10. *Dinopium* (Woodpecker).

7. Toes are four, two directed forward and two behind.
8. The tail is stiff and wedge-shaped.

Geographical distribution. *Dinopium* is distributed throughout India, Pakistan and Sri Lanka. *Dendrocopus mahrattensis* is also found in India. Its upper plumage has black and white spots on wings and tail feathers. Male has scarlet patches.

11. *Passer* (House Sparrow)

Classification: Same as that of Flamingo.

Order ... **Passeriformes** → Toes three in front and one behind, adapted for perching; beaks are adpated for cutting.

Genus ... *Passer*

Species ... *domesticus*

Habit and habitat. *Passer domesticus* is commensal on man, nesting in houses and commonly seen in humans habitat. They live in flocks in winter and feed on crop fields–ripening seeds of wheat and other cereals and also feed on insects injurious to crops.

Comments :

1. *Passer domesticus* (Fig. 14.11) is commonly known as **House Sparrow** and in Hindi it is called **Gauriyya**.
2. It is a small bird growing up to 10 to 12 cm in length.
3. Sexual dimorphism is exhibited. The upper surface is earthy-brown streaked with blackish and fulvous and underparts are whitish. The male has a black area on the throat and breast and female is ash white.
4. Head bears small eyes, nostrils, ear openings and short, conical beak adapted for seed eating.
5. Four clawed toes in each leg, three directed forward and one backwards. Feet adapted for perching.
6. Nest is a collection of straw and rubbish stuffed into a hole in a wall. Nesting throughout the year.
7. These are useful to agriculture by destroying large number of insect pests.

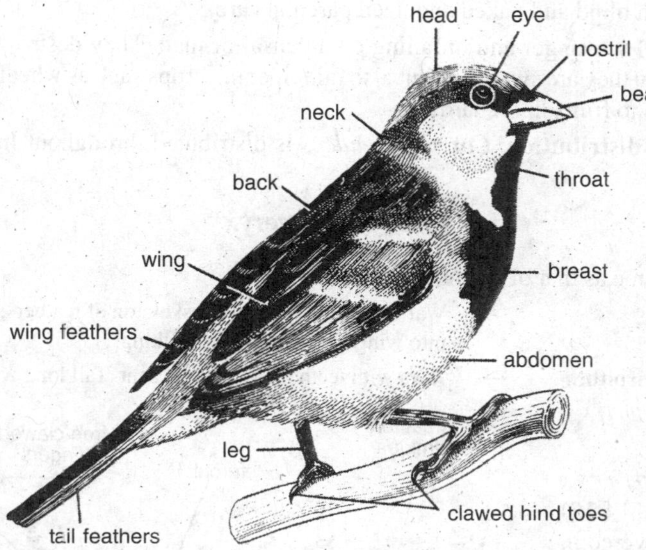

Fig. 14.11. *Passer domesticus* (House Sparrow).

Geographical distribution. *Passer domesticus* is distributed world wide except the Andmans and Nicobars. Eocene to Recent.

12. *Corvus* (Crow)

Classification: Same as that of Flamingo.

Order ... **Passeriformes** → Characters same as that of *Passer.*

Genus ... *Corvus*

Species ... *splendens*

Habit and habitat. *Corvus splendens* is the most familiar bird of villages and towns and is very smart and intelligent. Omnivorous. It eats almost anything, dead rat, kitchen refuse, fish, eggs, locusts, termites, fruits and grain, etc.

Comments :

1. *Corvus splendens* (Fig. 14.12) is commonly known an **House Crow** and in Hindi it is called **Kowwa** or **Cag**.

2. It grows in length from 32 to 42 cm. Its plumage colour is glossy black and the neck is grey.

3. Head bears large eyes, nostrils, ear openings and elongated, stout black coloured beak adapted for tearing and cutting.

4. Each leg bears 4 clawed toes, three directed forwards and one backwards, adapted for perching.

Fig. 14.12. *Corvus* (Crow).

5. Nest is a platform of twigs with a cup-like depression lined with tow, coir-fibre, etc.

6. Youngs born blind and naked and need parental care.

7. It is a useful scavanger and unfailing commensal of man. They destroy locusts and other injurious insects when they are swarming but also raid ripening crops such as wheat and maize and do considerable damage to fruits in orchards.

Geographical distribution. *Corvus splendens* is distributed throughout India, Pakistan, Sri Lanka and Myanmar.

13. *Archaeopteryx*

Classification : Same as that of Flamingo up to superclass.

Class ... **Aves** → Warm blooded with an exoskeleton of feathers; forelimbs modified into wings; hindlimbs for walking.

Subclass ... **Archaeornithes** → Jaws with teeth and wings present. Tail long with feathers.

Genus ... *Archaeopteryx*

Comments :

1. *Archaeopteryx lithographica* (Fig. 14.13) fossil bird was discovered in 1861 from Bavaria (Germany), belonged to Jurassiac period (140 million years ago).

2. Body was divisible into head, neck, breast, abdomen and elongated tail with rectrices.

3. Head was provided with eyes, nostrils, ear openings and long beak. Jaws were teethed like lizards.

4. Body was covered with scales. But forelimbs were modified into wings with remiges.

5. Legs were scaly and toes clawed one behind and 3 anterior.

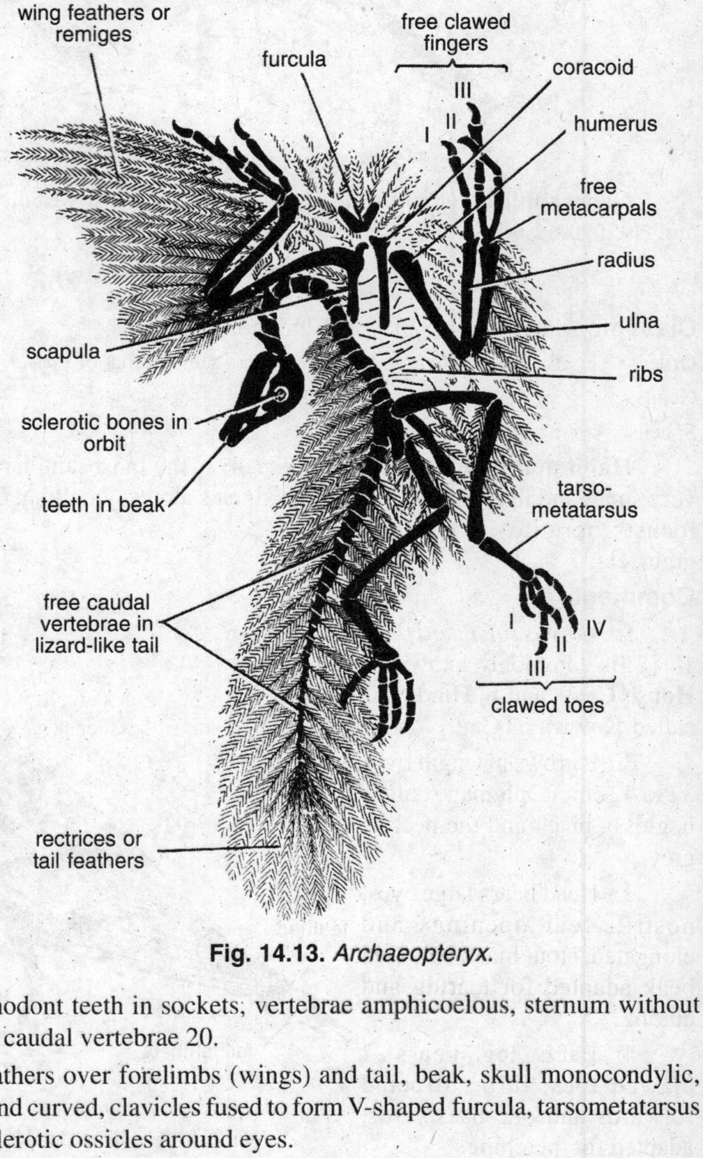

Fig. 14.13. *Archaeopteryx.*

Archaeopteryx had both reptilian and avian features.

Reptilian features were body covered with epidermal scales, brain simple with cylindrical cerebrum and small cerebellum, jaws with homodont teeth in sockets, vertebrae amphicoelous, sternum without keel, cervical vertebrae 9–10 and caudal vertebrae 20.

Avian characters were feathers over forelimbs (wings) and tail, beak, skull monocondylic, bones spongy, scapula elongated and curved, clavicles fused to form V-shaped furcula, tarsometatarsus present, hallux opposable, and sclerotic ossicles around eyes.

Thus, *Archaeopteryx* is a connecting link between reptiles and birds and provides proof of origin of bird.

TYPES OF FEATHERS IN BIRDS

1. Typical Quill Feather

Comments :

1. A typical quill feather (Fig. 14.14) has a strong **shaft** or **rachis** supporting the vane and characterised by the presence of interlocking arrangement of barbs and barbules.

2. It consists of a small proximal **calamus** or **quill** and an upper distal portion the **shaft** or **rachis**.

3. The calamus is hollow having a small aperture at its lower end the **inferior umbilicus** and another aperture, the **superior umbilicus** at the junction of calamus and shaft.

4. A small tufts of soft feathers, the **aftershaft**, near the superior umbilicus.

5. Shaft or rachis is solid and opaque, having the **ventral umbilical groove** running along the inner surface of the rachis throughout its length.

6. The **vane** is made up of two unequal lateral halves with its distal end narrower than its proximal end.

7. Each lateral half of the vane consists of numerous narrow, parallel and closely spaced thread-like structures the **barbs** with interlocking barbules.

8. The quill feathers are helpful in flight and are called **remiges** and over tail are called **rectrices.**

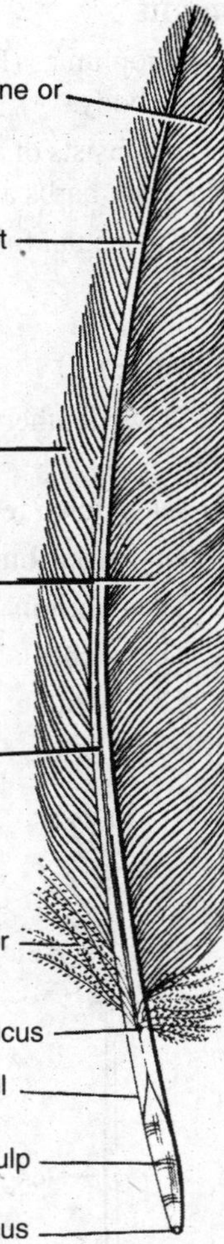

Fig. 14.14. Typical quill feather.

2. Contour Feather

Comments :

1. Contour feathers (Fig. 14.15) form the general covering of the body.

2. These are small feathers having poorly developed barbules which can be easily isolated.

3. The **calamus** is hollow and very small in comparison to that of the quill feather.

4. **Inferior umbilicus** lies at the proximal end of calamus.

5. The **rachis** is well developed but does not extend up to the distal extremity of the vane.

6. Both the halves of the vane are more or less equal in size.

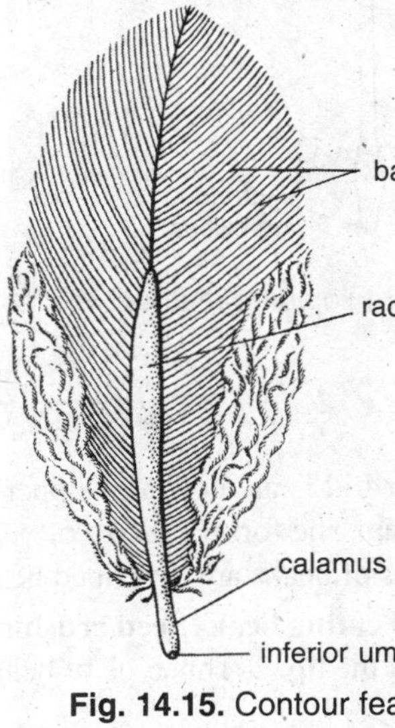

Fig. 14.15. Contour feather

3. Filoplume Feather

Comments :

1. Filoplumes (Fig. 14.16) are small, delicate and hair-like feathers found sparsely distributed all over the body.

2. It consists of a short **calamus** having **inferior umbilicus** at its proximal end and thread-like long few weak barbs and barbules at its free tip.

3. The actual function of these feathers is not known.

4. Down Feather

Comments :

1. Down feathers (Fig. 14.17) are very small, soft and wooly feathers found all over the body in young birds.

2. Each down feather consists of a calamus and a few long barbs and rudimentary barbules.

3. Numerous **barbs** arise as a tuft from the top of the short calamus.

4. The **calamus** is short having a **inferior umbilicus** at its proximal end.

5. The **rachis** is entirely absent.

6. These feathers conserve body heat in young birds and are replaced by contour feathers in most birds.

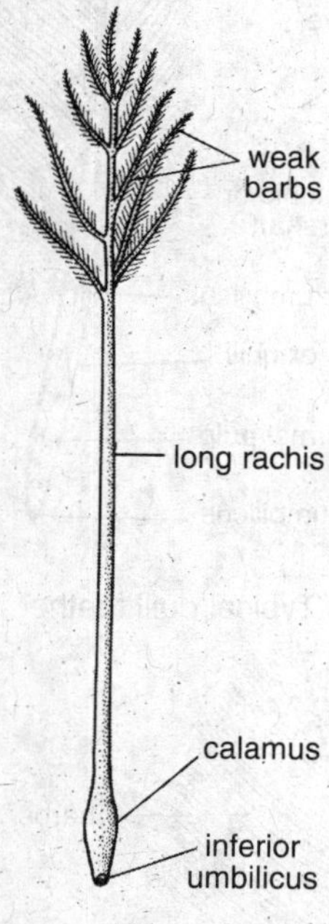

Fig. 14.16. Filoplume feather.

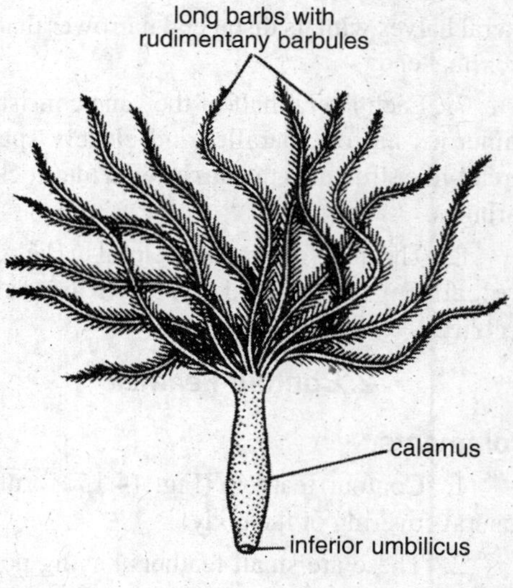

Fig. 14.17. Down feather.

TYPES OF BEAKS IN BIRDS

Beak or **bill** is an elongated upper and lower jaws and are covered by a horny sheath called **rhamphotheca.** The form and size of beak in various birds is related to their feeding habits. Some common types of beaks are described here.

1. Seed eating beak (Seed crushing break). 1. The seed eating beaks are short, stout, conical and pointed at the tip. 2. This type of beaks are characteristic of small seed eating or graminivorous birds such as **sparrows**, **finches** and **cardinals**, etc. 3. The weaker beaks are used to pierce the small seeds, while powerful beaks crush large, hard-shelled seeds, fruit stones, etc.

2. Cutting and biting beak. 1. The cutting and biting type beaks are long and strong and provided with sharp horny edges. 2. This type of beaks can be used for various purposes. 3. These are found in **crows** (*Corvus*) and **rovens**, etc.

3. Fruit eating beak. 1. The fruit eating beaks are large, sharp, powerful and hooked. 2. The upper beak is movable on the skull and is well adapted for tearing fruits and gnawing hard nuts and seeds. 3. Such beaks are found in **parrots** and **cockatoos**. In horn bills beak is large and heavy having cellular structure in its inner region that act as resonators to produce loud sound.

4. Piercing and tearing beak. 1. The piercing and biting beaks are short, pointed, sharp-edged and hooked at the tip and are adapted for piercing and tearing the flesh into pieces. 2. They are operated by well developed mandibular muscles. 3, These are commonly found in carnivorous birds such as **vultures**, **hawks**, **eagles**, **kites** and **owls**.

5. Insectivorous beak. 1. The insectivorous beaks are found in **swifts**, **swallows**, **flycatchers**. **night jars** and **hoopoe**, etc. 2. In **hoopoe** the beak is long, slender and slightly curved, adapted for turning the leaves and probing into soil for insect larvae, pupae and mites, etc. In swallows, frog-mouth and swifts, the beak is small, wide and delicate to prey upon flying insects. In fly catchers, the beak is small strong with notched mandibles at the tip end. Rictal bristles are also present at the base of the beak.

6. Fish catching beak 1. The fish catching beaks are long powerful and sharply pointed found in **storks**, **herons**, **kingfishers**, etc. 2. In **cormorant** the beak is slightly curved at the tip and have lateral tooth-like processes adapted for capturing the fish. In snake birds (Indian darters), these lateral tooth-like processes over the beaks are in the form of needle-like processes. 3. In **kingfisher** the beak is long powerful and sharply pointed to capture fish, frogs, tadpoles, molluscs and other aquatic animals.

7. Wood chiselling beak. 1. The wood chiselling beaks are long, straight, stout and chisel-like and are strongly articulated with the skull. 2. The skull bones are also thick and shock absorbent, Neck muscles are also strong. 3. This type of beaks are used for chiselling and drilling into bark or wood of trees for the prey, *e.g.*, insects and their larvae and for nest construction. 4. It is commonly found in **woodpeckers**.

8. Mud probing beak. 1. This type of beaks are very long, slender and slightly curved to probe the mud deep for searching the aquatic worms and larvae. 2. These are commonly found in **stilts, snipes, sand-pipers, jacanas, lapwings and curlews**, etc.

9. Mud and water straining beak. 1. This type of beaks are broad and flat, and the margins of the jaws are provided with transverse lamellae or horny serrations making it an efficient filter or sieve. 2. Because of the presence of transverse lamellae on the margin of jaws, the mud and water pass out leaving the food into the mouth. 3. This type of beaks are commonly found in **ducks**, **teals**, **geese** and **flamingos**. In flamingo, the beak is distally decurved. The two halves of lower jaw are enlarged to form a wide cavity that is covered over by upper jaw.

10. Flower probing beak. 1. This type of beaks are long, narrow and pointed for probing flowers and for sucking honey. 2. It is found in **humming birds**, which become stationary in air over the flowers for sucking honey. Their beaks are slightly decurved adapted for such function.

11. Pouched beak. 1. It is found in **pelecans**. 2. The beak is large having a large gular pouch. 3. The gular pouch is an extension of skin and attached to the mandible for storing engulfed fish.

12. Spatulate beak. 1. It is a characteristic of **spoon-bill**. 2. It is specially flattened throughout the length but ends in a broad spatulate or spoon-like expansion. 3. The spatulate expansion is used for dabbling in mud and water in search of fish, tadpoles, worms,insects and other small aquatic animals.

Fig. 14.18. Types of beaks in birds.

TYPES OF FEET OR CLAWS IN BIRDS

1. Running or cursorial feet. 1. In running birds, the legs are powerful and number of toes are reduced. 2. The hind toe is elevated, reduced or absent.3. Ratites, *e.g.*, **Rhea**, **Emu**, and **Cassowary** and **coursers**, **bustards** are short toed birds and having 3 toes directed forward. In **ostrich** the legs have only two toes, *i.e.*, 3rd or middle toe and 4th or outer toe is smaller and devoid of nail.

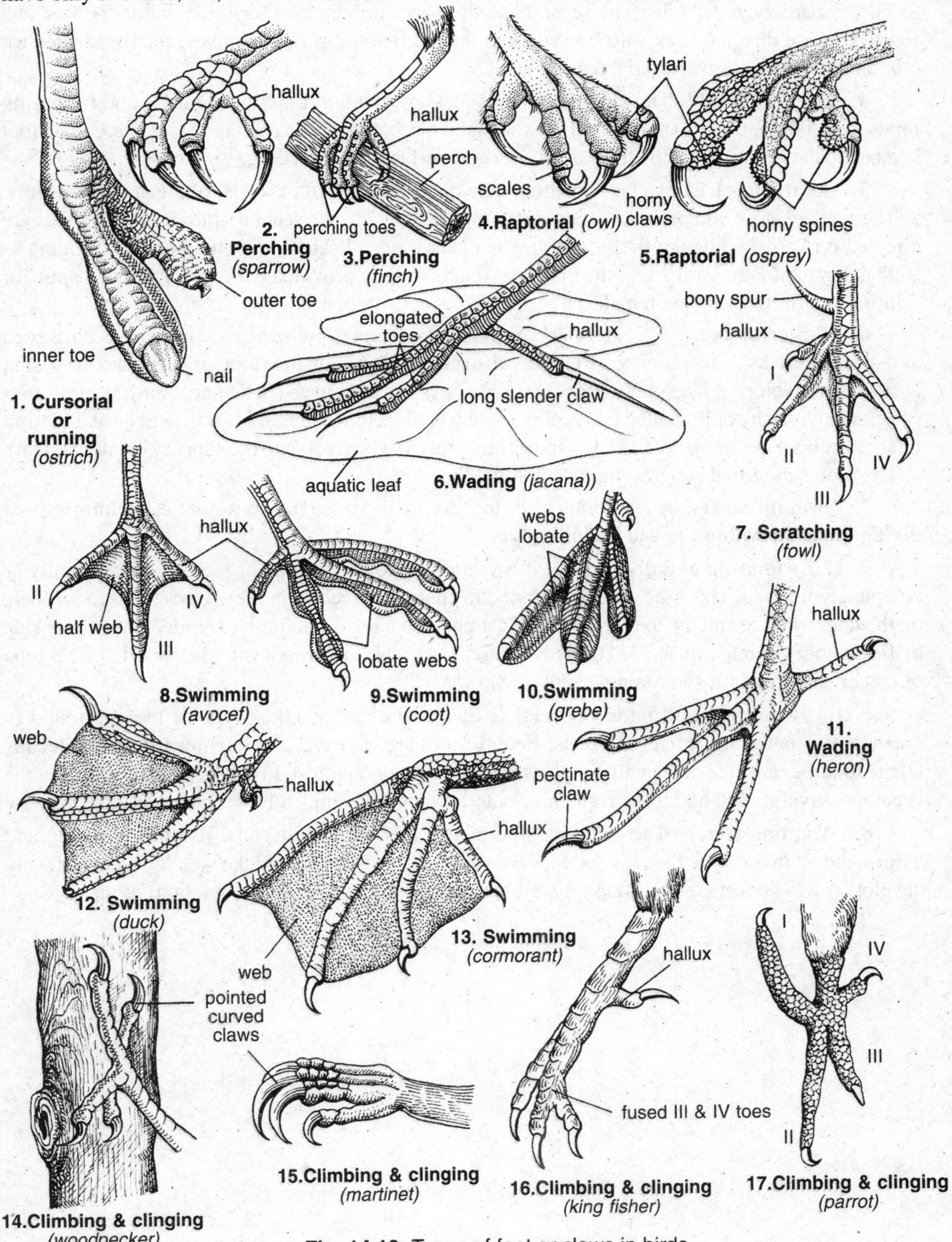

Fig. 14.19. Types of feet or claws in birds.

2. Perching feet. 1. The perching type of feet are found in majority of birds (passer birds), *e.g.*, **sparrows**, **crows**, **bulbuls**, **koels**, **mynas** and **robins**, etc. 2. In this type of foot there are four movables toes, three toes in front are slender and the hind or **hallux** is long, strong and sharply clawed. 3.Because of opposable toes these birds can fasten the foot to a branch of tree.

3. Climbing feet. 1. The climbing type of feet are found in **parrots**, **woodpeckers** and **hoopoes**. 2. The feet are zygotactylous. The second and third toes are directed forwards, while the first and fourth toes are directed backwards. 3. The foot is modified for grasping and especially adapted for climbing on vertical surfaces of trees and walls.

4. Clinging feet. 1. The clinging type of feet are found in **martinets**, **swifts** and **humming birds**. 2. All the four toes are directed forwards and have long, sharp and curved, claws. 3. The foot is modified for clinging on the trees or any vertical surface while feeding.

5. Scratching feet. 1. The scratching type of feet are found in **fowls, quails** and **pheasants**. 2. The feet are stout and provided with strong clawed toes. 3. The second, third and fourth toes are directed forwards, while the first toe is directed backwards. 4. The foot is modified for running as well as scratching the earth. 5. In the male, the foot is usually provided with a pointed bony **spur** for fighting and for treading the females.

6. Raptorial feet. 1. The raptorial type of feet are found in predatory or carnivorous birds such as **ospreys**, **hawks**, **kites**, **eagles**, **owls** and **vultures**, etc. 2. All the four toes are large and the hallux is strongly developed. The claws are large, strong, sharp and curved. The undersurface of all toes possess large fleshy bulbs called **tylari**. These are well developed in hawks. 3. In **osprey** and **ketupa** tylari are absent but sharp denticulated spines, are present which serve to the slippery creatures (fish). 4. The foot is modified for grasping and holding the prey.

7. Swimming feet. In swimming birds the feet are modified in two ways, *i.e.*, **swimming** and **diving type** and **swimming** and **paddling type**.

I. Swimming and diving feet. 1. Swimming and diving type of feet, toes are partially or completely webbed. This type of feet are found in **coots** and **grebes**. In these birds the web is lobate on the undersurface and the toes are free. 2. In grebes each toe is paddle-like bordered on either side by fleshy lobes of integument. 3. The nails are flat in a grebe and in coot slightly curved. 4. This type of feet are modified for swimming and diving in water.

II. Swimming and paddling feet. 1. Swimming and paddling type of feet are found in **cormorants**, **pelecans**, **ducks** and **teels**. Here the toes are webbed. 2. In **cormorants** and **pelecans** all the four toes are connected with tough web all along their length. 4. In **ducks** and **teals** the first toe is comparatively small and free, while three anterior toes are completely webbed.

8. Wadding feet. 1. The wadding type of feet are found in **herons**, **jacanas**, **lapwing** and **snipes**, etc. 2. In this type the legs and toes are very long and slender and the web is absent or feebly developed. 3. The feet are modified for walking over the aquatic leaves (vegetation) or marshes.

15
Chapter

CLASS–MAMMALIA

Mammals are covered with hair or fur and are warm-blooded (homoeothermous). The term mammal refers to mammary glands present in females for suckling youngs. In males, mammary glands remain undeveloped. The body insulation and separation of arterial and venous blood in the heart keep the body temperature constant. Like birds, rate of metabolism is also high and that needs more food. Cerebrum and cerebellum are highly developed and coordinate all the body activities.

Mammals live in all types of habitat like polar regions, (polar bear, arctic fox) tropical and subtropical regions, sea (walrus, seals, whales, sharks, etc.), forests, deserts and also air. Mammals also provide wool, meat and leather for human use. Horses, donkeys, camels are used in transportation, etc.

General Characters

1. Mammals are warm-blooded animals.
2. The skin is more or less covered with hairs (except Cetacea).
3. Sudoriphorous (sweat) glands and sebaceous (oil) glands are present in the skin.
4. Mammary glands in females, whose function is that of nourishing the young, are also present.
5. External ears (pinnae) are present.
6. Limbs are either plantigrade or digitigrade or unguligrade.
7. Teeth are heterodont (differentiation into incisors, canines and molars), thecodont (embedded in the alveolar pockets of jaw) and diphyodont (only two sets of teeth, a milk set replaced by permanent set).
8. Skull with two occipital condyles formed entirely by the exocciptals.
9. Skull is without prefrontals, postfrontals, quadrato-jugal, supraorbitals and basipterygoids.
10. The lower jaw is composed of a single bone, on each side, the **dentary** which articulates with squamosal of its side.
11. Vertebrae are gastrocentrous composed of three pieces, the centrum, and two epiphyses.
12. With few exceptions mammals possess seven cervical vertebrae.
13. The ribs articulate with the vertebrae by two heads, capitulum and tuberculum.
14. The digits in the fore- and hindlimbs are never more than five, but often reduced.
15. Presence of a muscular diaphragm between thoracic and abdominal cavities.
16. Heart is four chambered with only a left aortic arch.
17. RBCs are non-nucleated except Camelidae.
18. Brain with four optic lobes.
19. The kidney is metanephros.
20. Penis is always present in male.
21. Viviparous. The young, except in monotremes, develops in the uterus for some time and born alive.

CLASS – MAMMALIA

Subclasses

- I Prototheria
  - Order
    - Monotremata

      Ex. 1. *Echidna* 2. *Ornithorhynchus*
- II Allotheria Extinct
- III Theria

Subclass I – THERIA

Infraclasses

- I Metatheria
  - Order
    - Marsupialia

      Ex. 1. *Didelphis*, 2. *Macropus*, 3. *Perameles*
- II Eutheria

Infraclass II–EUTHERIA

Orders

1. Insectivora
   Ex. 1. Moles
   2. Shrews
   3. Hedgehogs
2. Dermoptera
   Ex. *Galeopithecus*
3. Chiroptera
4. Primates
5. Edentata
   Ex. 1. Anteaters
   2. Armadillos
   3. Sloths
6. Pholidota
   Ex. *Manis* (Pangolin)
7. Lagomorpha
   Ex. 1. Rabbits
   2. Hares
   3. Pikas
8. Rodentia
   Ex. 1. Squirrel
   2. Rat
   3. *Hystrix.*
9. Cetacea
   Ex. 1. Whales
   2. Dolphins
10. Carnivora
11. Tubulidentata
    Ex. Aard-varks
12. Proboscidea
    Ex. *Elephas* (Elephants)
13. Hyracoidea
    Ex. *Procavia*
14. Sirenia
    Ex. Sea-cows
15. Perissodactyla
    Ex. 1. Horses
    2. Zebra
    3. Asses
16. Artiodactyla

Order–Chiroptera

Suborders

I
Megachiroptera
Ex. *Pteropus, Cynopterus*
(Flying-foxes)

II
Microchiroptera
Ex. Bats.

Order–Primates

Suborders

I
Lemuroidea
Ex. *Lemur*
(Tree shrew)

II
Tarsioidea
Ex. *Tarsius*

III
Anthropoidea
Ex. 1. Monkeys
2. Apes
3. Man

Order–Carnivora

Suborders

I
Fissipedia
Ex. 1. Lion, 2. Tiger
3. Leopard, 4. Mangoose
5. Fox

II
Pinnipedia
Ex. 1. Sea-lion
2. Walrus
3. Seal

Note: The classification is adopted from **Storer** and **Usinger** (1957).

Subclass I – PROTOTHERIA

1. Primitive, reptile-like mammal. No external pinna on the ear.
2. Teeth are present only in young, adults and with horny beak.
3. A cloaca is present into which ureters and urinogenital sinus open.
4. Mammary glands are without nipples.
5. In the skull, tympanic bulla and lacrimals are absent; the jugal is reduced or absent.
6. The pectoral girdle possesses large coracoid bones and interclavicle.
7. The pelvic girdle possesses epipubic bones extending from the pelvis.
8. Vertebrae are without epiphyses.
9. The ribs have only a single head, the tuberculum being absent. The cervical ribs are present.
10. There is no corpus callosum in the brain.
11. The cochlea is a simple process of the sacculus and not coiled,
12. Testes are abdominal.
13. Oviparous.
14. Found in Australia, Tasmania and New Guinea.

Order 1. Monotremata

The general characters are same as those of Prototheria.

Examples : 1. *Echidna (*Spiny anteater*),* 2. *Ornithorhynchus* (*Platypus* or duckbill).

Subclass II – THERIA

1. Includes marsupial and placental mammals.
2. Ear usually with external pinna.
3. Teeth are usually present both in youngs and adults.
4. Cloaca is usually absent.
5. Mammary glands with nipples.
6. Testes usually in the scrotal sac.
7. Vasa deferentia and bladder opening through a common urethra in penis.
8. Ureters open into the base of the bladder.
9. Oviducts open into a vagina.
10. Viviparous.

Infraclass I – METATHERIA

1. Marsupium or brood pouch is present in the females.
2. Mammary glands are sebaceous and bearing nipples.
3. Epipubic (marsupial) bones are usually present and attached with the pubis.
4. Separate coracoid and interclavicle are absent.
5. Ribs bear two heads, tuberculum and capitulum.
6. Vertebrae with epiphyses.
7. Corpus callosum is feebly developed or absent.
8. Vagina and uterus are double (didelphic condition).
9. Viviparous.
10. Placenta is usually absent.

Order 1. Marsupialia

The general characters are same as that of Metatheria.

Examples : 1. *Didelphis* (Opossum), 2. *Dasyurus* (Tiger-cat), 3. *Caenolestes,* 4. *Notoryctes* (Marsupial mole), 5. *Parameles* (Bandicoot), 6. *Macropus* (Kangaroo).

Infraclass II – EUTHERIA

1. Marsupium (marsupial pouch) is entirely absent.
2. Epipubic bones are absent.
3. Cloaca is absent.
4. Corpus callosum is absent.
5. Mammary glands are well developed with nipples.
6. Ribs bear two heads, tuberculum and capitulum.
7. Urinogenital organs open independently of the rectum.
8. Testes are usually contained in scrotal sacs.
9. Vagina is single.
10. Viviparous.

11. The young always nourished for a considerable time in the uterus by means of allantoic placenta and born in a relatively advanced state.

Order 1. Insectivora

1. Small and furry animals.
2. Snout usually long and tapering.
3. Feet usually five toed with claws, inner toe not opposable.
4. Teeth sharp and pointed. Dental formula : *i* 3/3, *c* 1/1, *p* 4/4, *m* 3/3.
5. Placenta is discoidal.
6. Nocturnal and terrestrial.

Examples : 1. *Erinaceus,* 2. *Paraechinus* (Hedgehog), 3. *Scapanus,* 4. *Talpa* (Mole), 5. *Sorex,* 6. *Echinosorex* (Shrews).

Order 2. Dermoptera

1. Commonly known as "flying lemurs".
2. Membranes stretched between fore-and hindlimbs and also between the head and the forelimb and between the tail and hindlimbs.
3. Incisor teeth are 2/3.
4. Noctural in habit.

Example : *Galeopithecus.*

Order 3. Chiroptera

1. Commonly known as "true flying mammals.
2. Forelimbs are modified for flight. Second and fifth digits are greatly elongated, supporting the wing or flight membrane.
3. Hindlimbs are weak and have five clawed digits. Claws are sharp and curved.
4. Eyes are small and vision is weak.
5. Ears have large pinnae.
6. Teeth are sharp. Dental formula: *i* 2/3, *c* 1/1, *p* 3/3, *m* 3/3.
7. Sternum has a keel.
8. Mostly nocturnal in habit.

Suborder I – Megachiroptera

1. Large fruit-eating bats and also known as flying foxes.
2. Snout is elongated and without nose leaf.
3. The thumb and second digits are clawed.
4. Tail is absent.
5. Dental formula; *i* 2/2, *c* 1/1, *p* 3/3, *m* 2/3.
6. They live in flocks and sleep during the day on the branches of the tree, hanging head downward with wings folded.
7. Food mainly comprises fruits, *e.g.*, figs and guava.

Examples : 1. *Pteropus,* 2. *Xantharpyia,* 3. *Cynopterus.*

Subroder II – Microchiroptera

1. Small insectivorous bats.
2. Snout short usually with nose leaf.
3. Only thumb is clawed.

4. Tail, if present, is attached with the wing.
5. Nocturnal and usually gregarious in habit.
6. During the day, they hang by hind claws on the branches of trees or crevices of rocks, in caves or buildings keeping the head downwards.

Examples : 1. *Magaderma,* 2. *Rhinolophus,* 3. *Desmodus,* 4. *Eptesicus* (Brown bat), 5. *Myotis,* 6. *Vespertilo.*

Order 4. Primates

1. Completely hairy and generally arboreal mammals.
2. Hands and feet are more or less prehensile.
3. Five digits on each hand and foot provided with flat nails instead of claws.
4. Thumb and great toe are shorter and opposable to the other digits.
5. Mode of walking is plantigrade.
6. Orbits are directed forwards and surrounded by bony ring.
7. Clavicles are always present.
8. The femur has no third trochanter
9. Testes are enclosed in the scrotum and the penis is pendent.
10. Two pectoral mammae with teats are usually present.
11. Cerebrum is usually large and well convoluted
12. Placenta discoidal or metadiscoidal and haemochorial.

Suborder I – Lemuroidea

1. Exclusively arboreal and nocturnal.
2. Head with a sharp muzzle.
3. The long tail is never prehensile.
4. Forelimbs are shorter than hindlimbs.
5. The second toe or finger with claw but the rest of the fingers have flat nails.
6. Orbit is not surrounded by bone behind but opens into the temporal fossa.
7. Pectoral mammae are present.

Examples : 1. *Lemur,* 2. *Indris,* 3. *Chirogale,* 4. *Chiromys* (Aye-aye) 5. *Loris.* 6. *Tupaia.*

Suborder II – Tarsioidea

1. Eyes are extraordinarily developed.
2. Ears are large.
3. The limbs are extremely long, the hindlimbs longer than the forelimbs.
4. The digits have adhesive pads and flat nails except the second and third toes which have claws.
5. The tail is longer than the body with a tuft at the tip.
6. Nocturnal, arboreal and insectivorous.

Example : *Tarsius spectrum.*

Suborder III – Anthropoidea

1. Teats are always restricted to the thoracic region.
2. The orbit is completely closed behind from the lateral fossa.
3. Teeth are 32 to 36.
4. Hands and feet are prehensile except in man.

5. Fingers and toes have flat nails except in few forms.
6. The cerebral hemispheres are more highly developed and covering the cerebellum.
7. Terrestrial or arboreal and diurnal.

Examples : 1. Monkeys, 2. Apes, 3. Man.

Order 5. Edentata

1. Teeth are present or absent, if present, incisors and canines are always absent.
2. Feet have well developed claws.
3. The testes are abdominal.
4. The clavicle is always present.
5. Coracoid fused with acromian process.
6. Ischium unites with sacrum.
7. Lumbar vertebrae have extra articulating surfaces.

Examples : 1. Sloth, 2. Armadillos, 3. American ant-eaters.

Order 6. Pholidota

1. Body covered by large over-lapping horny plates.
2. Hairs are found between the plates.
3. The snout is elongated
4. Teeth are entirely absent.
5. Tongue is long, sticky and protrusible.
6. Ears are reduced.
7. The limbs are short each bears five digits. Forelimbs have well developed claws.
8. Nocturnal and burrowing in habits.

Examples : *Manis* (Pangolin or scaly ant-eater).

Order 7. Lagomorpha

1. Small to moderate sized mammals.
2. Two pairs of incisors are present in the upper jaw, Canines are absent.
3. Tibia is fused with the fibula.
4. Testes are always external in the scrotum.
5. Tail is short.
6. Soles of feet are hairy.
7. Toes bear claws.

Examples : Rabbits and Hares.

Order 8. Rodentia

1. Usually small mammals.
2. Limbs usually with five toes and claws.
3. The incisors, single pair, are long chisel-like and rootless and persistently growing with enamel confined chiefly to the anterior end.
4 Canines are entirely absent.
5. Space known as diastemma is present between the incisors and molars.
6. Testes are usually abdominal.
7. Placenta is discoidal and haemochorial.

Examples : 1. *Funambulus* (Squirrel), 2. *Rattus rattus* (House rat), 3. *Hystrix* (Porcupine), 4. Guinea pig (*Castor*), etc.

Order 9. Cetacea

1. Medium sized to very large aquatic mammals.
2. Body is usually spindle-shaped and fish-like.
3. The hairy covering of skin is reduced to few bristles on the muzzle.
4. Head long often pointed without neck.
5. Eyes are minute.
6. Nostrils on the top of the head and ear openings are minute.
7. Forelimbs are modified as paddles, digits are enclosed in a common integument devoid of claws.
8. Hindlimbs are entirely absent.
9. Tail is long ending in two broad transverse fleshy-flukes and notched in the mid-line.
10. Fleshy dorsal fin is present in some forms.
11. Dorsal and tail fins have no skeletal elements.
12. A thick layer of fat (blubber) is present beneath the skin.
13. Bones of skull are spongy and contain fat.
14. Stomach is complex and consists of three or four chambers.
15. Testes are abdominal.
16. Placenta is non-deciduate and diffuse.

Examples : Whales, Dolphins, Porpoise (*Phocaena*).

Order 10. Carnivora

1. Small to large quadrupeds, terrestrial, arboreal or aquatic.
2. Teeth generally have sharp and cutting edges and the canines are well developed. The incisors are small and always 3 on each side in each jaw.
3. The toes or digits never less than four and they all have strong and sharp claws.
4. The temporal fossa is open behind.
5. Tympanic bulla are large and rounded.
6. The clavicles are incomplete or absent.
7. In the hand scaphoid and lunar bones are always fused.
8. The stomach is always simple, caecum, if present, is always small.
9. The brain is well developed and the hemispheres are well convulated.
10. Mammae are abdominal.
11. Placenta is deciduate and zonary.

Suborder 1 – Fissipedia

1. The incisors are always six in each jaw and are feebly developed in many cases.
2. The canines are always very large and strong.
3. The last premolar in the lower jaw known as **carnassial** or **sctorial** teeth and the last two molars are crushing teeth.
4. Tympanic bulla is often very inflated.
5. The lower jaw has a high coronoid process and condyle is transversely elongated.
6. The clavicle is always small and some times quite rudimentary or even absent.
7. The digits are mostly five and never less than four. All the digits are separate and have claws.

Examples : 1. *Panthera leo* (Lion), 2. *Panthera tigris* (Tiger), 3. *Prionalurus bengalensis* (Leopard cat). 4. *Acinonyx* (Hunting leopard), 5. Civet cat, 6. *Herpestes* (Mangoose), 7. *Hyaena,* 8. *Canis (*Wolf*)*, 9. *Vulpes* (Fox), 10. *Lutra (*Otter*)*, *etc.*

Suborder II – Pinnipedia

1. Marine carnivores.
2. The greater part of the limbs are enclosed within the body skin and hands and feet are fully webbed.
3. The claws are reduced and the digits are increased in number.
4. Tail is very short and useful in swimming.
5. External ear small or degenerated.

Examples : 1. *Otario jubata* (Sea-lion), 2. *Odobenus* (Walrus), 3. *Phoca* (Seal).

Order 11. Tubulidentata

1. Body stout, somewhat pig-like.
2. Skin is very thick and covered with sparse hairs.
3. Snout is long and tubular with round nostrils at the tip.
4. Ears are long, erect and pointed.
5. Tongue is slender and protrusible.
6. Milk teeth numerous and permanent teeth fewer.
7. Digits 4 or 5 with heavy claws.
8. Placenta is zonary.

Example : *Orycteropus* (Aard-vark).

Order 12. Proboscidea

1. Largest and highly specialised terrestrial mammals.
2. The skin is thick and sparsely covered with hairs.
3. Long proboscis or trunk which is an elongation of the nose and upper lip, bearing the nostrils at the tip.
4. Eyes are small and ears are large.
5. The legs are large and pillar-like and bear five functional digits on both fore-and-hind-limbs.
6. The incisor teeth of the upper jaw developed into great tusks.
7. Canines are altogether absent. Molars are lophodont.
8. Dental formula : *i* 1/0, *c%, pm 3/3, m 3/3.*
9. Clavicle is absent.
10. The stomach is simple and the intestine is provided with caecum.
11. The cerebral hemispheres are much convoluted but they do not cover the cerebrum.
12. Testes are abdominal in position.
13. Two pectoral mammae.
14. Placenta non-deciduate and zonary.

Example : *Elephas* (Elephant).

Order 13. Hyracoidea

1. Small rabbit-like animals with a split snout.
2. Ears are short and the tail is much reduced.
3. Dentral formula : *i 1/2, c 1/0, pm 4, m 3/3.*
4. Four digits in the forelimbs and three in hindlimbs.
5. Clavicles are absent.
6. Dorsal gland is present.

7. Testes are abdominal.
8. Mammae are six pairs, four pairs inguinal and two pairs axillary.
9. Placentra is zonary.

Example : *Procavia* or *Hyrax.*

Order 14. Sirenia

1. Aquatic mammals commonly known as **sea cows**.
2. Body stream-lined or spindle-shaped.
3. Hindlimbs are absent.
4. Forelimbs modified as paddles.
5. Tail flattened with lateral flukes.
6. Nostrils are on the upper surface of the snout.
7. Clavicles are absent.
8. Stomach complex, consisting of several chambers.
9. Testes are abdominal.
10. Two pectoral mammae are present.
11. Placenta is non-deciduate and zonary.

Examples : 1. *Halicore dugong* (Dugong), 2. *Trichechus* (Manatees).

Order 15. Perissodactyla

1. Large sized hoofed mammals.
2. The middle digit of fore-and-hindlimbs is pre-eminent and carries most of the weight.
3. Teeth are lophodont.
4. Dorso-lumber vertebrae are 23.
5. Stomach is simple, caecum is large and sacculated.
6. Mammae are inguinal.
7. The brain is well convoluted.
8. Placentra is diffuse and epitheliochorial.

Examples : 1. *Equus* (Horse), 2. *Zebra.* 3. *Tapirus* (Tapir), 4. *Rhinoceros.*

Order 16. Artiodactyla

1. Terrestrial and semi-aquatic mammals.
2. Fore-and hindlimbs bear two rarely four digits.
3. Teeth are **selenodont** or **bunodont**.
4. Dorso-lumber vertebrae are 19.
5. Stomach is complicated and the coecum is small.
6. Mammae are few and inguinal or many and abdominal.
7. Placentra is diffuse or cotyledonary.

Examples : 1. *Hippopotamus* (Horse of river), 2. *Sus* (Pig), 3. *Dicotyles* (Peccaries), 4. *Camelus* (Camel), 5. Deer, 6. Girraffe, 7. Antelope, 8. Cow, 9. Buffalo, 10. Sheep, 11. Goat, etc.

MUSEUM SPECIMENS OF MAMMALIA

1. *Echidna (Tachyglossus)* (Spiny Anteater)

Classification :

Phylum	...	**Chordata**	→ Dorsal tubular nerve cord, notochord, and paired gill-slits.
Group	...	**Craniata**	→ Definite head cranium with brain
Subphylum	...	**Vertebrata**	→ Notochord is replaced by vertebral column; two pairs of appendages; circulatory system closed; hepatic portal system present; blood red containing R.B.Cs.
Division	...	**Gnathostomata**	→ Jaws and paired appendages are present.
Superclass	...	**Tetrapoda**	→ Paired limbs, lungs, cornified skin and bony skeleton
Class	...	**Mammalia**	→ Warm blooded; skin covered with hairs; mammary glands in females; pinnae present; teeth heterodont, thecodont, and diphyodont; skull with two occipital condyles; vertebrae gastrocentrous; limbs either plantigrade, digitigrade or unguligrade; presence of muscular diaphragm; heart four chambered; R.B.Cs. non-nucleated; kidneys metanephros; penis always present; viviparous.
Subclass	...	**Prototheria**	} → Pinna absent; teeth only in young, adults with horny beak; cloaca having the opening of ureters and urinogenital sinus; mammary glands without nipples; vertebrae without epiphyses; skull without tympanic bulla and lacrimals; corpus callosum in the brain; testes abdominal; females oviparous.
Orders	...	**Monotremata**	
Genus	...	*Echidna*	
Species	...	*aculeata*	

Habit and habitat. *Echidna* is terrestrial, burrowing and nocturnal. It feeds on ants by its long protrusible tongue.

Comments :

1. *Echidna aculeata* (Fig. 15.1) is commonly called **spiny-ant eater**. It is about 42 cm long. Body divisible into head, neck and trunk.

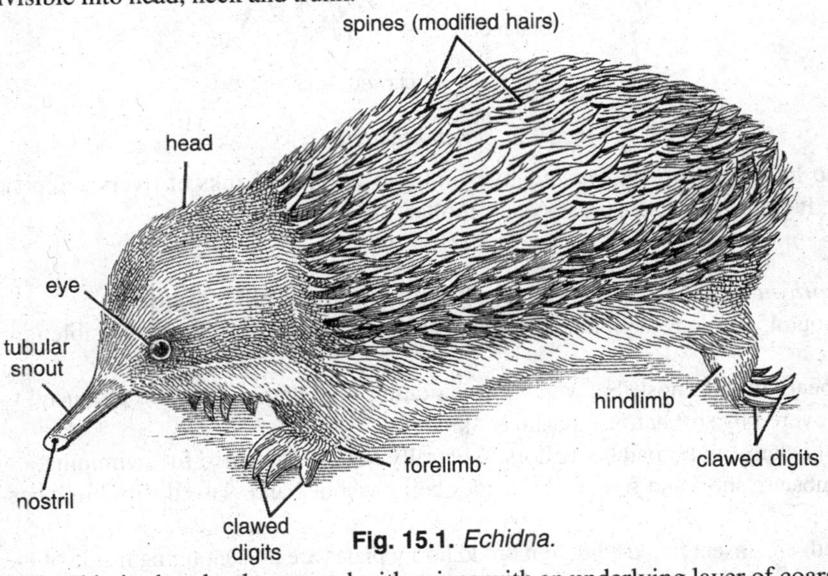

Fig. 15.1. *Echidna.*

2. The skin is abundantly covered with spines with an underlying layer of coarse hairs.

3. Head is small and projected into long and tapering, snout. The eyes are small and external ear or pinna is absent.

4. The tail is rudimentary.

5. Reptile-like five clawed digits are present on each limb. The claws are very long and powerful and are used for tearing down ant-hills or termite nests and for making burrows.

6. Second digit of hindlimb has a long and curved toilet claw to clean spines.

7. Male and female both possess mammary glands secreting milk to feed youngs. Teats are absent.

8. The tongue is extremely long and protrusible and is covered with a sticky salivary secretion for holding the ants. Posterior region of tongue is serrated to grind insects.

9. Teeth are absent in adults.

10. A small spur connected with a peculiar gland is present on the heal.

11. *Echidna* rolls its body to erect its spines to defend itself.

12. It is a connecting link between reptiles and mammals having characters of both the classes.

Geographical distribution. *Echidna aculeata* is found in New Guinea, Tasmania and Australia.

2. *Ornithorhynchus* (Duck-bill Platypus)

Classification : Same as that of *Echidna.*

Genus ... *Ornithorhynchus*

Species ... *anatinus*

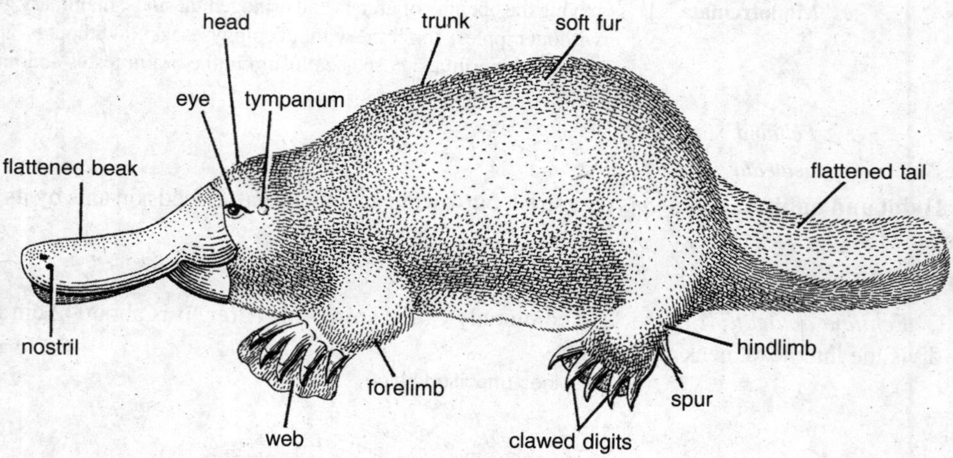

Fig. 15.2. *Ornithorhynchus.*

Habit and habitat. *Ornithorhynchus* lives in burrows on the banks of rivers and ponds and aquatic in habit. It feeds on worms, crustaceans, molluscan shells, etc.

Comments :

1. *Ornithorhynchus* (Fig. 15.2) is commonly known as **duck-bill platypus**.

2. It is about 45 cm long with a heavy coat of soft dark brown fur. Body divisible into head, trunk and tail.

3. Head bears mouth, nostrils, eyes with nictitating membrane. Jaws are produced to form flattened beak covered by soft hairless sensitive skin.

4. The tail is large and broad being dorso-ventrally flattened adapted for swimming.

5. The limbs are short and five toed, the toes being webbed and clawed. Hindlimb has horny spur.

6. The teeth are absent in the adult, but broad horny plates are present lining inside of the bill.

7. External ear-lobe or pinna is absent.

8. The male has a horny spur on the heal.

9. Cloaca present. Testes abdominal and penis conducts only sperms. Uterus or vagina absent.

10. Coracoid and precoracoid are present and interclavicle is T-shaped.

11. Eggs 1–3 are laid in nest in spring by the female. Youngs hatched and nursed by the female with its milk secreted by scattered milk glands present over the abdomen.

12. *Ornithorhynchus* possesses characters of reptiles and mammals and certain intermediate characters are also present.

(i) Reptilian characters are urinogenital system and precoracoids, and pinna and corpus callosum absent.

(ii) Mammalian characters are presence of hairs, diaphragm, 4-chambered heart, 3-ear ossicles, etc.

(iii) Intermediate characters are mammary glands without nipples, acromian spines in the scapulae and body temperature 25–28°C.

Geographical distribution. *Ornithorhynchus* is found in Eastern and Southern part of Australia and Tasmania.Triassic to Recent.

3. Macropus (Kangaroo)

Classification: Same as that of *Echidna*.

Subclass ... **Theria** → Placental mammals; ear usually with external pinna; teeth present both in youngs and adults; cloaca absent; mammary glands with nipples; testes in scrotal sacs; females viviparous.

Infraclass ... **Metatheria**
Order ... **Marsupialia** } → Brood pouch present in females; mammary glands sebaceous and with nipples; vertebrae with epiphyses, corpus callosum feebly developed or absent; vagina and uterus double (didelphic condition); viviparous; placenta usually absent.

Genus ... *Macropus*

Habit and habitat. *Macropus* is terrestrial, gregarious, herbivorous and browsing in habit. They cause much damage to crops.

Comments :

1. *Macropus* (Fig. 15.3) is commonly called **Kangaroo.** It is upto 2 metres while standing. Male is larger than the female. Male weighs about 100 kg and female 32 kg.

2. Body divisible into head, neck, thorax, abdomen and tail. Tail is powerful , non-prehensile and used as third leg in locomotion and at rest.

3. Head is small having mouth, nostrils and large eyes and pinnae.

4. The forelimbs are small having 5-clawed digits and hindlimbs and feet very long and powerful for leaping.

5. Hindlimbs have four digits, hallux being absent, second and third metatarsals and digits are thin and small and are united together by integument. The fourth metatarsal and digits are very large with a strong claw called (toilet claw) which is used in fighting, the fifth toe is small.

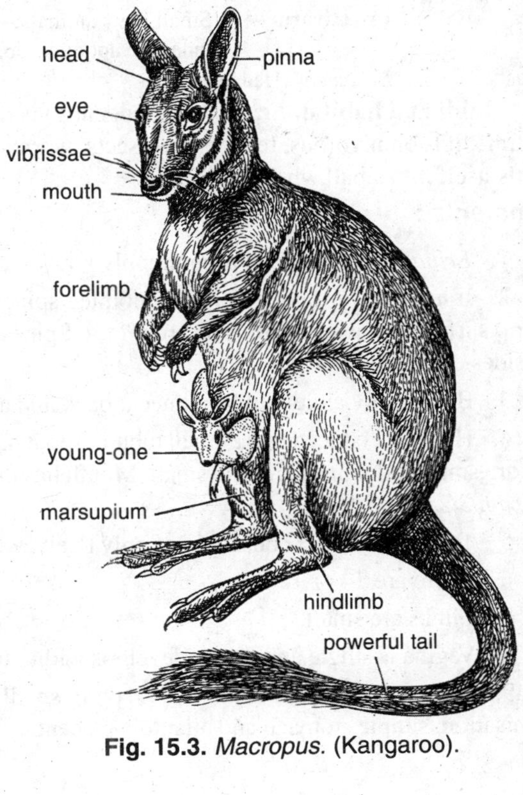

Fig. 15.3. *Macropus.* (Kangaroo).

6. Teeth are numerous with 5 incisors in each upper jaw and 3 in the lower jaw, there are 3 premolars and 4 molars in each half of each jaw; cheek teeth have grinding surfaces.

7. Locomotion is bipedal. It moves rapidly by long erect springs of over 8 metres by jumping, the tail is used for balancing.

8. Vertebrae have prominent metapophyses and anapophyses. There are no cervical ribs, sacrum is formed of a single vertebrae.

9. Pectoral girdle has large scapula with a spine, coracoid is reduced, clavicles are large, but there is no interclavicle.

10. The young develops in the uterus with only a yolk-sac placenta, they are born immature and naked, grub-like 2.5 cm long, they are transferred into a large masupium or brood pouch of the mother.

11. In the marsupium the young are fed on milk from true mammary glands having nipples. The young venture out when they are four months old.

Geographical distribution. *Macropus* is found in Australia, Tasmania and New Zealand in grassy plains.

4. *Erinaceus* (Hedgehog)

Classification : Same as that of *Echidna.*

Subclass	**Theria** →	Placental mammals; ear usually with external pinna; teeth present both in youngs and adults; cloaca absent; mammary glands with nipples; testes in scrotal sacs; females viviparous.
Infraclass	**Eutheria** →	Marsupial pouch absent;mammary glands well developed with nipples; cloaca absent; corpus callosum absent; vagina single; young always nourished for a considerable time in the uterus by allantoic placenta and born in a relatively advanced state.
Order ...	**Insectivora** →	Small furry animals; snout long and tapering; feet five toed with claws; inner toe not opposable; teeth sharp and pointed; placenta discoidal.
Genus ...	*Erinaceus* (Hedgehog)	

Habit and habitat. *Erinaceus* (Hedgehog) is nocturnal, inhabiting in holes and bushes in the day time. It is omnivorous, feeding on insects, worms, slugs, roots and fruits. It hibernates in winter. It rolls itself into a ball when alarmed.

Comments :

1. *Erinaceus* (Fig. 15.4) is commonly known as **Hedgehog**.

2. It has a covering of sharp and pointed spines on the back and sides of the body, while the ventral surface is covered with soft brown fur. Spines are moved by muscle bands found over neck and sides of the body.

3. Body is divisible into head, neck, back and abdomen.

4. Head is produced into a small tubular snout having nostrils at the tip. Mouth is small, tubular, eyes are small and pinnae are also small. Mouth has thirty six teeth. The dental formula being : *i 3/2, c1/1, pm 3/2, m 3/3*.

5. Tail is very short consisting of only twelve vertebrae.

6. There are 14 or 15 ribs.

7. Limbs are small.

8. Vagina is single and foetus develops within uterus attached by placenta.

9. They are primitive eutherians, having small cranial cavity containing brain of low grade organisation, simple stomach and allantoic placenta.

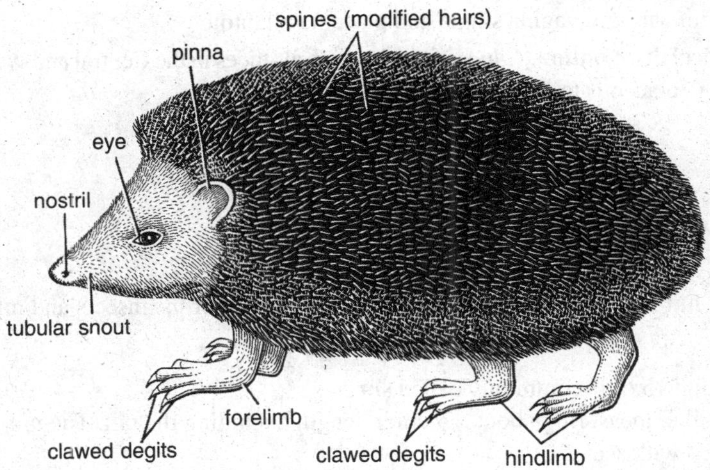

Fig. 15.4. *Erinaceus* (Hedgehog).

Geographical distribution. Hedgehogs are widely distributed in Palearctic, Oriental and Ethiopian regions. *Erinaceus* is found in North India, while *Paraechinus* is the common hedgehog found in the plains of South India. Cretaceous to Recent.

5. *Talpa* (Mole)

Classification : Same as that of *Erinaceus*.

Genus ... *Talpa* (Mole)

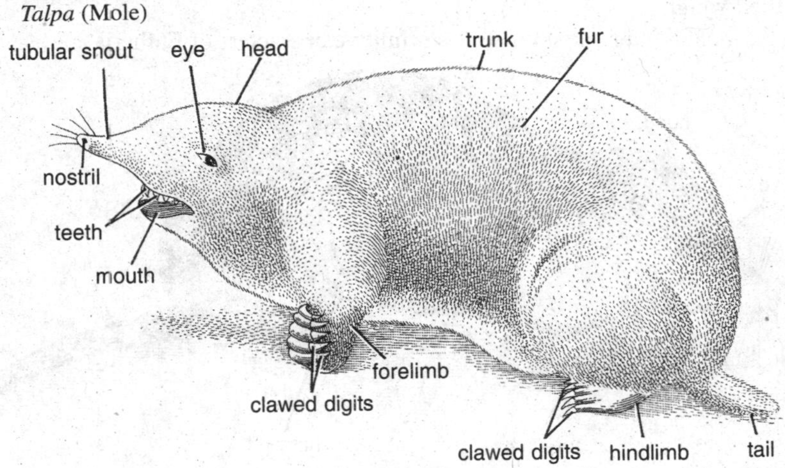

Fig. 15.5. *Talpa* (Mole).

Habit and habitat. *Talpa* (Mole) is burrowing, lives underground in burrows excavated by itself. Food chiefly consists of earthworms. etc.

Comments :

1. *Talpa* (Fig. 15.5) is commonly known as **short-tailed mole**.
2. The body is cylindrical covered with soft silky velvet black, silvery grey fur.
3. Head is short and conical with elongated snout bearing nostrils. Eyes are very small and covered with skin and pinnae (external ears) are entirely absent.
4. Tail is short, sensory and hairless.
5. Forelimbs are well developed and broad and provided with claws for digging, hindlimbs are large and broad with broad nail digits.
6. There are 40 teeth.

7. Marsupium absent, vagina single and placenta allantoic.

Geographical distribution. *Talpa* is found in high altitudes in the Central and Western Himalayas and Assam hills. Cretaceous to Recent.

6. *Sorex* (Shrew)

Classification : Same as that of *Erinaceus.*

Order ... **Insectivora**

Genus ... *Sorex*

Habit and habitat. *Sorex* lives in tunnels and feeds on worms, insects and mice.

Comments :

1. *Sorex* (Fig. 15.6) is commonly called **shrew**.
2. It is rat-like, measuring about 7.5 cm in length excluding the tail. The tail is about 5 cm in length and covered with scales.
3. Body is covered with a reddish-brown and pale silky-fur.
4. Head contains elongated snout with vibrisae, small uncovered eyes and ears are absent.
5. Limbs have five clawed-digits.
6. Mammary glands with teats are present.
7. First pair of incisors is long, lower canines are absent, all teeth are red. Dental formula is *i 3/2, c 1/0, pm 3/1, m 3/3 = 32.*
8. Shrews have a higher rate of metabolism than other mammals, they cannot live without food for more than ten hours. Because of their great expenditure of energy, shrews daily consume twice as much food as their weight.

Like *Erinaceus* and mole it possesses primitive characters of Eutheria.

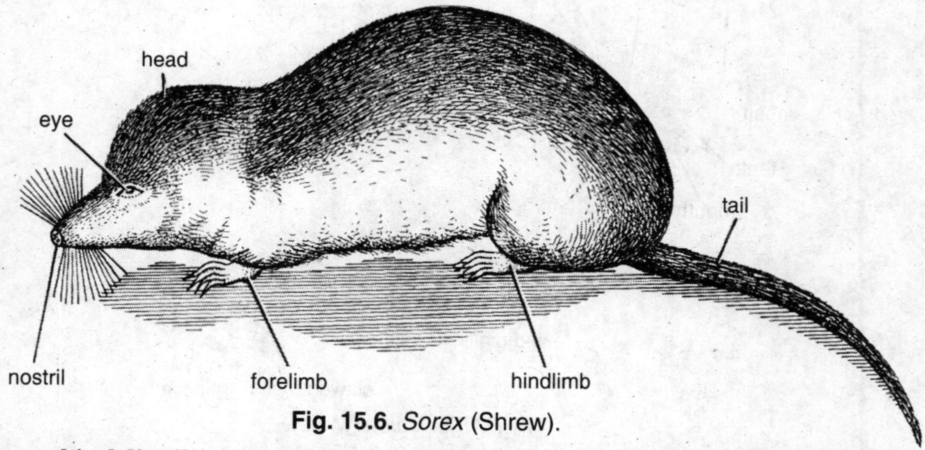

Fig. 15.6. *Sorex* (Shrew).

Geographical distribution. Shrews are found in Asia. Europe and North America. The Indian shrew belongs to genus *Soriculus.* Miocene to Recent.

7. *Pteropus* (Flying-fox)

Classification : Same as that of *Erinaceus.*

Order ... **Chiroptera** → Large fruit-eating bats; forelimbs modified for flight; second and fifth digits greatly elongated, supporting the wing or flight membrane; hindlimbs weak, have five clawed digits; eyes small and vision weak; ears with large pinnae, teeth sharp; sternum has a keel; nocturnal.

Suborder ... **Magachiroptera** → Large fruit-eating bats; snout elongated without nose leaf; thumb and second digits are clawed; tail absent; food fruits.

Genus ... *Pteropus* (Flying-fox)

Habit and habitat. *Pteropus* is arboreal, aerial and nocturnal and lives in groups. It feeds on fruits (frugivorous).

Comments :

1. Pteropus (Fig. 15.7) is commonly known as **flying-fox** and **Chamgadar** in Hindi.

2. It is a large bat with a wing-spread over one and a half metre though the body is only about 30 cm in length.

3. The body is covered with brown fur and shoulders are golden yellow.

4. Head is small with long snout, nostrils, large eyes, and pinnae. Nose leaf absent.

5. Tail is very small.

6. Forelimbs are modified into wings formed by a fold of skin (patagium) supported by long fingers 2nd to 5th. The thumb and second digits are clawed.

7. Hindlimbs and tail are also included in patagium. Hindlimbs bear sharp curved clawed digits.

8. Dental formula is *i 2/2, c 1/1, pm 3/3, m 2/3*. The molars are not tubercular but marked with a longitudinal groove.

9. They sleep with head hanging downwards and wings remain folded around the body.

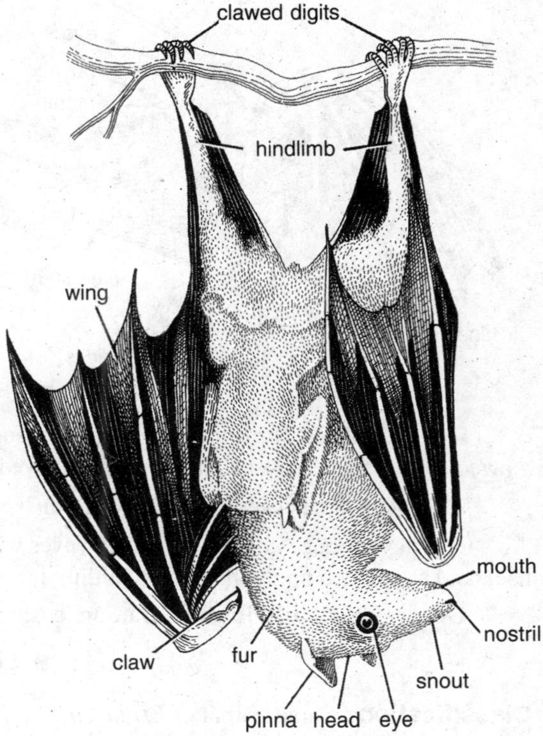

Fig. 15.7. *Pieropus* (Flying-fox).

10. Bats are the only flying mammals and have acute sonar or echo-apparatus like radar. It produces ultrasonic sound waves from vocal cords that strike with objects (wire, etc.) and reflected back to the bat for hearing.

Geographical distribution. *Pteropus* occurs in India, Sri Lanka, Australia, Africa, and Madgascar. Pliocene to Recent.

8. Bat

Classification : Same as that of *Pteropus* (flying-fox).

Suborder ... **Microchiroptera** → Small insectivorous bats; snout short usually with nose leaf; only thumb is clawed; tail if present, attached with the wing; molars with cusped crowns with transverse grooves; nocturnal and gregarious.

Type ... **Bat**

Habit and habitat. Bats are aerial, nocturnal, sleeping during daytime in caves or dark sheltered places and come out during night. Gregarious and insectivorous, occasionally frugivorous or blood sucking (sanguivorous).

Comments :

1. Bat (Fig. 15. 8) is small in size and body is covered with soft dark brown fur.

2. Head small with short snout with or without nose leaf, nostrils, small eyes with weak visual rods and large pinnae having extra lobes.

3. Tail is short and included in the inter-femoral membrane provided with a distinct flap.

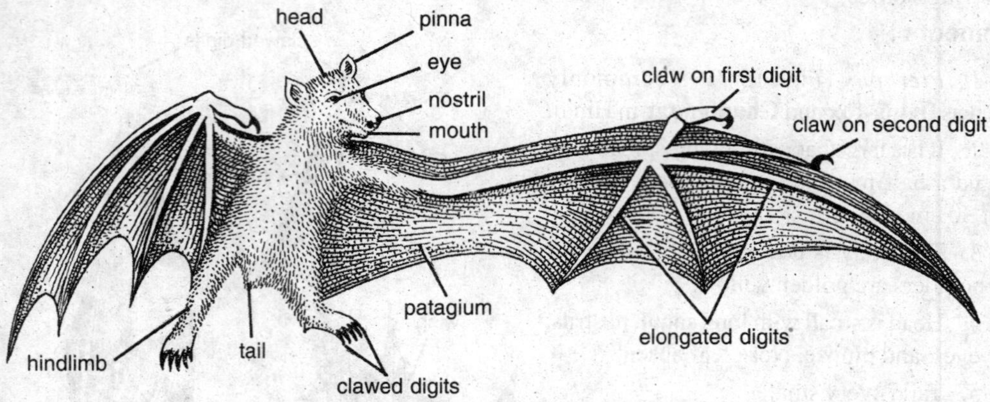

Fig. 15.8. Bat.

4. Forelimbs with 2 fingers are elongated supporting the patagium (a fold of skin). First digit is clawed. Hindlimbs are weak and have five clawed digits.

5. Molars have cusped crowns with transverse grooves like "W". Dentition 2,1, 2, ¾, 1, 3,3.

6. They produce ultra-sonic sound waves which after striking on the solid objects are reflected back and are picked up by their ears; guiding in their flight.

Geographical distribution. Bats are cosmopolitan in distribution.

9. *Loris*

Classification : Same as that of *Erinaceus.*

Order ... **Primates** → Hairy and arboreal; hands and feet prehensile; five digits on each hand and foot with flat nails; thumb and great-toe shorter and opposable to other digits plantigrade; orbits directed forwards; two pectoral mammae; placenta discoidal.

Suborder ... **Lemuroidea** → Arboreal and nocturnal; head with a sharp muzzle, long tail never prehensile.

Genus ... *Loris*

Habit and habitat. *Loris* is a primitive, nocturnal and arboreal primate. It feeds on fruits or small animals.

Comments :

1. The body of *Loris* (Fig. 15.9) is covered with thick and wooly brownish fur having a silvery look.

2. The head is fox-like and is produced into a pointed muzzle (snout) having aperture like nostrils, closely placed bulging eyes, and short pinnae.

3. Eye orbits are forwardly directed.

4. The tail is long but not prehensile.

5. Limbs are long. The digits have flat nails, except the second digit of the pes which have a sharp toilet claw. The fourth digit is the longest.

6. The lips are adherent to the gums and are not protrusible.

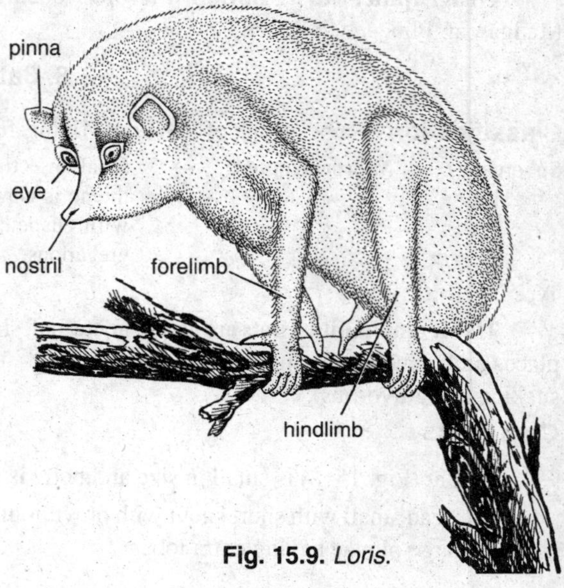

Fig. 15.9. *Loris.*

7. Dentition is primitive and they have 36 teeth.
8. Cerebral hemispheres are small and do not cover the cerebellum.
9. They exhibit remarkably slow and deliberate movements and often hanging upside down.
10. Lorises also show some features of higher primates, for example, fused tympanic ring with bulla, rounder brain and also have certain early characters such as a transverse fold of skin over abdomen which is considered the remains of marsupium.

Geographical distribution. Lorises are found outside Madagascar specially in India and Sri Lanka. Pliocene to Recent.

10. *Manis*

Classification : Same as that of *Erinaceus.*

Order ... **Pholidota** → Body covered with large over-lapping horny scales; hairs found between scales; snout elongated; teeth entirely absent; tongue long, sticky and protrusible; ears reduced; limbs short each with 5 digits, forelimb with well developed claws.

Genus ... *Manis*

Habit and habitat. *Manis* is a nocturnal and burrowing animal feeding on ants and termites.

Comments :

1. *Manis* (Fig. 15.10) is commonly known as **scaly-ant eater** or **pangolin**.
2. It is fairl, large animal, measuring up to 2 metres in length including tail.
3. Entire body is covered by imbricate thick epidermal scales arranged in longitudinal rows and are yellowish brown or clay coloured.
4. Scattered coarse hairs occur between the scales, and ventral body surface is covered with hairs.
5. Head is small with pointed snout bearing nostrils, eyes and ears are also small. The teeth are entirely absent.
6. Tail is long, broad and prehensile and covered with thick scales.
7. The limbs are short and bear five digits. Forelimbs have well developed curved claws.
8. The tongue is long, sticky and protrusible.
9. At the time of walking the claws of forelimbs are bent under the soles and the hindlimbs are plantigrade.

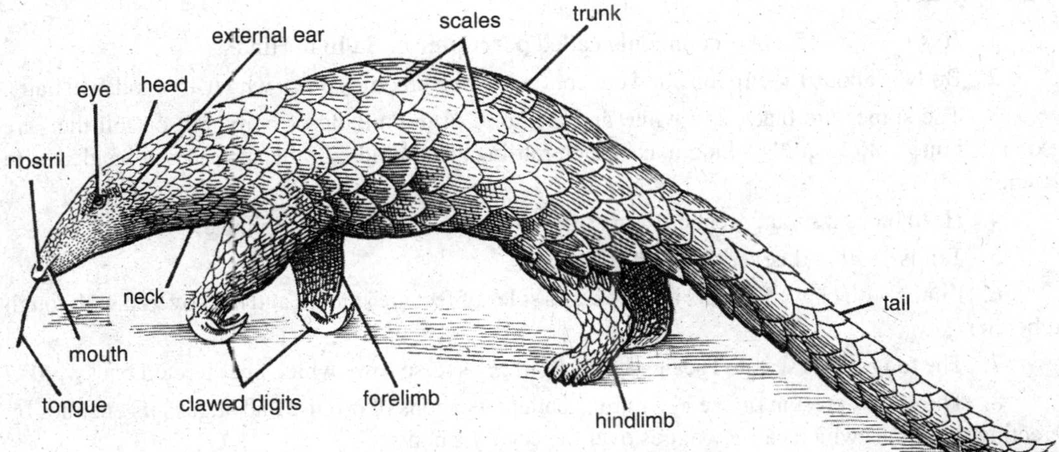

Fig. 15.10. *Manis* (Scaly ant-eater).

10. Its pyloric stomach is like gizzard of birds contains small pebbles or stones for grinding the food. It is, thus, called bajrokit or stone-eater.

Geographical distribution. *Manis or* Pangolins are found in Africa and Asia. *Manis crassicaudata* is found in the jungles of South India. *M. pentadactyla* and *M. aurita* are found in Hiamalayas, Assam, Sikkim and Nepal.

11. *Hystrix* (Porcupine)

Classification : Same as that of *Erinaceus.*

Order ... **Rodentia** → Limbs usually with five toes and claws; long incisors single pair, chisel-like and rootless, canines entirely absent; diastemma present between incisors and molars; testes usually abdominal; placenta discoidal and haemochorial.

Genus ... *Hystrix*

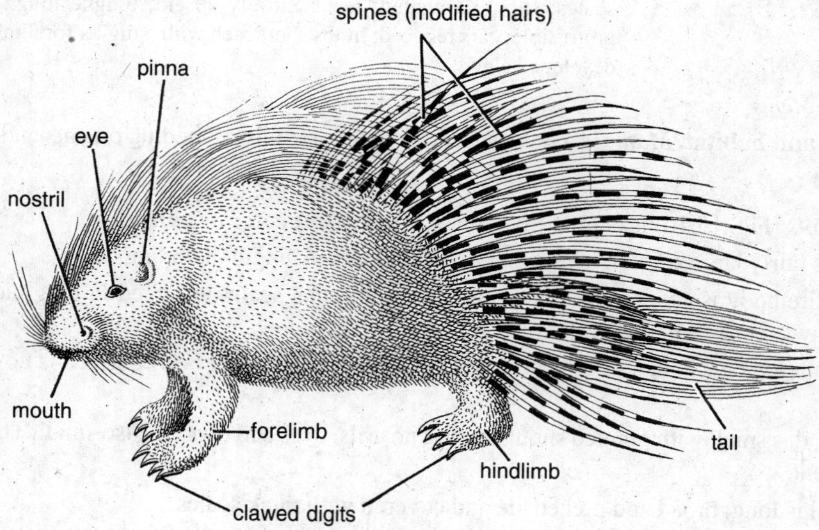

Fig. 15.11. *Hystrix* (Procupine).

Habit and habitat. H*ystrix* is commonly found near river banks, nallahs and bunds.It is nocturnal and herbivorous.

Comments :

1. *Hystrix* (Fig. 15.11) is commonly called **porcupine** or **Saihi** in Hindi.
2. Body is about 1 metre long and covered with long erectile spines which are modified hairs.
3. The spines are black and white and are solid in the middle of body, but on tail they are expanded into hollow quills which make much rattling. The middle of the spine being banded with black.
4. Head bears a snout, great crest of stiff long hairs, small eyes and pinnae.
5. Tail is short and never prehensile.
6. Limbs with clawed fingers and toes, and soles of feet are smooth and not covered with rough tubercles.
7. The tongue has serrated scales arranged in transverse rows which are directed backwards.
8. The spines present on the body are excellent weapons of offense and attacks its enemies by erecting the spines which cause wounds over the enemy's body.

Geographical distribution. *Hystrix* (Porcupine) is found from East Indies to Africa, Europe and Asia. *Hystrix leucura* and *H. bengalensis* are found in India. Eocene to Recent.

12. *Funambulus* (Squirrel)

Classification : Same as that of *Erinaceus*.

Order ... **Rodentia** → Characters same as that of *Hystrix*.

Genus ... *Funambulus*

Habit and habitat. *Funambulus* (Squirrel) is arboreal as well as ground dwelling and fast runner. It feeds on seeds, nuts and fruits. It makes nest of twigs and leaves on trees.

Comments:

1. *Funambulus* (Fig. 15.12) is commonly called **squirrel** or **gilhari** in Hindi.

2. Body is elongated and covered with soft white or greyish fur. Five longitudinal stripes of dark colour are present on the back.

3. Head with snout bearing vibrissae, eyes and pinnae are large. The eyes and pinnae or ears are large.

4. Tail is long and bushy.

5. The forelimbs have an inconspicuous thumb and hindlimbs have four clawed digits. The soles of forelimbs are naked and those of hindlimbs are hairy.

6. Incisors chisel-like and exposed, canines absent and cheek teeth are of equal size.

7. Squirrels are used in experiments in laboratories.

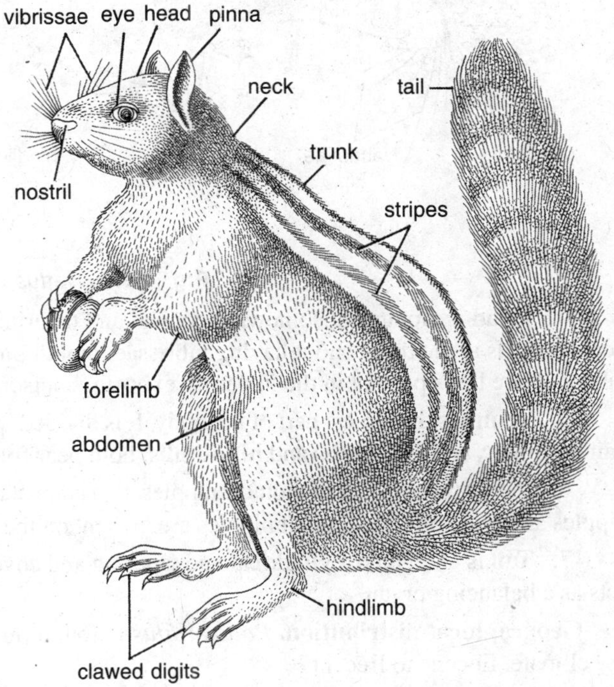

Fig. 15.12. *Funambulus* (Squirrel).

Geographical distribution. Squirrels are world wide in distribution with the exception of Australian region and islands of Madagascar. Eocene to Recent.

13. *Rattus* (Common Black Rat)

Classification : Same as that of *Erinaceus*.

Order ... **Rodentia** → Characters same as that of *Hystrix*.

Genus ... *Rattus*

Species ... *rattus*

Habit and habitat. The rat is chiefly nocturnal in habit and thrives best in warm climate. It inhabits holes and burrows in the houses and cultivated fields. It feeds on various kinds of grain and seeds. Viviparous. Rat also acts as a carrier of diseases like plague and typhus fever.

Comments :

1. *Rattus rattus* (Fig. 15.13) is commonly called **common black rat**.

2. The body is slender with thick-set fine greyish or brownish fur mixed with stiff hairs. Rat measures about 25cm including tail.

3. Body is divisible into head, neck, trunk and tail.

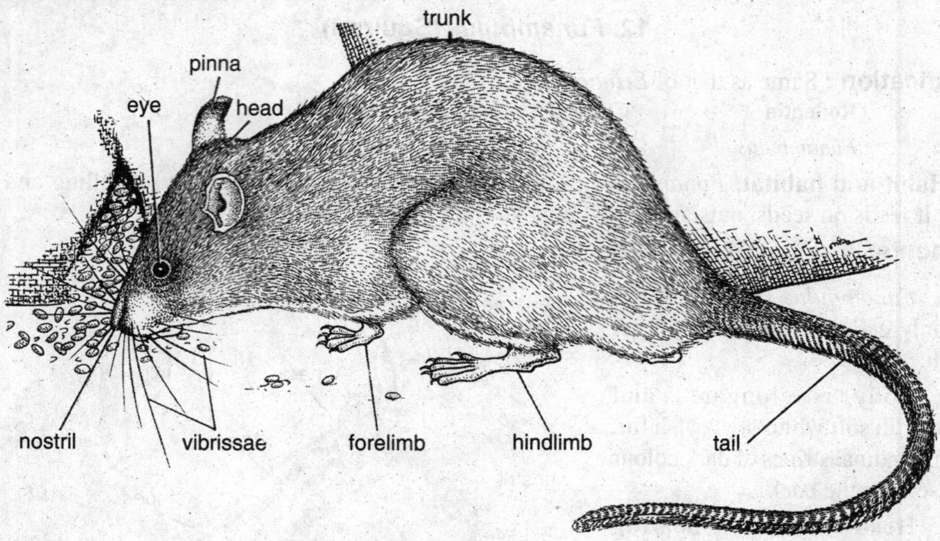

Fig. 15.13. *Rattus rattus.*

4. **Head** is elongated, somewhat conical and tapering anteriorly to a terminal nose. The head bears nostrils at the tip of snout having vibrissae, mouth bounded by upper and lower lips, eyes and pinnae. Hare lip is present in the upper lip exposing incisors.

5. **Trunk** is the largest part of the body. It is divided into **thorax** or **abdomen**. Trunk bears two pairs of limbs, *viz.,* **forelimbs** and **hindlimbs**, both bear five clawed digits.

6. In females one pair of thoracic nipples, two pairs of axillary nipples and one pair of abdominal nipples and two pairs of inguinal nipples are present on the ventral surface of the body.

7. **Tail** is elongated, cylindrical and tapering and covered with overlapping vestigial scales. It acts as a balancing organ.

Geographical distribution. *Rattus rattus* is found throughout the world. It prefers warm and dry climate. Eocene to Recent.

14. *Oryctolagus* (Rabbit)

Classification : Same as that of *Erinaceus* up to Infraclass.

Order .... **Lagomorpha** → Upper incisors 2 pairs and diastemma present.
Genus ... *Oryctolagus*
Species ... *cuniculus*

Habit and habitat. *Oryctolagus* inhabits fields, grassy lands, and woodlands. It is gregarious, crepuscular (come out of burrows in twilight for feeding), coprophagous (eats its soft stool for getting maximum nourishment) and polygamous.

Comments :

1. *Oryctolagus* (Fig. 15.14) is commonly called **rabbit**.
2. Body is divisible into head, neck, thorax, abdomen and short and bushy tail.
3. Body fur colour is white, black, brown or spotted.
4. Head has wide mouth with harelips, vibrissae, nostrils, eyes and large ears.
5. Forelimbs are used for burrowing and hindlimbs for leaping. Hindlimbs are stronger and longer.
6. Males have skin covered penis and females have clitoris (reduced nonfunctional penis).

Geographical distribution. Worldwide in distribution.

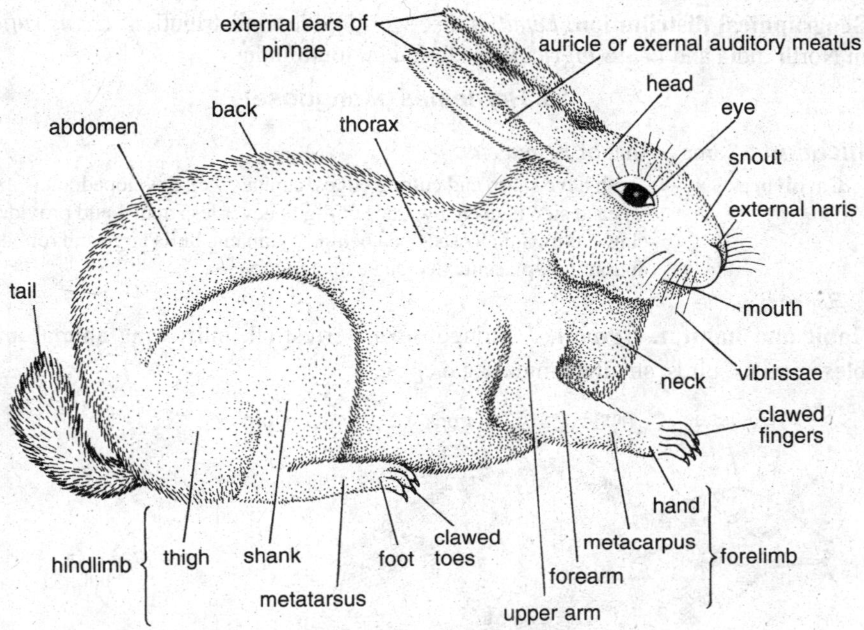

Fig. 15.14. *Oryctolagus* (Rabbit).

15. *Lepus* (Hare)

Classification: Same as that of *Oryctolagus*.

Habit and habitat. *Lepus* inhabits fields, grasslands and woodlands. It is gregarious, crepuscular (coming out of burrows for feeding during twilight), coprophagus (eating again their soft stool for maximum nourishment). Polygamous, nocturnal and living in temporary shelters.

Comments :

1. *Lepus* (Fig. 15.15) is larger in size (50–70 cm) than the rabbit.
2. Its body colour is brown dorsally and white on lateral and ventral sides.
3. Body divisible into head, neck, trunk, thorax, abdomen and short and bushy tail.
4. Head bears wide mouth with harelips, vibrissae, nostrils, eyes and longer erect ears. Ears have external auditory meatus.
5. Forelimbs and hindlimbs are more or less equal in size.
6. It runs at a fast speed of about 60 km per hour.
7. *Lepus* is wild animal and it cannot be domesticated.

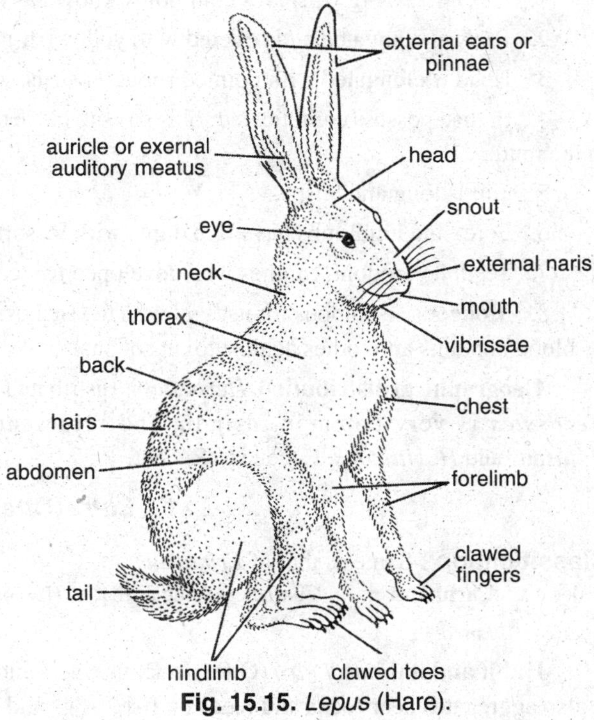

Fig. 15.15. *Lepus* (Hare).

Geographical distribution. *Lepus* is also world wide in distribution. *Lepus ruficaudatus* is found in North India and *Lepus nigricollis* is found in South India.

16. *Herpestes* (Mongoose)

Classification : Same as that of *Erinaceus*.

Order ... **Carnivora** → Teeth have sharp and cutting edges, canines well developed, incisors small and 3 on each side in each jaw; toes or digits never less than 4 and provided with sharp claws; tympanic fossae open behind; tympanic bulla large and rounded, mammae abdominal; placenta deciduate and zonary.

Genus ... *Hepestes*

Habit and habitat. *Herpestes* (Mongoose) is terrestrial, burrowing animal and feeds on vegetables, reptiles, birds, small mammals and eggs.

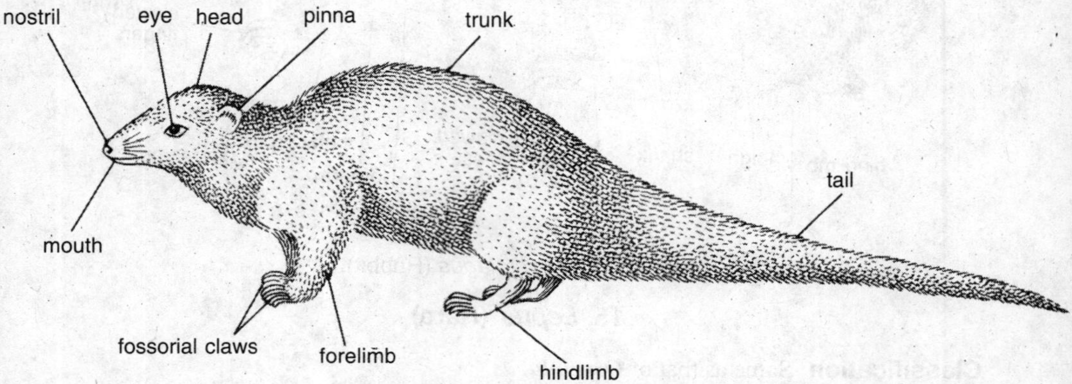

Fig. 15.16. *Herpestes* (Mongoose).

Comments :

1. *Herpestes* (Fig. 15.16) is commonly known as **mongoose** or **neola** in Hindi.
2. Body is elongated and covered with yellowish-grey fur.
3. Head is elongated with pointed snout, nostrils, vibrissae, small eyes and pinnae.
4. Pinnae possess complicated folds so as to close the ear openings when the animal is burried in the mud.
5. Tail is long and bushy.
6. Fore- and hindlimbs have five digits with fossorial claws.
7. Teeth heterodont, canines well developed for tearing flesh.
8. Mongoose is famous for its fight with the snakes. It kills the snakes and feeds on their blood. Its blood contains antibodies against toxins of snakes.

Geographical distribution. *Herpestes* is distributed between the Ethiopian and Oriental regions. *Herpestes* is very common in India and represented by *H. bengalensis, H. edwardsii, H. fuscus* and *H. vitticolis.* Eocene to Recent.

17. *Lutra* (Otter)

Classification : Same as that of *Erinaceus*.

Order ... **Carnivora** → Characters same as that of *Herpestes.*

Genus ... *Lutra*

Habit and habitat. *Lutra* (Otter) inhabits rivers and lakes of India, and adapted to aquatic life. It also aggregates near rocks and feeds on fish, frogs and crabs.

Comments :

1. *Lutra* (Fig. 15.17) is commonly called **otter** and **udbilao** in Hindi.

2. Body is slender covered with short dense oily fur.

3. Head is cat-like and bears mouth, nostrils vibrissae, small eyes and external ears.

4. Webbed digits in both limbs for swimming.

5. Tail is somewhat flattened.

6. Carnassial teeth are well developed, last upper molar is expanded on its inner side.

Geographical distribution. *Lutra* is found throughout World. Eocene to Recent.

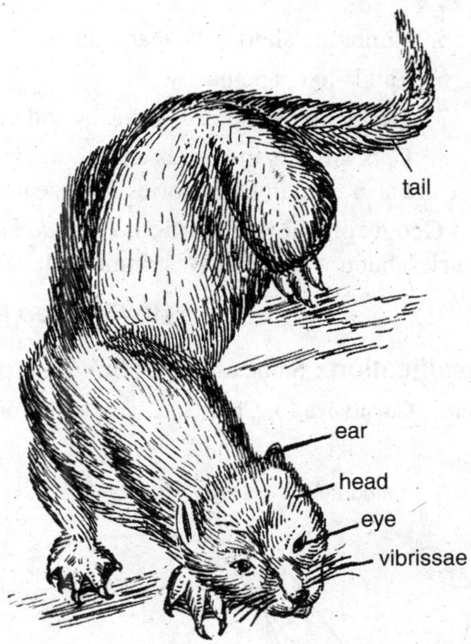

Fig. 15.17. *Lutra (otter)*

18. *Felis domesticus* (Cat)

Classification: Same as that of *Erinaceus* up to Infra class.

Order ... **Carnivora** → Characters same as that of *Herpestes*.

Genus ... *Felis*

Species ... *domesticus*

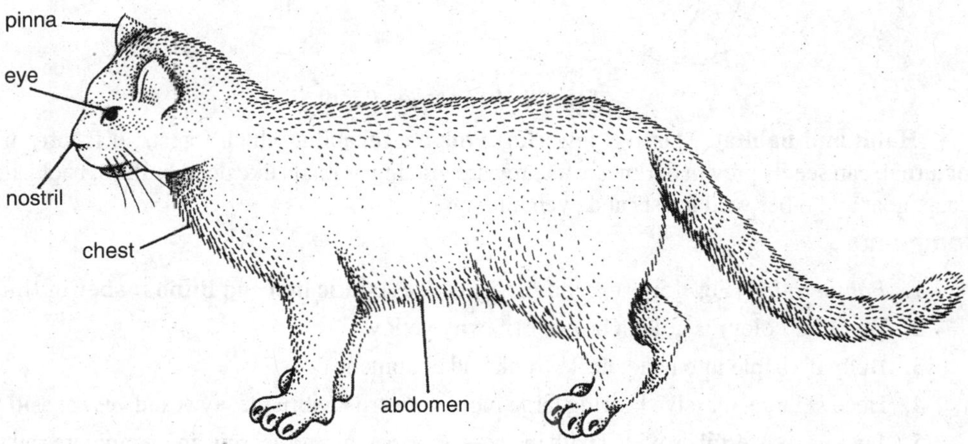

Fig. 15.18. *Felis domesticus* (Cat).

Habit and habitat. Solitary, nocturnal but also visible during day time and lives close to human habitat. It may be domesticated. It feeds on small mammals (rats, mice, etc) and birds.

Comments :

1. *Felis domesticus* (Fig. 15.18) is commonly called **Cat** and **Billi** in Hindi.
2. Body fur colour is variable. It may be dark brown, fawn, yellow and black.
3. Body divisible into head, neck, trunk, abdomen and a tail.
4. Head with wide mouth, nostrils, vibrissae, bulging eyes capable to see at night and external ears.

5. Limbs are short with sharp claws which are retractable into protective sheath.

6. Tail is long and bushy.

7. Gestation period is of 8 weeks and gives birth to 2 to 4 kittens.

8. Cats kill rats in houses and godowns, etc., and, thus, acts as agents for biological control. These are also used in laboratories for researches.

Geographical distribution. *Felis domesticus* (Cat) is distributed in tropical and subtropical countries, but not found in West Indies, Madagascar and Australia.

19. *Felis leo (Panthera leo)* (Lion)

Classification : Same as that of *Erinaceus* up to Infraclass.

Order ... **Carnivora** → Characters same as that of *Herpestes.*

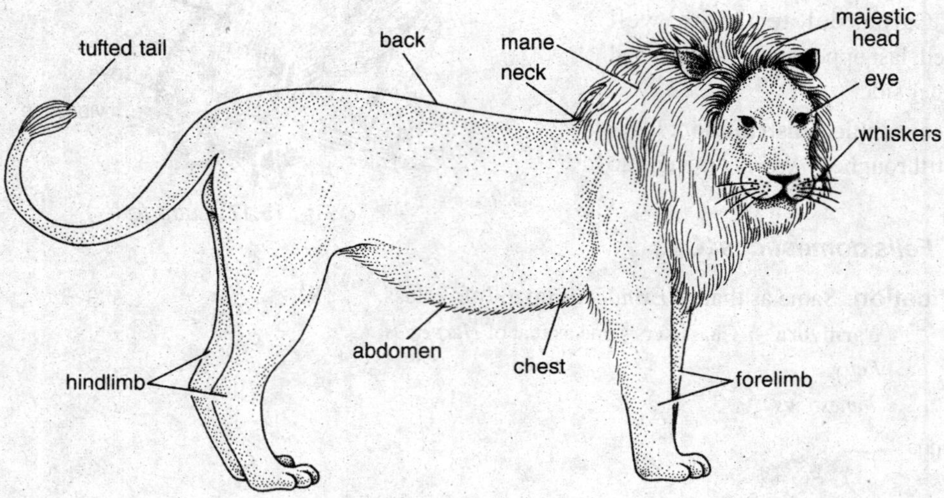

Fig. 15.19. *Felis leo* (Lion).

Habit and habitat. Massive powerful carnivore inhabiting thick forests. It is terrestrial and nocturnal, can see its prey in darkness. Its prey are grazing animals like deer, buffalo, neel gai, zebra, sheep, goat and other wild beasts and even humans.

Comments :

1. *Panthera leo* (Fig. 15.19) is commonly called **Asiatic lion** and **Bubbar sher** in Hindi.

2. Body fur colour is golden brown or fawny yellow.

3. Body divisible into head, neck, trunk and abdomen.

4. Head is large, massive bearing large mouth, nostrils, vibrissae, eyes and very sensitive ears,

5. Jaws are powerful provided with incisors, long canines for tearing and premolars and molars.

6. Tail is long having tuft of hairs at the tip end which conceal sharp horny tail spur.

7. Limbs are well developed having thick pads below sharp clawed digits. Claws are retractile into sheath.

8. Males are larger than females measuring 3.5 feet tall and 225 kg in weight. Male has a ruff of long hairs over neck and shoulders.

9. Male has mane. Mane is darker than body covering head and shoulder.

10. Female gives birth to 2–6 cubs and these are cared and trained up to 9 months for hunting.

Geographical distribution. Found in Asia, India (Gir forest of Gujarat), Africa (South of Sahara desert). These are kept in zoo for public seeing.

Felis tigris (tiger) is also found in Asian and African jungles. Its fur colour is orange-yellow, striped black. Its male has no long hairs over neck and shoulders. It is the national animal of India.

20. *Canis familiaris* (Dog)

Classification : Same as that of *Erinaceus.*

Order ... **Carnivora** → Characters same as that of *Herpestes.*

Habit and habitat. They live around human habitat and wild dogs are also found in jungles in groups. They are also domesticated.

Comments :

1. *Canis familiaris* (Fig. 15.20) is commonly called **dog** and **Kutta** in Hindi.
2. Body fur colour and size is variable.
3. Body is divided into head, neck, trunk and abdomen.
4. Head is small with large mouth gape, nostrils, vibrissae, eyes and ears.
5. Limbs are of equal size and are adapted for fast running. Digits clawed and provided with small pads beneath.
6. Tongue is protruding. Respiration is fast with protruded tongue in summers.
7. Male and female both come into heat once or twice a year. Male becomes more ferocious during this period.
8. Dogs are domesticated as watch dogs. They are also used in tracing criminals because they have a strong power of smell. Dogs are also employed for various purposes by the army.
9. Dogs are carrier of a viral disease called **Rabies** or **Hydrophobia**. Infected dog runs away from water and if such dog bites a man, it transmits rabies virus. In such a case patient needs HDC vaccine (human diploid cell vaccine) for treatment. Rabies is a fat disease.

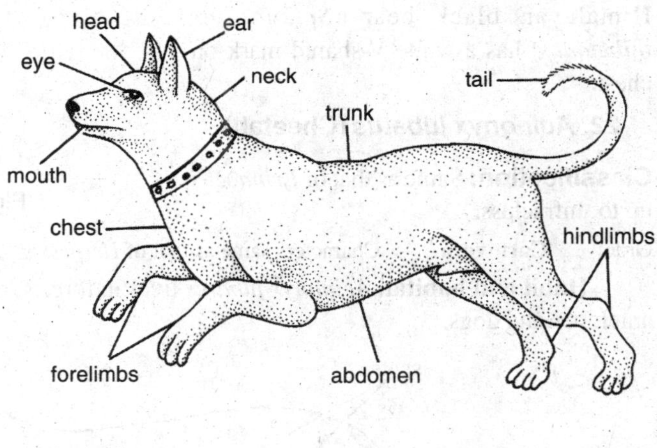

Fig. 15.20. *Canis familiaris* (common dog).

Geographical distribution. World wide in distribution.

21. *Melursus ursinus* (Sloth Bear)

Classification : Same as that of *Erinaceus* up to Infraclass.

Order ... **Carnivora** → Characters same as that of *Herpestes.*

Habit and habitat. They are found in thick forests. They feed on fruits, honey and insects and can climb trees.

Comments :

1. *Melursus ursinus* (Fig. 15.21) is commonly called **Sloth-bear**.
2. Body fur colour is variable: black (Indian bear), brown (European bear *Ursus arctus*) and white (polar bear *Thalarctos maritimus*). Body hairs are long and coarse.

3. Body heavily built and divisible into head, neck, trunk, abdomen and tail.

4. Head is large with protruding naked snout bearing mouth, nostrils and small eyes and ears. Lower lip is elongated.

5. Limbs are strong and well developed with large clawed plantigrade feet.

6. Tail is very short.

7. Canines are not so long as found in lion.

8. Males are larger than females.

9. Female breeds every second or third year and gives birth to 2–4 small cubs.

Geographical distribution. Distributed in tropical and temperate forests of India. Himalayan black bear (*Sclenarctos thibetanus*) has a white V-shared mark on chest.

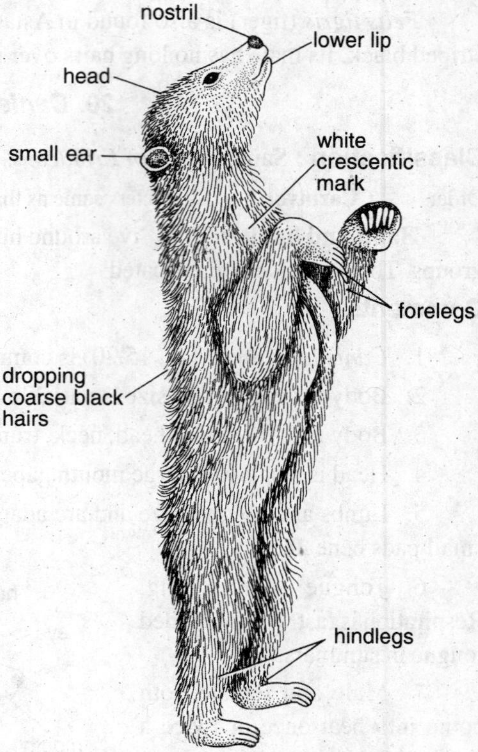

Fig. 15.21. *Melursus ursinus.*

22. *Acinonyx jubatus* (Cheetah)

Classification: Same as that of *Erinaceus* up to Infraclass.

Order ... **Carnivora** → Characters same as that of *Herpestes*.

Habit and habitat. *Acinonyx jubatus* lives in forests, grossy woody lands and feeds on birds, antelopes and dogs.

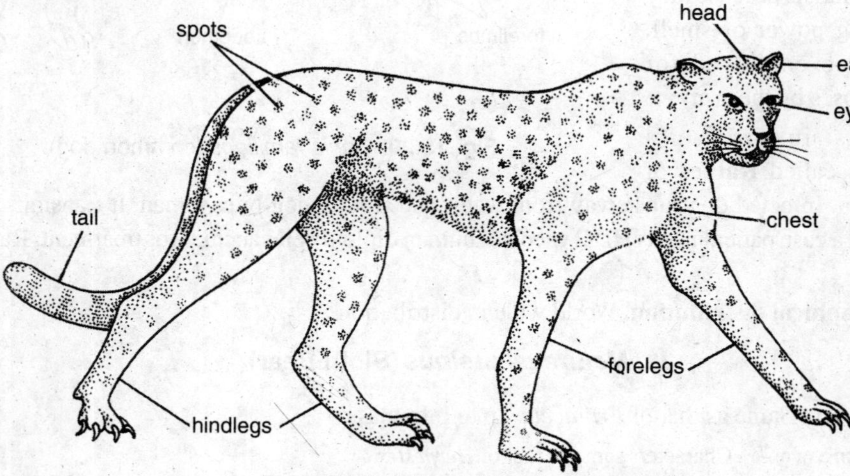

Fig. 15.22. *Acinonyx jubatus* (Cheetah).

Comments :

1. *Acinonyx jubatus* (Fig. 15.22) is commonly called **Cheetah** or **hunting leopard**.
2. Body fur colour is greyish with black spots over the entire body.
3. Body dog-like measuring 1.5 metres long and divisible into head, neck, trunk and abdomen.
4. Head is small having distinctive black stripe from inner conrer of eyes to the upper lip.

5. Head also bears wide mouth, nostrils, vibrissae, small eyes and ears.

6. Limbs are larger and slender for fast running and have clawed and padded feet.

7. Males are slightly larger than females.

8. Gestation period is 84 to 95 days. Female gives birth to 2–4 cubs which are born with silvery mane which disappears when they are 2–5 months old.

9. It is the fastest running land animal reaching a top speed of 110 km per hour.

10. It hunts in groups. Easily tamed, they make docile pets.

Geographical distribution. *Acinonyx jubatus* (Cheetah) inhabits grass lands and semideserts of Africa and South West Asia. Cheetah became extinct in India early in the present century.

23. *Hyaena striata* (Hyaena)

Classification: Same as that of *Erinaceus* up to Infraclass.

Order ... **Carnivora** →Characters same as those of *Herpestes*.

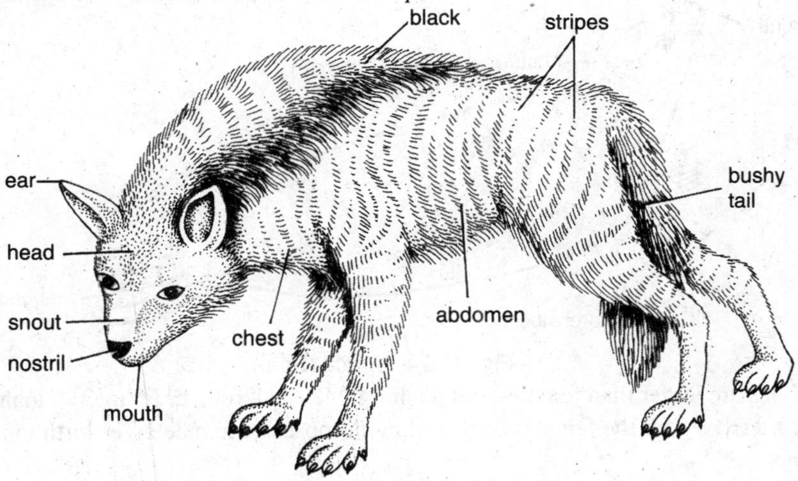

Fig. 15.23. *Hyaena striata* (Hyaena).

Habit and habitat. *Hyaena striata* is nocturnal inhabiting forests and lives in burrows. It is a scavanger feeding chiefly on carrion. It often robs graves regions where it is found. It also produces sound similar to cracking human laugh.

Comments :

1. *Hyaena striata* (Fig. 15.23) is commonly called **Striped hyaena** and **Lakad baggha** in Hindi.

2. It looks like a dog and measures 1.5 metres in length. Its body colour is greyish having dark stripes.

3. Body is divisible into head, neck, trunk, chest and abdomen and bushy tail.

4. Head is small with large mouth, nostrils, small eyes and large ears.

5. Cheek teeth are well developed for crushing.

6. Limbs have 4 digits, with non-retractile claws.

7. Forelimbs are stronger than the hindlimbs.

8. Paws of forelimbs are used for digging human cave for eating freshly burried body.

Geographical distribution. *Hyaena striata* is found in India and *Crocuta* (spotted or laughing hyaena) is found in African plains.

24. *Phoca* (Seal)

Classification : Same as that of *Erinaceus* up to Infraclass.

Order ... **Carnivora** → Characters same as that of *Herpestes.*

Habit and habitat. *Phoca* is marine and comes ashore for breeding. Seals show a little territorial behaviour. They are usually gregarious.

Comments :

1. *Phoca* (Fig. 15.24) is commonly called **Seal.**
2. Body is streamlined, fish-like form.
3. Head bears mouth, nostrils, vibrissae, and eyes. External ears are absent. (But fur seals, *Callorhinus* and sealion *Zalophus* possess external ears).
4. Limbs are modified as flippers acting as oars for swimming. Feet are webbed. (Fur seals and sealion use their limbs for moving on land).

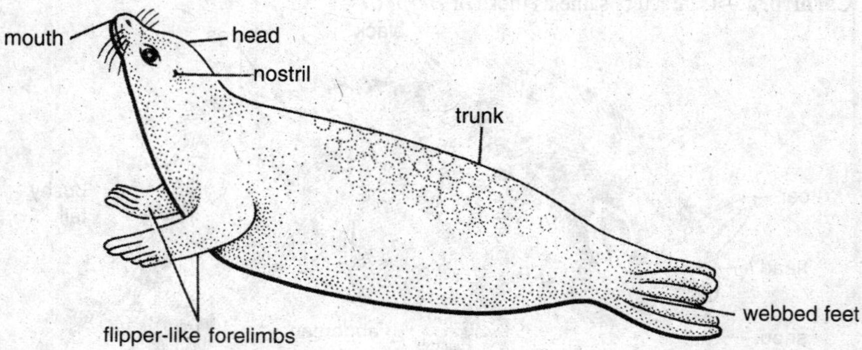

Fig. 15.24. *Phoca* (Seal).

5. Males are larger than females. Each adult male is a lord over 50 to 60 females. Adult male establishes a territory for the females for breeding purposes. A female gives birth to a single young seal.
6. Seals are hunted for oil producing blubber (fatty layer beneath skin) and skins.

Geographical distribution. *Phoca* is found in Arctic and Antarctic regions.

25. *Platanista gangetica* (Ganges Dolphin)

Classification : Same as that of *Erinaceus* up to Infraclass.

Order ... **Cetacea** → Fish-like mammals; a few bristles are present over snout (modified hairs).

Habit and habitat. *Platanista gangetica* lives in the Ganges, Brahmaputra and perhaps in the Indus rivers and their tributaries.

Comments :

1. *Platanista gangetica* (Fig. 15.25) is commonly called **Ganges dolphin** or **Susu**.

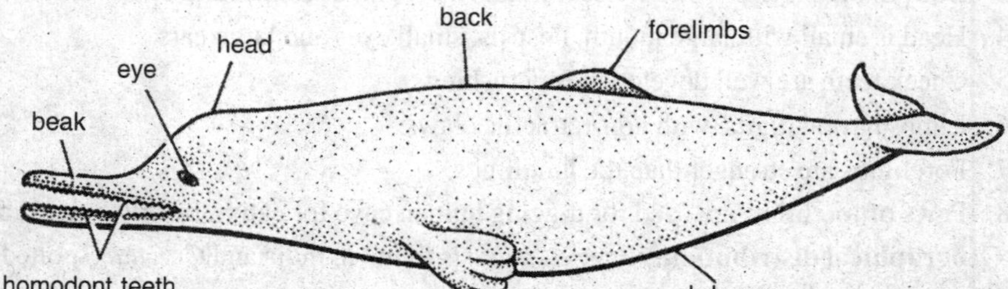

Fig. 15.25. *Platanista gangetica* (Ganges dolphin).

2. Body is 2 to 3 metres long, dark grey in colour and streamlined.
3. Head is beak-like with long mouth, small eyes and without external ears.
4. Jaws are projected as bird-like beaks containing up to 200 homodont sharp teeth in both the jaws.
5. Forelimbs are modified into broad flippers. Hindlimbs are absent.
6. Tail is fish-like divided into two broad horizontal fleshy flukes or lobes used for propulsion.
7. Ganges dolphin probes in mud for shell-fishes.

Geographical distribution. It lives in freshwater of Ganges and Brahmaputra and their tributaries. Remaining cetaceans are marine like common dolphin *Dolphinus* possesses neck, well developed eyes and a dorsal fin over the back; *Phocaena* (porpoise), and whales.

26. *Macaca* (Rhesus Monkey)

Classification: Same as in *Erinaceus* up to Infraclass.

Order	... **Primates**	→	Head moves readily on neck; innermost toe and thumb usually opposable.
Suborder	... **Anthropoidea**	→	Eyes directed forward; cerebrum extends over cerebellum.
Genus	... *Macaca*		
Species	... *mulata*		

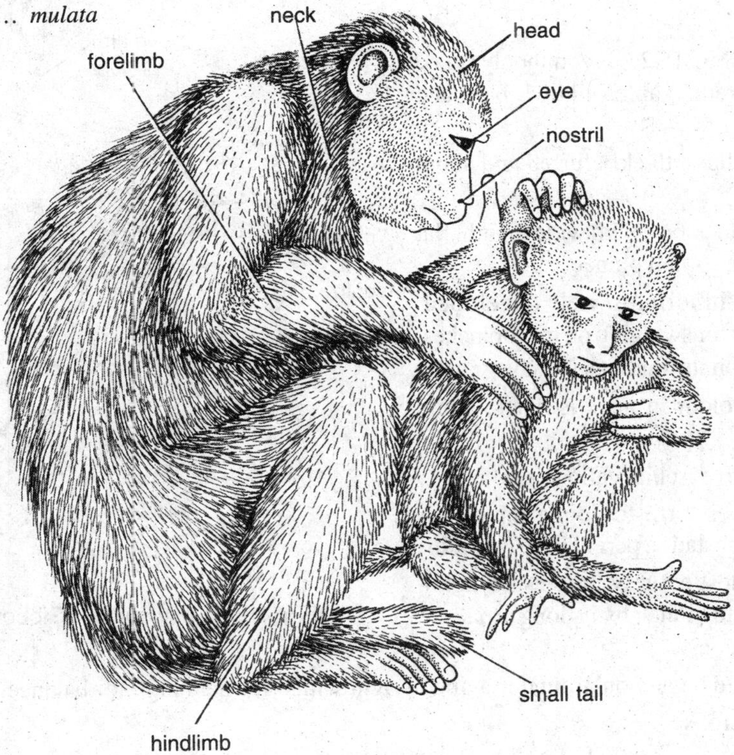

Fig. 15.26. *Macaca* (Rhesus monkey).

Habit and habitat. Arboreal, terrestrial and social animal.

Comments :

1. *Macaca mulata* (15.26) is commonly known as **Rhesus monkey** and **Bandar** in Hindi.
2. Body is covered with light brown or golden hairs.
3. Body is divided into head, trunk, abdomen and tail.
4. Head bears wide mouth, downward directed nostrils, eyes and external ears. Cranium is large.
5. Facial muscles are protrusible which permit facial expression.

6. Cheek are internally pouched for food storage. Teeth heterodont.

7. Limbs are well developed for tree dwelling and digits are nailed.

8. Buttocks have two red coloured naked tuberosities.

9. Monkeys can be easily trained for dancing, etc., and are used to amuse children. Rh blood factor was first discovered in monkeys. *Samiri* (squirrel monkey), *Ateles* (spider monkey), *Callithris* (marmoset) and *Presbystis* (langur) are also other monkeys. Langur is greyish coloured having elongated limbs and tail.

10. *Macaca* are used in laboratories for research purposes.

Geographical distribution. These are old world monkeys found in Asia, China, India and Vietnam.

27. *Hylobates* (Gibbon)

Classification : Same as that of Rhesus monkey.

Habit and habitat. Arboreal, territorial and rarely descend to ground.

Comments :

1. *Hylobates* (Fig. 15.27) is commonly called **Gibbon**. Its call is remarkable – howl, bark and cry.

2. Body is tailless, thickly furred and about 1 metres in height.

3. Head has black face with wide mouth, flat nose, sunken eyes and small pinnae.

4. Fore- and hindlimbs well developed. Forelimbs are longer for swinging on tree branches. On ground it walks on its hindlegs slowly.

5. Body fur colour in male is black, while female is yellowish.

6. Buttocks have ischial callosities.

7. Female gives birth to young ones after every two years and gestation period is of 7 months.

8. It lives almost entirely in trees swinging rapidly from branch to branch by its long arms and famous for its acrobatics.

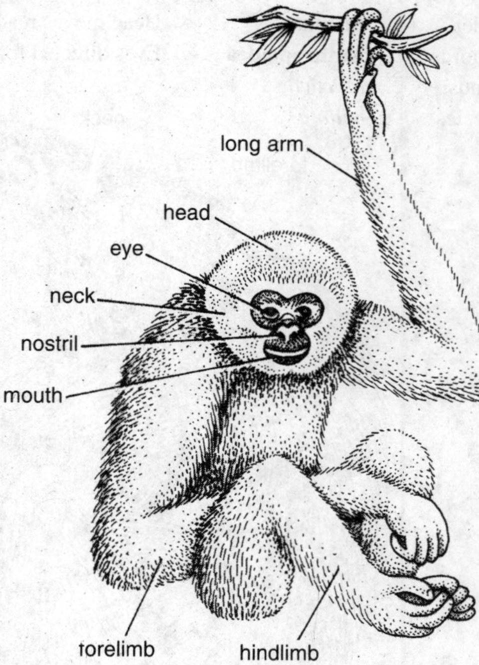

Fig. 15.27. *Hylobates* (Gibbon).

9. It has a bipedal gaite on ground moving upright with arms held high for balance.

10. It is frugivorous.

11. Gibbons live in pairs and make loud territorial calls.

Geographical distribution. *Hylobates* (Gibbon) is found in rain forests in South East Asia, Assam, and Indonesia.

28. *Pan triglodytes* (Chimpanzee)

Classification : Same as that of Rhesus monkey.

Habit and habitat. *Pan triglodytes* is arboreal but mostly lives on trees and on ground and feeds on buds of fruits (frugivorous). It is quite intelligent primate.

Comments :

1. *Pan triglodytes* (Fig. 15.28) is commonly called **Chimpanzee**.

2. Body except face and limbs is covered over by thick black hairs, while back has grey hairs.

3. It is about 1.2 metres in height and weighs 150–175 pounds.

4. Head with protruded snout, wide mouth with bulging lips, wide flat nose, sunken eyes and large ears.

5. Limbs are well developed. Forelimbs are longer. It walks on knuckles on ground.

6. Male has goitre-like throat sac and two fatty swellings in cheeks.

7. Female gives birth to a single young and suckles for four years. Gestation period is of 225 days.

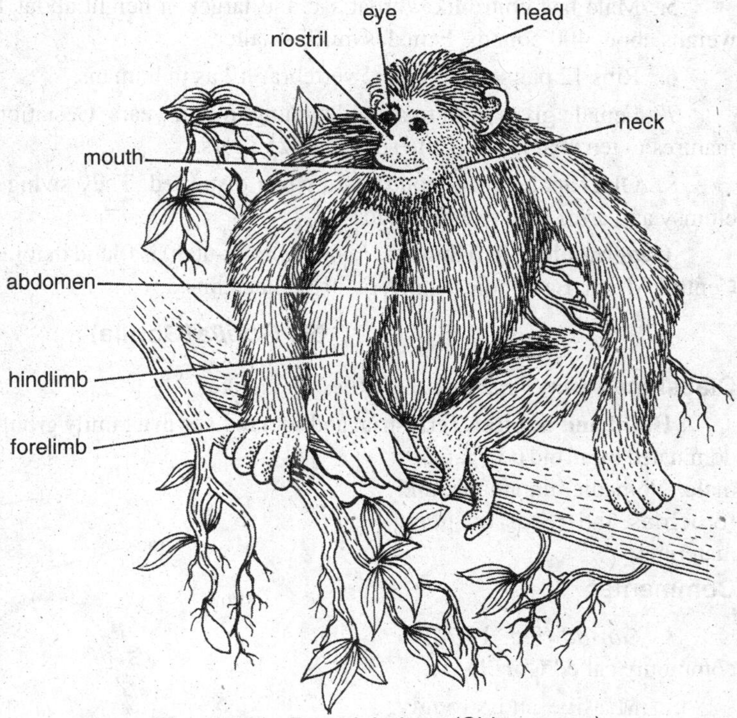

Fig. 15.28. *Pan triglodytes* (Chimpanzee).

8. Chimpanzee is most intelligent of all apes and possesses high level of curiosity. It is more like man than any other living mammal.

9. It is easily tamed when young and trained to perform simple acts.

Geographical distribution. *Pan triglodytes* (Chimpanzee) is found in rain forests of West Africa.

29. *Pongo* (Orang-utan)

Classification: Same as that of Rhesus monkey.

Habit and habitat. *Pongo* (Orang-utan) is arboreal, frugivorous and lives singly or in pairs or in groups of four. It swings over tree branches. It can be tamed.

Comments :

1. *Pongo* (Fig. 15.29) is commonly called **Orang-utan**.

2. Body is covered by reddish brown fur which becomes darker in older males. Entire body except fingers, face and toes are covered by hairs.

3. Head with high forehead, wide mouth with bulging lips, nostrils large, sunken eyes and small ears.

4. Arms are exceptionally long and legs are comparatively short.

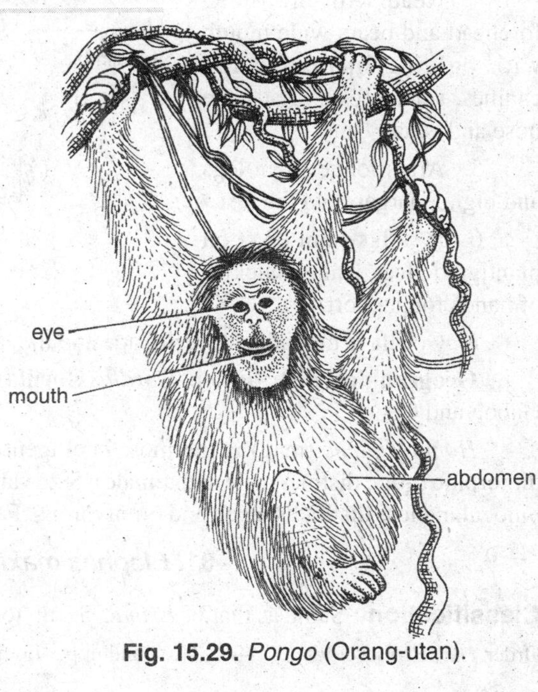

Fig. 15.29. *Pongo* (Orang-utan).

5. Male has goitre-like throat sac and larger in height about 1.5 metre than the female and weighs about 400 pounds. Females much smaller.

6. Ribs 12 pairs and thoracic vertebrae 12 as in humans.

7. Female gives birth to a single young in four years. Gestation period is of 275 days. Young matures in ten years and its life span is of 25 years.

8. Older males make loud roars when disturbed. They swing from branch to branch. Gaite clumsy and arms held over its head.

Geographical distribution. *Pongo* (Orang-utan) is found in forests of Borneo, Sumatra, outside Central Africa, East Indies and West of Wallace line.

30. *Gorilla* (Gorilla)

Classification : Same as that of Rhessus monkey.

Habit and habitat. Terrestrial and social, lives in a family group of 30 headed by an adult male dominated by an old polygamous male. Nomadic, untamable and ferocious and feeds on plant materials.

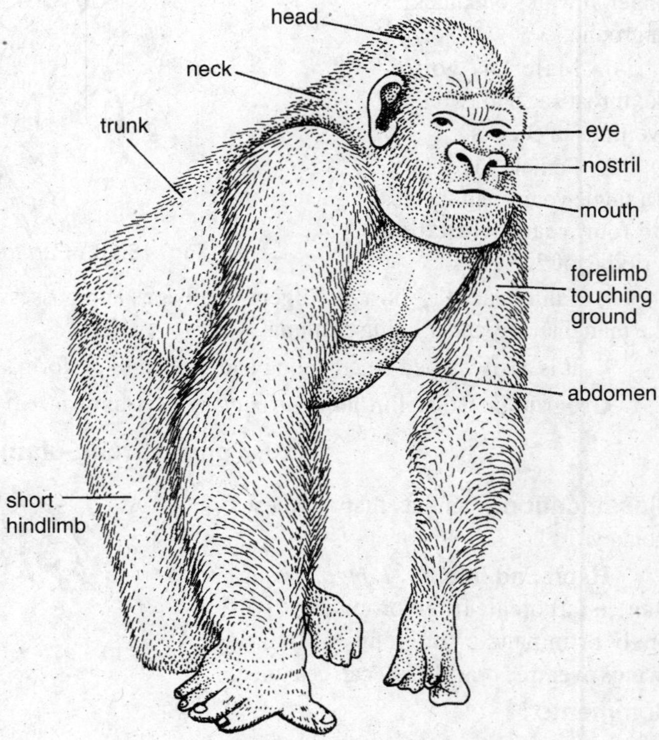

Fig. 15.30. *Gorilla* (Gorilla).

Comments :

1. *Gorilla* (Fig. 15.30) is commonly called **Gorilla**.

2. Massive and strongly built body is 1.8 metres tall, 250 kg in weight.

3. Body of mountain gorilla is covered by dense black hairs and lowland gorilla is of blackish brown colour.

4. Head with protruded forehead and bears wide mouth with bulging lips and large canines, sunken eyes, flat wide nose and small ears.

5. Arms longer than legs and big toe larger than the rest.

6. It walks semi-erect on plantigrade feet, using knuckles of hands for support.

7. Gorilla is the most powerful ape and the largest living primate.

Geographical distribution. *Gorilla* (**Gorilla**) is found in Equatorial African forests of Cameron Gabon and Congo.

Homo sapiens (man) is the most intelligent primate having large cranial capacity. Sexually dimorphic (quite distinct male and female). Size varies from 5 to 6 feet or above. Posture erect, vision binocular, clear distinct speech and omnivorous. Female gestates for 9 months.

31. *Elephas maximus* (Elephant)

Classification : Same as that of *Erinaceus* up to Infraclass.

Order ... **Proboscidea** → Nose and upper lip modified as elongated, flexible proboscis or trunk.

Habit and habitat. *Elephas maximus* is gregarious lives in small or large hordes. It feeds on tree leaves, tall grasses, bamboos and sugar canes. It inhabits semi-arid bushy fields and forests from coastal plains to high mountain forests and cool Alpine meadows, and lives near water bodies.

Comments :

1. *Elephas maximus* (Fig. 15.31) is commonly called **Asiatic elephant** or **Hathi** in Hindi.
2. Body is massive, divided into head, short mobile neck, huge trunk, thorax, abdomen and tail. It measures 3 metres tall at shoulder and weigh over 6 tons.
3. Hairs over the body are sparse.
4. Head is large, small eyes and broad ears.
5. Nose and upper lip long forming proboscis or trunk having nasal passage within and nostrils at the tip end. It is used to lift food and water into the mouth.
6. Limbs heavy pillar-like, 5 nails on each foot. Tail is small with bunch of hair at the tip end.
7. In male two upper incisors are elongated as tusks which may grow 2 metres or more. Dental formula is 1003/0003 = 14. Molar teeth laphodont, *i.e.*, enamel and dentive have intricate foldings for grinding food.

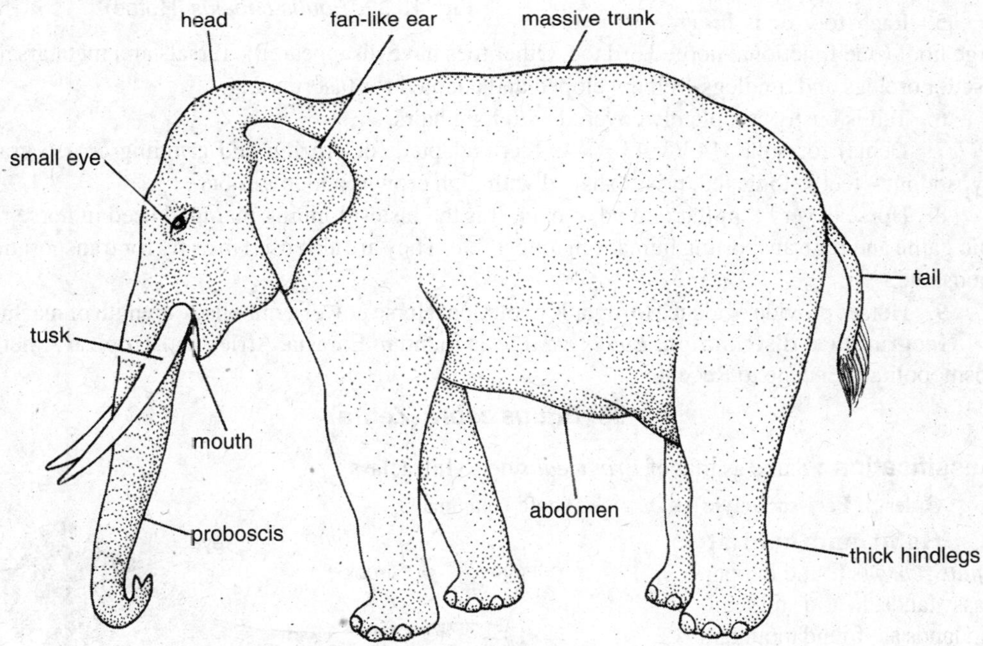

Fig. 15.31. *Elephas maximus* (Asiatic elephant).

8. Gestation period is of 22 months. Female gives birth to a single calf.
9. Elephants have long been domesticated and trained to do heavy work.
10. Elephants' tusk is used to make toys and ornaments. Elephants are used to transport logs in forests and in temples also during festivals. Elephants provide evidence for organic evolution.

Geographical distribution. *Elephas maximus* is found in South Asian countries such as India, Sri Lanka and Myanmar. Eocene to Recent.

32. *Equus cabalus* (Horse)

Classification : Same as that of *Erinaceus* up to Infraclass.

Order ... **Perissodactyla** → Odd toed hoofed mammals.

Habit and habitat. Terrestrial, lives in herds in open plains and grassy lands. Wild Przewalski's horses live in deserts of Mongolia.

Comments :

1. *Equus cavalus* (Fig. 15.32) is commonly called **Horse or Ghora** in Hindi.

2. Body is massively built and strong. Colour is variable. Body is divided into head, neck, trunk, abdomen and tail.

3. Head is small with wide mouth, nostrils, eyes with eye lashes and ears. Jaws elongated typically horse-like.

4. Neck is long and mobile. Mane (hairs over back of neck) is longer and pendent.

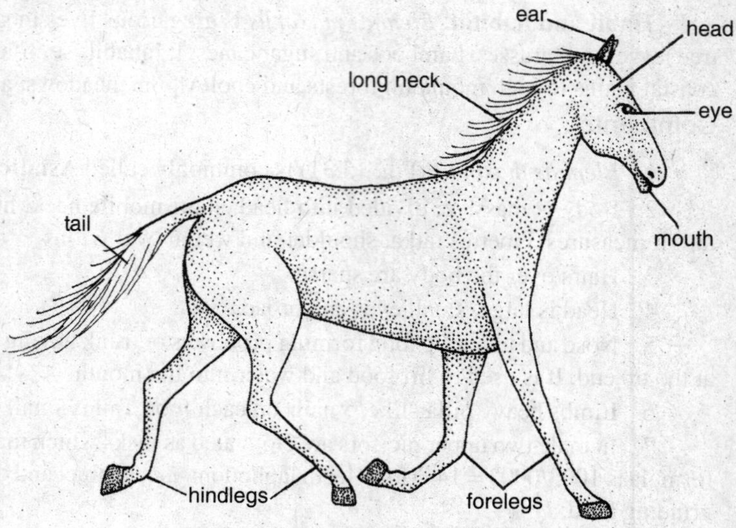

Fig. 15.32. *Equus cabalus* (Horse).

5. Each foot ends in a large hoof (one functional hoofed 3rd toe. Other toes have disappeared). Tarsals and metatarsals are fused. Forelegs and hindlegs well developed and adapted for fast running.

6. Tail is bushy completely covered with long hairs.

7. Dental formula 3143/3143 = 44. Teeth adapted for grazing and grinding coarse grasses, (hypsodont – teeth elongated, prism-shaped with high crowns and low roots).

8. Horse is very familiar animal to man. It is the fastest running mammal used in horse races, polo game and for transporting humans in tongas, etc. They are also used in army for transporation of goods, etc.

9. Horses provide evidence to organic evolution. Horse power symbolises strength of machinery.

Geographical distribution. *Equus cabalus* is found in Eurasia, Africa and Tropical America. Cosmopolitan. Eocene to Recent.

33. *Equus zebra* (Zebra)

Classification : Same as that of *Erinaceaus* up to Infraclass.

Order ... **Perissodactyla** → Odd toed hoofed mammals.

Habit and habitat. *Equus zebra* is found in open grassy lands. In arid and semi-arid lands are found mountain zebras. Zebra is also kept in zoological gardens. They are seldom domesticated.

Comments :

1. *Equus zebra* (Fig. 15.33) is commonly called **zebra**.

2. Body size is like that of ass. Body is elegantly built.

3. Colour of body is white or buff and fully stripped with black or brown. These coloured stripes are present all over body.

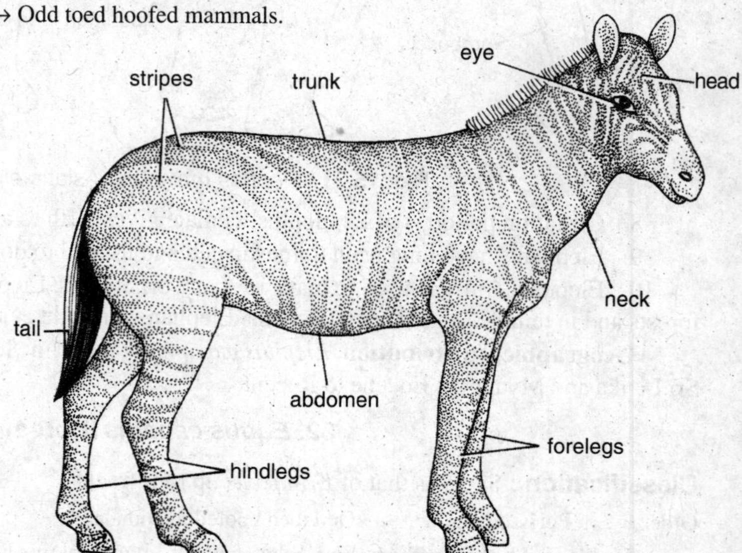

Fig. 15.33. *Equus zebra.*

4. Head bears mouth, nostrils, eyes and ears.
5. Jaws are elongated like horse and teeth are adapted for grazing.
6. Tail is bushy like horse.
7. Legs end in a large hoof (only third digit is present) for fast running.
8. Mane is like horse but smaller in size of hairs.
9. Gestation period is 11–12 months. Only one young is born.
10. For safety, zebras rely on their keen eye-sight and speed, but they can use their hooves and teeth for defence.

Geographical distribution. Zebras are found in open grassy lands of Africa.

34. *Equus hemionus* (Ass or Donkey)

Classification: Same as that of horse.

Habit and habitat. *Equus hemionus* is found grazing in open grass lands. Wild ass lives on high plateaus.

Comments :

1. *Equus hemionus* (Fig. 15.34) is commonly called **Asiatic wild ass**. They are on the verge of extinction. Their domesticated variety is known as **donkey**.
2. It is smaller than the horse and zebra.
3. Its body colour is sandy or white.
4. Head is relatively larger, mane shorter and erect.
5. Head bears mouth, nostrils, eyes with eye lashes and longer ears.
6. Tail's lower part is only covered by hairs.

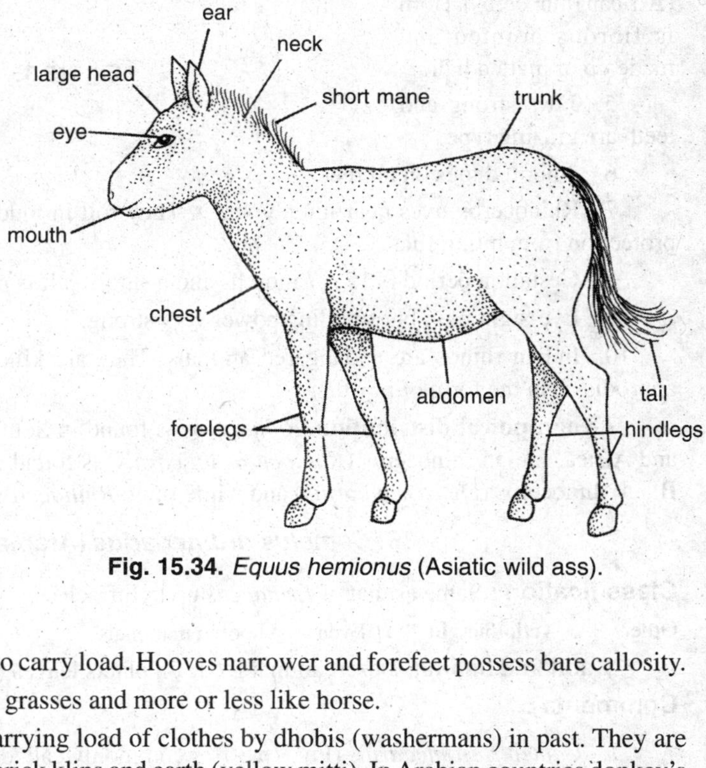

Fig. 15.34. *Equus hemionus* (Asiatic wild ass).

7. Legs are strong and stout to carry load. Hooves narrower and forefeet possess bare callosity.
8. Teeth adapted for grazing grasses and more or less like horse.
10. Donkeys were used for carrying load of clothes by dhobis (washermans) in past. They are also used for carrying bricks from brick klins and earth (yellow mitti). In Arabian countries donkey's race is also performed.

Geographical distribution. In India *Equus hemionus* is found in Rann of Kutch (Gujarat), Mongolia and Syria.

35. *Rhinoceros*

Classification : Same as that of horse.

Habit and habitat. *Rhinoceros* lives in grasslands, forests in the foothills of Himalayas and feed mainly on grasses and succulent vegetation of swamps. It is solitary, swift and agile.

Comments :

1. *Rhinoceros* (Fig. 15.35) is commonly called **Rhinoceros** or **Genda** in Hindi.

2. Body is large (2 metres in height and 2.5 tons in weight), heavily built and thick armour-like folded skin with numerous tubercles. Skin is usually hairless.

3. Head bears large mouth with prehensile upper lip, nostrils, small eyes and ears.

4. Snout bears a single horn (Indian rhinoceros) or two horns (African rhinoceros). Horn is fibrous pointed and made up of matted hairs.

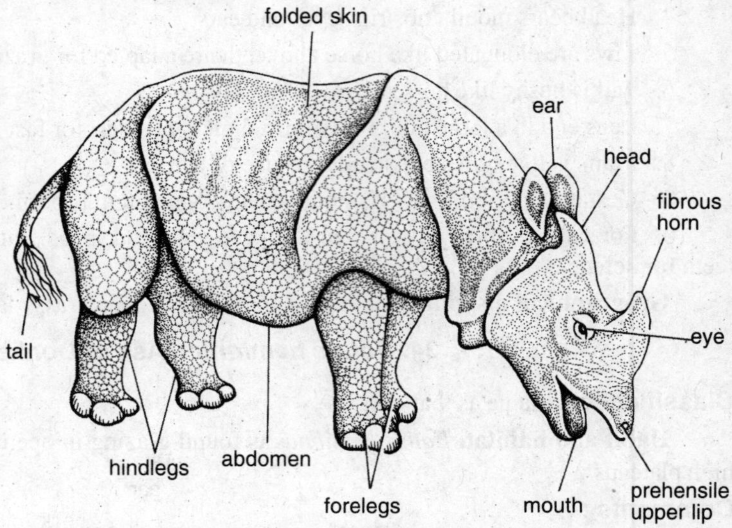

Fig. 15.35. *Indian rhinoceros.*

5. Jaws strong and teeth are grazing type.

6. 3 hoofed toes on all feet.

7. Rhinoceros lives near muddy water. They roll in mud for cooling their skin and also for protection from biting flies.

8. Gestation period is 15–17 months and a single calf is produced.

9. Eye-sight poor but smelling power very strong.

10. Indian rhinos are endangered animals. They are killed for their horns which is used as aphrodiasic in the form of powder.

Geographical distribution. *Rhinoceros* is found in South Asia, India, Nepal, Java, Sumatra and Africa. Indian rhinoceros (*Rhinoceros unicornis)* is found in North Bengal, Assam and Nepal. Black rhinoceros (*Diceros bicornis*) and white rhino *Rhinoceros simus* are found in Africa.

36. *Camelus dromedarius* (Arabian Camel)

Classification : Same as that of *Erinaceus* up to Infraclass.

Order … **Artiodactyla** → Even toed hoofed mammals.

Habit and habitat. Desert animal feeds on plants leaves (herbivorous).

Comments :

1. *Camelus dromedarius* (Fig. 15.36) is commonly called **Camel** or **Oont** in Hindi.

2. Camels have attained highest height.

3. Body divisible into small head, thin and long upright neck, trunk (back) humped, abdomen and tail having hairs on its lower half.

4. Head bears large mouth with powerful jaws and clefted upper lip, muscular nostrils, eyes with long eye lashes and hairy small ears.

5. Limbs elongated and well developed with two toes without hoofs, supported by leathery pads below for moving over sand.

6. Stomach is a complex organ having 4 -chambers and enough space for storing water. They chew their cud like cows and buffaloes, and can carry a reserve water for 2 weeks.

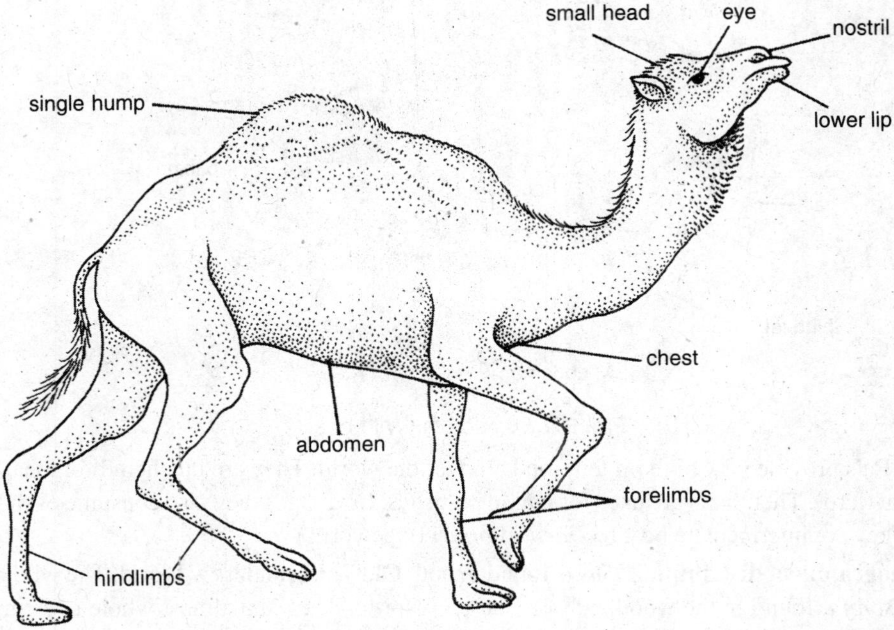

Fig. 15.36. *Camelus dromedarius* (Camel).

7. Camels are characterised by having one or two humps over back. Arabian camel has single hump, while bactrian camel has two humps.

8. Camels are useful domesticated animals used to transport loads. They are called ships of the desert, used to carry loads. Border security forces (BSF) use camels in desert areas of Rajasthan to protect border areas from illegal infiltration, etc. Camel race is famous in Arabian countries. Camels also provide wool and their hairs are used to prepare brushes.

Geographical distribution. *Camelus dromedarius* is an old world animal found in South West Asia and North Africa. *Camelus bactrianus* (Bactrian camel) is inhabitant of cold rocky deserts of Central Asia. (Gobi desert).

37. *Sus scrofa* (Wild Boar)

Classification : Same as that of Camel.

Habit and habitat. *Sus scrofa* is found in herds in plains, deserts and rain forests. *Sus* is domesticated and feeds on both plants and animals and even human faeces. They do not ruminate.

Commets :

1. *Sus scrofa* (Fig. 15.37) is commonly called **Wild boar** or **Pig** and **Suar** in Hindi.
2. Skin thick having bristle-like hairs.
3. Body is stout and covered with short stiff hairs, Male is larger than the female.
4. Head is long with mobile disc-like snout and bears mouth, nostrils at the tip of snout, small eyes and ears.
5. Canines of both the jaws are curved upwards. Dental formula 3143/3143 = 44.
6. Legs short, stout and four toed (four toes on each foot) but only 3rd and 4th toes reach the ground.
7. Tail is short with hairs at its tip end. In domestic pig tail is lifted and coiled.
8. Stomach is simple and 2-chambered. Food is both plant and animal. They do not ruminate.
9. Gestation period is of four months and gives birth to 4–8 piglets.

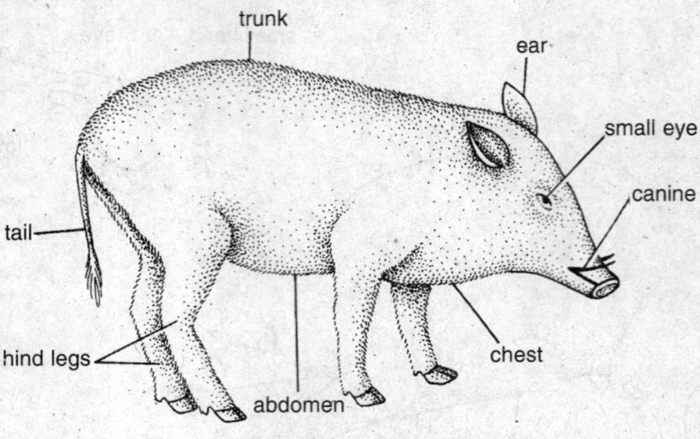

Fig. 37. *Sus scrofa* (Wild boar).

10. Pigs provide very high protein food all over the world. They are butchered for meat (pork) and for fat (lard). Their hairs are used for making brushes. In villages they also consume even human faeces. They are intermediate host for *Taenia solium* (tapeworm).

Geographical distribution. *Sus* is found in both Old World and New World. The wild boar or pig, *Sus scrofa*, found in the woodlands of Europe, North Africa and almost whole of Asia, is the ancestor of all domesticated pigs.

38. *Giraffa camelopardalis* (Giraffe)

Classification : Same as that of Camel.

Habit and habitat. *Giraffa camelopardalis* (Giraffe) is found in woody grasslands.. It feeds on tree leaves and is ruminant.

Comments:

1. *Giraffa camelopardalis* (Fig. 15.38) is commonly called **Giraffe.**

2. It is the tallest animal. Male is 4–6 metres in height, while female is 3.5 – 4.5 metres in height.

3. Body colour is yellow having brownish spots all over the body.

4. Head is small with mouth, nostrils at the tip of snout, eyes with eyelashes, ears and 3–5 small bony horns covered with velvety skin over the top of head.

5. Mouth is provided with leathery lips. It has long prehensile tongue and conical jaws. Its prehensile tongue can extend up to 50 cm to eat leaves of trees.

6. Neck with short erect mane is very long for eating leaves of tall trees. Cervical vertebrae 7.

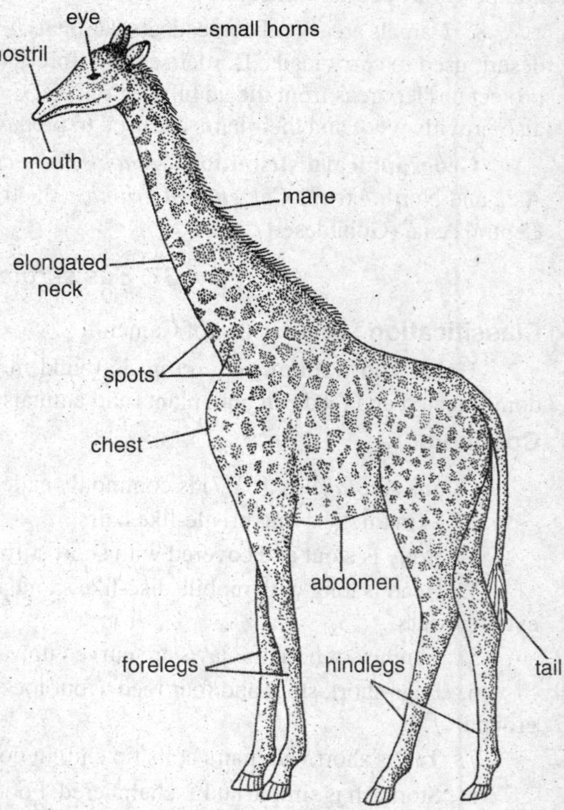

Fig. 15.38. *Giraffa camelopardalis* (Giraffe).

7. Forelegs longer than the hindlegs and both are 2-toed. It runs very fast to outrun most of its enemies.

8. They can live for longer period without water but when available they drink water regularly.

9. It cannot make sound.

Lamarck gave his theory of organic evolution on the basis of Girrafe's evolution.

Geographical distribution. *Giraffa camelopardalis* is found in East Africa.

39. *Ovibos moschatus* (Musk Ox)

Classification : Same as that of Camel.

Habit and habitat. Gregarious, ruminant and inhabitant of colder regions. It eats grasses hedges and browse on willows, lichens, mosses and bark.

Comments :

1. *Ovibos moschatus* (Fig. 15.39) is commonly called **Musk ox**.

2. It measures about 1.5 metres in height at shoulder. Its body weight is about 300 to 400 kg.

3. Body is covered with thick and shaggy coat of long hairs hanging downward the body up to hooves. Tail is covered under the coat. The coat protects the body from cold.

4. Head is large having mouth, nostrils, eyes and a pair of curved pointed massive horns in both the sexes.

5. Both pairs of limbs are stout with broad hooves for movement on snow.

6. Female gives birth to a single calf. They protect their calves from wolves by making a ring around them.

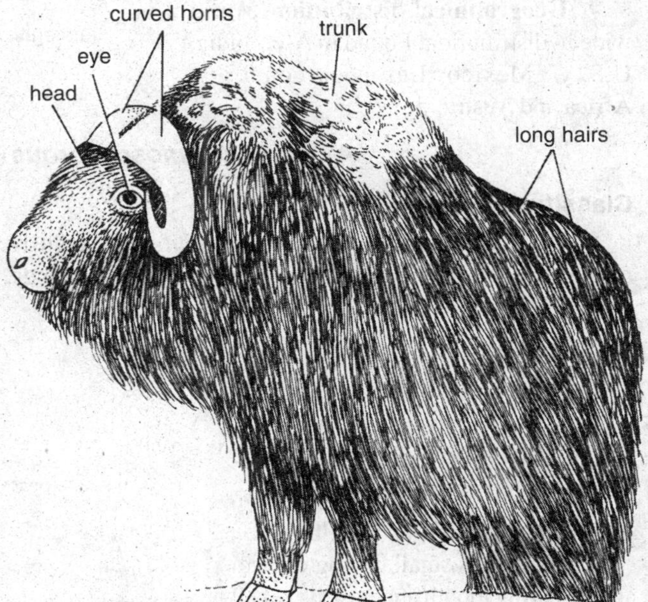

Fig. 15.39. *Ovibos moschatus* (Musk ox).

7. They are killed in hundreds for fur trade and for preparing sports material due to which their population is declining.

Geographical distribution. *Ovibos moschatus* lives in Tundra and North Asia.

40. *Ovis* (Sheep)

Classification : Same as that of Camel.

Habit and habitat. *Ovis* (sheep) is gregarious lives in herds, ruminant and feeds on grasses and tree and shrub leaves.

Comments :

1. *Ovis* (Fig. 15.40) is commonly called **Sheep** and **Bher** in Hindi, measuring about 1 metre inches in height.

2. Body covered with wooly fur. The colour of fur is generally white and is of high economic value.

3. Head bears mouth, nostrils, eyes, ears and a pair of ringed, hollow and backwardly curved horns. Teeth are adapted for grazing.

4. Dentition 0033/3133 = 32.

5. Limbs are short having divided hooves.

6. Female gives birth to a single lamb.

7. They carry a number of protozoans and helminth parasites.

8. Sheep provides us milk, mutton, wool, leather, etc., and, hence, they are domesticated in huge numbers. In India, these are poor men's animals and are grazed in herds protected by stout dogs.

Geographical distribution. World wide in distribution. Found in Asia, India, U.S.A., Mexico, European countries, Africa and Australia.

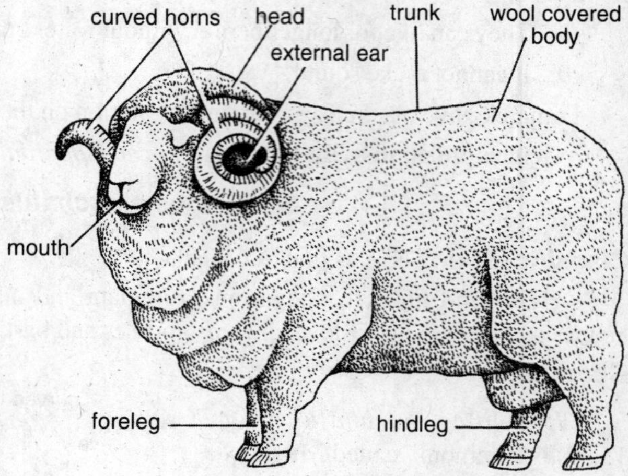

Fig. 15.40. *Ovis* (sheep).

41. *Moschus moschiferous* (Musk Deer)

Classification : Same as that of Camel.

Habit and habitat. *Moschus moschiferous* is elusive, solitary and nocturnal, generally lives near water. It lives in high mountain forests and bushy lands. It is a shy animal, becomes active mainly at dusk and dawn.

Comments :

1. *Moschus moschiferous* (Fig. 15.41) is commonly called **Musk deer** and **Kasturi hiran** in Hindi.

2. Body colour is golden yellow, measuring about 50 cm in height.

3. Head is small having conical snout bearing mouth and nostrils, small eyes and erect ears.

4. Limbs well developed and hoofed even toed.

5. Ruminants and chew their cud.

6. Upper canines in male are highly developed forming tusks up to 7.5 cm long which protrude below jaw and are used for fighting.

7. Adult male possesses an **abdominal musk gland** that secretes **kasturi** (waxy substance). It is used in perfumes and medicines.

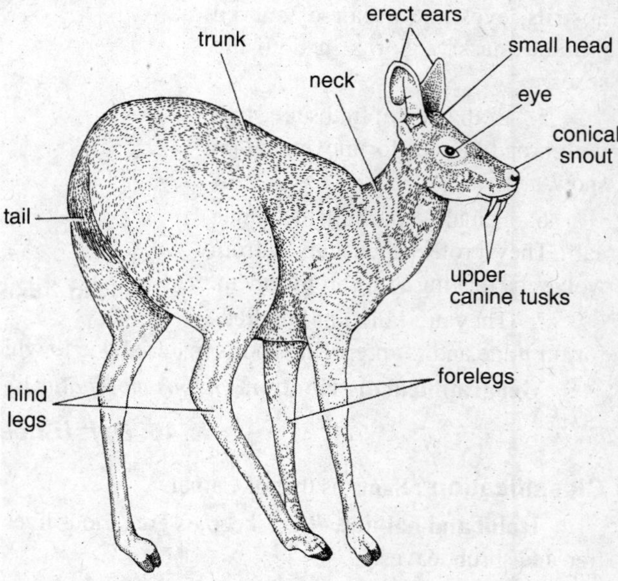

Fig. 15.41. *Moschus moschiferous* (Musk dear)

8. Musk deers are hunted for kasturi which is a precious product. Musk deers belong to threatened species in India.

Geographical distribution. *Moschus moschiferous* (**Musk deer**) is found in equatorial rain forests of Africa, Malaysia and India (in Nanda Devi mountain region).

16
Chapter

SLIDES OF PROTOCHORDATA

1. T.S. of Proboscis of *Balanoglossus*

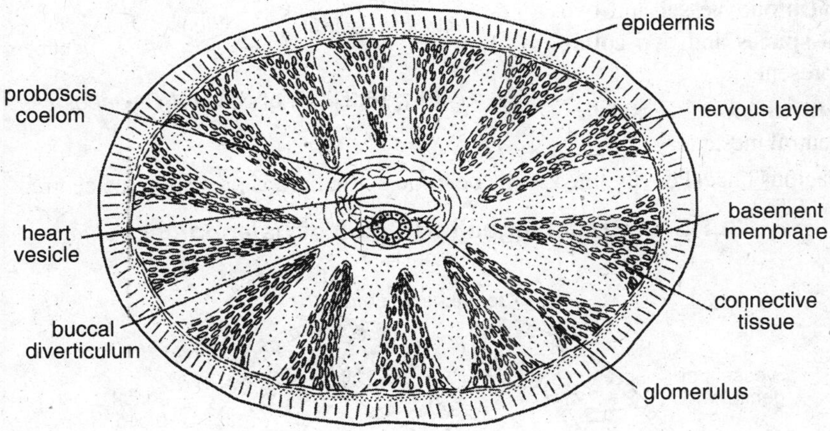

Fig. 16.1. T.S. of proboscis of *Balanoglossus*.

Comments :

Transverse section passing through the **proboscis** of *Balanoglossus* (Fig. 16.1) shows the following structures :

1. Body wall is composed of single layer of epidermis. Dermis is absent.
2. **Nervous layer** intra-epidermal lying beneath the epidermis and bounded internally by the basement membrane.
3. **Proboscis coelom** is very much reduced due to the development of radiating bands of connective tissue and muscle fibres.
4. In the centre of the section there is a **central complex** (basal organ) consisting of **buccal diverticulum**, **central sinus**, **heart vesicle** and **glomerulus**.

2. T.S. of Collar of *Balanoglossus*

Comments :

Transverse section through the **collar** of *Balanoglossus* (Fig. 16.2) shows the following structures :

1. Body wall is composed of **epidermis** provided with blood plexus on its inner side. Dermis absent.

2. Below the epidermis lies the **collar coelom** incompletely divided by the dorsal and ventral mesenteries.

3. The collar coelom is filled by the musculature and connective tissue.

4. In the dorsal portion a **collar cord**, **dorsal blood vessel** and two **perihaemal spaces** and two **collar canals** are present.

5. **Ventral blood vessel** lies above the ventral mesentery.

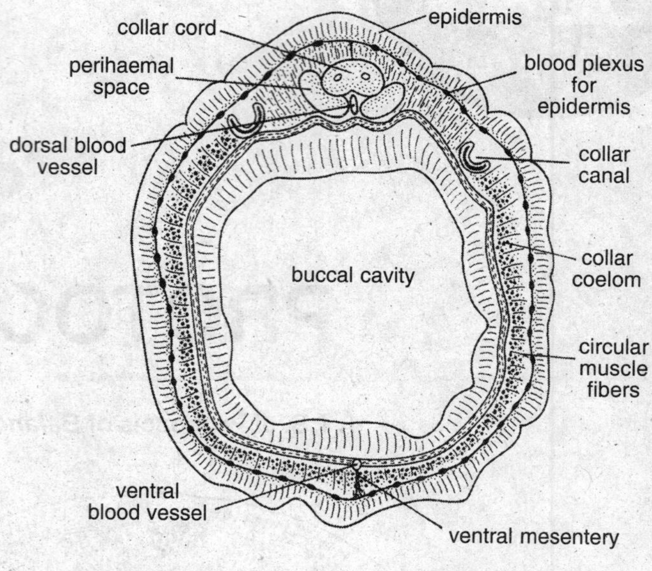

Fig. 16.2. T.S. of collar of *Balanoglossus.*

6. Spacious buccal cavity lined with coelomic epithelium is present in the centre.

3. T.S. of Branchiogenital Region of *Balanoglossus*

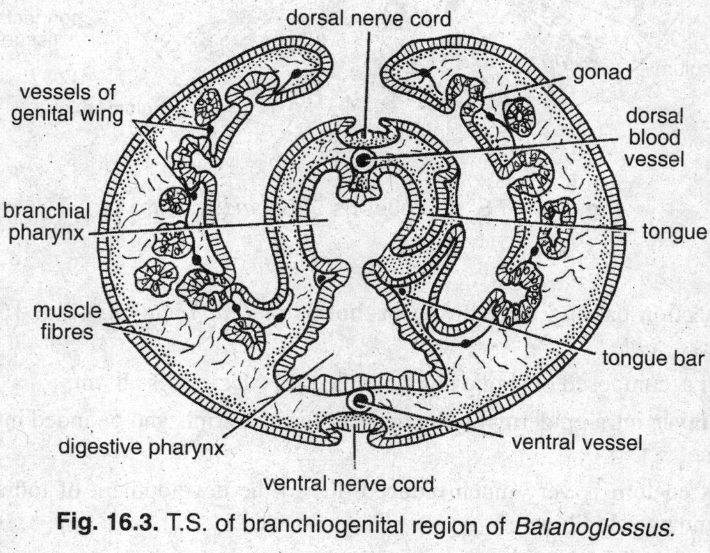

Fig. 16.3. T.S. of branchiogenital region of *Balanoglossus.*

Comments :

Transverse section of *Balanoglossus* passing through the **branchiogenital region** (Fig. 16.3) shows the following important structures :

1. Outer layer is of skin consisting of only epidermis and dermis is absent. Beneath epidermis lies the musculature and connective tissue.

2. A pair of genital wings is present which contain numerous gonads opening outside through the gonopores and blood vessels.

3. In the centre, pharynx is present which is divided by lateral constrictions into an upper **branchial** or **respiratory pharynx** for respiration and lower **digestive-pharynx**.

4. The **branchial** or **respiratory pharynx** communicates to the outer side by lateral branchial aperture.

5. **Dorsal blood vessel** and **dorsal nerve cord** lie above tne branchial pharynx and **ventral blood vessel** and **ventral nerve cord** lie below the digestive-pharynx.

4. T.S. of Post-hepatic Intestine of *Balanoglossus*

Comments :

Transverse section of *Balanoglossus* through the **post-hepatic intestine** (Fig. 16.4) reveals the following structure :

1. Epidermis is the outermost layer and consists of single layer of cells : goblet cells and neurosensory cells.

2. On the dorsal side **dorsal nerve cord** and **dorsal blood vessel** are seen.

3. On the ventral side **ventral blood vessel** and **ventral nerve cord** are present.

4. In the centre a large lumen of intestine having a pair of dorso-lateral **ciliated grooves** are present. Intestine is lined by a single layer of flattened epithelial cells.

5. Several layers of **longitudinal muscle fibres** are present between the epidermis and the wall of the intestine.

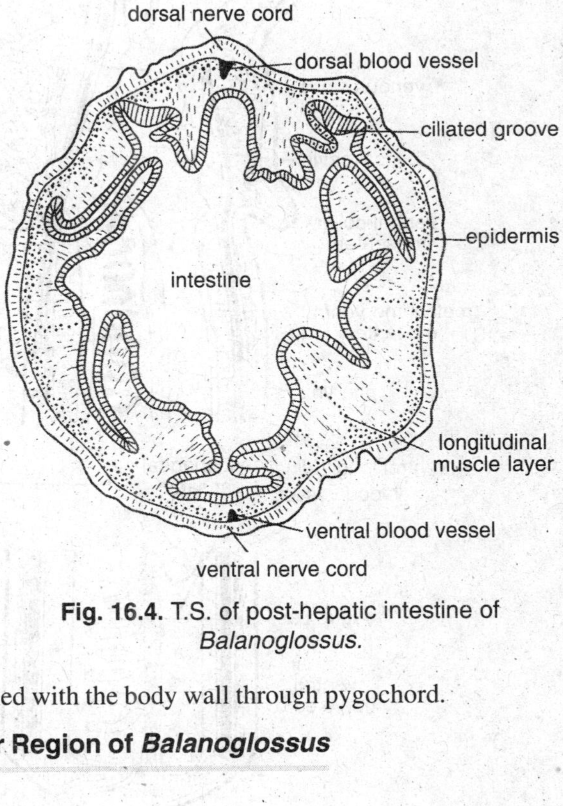

Fig. 16.4. T.S. of post-hepatic intestine of *Balanoglossus.*

6. On the ventral side intestine is attached with the body wall through pygochord.

5. M.L.S. of Anterior Region of *Balanoglossus*

Comments:

Median longitudinal section of **anterior region** of *Balanoglossus* (Fig. 16.5) shows the following important structures :

1. The body wall composed of single layer of **epidermis**. Dermis is absent.

2. Nervous layer is **intra-epidermal**, *i.e.*, lying below the epidermis and bounded internally by basement membrane.

3. The section comprises **proboscis**, **collar** and **anterior region** of trunk having **pharynx**.

4. **Proboscis** is conical with narrow posterior proboscis stalk connected with the collar.

5. The **proboscis coelom** is reduced due to the presence of muscles (longitudinal and circular) and contains the central complex.

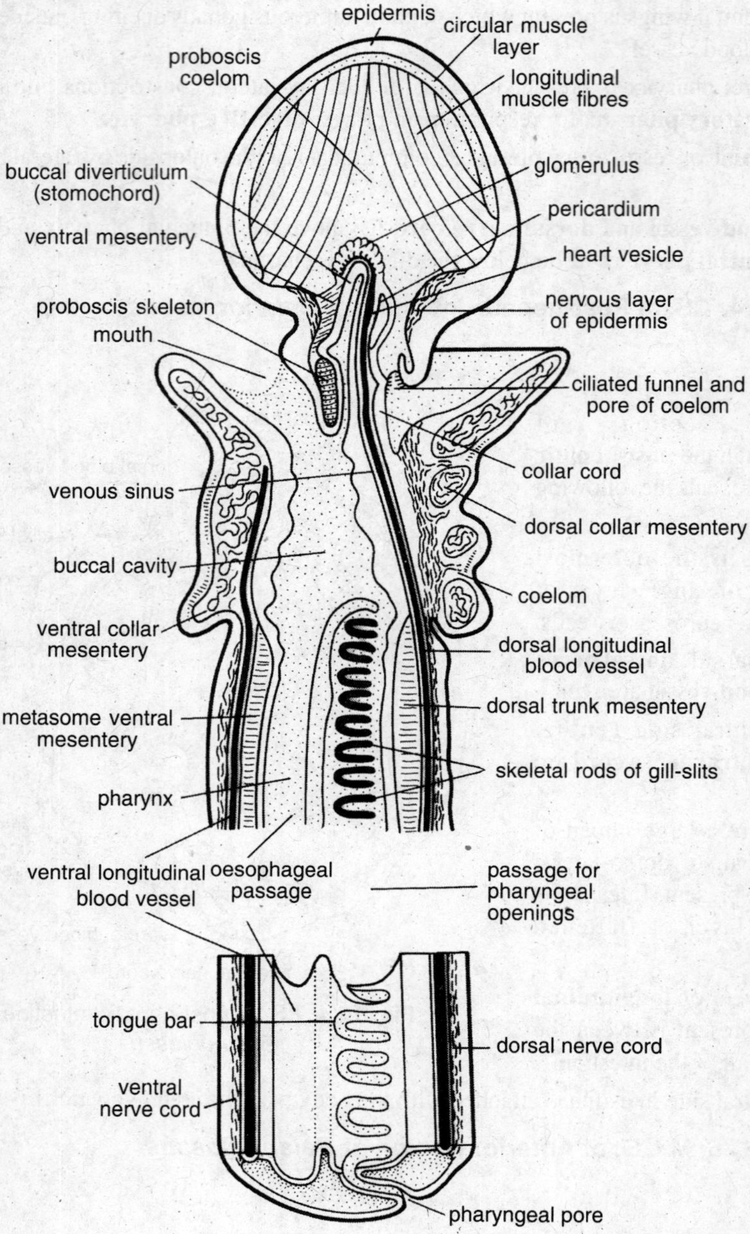

Fig. 16.5. M.L.S. of anterior region of *Balanoglossus*.

6. The **central complex** comprises **buccal diverticulum**, **central sinus**, **heart vesicle** and **glomerulus**.

7. In the posterior part of the proboscis at the base of buccal diverticulum is present the **proboscis skeleton**.

8. **Mouth** is ventral and lies in a groove at the junction of proboscis and collar. It leads into the buccal cavity confined to the collar region.

9. The **collar** contains **buccal cavity**, **venous sinus** (containing dorsal longitudinal blood vessel), **collar cord**, **dorsal** and **ventral collar mesenteries** and **coelom**.

10. The anterior trunk region contains **pharynx** which is divided into dorsal respiratory-pharynx perforated by numerous U-shaped gill-slits having U-shaped skeleton rods (tongue bars) and ventral digestive pharynx.

11. The trunk region also shows **dorsal** and **ventral nerve cords**, **dorsal** and **ventral trunk mesenteries** and **oesophageal passage**.

12. The pharyngeal pore is also seen at the posterior end.

6. Tornaria Larva (Whole mount)

Comments :

This is the W.M. of **tornaria larva** of *Balanoglossus* (Fig. 16.6). It shows the following characteristic features :

1. The full grown **tornaria larva** is usually ovoid in shape and transparent, measuring up to 3 mm in size.

2. It is provided with two ciliated bands encircling the body : **A pre-oral ciliated band** lies anterior to mouth and forms a **post-oral loop** behind the mouth, and a **circum-anal ring** or **posterior ciliated band** lies in front of the anus.

3. At its anterior end is a thickened plate the **apical plate** which bears a tuft of cilia and a pair of pigmented **eye-spots**.

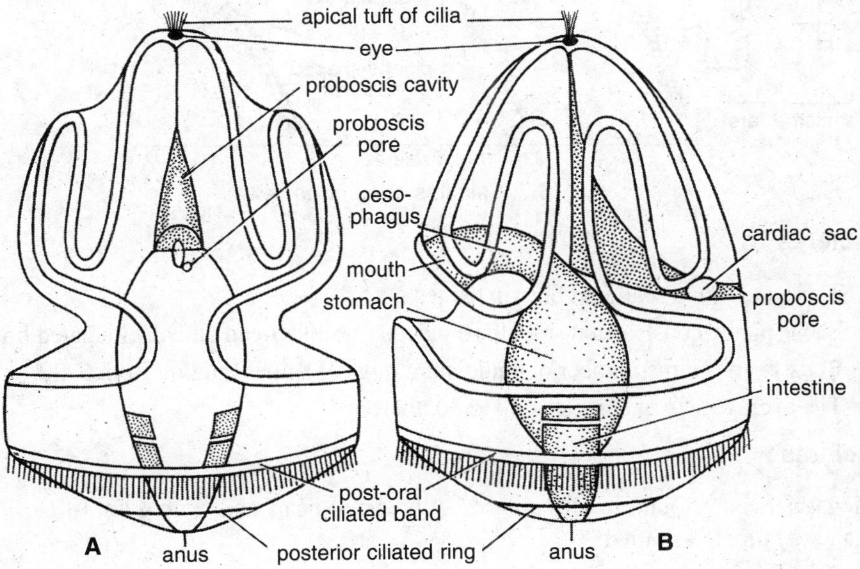

Fig. 16.6. Tornaria larva of *Balanoglossus*. A – Dorsal view; B – Ventral view.

4. The circum oral band cilia are short and collect food. The circum-anal ring or telotroch have long and powerful cilia and is the chief locomotory organ of the tornaria.

5. **Mouth** is situated on the ventral surface, while the **anus** lies at the posterior terminal end. The alimentary canal is simple divided into **oesophagus**, **stomach** and **intestine**.

6. A thin walled **water sac** (protocoel) opens out side through the dorsal pore, the **hydropore**.

7. The tornaria larva resembles the echinoderm larvae.

8. After metamorphosing it attains the adult form.

7. Spicules of *Herdmania*

Comments :

Spicules of *Herdmania* (Fig. 16.7) are calcareous and are of two types found embedded in the test, *viz.*, very small known as **microscleres** and larger ones are known as **megascleres**. All the spicules have rings of minute spines all directed in the same direction.

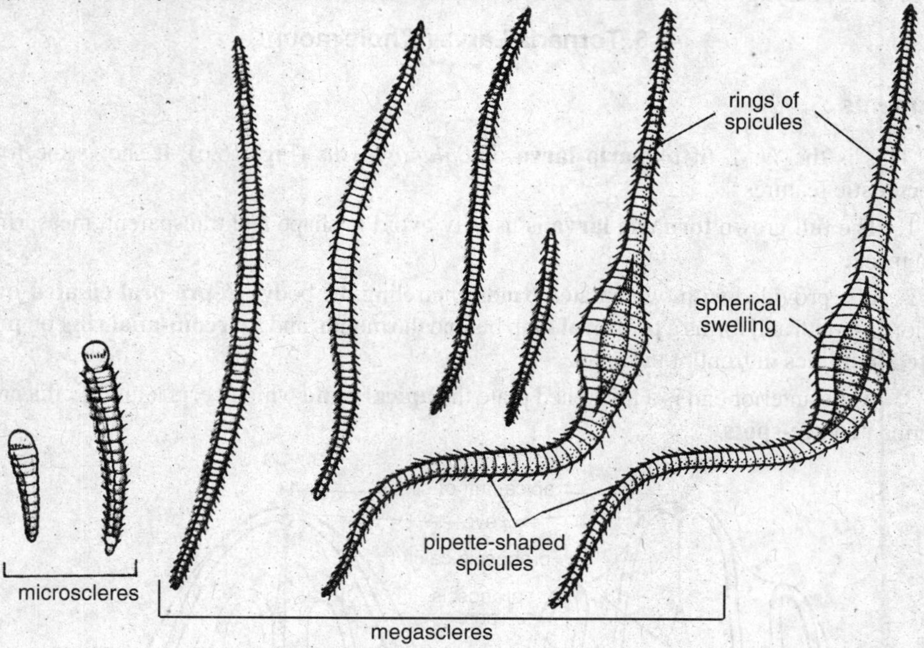

16.7. Spicules of *Herdmania.*

Microscleres :

1. These are found scattered only in the test.
2. Each spicule consists of a smooth rounded knob-like **head** and an elongated **body**.
3. Body bears a large number of equidistant rings of spines ranging from 5–20.
4. The average size of spicule is 40 to 80 microns.

Megascleres :

Megascleres are found in all parts of body except heart and of two kinds, *i.e.*, *(i)* **spindle-shaped** and *(ii)* **pipette-shaped**.

(i) **Spindle-shaped spicules :**

1. These are always found enclosed in a connective tissue sheath and lie scattered throughout the body in most of its tissues.
2. Each spicule has 20–60 equidistant a rings of spines, pointing in the same direction.
3. The average size is 1·5 – 2·5 mm.

(ii) **Pipette-shaped spicules :**

1. These spicules are larger ranging up to 3–6 mm in size.
2. The characteristic feature of each spicule is the presence of large spherical swelling in the middle. They are of straight and curved type and pointed at the end.
3. They have a large number of rings of spines.

8. Tadpole Larva of *Herdmania*

Comments :

Larva of *Herdmania* is called **tadpole larva**. It leads a free swimming life for about 3 to 4 hours after hatching and then undergoes retrogressive metamorphosis.

1. Tadpole larva of *Herdmania* (Fig. 16.8) is transparent, minute about 1.2 mm long and highly motile.

2. **Body** is covered by a thin test and divided into an anterior oval trunk (head) and a long posterior tail.

3. **Trunk** bears three anterior adhesive papillae for attachment during metamorphosis.

4. (*i*) Trunk internally has a dorsal oval and hollow brain (sensory vesicle) containing two ocelli and an otocyst.

(*ii*) Behind sensory vesicle is present visceral ganglion extending into the tail as nerve cord.

(*iii*) Alimentary canal is rudimentary having antero-dorsal mouth, a large sac-like pharynx bearing gill-slits and enclosed with in atrium, intestine and rectum. Rectum opens into the atrium.

(*iv*) Atrium opens outside through atriopore.

(*v*) Below the pharynx lies the heart-enclosed within pericardium.

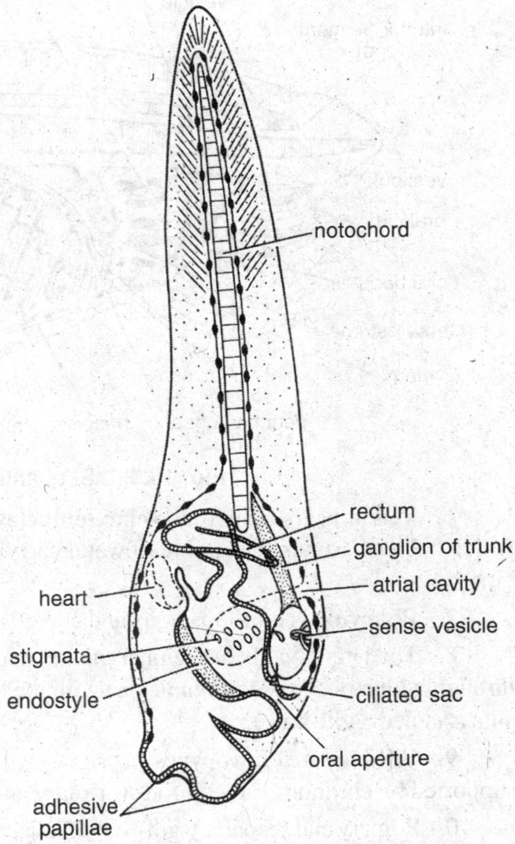

Fig. 16.8. Free-swimming larva of *Herdmania*.

5. **Tail** is laterally compressed and fringed with tail fin. Tail possesses dorsal nerve cord, notochord and muscle bands. Tail is its locomotor organ.

9. L.S. of Anterior Part of *Amphioxus*

Comments :

Longitudinal section of **anterior part** of *Amphioxus* (Fig. 16.9) shows the following important structures :

1. Body is provided with dorsal fin having dorsal fin-rays along the dorsal surface of the body.

2. **Notochord** extends from the anterior end to the posterior end of the body.

3. **Nerve cord** lies just above the notochord and extends from the anterior end to the posterior end.

4. **Oral hood** is ventral and provided with a number of **oral cirri** which help during feeding by turning inwards to prevent sand particles from passing into the buccal cavity.

5. At the posterior wall of the buccal cavity lies a vertical transverse partition called **velum** perforated by **enterostome**.

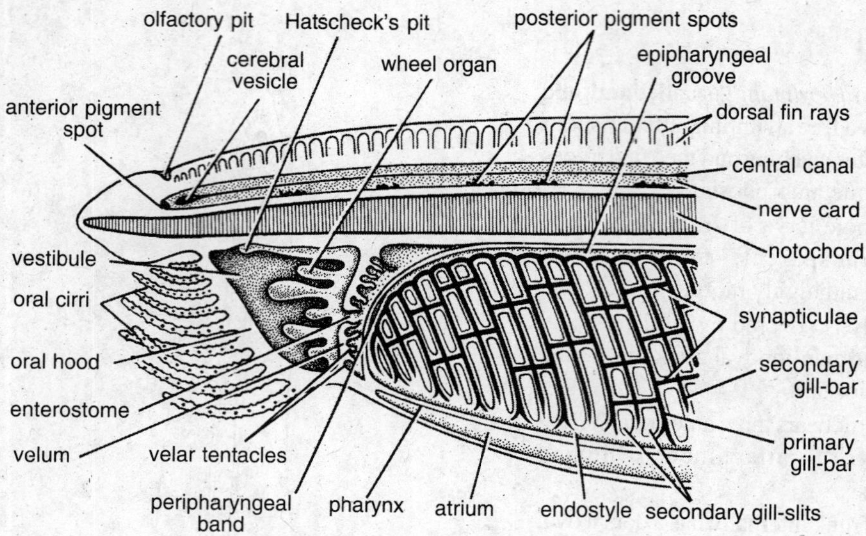

Fig. 16.9. L.S. of anterior part of *Amphioxus.*

6. Velum bears numerous **velar tentacles**. In the buccal cavity lies a peculiar **wheel organ** which helps in driving the current of water carrying food particles into the mouth surrounded by oral cirri.

7. **Pharynx** is quite spacious and extends from the velum to the beginning of the intestine.

8. The lateral walls of the pharynx are perforated by a longitudinal series of vertical gill-slits through which pharynx communicates with the atrial cavity. The pharyngeal wall between two gill-slits is called a **gill-bar**.

9. Gill-bars are of two types : a primary gill-bar and a secondary gill-bar, both alternate and are supported by chitinous skeletal rods (primary and secondary gill-rods).

10. Primary and secondary gill-rods of adjacent gill-bars are connected by transverse synapticulae.

11. Pharynx also have a dorsal epipharyngeal groove and a ventral ciliated and glandular endostyle.

10. T.S. of *Amphioxus* through Oral Hood

Comments :

Transverse section of *Amphioxus* passing through the region of **oral hood** (Fig. 16.10) shows the following details :

1. Body wall comprises single layer of **epidermis** lined internally by basement membrane, compact and fibrous cutis and spongy subcutis. Dermis absent.

2. On the dorsal surface **dorsal fin** having the **dorsal fin-ray** is present.

3. **Myotomes** separated by **myocommata** are present on both the dorso-lateral sides of the body enclosing the nerve cord and notochord.

4. Dorsal tubular and hollow **nerve cord** lies below the dorsal fin. Ocelli (eye spots) are present on the ventro-lateral sides of nerve cord.

5. **Notochord** composed of vacuolated cells is enclosed in the notochordal sheath and lies below the nerve cord.

6. The oral hood encloses a large **buccal cavity**. In its dorsal wall has a groove called **Hatscheck's groove** which is sensory in nature.

7. In the buccal cavity several sections of the buccal cirri are seen which actually surround the mouth and arise from the margin of oral hood.

11. T.S. of *Amphioxus* through Anterior Pharynx

Comments :

Transverse section of *Amphioxus* passing through the **anterior region of pharynx** (Fig. 16.11) shows the following important structures :

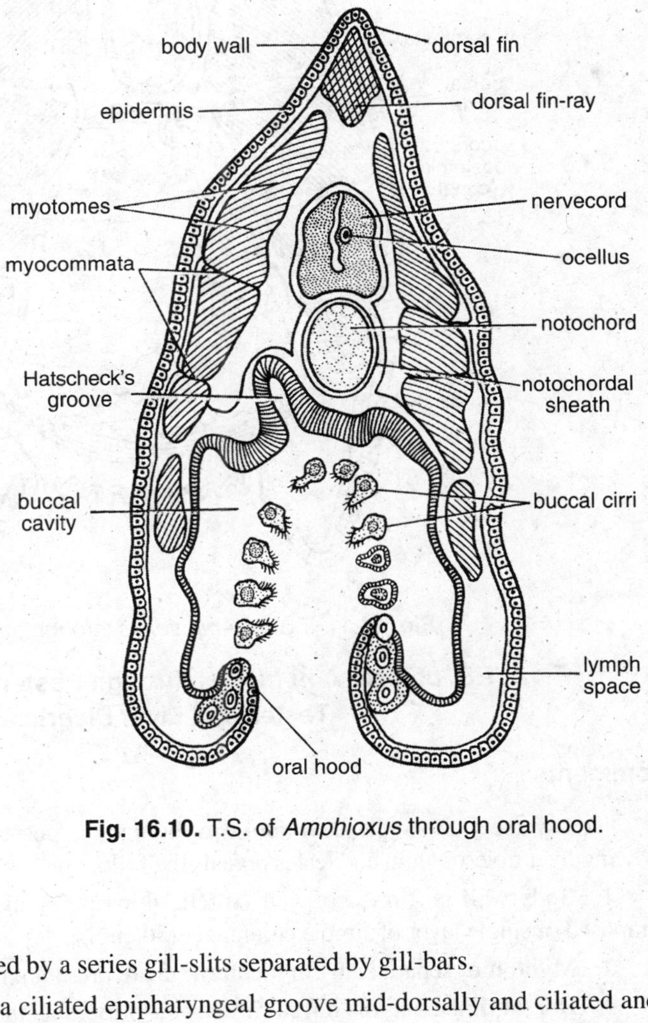

Fig. 16.10. T.S. of *Amphioxus* through oral hood.

1. Body wall is formed of thin **cuticle**, **epidermis, cutis** and **subcutis**.

2. **Myotomes** (muscles) enclosed within **myocommata** of both the sides alternate with each other.

3. **Dorsal fin** containing the **dorsal fin-rays** lies at the dorsal surface.

4. Beneath the dorsal fin is present the dorsal tubular **nerve cord**, **notochord** comprising of vacuolated cells surrounded by notochordal sheath.

5. **Pharynx** lying beneath the notochord is quite spacious and surrounded ventro-laterally by the atrial cavity.

6. Pharynx is laterally perforated by a series gill-slits separated by gill-bars.

7. Pharynx internally contains a ciliated epipharyngeal groove mid-dorsally and ciliated and glandular endostyle or hyopharyngeal groove mid-ventrally.

8. Ciliated epipharyngeal groove push the food laden mucous cord into oesophagus.

9. Other structures present are the **dorsal aortae**, one on either side of the epipharyngeal groove, coelomic spaces, gill-arches, ventral vessel within endostylar coelom and transverse muscle fibres.

10. The **metapleural** folds having cavities are present on the ventro-lateral sides of the anterior two-third of body.

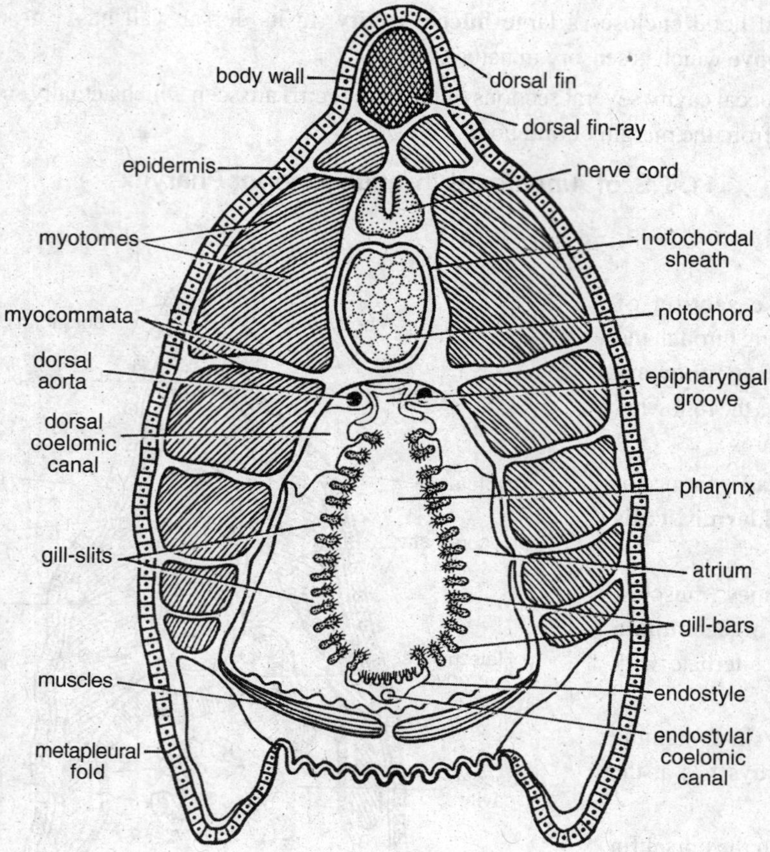

Fig. 16.11. T.S. of *Amphioxus* through anterior pharynx.

12. T.S. of Male *Amphioxus* through Posterior Pharynx showing Testes and Liver Diverticulum

Comments :

Transverse section of male *Amphioxus* through the **posterior region of pharynx** (Fig. 16.12) showing liver diverticulum and testes reveals the following structures :

1. Body wall is formed of thin **cuticle**, thin **epidermis**, **cutis** and **subcutis**. Epidermis is composed of single layer of simple columnar epithelium.
2. **Myotomes** separated by **myocommata** are present on both the lateral sides of the body.
3. **Dorsal fin** having the **dorsal fin-rays** is present on the dorsal surface.
4. Beneath the dorsal fin is present hollow **nerve cord**, **dorsal fin-ray** and **notochord** composed of vacuolated cells and surrounded by **notochordal sheath**.
5. Beneath the notochord lies a spacious **pharynx** perforated by numerous **gill-slits** separated by gill-bars containing gill-arches.
6. In the mid-dorsal line of pharynx lies a ciliated **epipharyngeal groove**, while in the mid-ventral line is present a glandular ciliated **endostyle**.
7. The **atrium** is present around the pharynx ventro-laterally.

8. **Coelom** appears as dorsal coelomic canals on either side of the epipharyngeal groove. Parts of coelom are also present in the endostyle, in metapleural folds and around the liver diverticulum and gonads.

9. The **liver diverticulum** surrounded by coelom lies on the right side of the pharynx.

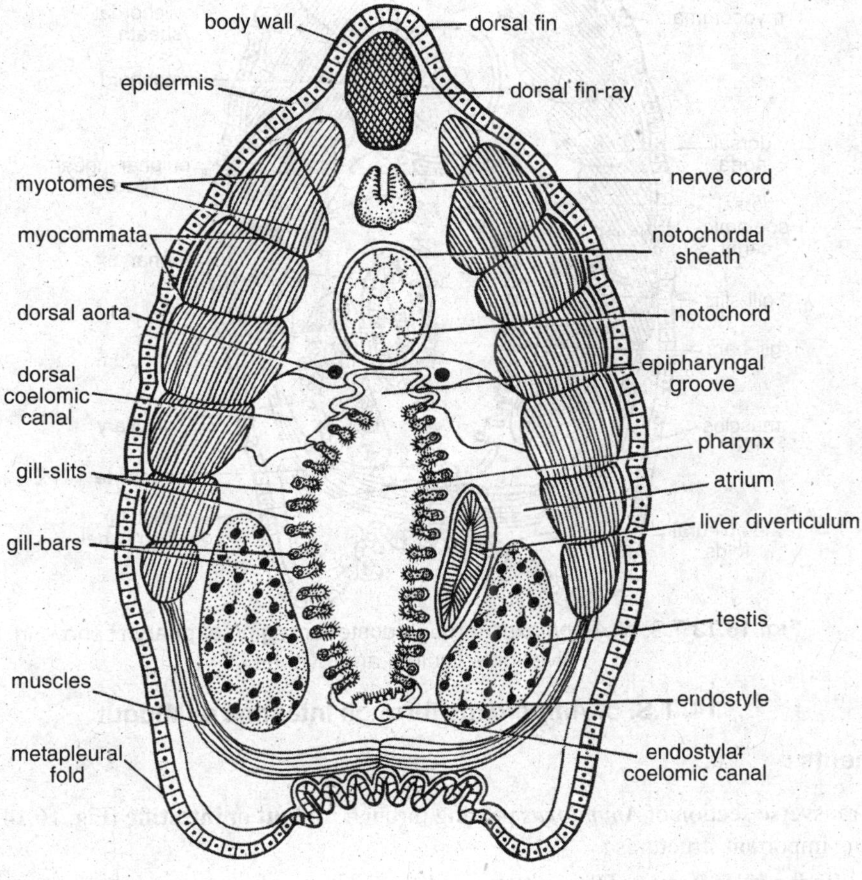

Fig. 16.12. T.S. of *Amphioxus* through posterior pharynx showing testes and liver diverticulum.

10. The **testes**, one pair in each segment bulging in the atrial cavity on both sides of the pharynx. Spermatozoa are visible with in each testis in the form of streaks.

11. **Metapleural folds** are present on both the ventro-lateral sides of the body.

13. T.S. of Female *Amphioxus* through Posterior Pharynx showing Ovaries and Liver-diverticulum

Comments :

1 – 9 points are the same as in the previous section.

10. The **ovaries**, one pair in each segment bulging in the atrium on both sides of the pharynx. The ovaries contain several rounded ova.

11. **Metapleural folds** are present on both the ventro-lateral sides of the body up to atriopore.

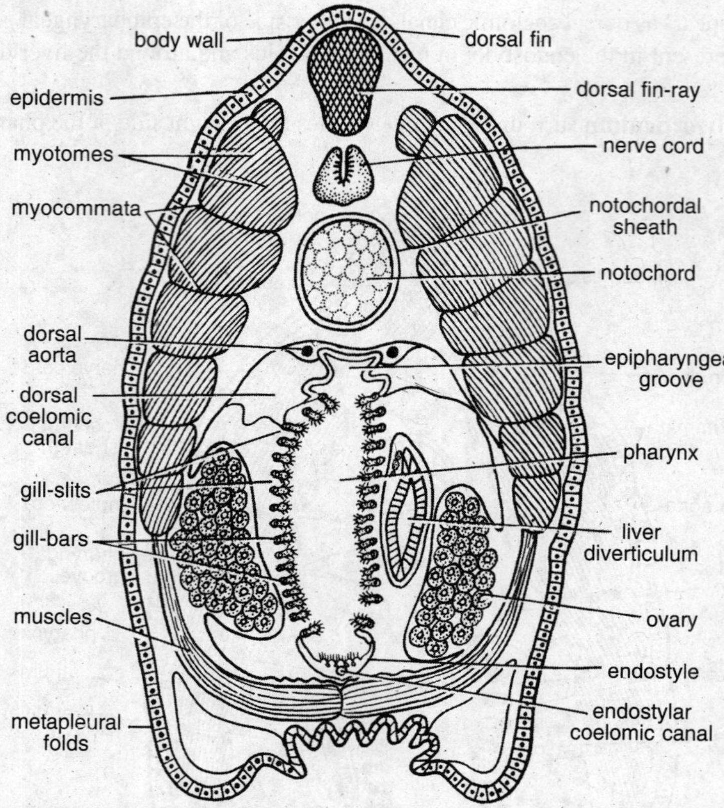

Fig. 16.13 T.S. of *Amphioxus* through posterior region of pharynx showing liver-diverticulum and ovaries.

14. T.S. of *Amphioxus* through Intestine or Midgut

Comments :

Transverse section of *Amphioxus* passing through **midgut** or **intestine** (Fig. 16.14) shows the following important structures :

1. Body wall is formed of thin **cuticle**, **epidermis** is composed of single layer of simple columnar epithelium, **cutis** and **subcutis**.
2. **Dorsal fin** having **dorsal fin-ray** lies on the dorsal surface.
3. **Myotomes** (longitudinal muscles) surrounded by **myocommata** of both the sides alternate with each other.
4. Beneath the dorsal fin lies the **nerve cord** and **notochord** formed of vacuolated cells and surrounded by **notochordal sheath**.
5. Single median **dorsal aorta** is seen ventral to the notochord.
6. The oval **midgut** or **intestine** composed of single layer of ciliated columnar epithelium lies in the coelomic cavity in the centre of the section.
7. **Coelom** is very well-developed and surrounds the intestine.
8. The **atrium** is well-developed lying on the ventro-lateral sides of the coelom and intestine.
9. The floor of atrial cavity contains several **renal papillae**. Transverse muscles are present in between two metapleural folds.
10. The **metapleural folds** are present on the ventrolateral sides of the anterior two-third part of the body.

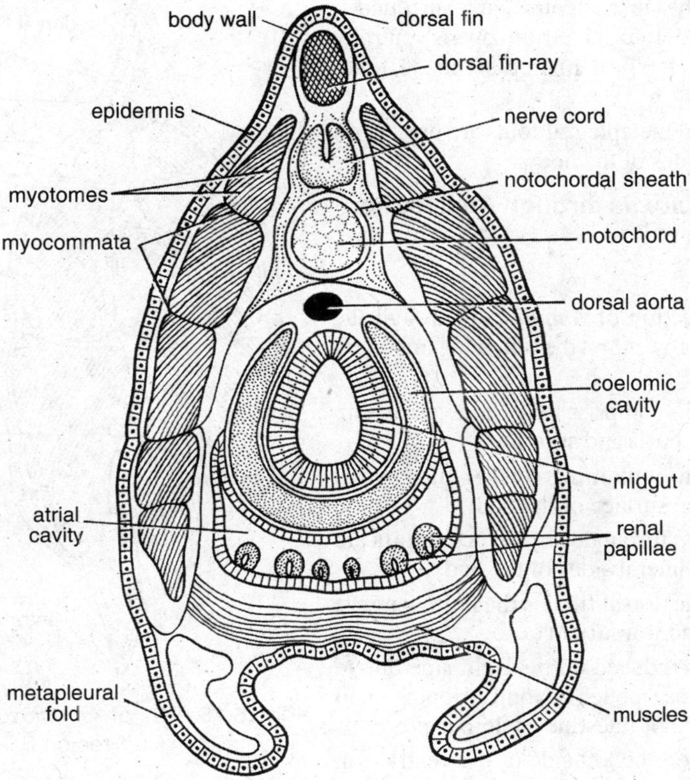

Fig. 16.14. T.S. of *Amphioxus* through intestine.

15. T.S. of *Amphioxus* through Atriopore

Comments :

Transverse section of *Amphioxus* through the atriopore (Fig. 16.15) shows the following important details :

1. Body wall is formed of thin **cuticle**, **epidermis** composed of single layer of simple columnar epithelium, **cutis** and **subcutis**.

2. **Myotomes** enclosed within **myocommata** of both the sides alternate with each other.

3. **Dorsal fin** having the **dorsal fin-rays** is present on the dorsal surface of the section.

4. Beneath the dorsal fin lies hollow **nerve cord**, **notochord** surrounded by **notochordal sheath** and single median **dorsal aorta.**

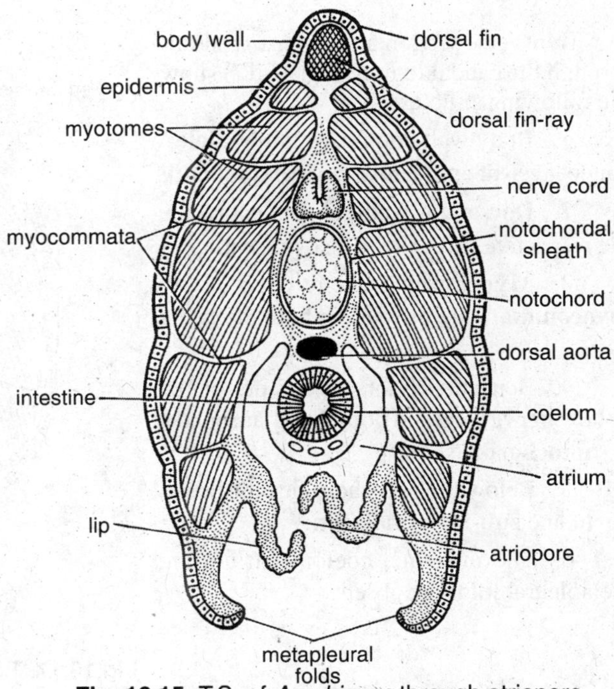

Fig. 16.15. T.S. of *Amphioxus* through atriopore.

5. **Intestine** lies in the centre and is surrounded ventrolaterally by coelom and atrium opens ventrally by an **atriopore**. Atrium protects the delicate pharyngeal region.

6. Two lateral metapleural folds are present on the ventro-lateral sides of the body.

16. T.S. of *Amphioxus* through Anal Region

Comments :

Transverse section of *Amphioxus* through the anal region (Fig. 16.16) shows the following important structures:

1. Body wall comprises single layer of thin **cuticle**, **epidermis**, **cutis** and **subcutis**.

2. **Dorsal fin** containing **dorsal fin-rays** is present on the dorsal surface of the body.

3. **Myotomes** surrounded by **myocommata** are present on both the lateral sides of the body.

4. Beneath the dorsal fin lies the hollow **nerve cord**, **notochord** and **dorsal aorta**.

5. **Coelom** is reduced on the right side due to extension of atrium as a pouch beyond atriopore up to anus. On the left side of intestine coelom is present.

6. **Anus** lies on the left side of the ventral fin. **Ventral fin** is well developed. Metapleural folds are absent beyond atriopore.

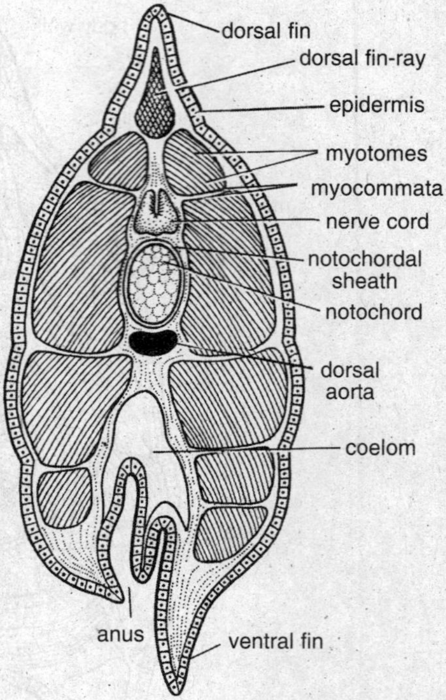

Fig 16.16. T.S., of *Amphioxus* through anal region.

17. T.S. of *Amphioxus* through Caudal Region

Comments :

Transverse section of *Amphioxus* passing through the caudal region (Fig. 16.17) shows the following structures :

1. Body wall comprises thin **cuticle**, a single layer of **epidermis**, **cutis** and **subcutis**.

2. **Dorsal** and **ventral fins** containing the respective fin-rays are also present.

3. **Myotomes** enclosed by **myocommata** are present on both the lateral sides.

4. Beneath the dorsal fin is present the hollow **nerve cord** and **notochord** surrounded by **notochordal sheath**.

5. Below the notochord lies the **dorsal aorta** and **sub-intestinal vein**.

6. The intestine, coelom, atrium and metapleural folds are absent.

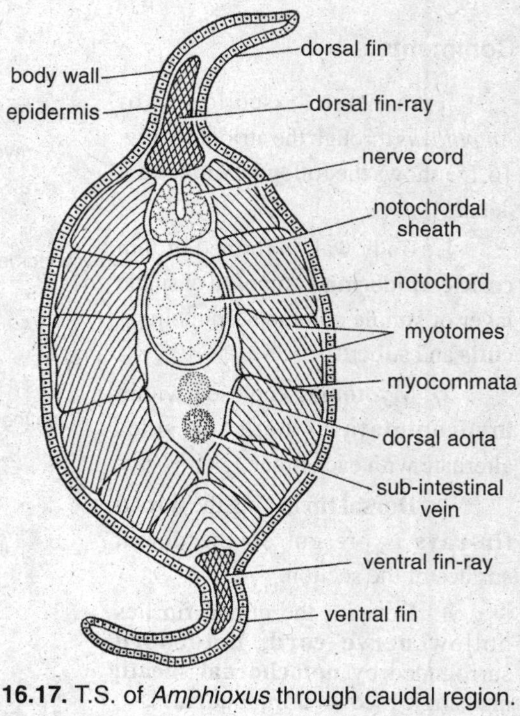

Fig 16.17. T.S. of *Amphioxus* through caudal region.

18. Velum of *Amphioxus* (Whole mount)

Comments :

1. **Velum** of *Amphioxus* (Fig. 16.18) is a vertical membranous partition at the hinder wall of the vestibule or buccal cavity.

2. It is composed of velar-ring with sphincter and velar-tentacles.

3. The **sphincter** composed of circular muscle fibres and controls the opening and closing of enterostome.

4. The **velar-tentacles** having **sensory papillae** are 10–12 in number arising from the posterior border of velum and projecting into the enterostome.

5. The **enterostome** is the opening in the centre of velum.

6. The velar tentacles over enterostome form a sieve and filter the food particles.

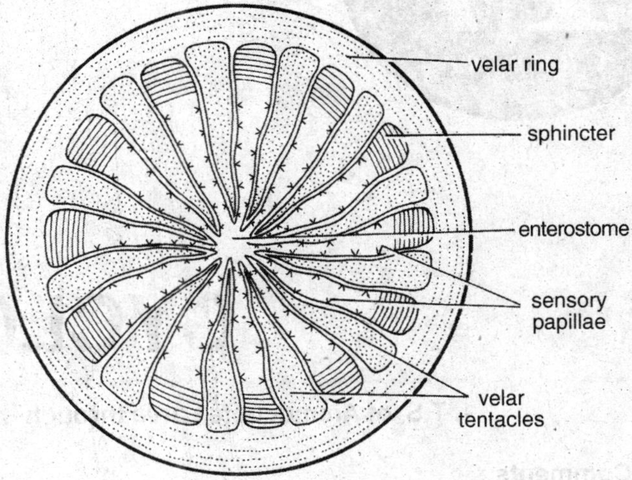

Fig. 16.18. Velum of *Amphioxus.*

19. Oral hood of *Amphioxus* (Whole mount)

Comments :

1. The membranous **oral hood** (Fig. 16.19) encloses the buccal cavity or vestibule of *Amphioxus.*

2. Anterior ventro-lateral margin of oral hood bears oral cirri (10–11 pairs).

3. Each oral cirrus contains a gelatinous **skeletal-rod** and bears small **sensory papillae**.

4. Oral cirri form a sieve over the mouth to prevent entry of larger food particles into the vestibule.

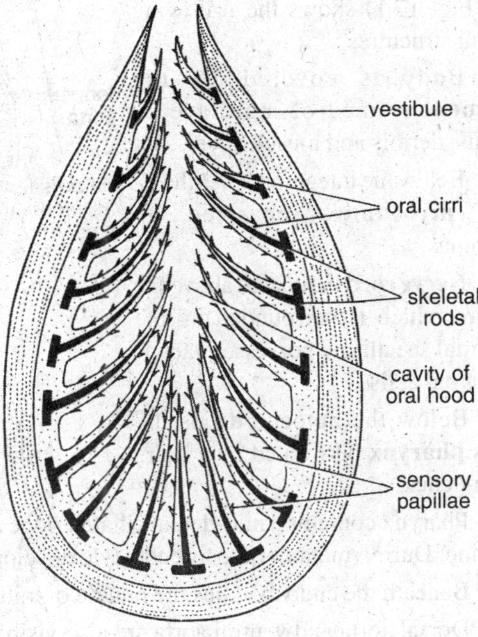

Fig. 16.19. Oral hood of *Amphioxus.*

17

Chapter

SLIDES OF CYCLOSTOMATA

1. T.S. of Ammocoete Larva through the Pharyngeal Region

Comments :

Transverse section of **ammocoete larva** passing through the pharyngeal region (Fig. 17.1) shows the following important structures :

1. Body is covered by the **integument** formed of multilayered epidermis, dermis and muscle layer.

2. Below the integument lies the E-shaped **myotomes** separated by myocommas.

3. **Nerve cord** is situated above the notochord which is surrounded by the notochordal sheath and possesses vacuolated chordal cells.

4. Below the notochord lies the spacious **pharynx** perforated by 7 pairs of gill-pouches.

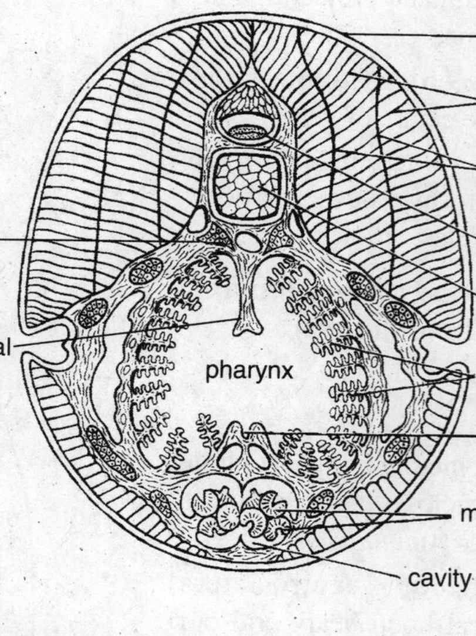

Fig 17.1. T.S. of **ammocoete larva** through the pharyngeal region.

5. Pharynx contains a mid-dorsal ciliated ridge and ciliated and glandular endostyle in the mid-ventral line. During metamorphosis endostyle develops thyroid gland.

6. Beneath the endostyle lies the bundle of mid-ventral muscles.

7. Dorsal aorta and ventral aorta are also visible.

8. Pharynx also contains velum formed of a pair of muscular flaps. It regulates entry of water current into the pharynx.

2. T.S. of Ammocoete Larva through the Liver

Comments :

Transverse section of **ammocoete larva** passing through the gut and liver (Fig. 17.2) shows the following important structures :

1. **Integument** formed of epidermis and dermis forms the external covering.

2. Below the integument lies the E-shaped **myotomes** separated by myocommas.

3. **Nerve cord** lies above the notochord which is surrounded by notochordal sheath and have vacuolated chordal cells.

4. **Dorsal aorta**, **cardinal veins, ventral aorta** and **pronephros** are seen below the notochord.

5. In the ventral part the section, **liver** around **gut** (intestine) is visible.

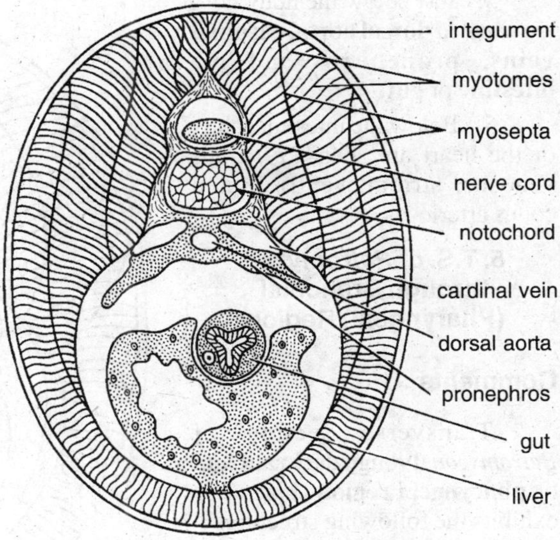

Fig 17.2. T.S. of **ammocoete larva** through the gut and liver.

3. T.S. of Ammocoete Larva through the Intestine or Gut

Comments :

Transverse section of **ammocoete larva** passing through the intestine or gut (Fig. 17.3) reveals the following important structures :

1. **Integument** formed of epidermis and dermis forms the external covering.

2. Just below the integument lies the E-shaped **myotomes** separated by myocommas.

3. **Nerve cord** lies above the notochord that is surrounded by the notochordal sheath and possesses vacuolated chordal cells.

4. The **dorsal aorta**, **cardinal vein** and **ventral aorta** are also visible.

5. Section of **intestine** lies below the cardinal vein and has typhlosole. Intestine is surrounded by coelom.

6. On one side of the intestine is seen the **pronephros**.

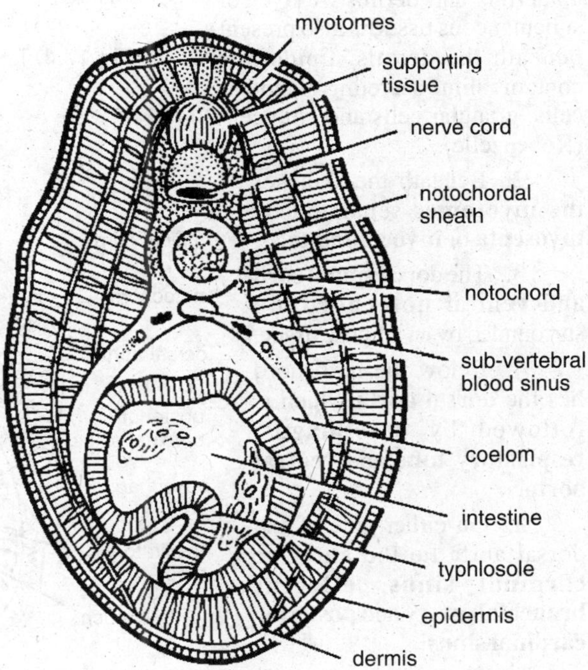

Fig. 17.3. T.S. of **ammocoete larva** through the intestine.

4. T.S. of Ammocoete Larva through the Heart

Comments :

Transverse section of **ammocoete larva** passing through the **heart** (Fig. 17.4) reveals the following important structures :

1. External covering is formed by the **integument** composed of epidermis and dermis.

2. Below the integument lies the E-shaped **myotomes** separated by myocommas.

3. **Nerve cord** is situated just above the notochord that is surrounded by the notochordal sheath and possesses vacuolated chordal cells.

4. Just below the notochord are situated **dorsal aorta**, **cardinal veins**, **pronephros** and the **intestine** or **gut**.

5. Below the intestine, parts of the heart are seen, *i.e.*, sinus venosus, atrium, ventricle and conus arteriosus.

Fig 17.4. T.S. of **ammocoete larva** through the heart.

5. T.S. of *Petromyzon* through Branchial (Pharyngeal) Region

Comments :

Transverse section of *Petromyzon* through the **branchial** or **pharyngeal region** (Fig. 17.5) exhibits the following structures :

1. The outermost covering of the body is skin formed of stratified squamous epithelial epidermis and dermis. A layer of subcutaneous tissue is also present beneath the dermis. Epidermis contains slime secreting glandular cells, granular cells and club cells (Kobenzelle).

2. Beneath the skin lies the **myotomes** separated by **myosepta** or **myocommata**.

3. The dorsal **spinal cord** and ventral **notochord** are surrounded by an adipose tissue.

4. Below the notochord lies the **dorsal aorta** which is followed by **oesophagus**, **respiratory tube** and **ventral aorta**.

5. On either side of the dorsal aorta lie the **anterior cardinal sinus**, **efferent branchial artery** and **posterior cardinal sinus**.

6. The ventral half portion of the body is occupied by **coelomic cavity**.

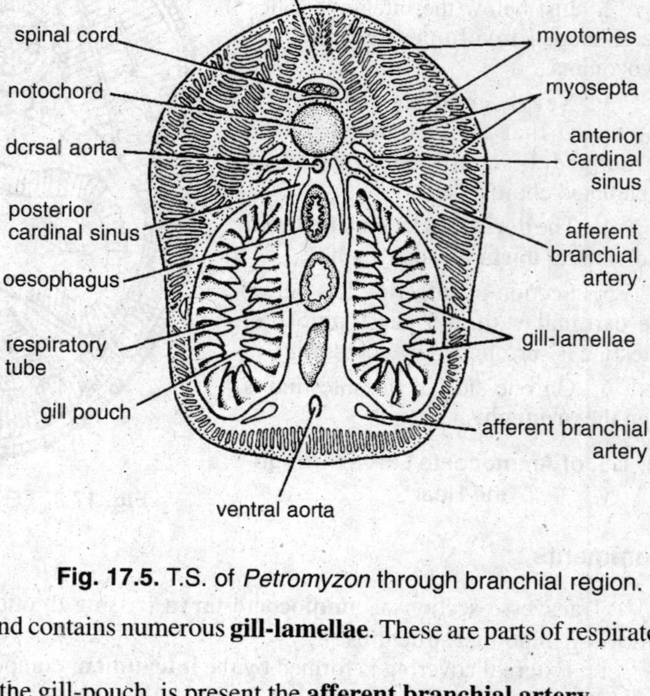

Fig. 17.5. T.S. of *Petromyzon* through branchial region.

7. Two **gill-pouches**, one on each side of the coelom, are present. Each gill-pouch is hollow and contains numerous **gill-lamellae**. These are parts of respiratory pharynx.

8. On either side, just below the gill-pouch, is present the **afferent branchial artery**.

6. T.S. of *Petromyzon* through Posterior Trunk Region

Comments :

Transverse section of *Petromyzon* through the **posterior trunk region** (Fig. 17.6) reveals the following structures :

1. The outermost covering of the body is skin formed of epidermis, dermis and a layer of subcutaneous tissue.

2. Beneath the skin lies the E-shaped **myotomes** separated by **myosepta** or **myocommata**.

3. **Dorsal fin** is supported by cartilaginous fin-rays arranged in a single series. Their bases are fused with the membranous sheath surrounding the neural tube and notochord.

4. **Fin muscles** are also present at the base of fin.

5. In the mid-dorsal line is seen a **spinal cord** surrounded by adipose tissue and **notochord** below.

6. **Dorsal aorta** and **posterior cardinal sinus** are also visible beneath the notochord.

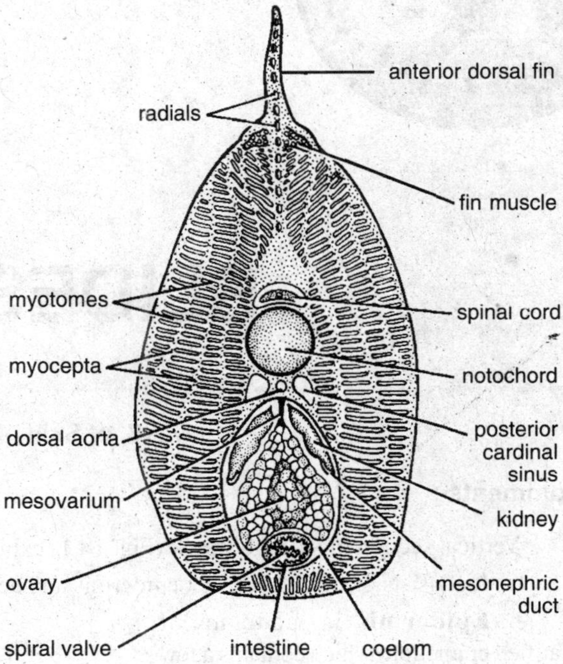

Fig. 17.6 .T.S. of *Petromyzon* through posterior trunk region.

7. In the coelomic cavity lies a pair of **kidneys** showing the **mesonephric duct**, **ovary** and **intestine** with **spiral valve**.

7. T.S. of *Petromyzon* through the Tail (Caudal) Region

Comments :

Transverse section of *Petromyzon* through the **tail region** (Fig. 17.7) shows the following structures :

1. The outermost covering of the body is skin composed of epidermis, dermis and a subcutaneous layer. Epidermis contains slime secreting gland cells, granular cells and club cells. Dermis is composed of collagen and elastin fibres, squamous epithelial cells and mucous cells.

2. Beneath the skin are present **myotomes** separated by **myosepta** or **myocommata**.

3. Dorsal lobe **of caudal fin** and ventral lobe of **caudal fin** are supported by cartilaginous **radials** arranged in a single row.

4. **Fin muscles** are present beneath the caudal fins.

5. In the centre of the section is present **spinal cord** above and **notochord** below.

6. Just below the notochord lies a **caudal artery** and a **caudal vein**.

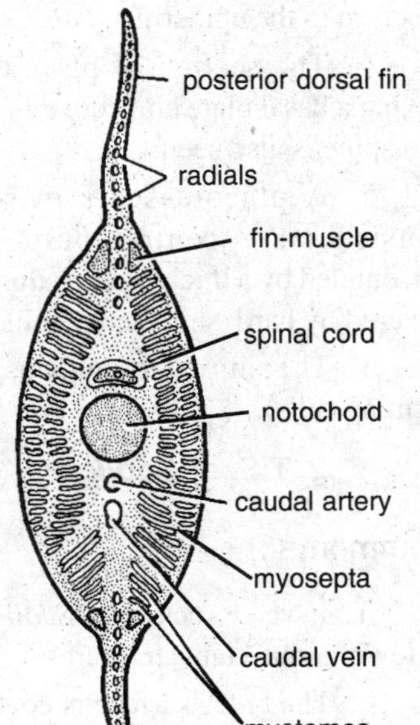

Fig. 17.7 .T.S. of *Petromyzon* through tail region.

18

Chapter

SLIDES OF FISHES

1. V.S. of Skin of Shark

Comments :

Vertical section of **skin of Shark** (Fig. 18.1) exhibits the following structures :

1. It consists of two layers, *i.e.*, epidermis and dermis or corium.

2. **Epidermis** is made up of stratified epithelium which contains a few mucous secreting gland cells.

3. Dermis is made up of areolar connective tissue mingled with smooth muscle fibres, blood capillaries, nerves and pigment cells (**melanophores**) which lie close to the epidermis.

4. The section of placoid scale having a basal plate embedded in dermis and spine is also seen.

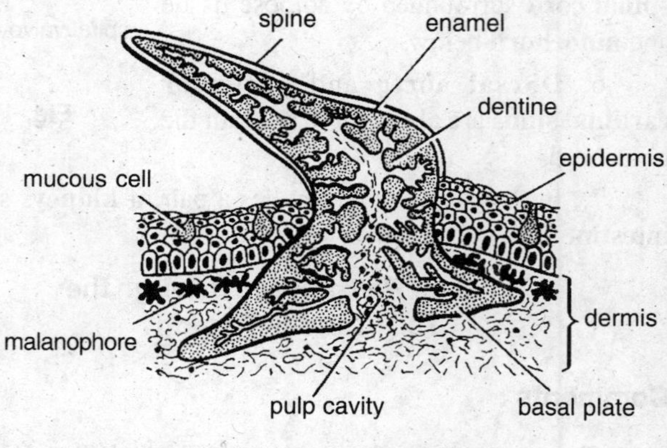

Fig 18.1. V.S. of skin of Shark.

5. A placoid scale in section consists of a central pulp cavity surrounded by a thick layer of **dentine** is covered by hard, shiny enamel like vitrodentine.

6. The pulp cavity contains blood vessels, lymph vessels, nerves and odontoblast cells (dentine forming cells).

2. T.S. of *Scoliodon* (Hand cut Section) through the Olfactory Region

Comments :

Transverse section of *Scoliodon* through the **region of olfactory sacs** (Fig. 18.2) reveals the following important structures :

1. The outer surface is covered by **integument** formed of outer epidermis and lower dermis bearing the placoid scales.

2. A pair of lateral olfactory sacs are distinct. These open externally by a pair of nostrils.

3. Each olfactory sac is covered externally by a thin membrane and is lodged into cartilaginous **olfactory capsule**.

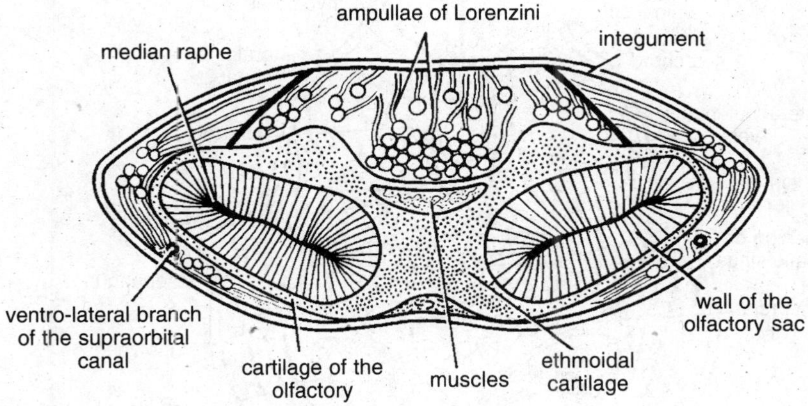

Fig. 18.2. T.S. of *Scoliodon* through olfactory region.

4. The mucous membrane lining its inner wall is thrown into a double series of closely set folds, the **Schneiderian folds** held in place by a median band (median raphe).

5. The epithelium of folds consists of olfactory receptor cells intermixed with stiff supporting cells.

6. The central portion of the section is occupied by **ethmoidal cartilage**.

7. Numerous **ampullae of Lorenzini** are present above and on the lateral sides of the ethmoid cartilage of olfactory capsules.

8. **Ventro-lateral branch of the supra orbital canal** is seen on both the lateral sides.

3. T.S. of *Scoliodon* (Hand cut section) through the Branchial Region

Comments :

Transverse section of *Scoliodon* through the **branchial region** (Fig. 18.3) shows the following structures :

1. The outer surface is covered by integument comprising the epidermis, dermis and dermal placoid scales.

2. Well-developed **myotomes** separated by **myocommata** are seen in the dorsal half portion of the section.

3. **Myotomes** have concentrically arranged muscle fibres of striated type and **myocommata** composed of tough connective tissue.

4. In mid-dorsal line in between myotomes of both sides lies the cartilaginous **vertebra** having the spinal cord in the neural canal.

5. Just below the vertebra lies the **dorsal aorta** and thick-walled **oesophagus**.

6. Below the oesophagus lies thin-walled **pericardial cavity** containing the heart.

7. **Coracoid cartilages** give support to the dorsal pericardium.

8. On either side of the pericardial cavity and coracoid cartilages are seen obliquely placed gill-pouches having gill-lamellae.

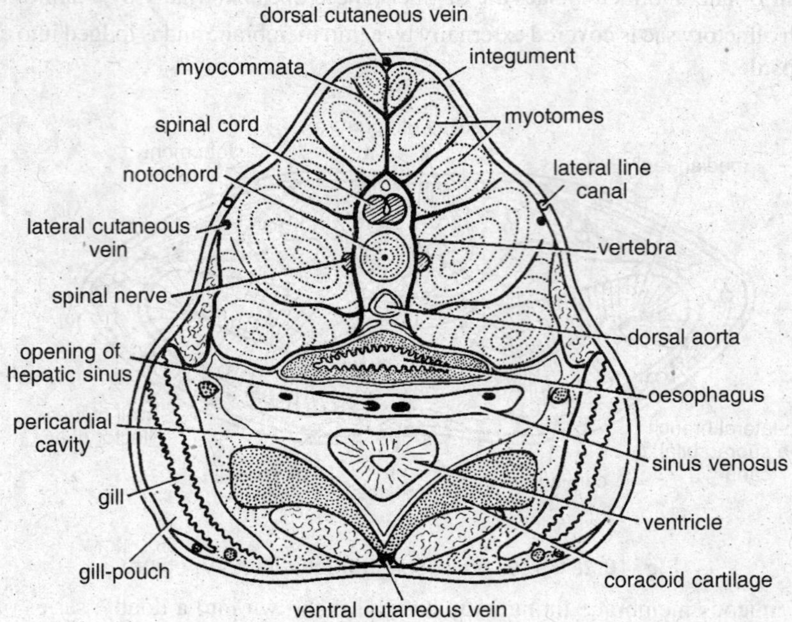

Fig. 18.3. T.S. of *Scoliodon* through branchial region.

4. T.S. of *Scoliodon* (Hand cut section) through the Stomach and Liver

Comments :

Transverse section of *Scoliodon* passing through the **stomach** and **liver** (Fig. 18.4) exhibits the following important structures :

1. The outer surface of the body is covered by the **integument** comprising epidermis, dermis and projecting dermal placoid scales.

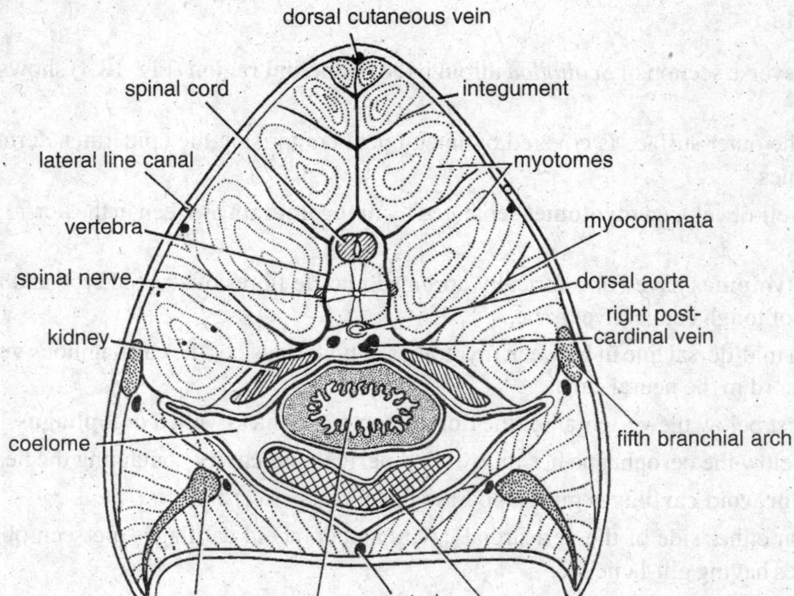

Fig. 18.4. T.S. of *Scoliodon* through the stomach and liver.

2. Well-developed **myotomes** formed of concentrically arranged striated muscle fibres separated by **myocommata** are seen in the dorsal half portion.

3. In mid-dorsal line in between myotomes of both sides lies the **vertebra** enclosing the **spinal cord**. Just below the vertebra lies the **dorsal aorta**.

4. A spacious coelomic cavity lined by coelomic epithelium contains the sections of stomach and bilobed liver.

5. Two mesonephric **kidneys** outside the coelom and posterior cardinal veins are seen.

6. Cartilages of pectoral fins are seen on both the ventro-lateral sides.

5. T.S. of *Scoliodon* (Hand cut section) through the Intestine

Comments :

Transverse section of *Scoliodon* passing through the **intestine** (Fig. 18.5) shows the following important structures :

1. The outer surface is covered by the **integument** composed of epidermis and dermis and dermal placoid scales.

2. **Myotomes** contain concentrically arranged striated muscle fibres and separated by the **myocommata** of tough connective tissue.

3. In between the myotomes, in the median line lies the **vertebra** enclosing the **spinal cord**. Notochord persists in the centrum of vertebra.

4. **Dorsal aorta** and lateral **cardinal veins** lie below the vertebra.

5. Below the dorsal aorta lies the **mesonephric kidney**.

6. Just below the kidney lies the **seminal vesicles** in male and **uterus** in female and **intestine** showing scroll valve in the centre below the seminal vesicles. It looks like a watch spring.

7. In male on either side of the intestine lies testes and in female ovaries.

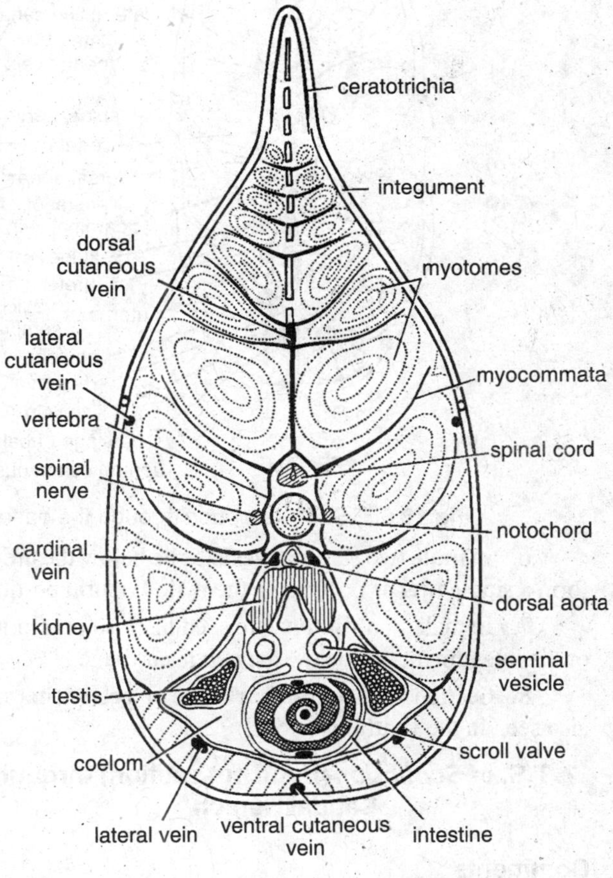

Fig 18.5. T.S. of *Scoliodon* through intestine.

6. T.S. of *Scoliodon* (Hand cut section) through the Base of Pelvic Fin of Male and Female

Comments :

Transverse section of *Scoliodon* passing through the **pelvic region** of **male** and **female** specimens (Fig. 18.6) exhibits the following structures :

1. The outer surface is covered by the **integument** composed of epidermis, dermis and projected dermal placoid scales.

2. The **myotomes** contain concentrically arranged muscle fibres and are separated by the **myocommata** of tough connective tissue.

3. In between the lateral myotomes lies the **vertebra** enclosing the **spinal cord**. Notochord is persistent in the centrum.

4. Below the vertebra lies the dorsal aorta and two **posterior cardinal** veins.

5. **Kidneys** and **ureters** also lie below the posterior cardinal veins.

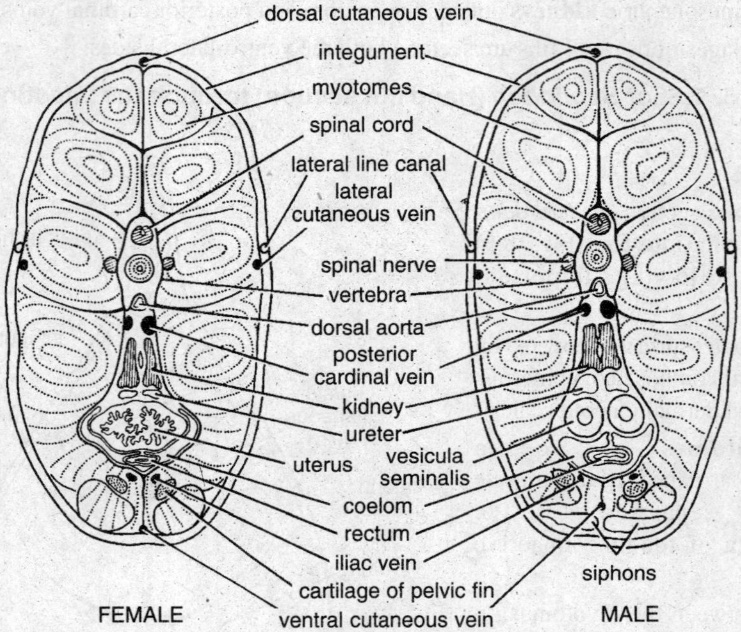

Fig 18.6. T.S. of *Scoliodon* through the base of pelvic fin of male and female.

6. In **male** two **seminal vesicles** lie below the ureters and in **female** two uteri fused together to form a single **uterus** occupying the major ventral portion.

7. In male two **siphons** are seen in close proximity of the integument just below the pelvic cartilage.

8. Sections of **rectum**, **cartilage of pelvic fins** are also seen in the ventral portion.

7. T.S. of *Scoliodon* (Hand cut section) through Caudal Region

Comments :

Transverse section of *Scoliodon* passing through the **caudal region** (Fig. 18.7) shows the following details :

1. The outer surface is covered by the **integument** composed of epidermis, dermis and projected dermal placoid scales.

2. The **myotomes** formed of concentrically arranged striated muscle fibres and are separated by the **myocommata**.

3. In between the myotomes, in the median line lies the **vertebra** which is enclosing the **spinal cord** and

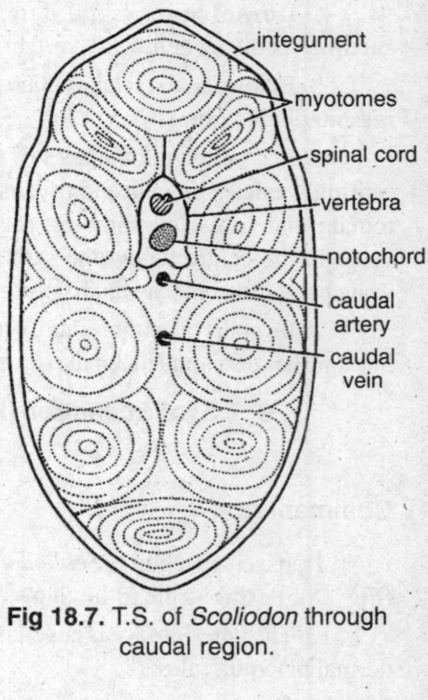

Fig 18.7. T.S. of *Scoliodon* through caudal region.

notochord persistent in the centrum.

4. Below the vertebra lies the **caudal artery** and **caudal vein** respectively.

5. None of the body organs are present in the caudal region.

8. Ampulla of Lorenzini

Comments :

1. Each ampulla (Fig. 18.8) is made up of **ampullary sac** and an elongated tubule opening over the head through a minute aperture.

2. Each ampullary sac consists of eight to nine radially dilated chambers arranged around a central core, the **centrum**.

3. Long **tubule** of ampullary sac is filled up with mucus and opens at the surface of the head through a minute pore known as **external aperture**.

4. Each ampulla is innervated by the nerve branches of VIIth cranial nerve.

5. The ampullae lie together in clusters like bunches of grapes in the snout of dogfish.

6. The ampullae of Lorenzini are **thermoreceptors** (temperature receptors).

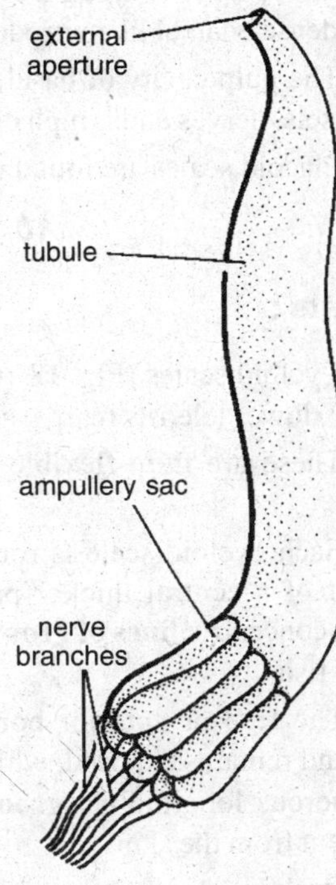

Fig 18.8. Ampulla of Lorenzini.

9. Placoid Scale of *Scoliodon* (Whole mount)

Comments :

1. The placoid scales are arranged in regular oblique rows covering the entire surface of the body and form the exoskeleton of the shark (dogfish).

2. Placoid scales are small pointed and tri-radiate denticles found embedded in the dermal layers of the skin.

3. A typical placoid scale consists of a diamond-shaped or rhomboidal **basal plate** having an **opening of the pulp cavity** and flat **trident spine**.

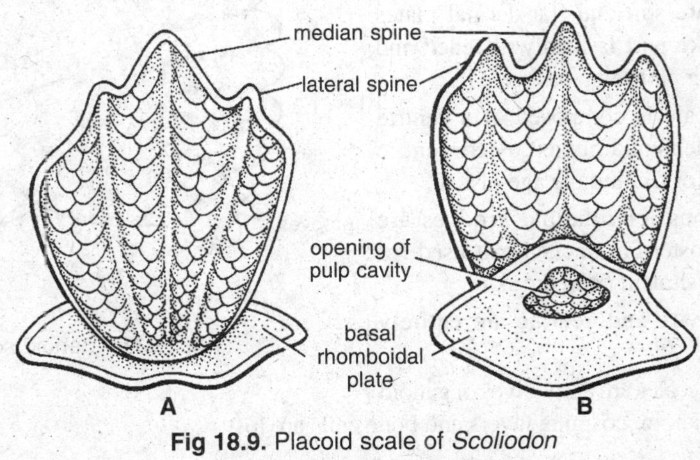

Fig 18.9. Placoid scale of *Scoliodon*
A-Dorsal view; B-Ventral view.

4. The basal plate remains embedded in the dermis held by Sharpey's and other connective tissue fibres and formed of a trabecular calcified tissue, the **cement**.

5. The spine is composed of a hard calcareous substance, the **dentine** coated externally with a hard and dense enamel like vitrodentine.

6. The pulp cavity of basal plate and spines contain dentine forming cells the odontoblasts, blood vessels, nerves and lymph channels.

7. Placoid scales are found in sharks, rays and skates.

10. Cycloid Scale (Whole mount)

Comments :

1. Cycloid scales (Fig. 18.10) are found in *Amia* (Holostei), lower teleosts (carps, cods, etc) and dipnoi.

2. These are thin, flexible transluscent dermal plates.

3. Each cycloid scale is roughly circular and is composed of a central thicker part the **nucleus** and numerous concentric **lines of growth** used to determine the age of fish.

4. The free or anterior border is more or less rounded and remains exposed, while the posterior part have numerous longitudinal **grooves** for sucking the nourishment from the skin.

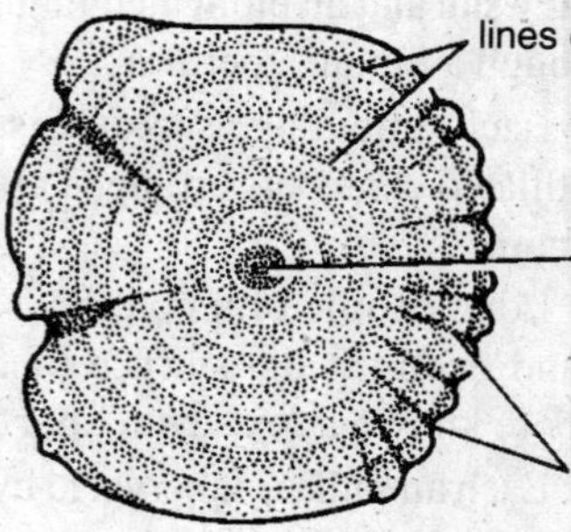

Fig 18.10. Cycloid scale.

5. Cycloid scale is composed of a thin upper layer of bone and lower layer of fibrous connective tissue. Pulp cavity is absent.

6. Cycloid scales are derivatives of the ganoid scales in which ganoin and cosmine layers and bone cells are lost.

11. Ctenoid Scale (Whole mount)

Comments :

1. Ctenoid scales (Fig. 18.11) are commonly found in higher teleosts like perch, sunfish, etc.

2. These are soft and flat dermal plates whose groove-like part is firmly attached and hind teeth bearing part remains free.

3. Each scale is composed of a central thicker part **nucleus** and numerous concentric **lines of growth** for finding the age of fish.

4. Numerous longitudinal grooves are present on the posterior attached end used for sucking the nourishment from the skin.

5. Pulp cavity and dentine are entirely absent.

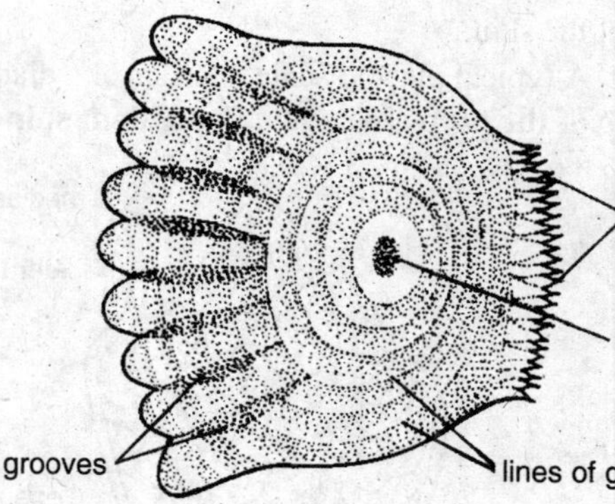

Fig 18.11. Ctenoid scale.

6. Ctenoid scales are derivatives of ganoid scales in which ganoin, cosmine layers and bone cells are lost.

12. Ganoid Scale (Whole mount)

Comments :

1. Ganoid or rhomboid scales (Fig. 18.12 and 18.13) were the characteristic of primitive Actinoptergyii (Chondrostei such as *Polypterus, Acipenser* and Holostei like *Lepidosteus*. These scales remain closely fitted side-by-side like tiles.

2. Each scale has bony base, **isopedine** middle dentile-like cosmine externally covered by shining material the **ganoin**. Such scale is called palaeoniscoid ganoid scale.

These scales fit together and arranged in diagonal rows.

Fig. 18.12. Ganoid scales.

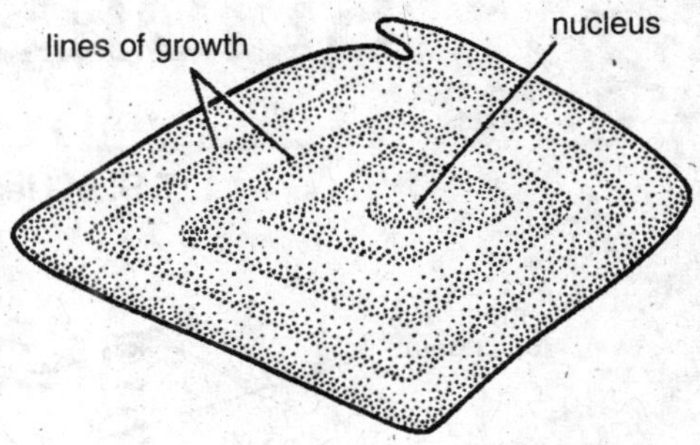

Fig. 18.13. Rhomboid scale.

13. Ganoid Scale (Whole mount)

1. Rhomboid scales of *Lepidosteus* (gar-pike) are called Lepidosteoid ganoid scale.
2. Such scale has only two layers : outer ganoin and inner isopedine.
3. Each scale possesses lines of growth and a nucleus.

Chapter 19

SLIDES OF AMPHIBIA

1. T.S. of Head Region of Frog

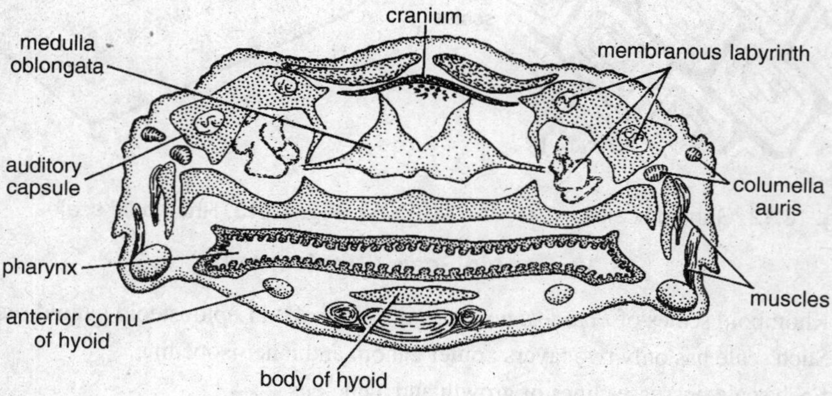

Fig. 19.1. T.S. of head region of frog.

Comments :

Transverse section of the **head region** of **frog** (Fig. 19.1) passing through the auditory capsule shows the following important structures :

1. The outermost covering is skin devoid of exoskeleton. The skin comprises multilayered epidermis and dermis formed of connective tissue.
2. Below the skin is seen the cranium enclosing the **medulla oblongata** part of brain and on lateral sides, the auditory capsules containing the parts of **membranous labyrinth** and **columella auris** part of ear.
3. Beneath the cranium is present a horizontal space, the **pharynx**.
4. Just below the pharynx lies the body of cartilaginous **hyoid apparatus** in the middle and also the **anterior cornu** of hyoid on either side.
5. On both the lateral sides and just below the hyoid apparatus muscles are seen.

2. V.S. of Skin of Frog

Comments :

Vertical section of **skin** of **frog** (Fig. 19.2) shows the following histological details :

1. The skin or integument consists of two distinct layers, the outer ectodermal **epidermis** and inner mesodermal **dermis**.

2. **Epidermis** is made up of outer **stratum corneum** of flattened dead horny cells arranged in one or two layers which is cast off as squamous epithelium and an inner most layer of **stratum Malpighi** placed over basement membrane.

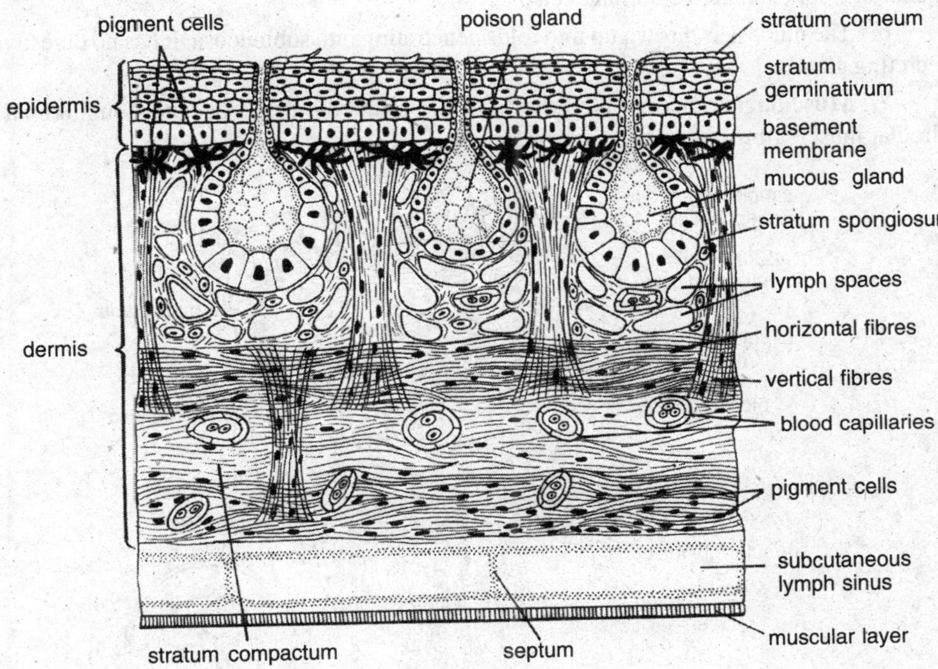

Fig. 19.2. V.S. of skin of frog.

3. **Dermis** consists largely of connective tissue which is differentiated into two distinct layers, *i.e.*, outer **spongy layer** and inner **compact layer**.

(*i*) **Spongy layer** is composed of loose connective tissue and contains **mucuous glands**, **blood vessels**, **nerve fibres**, **lymph spaces** and **melanophores**, etc.

(*ii*) **Mucous glands** are flask-shaped, opening on the surface of the skin. They secrete mucus and make the body slippery. They originate from the stratum Malpighi and their body lies in dermis.

(*iii*) **Melanophores** are pigment cells present in the upper layer of dermis and is ectodermal in origin. It is responsible for the change of colours of skin.

4. **Compact layer** is made up of dense fibrous connective tissue having horizontal and vertical strands.

Functions : The skin of frog serves several functions :

(*i*) Protection of body; (*ii*) Production of mucus to keep the skin moist and slippery to protect it from enemies; (*iii*) Respiratory, allows exchange of O_2 and CO_2 ; (*iv*) Excretory, and (*v*) Sensory.

3. T.S. of Oesophagus of Frog

Comments :

Transverse section of **oesophagus** of **frog** (Fig. 19.3) shows the following histological structures :

1. It is composed of usual four layers, *i.e.*, **serosa, muscularis, submucosa** and **mucosa.**

2. **Serosa** is not found in oesophagus because it lies outside the coelom. Here outermost layer is of a thin fibrous connective tissue.

3. **Muscularis coat**, the muscle fibres of which are arranged in two layers : outer thin **longitudinal muscle layer** and inner thick **circular muscle layer**.

4. **Submucosa** lies below the muscular layers. It is composed of thick layer of loose connective tissue having blood vessels, lymph vessels, nerves, etc.

5. **Mucosa** consists of simple columnar epithelium containing tubular goblet cells secreting mucus and some ciliated columnar cells.

6. The mucosa is thrown up into folds penetrating into submucosa. It has no digestive enzymes secreting glands.

7. **Muscularis mocosae** is in the form of scattered patches of outer longitudinal and inner circular muscle fibres lying below the submucosa.

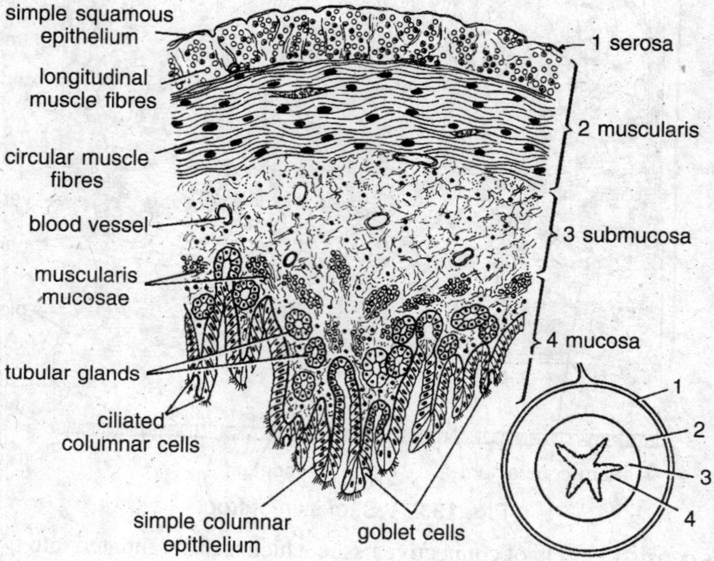

Fig. 19.3. T.S. of oesophagus of frog.

4. T.S. of Stomach of Frog

Comments :

Transverse section of **stomach** of **frog** (Fig. 19.4) show the following histological details :

1. It is composed of four layers, *i.e.*, **serosa**, **muscularis**, **submucosa** and **mucosa**.

2. **Serosa** is thin and made up of an outer layer of flattened mesothelial cells and an inner very thin connective tissue layer (subserosa).

3. **Muscularis** consists of a thick and prominent layer of inner circular and outer longitudinal smooth muscle fibres.

4. **Submucosa** lies below the circular muscle fibres and made up of loose connective tissue containing blood vessels, lymph vessels and nerves.

5. **Mucosa** is the innermost layer and is variously folded forming pits and different types of glands. It is formed of three layers :

(*i*) **Columnar epithelium** layer is the innermost, supported on thin basement membrane ; (*ii*) **Lamina propria** is a thin connective tissue layer; (*iii*) **Muscularis mucosae** is a narrow layer of inner circular and outer longitudinal muscle fibres located below the submucosa.

6. **Mucosa** is thrown into numerous folds or **villi** projecting into the lumen. The villi increase the maximum absorptive surface.

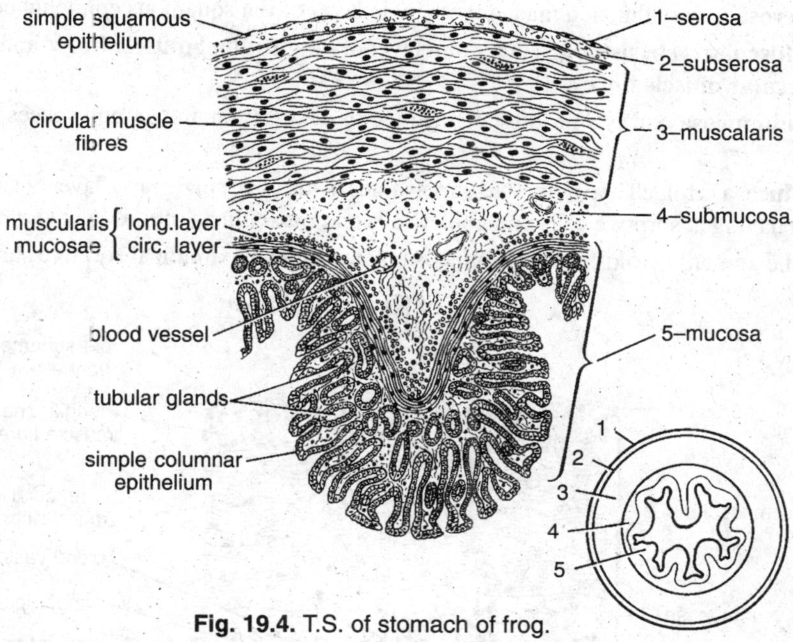

Fig. 19.4. T.S. of stomach of frog.

7. The tubular gastric glands open into the lumen of stomach through narrow ducts and secrete pepsin and HCl.

5. T.S. of Intestine of Frog

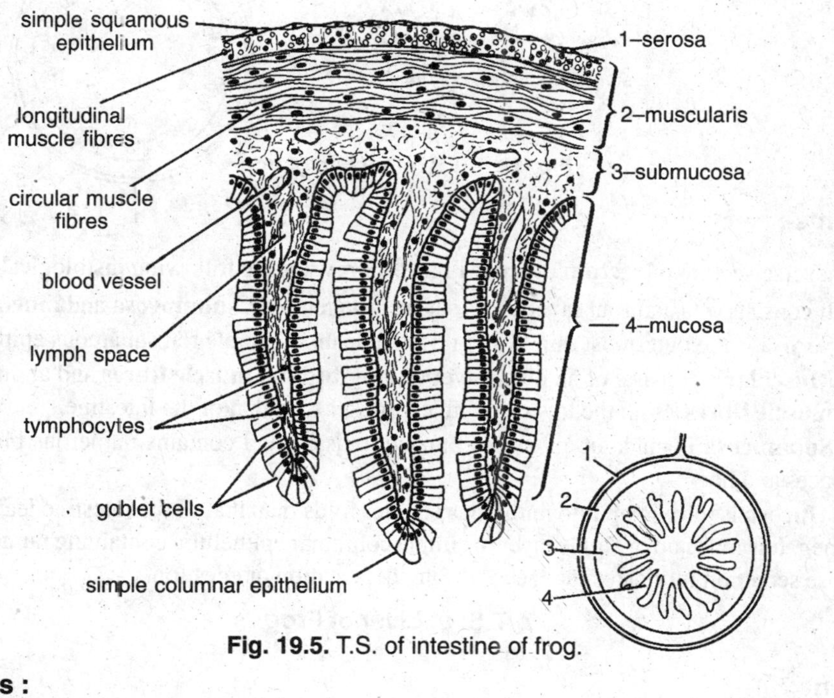

Fig. 19.5. T.S. of intestine of frog.

Comments :

Transverse section of **intestine (ileum)** of **frog** (Fig. 19.5) shows the following histological details :

1. It consists of the usual four layers, *viz.*, **serosa, muscularis, submucosa** and **mucosa**.

2. **Serosa** is very thin and made up of single layer of flat squamous epithelial cells.

3. **Muscularis** consists of an outer thin layer of **longitudinal muscle fibres** and an inner thick layer of **circular muscle fibres**.

4. **Submucosa** consists of loose connective tissue and contains lymph spaces, blood vessels and nerves.

5. **Mucosa** is folded into numerous simple folds made up of single inner layer of simple columnar epithelium having absorptive cells and goblet cells and **muscularis mucosae** is very thin layer.

6. The numerous folds or villi of mucosa increase the maximum absorptive surface.

6. T.S. of Rectum of Frog

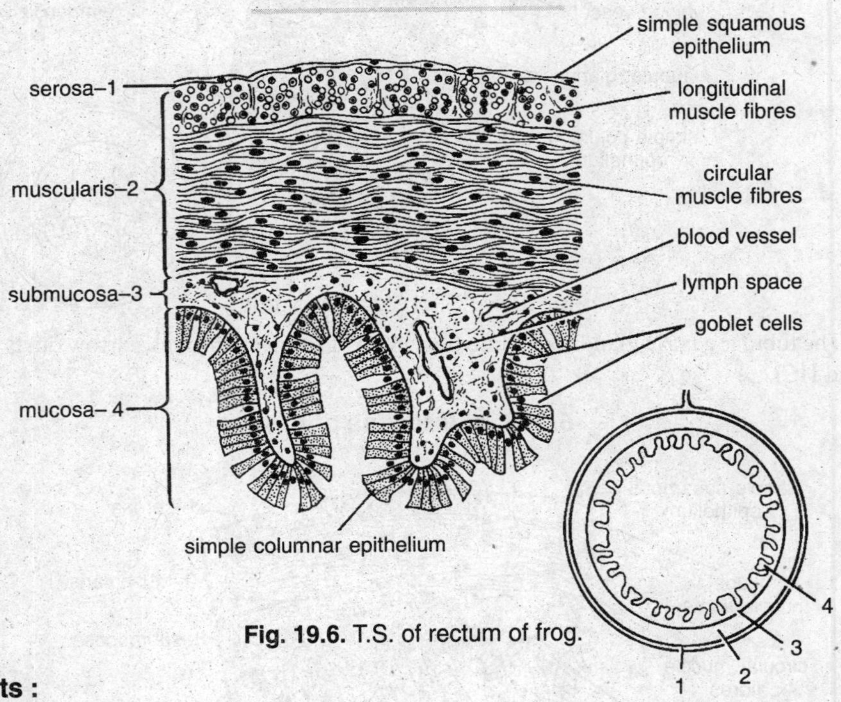

Fig. 19.6. T.S. of rectum of frog.

Comments :

Tansverse section of **rectum** of **frog** (Fig. 19.6) reveals the following histological details :

1. It consists of usual four layers, *i.e.*, **serosa**, **muscularis**, **submucosa** and **mucosa**.

2. **Serosa** is the outermost thin layer and is made up of simple flat squamous epithelium.

3. **Muscularis** consists of an outer layer of **longitudinal muscle fibres** and an inner layer of **circular muscle fibres**. Both the layers are much thicker than that of the intestine.

4. **Submucosa** is made up of loose connective tissue and contains numerous blood vessels, lymph spaces and nerves.

5. **Mucosa** is thrown up into numerous smaller folds than that of the intestine leaving a wider rectal lumen. It is made up of single layer of simple columnar epithelium containing numerous goblet cells whose secretion lubricates the faeces during defaecation or egestion.

7. T.S. of Liver of Frog

Comment :

Transverse section of **liver** of **frog** (Fig. 19.7) exhibits the following histological details :

1. It consists of a large number of small rounded **hepatic acini** or lobules formed of large polygonal and cuboidal hepatic cells.

2. Hepatic cells are arranged in solid cords or trabeculae.

3. Interspaces between lobules have connective tissue, hepatic ducts, blood sinuses, and blood vessels.

4. Each hepatic cell contains a prominent nucleus and granular cytoplasm which indicates its secretory nature.

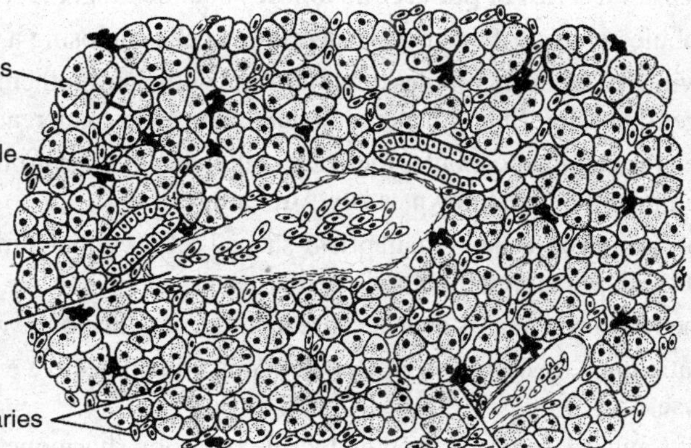

Fig. 19.7. T.S. of liver of frog.

5. Adjacent liver cells enclose fine wall-less bile canaliculi (bile capillaries) into which bile is secreted by liver cells.

6. **Bile canaliculi** unite to form the **bile ductules** and these in turn unite to form **bile duct** which takes bile into gall bladder.

7. Wall-less sinusoids or blood spaces are seen between hepatic cords which receive blood from branches of hepatic portal vein and hepatic artery.

8. Sinusoids are also lined by semi-irregular phagocytic cells called Kupffer cells.

8. T.S. of Pancreas of Frog

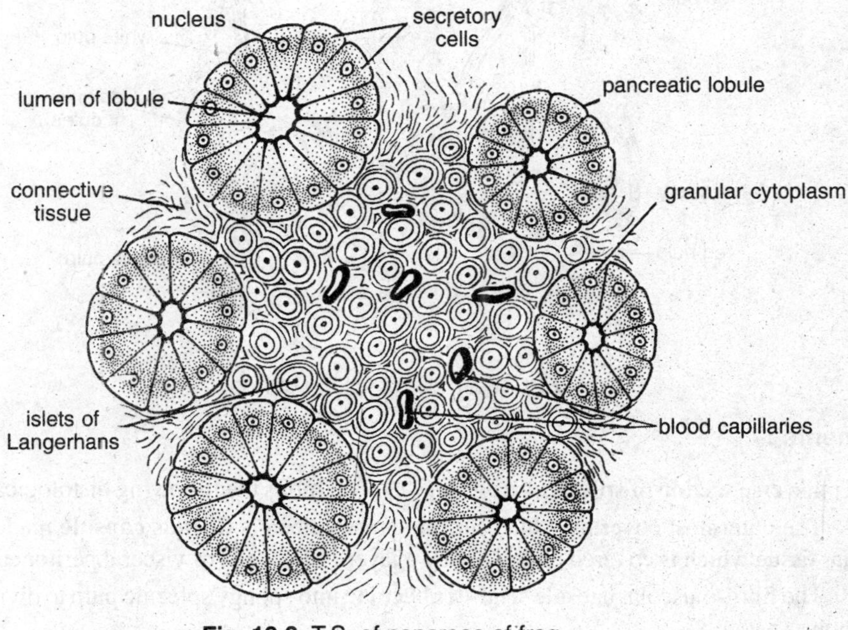

Fig. 19.8. T.S. of pancreas of frog.

Comments :

Transverse section of **pancreas** of **frog** (Fig. 19.8) shows the following histological details :

1. It is a much branched and grape-bunch-like (racemose) gland externally covered by a thin loose connective tissue sheath.
2. It consists of a series of **pancreatic lobules** bounded together by connective tissue.
3. Each lobule consists of a cluster of **secretory cells** enclosing a very narrow central lumen.
4. The cavities of adjacent lobules communicate with one another through fine ductules which join to form larger ducts and finally pancreatic duct that carries pancreatic juice to the duodenum.
5. Between the pancreatic lobules there are scattered, compact groups of cells in the connective tissue which stain paler and are known as **islets of Langerhans**.
6. **Islets of Langerhans** have small prism-shaped cells without lumen. These are of three types, alpha, beta and D cells separated by blood capillaries.
7. The function of pancreas is dual because it produces **exocrine** as well as **endocrine** secretions.
8. Pancreatic juice (exocrine secretion) contains three digestive enzymes such as **amylopsin**, **trypsin** and **lipase**.
9. Islets of Langerhans produce an endocrine secretion (hormone) **glucagon** from alpha cells and **insulin** from beta cells. Glucagon increases concentration of sugar in blood and insulin regulates sugar in blood.

9. T. S. of Spleen of Frog

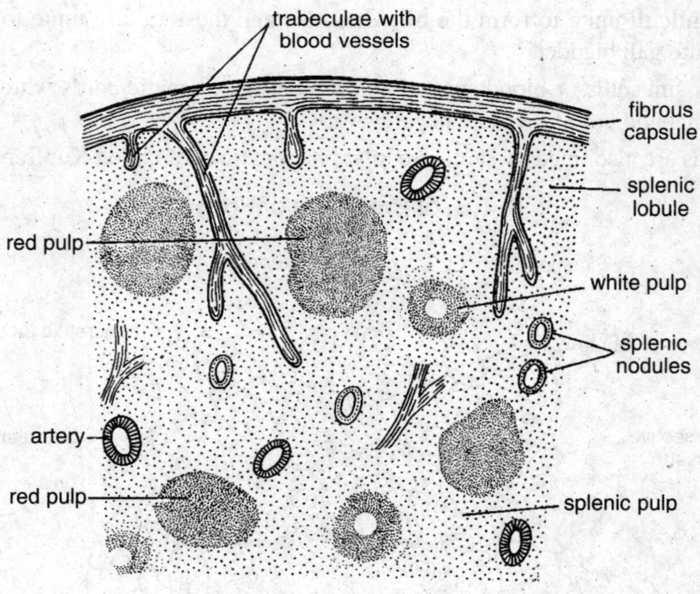

Fig. 19.9. T.S. of spleen of frog.

Comments :

Transverse section of **spleen** of **frog** (Fig. 19.9) exhibits the following histological structures :

1. The outermost covering which surrounds the spleen is **fibrous capsule** made up of fibro-muscular tissue which is covered externally by a layer of flat cells of visceral peritoneum.
2. The fibro-muscular capsule sends **trabeculae** into spongy spleenic pulp to divide the spleen into lobules.

3. Spleenic pulp contains red and white pulps, arteries, veins, and nerves.

4. **Red pulp** forms the bulk and contains mostly red blood cells. Red pulp also contains irregular-shaped pulp cords, formed of red and white blood cells, reticular cells, macrophages and splenocytes.

5. **White pulp** is formed of scattered patches of spleenic nodules within red pulp.

6. **White pulp** is formed of accumulations of lymphocytes histiocytes and reticular cells.

7. The spleen has several functions which are as follows :

(*i*) Its macrophages destroy the worn out old erythrocytes.

(*ii*) It forms the leucocytes.

(*iii*) It produces a secretion which activates the protein enzymes of the pancreas.

(*iv*) Its lymphocytes also produce the antibody.

10. T.S. of Lung of Frog

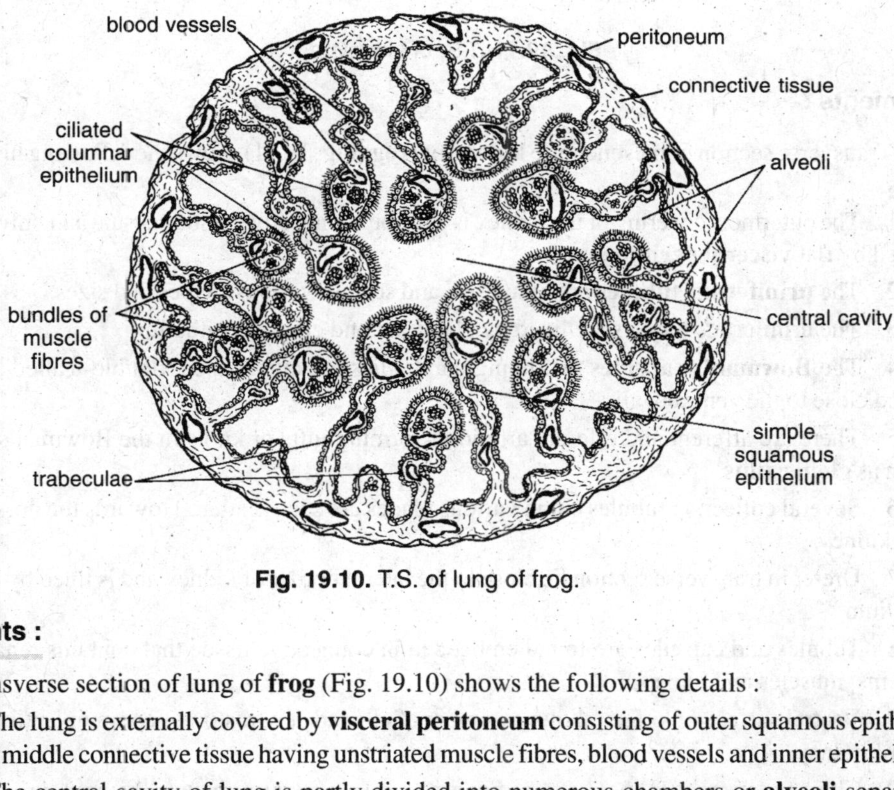

Fig. 19.10. T.S. of lung of frog.

Comments :

Transverse section of lung of **frog** (Fig. 19.10) shows the following details :

1. The lung is externally covered by **visceral peritoneum** consisting of outer squamous epithelial cells layer, middle connective tissue having unstriated muscle fibres, blood vessels and inner epithelium.

2. The central cavity of lung is partly divided into numerous chambers or **alveoli** separated from each other by septa or **trabeculae** which are projections of wall of lung.

3. The trabeculae are lined by thin, flattened simple squamous epithelium secreting mucus and the inner edges of septa are lined by a ciliated columnar epithelium.

4. The walls of the trabeculae are richly supplied with blood capillaries.

5. The respiratory surface is increased by the presence of trabeculae and alveoli.

6. In the lung the blood is separated from the air contained in the alveoli by two thin membranes, namely the exeedingly thin epithelium of the trabeculae and the thin walls of the capillaries.

7. Within the lung the oxygen of the air combines with blood haemoglobin of capillaries forming oxyhaemoglobin and transported to body tissues for cellular respiration.

11. T.S. of Kidney of Frog

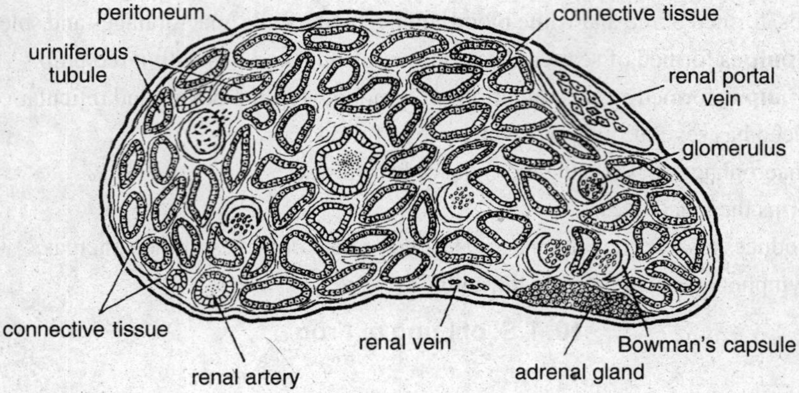

Fig. 19.11. T.S. of kidney of frog.

Comments :

Transverse section of mesonephric **kidney** of **frog** (Fig. 19.11) shows the following histological details :

1. The outermost covering of the kidney is a fibrous connective tissue capsule and only ventrally covered by flat **visceral peritoneum**.
2. The **uriniferous tubules** are numerous and seen in various shapes and sizes.
3. The uriniferous tubules are lined by glandular and ciliated epithelium.
4. The **Bowman's capsules** or Malpighian bodies are cup-shaped, double-walled structures situated close to the ventral wall.
5. There are afferent and efferent arterioles forming tufts or knots in the Bowman's capsules known as **glomerulus**.
6. Several collecting tubules cut in various planes are seen scattered towards the dorsal border of the kidney.
7. Ureter in transverse section lie close to the lateral border of kidney and is lined by columnar epithelium.
8. Tubules and capsules are found embedded in connective tissue that contains renal arteries and veins, muscles and nerves.
9. Ventral surface of kidney shows a number of ciliated nephrostomes communicating with the veins of kidney.
10. Chief function of kidney is to remove non-gaseous waste matter like urea, uric acid, and certain salts (phosphates and sulphates) that are brought to them from different parts of the body.

12. T.S. of Testis of Frog

Comments :

Transverse section of **testis** of **frog** (Fig. 19.12) shows the following structures :

1. The outer covering is the **visceral peritoneum** called **mesorchium**.
2. Numerous microscopic **seminiferous tubules** are held together by the connective tissue.
3. Each seminiferous tubule is blind at its inner end and is lined with stratified germinal epithelium whose cells undergo **spermatogenesis** to produce spermatozoa.

4. Bundles of spermatozoa are seen in the lumen of the mature seminiferous tubules.

5. Small and rounded **spermatogonia**, largest primary **spermatocytes**, half-sized secondary **spermatocytes** and smallest **spermatids** border the central lumen are also seen.

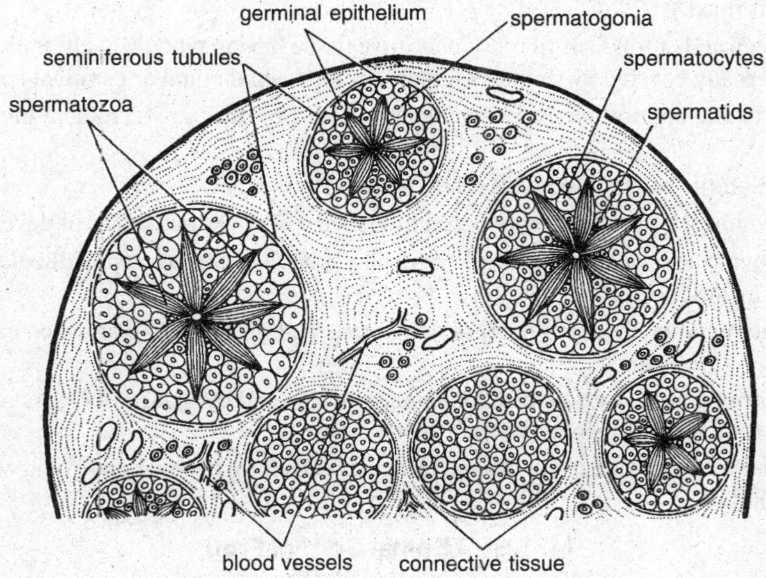

Fig. 19.12. T.S. of testis of frog.

6. The connective tissue contains small patches of **interstitial cells** which secrete hormone **testosterone** responsible for the appearance of the secondary sexual characters.

7. **Sertoli cells** in frog are absent.

13. T.S. of Ovary of Frog

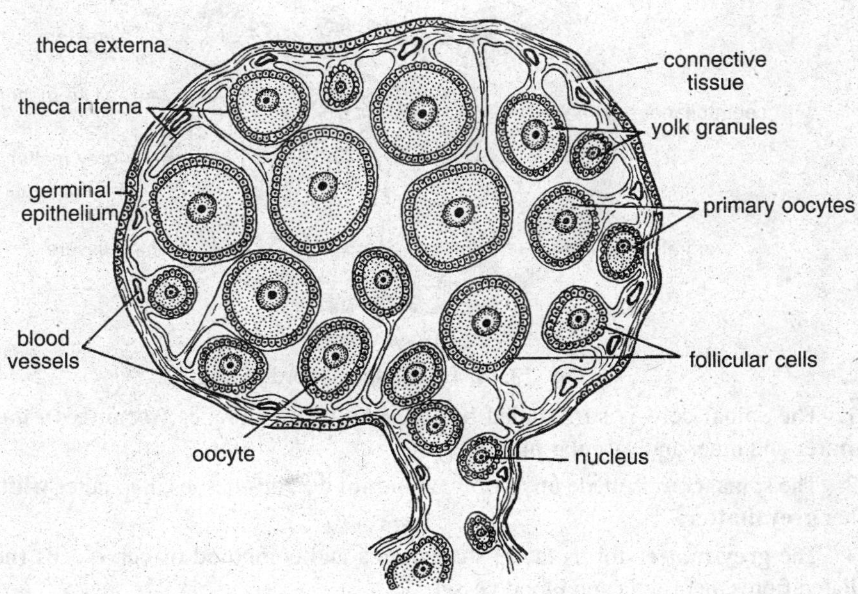

Fig. 19.13. T.S. of ovary of frog.

Comments :

Transverse section of **ovary** of **frog** (Fig. 19.13) exhibits the following structures :

1. The irregular hollow sac-like ovary consists of a number of hollow lobes or **lobules** in which ova are formed.

2. Its thin wall is formed of fibrous connective tissue having blood vessels, muscle fibres and nerves and externally covered by simple columnar coelomic epithelium or germinal epithelium.

3. Each lobule and follicle is externally surrounded by **theca externa** and internally **theca interna**.

4. Several follicles of various sizes are connected to the theca externa.

5. In the lumen of each lobule are present rounded ova in various stages of development.

6. Each ovum has a nucleus surrounded by yolky granular cytoplasm and **follicular cells** which develop from smallest oogonia.

7. The germinal epithelium undergoes oogenesis and give rise to ova which can be seen in various stages of development.

8. Fully formed ova are rounded containing nucleus and cytoplasm with granular yolk and surrounded by follicular epithelium (vitelline membrane).

9. The theca externa, theca interna and follicular cells form the ovarian stroma which secretes ovarian hormones.

14. T.S. of Spinal Cord of Frog

Comments :

Transverse section of **spinal cord** of **frog** (Fig. 19.14) shows the following histological structures :

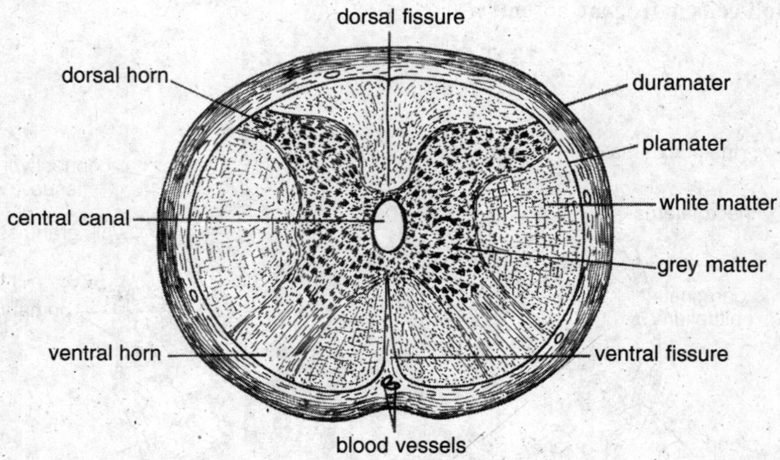

Fig. 19.14. T.S. of spinal cord of frog.

1. The spinal cord is surrounded by two fibrous membranes (meninges), *i.e.*, outer thick **duramater** and inner and vascular **piameter**.

2. The spinal cord is made up of of two kinds of nervous tissues, *i.e.*, outer **white matter** and the inner **grey matter**.

3. The **grey matter** forms the H-shaped area and composed of nerve cells (neurons), non-medullated fibres, neuroglia and blood vessels.

4. In the centre of grey matter is a small canal known as **central canal** lined by single layer of columnar epithelium (ependyma). Its function is presumably to circulate the cerebro-spinal fluid.

5. Grey matter is projected at the corners into **dorsal** and **ventral horns** to which spinal nerves are attached.

6. Surrounding the grey matter is the **white matter**. It consists of medullated nerve fibres.

7. Two deep clefts or grooves dorsal and ventral median fissures divide spinal cord into right and left lateral halves.

15. T.S. of Bone of Frog

Comments :

Transverse section of limb bone of frog (Fig. 19.15) shows the following structures :

1. The bone is a hard connective tissue.

2. Its outermost **periosteum layer** is formed of an outer layer of **dense connective tissue** having blood vessels, lymph vessels and nerves and an inner layer of bone forming cells called **outer osteoblasts**.

3. In the centre of bone is present a large **marrow cavity** filled with soft fatty tissue and blood vessels, called **bone marrow**.

4. Bone marrow is lined by delicate connective tissue layer called **endosteum** that contains inner layer of bone forming **osteoblasts**.

5. In between periosteum and endosteum matrix is arranged in thin concentric layers or **lamellae** around bone marrow.

6. In between adjacent lamellae are present numerous small cavities or **lacunae** occupied by **osteocytes** (bone cells) which give off fine branching protoplasmic processes which unite with neighbouring cells.

7. Each lacuna gives off several branching tubules (**canaliculi**) extending in all the directions and opening into canaliculi of neighbouring lacunae.

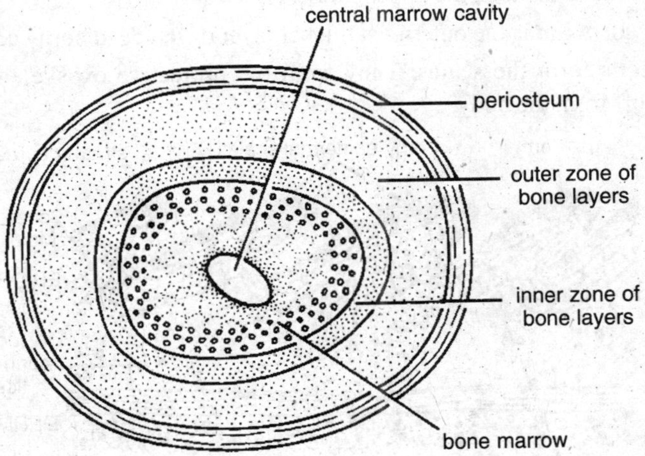

Fig. 19.15. T.S. of limb bone of frog.

Chapter 20

SLIDES OF REPTILIA

1. V.S. Skin of a Lizard

Comments :

Vertical section of skin of lizard (Fig. 20.1) shows the following structures :

1. Skin of lizards is scaly and have no cutaneous glands.
2. Skin is composed of two layers : outer **epidermis** and inner **dermis**.
3. Epidermis is thin, stratified and ecotodermal in origin.
4. **Epidermis** has two layers : innermost single layered **stratum germinativum** whose cells actively divide and move towards the upper surface.
5. **Stratum corneum** is the outermost thicker layer of flat dead horny cells.
6. **Horny cells** form the scales. They form the protective exoskeleton of lizard and are periodically shed off from the body.

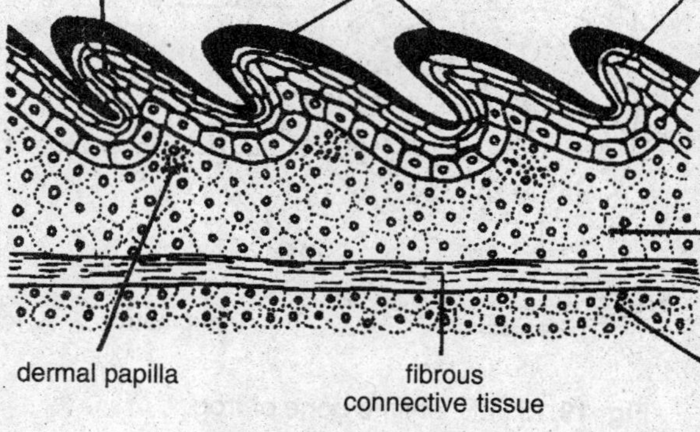

Fig. 20.1. V.S. of skin of a lizard.

7. **Dermis** is a thick layer of fibrous connective tissue and contains blood vessels, nerves, muscle fibres and chromatophores.
8. Integumentary glands are absent except femoral glands over thigh.

2. V.S. of Lung of Lizard

Comments :

Vertical section of lung of lizard (Fig. 20.2) shows the following structures :

1. Lungs in lizard are thin-walled, elastic and hollow sacs.
2. Lung is formed of outer **serosa** and inner wall is raised into a network of septa forming **alveoli** or **air sacs**.
3. Lung wall is richly vascularised.
4. Anterior part of each lung is more sacculated, thicker and more vascularised and, thus, is more respiratory.
5. Posterior part of each lung acts as a reservoir for the residual air.

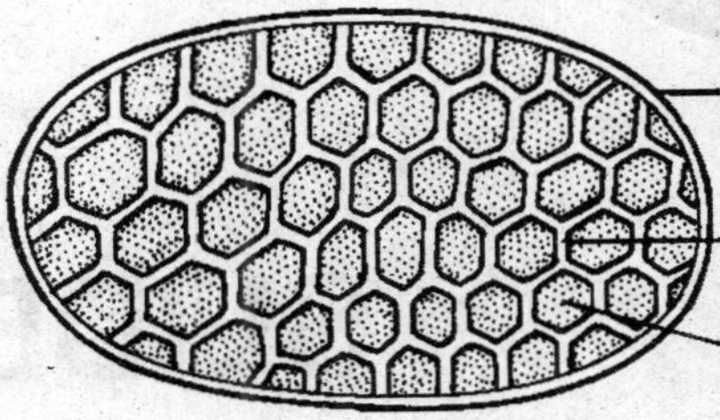

Fig. 20.2. V.S. of lung of a lizard.

6. Bronchus after entering into lung does not divide into bronchioles, but directly forms alveoli.
7. Pulmonary artery brings deoxygenated blood into the lungs and pulmonary vein carries oxygenated blood into the heart.

21
Chapter

SLIDES OF BIRDS

1. V.S. of Skin of a Bird

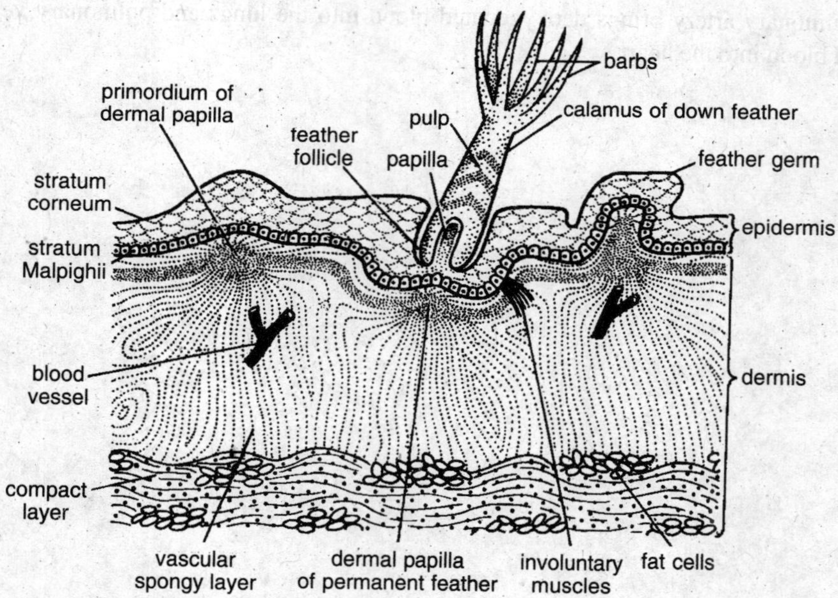

Fig. 21.1. V.S. of skin of a bird.

Comments :

Vertical section of **skin** of a **bird** (Fig. 21.1) shows the following details :

1. It consists of two layers, outer **epidermis** and inner **dermis**.
2. **Epidermis** is composed of several layers, *i.e.*, epitrichium, stratum corneum and stratum Malpighii.
3. **Epitrichium** is the outermost layer and consists of a single layer of flattened delicate cells.
4. **Stratum corneum** is the middle layer of flat cornified cells. It is horny and protective.
5. **Stratum Malpighii** (stratum germinativum) is the inner layer composed of large cylindrical actively dividing cells.

6. Feather papilla, calamus of down feather along with barbs are also seen in the epidermal layer.

7. **Dermis** is composed of upper broad vascular **spongy layer** and lower narrow **compact layer**.

8. In the spongy layer dermal papilla of permanent feather, feather follicle, pulp, involuntary muscles and blood vessels are seen.

9. The innermost compact layer have fat cells.

2. L.S. of Syrinx of a Bird

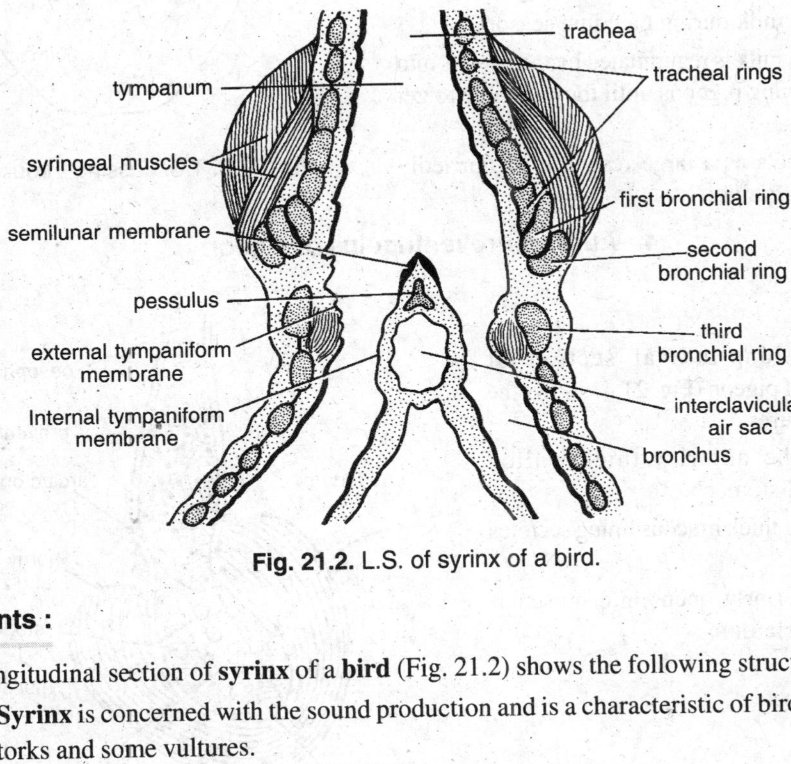

Fig. 21.2. L.S. of syrinx of a bird.

Comments :

Longitudinal section of **syrinx** of a **bird** (Fig. 21.2) shows the following structures :

1. **Syrinx** is concerned with the sound production and is a characteristic of birds, but absent in ostrich, storks and some vultures.

2. It consists of an expanded chamber called **tympanum**.

3. Its walls are supported by the last three or four rings of trachea and first half ring of each bronchus.

4. **Pessulus** is a bony ridge which extends dorso-ventrally at the junction of two bronchi and gives support to a small vibratory crescentic **mucous membrane**.

5. The mucous membrane of each bronchus forms an **external tympaniform membrane** and an **internal tympaniform membrane** associated with the respective wall of the bronchus.

6. These tympaniform membranes are controlled and regulated by a pair of **intrinsic syringeal muscles** and a pair of **sternotracheal muscles**.

7. The voice is produced by the vibrations of the tympaniform membranes as the air expelled from lungs passes between them during expiration and pitch of the voice is altered by changes in the tension of these membranes controlled by above given muscles.

3. L.S. of Crop of Pigeon

Comments :

Longitudinal section of **crop** of **pigeon** (Fig. 21.3) shows the following structures :

1. Crop is a thin-walled, bilobed elastic sac.
2. Crop is internally lined by the epithelial cells layer that also contains mucus secreting gland cells.
3. Epithelial cells in both séxes degenerate and produce pigeon milk during breeding season.
4. Pigeon milk is regurgitated by the parents into the mouth of young pigeons until they are able to get food themselves.
5. Crop acts as a large reservoir of hurriedly swallowed food grains.

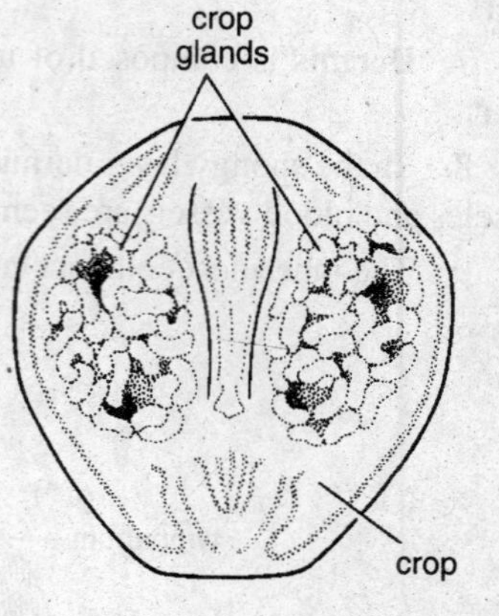

Fig. 21.3. T.S. passing through crop.

4. V.L.S. of Proventriculus of Pigeon

Comments :

Vertical longitudinal section of proventriculus of pigeon (Fig. 21.4) shows the following structures :

1. It is the anterior thick-walled glandular part of stomach.
2. Its inner thick mucous lining secretes the gastric juice.
3. It posteriorly opens into muscular **gizzard** or **ventriculus**.

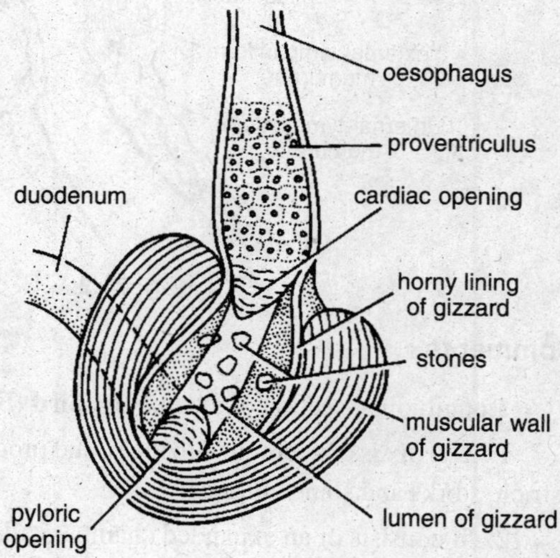

Fig. 21.4. V.L.S. of proventriculus of pigeon.

22

Chapter

SLIDES OF MAMMALIA

1. V.S. of Skin of a Mammal

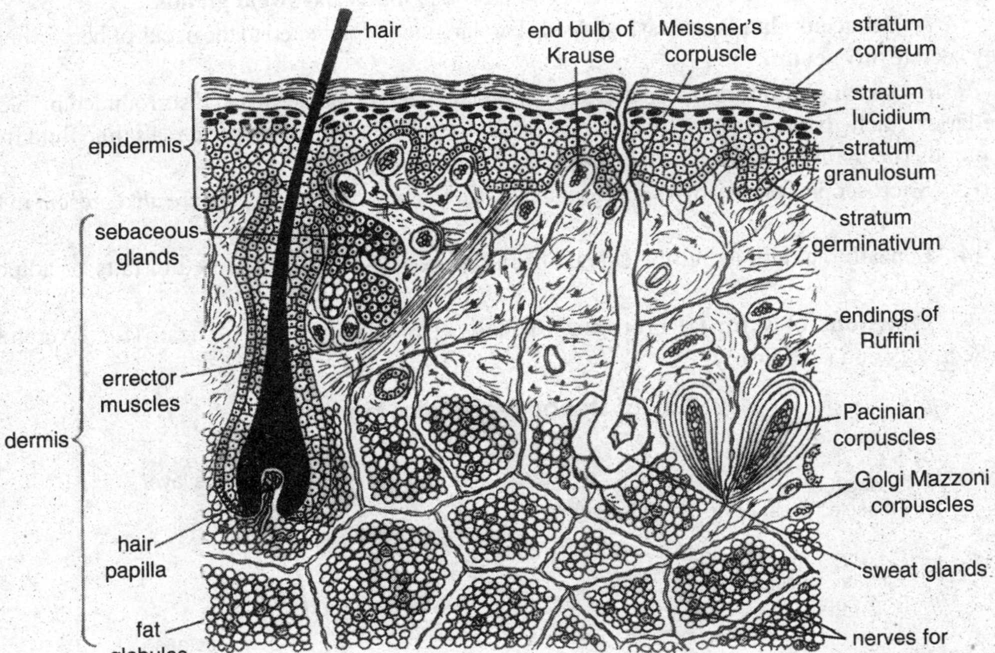

Fig 22.1. V..S of skin of mammal.

Comments :

Vertical section of **skin** of a **mammal** (Fig. 22.1) shows the following histological structures :

1. **Skin** is composed of two layers, *i.e.,* outer layer **epidermis** and inner layer **dermis**.

2. **Epidermis** contains no blood vessels and comprises four layers, namely outermost **stratum corneum**, next **stratum lucidium, stratum granulosum** and innermost layer is **stratum germinativum**.

3. **Stratum corneum** consists of dead horny cells which are periodically moulted. Its example is dandruff.

4. **Stratum lucidum** is found only in palms and soles. It is a hyaline non-stainable layer.

5. **Stratum granulosum** is made up of granular cells containing keratohyaline granules. This layer is also found in palms and soles.

6. **Stratum germinativum** is the innermost columnar cells layer of epidermis based over basement membrane. Its cells are continually divided and pushed upward. All the glands and keratin structures are formed from this layer.

7. **Dermis** consists of dense fibrous connective tissue containing muscle fibres, blood vessels, lymph vessels, pigment cells, nerves and glands.

8. Mammalian skin is characterised by the presence of **hairs** and **glands** whose bases are present in dermis and penetrate the epidermis above. These are epidermal structures.

9. **Hair** is present within a hair follicle which is invagination of Malpighian layer into the dermis, and supported by fibres.

10. Base of hair follicle is dilated and evaginated to form hair papilla supplied by blood capillaries and nerves.

11. Unstriated muscle fibres connecting the hair with the epidermis move the hair involuntarily and are known as errector muscles.

12. The glands are of two types namely **sebaceous glands** and **sweat glands**.

(*i*) **Sebaceus glands** are flask-shaped alveolar glands connected to the necks of hair follicles. They secrete oily **sebum**.

(*ii*) **Sweat glands** are coiled and tubular and lie deep in the dermis and surrounded by blood capillaries. Each gland opens on the surface through a long coil duct and separates a saline fluid from the blood. The main function of sweat glands is temperatures regulation of body.

13. Microscopic sensory organs of touch, pressure, temperature and pain are also present in the dermis.

14. Beneath the dermis lies the subcutaneous tissue in which fat is stored in fatty or adipose cells.

Functions of skin: (*i*) Protection of body, (*ii*) Conservation of heat, (*iii*) Excretory, (*iv*) Secretory and (*v*) Sensory.

2. T.S. of Oesophagus of a Mammal

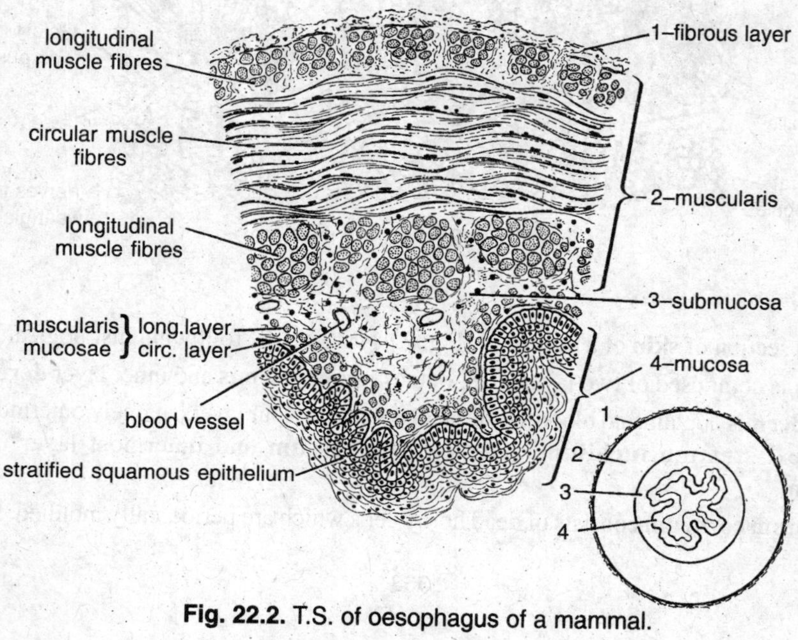

Fig. 22.2. T.S. of oesophagus of a mammal.

Comments :

Transverse section of **oesophagus** of a **mammal** (Fig. 22.2) is composed of four layers : serosa, muscular coat, submucosa and mucosa.

1. Outermost layer is composed of a thin layer of fibrous connective tissue known as **tunica adventitia**.
2. The **muscularis** consists of two layers of muscle fibres : outer is **longitudinal muscle fibres** and the inner layer is of **circular muscle fibres (smooth muscles).**
3. The **submucosa** is composed of connective tissue containing blood vessels, lymph vessels and nerves.
4. **Muscularis mucosae** is also present and mainly composed of longitudinal muscle fibres.
5. **Mucosa** is the innermost stratified columnar epithelium and is folded.
6. **Mucosa** is formed of usual three layers : innermost simple columnar epithelium, lamina propria of loose connective tissue and muscularis mucosae described above. Lamina propria papillae project into columnar epithelium.

3. T.S. Stomach of a Mammal

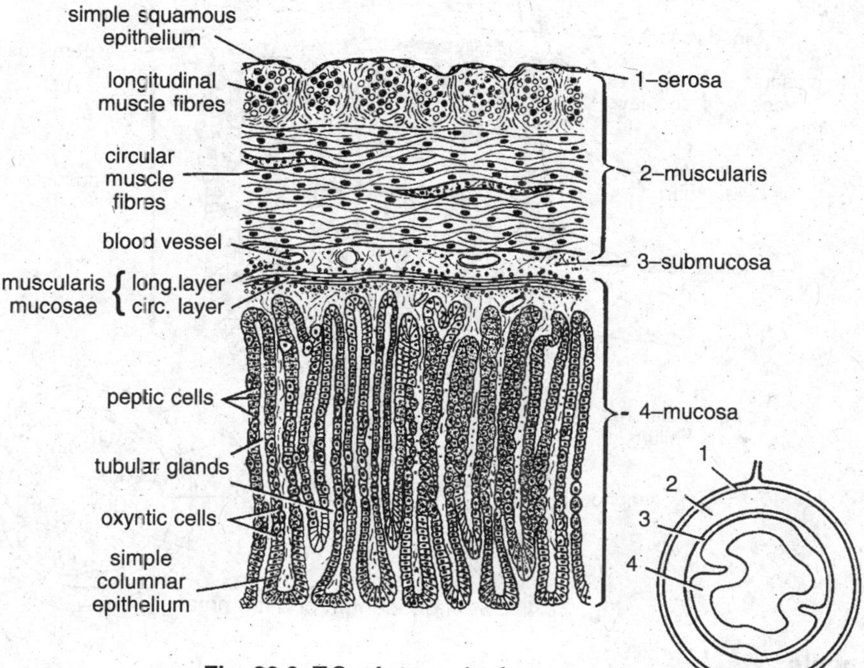

Fig. 22.3. T.S. of stomach of a mammal.

Comments :

Transverse section of **stomach** of a **mammal** (Fig. 22.3) show the following four layers : serosa, muscularis, submucosa and mucosa.

1. **Serosa** is the outermost layer composed of simple squamous epithelium.
2. **Muscularis** consists of three layers : the outer layer is composed of **longitudinal muscle fibres**, middle layer is composed of circular muscle fibres and inner is of oblique muscle fibres. Oblique muscles are best developed in cardiac stomach, Longitudinal and circular muscles become thicker towards pylorus.

3. **Submucosa** is made up of connective tissue containing blood vessels, lymph vessels and nerves. Meissner's nerve plexus is also present in the submucosa.

4. **Muscularis mucosae** is a thin layer of outer longitudinal and inner circular muscle fibres present in between submucosa and mucosa.

5. **Mucosa** consists of **simple columnar epithelium** which is invaginated at close intervals into lamina propria forming gastric pits which lead into simple or branched tubular gastric glands.

6. The **gastric glands** composed of three types of cells namely **peptic cells**, **oxyntic cells**, and **mucous cells**. **Peptic cells** secrete pepsin, **oxyntic cells** (parietal cells) secrete HCl, and **mucous cells** secrete mucus.

7. Villi are not found in stomach.

4. T.S. of Duodenum of a Mammal

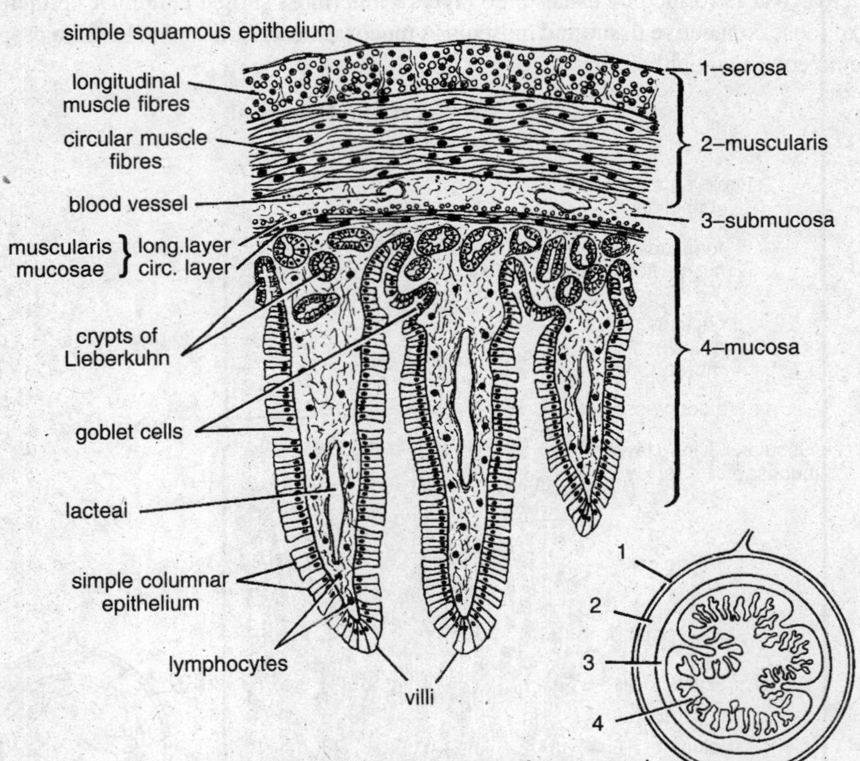

Fig. 22.4. T.S. of duodenum of a mammal.

Comments :

Transverse section of **duodenum** of a **mammal** (Fig. 22.4) shows the usual four layers :

1. Outermost layer is **serosa** which usually consists of a simple squamous epithelium.

2. **Muscularis** consists of outer **longitudinal muscle fibres** and inner thicker **circular muscle fibres**.

3. **Submucosa** consists of connective tissue holding blood vessels, lymph vessels, and nerves It is raised up into low circular folds, **plicae circulares**.

4. **Muscularis mucosae** is thin and double layered consisting of outer layer of longitudinal fibres and inner layer of circular fibres. It is a reduced layer.

5. **Mucosa** is thrown into numerous large and small finger-like folds called **villi** which are covered by **simple columnar epithelium** having scattered **goblet cells**, and **absorptive cells**.

6. Each villus contains connective tissue, unstriated muscle fibres, **lymphocytes**, **blood vessels** and a **lacteal** surrounded by capillary network.

7. At the base of the villi are present simple tubular pits of intestinal glands called **crypts of Lieberkuhn**. These secrete digestive enzymes (**succus entericus**).

8. Within submucosa are present racemose **Brunner's glands**. (mucous glands). They secrete mucus and pour into crypts. Brunner's glands are only found in duodenum.

5. T.S. of Liver of a Mammal

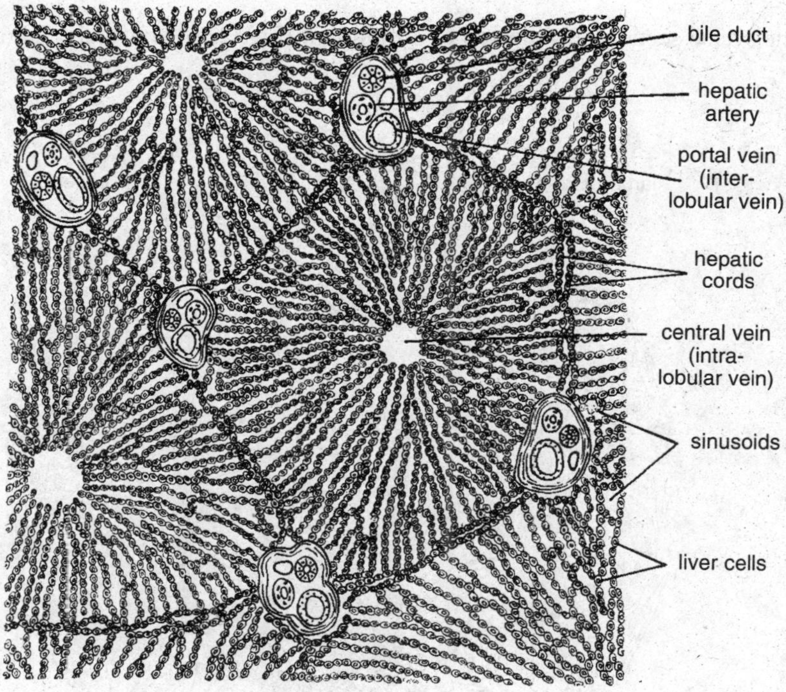

Fig. 22.5 .T.S. of liver of a mammal.

Comments :

Transverse section of **liver** of a **mammal** (Fig. 22.5) shows the following histological structures :

1. Liver is a spongy racemose gland externally covered by fibro-elastic serous coat (liver capsule). Each liver lobe is composed of numerous small polygonal **lobules** separated by a thin layer of connective tissue septa or **Glissons' capsule**. Each lobule contains a **central vein** (intra-lobular vein) in the centre and **portal canals** at the corners between adjacent lobules.

2. Each portal canal consists of connective tissue strand and contains a branch of **portal vein** (inter-lobular vein), **hepatic artery**, **bile duct**, **lymph vessel** and **nerves.**

3. The liver cells are cuboidal or polyhedral and arranged in one or two celled thick anastomosing long chains extending radially from the central vein to the periphery of the lobule.

4. Each liver cell has granular cytoplasm and a prominent nucleus and deposits of glycogen and lipids.

5. Between hepatic chains are present narrow irregular spaces (**lacunae**) through which run blood capillaries called **hepatic sinusoids**. Capillaries have discontinuous walls of endothelial cells, some of these cells become **Kupffer cells**.

6. Around liver cells is present meshwork of fine intercellular channels, the **bile canaliculi** or **bile capillaries** into which bile is secreted by liver cells.

7. Liver secretes bile that contains bile salts, pigments and toxins.
8. Liver has several functions which are as follows :
   (*i*) It produces bile which plays an important role in the digestion of food especially fat,
   (*ii*) It stores the soluble products of digestion and metabolise them for assimilation,
   (*iii*) Oxidation of sugar takes place in it, (*iv*) Toxic substances are detoxicated in the liver.

6. T.S. of Pancreas of a Mammal

Comments :

Transverse section of **pancreas** of a **mammal** (Fig. 22.6) shows the following histological details :

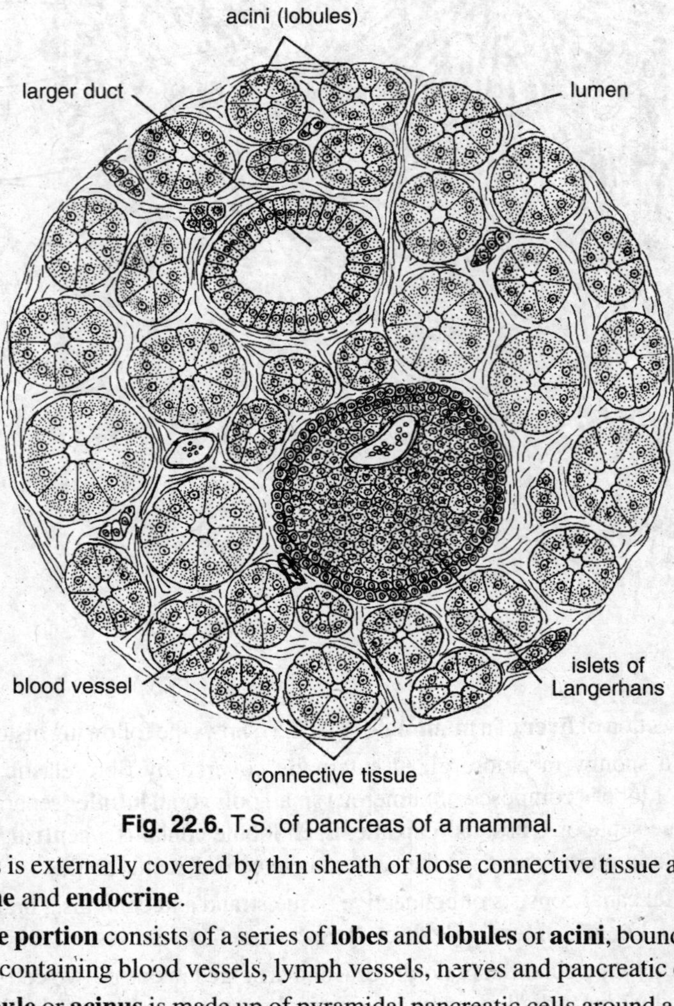

Fig. 22.6. T.S. of pancreas of a mammal.

1. Pancreas is externally covered by thin sheath of loose connective tissue and consists of two portions : **exocrine** and **endocrine**.

2. **Exocrine portion** consists of a series of **lobes** and **lobules** or **acini**, bound together by loose connective tissue containing blood vessels, lymph vessels, nerves and pancreatic ducts.

3. Each **lobule** or **acinus** is made up of pyramidal pancreatic cells around a central lumen and having granular cytoplasm and prominent nuclei.

4. The lobules open into small ductules which join large ducts and eventually the main pancreatic duct.

5. **Exocrine portion** produces pancreatic juice which contains **trypsin**, **amylase** and **lipase** enzymes.

6. **Endocrine portion** is composed of **islets of Langerhans** found between the acini in the connective tissue.

7. Islets of Langerhans are compact masses of cells and secrete two hormones namely **insulin** and **glucagon**.

8. Insulin is said to be produced by the **beta cells** of islets of Langerhans. It regulates the amount of sugar contents of the blood. Its deficiency causes **diabetes mellitus**.

9. **Glucagon** is said to be produced by **alpha cells**. It increases the sugar contents of the blood and its deficiency causes **hypoglycemia**.

7. T.S. of Spleen of a Mammal

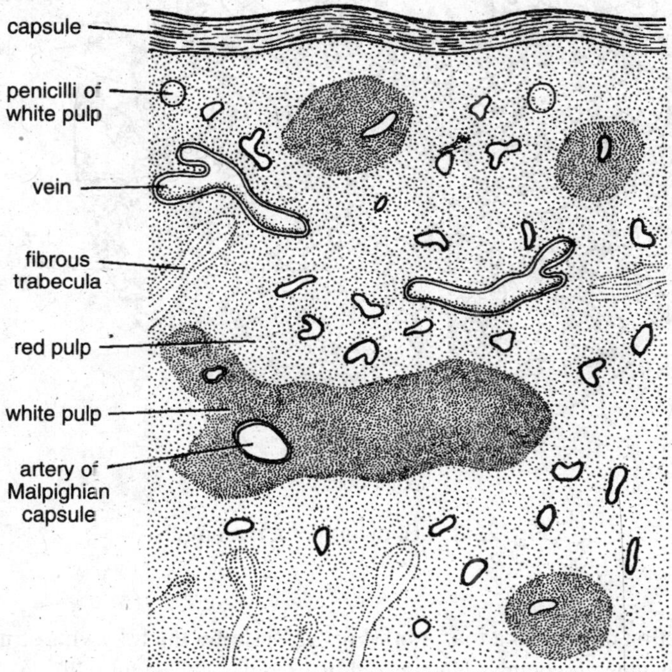

Fig. 22.7. T.S. of spleen of a mammal.

Comments :

Transverse section of **spleen** of a **mammal** (Fig. 22.7) shows the following structures :

1. The spleen is the largest lymphoid gland and is surrounded by a thin fibromuscular capsule which is externally covered by visceral peritoneum.

2. From the capsule arises branching **trabeculae** which penetrate into the spongy or spleenic pulp of the spleen.

3. The spleenic pulp is distinguished into **white pulp** and **red pulp**.

4. The spleenic pulp is supported by reticular tissue and divided into lobules.

5. The **red pulp** forms the bulk and contains **erythrocytes** which give red colour to it.

6. Also contains pulp cords formed of erythrocytes, leucocytes, reticular cells, **macrophages** or **phagocytes** and venous sinuses are also present.

7. **White pulp** is darker than red pulp in stained preparation and is formed of lymphocytes, macrophages, reticular cells and eccentric artery or arteriole.

8. The spleen is very important organ of the body and serves the following functions : (*i*) Produces erythrocytes and lymphocytes. (*ii*) Eats foreign bodies, bacteria and dead cells. (*iii*) Produces antibodies which prevent the entrance of bacteria. (*iv*) It acts as a filtering organ of the blood.

8. T.S. of Lung of a Mammal

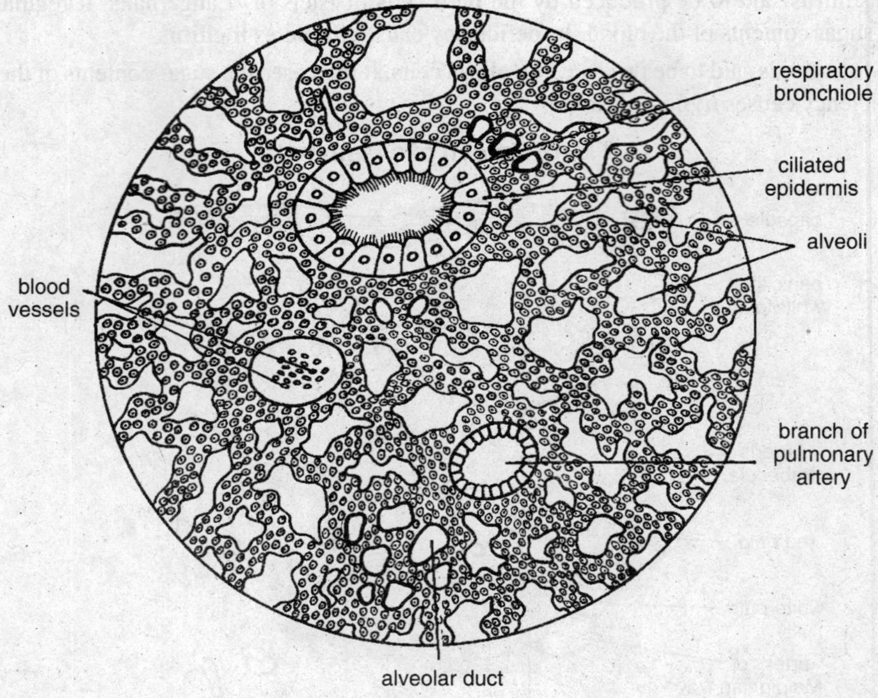

Fig. 22.8. T.S. of lung of a mammal.

Comments :

Transverse section of **lung** of a **mammal** (Fig. 22.8) shows the following structures :

1. Mammalian lung is spongy, elastic, richly vascular and externally covered by visceral peritoneum.
2. Histologically it consists of numerous **alveoli** formed by the repeated divisions of bronchioles which terminate into **air sacs** and these bear hemispherical hollow **alveoli**.
3. **Alveoli** present a honey combed appearance.
4. Walls of air sacs and alveoli are very thin made up of a single layer of squamous epithelium and covered by network of capillaries of pulmonary artery and vein.
5. Alveoli are the final structural and physiological units of lung where exchange of gases takes place.
6. Bronchioles are lined with ciliated columnar epithelium, placed on basement membrane.
7. Wall of bronchiole is made of elastic connective tissue and circular muscle fibres.
8. Route of air during respiration is as follows : air → trachea → bronchus → bronchioles → respiratory bronchioles → alveolar ducts → alveoli → gaseous exchange takes place and CO_2 is then taken out.

9. L.S. of Kidney of a Mammal

Comments :

Longitudinal section of metanephric **kidney** of a **mammal** (Fig. 22.9) shows the following histological details :

1. The bean-shaped kidney is surrounded by a **capsule** of tough fibrous connective tissue and covered by peritoneum only on its ventral surface.

2. The glandular part of the kidney composed of outer striated **cortex** and inner deeply strained **medulla**.

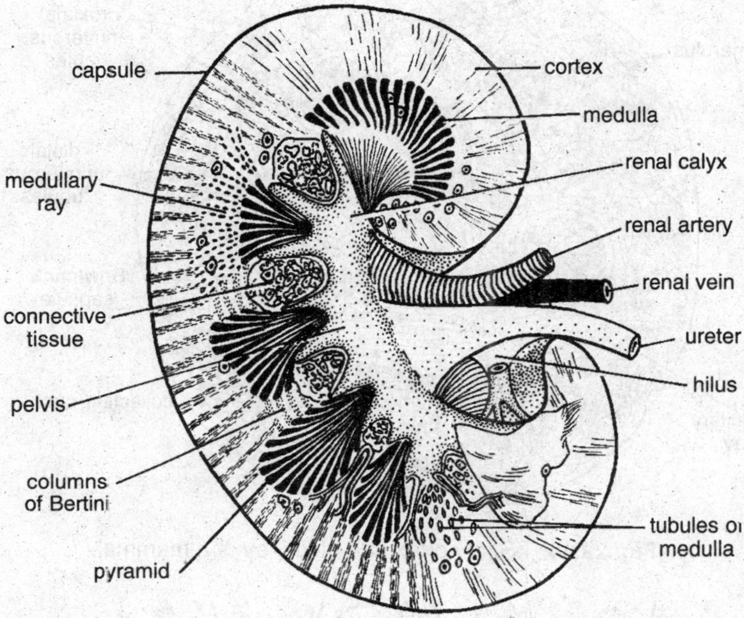

Fig. 22.9. L.S. of kidney of a mammal.

3. **Cortex** contains convoluted portions of **uriniferous tubules**, **Malpighian capsules** (**Bowman's capsules** and **glomerulus**).

4. **Medulla** is composed of several conical renal **pyramids** whose apices project as renal papillae into pelvis called **calyces**, **medullary rays**, **columns of Bertini** (extensions of cortex between pyramids), **tubules of medulla** and connective tissue.

5. **Hilus** is the depression in the middle of the inner concave region of the kidney. Ureter and renal vein leave the kidney from the hilus or hilum region.

6. **Renal artery** and a nerve enter the kidney at the hilus.

7. **Renal pelvis** is the proximal expanded end of ureter.

10. T.S. of Cortex Part of Kidney of a Mammal

Comments :

Transverse section of **cortex part** of **kidney** (Fig. 22.10 and 22.11) shows the following histological details :

1. Cut ends of large number of coiled proximal and distal **uriniferous tubules** of various shapes and sizes are present.

2. The **proximal** and **distal uriniferous tubules** are lined by cuboidal epithelium which is ciliated at intervals.

3. Cortex also contains sections of Malpighian capsules (Bowman's capsule and glomerulus).

4. **Glomerulus** is formed by the network of afferent and efferent arterioles.

5. The inner lining (squamous epithelium) of Malpighian's capsule filters waste products dissolved in water and converts it into urine which passes out through the uriniferous tubules.

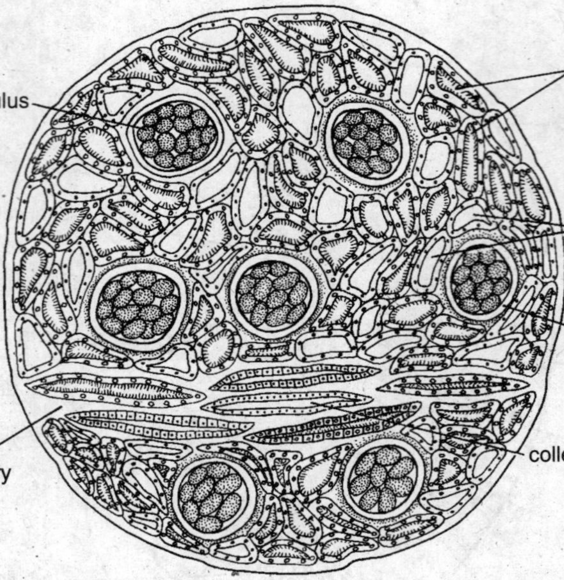

Fig. 22.10. T.S. of cortex part of kidney of a mammal.

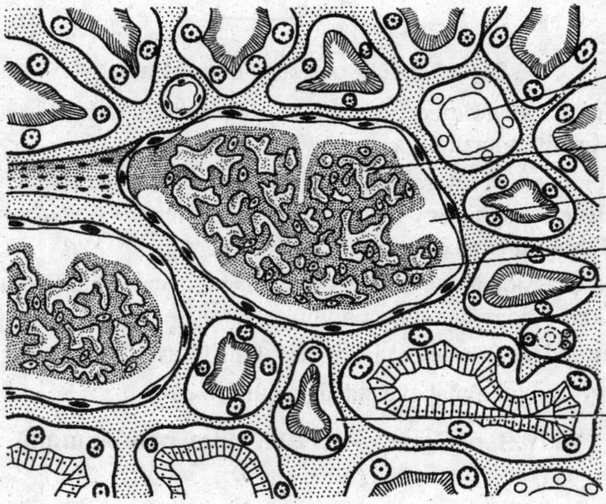

Fig. 22.11. T.S. of cortex part of kidney of a mammal (highly magnified).

11. T.S. of Medulla Part of Kidney of a Mammal

Comments :

Transverse section of **medulla part of kidney** of a **mammal** (Fig. 22.12) shows the following histological details :

1. The medulla part of the kidney is characterised by the absence of numerous convoluted parts of uriniferous tubules, Bowman's capsules and glomerulus.
2. Cut ends of **thin** and **thick** segments of **Henle's loop** are present in the section.
3. Numerous large collecting tubules and ducts are present.
4. Blood vessels, capillaries and nerves are present in the connective tissue.
5. Medullary rays are also present in the medulla.

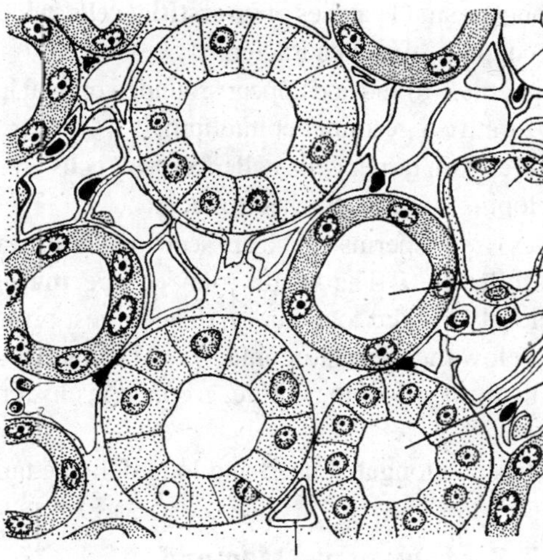

Fig. 22.12. T.S. of medulla part of kidney of a mammal.

6. The main function of medulla part of kidney is to collect the urine from various ducts coming from cortex part and ultimately release it into the pelvis, proximal swollen part of ureter.

12. T.S. of Testis of a Mammal

Fig. 22.13. T.S. of testis of a mammal.

Comments :

Transverse section of **testis** of a **mammal** (Fig. 22.13) shows the following details :

1. The testis is some what rounded or oval in shape and externally surrounded by visceral peritoneum followed by a layer of fibrous connective tissue, the **tunica albuginea**.

2. Histologically each testis is composed of a mass of coiled microscopic **seminiferous tubules**, held together by connective tissue that contains blood vessels, lymph vessels, nerves and muscle fibres.

3. The connective tissue also contains small patches of **interstitial cells**, which secrete a hormone **testosterone** responsible for male secondary sexual characters.

5. Each seminiferous tubule is internally blind and appears rounded or oval in section surrounded by basement membrane and lined by stratified germinal epithelium.

6. In between the germinal cells certain larger cells called **Sertoli cells** are usually seen. These cells supply nourishment to the developing sperms attached to them.

7. The germinal epithelium gives rise to sperms which are seen in various stages of development in a seminiferous tubule and are as follows: (*i*) The small, rounded **spermatogonia** lie along the periphery of the tubule and appear closely packed together. (*ii*) The largest primary **spermatocytes** and half sized spermatocytes lie just below the spermatogonia. (*iii*) The smallest **spermatids** border the lumen. (*iv*) The **spermatozoa** lie in the cavity of the tubule, grouped in clusters and are connected with Sertoli cells.

8. A **spermatozoon** or sperm has an elongated head and long delicate tail. Its nucleus lies in the head.

13. T.S. of Ovary of a Mammal

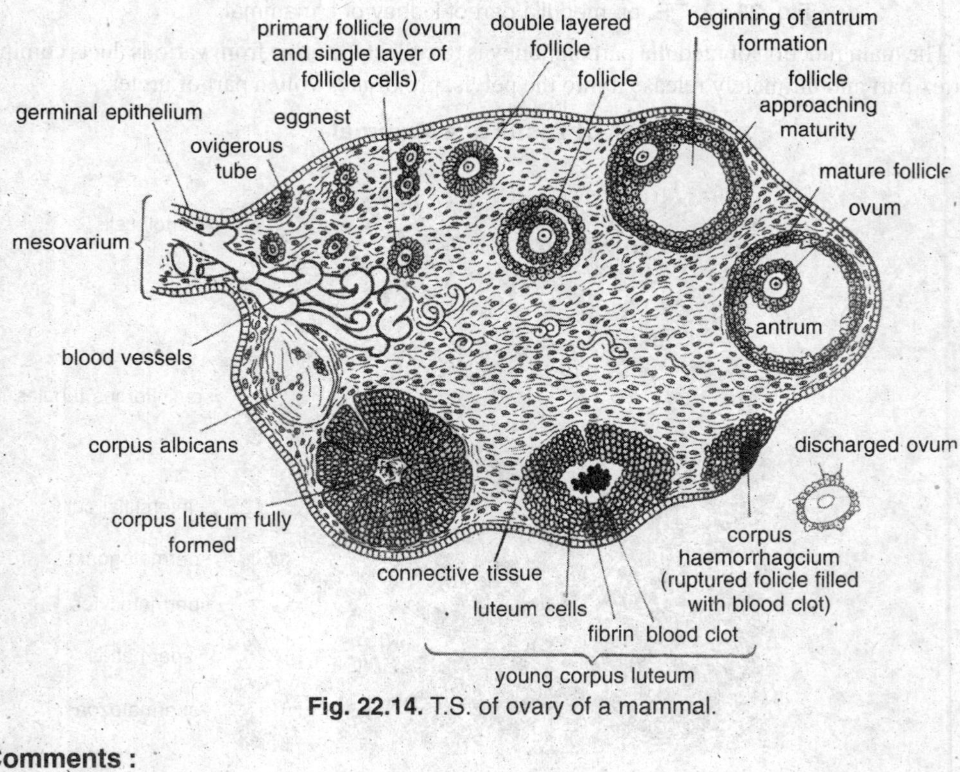

Fig. 22.14. T.S. of ovary of a mammal.

Comments :

Transverse section of **ovary** of a **mammal** (Fig. 22.14) shows the following important structures :

1. The ovary is solid and oval in shape and is invested by peritoneum of cubical cells that forms **germinal epithelium** beneath which is present the dense connective tissue, the **tunica albuginea.**

2. It consists of mass of connective tissue fibres and spindle-shaped cells, forming the **stroma**, that is differentiated into a broad peripheral **cortex** and narrow central **medulla**.

3. Lying in the cortex are ovarian follicles, corpora lutea, corpora albicans and interstitial cells. Medulla contains only blood vessels.

4 In the section **egg nest, primary follicle** (central oocyte and single layer of follicle cells), **double layered follicle, follicle approaching maturity** showing **antrum formation, mature follicle**

or **Graafian follicle**, **ruptured follicle**, **young** and **fully formed corpus luteum** and **corpus albicans** are seen.

5. **Primary follicles** arise from ingrowth of the germinal epithelium into the cortex and each contains a central oocyte surrounded by a single layer of follicle cells.

6. **Secondary follicles** contain a non-cellular transparent area around the oocyte called **zona pellucida**, several layers of follicle cells and a thin connective tissue layer, called **theca** having blood capillaries.

7. **Mature** or **Graafian follicle** contains a large central follicular cavity, **antrum** externally lined by follicular cells, **membrana granulosa**. Mass of cells surrounding the primary oocyte and attaching it to the follicle wall is called **discuss proligerous** or **cumulus**. Inner cells of cumulus around zona pellucida form the **corona radiata**.

8. **Corpus luteum** is a ruptured follicle having a solid mass of cells. It secretes a hormone, **progesterone**.

9. **Corpus albicans** is a scar of white tissue formed by the degenerated corpus luteum in case of non-fertilisation of ovum.

10. **Interstitial cells** in cortex also produce sex hormones.

14. T.S. of Penis of a Mammal

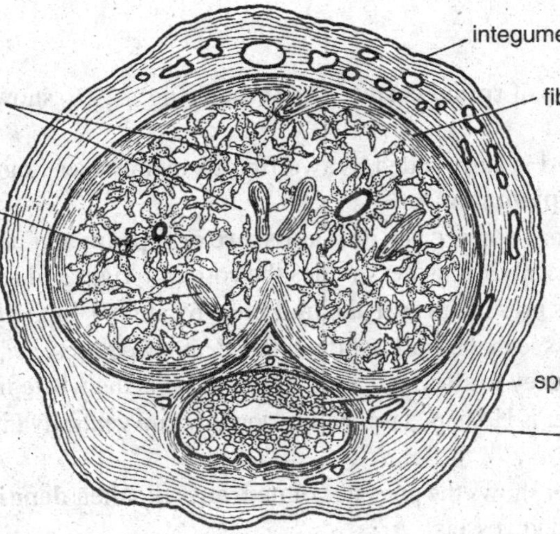

Fig. 22.15. T.S. of penis of a mammal.

Comments :

Transverse section of **penis** of a **mammal** (Fig. 22. 15) shows the following histological details :

1. The outermost layer is **integument** or skin which surrounds the penis and composed of circular muscle fibres.

2. The penis consists of three longitudinal sets of spongy erectile tissue namely : (*i*) two **corpora cavernosa** and, (*ii*) one **corpus spongiosum**.

3. The **corpus spongiosum** surrounds the **urethra**.

4. Two **corpora cavernosa** lie dorsal to spongiosum.

5. The corpora cavernosa and spongiosum are spongy in structure and contains several blood spaces.

6. The **corpus spongiosum** forms the tip of the penis called **glans penis** covered by prepuce (part of skin).

7. The corpora cavernosa and corpus spongiosum are surrounded by a thick fibrous membrane, the **tunica albuginea**.

8. The **tunica albuginea** projects into corpora cavernosa and forms trabeculae.

15. T.S. of Spinal Cord of a Mammal

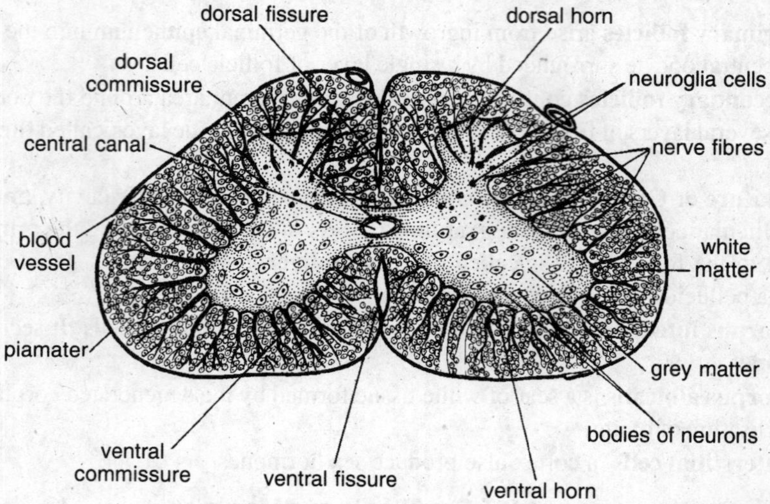

Fig. 22.16. T.S. of spinal cord of a mammal.

Comments :

Transverse section of **spinal cord** of a **mammal** (Fig. 22.16) shows the following histological structures :

1. The spinal cord is surrounded by two fibrous membranes : outer thick **duramater** and an inner vascular and thin **piamater**.
2. In the mid-dorsal surface is a **dorsal fissure** or **septum** and in the mid-ventral surface is a **ventral fissure** which is slightly wider divides the cord into right and left lateral halves.
3. In the centre is present a **central canal** lined by simple epithelial cells.
4. The nervous tissue of the cord is differentiated into two zones, *i.e.*, the central zone surrounding the central canal is the **grey matter** and a peripheral zone is the **white matter**.
5. The grey matter is H-shaped projecting dorsally and ventrally into **dorsal horns** and **ventral horns**.
6. The grey matter shows the presence of **neurons** with their **dendrons**, non-medullated fibres, **neuroglial cells** and blood vessels.
7. The white matter is composed of only medullated nerve fibres supported by prolongations of the neuroglia.
8. The bands of fibres which extend transversely, one dorsal and other ventral to the central canal, are known as **dorsal** and **ventral commissures** respectively.
9. Spinal cord conducts impulses to and from brain and also controls reflex activities.

16. V.L.S. of Anterior Lobe of Pituitary Gland of a Mammal

Comments :

Vertical longitudinal section of **anterior lobe** of **pituitary gland** of a mammal (Fig. 22.17) shows the following histological structures :

1. The **pituitary gland** is more or less rounded in shape and occurs at the base of brain just beneath the diencephalon.
2. It is composed of three lobes namely, **anterior lobe**, **intermediate lobe** and **posterior lobe**.

3. The **anterior lobe** forms the largest part of pituitary gland formed of three distinct kinds of cells differing in their staining reactions : basophil cells, acidophil cells and chromophobes.

4. Usually on the outerside are the **basophil cells** which are stained by basic stains.

5. In the centre are found **acidophil** or **oxyphil cells** which take acidic stains.

6. The **chromophobe** cells are indifferent to either basic or acid stains. They are found scattered throughout the anterior lobe.

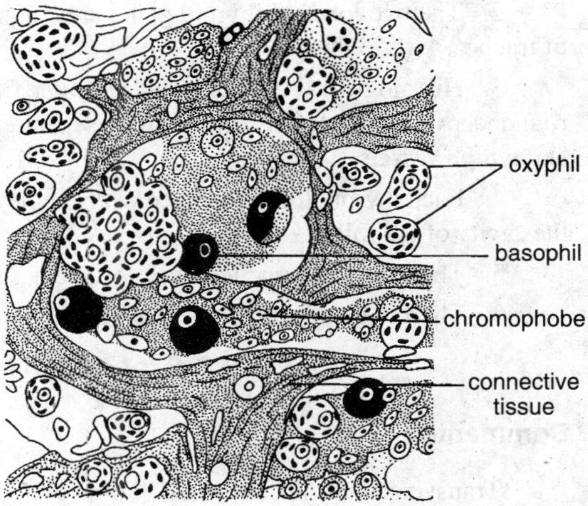

Fig. 22.17. V.L.S. of anterior lobe of pituitary gland of a mammal.

7. **Anterior lobe** produces many hormones, namely **somatotrophic hormone**, **thyrotrophic hormone**, **adrenocorticotrophic hormone**, **gonadotrophic hormone** and, thus, controls growth, development of sex glands as well as the activities of thyroid, parathyroid and adrenal glands.

8. **Intermediate lobe** is composed of cell cords with colloid filled follicles. It produces an **intermedin hormone**.

9. **Posterior lobe** is composed of neuroglial cells, connective tissue fibres and blood vessels. It produces **pituitrin**, **vasopressin** and **oxytocin** hormones.

10. The pituitary gland is an endocrine gland of utmost importance to organisms. It is called master gland.

17. V.S. of Thyroid Gland of a Mammal

Comments :

Vertical section of **thyroid gland** of a **mammal** (Fig. 22.18) shows the following histological details :

1. The thyroid gland of mammal is bilobed and lies on the ventro-lateral surface of the larynx.

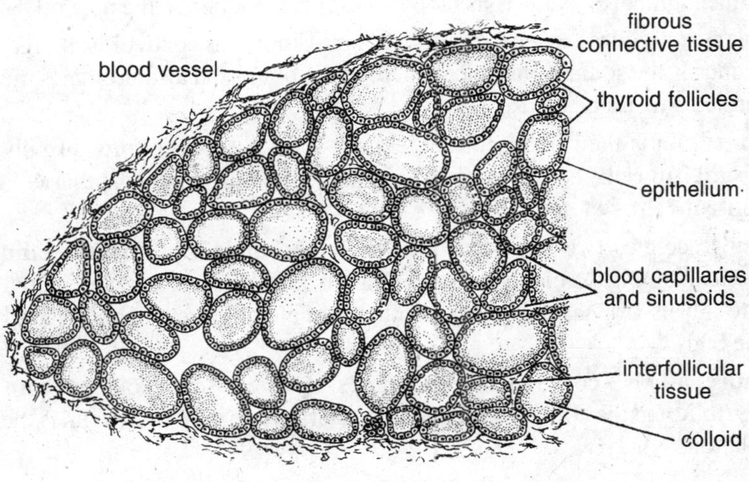

Fig. 22.18. V.S. of thyroid gland of a mammal.

2. The two lobes of thyroid gland are connected by a narrow **isthumus** across ventral surface of trachea.

3. Histologically it consists of an outer fibrous connective tissue capsule and a number of rounded, oval or oblong **thyroid follicles** separated by inter-follicular connective tissue having blood and lymph vessels.

4. Each thyroid follicle is lined with simple cuboidal epithelium which pour their secretion into the cavity of the follicle.

5. The cavity or lumen of each follicle is filled with **colloid**.

6. Thyroid gland secretes **thyroxine hormone** which controls the entire metabolism of animals.

18. T.S. of Adrenal Gland of a Mammal

Comments :

Transverse section of **adrenal gland** of a **mammal** (Fig. 22.19) shows the following histological structures :

1. The adrenal gland is an endorcrine gland, composed of two distinct parts, *i.e.*, outer **mesodermal cortex** and inner **nervous medulla** surrounded by the **capsule** composed of fibrous connective tissue containing blood vessels and nerves.

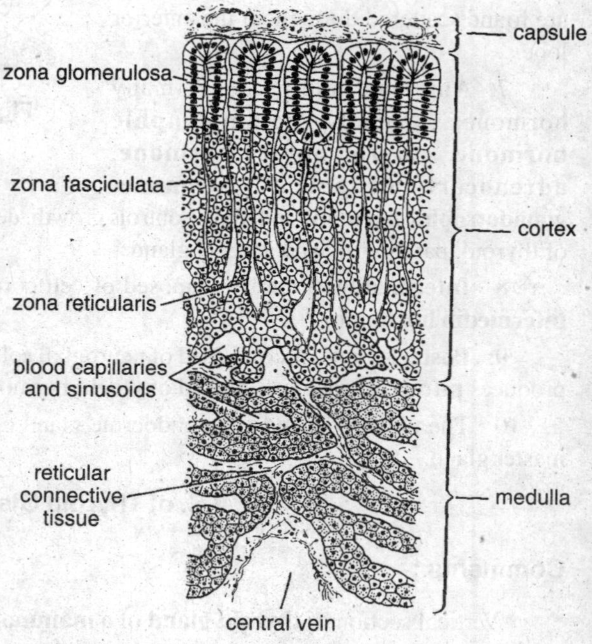

Fig. 22.19. T.S. of adrenal gland of a mammal.

2. The **cortex** lies next to the capsule and is differentiated into three zones namely **zona glomerulosa**, **zona fasciculata** and **zona reticularis**. (*i*) **Zona glomerulosa** is made up of columnar cells arranged in oval groups which resemble either closed or open vesicles. (*ii*) **Zona fasciculata** is thickest and consists of columns of large polyhedral cells arranged radially in double rows. (*iii*) **Zona reticularis** consists of anastomosing networks of columnar cells containing pigment granules. Numerous blood sinusoids are found in the networks.

3. The **cortex** produces several steroid hormones known as **corticoids**. It regulates the general metabolism, controls the sodium chloride content of the blood and also promotes the breaking down of the tissue proteins to amino acids.

4. The **medulla** is the central portion, consists of networks or cords of polygonal cells and clusters of **chromaffin cells**, networks of cells contain numerous blood capillaries, sinusoids and in the centre a **central vein**.

5. Medulla secretes two hormones known as **adrenalin** and **noradrenalin**. **Adrenalin** is responsible for maintaining the blood pressure, dilation of vessels and muscles, increasing the general metabolism rate and also for hastening the coagulation of blood. Adrenalin controls all the involuntary activities of the body.

6. **Noradrenalin** has constrictor effect, causes a greater rise in blood pressure than adrenalin. It raises both systolic and diastolic pressures. The main function of noradrenalin is the normal control of blood circulation.

23
Chapter

HISTOLOGY OF SOME VERTEBRATE TISSUES

1. Simple Squamous Epithelium

Comments :

1. Simple squamous epithelium (Fig. 23.1) consists of a single layer of flattened plate-like cells.

2. The cells are closely fit together by their edges and form a sort of mosaic.

3. The cells frequently have a roughly hexagonal, irregular and wavy outline.

4. Each cell contains a nucleus and cytoplasm.

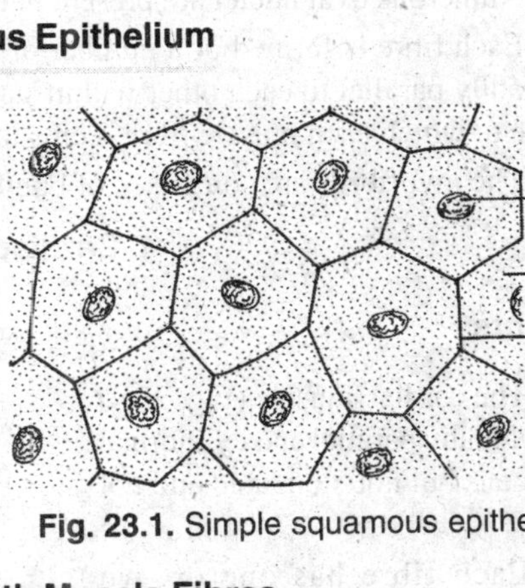

Fig. 23.1. Simple squamous epithelium.

2. Unstriated or Smooth Muscle Fibres

Comments :

1. The unstriated muscles (Fig. 23.2) consist of elongated cell or fibres bound together by delicate connective tissue.

2. Each muscle fibre is an elongated cell, pointed at both the ends and thickened out in the middle containing a nucleus.

3. Each cell is bounded by a thin plasmalemma, the **sarcolemma**.

4. A number of thread-like **myofibrillae** or **sarcostyles** extend lengthwise through the fibres.

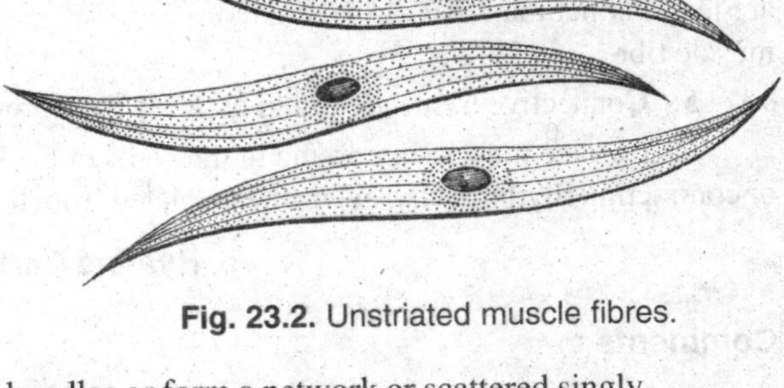

Fig. 23.2. Unstriated muscle fibres.

5. The fibres are either bound together by connective tissue forming small bundles or form a network or scattered singly.

6. Unstriated muscle fibres are found in the wall of the digestive tract, in the ducts of digestive glands and in the urinary bladder.

3. Striated or Skeletal Muscle Fibres

Comments :

1. The striated or voluntary muscle (Fig. 23.3) consists of many fibres or cells which extend parallel to one another but do not branch or anastomose and are bound together in bundles (fasciculi) by connective tissue.

2. Each muscle fibre is elongated and cylindrical and contains cross striations of light and dark colour and both are arranged alternately.

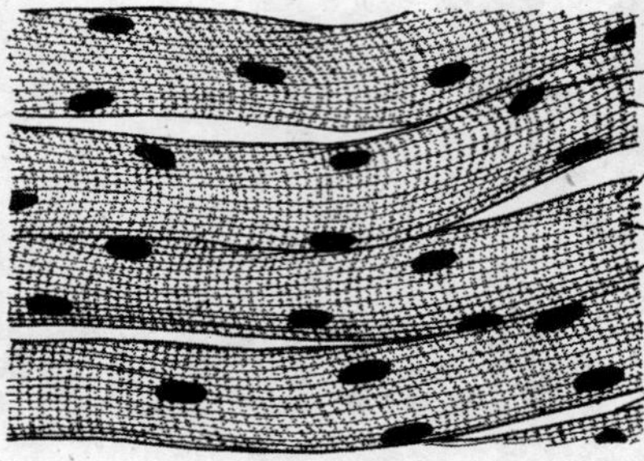

Fig. 23.3. Striated muscle fibres.

3. These striations are known as **light** and **dark bands**.

4. Each fibre is surrounded by a thin sheath called the **sarcolemma**.

5. Numerous oval nuclei are present in each fibre making it a **syncytium**.

6. Each fibre is formed of a number of cross striated **myofibrillae** or **sarcostyles** which run longitudinally parallel to each other within sarcoplasm. Each myofibril has alternate light and dark bands due to which each muscle fibre seems to be striated.

7. The myofibrils are composed of bundles of myosin and actin filaments.

4. Cardiac Muscle Fibres

Comments :

1. The cardiac muscle (Fig. 23.4) consists of many parallel and cylindrical fibres or cells but not elongated and not pointed.

2. Each fibre has one or two nucleus and covered by thin sarcolemma. The fibres branch and anastome.

3. The two connected fibres are separated from each other by a distinct **intercalated disc**.

4. Dark and light alternate bands are less distinct than that of the striated muscle fibres.

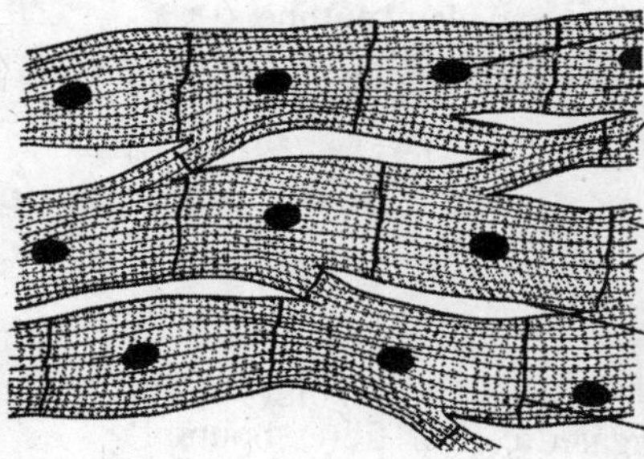

Fig. 23.4. Cardiac muscle fibres.

5. Connective tissue are found between the fibres.

6. Cardiac muscles are found in the walls of the heart of all vertebrates and have the property of contracting rhythmically, even when isolated from the body and has a very long refractory period.

5. Hyaline Cartilage

Comments :

1. Microscopically cartilage (Fig. 23.5) is a clear bluish coloured, translucent, glossy structure.

2. It is externally bounded by a fibrous membrane, the **perichondrium** having numerous blood vessels.

3. **Matrix** or ground substance is called **chondrin** which is produced by **chondrocytes** (mature cartilage cells) or **chondroblasts**.

4. Matrix contains the delicate network of collagen fibrils and groups of chondroblasts.

5. The chondroblasts secrete the chondrin lie in small fluid-filled spaces called **lacunae** which are surrounded by **capsule**.

6. Hyaline cartilage is found in sternal ribs, at the ends of long bones, trachea, bronchi, xiphisternum,larynx, nose and ears.

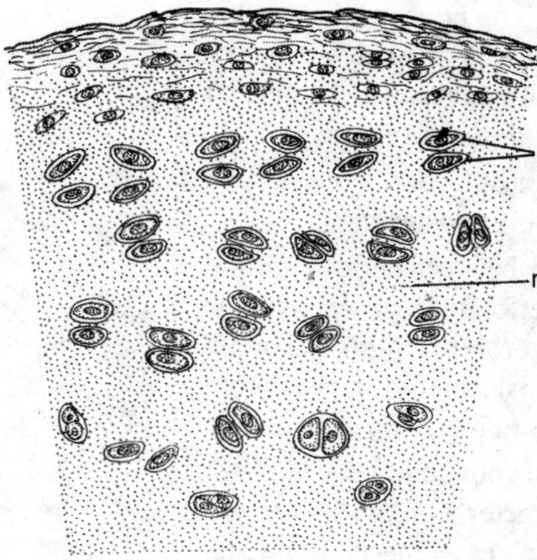

Fig. 23.5. Hyaline cartilage.

6. Elastic Cartilage

Comments :

1. Elastic cartilage (Fig. 23.6) is fundamentally similar to hyaline cartilage.

2. It is yellowish in colour and more flexible and elastic.

3. In its matrix are present numerous **yellow fibres** and rounded **chondrocytes** which lie with in the lacunae. These cells are scattered in the matrix in groups of two or singly.

4. It is found in mammals in pinna of ear, end of nose, in epiglottis, etc.

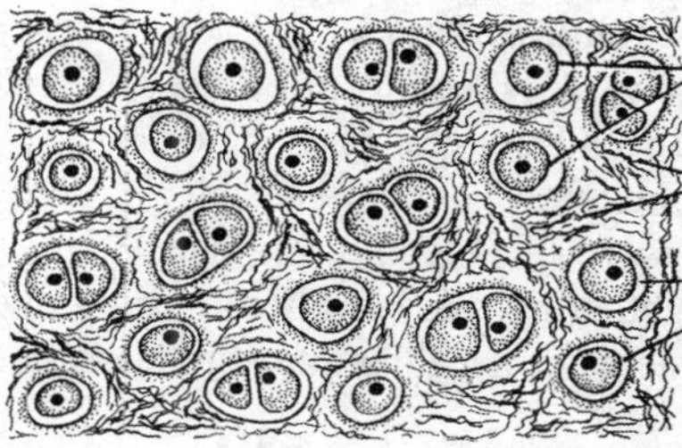

Fig. 23.6. Elastic cartilage

7. Fibro-cartilage

Comments :

1. The matrix of **fibro-cartilage** (Fig. 23.7) contains thick compact bundles of white fibres running longitudinally.

2. **Chondrocytes** are more or less oval and are enclosed within the lacunae.

3. It may be regarded as transitional type between hyaline cartilage and areolar tissue.

4. It is typically found in the inter-vertebral cartilaginous discs in mammals.

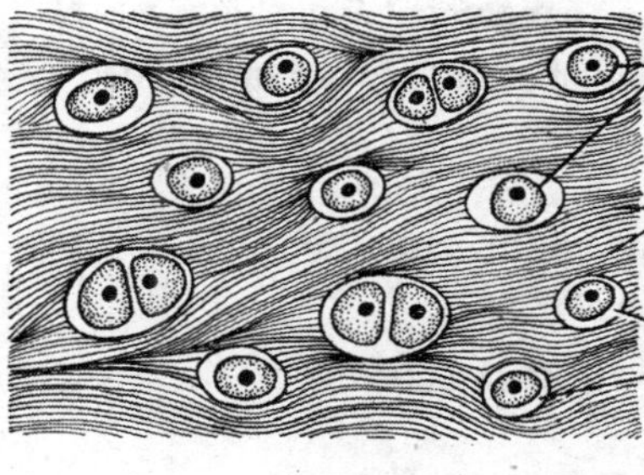

Fig. 23.7. Fibro-cartilage.

8. T.S. of Bone of Mammal

Comments :

Transverse section of **bone** (Fig. 23.8) exhibits the following important characters :

1. Bone is surrounded by thin connective tissue membrane called **periosteum**.

2. In the long bones of mammal is present a central **marrow cavity** containing **yellow marrow** in shafts and **red marrow** at the ends (epiphyses).

3. Innermost layer of bone around marrow cavity is **endosteum**.

4. Matrix between periosteum and endosteum is perforated by a series of fine channels, **Haversian canals** (Clopton Havers an English anatomist) which run almost paralled to long axis of the bone.

5. Haversian canals, (22–110 microns in diameter) contain an artery, a vein, lymphatic vessel, nerve fibres, fat cells, osteoblasts and small amount of areolar connective tissue.

6. Haversian canals are interconnected by **Volkman canals**.

7. **Haversian system :** (*i*) Each **Haversian canal** is surrounded by alternating concentric layers of bone or **lamellae** and **lacunae** having bone cells or **osteocytes**. (*ii*) The lacunae are connected together by fine **canaliculi** through which run processes of osteocytes. (*iii*) In between Haversian systems lie irregular interstitial lamellae and cells.

Fig. 23.8. T.S. of bone.

9. L.S. of Bone of Mammal

Comments :

Longitudinal section of bone (Fig. 23.9) reveals the following details :

1. **Haversian canals** are transversely connected with each other by **connecting canals** called **Volkman canals** containing blood vessels.

2. Volkman canals are not surrounded by concentric lamellae.

3. Osteocytes, bone lacunae, bone canaliculi and bone lamellae are arranged lengthwise.

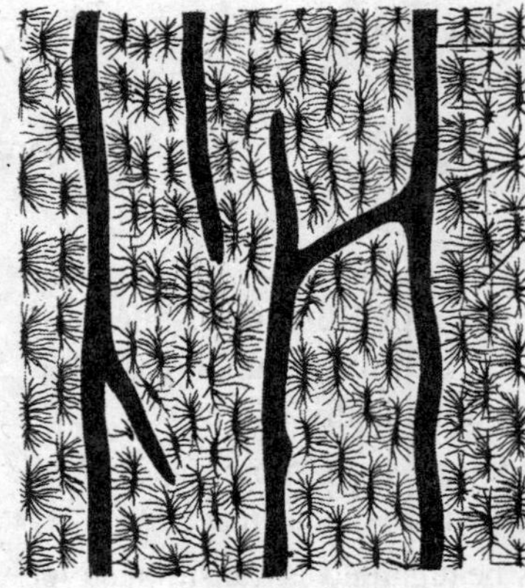

Fig. 23.9. L.S. of bone.

10. Blood of Frog

Comments :

The blood film of frog (Fig. 23.10) on the slide exhibits the following important structures :

1. It shows numerous **erythrocytes** (R.B.Cs) and **leucocytes** (W.B.Cs) of different shapes and sizes suspended in the plasma of blood.

2. The **erythrocytes** (R.B.Cs) are oval, biconvex and have a nucleus.

3. The **leucocytes** (W.B.Cs) are colourless, nucleated and are phagocytic in nature. Two main types of leucocytes are seen, *viz.*, (*i*) **polymorphonuclear leucocytes** are with lobulated nucleus and cytoplasm with granules and, hence are called **granulocytes**. (*ii*) **mononuclear leucocytes** lack cytoplasmic granules. **Lymphocytes** have large spherical nucleus and **monocytes** are larger with oval or kidney-shaped nucleus, (*iii*) **blood platelets** or **thrombocytes** are spindle-shaped and nucleated.

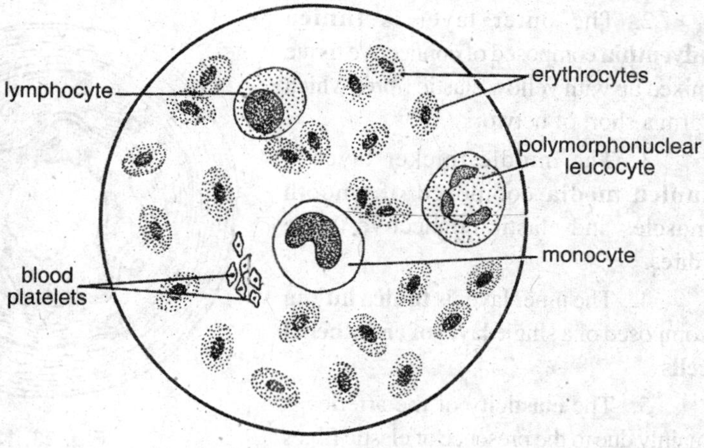

Fig. 23.10. Blood of frog.

11. Blood of Man (Mammal)

Comments :

The film of mammalian blood (Fig. 23.11) on the slide exhibits the following structures :

1. It shows numerous blood corpuscles of different shapes and sizes suspended in the **plasma**.

2. **Erythrocytes** (R.B.Cs) are round, non-nucleated and biconcave and in side view they appear biconvex. The size of erythrocytes is 7.74 to 8 μ in diameter.

3. **Leucocytes** (W.B.Cs) are colourless, nucleated and show amoeboid movement. Three types of leucocytes are seen in the slide.

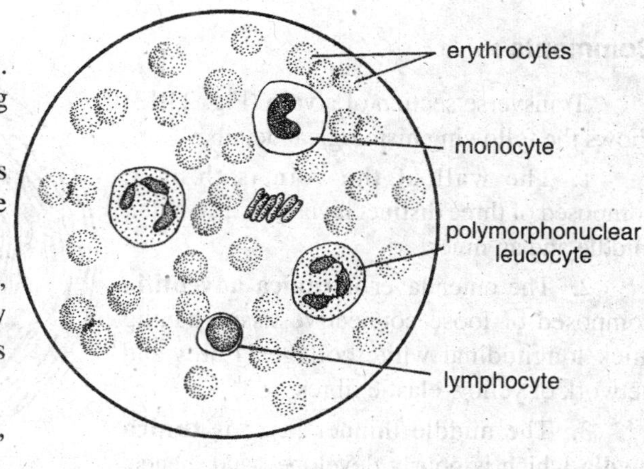

Fig. 23.11. Blood of man.

(*i*) **Polymorphonuclear leucocyte** or **granulocytes :** Their nucleus is divided into a number of segments (3–5) connected with one another by fine thread and cytoplasm is granular and stain well with most dyes (acid and basic).

(*ii*) **Agranulocytes** or mononuclear leucocytes lack granules. **Monocytes** are largest leucocytes and possess a horse-shoe-shaped or kidney-shaped nucleus.

(*iii*) **Lymphocytes** are small with large spherical nucleus and little cytoplasm.

4. **Platelets** are very small and non-nucleated.

5. The number of erythrocytes in normal adult man and woman is about four and half million or five million in 1 cubic millimetre of blood.

12. T.S. of Artery

Comments :

Transverse section of an **artery** (Fig. 13.12) shows the following histological details :

1. The wall of an artery is composed of three distinct layers or coats, an outer, middle and an inner.

2. The outer layer is **tunica adventitia** composed of connective tissue mixed up with yellow elastic fibres which form a short of network.

3. The middle thicker layer is **tunica media** composed of smooth muscles and elastic connective tissue fibres.

4. The inner layer is **tunica intima** composed of a single layer of **endothelial cells**.

5. The elasticity of the arteries is mainly due to the presence of elastic fibres in its walls.

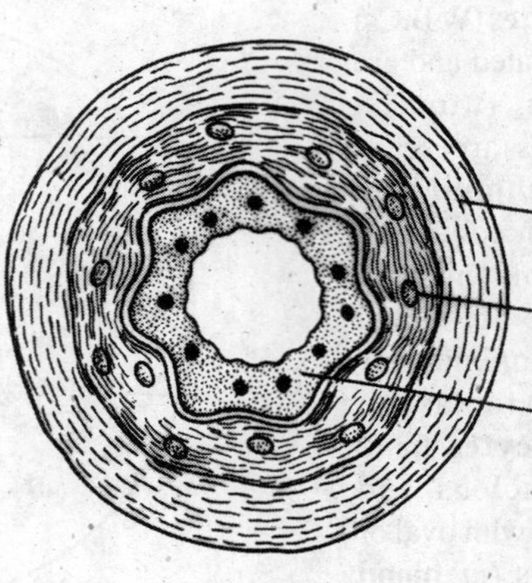

Fig. 23. 12. T.S. of an artery.

6. The middle layer contains sympathetic nerves that controls the flow of the blood.

13. T.S. of Vein

Comments :

Transverse section of a **vein** (Fig. 23.13) shows the following histological details :

1. The wall of the vein is thinner, composed of three distinct layers, *viz.*, an outer, middle and an inner.

2. The outer layer is **tunica adventitia** composed of loose connective tissue having thick longitudinal white collagen fibres and network of yellow elastic fibres.

3. The middle thinner layer is **tunica media** which is poorly developed and, hence, the veins are thin-walled. It consists of circular smooth muscle fibres which are separated by longitudinal fibres.

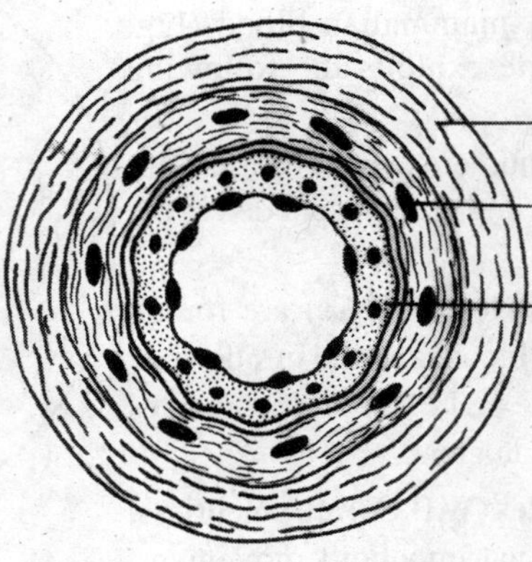

Fig. 23.13. T.S. of a vein

4. The inner layer is **tunica intima** composed of a single layer of squamous endothelial cells.

5. The wall of the veins is always softer, less elastic and thin walled in comparison of arteries.

6. The middle layer contains branches of sympathetic nerve which controls the flow of the blood.

14. Section of Taste Bud

Comments :

The section of **taste bud** (Fig. 13.14) shows the following structures :

1. Each taste bud is spherical or flask-shaped structure embedded in the stratified epithelium lining the buccal cavity or tongue.

2. It is composed of cluster of cells which are of two types namely: (i) **receptor** or **gustatory cells** or **sensory cells** and (ii) **supporting cells**.

3. **Gustatory** or **sensory cells** are ensheathed in supporting cells and are long and narrow and their free ends bear short, hair-like processes, while their basal ends bear nerve fibres.

4. **Supporting cells** are deeper in position and variable in shape and lie in between the gustatory or sensory cells to provide them support.

5. Taste buds are innervated by facial, glossopharyngeal and vagus nerves.

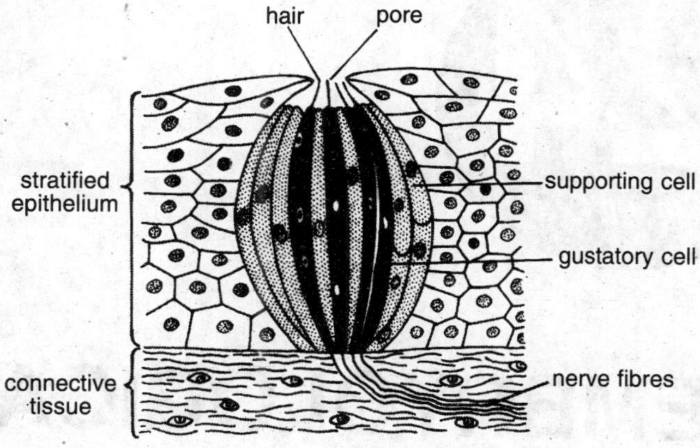

Fig. 23. 14. Section of a taste bud.

15. Medullated Nerve Fibres

Comments :

1. The medullated nerve fibres (Fig. 23.15) are white coloured due to the presence of myelin sheath and found in the cranial and spinal nerves of vertebrates.

2. Each fibre is composed of a nerve cell or **cyton** that gives off a number of short and branched dendrons and a long central **axon** surrounded by a **medullary sheath** or **myelin sheath**.

3. The medullary sheath is broken at definite intervals.

4. The axon and medullary sheath are surrounded by another sheath called **neurilemma** which is constricted at broken points of medullary sheath to such an extent that it touches the axon and forms **Ranvier's nodes**.

5. Nuclei are found in neurilemma.

6. The portion between adjacent Ranvier's nodes is known as neurilemma.

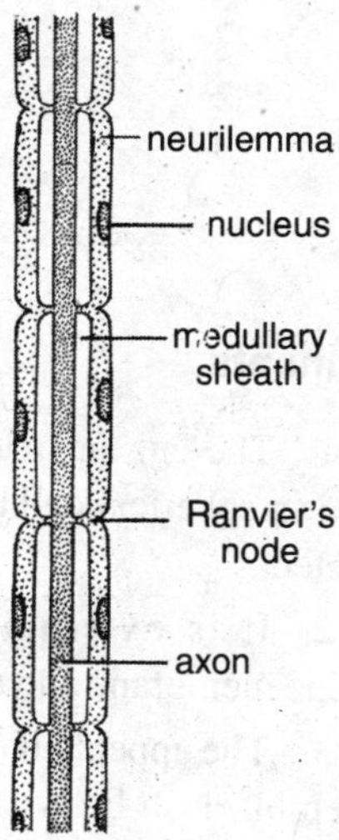

Fig. 23. 15. Medullated nerve fibre.

16. Non-medullated Nerve Fibres

Comments :

1. The non-medullated nerve fibres (Fig. 23.16) are grey in colour and found in sympathetic nerves.

2. Each fibre is composed of a nerve cell or **cyton** that gives off a number of short and branched dendrons and central long **axon** surrounded by the **neurilemma**.

3. The **neurilemma** is wavy sheath and contains nuclei.

4. The medullary sheath, Ranvier's nodes and internodes are entirely absent.

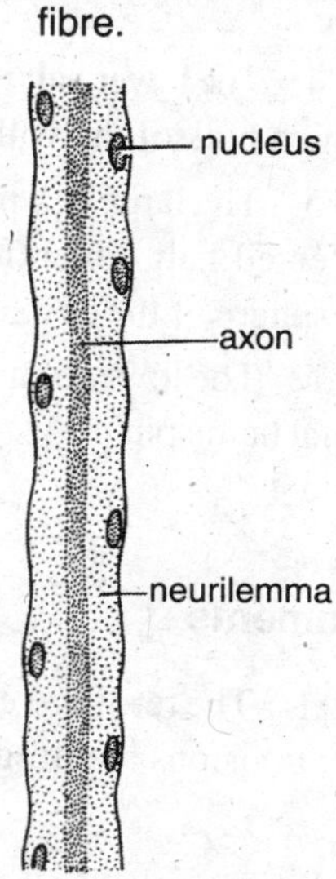

Fig. 23. 16. Non-medullated nerve fibre.

24

Chapter

EMBRYOLOGICAL SLIDES

DEVELOPMENT OF FROG

1. Unfertilized Egg of Frog

Comments :

1. The egg or ovum of frog (Fig. 24.1) is a spherical cell about 1.6 mm in diameter.

2. It is externally covered by vitelline membrane and albumen layer.

3. The upper half of egg cytoplasm is deep blackish brown in colour due to the presence of a superficial layer of pigment.

4. The lower half is lighter in colour in which bulk of the yolk is aggregated.

5. The large egg nucleus lies in the pigmented half which forms the animal hemisphere of the egg.

6. The lower lighter area forms the vegetal hemisphere.

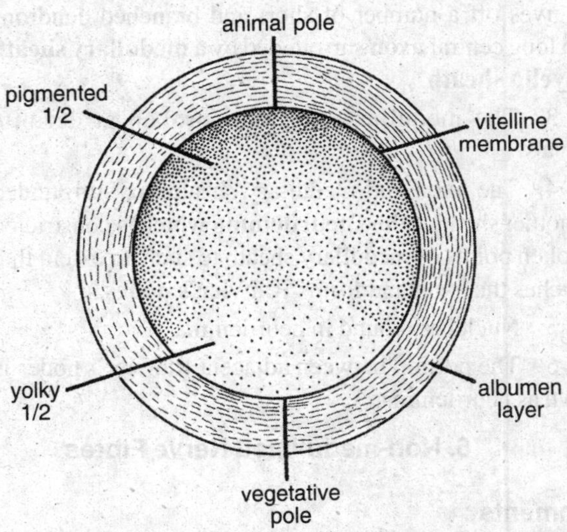

Fig. 24.1. Unfertilised egg of frog.

2. L.S. of Early Blastula of Frog

Comments :

1. The fertilised egg undergoes two vertical and third horizontal cleavage to form early blastula. Later divisions are irregular to form late blastula.

2. The section of blastula (Fig. 24.2) shows that it contains a cavity in the centre known as **blastocoel** surrounded by the blastomeres.

3. The blastomeres are distinguished into two types, *i.e.*, upper smaller pigmented **micromeres** and lower larger yolk-filled **megameres**.

4. Micromeres lie in the animal hemisphere and megameres are located in the vegetal hemisphere.

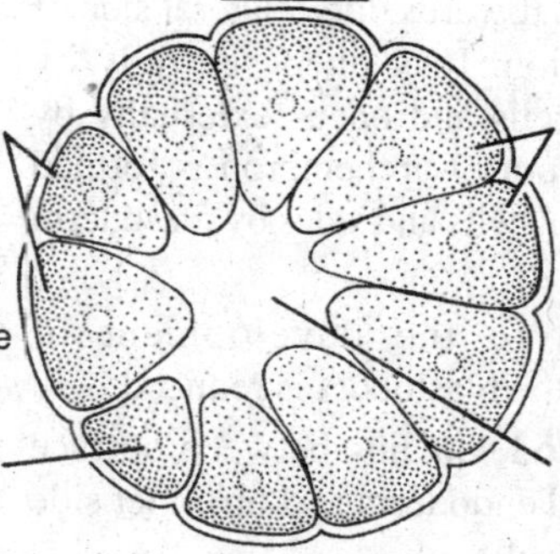

Fig. 24.2. L.S. of early blastula (8-celled) of frog.

3. V.S. of Early Gastrula of Frog

Comments :

The vertical section of **early gastrula** (Fig. 24.3) shows the following important structures :

1. Gastrulation is a mass movement of various cells or presumptive areas of blastula to form a two or three-layered embryo or gastrula.

2. Gastrulation takes place by **epiboly** (Mass movement of presumptive ectoderm cells of blastula covering the outer surface of embryo) and **emboly** (Mass movement of future endoderm and mesoderm cells of blastula in the interior of embryo).

3. **Notochord** and **neural plate** formed by ectoderm cells lie in the dorsal and dorso-lateral position respectively.

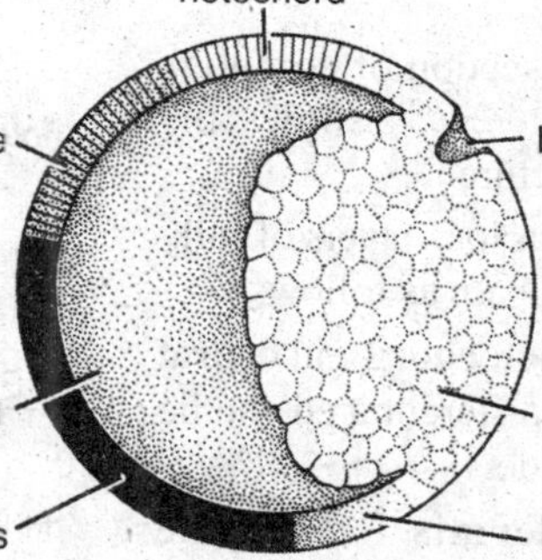

Fig. 24.3. V.S. of early gastrula of frog.

4. **Epidermis** is differentiated in the ventro-lateral position in animal pole.

5. **Blastocoel** is large central cavity within the gastrula and surrounded by a layer of cells.

6. **Blastopore** is a slight depression at grey crescent guarded by dorsal and ventral lips towards vegetative pole.

7. The **endoderm** is well developed and formed by cells of yolky vegetal pole.

8. A small **mesoderm** is also seen on the ventral side towards the posterior pole.

4. V.S. of Late Gastrula of Frog

Comments :

Vertical section of **late gastrula** of **frog** (Fig. 24.4) shows the following important structures:

1. Ecotodermal **notochord** and **neural plate** cells lie in the dorso-lateral position and **epidermis** is in the ventral portion towards the anterior side.

2. **Archenteron** is a newly formed cavity located towards dorso-lateral side of anterior portion. It is lined ventrally by endodermal cells and opens externally by blastopore. Roof of archenteron is formed by notochordal cells and sides by mesoderm cells.

3. Due to the development of archenteron, the blastocoel is reduced.

4. **Mesoderm** lies between surface epidermis and endoderm and on either side of notochord cells.

5. Gastrula has three germinal layers, *viz.*, **epidermis** (ectoderm), **mesoderm** and **endoderm** from which various organs are derived.

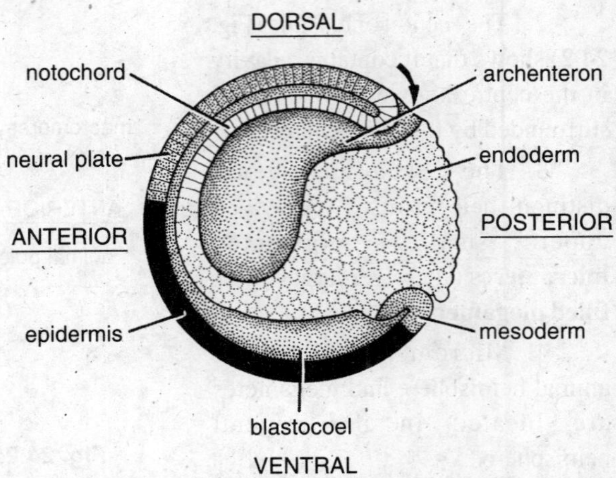

Fig. 24.4. V.S. of late gastrula of frog.

5. V.S. of Complete Gastrula

Comments :

Vertical section of **complete gastrula** (Fig. 24.5) shows the following structure :

1. The various layers of fully formed gastrula are as follows :

2. The **notochord** and **neural plate** are well differentiated and lie at the dorsal side.

3. **Epidermis** lies on the antero-ventral side.

4. **Archenteron** forms an extensive cavity and lies towards the dorsal side in the animal half and opens externally through blastopore. The **blastocoel** is obliterated.

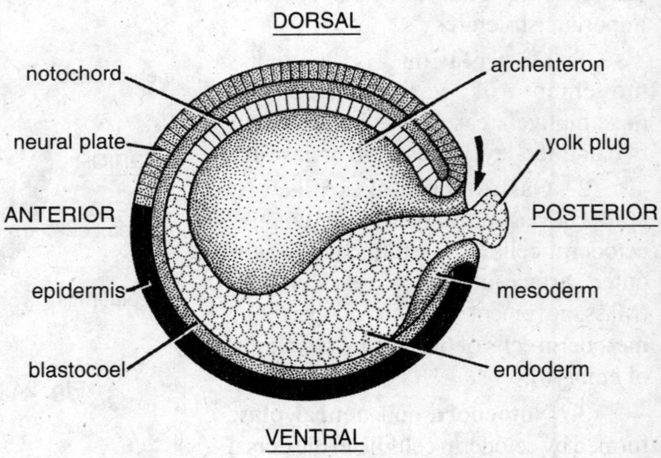

Fig. 24.5. V.S. of complete gastrula of frog.

5. Due to contraction of blastoporal lips some endodermal cells protrude through the blastopore as the **yolk plug**.

6. The **mesoderm** increases and lies in between epidermis and yolky endoderm and on either side of notochord.

6. Tadpole Larva of Frog

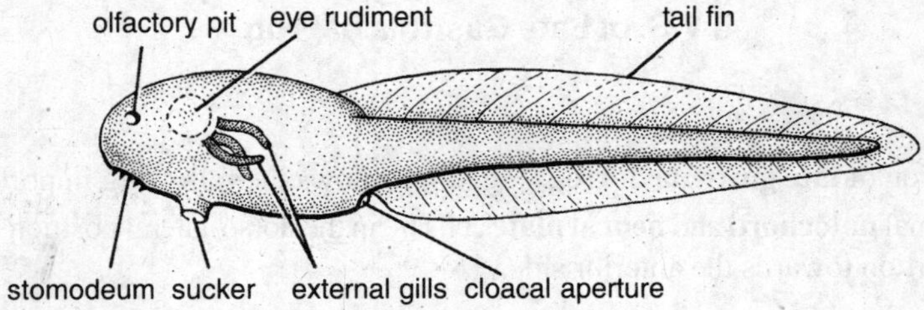

Fig. 24.6. Tadpole larva of frog.

Comments :

1. The chick embryo of 43 hours of incubation (Fig. 24.17) exhibits the **cranial flexion** and **torsion**.

2. The cephalic region of the embryo is twisted in such a manner that the left side comes to lie over the yolk and right side away from the yolk. It is cranial flexion.

3. Brain region is differentiated into **prosencephalon**, **mesencephalon**, and hindbrain constricted into **metencephalon** and **myelencephalon**.

4. **Auditory pits** are also appeared in the hindbrain.

5. **Heart** differentiated into **ventricular region**, **atrial region** and **sinus region**. **Truncus arteriosus** also develops.

6. **Omphalomesenteric veins**, **omphalomesenteric arteries** and **extra-embryonic vascular plexus** are well developed.

7. **19 pairs of somites** are formed.

8. **Primitive streak** diminishes to a small primitive plate.

9. **Body folds** (head and lateral) are clearly visible.

18. Chick Embryo-55 Hours (Whole mount)

Comments :

1. In chick embryo of 55 hours of incubation (Fig. 24.18), the head is free from the yolk .

2. The anterior half of the embryo bends downward (torsion) along with further cranial flexion.

3. The beginning of the caudal fold marks off the tail region of the embryo from the extra-embryonic membranes.

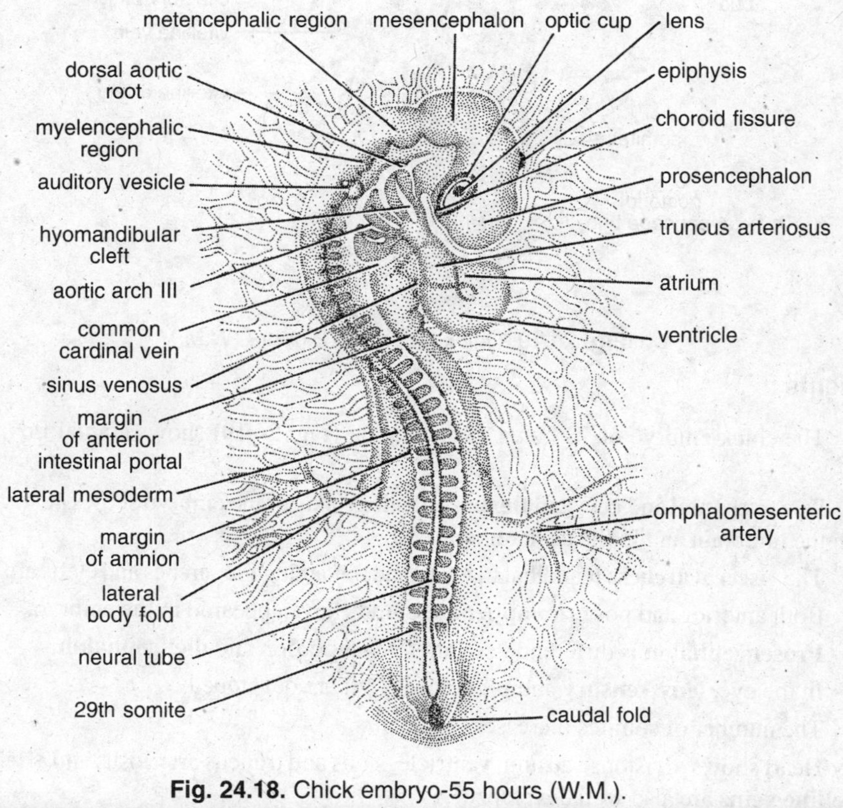

Fig. 24.18. Chick embryo-55 hours (W.M.).

4. **Optic cup** and **lens** and **auditory vesicles** are distinct.
5. **Epiphysis** is the evagination of dorsal wall of diencephaton.
6. **Heart** is differentiated into **atrium**, **ventricle**, **truncus arteriosus** and **sinus venosus**.
7. **Branchial arches**, **branchial clefts**, **dorsal aorta** and **aortic arches** have appeared.
8. The number of somites increases to 29 pairs.
9. Besides the above structures, **body folds** (head, lateral and caudal), **common cardinal vein**, and **omphalomesenteric artery** are also present.

19. Chick Embryo-72 Hours (Whole mount)

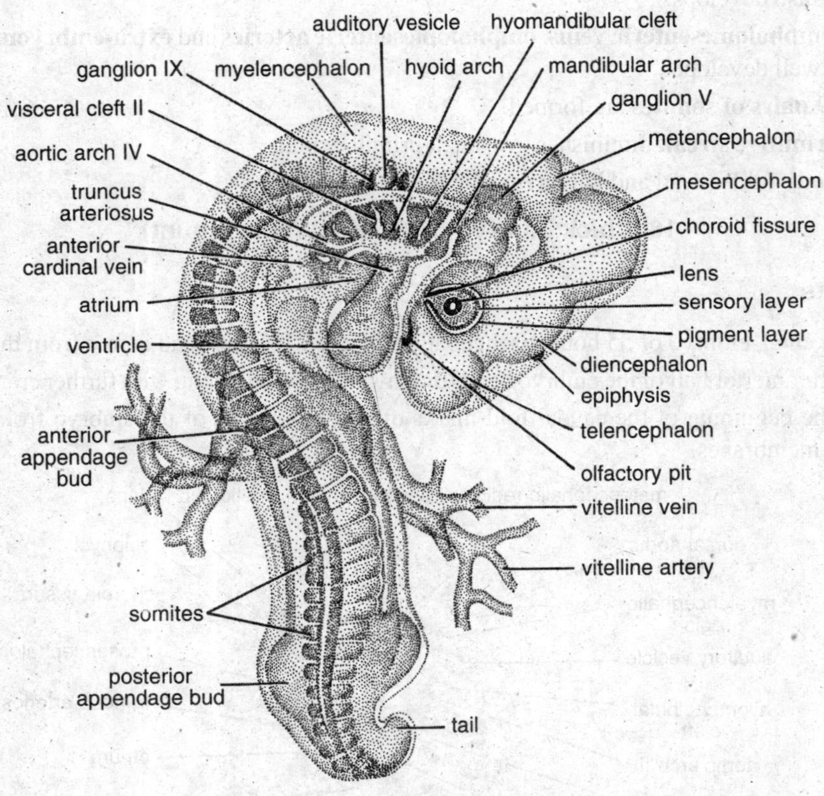

Fig. 24.19. Chick embryo -72 hours (W.M.).

Comments :

1. The chick embryo of 72 hours of incubation (Fig. 24.19) shows cranial flexion and body torsion.
2. Due to **cranial** and **cervical flexures**, the long axis of the embryo shows nearly right-angled bends in the midbrain and in the neck region.
3. The **visceral arches** : mandibular, hyoid and remaining four arches and clefts are conspicuous.
4. Both anterior and posterior **appendage buds** have appeared in the embryo.
5. **Prosencephalon** is differentiated into **telencephalon** and **diencephalon**.
6. In the eye, **lens**, **sensory** and **pigment layers** are developed.
7. The number of somites increases to 36 pairs.
8. Heart shows divisions : atrium, ventricle, sinus and truncus arteriosus and **vitelline arteries** and **vitelline veins** are also well developed.

9. **Nasal pits** appear as shallow depressions in the ectoderm of the rostral part of the head which overhangs the mouth region.

20. Chick Embryo-96 Hours (Whole mount)

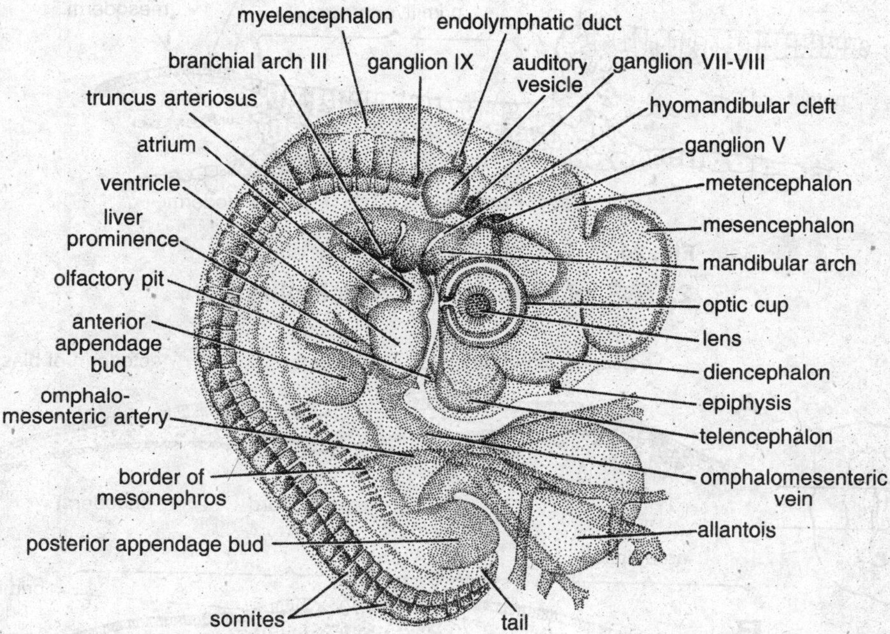

Fig. 24.20. Chick embryo-96 hours (W.M.).

Comments :

1. In the chick embryo of 96 hours of incubation, (Fig. 24.20), the entire body has been bent through 90 degrees and the embryo lies with its left side on the yolk.

2. At the end of 96 hours the body folds have been completed so that the embryo remains attached to the yolk only by a slender stalk.

3. The progressive increase in the cranial, cervical, dorsal and caudal flexures results in the bending of the embryo on itself so that it becomes C-shaped and its head and tail lie close together.

4. Five part of brain, optic cup with lens and auditory vesicle are clearly visible.

5. **Endolymphatic duct** arises from the auditory vesicle.

6. Divisions of heart and visceral arches along with **omphalomesenteric artery** and **omphalomesenteric vein** are clearly visible.

7. **Appendage buds** increase rapidly in size and become elongated.

8. The number of somites increases to 41 pairs.

9. **Allantois** attached with hindgut has also appeared.

21. L.S. of 17-Hours of Chick Embryo

Comments :

1. The section (Fig. 24.21) is just after the formation of primitive streak and migration of cells forming the notochord.

2. Primitive streak, primitive groove, Hensen's node and primitive pit are clearly seen.

3. Chorda-mesoderm cells from Hensen's node invaginate and migrate anteriorly to form notochord in between ectoderm and endoderm.

4. Mesoderm also extends laterally on either side of notochord and primitive streak.

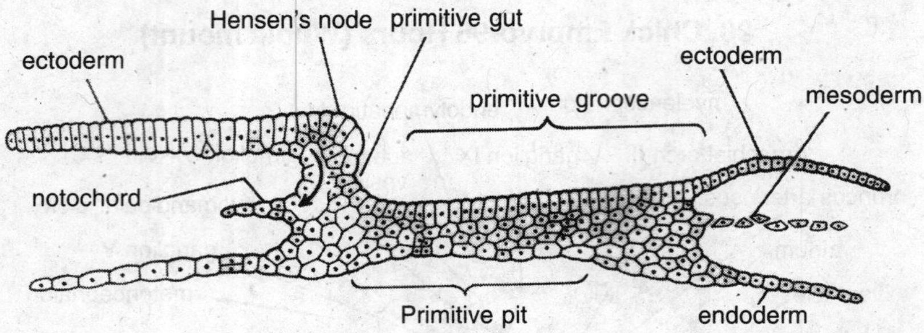

Fig. 24.21. L.S. of 17-hours of chick embryo

22. T.S. of 18-hours of Chick Embryo

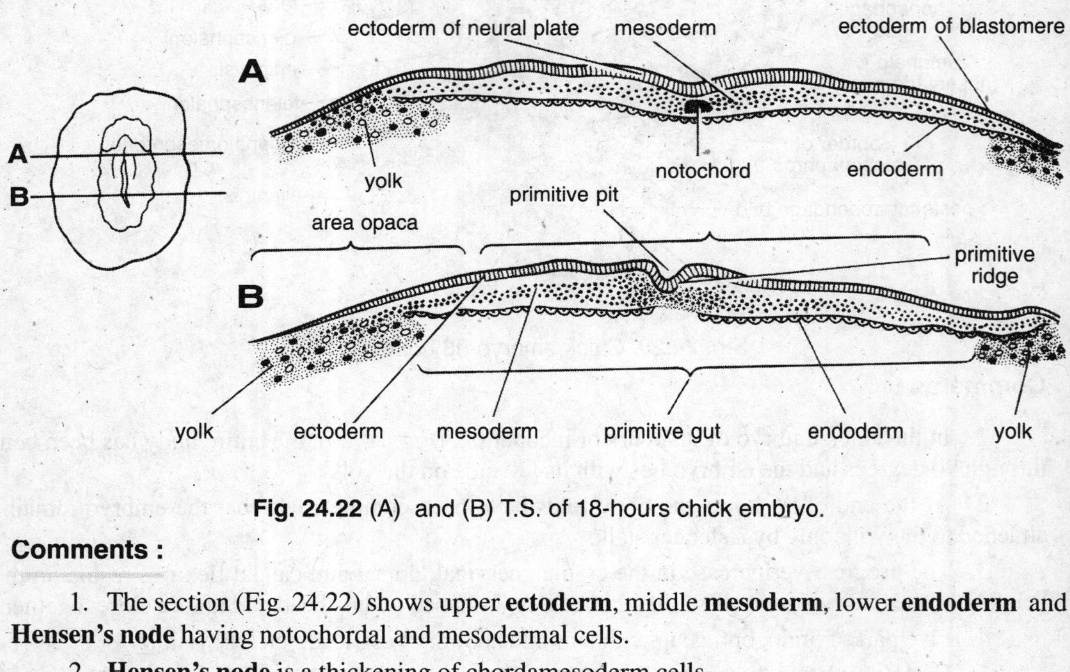

Fig. 24.22 (A) and (B) T.S. of 18-hours chick embryo.

Comments :

1. The section (Fig. 24.22) shows upper **ectoderm**, middle **mesoderm**, lower **endoderm** and **Hensen's node** having notochordal and mesodermal cells.
2. **Hensen's node** is a thickening of chordamesoderm cells.
3. Beneath endoderm over yolk lies primitive gut.

23. T.S. of 24-Hours Chick Embryo

(A) T.S. through Head Region

Comments :

1. The **neural plate** is folded forming neural groove and neural folds.
2. **Notochord** lies just below the neural groove.
3. Ectoderm of head, mesenchyme (loose mesoderm cells), ectoderm of blastoderm, mesoderm, endoderm of blastoderm, foregut and yolk are also seen in the section.

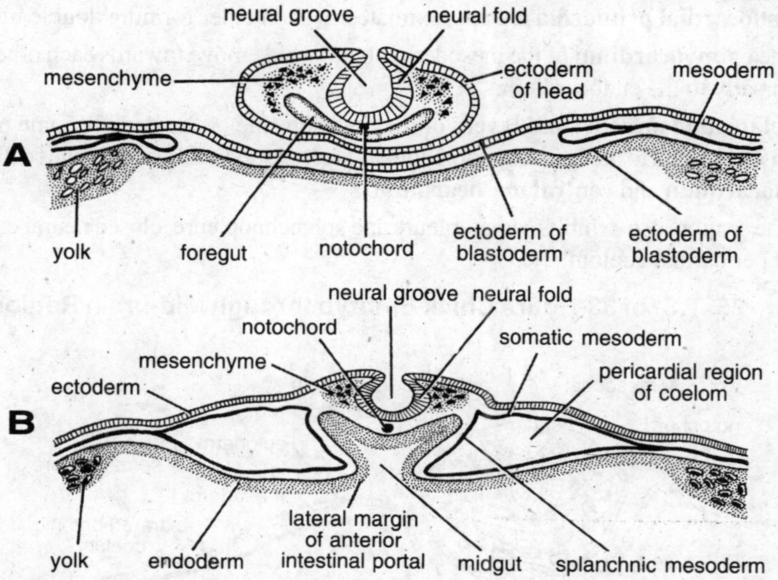

Fig. 24.23. T.S. of 24-hours chick embryo. A –T.S. through head; B–T.S. through mid-body

(B) T.S. through Mid Body

Comments :

1. The margins of the **neural folds** are less closely approximated to each other.
2. **Notochord** lies below the neural groove.
3. **Mesoderm** is differentiated into **somatic** and **splanchnic mesoderm** in between ectoderm and endoderm.
4. Endoderm lies over the yolk in the area opaca and is separate in area pellucida and folded to form the foregut.
5. Anterior intestinal portal is opening of foregut and posterior to it is the midgut.

24. T.S. of 28-Hours Chick Embryo through Pericardial Region

Comments :

1. This section (Fig. 24. 24) shows formation of heart from mesoderm.

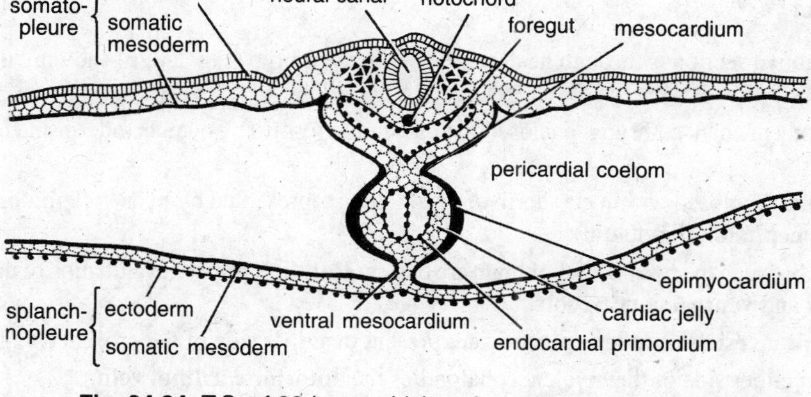

Fig. 24.24. T.S. of 28-hours chick embryo through pericardial region.

2. **Endocardial primordia** are approximated to each other forming double tubes.

3. The **epimyocardium** of the mesoderm of either side move towards each other first ventrally and then dorsally to the endocardium.

4. **Splanchnic mesodermal layers** of the opposite side of the body become opposed to each other, dorsal and ventral to the heart and form double-layered supporting membranes called respectively, **dorsal mesocardium** and **ventral mesocardium**.

5. The section also exhibits somatopleure and splanchnopleure, closed neural canal, notochord, foregut and pericardial coelom.

25. T.S. of 33-Hours Chick Embryo through Mid-brain Region

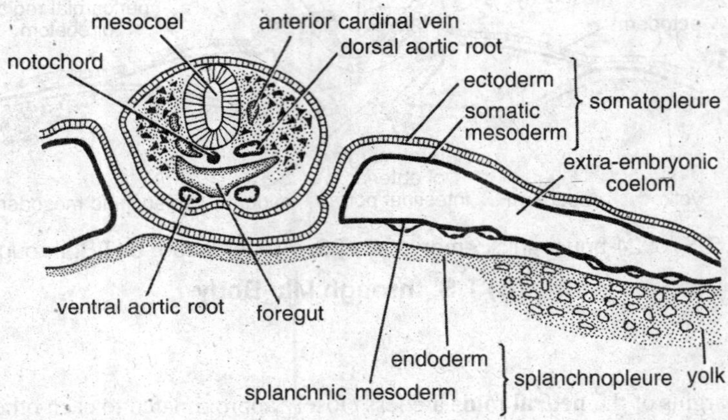

Fig. 24.25. T.S. of 33-hours chick embryo through mid-brain region.

Comments :

This section (Fig. 24.25) shows the following structures :

1. **Somatopleure** comprising of outer ectoderm and inner somatic **mesoderm**.
2. **Splanchnopleure** composed of inner endoderm and outer splanchnic mesoderm.
3. In the median line within head of embryo **mesocoel of mid brain**, **anterior cardinal veins**, **dorsal aortic roots**, **notochord**, **foregut** and **ventral aortic roots** are present.
4. **Extra-embryonic coelom** lies between the somatopleure and splanchnopleure.
5. **Yolk** is present on either side below the splanchnopleure.

26. T.S. of 48-hours Chick Embryo through Brain and Optic Vesicles

Comments :

The above section is through head region of chick embryo (Fig. 24.26) showing the following structures :

1. Body of chick embryo is enclosed within amnion formed by outer ectoderm and inner somatic mesoderm.
2. Brain enclosed within ectoderm of head fold is represented by **prosencephalon** (forebrain) and **myelencephalon** (hindbrain).
3. In between the two cavities of brain are present **Rathke's pouch** (evagination of diencephalon) **gut, dorsal** and **ventral aortic roots**.
4. **Optic vesicles** and **lens placode** are present on either side of the prosencephalon.
5. On either side of the myelencephalon lies the **anterior cardinal vein**.

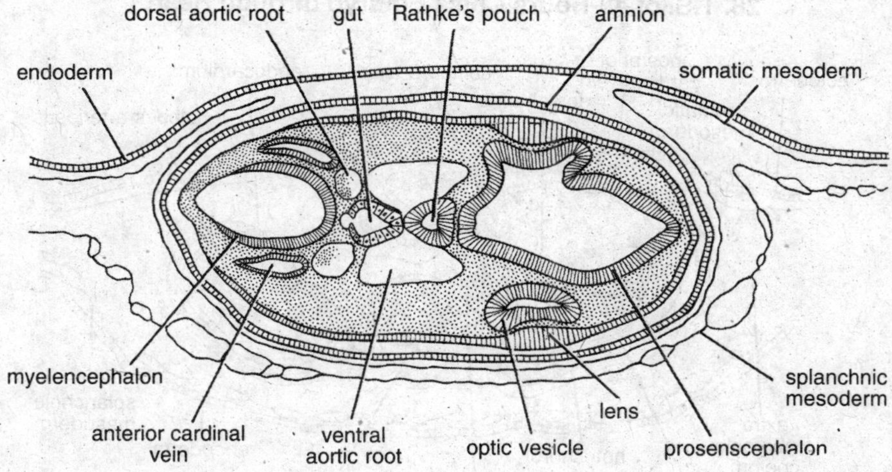

Fig. 24.26. T.S. of 48-hours chick embryo through brain and optic vesicles.

27. T.S. of 48-Hours Chick Embryo through Auditory Vesicles

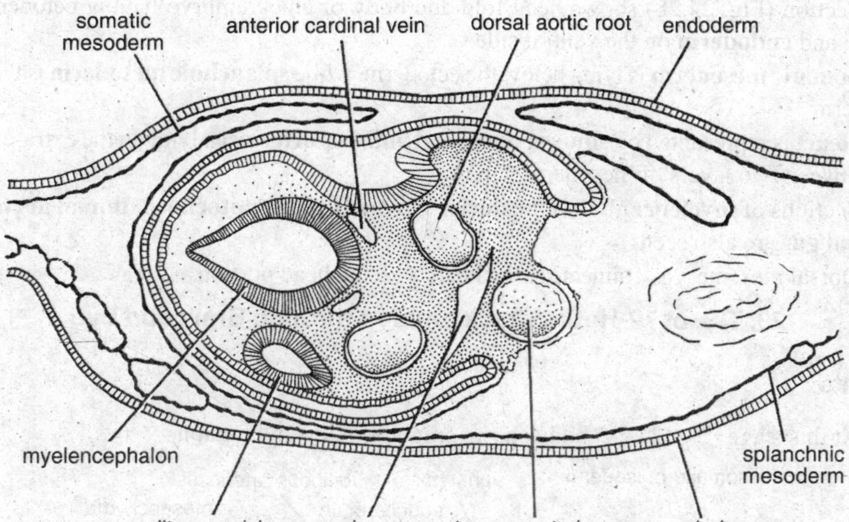

Fig. 24.27. T.S. of 48-hours chick embryo through auditory vesicles.

Comments :

1. Head fold of embryo is covered by ectoderm. Beneath it lies body **ectoderm** on the dorsal side and **endoderm** on the ventral side.

2. Below the body ectoderm is present **somatic mesoderm** and **splanchanic mesoderm** lies above the endoderm.

3. Head part of the embryo is enclosed within the amnion.

4. Brain is represented by **myelencephalon** showing **auditory vesicle** in the form of small cavity, one on either side of the myelencephalon.

5. Sections of anterior cardinal veins, dorsal aortic roots, pharynx and truncus arteriosus are also seen.

28. T.S. of 48-Hours Chick Embryo through Heart

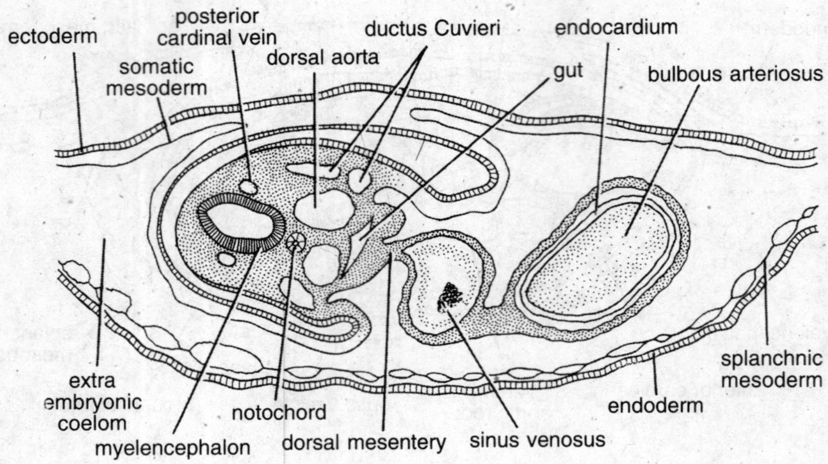

Fig. 24.28. T.S. of 48-hours chick embryo through heart.

Comments :

1. Section (Fig. 24.28) shows head fold and body of chick embryo having ectoderm on the dorsal side and **endoderm** on the ventral side.
2. **Somatic mesoderm** is lying below the ectoderm, while **splanchnic mesoderm** is lying above the endoderm.
3. **Heart** is represented by **sinus venosus** and **bulbous arteriosus** lying in the extra-embryonic coelom between two layers of mesoderm.
4. Sections of **myelencephalon**, **posterior cardinal veins**, **notochord**, **dorsal aorta**, **ductus Cuvieri** and **gut** are also seen.
5. Dorsal mesentery is connecting the heart with the head portion.

29. T.S. of 72-Hours Chick Embryo through Brain and Eye

Comments :

1. At this stage (Fig. 24.29) body shows cranial flexion and torsion.

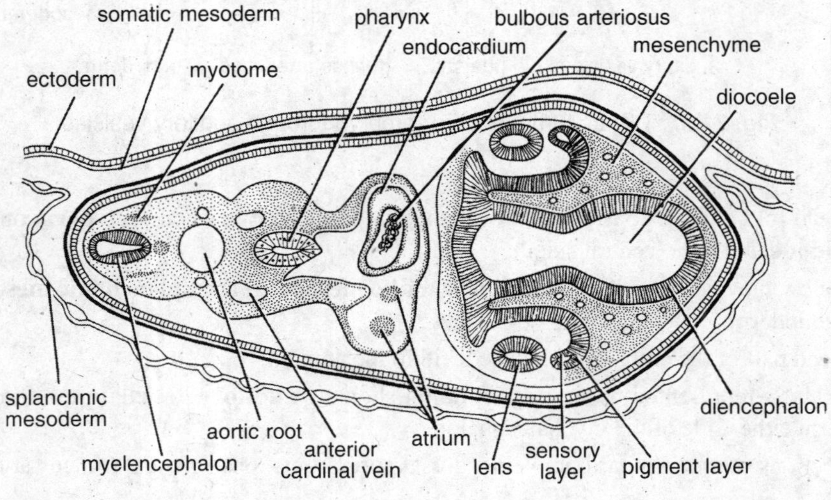

Fig. 24.29. T.S. of 72-hours chick embryo through brain and eye.

2. Head and body are enclosed within amnion (outer ectoderm and inner somatic mesoderm) and chorion (inner splanchnic mesoderm and outer endoderm).

3. Brain is represented by a large cavity **diencephalon** and a small cavity **myelencephalon**.

4. On either side of the prosencephalon is present an optic vesicle differentiated into **sensory** and **pigment layers** and a **lens**.

5. Sections of heart (**atrium** and **bulbous arteriosus**), **pharynx**, **anterior cardinal veins** and **aortic roots** are also present behind the brain.

30. T.S. of 72-hours Chick Embryo through Pharynx

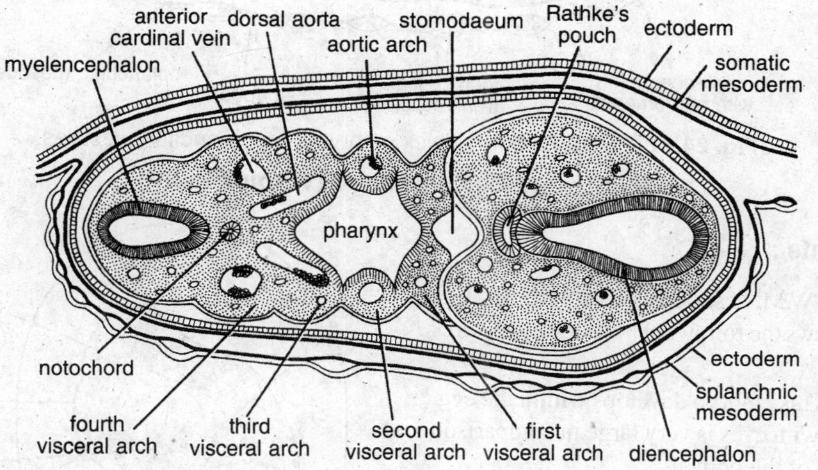

Fig. 24.30. T.S. of 72-hours chick embryo through pharynx.

Comments :

1. Body of chick embryo is enclosed within amnion and chorion.

2. Besides these ectoderm forms the outer layer of head and body of embryo.

3. Brain is represented by a large cavity the **diencephalon** and a small cavity the **myelencephalon**.

4. In between the diencephalon and myelencephalon in the head are present the sections of **Rathke's pouch** and **stomodaeum** and behind head lies **pharynx**, **dorsal** and **ventral aorta, anterior cardinal veins**, first, second, third and fourth **visceral arches** and **notochord** just below the brain.

31. T.S. of 72-Hours Chick Embryo through Auditory Vesicles

Comments :

1. Body of embryo is enclosed within amnion (ectoderm and somatic mesoderm) and chorion (splanchnic mesoderm and endoderm).

2. Head is also covered by ectoderm.

3. Brain is represented by **diencephalon** and **myelencephalon** with their cavities diocoel and IV ventricle.

4. **Auditory vesicles** lie on either side of the myelencephalon which are its evaginations.

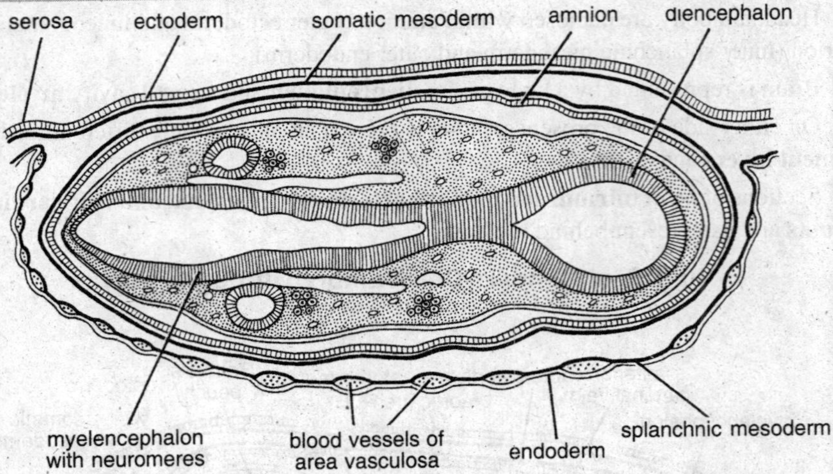

Fig. 24.31. T.S. of 72-hours chick embryo through auditory vesicles.

32. Six-day Chick Embryo (Whole mount)

Comments :

The W.M. of **six-day chick embryo** (Fig. 24.32) shows the following noteworthy features :

1. This embryo develops within the egg. The head with eyes is very large in comparison to the rest of the body.
2. The **ear** also begins to appear.
3. The beak begins to appear near the lower edge of the head.
4. Neck and trunk are also distinguishable.
5. Feather buds start appearing.
6. Developing toes are also marked out on the feet.

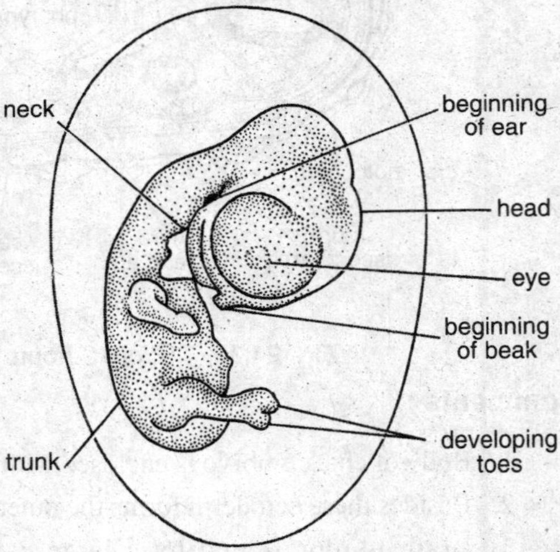

Fig. 24.32. Six-day chick embryo (whole mount).

33. Eight-day Chick Embryo (Whole mount)

Comments :

The W.M. of **eight-day chick embryo** (Fig. 24.33) exhibits the following important features :

1. The size of head reduces to some extent in comparison to the six-day chick embryo.
2. The developing beak is now distinct.
3. The eye is large and slightly more developed.
4. The ear-opening is also quite conspicuous.

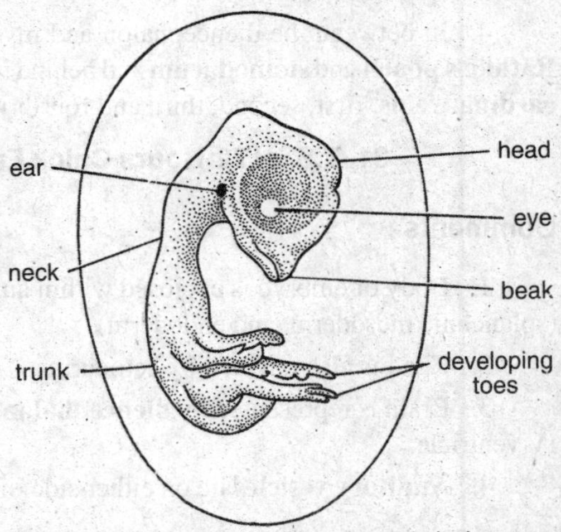

Fig. 24.33. Eight-day chick embryo (whole mount).

5. The feather buds become quite conspicuous.
6. The toes are becoming more clear.

34. Twelve-day chick embryo

Comments :

The W.M. of **twelve-day chick embryo** (Fig. 24. 34) shows the following important features :

1. Reduction in the size of the head takes place as compared to the earlier stage.
2. Eye-lids start developing to cover the eyes.
3. The developing beak shows the upper and lower jaws quite distinct.
4. The feather-buds become very prominent.
5. The legs with clawed toes are more or less formed. Forelimbs developing into wings.
6. The elongation in the size of feather buds gives the embryo a more bird-like appearance.

35. Eighteen-day Chick Embryo (Whole mount)

Comments :

The W.M. of **eighteen-day chick embryo** (Fig. 24.35) exihibits the following important features :

1 Young bird acquires features of a bird. Body is distinctly divided into head, neck and trunk.
2. The eye-lids are now completely developed to cover the eyes.
3. The beak is completely developed.
4. The down feathers start appearing.
5. The legs with clawed toes are quite distinct.
6. Wings are also developing.

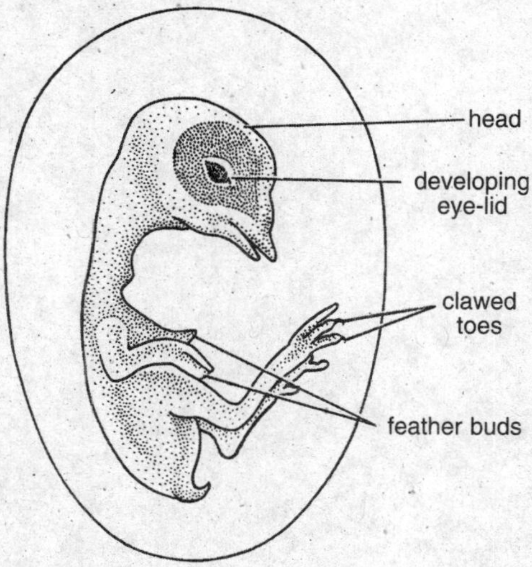

Fig. 24.34. Twelve-day chick embryo (whole mount).

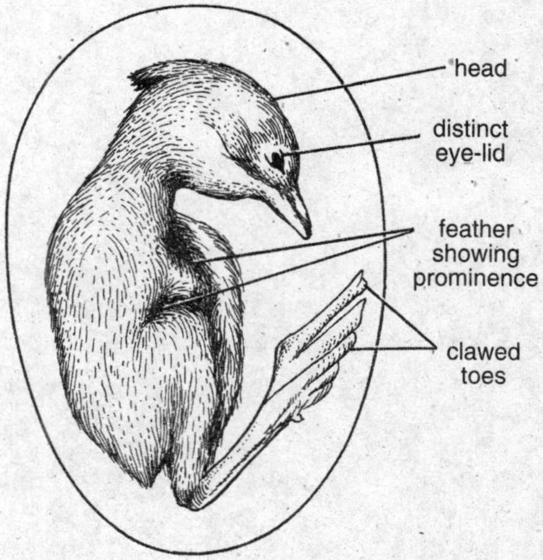

Fig. 24.35. Eighteen-day chick embryo (whole mount).

36. Twenty one-day Chick Embryo (Whole mount)

Comments :

The W.M. of **twenty-one day chick embryo** (Fig. 24.36) shows the following noteworthy features :

1. The body proportions are now completely distinct having head, neck and trunk.
2. Eyes are with eyelids.
3. Beak is developed.
4. Body except beak and toes are covered by tender down feathers (plumes).
5. Rudiments of flying feathers are distinct.
6. It is now almost ready to hatch with the help of calcareous egg tooth (caruncle) at the tip of upper beak.

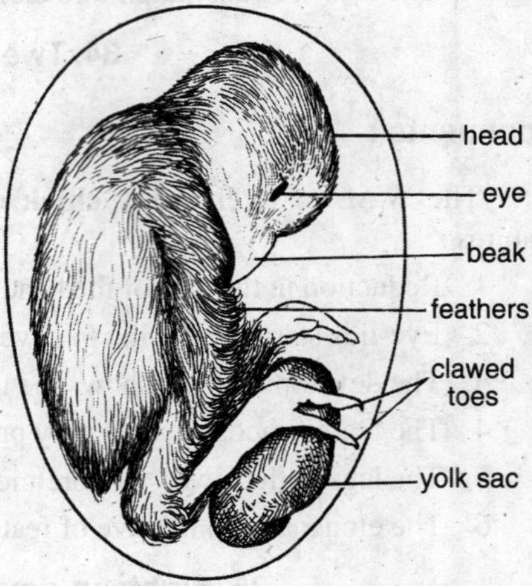

Fig. 24.36. Twenty one-day chick embryo (whole mount)

Chapter 25

OSTEOLOGY

The term **osteology** refers to the study of bones. The **endoskeleton** is the internal framework of the body that forms the main supporting framework of body. It primarily comprises **cartilages** and **bones**. The endoskeleton is mesodermal in origin.

The endoskeleton has been divided into two types :

1. Axial skeleton; 2. Appendicular skeleton.

1. Axial skeleton. The axial skeleton comprises those parts which lie along the median axis of the body. These are the **skull**, **vertebral column**, **sternum** and **ribs**.

2. Appendicular skeleton. The appendicular skeleton comprises those parts which lie on both sides of the axis. These are **girdles** and **limb bones**.

OSTEOLOGICAL TERMS

1. Exoskeleton. Skeletal structures found outside the body over the integument. Exoskeleton is dead or lifeless in invertebrates. It is often secreted on the surface.

Exoskeleton is either (1) **horny epidermal**, *i.e.*, derived from epidermis like reptilian scales, bird feathers, mammalian hairs, horns, nails and hoofs, etc., or (2) **dermal bony** derived from skin dermis like bony scales, plates (scutes), fin rays and antlers of fishes, reptiles and mammals.

2. Endoskeleton. Internal framework of cartilages or bones derived from mesoderm.

3. Axial skeleton. Endoskeleton found along the median axis of the body.

4. Appendicular skeleton. Endoskeleton attached to the axial skeleton of the body like girdles and limb bones.

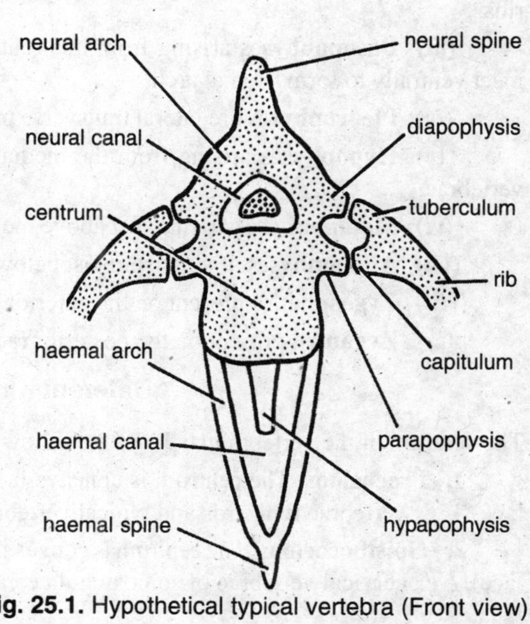

Fig. 25.1. Hypothetical typical vertebra (Front view).

5. Visceral skeleton. This is skeleton of pharyngeal wall giving support to gills. It partly develops from neural crest cells and splanchnic mesoderm.

6. Vertebral column. The notochord in fishes and higher vertebrates is surrounded by cartilaginous or bony rings called vertebrae. The notochord is constricted within each vertebra in fishes and aquatic amphibians and in tetrapoda it is obliterated.

Parts of a Typical Vertebra

The typical vertebra consists of the following parts :

1. Centrum is the solid cylindrical portion which encloses or replaces the embryonic notochord.

2. Neural arch arises from the dorso-lateral sides of the centrum and encloses the **neural canal**. It also forms the **neural spine** on its mid-dorsal side. Neural arches of successive vertebrae enclose the neural canal in which lies the spinal cord.

3. Haemal arch arises from the ventral side of the centrum and encloses the **haemal canal**. It also forms the **haemal spine** by meeting in the mid-line. Through haemal canal runs the caudal artery and vein.

4. Zygapophyses are the articular processes arise from neural arch by which vertebrae articulate with each other. These are of two types :

(*i*) **Prezygapophyses** are a pair of process on the anterior side of the neural arch. The articular surface is directed towards the upper side.

(*ii*) **Postzygapophyses** are a pair of processes present on the posterior surface of the neural arch. The articular surface is directed towards the lower side.

Zygapophyses are not found in fish vertebrae.

5. Transverse processes arising from the sides of the neural arch or from the lateral sides of the centrum. These are of the following types :

(*i*) **Diapophyses** arising laterally from the neural arch or centrum and articulates with dorsal head of ribs.

(*ii*) **Parapophyses** arising laterally from the centrum and articulates with ventral head of ribs.

(*iii*) **Basopophyses** arising from the ventro-lateral sides of the centrum or haemal arch or meet ventrally to form haemal arch.

(*iv*) **Pleurophyses** are lateral transverse processes arising from the centrum.

(*v*) **Hypophyses** arising from the median ventral surface of the centrum found in certain vertebrae.

(*vi*) **Metapophyses** arising just above the prezygapophyses.

(*vii*) **Anapophyses** arising from just below the postzygapophyses.

(*viii*) **Zygosphenes** present on the anterior face of the neural arch of snake's vertebra.

(*ix*) **Zygantra** present on the posterior face of the neural arch of snake's vertebra.

Different Kinds of Centra

The centrum of the vertebra may be of the following types :

1. Procoelous. The centrum is concave in front (anterior face) and convex behind (posterior face), *e.g.*, vertebrae in lizards and typical vertebrae of frog.

2. Opisthocoelous. The centrum is convex in front (anterior face) and concave behind (posterior face), *e.g.*, cervical vertebrae of some ungulates.

3. Heterocoelous or **saddle-shaped.** The anterior face of centrum is convex dorso-ventrally and concave from side to side. The posterior face being concave dorso-ventrally and convex from side to side, *e.g.*, vertebrae of birds.

4. Amphicoelous. Both the anterior and posterior surfaces of the centrum are concave, *e.g.*, in vertebrae of fishes tailed amphibians and VIIIth vertebra of frog.

5. Acoelous or amphiplatyan. Both the surfaces of the centrum are nearly flattened, *e.g.*, vertebrae of mammals.

6. Biconvex. Centrum is convex at both the ends, *e.g.*, IX vertebra of frog.

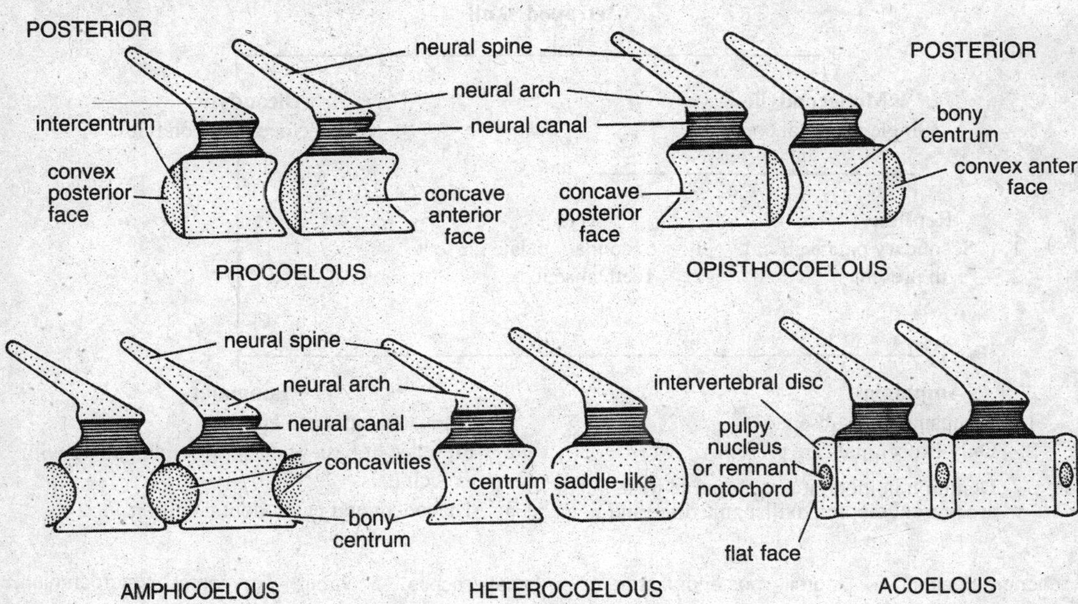

Fig. 25.2. Different types of centra.

JAW SUSPENSIONS (SUSPENSORIUM)

The method of jaw attachment or suspension with the skull is called jaw suspension or suspensorium. The various types of jaw suspension are as follows :

1. Autodiastylic. Jaws are attached with the cranium by anterior and posterior ligaments. **Hyoid** (II) **arch** is completely free and gives no support to the jaws. Example : Acanthodian fishes.

2. Amphistylic. Quadrate and otic processes of upper jaw are attached by ligaments to the skull. In the same way the upper end of hyomandibula of hyoid arch is also attached to the skull and both the jaws are suspended from its other end. Thus, in this type both the arches, mandibular (I) and hyoid (II) take part in jaws suspension with the skull. Example : Crossopterygii and some primitive sharks.

3. Hyostylic. Upper jaw (palatoquadrate bar) is loosely attached by ethmopalatine to cranium. Hyoid arch (hyomandibular part) binds the two jaws with the cranium. Example : Most elasmobranchs and all bony fishes.

4. Autostylic. Upper jaw (palatoquadrate bar) is closely bound to cranium and articular of lower jaw articulates with the quadrate of upper jaw. Example : Placoderms, chimaeras, lungfishes, amphibians, reptiles and birds.

Autostylic is further distinguished into three types :

(*i*) **Holostylic.** Upper jaw is fused with the skull and lower jaw is suspended from it. Hyoid arch is completely free. Example : Holocephali (chimaeras).

(*ii*) **Monimostylic.** Articular of lower jaw articulates with the quadrate and quadrate is immovably attached with the skull. Hyomandibular of hyoid arch forms columella. Example : Many tetrapods.

(*iii*) **Streptostylic.** In this type quadrate is movably attached with the skull. Examples : Lizards, snakes and birds.

5. Craniostylic. Upper jaw is fused with the cranium. Dentary of lower jaw articulates with the squamosal of skull. Hyomandibular forms stapes, articular and quadrate form malleus and incus. Example : Mammals.

Identification of Skull

Tetrapod Skull

Monocondylic
(Single occipital condyle)

Reptilia
1. Secondary palate absent.
2. Teeth present.

Aves
1. Secondary palate present.
2. Teeth absent.

Dicondylic
(Double occipital condyle)

Amphibia
1. Tympanic bulla absent.
2. Only exoccipitals are present.
3. Dentition homodont and absent in toad. Teeth are present on premaxillae, maxillae and vomers.

Mammalia
1. Tympanic bulla present.
2. Presence of basi- and supraoccipital along with exoccipitals.
3. Thecodont and heterodont dentition.

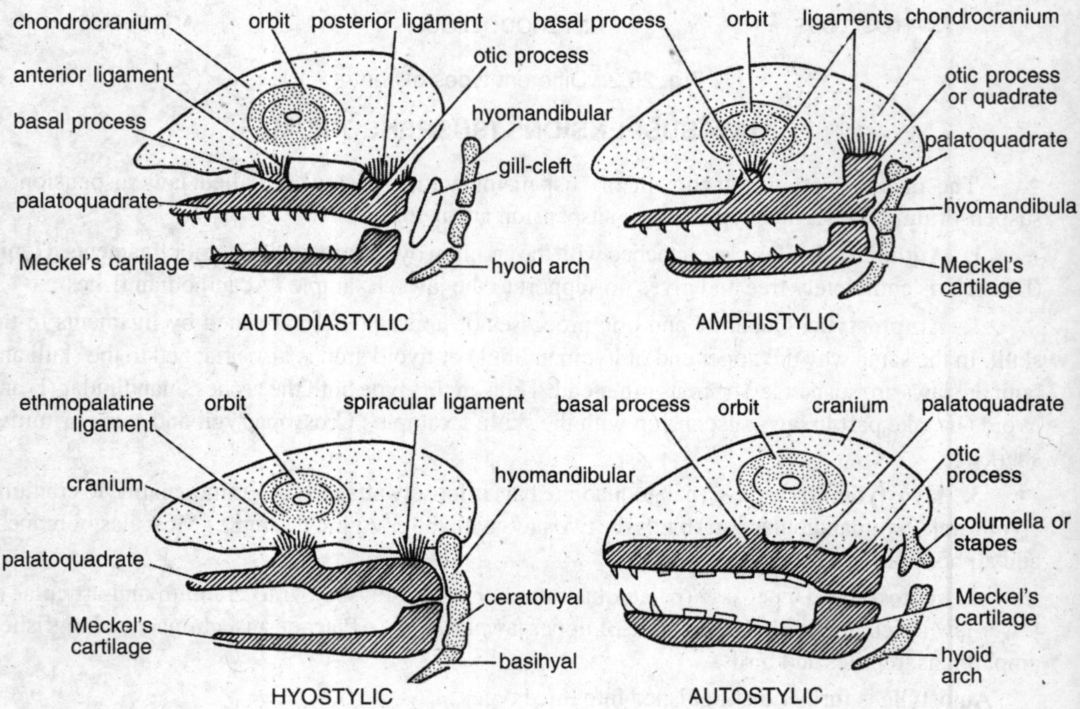

Fig. 25.3. Jaw suspension in vertebrates.

Chapter 26

SKELETON OF *SCOLIODON*

ENDOSKELETON

Endoskeleton of *Scoliodon* is completely cartilaginous, but basal plate of placoid scales is bony. Axial skeleton includes skull and vertebral column and appendicular skeleton consists of girdles and skeleton of fins.

Skull (Chondrocranium)

Skull consists of **cranium** and **sense capsules**, and **visceral skeleton** includes jaws and gill-arches.

(I) Cranium

1. Cranium encloses the brain.
2. It is formed of a single piece of cartilage.
3. With the cranium are fused the olfactory and auditory capsules.
4. Skull is perforated by formina for blood and nerves.
5. Its roof has two spaces or fontanelles.
6. Cranium has four regions : occipital, auditory, orbital and ethmoidal.

1. Occipital region. The posterior part of skull having a large hole, the **foramen magnum** is the occipital region. On either lateral side of foramen magnum is a small rounded occipital condyle for attachment with first vertebra. Outside of each condyle is present a foramen for X (vagus) nerve. Occipital crest is also present on the dorsal side of this region.

2. Auditory region. It is in front of occipital region and has two capsules one on either lateral side of the cranium. Internal ear is enclosed within each capsule. There is an oval depression on the dorsal side of cranium called **parietal fossa**. It contains two pairs of apertures : anterior smaller are the apertures of **endolymphatic ducts** and posterior larger openings are of **perilymphatic foramina**. Each auditory capsule shows three ridges indicating the position of semicircular canals.

3. Orbital region. It is the middle part lying in front of auditory region. On either lateral side is present a large hollow orbit in which lies the eye. Each orbit is bordered dorsally by a supraorbital ridge, ventrally by infra- or suborbital ridge, anteriorly by a preorbital process and posteriorly by a postorbital process. A number of foramina are present for nerves.

4. **Ethmoidal region.** It is composed of anterior part of cranium, rostrum and two lateral olfactory capsules, one on either side of cranium. Each olfactory capsule lodges olfactory sac and is opened ventrally. Both the capsules are separated from each other by a thin median vertical plate called **mesethmoid** (internasal septum).

Cranium roof has a depression called **anterior fontanellae** covered by a membrane.

Rostrum is formed of three rod-like **rostral cartilages** (one ventro-median and two dorso-lateral). All the three unite anteriorly.

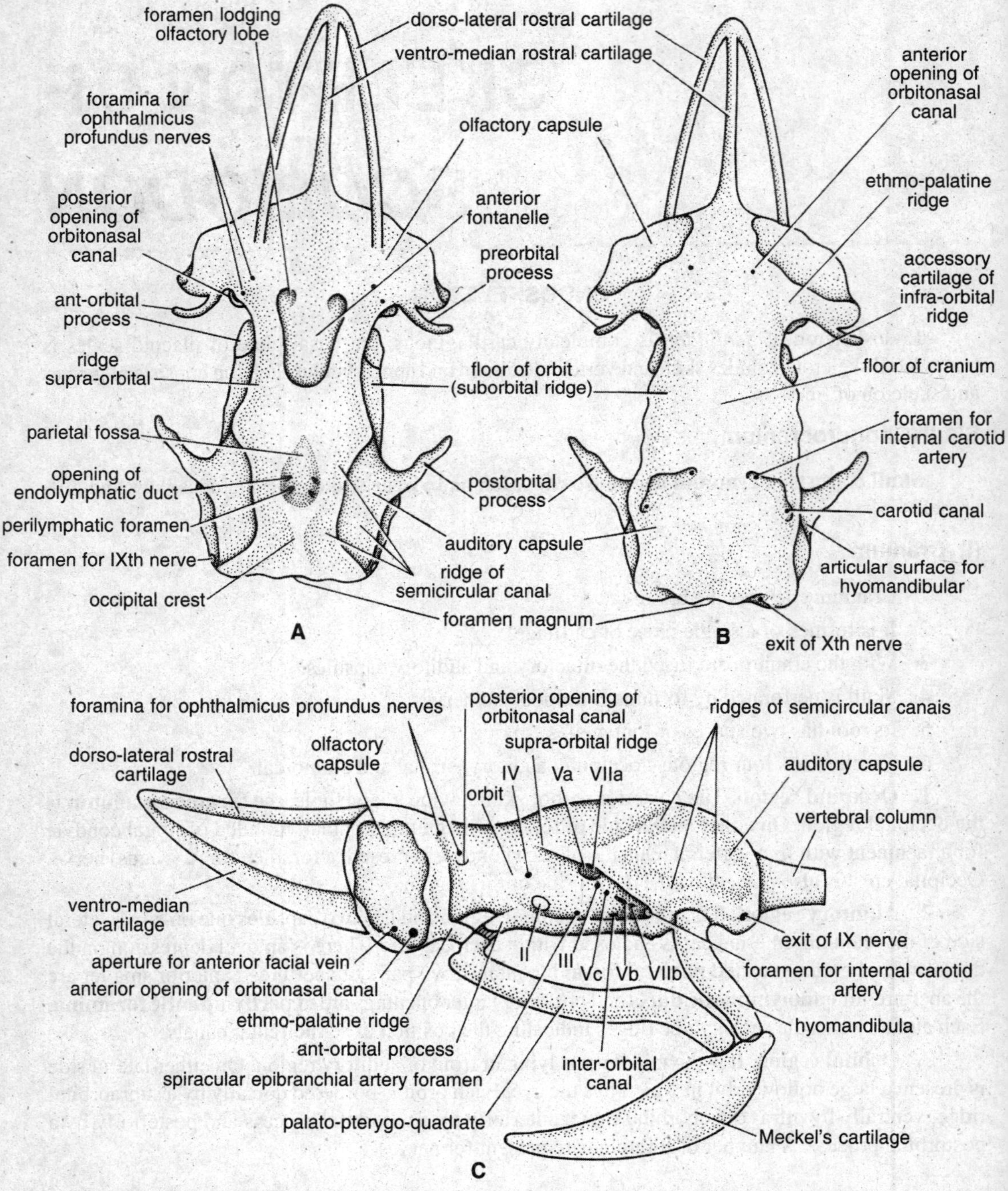

Fig. 26.1. *Scoliodon.* Skull A–Dorsal view; B–Ventral view; C–Lateral view with jaws.

II. Visceral Skeleton

It forms the jaws, skeleton of pharynx and a series of 7 pairs of U-shaped visceral arches encircling the buccal cavity and pharynx.

Seven visceral arches are as follows :

1. Mandibular arch (First arch). It is formed on either side of two segments : (*a*) Dorsal segment is called **palato-pterygo-quadrate** which forms the upper jaw. It is attached with the cranium by ligament. (*b*) Lower segment is called **Meckel's cartilage** which forms the lower jaw.

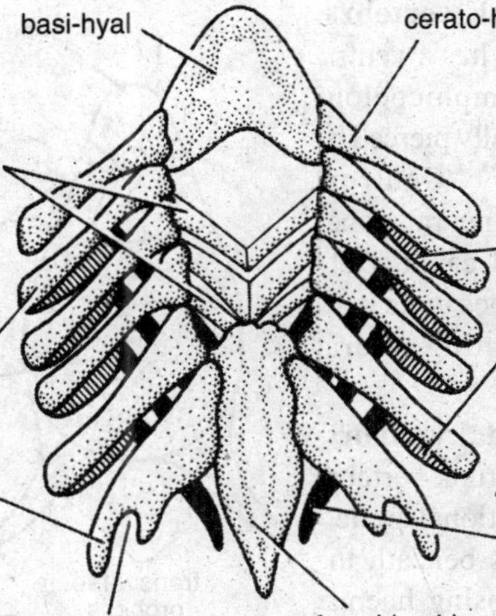

Fig. 26.2. *Scoliodon.* Visceral skeleton (ventral view).

Both the jaws are united by a hinge joint at the back of mouth.

2. Hyoid arch (Second arch). It consists of three cartilages: (*a*) Single median ventral **basihyal** giving support to buccal floor and tongue.(*b*) Lateral two **ceratohyals** lying in the buccal floor behind Meckel's cartilage. (*c*) Dorsal two **hyomandibula** whose lower ends connect with the posterior ends of upper and lower jaws and their upper ends join with cranium.

Thus, hyomandibula suspend the jaws with the cranium called **hyostylic**.

3. Branchial arches. These are five in number behind hyoid arch. These support the walls of pharynx and gills.

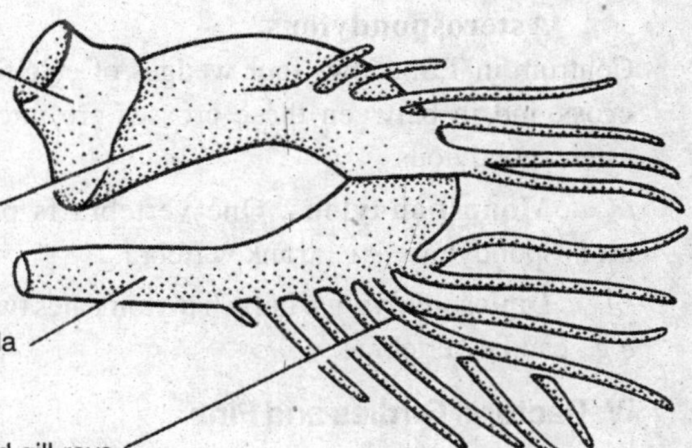

Fig. 26.3. *Scoliodon.* Skeleton of hyoid arch with gill-rays in ventral view.

Each branchial arch on either side consists of four rod-like parts joined with each other. These are :

(*a*) Dorsal pharyngobranchials;

(*b*) Lateral epibranchials having gill-rays;

(*c*) Lateral ceratobranchials bearing gill-rays;

(*d*) Ventral hypobranchials;

(*e*) Fused basibranchials of all five branchial arches.

All these five arches are joined with each other.

First 4 pairs of branchial arches possess at right angles four pairs of extrabranchial cartilages.

III. Vertebral Column

In adult *Scoliodon* vertebral column consists of about 130 vertebrae arranged in a longitudinal series divided into trunk and caudal regions. Vertebrae are amphicoelous.

1. Trunk vertebra
Each vertebra has a ventra cylindrical amphicoelou centrum centrally pierced b notochord. Centrur dorsally bears neural arche around neural canal. Bot the neural arches unite mid dorsally forming neura spine.

2. Caudal vertebra
It is also like trunk vertebr but it has additional arches haemal arches beneath th centrum enclosing haema canal which fuse in mid ventral aspect to form haemal spine.

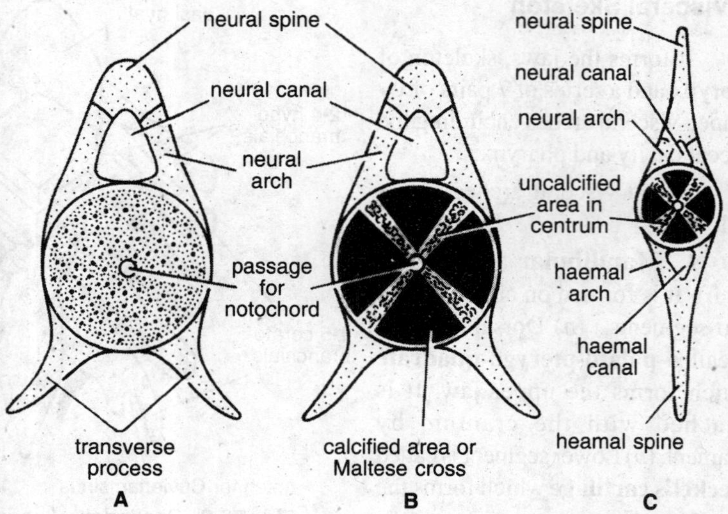

Fig. 26.4. *Scoliodon.* Vertebrae, A – Complete trunk vertebra in front view; B – Trunk vertebra in T.S; C – Caudal vertebra in T.S.

Extra Terms

Asterospondylous.
Centrum in T.S. shows four wedges of calcified fibrocartilage arranged in a cross called **Maltese cross** and in between these crosses are uncalcified areas. Such centrum or vertebra is called asterospondylous.

Monospondylous. One vertebra is present in each myotome. Such vertebra is called monospondylous, *e.g.*, trunk vertebra.

Diplospondylous. In each mytome lies two vertebrae, such vertebrae are called diplospondylous, *e.g.*, caudal vertebrae.

IV. Pectoral Girdles and Fins

Pectoral Girdle

1. Pectoral girdle (Fig. 26.4) is U-shaped composed of two right and left halves united mid-ventrally. It is not attached to the ventral column.

2. Each half has a dorsal rod-like scapula and vertebral flat coracoid.

3. Dorsal end of scapula is suprascapula.

4. Pectoral girdle is embedded in ventro-lateral body wall just posterior to gills in heart region.

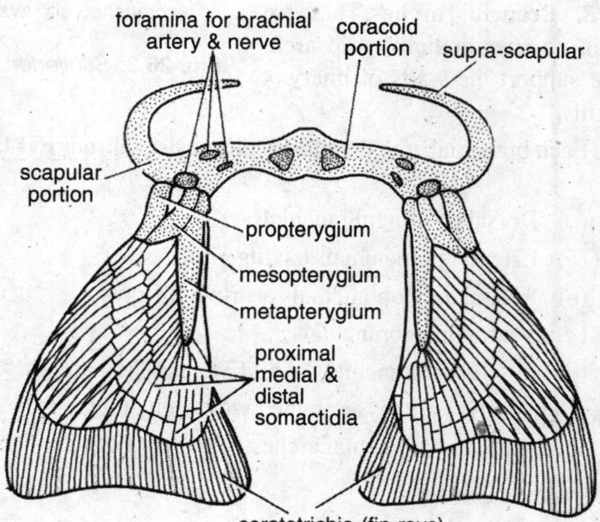

Fig. 26.5. *Scoliodon.* Pectoral girdle and fins. (Dorsal view).

Pectoral fins

1. Two pectoral fins are attached to pectoral girdle on either ventrolateral side.

2. Each pectoral fin is basally formed of 3 basal cartilages : propterygium, mesopterygium and metapterygium. These three cartilages articulate with the pectoral girdle at the junction of scapula and coracoid.

4. From these basal cartilages are attached numerous radial cartilages called **somactidia**. Each radial cartilage is formed of 2 or 3 somactidia (proximal, medial and distal).

5. Each distal somactidia bears two sets of long, thread-like, stiff and horny fin rays or **ceratotrichia** which support peripheral membranous part of the fin.

V. Pelvic Girdle and Fins

Pelvic Girdle

1. Pelvic girdle consists of a transverse flat rod called **ischiopubic bar**. It is not attached to the vertebral column.

2. It is embedded in the ventral abdominal wall in front of cloaca.

3. Ischiopubic bar on either side has a blunt iliac process above the acetabular facet.

Pelvic Fins

1. Each pelvic fin has a large and curved basal cartilage called **basipterygium** attached to the pelvic girdle at acetabular facet.

2. Basipterygium bears 15 or more slender radials or **somactidia**.

3. Somactidia bear numerous distal ceratotrichia (fin-rays).

Claspers

1. Claspers are found in male *Scoliodon* and are attached with the disal end of basipterygium.

2. Its skeleton has an elongated tubular **axial cartilage** which is dorsally grooved.

3. The proximal opening of the groove (**apopyle**) communicates with cloaca and glandular sac (siphon).

4. The distal opening of groove (hypopyle) opens on a sharp pointed **style** that is enclosed within a dorsal and a ventral **sheath plate**.

5. At the upper end of style is present a small serrated accessory cartilage.

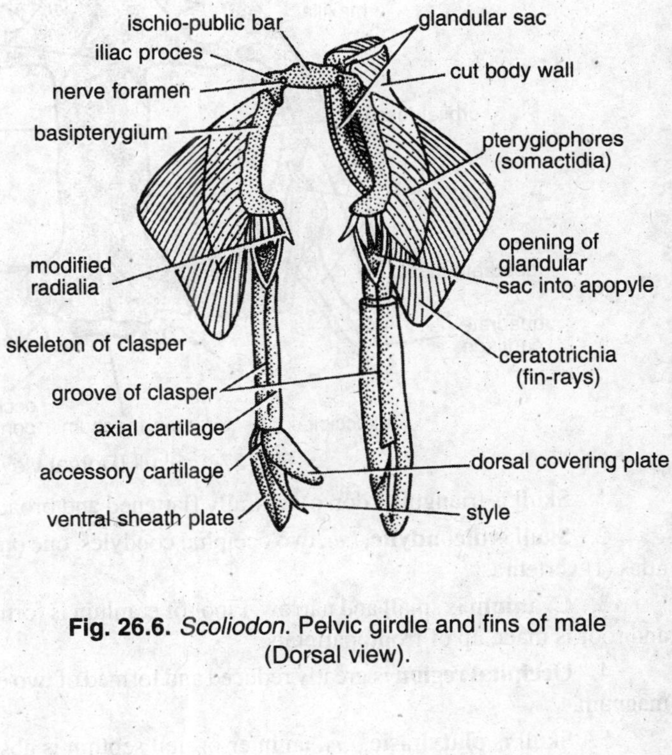

Fig. 26.6. *Scoliodon.* Pelvic girdle and fins of male (Dorsal view).

27
Chapter

SKELETON OF FROG

AXIAL SKELETON

Skull of Frog

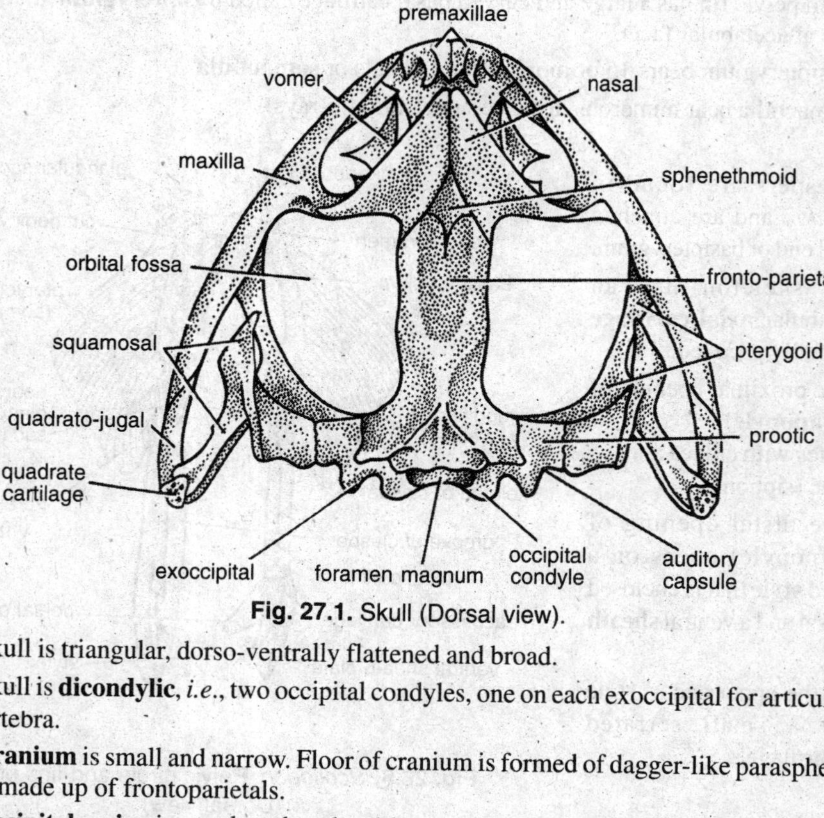

Fig. 27.1. Skull (Dorsal view).

1. Skull is triangular, dorso-ventrally flattened and broad.

2. Skull is **dicondylic**, *i.e.*, two occipital condyles, one on each exoccipital for articulating with atlas (I) vertebra.

3. **Cranium** is small and narrow. Floor of cranium is formed of dagger-like parasphenoid bone and roof is made up of frontoparietals.

4. **Occipital region** is greatly reduced and formed of two exoccipitals enclosing a large foramen magnum.

5. Skull is **platybasic**, *i.e.*, an inter-orbital septum is absent and the cranium extends beyond orbits.

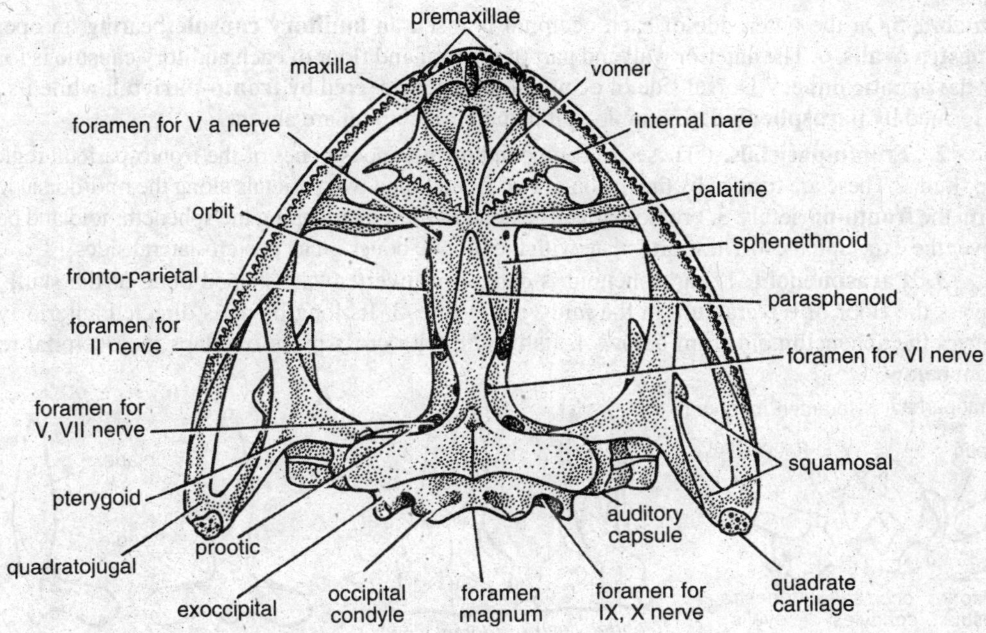

Fig. 27.2. Skull (Ventral view).

6. **Sphenethmoid** is tubular and forms the posterior wall of olfactory chambers and encircles the anterior end of cranial cavity. It is partly covered by the fronto-parietal and nasals above and parasphenoid below.

7. **Sense capsules.** A pair of **olfactory capsules** enclosing organs of smell and a pair of **auditory capsules**, lodging the organs of hearing and both are fused with the cranium. A pair of **optic capsules** enclosing the eyes are not fused with the skull.

8. **Olfactory capsules** are dorsally covered by a pair of large triangular **nasals** and floor is covered by a pair of small bones, the **vomers** bearing vomerine teeth. In front of each nasal lies a minute, irregular-shaped bone, **septomaxillary**.

9. Upper jaw consists of **premaxillae, maxillae, quadratojugals, quadrate cartilages** behind quadratojugals, pterygoids and palatines.

10. Lower jaw consists of mentomeckelian cartilage anterior to each dentary, **dentaries** and **angulosplenials**.

11. The suspensorium is **autostylic**, *i.e.*, the lower jaw is attached to the skull through rod-like quadrate cartilage.

12. Basisphenoids, alisphenoids, presphenoids, and supra-and basioccipitals are absent.

Bones Seen on the Dorsal Surface

Premaxillae, maxillae, quadratojugals, squamosals, septomaxillaries, nasals, fronto-parietals, prootics and exoccipitals.

Bones Seen on the Ventral Surface

Premaxillae, maxillae, quadratojugals, vomers, palatines, sphenethmoids, parasphenoid, pterygoids and exoccipitals.

DISARTICULATED BONES OF FROG

1. **Occipital segment.** 1. Forms the posteriormost part of the skull. 2. It is formed of two **exoccipitals** enclosing the foramen magnum through which spinal cord passes. 3. Each exoccipital bears an **occipital condyle** on its posterior surface. 4. Occipital condyles articulate with the atlas

vertebra. 5.On the outer side of each occipital is fused an **auditory capsule** bearing an opening, **fenestra ovalis**. 6. The anterior wall and part of the roof and floor of each auditory capsule is formed by the **prootic bone**. 7.Dorsal side of occipital region is covered by **fronto-parietal**, while its floor is formed by **parasphenoid**. 8. Supraocipital and basioccipital are absent.

2. Fronto-parietals. 1.These are compound membranous bones of the fronto-parietal region of the skull. 2.These are formed by the fusion of two frontals and two parietals along the mid-dorsal line to form the **fronto-parietals**. 3. Fronto-parietals extend in front overlapping the sphenethmoid and behind cover the exoccipitals. 4. These articulate with the prootic bones on its postero-lateral sides.

3. Parasphenoid. 1. Parasphenoid is dagger or inverted "⊥" shaped bone of the skull. 2. It covers the floor of the cranium in the mid-ventral line. 3. Its long blade is directed anteriorly and covers the sphenethmoid from below. 4. Its handle with cross-piece overlaps the occipital region from below.

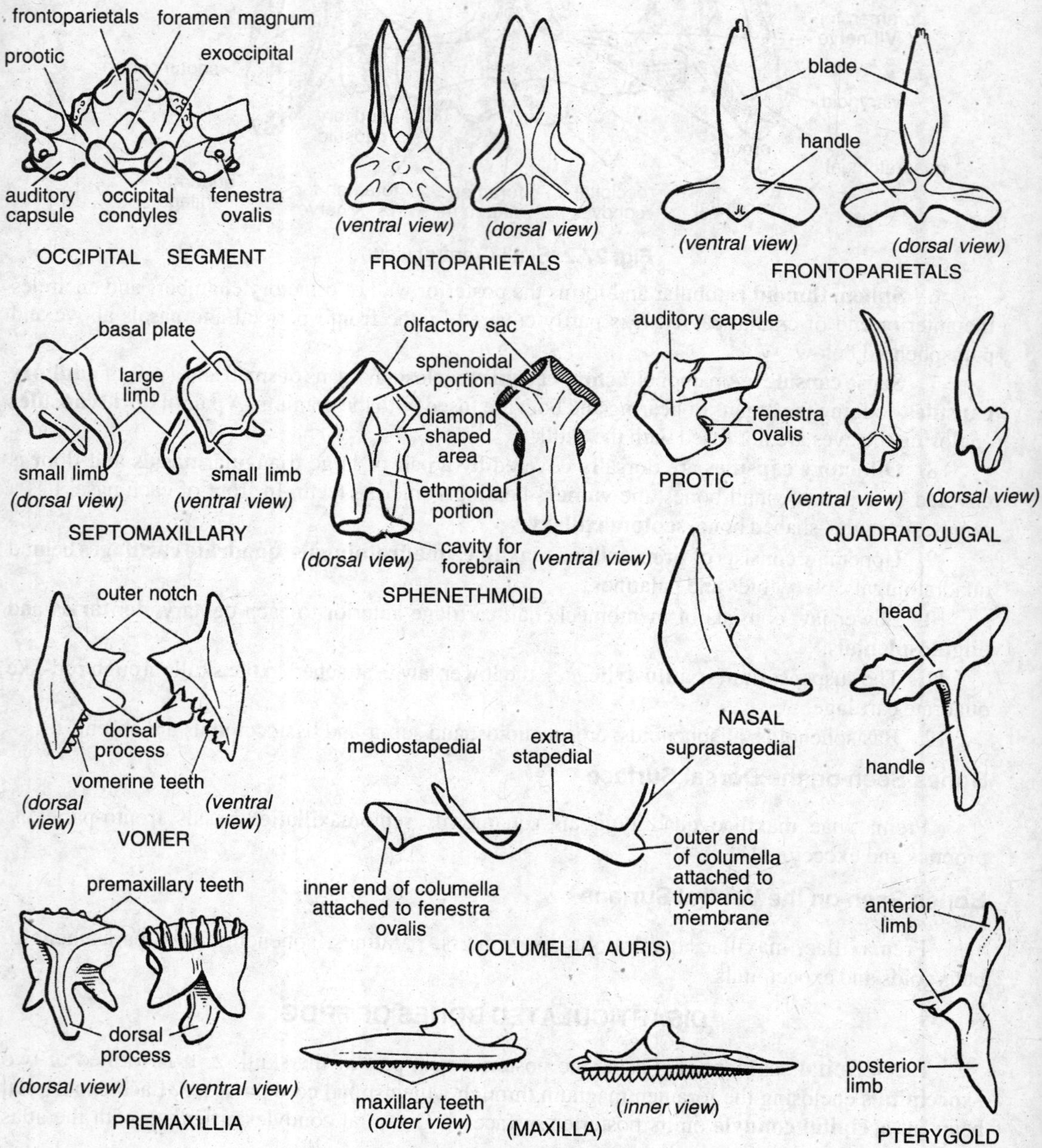

Fig. 27.3. Disarticulated skull bones of frog.

4. Sphenethmoid. 1. Sphenethmoid is hollow tubular bone that encircles the anterior end of cranial cavity and forms the posterior wall of nasal cavity. 2. It is partly covered by the fronto-parietals and nasals above leaving small diamond-shaped area, and parasphenoid below. 3. It is divisible by a transverse partition into an anterior **ethmoidal portion** and a posterior **sphenoidal portion**. 4. **Ethmoidal portion** is divided by a longitudinal septum into right and left portion, each enclosing the olfactory sac. 5. **Sphenoidal portion** encloses the forebrain.

5. Premaxilla. 1. Premaxilla is the anteriormost bone of the upper jaw. 2. It is a small irregular bone in the anterior portion of the snout and meet with its counterpart in the middle line. 3. It bears two rows of conical teeth on its ventral surface along the anterior end. 4. Dorsally it gives off a posteriorly directed **dorsal process** that forms part of the inner boundary of external nostril. 5. On its posterior side, premaxilla meets the maxilla of its side.

6. Maxilla. 1. It is a long, thin and slightly curved bone, forming the greater portion of the outer margin of the upper jaw. 2. It bears numerous minute, sharp, pointed and backwardly directed homodont teeth along its whole length. 3. It articulates in front with premaxilla and behind with quadratojugal. 4. On its inner side in the middle of its length, it articulates with the palatine and the anterior limb of pterygoid.

7. Quadrato-jugal. 1. Quadrato-jugal is small, slender and coma-shaped bone. 2. It forms the posterior portion of the outer margin of the upper jaw. 3. It articulates anteriorly with maxilla and posteriorly with quadrate cartilage. **Quadrate cartilage** is a small thin rod completely covered by pterygoid and squamosal.

8. Nasal. 1. Nasal is large, flat and triangular membranous bone covering the olfactory capsule. 2. Both the nasals unite in the median line. 3. Its anterior end reaches up to the dorsal process of the premaxilla and partly forms the boundary of external nares. 4. Its outer process unites with the maxilla, while its posterior process partly overlaps the ethmoidal portion of sphenethmoid.

9. Septomaxillary. 1.Septomaxillary is a very small irregular bone lying close to anterior process of nasal. 2. It consists of a basal plate and a pair of backwardly directed small and larger processes.

10. Vomer. 1. Vomer is somewhat triangular bone forming the floor of olfactory capsule. 2. The posterolateral margin bears a few pointed **vomerine teeth**. 3. It forms inner margin of the posterior nares.

11. Palatine. 1. Palatine is a slender, rod-like bone lying just in front of the orbit on the ventral surface of the skull. 2. It connects the anterior side of cranium with the middle of maxilla.

12. Squamosal. 1. Squamosal is "T" or hammer-shaped bone lying on the dorso-lateral side of the posterior end of cranium above the pterygoid. 2. It consists of a backwardly directed **handle** and a forwardly directed **head**. 3. **Handle** articulates to the quadrate cartilage and the **head** articulates to the auditory capsule and prootic.

13. Pterygoid. 1. Pterygoid is a triradiate or Y-shaped bone lying ventral to the squamosal towards the posterior end of cranium. 2. Its anterior limb articulates with the maxilla and outer end of palatine. 3. Its inner shorter limb joins with the parasphenoid and auditory capsule. 4. Its posterior limb meets the quadratojugal and quadrate cartilage. 5. Pterygoid also forms the posterior ventral boundary of the orbit.

14. Lower jaw. 1. The lower jaw of frog is

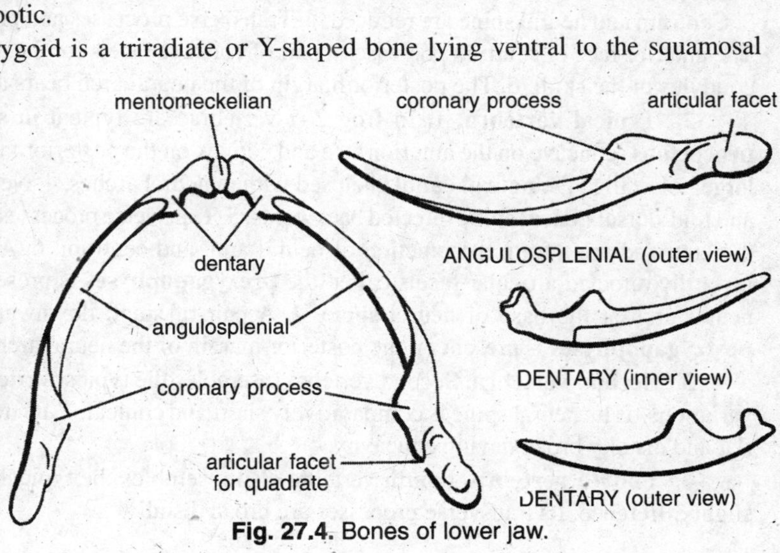

Fig. 27.4. Bones of lower jaw.

composed of two halves or rami united anteriorly by a ligament. 2. Each half or ramus consists of a core of Meckel's cartilage surrounded by three bones, *viz.*, **mentomeckelian**, **dentary** and **angulosplenial**. 3. **Mentomeckelian** is a small cartilaginous bone, present at the anterior symphysis of two rami. It is an ossification of anterior end of Meckel's cartilage. 4. Teeth are entirely absent.

15. Dentary. 1. It is a small, flat, dagger-shaped bone of the lower jaw. 2. It covers the outer anterior surface of Meckel's cartilage. 3. Anteriorly it unites with the mentomeckelian, while its posterior part is closely applied to the outer side of the angulosplenial.

16. Angulosplenial. 1. It is a long curved bone, constituting most of the inner and posterior portion of each ramus of the lower jaw. 2. Its anterior end is tapering, while its posterior end bears a knob like **articular surface** for the articulation with the quadrate cartilage. 3. Just in front of the articular surface is present a protuberance the **coronary process**.

17. Hyoid apparatus. 1. The hyoid apparatus is a cartilaginous plate lying below the tongue in the floor of buccal cavity. 2. It provides support to the tongue. 3.
It is composed of three parts, *viz.*, body of hyoid, anterior cornua and posterior cornua. 4. **Body of hyoid** is a thin, membranous and square-shaped plate bearing processes at each angle of the plate. 5. From the antero-lateral ends of the body of hyoid arise a pair of long, slender and curved cartilaginous process, **anterior cornua**, or horn attached with auditory capsule. 6. From the posterior side of the body of hyoid arise a pair of short and stout bony processes the **posterior cornua** or **thyrohyals.** 7. Posterior cornua supports the laryngo-tracheal chamber.

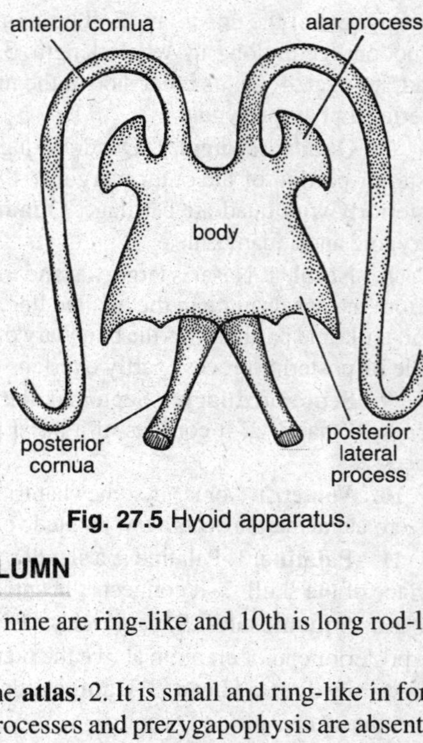

Fig. 27.5 Hyoid apparatus.

VERTEBRAL COLUMN

Vertebral column is formed of ten vertebrae : First nine are ring-like and 10th is long rod-like urostyle.

1. Atlas vertebra. 1. The first vertebra is called the **atlas**. 2. It is small and ring-like in form. 3. Centrum and neural spine are reduced. 4. Transverse processes and prezygapophysis are absent. 5. The anterior face of centrum possesses a pair of concave facets for the articulation with the occipital condyles of the skull. 6. The posterior margin of the neural arch bears a pair of postzygapophyses.

2. Typical vertebra. 1. In frog 2–7 vertebrae are typical in structure. 2. The centrum is **procoelous** (concave on the anterior face and convex on the posterior face). 3. It is ring-like having a large hole called the **neural canal** enclosed with in neural arches. 4. Neural arch bears a small, blunt and mid-dorsal **neural spine** directed backwards. 5. Transverse processes are long, tapering, outwardly directed and arise from the junction of neural arch and centrum. 6. A pair of small upwardly and inwardly directed articular facets called the **prezygapophyses** is present on the anterior margin of neural arch at the base of neural spine. 7. A pair of small downwardly and outwardly directed **postzygapophyses** is present on the posterior margin of the neural arch.

3. Second vertebra. Second vertebra resembles the typical vertebra in structure except slight variations. 1. Its neural spine is comparatively short and conical. 2. Its transverse processes are small, flat and distally broad having concavity.

4. Fourth vertebra. Fourth vertebra also resembles the typical vertebra in structure except slight difference. Its transverse processes are broad distally.

5. Eighth vertebra. 1. The centrum is **amphicoelous** (biconcave on both the sides). 2. The anterior concavity receives the posterior convexity of seventh vertebra. 3. The posterior concavity receives the anterior convexity of ninth vertebra. 4. Transverse processes are long, slender and outwardly directed. 5. Prezygapophyses and postzygapophyses are present on the anterior and posterior margins of neural arch respectively.

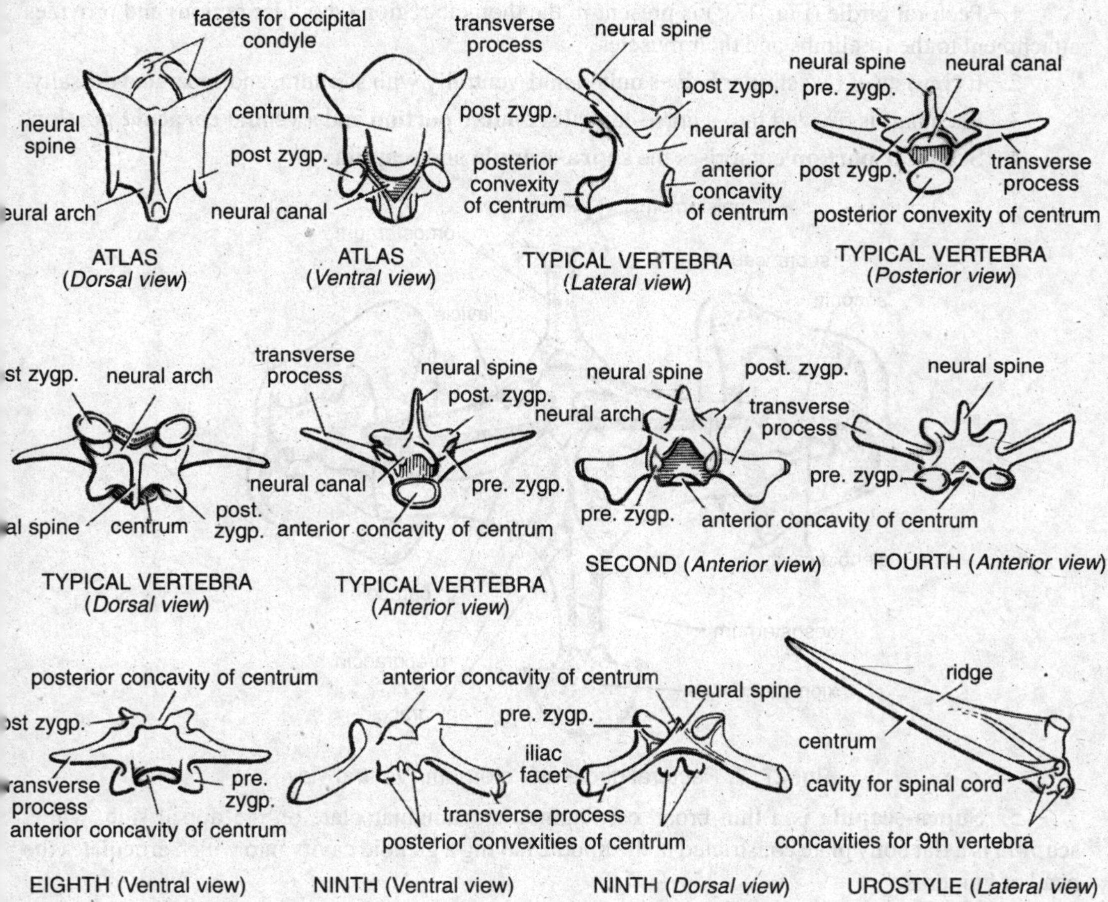

Fig. 27.6. Vertebrae of frog.

6. Ninth vertebra. 1. Ninth vertebra is also known as sacral vertebra. 2. The centrum is **biconvex**, *i.e.*, convex on both the sides (bearing one convexity anteriorly and two convexities posteriorly). 3. The anterior convexity fits into the posterior concavity of eighth vertebra. 4. The posterior convexities fit into the anterior concavities of urostyle. 5. Transverse processes are large cylindrical, stout and backwardly directed having facets at the tip for the articulation of ilium bone of pelvic girdle. 6. Neural spine is inconspicous, *i.e.*, greatly reduced. 7. Prezygapophyses are well developed along the anterior end of neural arch, while the postzygapophyses are entirely absent.

7. Urostyle. 1. Urostyle is the Xth vertebra representing the caudal region in the frog. 2. It is long and triangular with pointed apex directed backwards. 3. Centrum is long, rod-like with broad anterior end bearing two concavities to receive the convexities of IXth vertebra. 4. Dorsally it is raised into a vertical ridge gradually tapering posteriorly and has a narrow canal for the spinal cord. 5. Transverse processes, pre-and postzygapophyses are entirely absent.

APPENDICULAR SKELETON

GIRDLES

1. Pectoral Girdle and Sternum

1. Pectoral girdle (Fig. 27.7) is present in the thoracic region (shoulder region) and provides attachment to the forelimbs and their muscles.

2. It consists of two similar halves united mid-ventrally with sternum, and separated dorsally.

3. Each half is divided into a dorso-lateral **scapular portion** and a ventral **coracoid portion**.

4. **Scapular portion** comprises the **supra-scapula** and **scapula**.

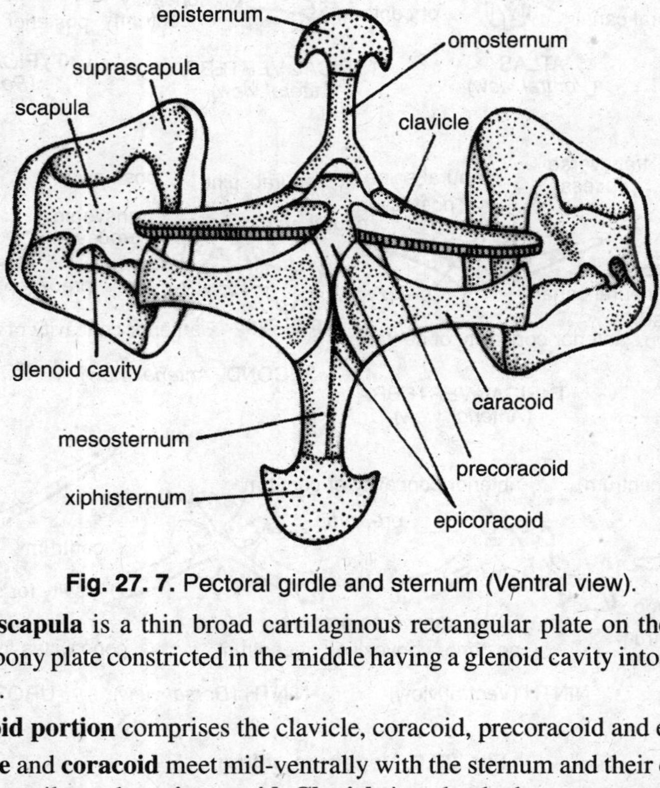

Fig. 27. 7. Pectoral girdle and sternum (Ventral view).

5. **Supra-scapula** is a thin broad cartilaginous rectangular plate on the dorsal side, while. **scapula** is a flat bony plate constricted in the middle having a glenoid cavity into which articulates the head of humerus.

6. **Coracoid portion** comprises the clavicle, coracoid, precoracoid and epicoracoid.

7. **Clavicle** and **coracoid** meet mid-ventrally with the sternum and their counterparts of other side by a strip of cartilage–the **epicoracoid**. **Clavicle** is a slender bone separated from coracoid by a wide gap.

8. **Sternum** lies in the mid-ventral line between two halves of pectoral girdle. It consists of anterior most cartilaginous disc **episternum** and bony rod **omosternum** connecting it with clavicles, **mesosternum** is rod-like behind epicoracoids and **xiphisternum** is the terminal broad cartilaginous plate.

2. Pelvic Girdle

1. Pelvic girdle (Fig. 27.8) lies in the posterior region of the trunk and gives support to the hindlimbs.

2. It is V-shaped and composed of two similar halves each of which is known as **os-innominatum**.

3. Each os-innominatum is composed of three bones–silium, pubis and ischium, all of which share in the formation of median disc and acetabulum.

4. **Ilium** is greatly elongated and forms the major part of each os-innominatum and it runs forwards to meet the transverse process of the ninth vertebra and bears a prominent dorsal vertical ridge the **iliac crest**.

5. **Pubis** is a much reduced triangular piece of calcified cartilage forming ventral part of the disc.

6. **Ischium** is larger and forms the posterior part of the disc.

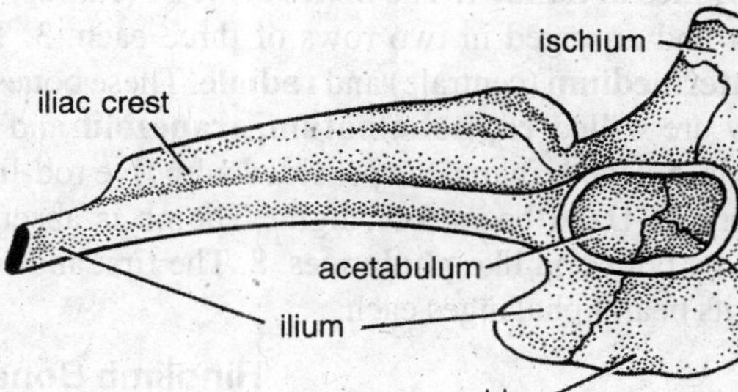

Fig. 27. 8. Pelvic girdle (Lateral view).

7. The disc formed by the union of three bones contains on either side a cup-shaped cavity the **acetabulum** for the articulation of the head of humerus.

FORELIMB BONES

1. Humerus. 1. It is the bone of upper arm of forelimb. 2. It is a short, stout and cylindrical bone with a slightly curved shaft. 3. Its proximal end is known as **head** which fits into the glenoid cavity of pectoral girdle. Head is covered with calcified cartilage. 4. The antero-ventral ridge below the head is known as **deltoid ridge** to which muscles are attached. 5. The distal end forms a rounded **trochlea** with a condylar ridge on either side. 6. Trochlea articulates with the groove of radio-ulna.

2. Radio-ulna. It is a compound bone of forearm of forelimb 2. It is formed by the fusion of **radius** and **ulna** bones. 3. Its proximal end has a concavity for the trochlea of humerus and further projects into an **olecranon process** forming elbow joint. 4. The distal portion of radio-ulna is somewhat flat having a groove divisible into an anterior (radial) part and a posterior (ulnare) part having facets for the proximal row of carpals.

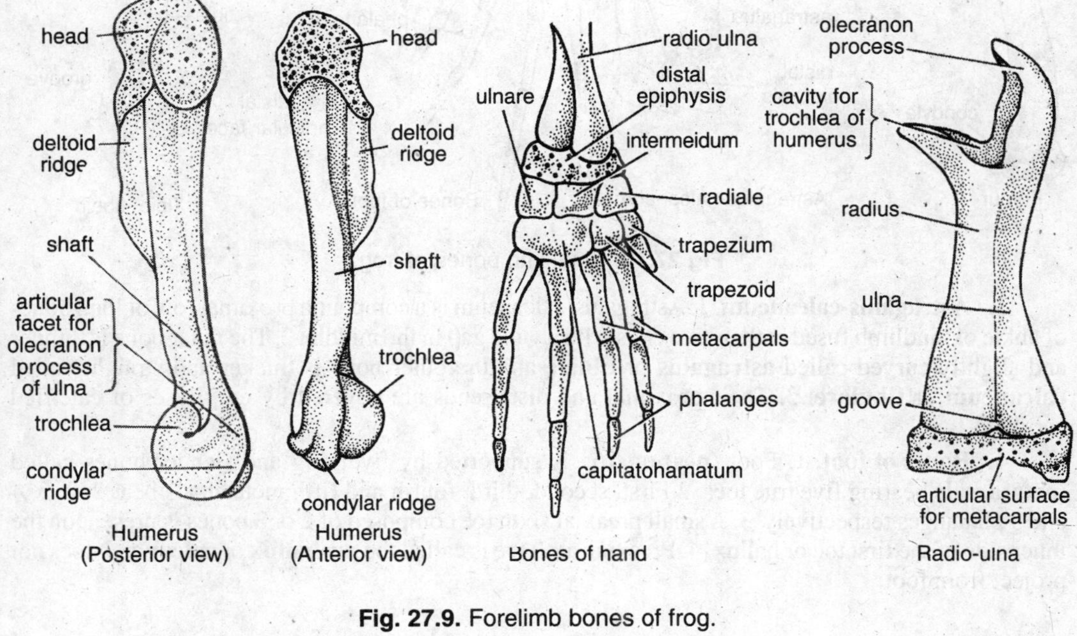

Fig. 27.9. Forelimb bones of frog.

3. Bones of hand. 1. The bones of **wrist** (carpus) are called **carpals**. 2. Carpal bones are six in number and arranged in two rows of three each. 3. The bones of the proximal row are called **ulnare**, **intermedium** (centrale) and **radiale**. These bones articulate with radio-ulna. 4. The bones of distal row are called **capitohamatum**, **trapezoid** and **trapezium**. These bones articulate with metacarpals. 5. Hand (manus) is provided with five rod-like slender metacarpals. First metacarpal is rudimentary. 6. Hand bears four digits. Thumb is absent. 7. The remaining four metacarpals are supported by bony rod like **phalanges**. 8. The first and second digits bear phalanges, and third and fourth digits bear 3 phalanges each.

Hindlimb Bones

1. Femur. 1. **Femur** is a long slender bone of thigh region of hindlimb. Its shaft is slightly curved. 2. The proximal swollen end is called **head** that fits into the acetabulum of pelvic girdle forming ball and socket joint. 3. The distal end forms a condyle which articulates with tibio-fibula. 4. Head and condyle are covered by calcified cartilage.

2. Tibio-fibula. 1. Tibio-fibula is a large compound bone of the shank region of hindlimb. 2. It is formed by the fusion of inner tibia and outer fibula bones forming a single bone called the **tibio-fibula**. 3. The proximal and distal ends are covered by cartilage. 4. Near the proximal end tibia bears a **cnemial** or **tibial crest**. 5. The proximal end articulates with femur and distal end articulates with **astragalus-calcaneum**.

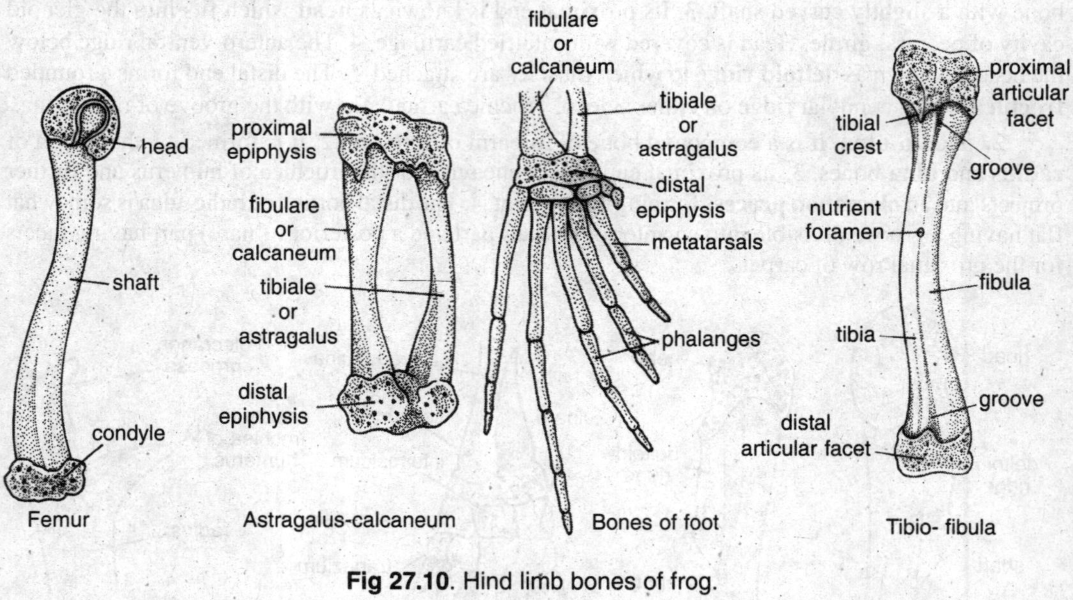

Fig 27.10. Hind limb bones of frog.

3. Astragalus-calcaneum. 1. Astragalus-calcaneum is a compound proximal row of long bones of ankle of hindlimb fused at their two ends with a wide gap in the middle. 2. The inner bone is thinner and slightly curved called **astragalus** or **tibiale** and the outer bone is thicker and straight called **calcaneum** or **fibulare**. 3. Their proximal and distal ends are covered by **epiphyses** of calcified cartilage.

4. Bones of foot. 1. Foot **(pes)** of frog is supported by five long and slender bones called **metatarsals** bearing five true toes. 2.First, second, third, fourth and fifth metatarsals bear 2, 2, 3, 4 and 5 phalanges respectively. 3. A small preaxial sixth toe composed of 2 or 3 bones is present on the inner side of the first toe or hallux. 4. Preaxial sixth toe is called the **prehallux** or **calcar** and does not project from foot.

28

Chapter

SKELETON OF *VARANUS*

AXIAL SKELETON

1. Skull

1. Skull is compact and more or less elongated having narrow anterior end.
2. It is **monocondylic**, *i.e.*, single condyle is borne by basioccipital.
3. It is **tropibasic**, *i.e.*, the inter-orbital septum is present between two orbits.
4. **Skull** has three fossae on either side, supraorbital, suborbital and supratemporal fossae.
5. Suspensorium is **autostylic**, *i.e.*, lower jaw is attached with the quadrate bone of the skull.
6. **Auditory region** has three bones **prootic**, **epiotic** and **opisthotic**. Opisthotic fused with exoccipital, epiotic with supraoccipital and prootic is free.
7. Two parietals are fused forming a single bone, having a parietal foramen between the two.
8. Two premaxillae are also fused into a single bone.
9. Prefrontal, supraorbital, and postorbital bones are present.
10. Alisphenoid, orbitosphenoid and presphenoid bones are absent (present in rabbit skull).
11. Lower jaw consists of six bones.
12. Teeth are present on premaxillae, maxillae and dentaries and are **homodont**, **polyphyodont** and **pleurodont** type.

Parts of the skull. Cranium has three regions : occipital, parietal and frontal.

1. Occipital Segment

1. Occipital segment is the posteriormost part of the cranium, enclosing the foramen magnum. 2. It consists of four fused bones, *i.e.*, the **supraoccipital, basioccipital** and paired **exoccipitals**. 3. **Supraoccipital** is more or less rectangular forming the roof of the cranium. It articulates anteriorly with the parietals and prootics. 4. **Basioccipital** forms the floor of the cranium and posteriorly bears the single occipital condyle. It articulates anteriorly with basisphenoid laterally with exoccipitals. 5. **Exoccipitals** surround the foramen magnum from its lateral sides. Each exoccipital is produced laterally into a **paroccipital process** articulating with the supratemporal. 6. Each exoccipital has a combined jugular and condylar foramen for the exit of X, XI and XII cranial nerves vein and artery.

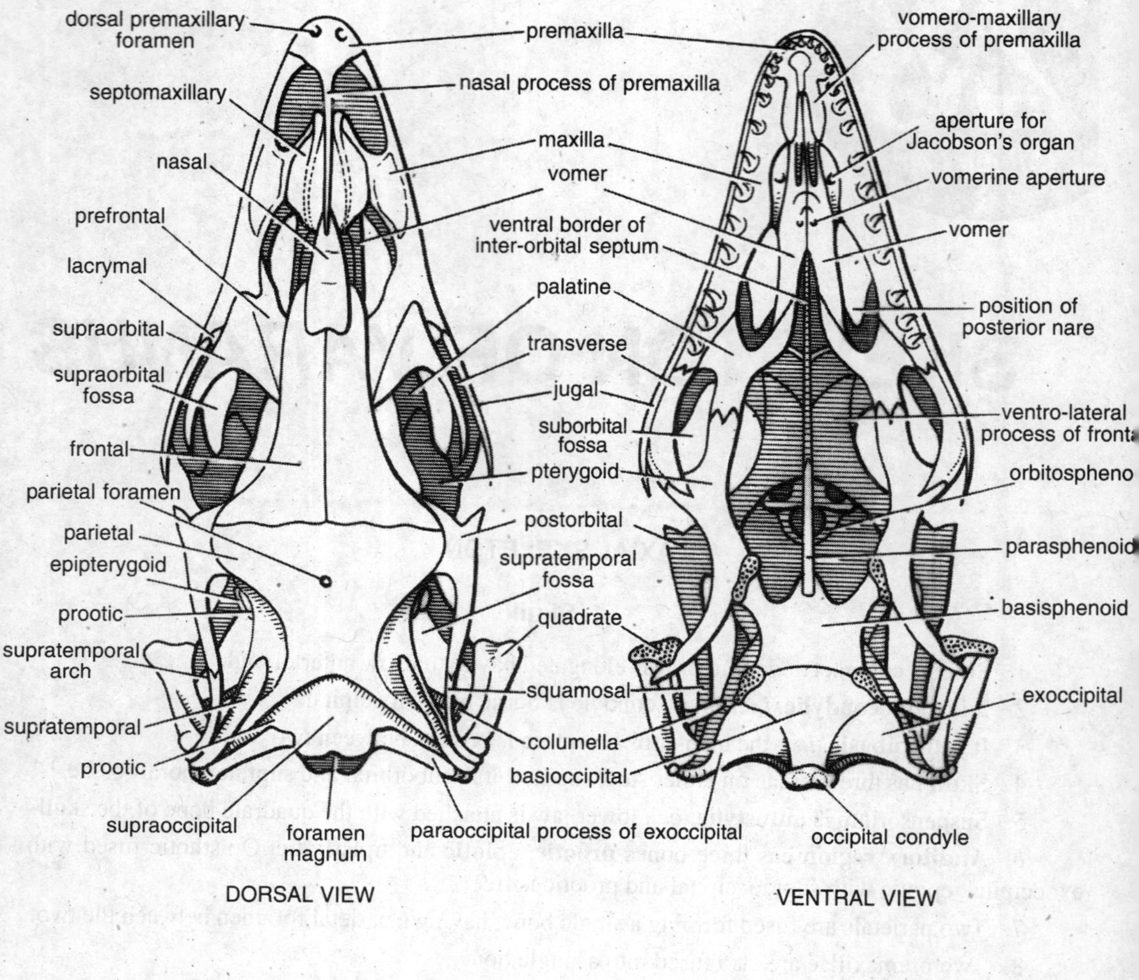

Fig. 28.1. Skull of *Varanus* in dorsal, ventral and lateral views.

2. Parietal Region

1. Parietals. 1.Two parietals fused to form a single bone, forming the roof of cranium. 2. It is broad in front and narrow behind and has a parietal foramen in its centre. 3. It postero-laterally is produced behind into **supratemporal processes**, each of which articulates with the quadrate, squamosal, supratemporal and paroccipital processes of exoccipital of its side. 4. Parietals articulate anteriorly with the frontals, posteriorly with the supraoccipital, on the sides with the post-orbitals.

2. Basisphenoid. 1. Basisphenoid is roughly rectangular and forms the floor of the cranium in the parietal region. 2. On its antero-lateral sides it gives off a pair of **basipterygoid processes** which unite with the pterygoids. 3. At the anterior margin it bears a median knob-like process which receives the hinder end of slender parasphenoid. 4. It articulates posteriorly with the basioccipitals.

3. Frontal Region

It is formed of three bones : a pair of frontals and a parasphenoid.

1. Frontals. 1. It is formed by the fusion of two frontal bones in the mid-line forming the roof of the frontal region of the cranium. 2. Each frontal is narrow in front and broad behind. 3. The frontals posteriorly join with the parietals by a **coronal suture**. 4. Frontals articulate in front with nasals, antero-laterally with prefrontals, postero-laterally with postorbitals, behind with parietals and below with parasphenoid.

2. Parasphenoid. It is a large rod-like bone present at midventral line in front of basisphenoid.

4. Sense Capsules (Olfactory Capsules)

1. Nasals. Two nasals fused in the mid-line into a single compound bone form the roof of the olfactory capsule. 2. Each nasal is flat and triangular bone narrow anteriorly and broad behind. 3. Nasals articulate in front with the nasal premaxillary process and two septomaxillaries and behind with frontals.

2. Vomer. 1. Vomer lies on the floor of olfactory capsule. 2. It is a rod-shaped slender bone uniting with its counterpart in the mid-line only in its anterior half. 3. It forms the inner margin of the posterior nare and is perforated by a vomerine aperture in its centre. 4. It articulates with premaxilla and maxilla in front and with palatine behind.

3. Septomaxillary. It is an irregular flat bone present on outer side of nasal process of premaxilla.

4. Orbits. Both the lateral orbits are separated from each other by a thin vertical interorbital septum. Each orbit is associated with the following bones :

1. Prefrontal. 1. It is a small, more or less triangular bone bearing a deep cup-like concavity on its ventral side. 2. It lies obliquely between the frontal and maxilla. 3. It forms the anterior boundary of orbit and articulates with supra-orbital posteriorly and with lacrymal ventrally.

2. Postorbital or postfrontal. 1. It forms the postero-dorsal boundary or orbit. 2. It is an irregular bone producing four processes. 3. In between the two inner processes is present a deep notch which receives the parts of parietal, and frontal. 4. The posterior longest process articulates with the squamosal to form supratemporal arcade.

3. Supraorbital. 1. It is somewhat triangular bone overhanging the anterior portion of the orbit. 2. Its broad base articulates in front with the prefrontal, while its free pointed apex is directed posteriorly towards postorbital.

4. Lacrymal. 1.It is a small square-shaped bone lying within orbit. 2. It contains a small aperture for lacrymal duct and articulates in front with maxilla, behind with jugal and above with prefrontal.

SUSPENSORIUM

It includes **quadrates** and **squamosals** which suspend lower jaw from cranium.

1. Quadrate. 1. It is a small rod-like bone of the suspensorium. 2. It lies obliquely in the posterolateral side of the cranium. 3. Posteriorly it articulates with the squamosal, supra-temporal, supra-temporal process of the parietal and paroccipital process of exoccipital. 4. Anteriorly it articulates with the quadrate process of pterygoid and the articular bone of the lower jaw.

6. Bones of Upper Jaw

Upper jaw is fused with the cranium and includes premaxilla, maxilla, jugal, septomaxillary, palatine, pterygoid, transverse and epipterygoid.

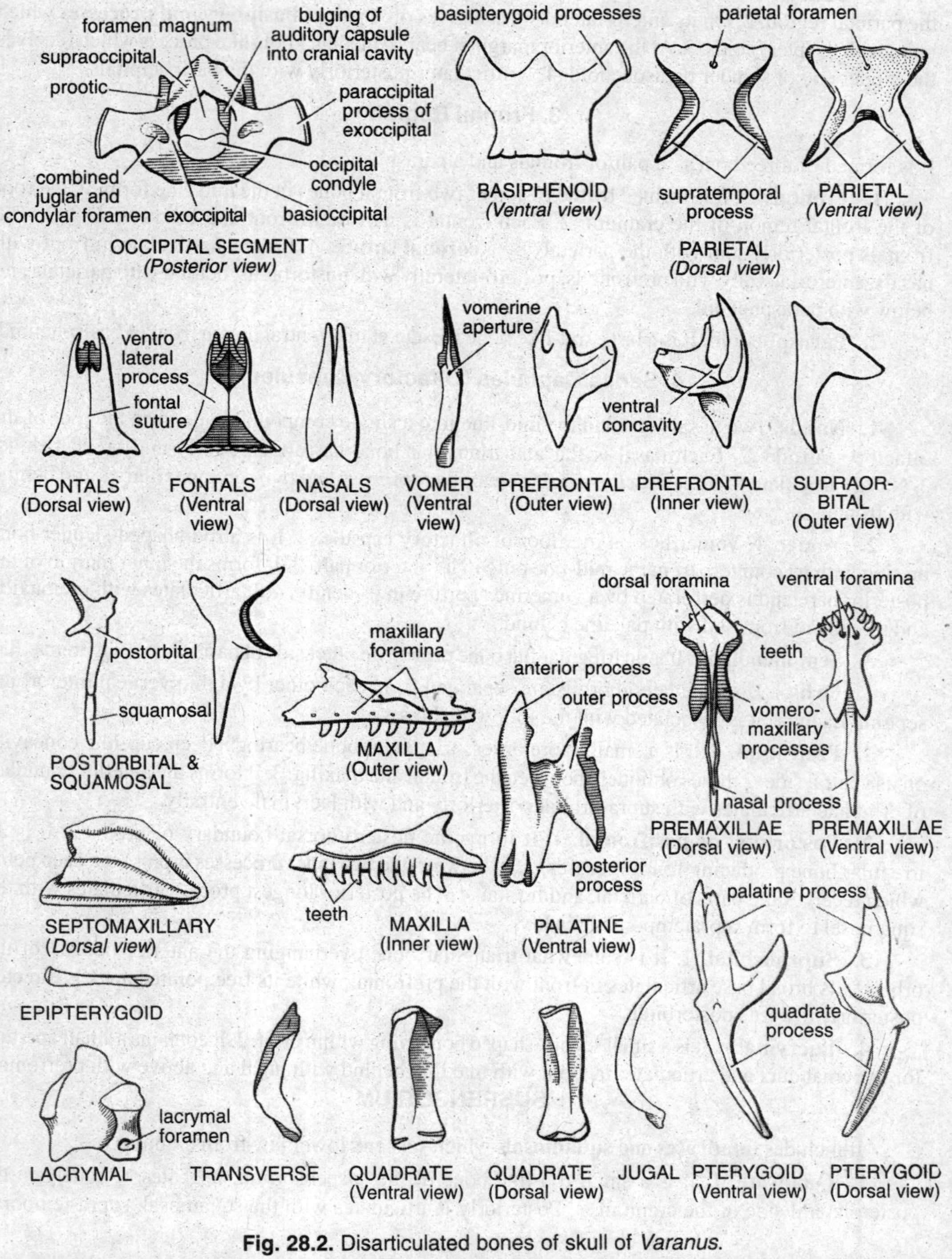

Fig. 28.2. Disarticulated bones of skull of *Varanus*.

1. Premaxillae. 1.Two premaxillae are fused in the midline into a single bone and forms the anterior limit of snout. 2. Posteriorly it is produced into a long narrow median nasal process which is lodged into the anterior groove of nasals. Nasal process lies between external nares and septomaxillaries. 3. A pair of wing-like ventral vomero-maxillary processes articulate with the vomers. 4. On its ventral surface it possesses 6–8 small conical teeth along its margins. 5. Dorsally it bears two premaxillary foramina. 6. Premaxillae articulate on either side with maxillae.

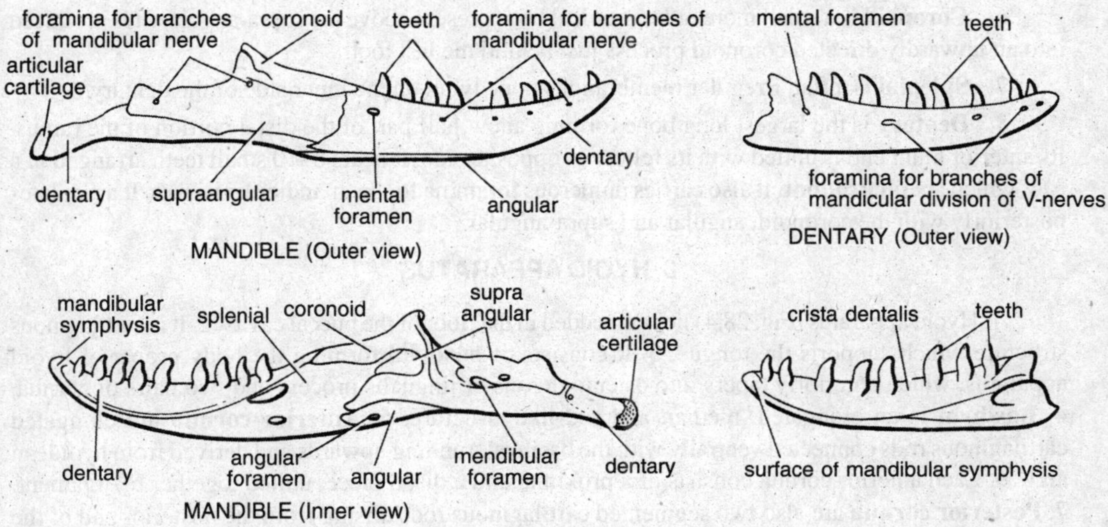

Fig. 28. 3. Lower jaw (Mandible) of *Varanus*.

2. Maxilla. 1. Maxilla forms the anterior **half of** upper half jaw. 2. It is a long, irregular bone lying on either side behind the premaxilla. 3. Outer surface of each maxilla is perforated by a number of maxillary foramina. 4. The main body of each maxilla is known as alveolar part which bears a row of 8–10 teeth along its ventral margin. 5. Teeth are small, conical and **pleurodont** type. 6. Palatine process present on its inner side is poorly developed. 7. Maxilla articulates in front with premaxilla, septomaxillary and vomer, and posteriorly with the palatine, jugal and transverse. Ascending process of maxilla joins with prefrontals, nasal and lacrymal.

3. Palatine. 1. Palatine is a small, flat and more or less irregular bone and forms the roof of the buccal cavity. 2. It is produced into three processes. Anterior longer process joins with vomer, posterior smaller process with pterygoid and outer process with maxilla and transverse. 3. It also forms the posterior boundary of internal nare.

4. Pterygoid. 1. It is a large enlongated and irregular bone found in the roof of mouth. 2. It has an anterior palatine process and a posterior quadrate process. 3. It articulates anteriorly with palatine on the inner side and transverse on the outer side, and posteriorly with basipterygoid process of basisphenoid on the inner side and with quadrate on the outer side.

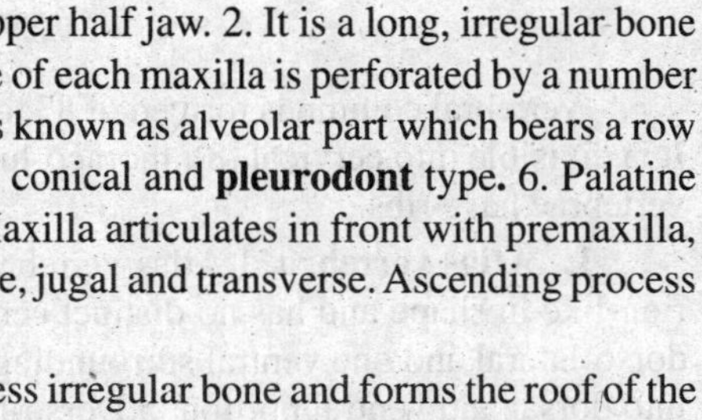

Fig. 28.4. Hyoid apparatus of *Varanus.*

7. Lower Jaw (Mandible)

1. The lower jaw or mandible consists of two similar half or rami meeting anteriorly.

2. Each ramus is composed of six bones, *i.e.*, articular, angular, supra-angular, coronoid, splenial and dentary around an axial Meckel's cartilage.

3. **Articular** is the posteriormost bone of the ramus. On its dorsal surface it bears an articular surface for quadrate and extend posteriorly into an articular cartilage.

4. **Angular** is small splint-like bone fitted in between the dentary and articular and perforated by an angular foramen.

5. **Supra-angular** is a flat, elongated and more or less rectangular bone lying in the middle of the ramus. It carries a pair of mandibular formina for the mandibular nerves.

6. **Coronoid** is a small more or less conical bone present above the supra-angular. It is produced into an upwardly directed coronoid process just behind the last tooth.

7. **Splenial** is a flat, irregular membranous bone lying on the inner side of the dentary.

8. **Dentary** is the largest long bone forming about half part of the distal portion of the ramus. Its anterior blunt end is united with its fellow of opposite side. It bears 8–10 small teeth arranged in a row, which are **pleurodont**. It also carries numerous foramina for the mandibular nerves. It articulates posteriorly with the coronoid, angular and supra-angular.

8. HYOID APPARATUS

1. Hyoid apparatus (Fig. 28.4) lies embedded in the floor of the buccal cavity. 2. It is cartilaginous structure which supports the tongue. 3. It consists of **basihyal** forming the body proper of hyoid apparatus, which anteriorly tapers into os entoglossus or lingualis process and two pairs of cornua. 4. **Basihyal** is an elongated, median and rod-like structure. 5. **Anterior cornua** are elongated cartilaginous rods connected ventrally with the basihyal, running upwards and derived from hyoidean arch. 6. Each anterior cornua consists of a proximal and a distal piece, united together by ligament. 7. **Posterior cornua** are also two segmented cartilaginous rods arising from the posterior end of the basihyal and running towards the hind end. These are derived from branchial arches.

VERTEBRAL COLUMN

Vertebral column is formed of 83 to 85 vertebrae having no epiphyses. Centrum is procoelous. It is divisible into cervical (8), thoraco-lumbar (22), sacral (2) and caudal regions (51–53). Thoracic vertebrae have ribs.

1. Atlas vertebra. 1.Atlas vertebra is the first vertebra of cervical region. 2. It is small and ring-like in shape and has no distinct centrum. 3. It is composed of three distinct bony pieces, two dorso-lateral and one ventral surrounding the neural canal. 4. Neural canal is divided by a ligament into dorsal and ventral portion. 5.Dorso-lateral pieces have a space filled by a membrane at the point of their union in the mid-dorsal line. 6. Along the anterior face of ventral piece is present a cavity for the articulation of occipital condyle of skull and on its posterior side present an articular facet for odontoid process of axis (2nd) vertebra. Ventral piece has also a small ventral spine. 7. Pre-and postzygapophyses and transverse processes are entirely absent.

2. Axis vertebra. 1.Axis is the second cervical vertebra. 2. It resembles the typical cervical vertebra in structure but it is slightly larger. 3. Neural spine forms a vertical crest or ridge. 4. **Centrum** anteriorly bears an odontoid process and beneath it is present a posteriorly directed spine. 5. A **hypapophysis** is also present on the postero-ventral surface of the centrum. 7. Prezygapophyses are rudimentary and postzygapophyses are well developed.

3. Typical cervical vertebra. 1. Centrum is stout, elongated and **procoelous** (concave in front and convex behind). 2.Neural spine is vertical and crest-like. 3. Neural arches bear anterior prezygapophyses directed upwards and posterior postzygapophyses directed downwards. 4. **Centrum** on the postero-ventral surface carries a backwardly directed **hypapophysis**. 5. Centrum of 4 to 8 cervicals on either antero-lateral surfaces have articular surfaces for cervical ribs.

4. Thoraco-lumbar vertebra. 1. Thoraco-lumbar vertebra is similar to cervical vertebra but it is slightly larger in size. 2. Centrum is strongly **procoelous.** 3. Neural spine is vertical crest-like and low. 4. Pre-and postzygapophyses are well formed. 5. Hypapophysis is absent. 6. On either side at the junction of neural arch and centrum is present a capitular facet for the articulation with thoracic rib.

5. First sacral vertebra. 1.It is stout and strong vertebra giving support to the pelvic girdle. 2. **Centrum** is short and strongly **procoelous** (concave in front and convex behind). 3. The neural spine is a low crest. 4. Pre-and postzygapophyses are well-developed. 5. Transverse processes are strong, expanded at the tips and are also notched to receive the ilia bones of pelvic girdle. 6. Hypapophysis is absent.

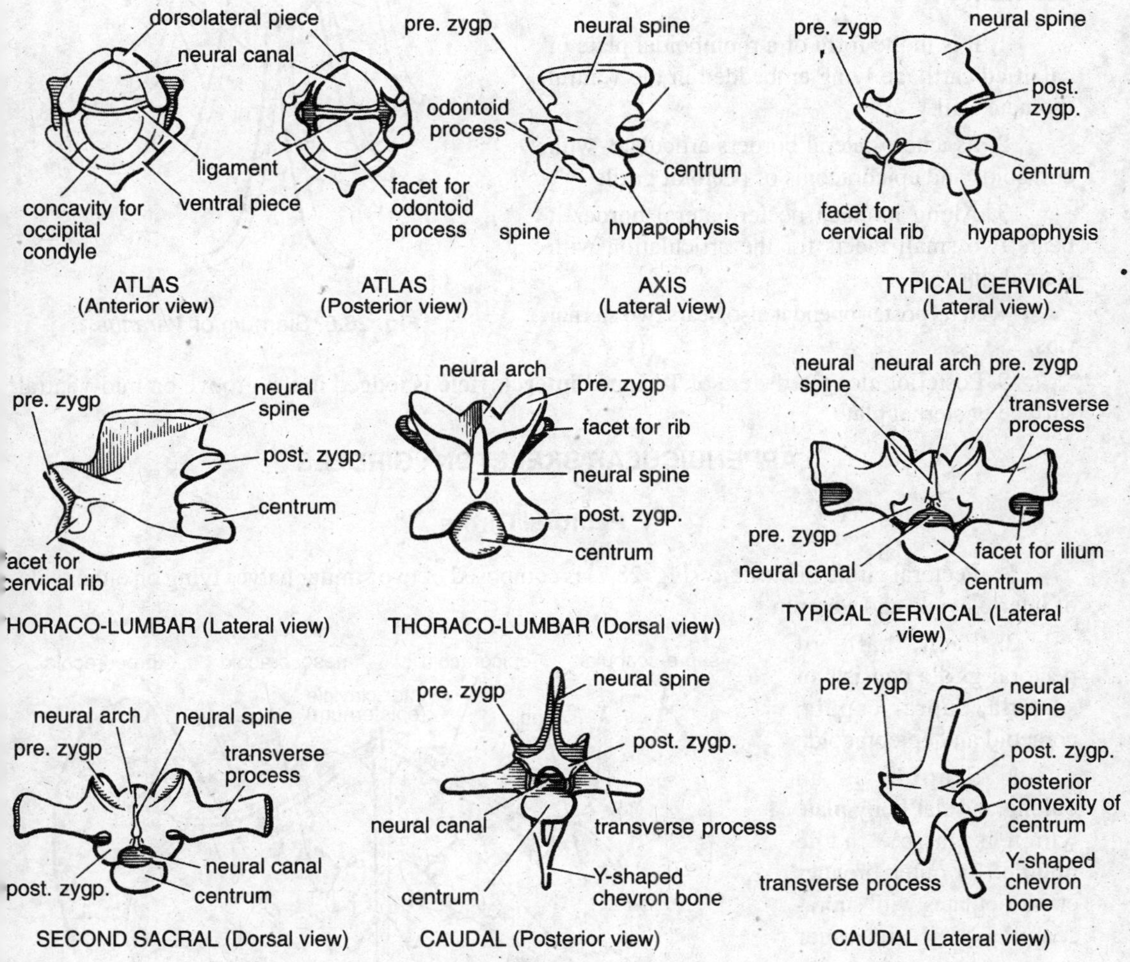

Fig. 28.5. Vertebrae of *Varanus.*

6. Second sacral vertebra. 1.Second sacral vertebra is similar to the first sacral in structure except only one difference, *i.e.*, transverse processes are not notched at tips. 2. **Centrum** is short and strongly **procoelous**. 3. Neural spine is a low crest. 4. Pre- and postzygapophyses are well-developed. 5. Transverse processes are strong and expanded at the tips but are not notched.

7. Anterior caudal vertebra (Typical). Numerous caudal vertebrae gradually reduce in size behind. 1.**Centrum** is longer and strongly **procoelous**. 2. **Neural spine** is long, pointed and backwardly directed. 3. Pre-and postzygapopyhses are well-developed. Transverse processes are slender and downwardly directed. 4. The most characteristic feature of the vertebra is the presence of Y-shaped **chevron bone** attached to ventral surface of the centrum. Centurm bears ventrally a pair of small nodules for the attachment of chevron bone. 5. Chevron bone encloses the haemal canal for the passage of artery and vein.

8. Posterior caudal vertebrae. Posterior caudal vertebrae resemble the anterior caudal vertebrae in structure except the absence of **chevron bone**. They become gradually smaller in size towards the posterior end. Near the tail end vertebrae are merely represented by a rod-like centrum, all parts of vertebrae get reduced.

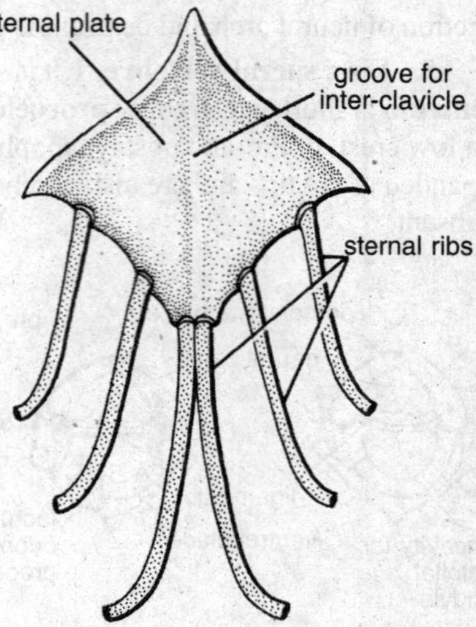

Fig. 28.6. Sternum of *Varanus.*

Sternum

1. It is in the form of a rhomboidal plate of calcified cartilage lying embedded in the ventral thoracic wall.

2. Its antero-lateral borders articulate with coracoids and epicoracoids of pectoral girdle.

3. Along its each posterolateral border it bears two small facets for the articulation with sternal ribs.

4. At its posterior end it also bears two sternal ribs.

5. Posterior elongated stem of T-shaped **interclavicle** is lodged in the groove on mid-ventral surface of sternal plate.

APPENDICULAR SKELETON : GIRDLES

1. Pectoral Girdle

1. Pectoral girdle of *Varanus* (Fig. 28.7) is composed of two similar halves lying on either side of inter-clavicle and sternum.

2. Each half of pectoral girdle consists of scapula, supra-scapula, coracoid and epicoracoid.

3. **Scapula** is an oblong and flat bony plate which is narrow in the middle. Its outer broader end articulates with supra-scapula, while its inner narrow end unites with coracoid.

4. **Supra-scapula** is more or less rectangular thin plate of calcified cartilage. It articulates ventrally or proximally with scapula and its distal or dorsal border is free and slightly curved inward.

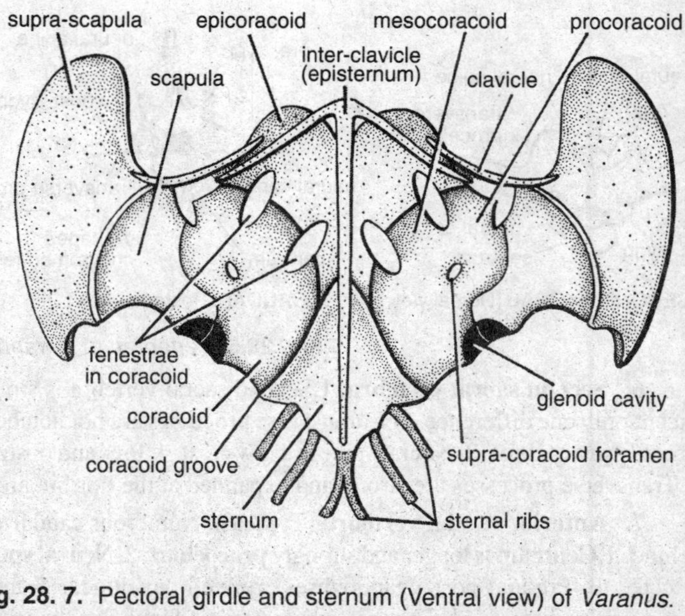

Fig. 28. 7. Pectoral girdle and sternum (Ventral view) of *Varanus.*

5. **Coracoid** is large, flat, fenestrated partly bony and partly cartilaginous. Two large fenestrae divide the bony coracoid part into outer procoracoid, middle mesocoracoid and inner coracoid proper that articulates with antero-lateral border of sternum.

6. **Epicoracoid** is cartilaginous and irregular-shaped arising from the anterior inner part of coracoid above the fenestrae.

7. Epicoracoids of two sides abut medially with posterior stem of interclavicle.

8. At the joint of scapula and coracoid along the ventral side is present a glenoid cavity for the articulation of head of humerus.

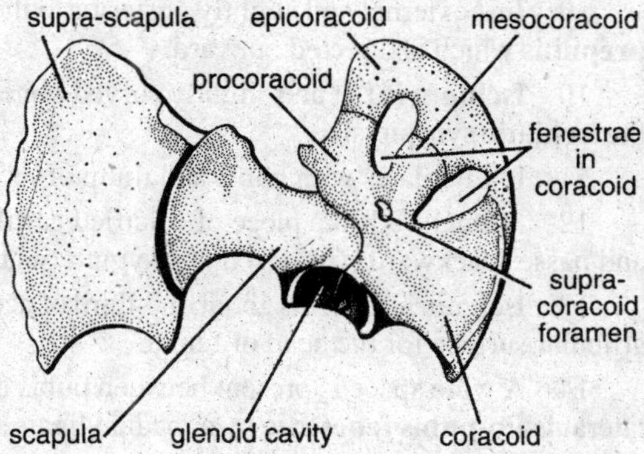

Fig. 28.8. One-half of pectoral girdle of *Varanus.*

Interclavicle and clavicle (Episternum). 1. **Episternum** is T-shaped present in between the two halves of pectoral girdle. 2. Anterior lateral limbs of interclavicle are lying anterior to the scapulae. 3. Its posterior stem is closely applied to the ventral surface of sternum. 4. **Clavicle** is a small, narrow, flat and curved dermal bone lying attached to the anterior side of the lateral limbs of the interclavicle. 5. Its outer end joins with the junction of scapula and supra-scapula. But its inner end does not unite with its fellow of opposite side in the middle.

2. Pelvic Girdle

1. Pelvic girdle (Fig. 28.10) consists of two similar halves (**ossa innominata**), meeting in the mid-line by a vertical ligament.

2. Each half is composed of ilium, pubis and ischium which are not fused with each other.

3. **Ilium** is strong, compressed rod-like bone directed upwards and backwards to articulate with the sacral vertebra.

4. On outer side it is produced into a knob-like **pre-acetabular process** in front of acetabulum and also contributes in the formation of about one third part of acetabulum.

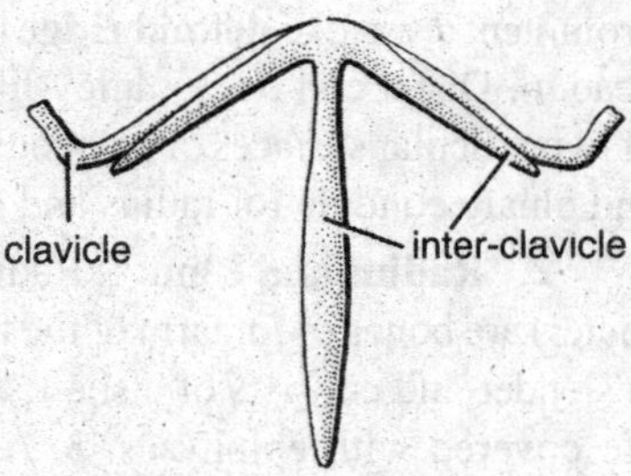

Fig. 28.9. Interclavicle and clavicle of *Varanus.*

5. **Pubis** is a flat and slightly curved bone passes downwards and forwards to meet in front with its fellow of the other side in mid-ventral line at pubic symphysis.

6. Between the anterior ends of two pubes in front of symphysis is present a nodule of calcified cartilage, the **epipubis**.

7. Pubis also contributes to about one-third of the acetabulum.

8. On the posterior end near its fusion with ilium and ischium it has an oval foramen for the obturator nerve.

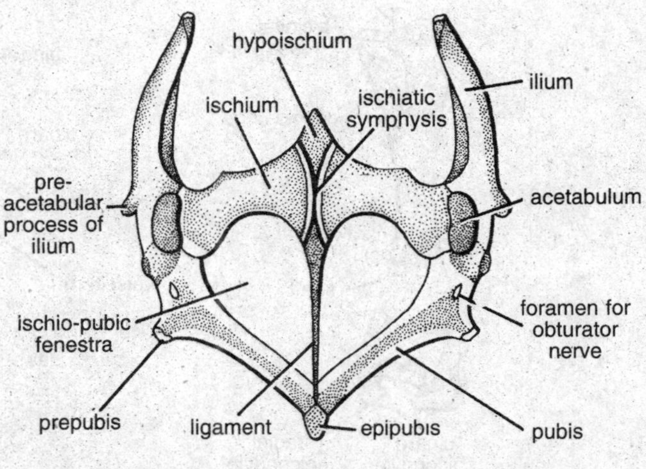

Fig. 28.10. Pelvic girdle (Ventral view) of *Varanus.*

9. Just external and slightly posterior to the above foramen arises a small rod-like process the **prepubis** which is directed outward.

10. **Ischium** is flat and slightly curved bone, runs inwards to meet its fellow of the other side at the **ischiatic symphysis**.

11. It articulates with pubis and ilium of its side.

12. A small rod-like piece of calcified cartilage **hypoischium** is present in between two ischia and passes backwards from ischiatic symphysis to give support to ventral wall of cloaca.

13. Ischium also forms about one-third of acetabulum which is a cup-shaped cavity that provides articular surface for the head of femur.

14. A wide space is present between pubis and ischium of both sides which is divided into two lateral **ishio-pubis fenestrae** by a median ligament.

LIMB BONES

Forelimb Bones

1. **Humerus.** Humerus (Fig. 28.12) is the bone of the upper arm of forelimb. 2. It has an elongated shaft, whose both ends are expanded and covered by epiphyses of calcified cartilage. 3. Proximal end bears rounded head which fits into the glenoid cavity of pectoral girdle. 4. A **bicipital fossa** is enclosed between medial process and head at proximal end. 5. A prominent crest-like **deltoid ridge** is present below the head. 6. Distal end has a pulley-like trochlea formed of two articular surfaces or epicondyles: radial condyle and ulnare condyle for radius and ulna respectively.

2. Radius and Ulna. 1. Radius (inner) and ulna (outer) are bones of forearm of the forelimb. 2. **Radius** is slender and consists of a shaft, both ends of which are covered with epiphyses. 3. At the distal end of

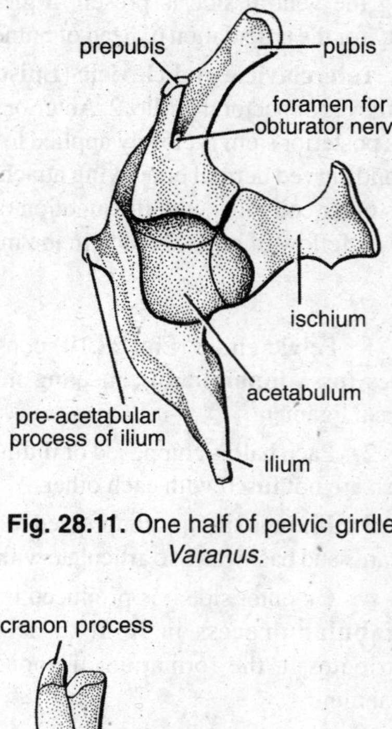

Fig. 28.11. One half of pelvic girdle of *Varanus*.

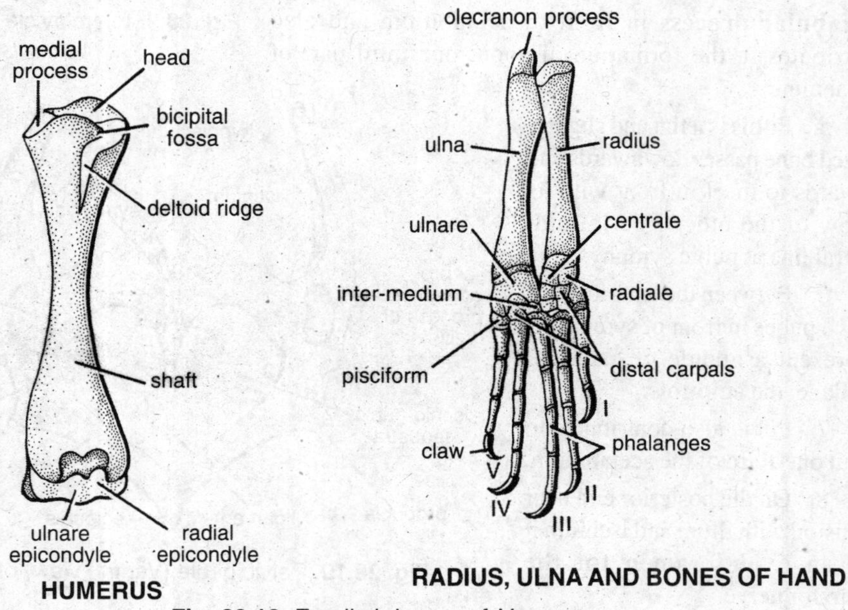

Fig. 28.12. Forelimb bones of *Varanus*.

radius is present a concave articular facet for carpus and preaxial styloid process. 4. **Ulna** is relatively stouter. Its proximal end is produced into an upwardly directed **olecranon process**, whereas distally it bears a convex articular surface for carpus.

3. Bones of forefoot. 1. Carpus (wrist) consists of ten rounded polyhedral carpals arranged in two rows 2. **Proximal row** contains three carpals–radiale, ulnare and intermedium. 3. **Distal** row has five small carpals. 4. **Central** lies in between two rows of carpals. 5. Besides the above a **pisciform** is attached to the distal epiphysis of ulna on its post-axial side. 6. **Manus** (hand) has five long **metacarpals** which bear five digits having 2, 3, 4, 5 and 3 phalanges with terminal horny claw.

Hindlimb Bones

1. Femur. 1. Femur is stout bone of thigh of hindlimb having an elongated shaft and two epiphyses at both ends. 2. Proximal end has a rounded head which fits into acetabulum of pelvic girdle. 3. Proximal end also bears two tuberosities–greater trochanter at postaxial side and lesser trochanter at preaxial side. 4. Distal end is pulley-shaped having two condyles for articulation with tibia and fibula.

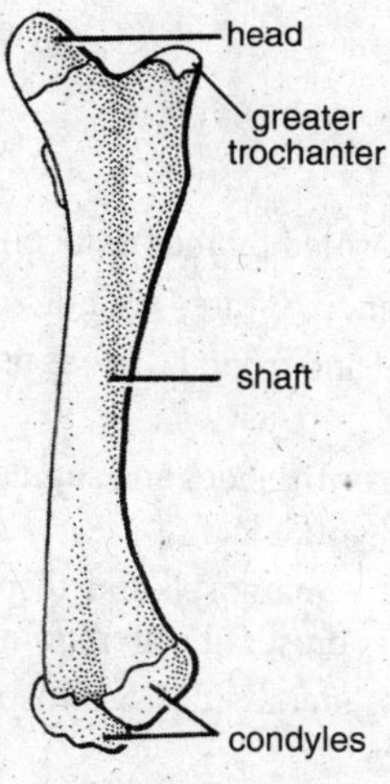

FEMUR

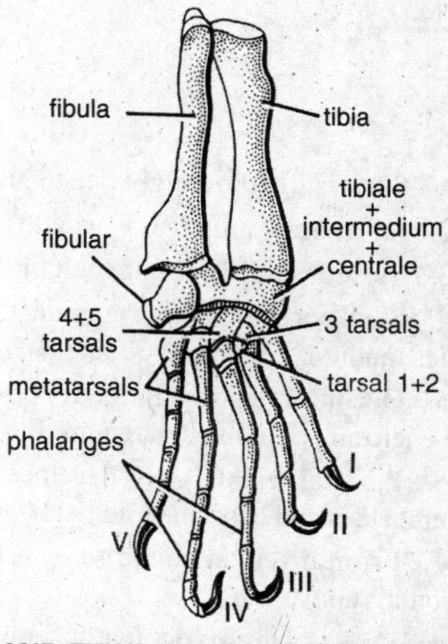

TIBIA AND FIBULA AND BONES OF FOOT

Fig. 28.13. Hindlimb bones of *Varanus*.

2. Tibia and fibula. 1. Tibia and fibula are the bones of shank of hindlimb. 2. **Tibia** is a stout bone and bears a longitudinal ridge, **cnemial crest** on the dorsal (anterior) surface. 3. Proximal end of tibia also bears two concave facets for articulation with condyles of femur. 4. **Fibula** is relatively slender bone proximally articulating with femur. 5. Distal ends of both tibia and fibula provide surface for the articulation with tarsals.

3. Bones of hindfoot. 1.Tarsus contains five **tarsal bones**, two in the proximal row and three in distal row. 2.Two tarsals of proximal row are suturally fused and articulate with tibia and fibula. 3. Distal row consists of three cuboid tarsals articulating with five metatarsals of foot (pes). 4.The number of phalanges in five toes is variable; first has 2, second has 3, third has 4, fourth has 5 and fifth has 3 phalanges. Terminal phalanx of each toe is clawed.

29

Chapter

SKELETON OF FOWL

AXIAL SKELETON

1. Skull

1. Skull is light in weight due to pneumatic (spongy) bones.
2. It is **monocondylic**, *i.e.*, single occipital condyle is present formed by the basioccipital.
3. Most of skull bones of adult bird are intimately fused and, hence, sutures are not visible.
4. Skull consists of large cranium that includes occipital, parietal and frontal regions and sense capsules (auditory, optic and olfactory).
5. **Occipital region** is formed of basioccipital below, exoccipitals on the sides and supraoccipital above enclosing a foramen magnum. Basioccipital forms the occipital condyle.
6. **Parietal region** is well developed having fused parietals above, squamosals and alisphenoids on lateral sides and basisphenoid below, and below the basisphenoid is present basitemporal.
7. **Frontal region** comprises presphenoid on ventral sides, on lateral sides orbitosphenoids and frontals above.
8. Skull is **tropibasic**, that is, cranium does not extend in front into orbital region. Both the orbits are separate from each other by interorbital septum formed mainly by mesethmoid together with presphenoid, parasphenoid, rostrum and orbitosphenoids.
9. **Auditory capsule** is composed mainly of prootic, lying on either lateral side of occipital segment. Tympanic cavity contains columella (ear ossicle) whose inner slender rod-like bone, **columella auris** and outer triradiate part is called **extra columella**.
10. **Olfactory capsules** are smaller in size, lie in front of cranium and composed of nasals forming roof of nasal chambers, vomers below, and poorly developed ectoethmoids (turbinals).
11. Facial portion is prolonged forwards to form a long pointed toothless beak.
12. **Upper jaw** is composed of paired premaxilla, maxilla, jugal and quadratojugal.
13. Palate **schizognathus** type formed by vomers, palatines, pterygoids and palatal processes of maxilla. Vomers are short, thus, palatines meet in the middle.
14. Suspensorium is **autostylic**.
15. **Lower jaw** articulates with the upper jaw by distinct movable quadrate bones.
16. Lower jaw consists of five paired bones, *i.e.*, articular, angular, supra-angular, dentary and splenial.

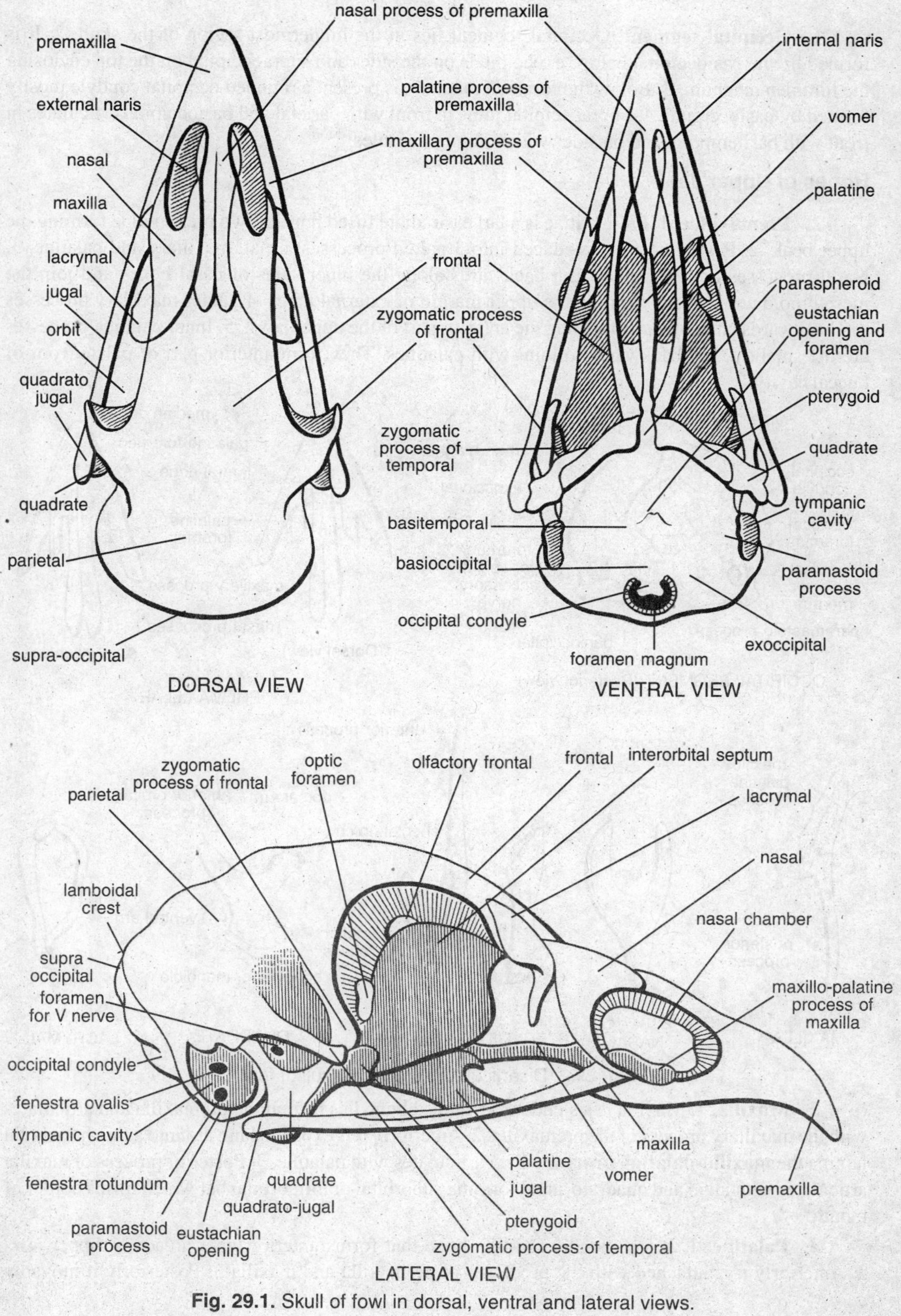

Fig. 29.1. Skull of fowl in dorsal, ventral and lateral views.

DISARTICULATED SKULL BONES OF FOWL

1. Occipital segment. Occipital segment lies in the hindermost region of the skull, 2. It is formed by the basioccipital below, exoccipitals on the sides and supraoccipital on the top enclosing the foramen magnum, 3. Below the foramen magnum is present a rounded occipital condyle mostly formed by basioccipital. 4. Supraoccipital joins in front with parietal and basioccipital articulates in front with basitemporal and on sides with auditory capsules.

Bones of Upper Jaw

2. Premaxillae. 1. Premaxillae is a large triradiate fused bone of two premaxillae forming the upper beak. 2. Posteriorly it is produced into 3 paired processes: nasal, maxillary and palatine. 3. Nasal processes are the longest run backwards along the inner sides of nasal bones and join the mesethmoid and frontals forming the upper margin of external nares. 4. Outer maxillary processes run backwards and outwards forming the anterior part of the upper jaws. 5. Inner palatine processes are the smallest, run backwards to join with palatines. They form anterior part of palate (roof of buccal cavity).

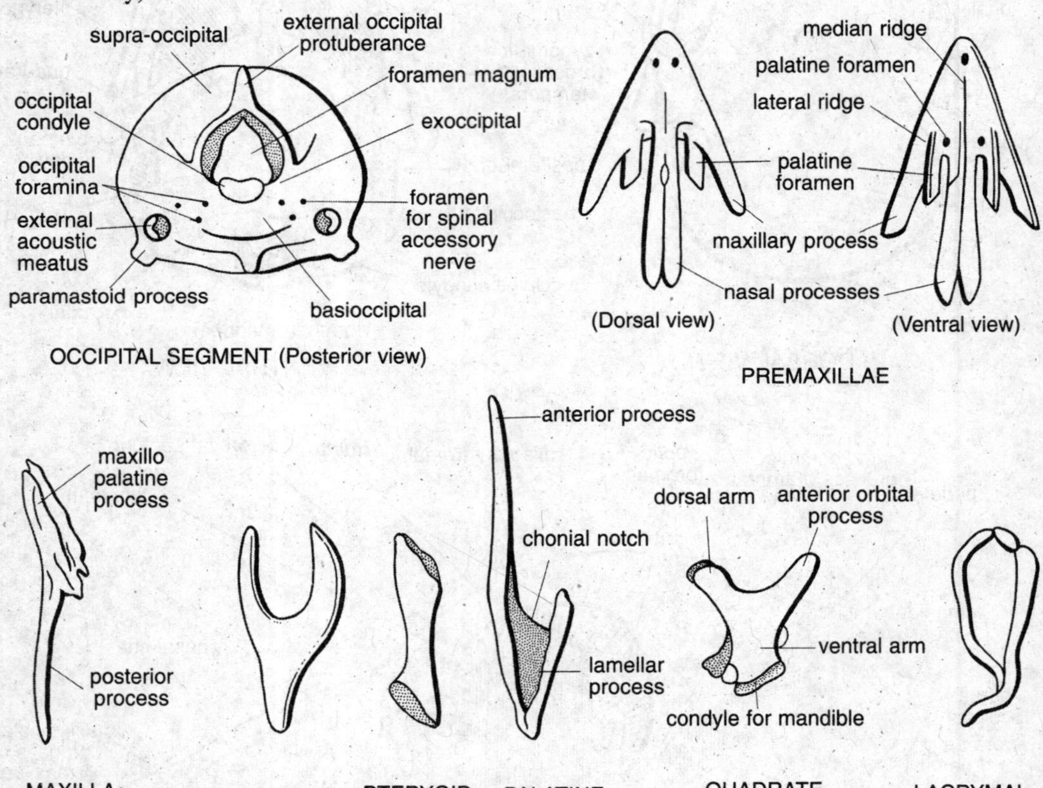

Fig. 29.2. Disarticulated bones of skull of fowl.

3. Maxilla. 1. Maxilla is a slender rod-shaped bone lies behind the premaxilla and articulates with the maxillary process of the premaxilla, 2. Anteriorly it is expanded inward into a thin horizontal lamina the **maxillo-palatine process** which articulates with palatine, 3. Posterior process of maxilla articulates with jugal and quadrato-jugal forming suborbital or infraorbital bar which forms border of mouth.

4. Palatine. 1. It is a slender rod-like bone that forms anterior inner arcade of upper jaws 2. Anteriorly it is attached with the processes of premaxilla and maxilla 3. Posteriorly it movably articulates with pterygoid. Its posterior end expands inward into a broad lamella which articulates with parasphenoidal rostrum.

5. Pterygoid. It is a short and stout rod-shaped bone, lying obliquely behind the palatine and forms posterior roof of mouth cavity. 2. It articulates behind with the inner surface of quadrate and in front with parasphenoid rostrum and palatines.

6. Quadrate. 1. It is a stout and triradiate bone located in front of tympanic cavity and forms suspensorium with squamosal. 2. Its dorsal arm movably articulates with squamosal between the base of its zygomatic process and tympanic cavity. 3. Its ventral arm is flattened and forms condyle to articulate with lower jaw (mandible). 4. Its anterior orbital arm runs anteriorly parallel to pterygoid and remains free.

7. Lacrymal. 1. It is the bone of orbital capsule and forms the anterior boundary of orbit. 2. It is produced below into a slender curved process. 3. It bears a characteristic lacrymal foramen.

8. Nasal. 1.It is the bone of olfactory capsule and both nasals cover the roof and sides of nasal chambers (olfactory capsule). 2.It is a thin more or less triangular bone having three processes. 3. It gives posterior process joins with the frontal and lacrymal, while two anterior processes with a notch form the posterior boundary of the external nare and articulate with the process of premaxillae.

Lower Jaw (Mandible)

1. Lower jaw (mandible) consists of two long laterally compressed rami firmly united in front articulate behind with quadrate. 2. Each ramus is composed of five fused bones, *i.e.*, one replacing bone **articular** and four investing bones **angular**, **supraangular**, **dentary** and **splenial** which develop around the Meckel's cartilage. 3. **Articular** forms the posterior expanded part of each ramus. On dorsal surface it bears concave articular surface for the quadrate. Posteriorly it is produced into medial and angular processes. 4. **Angular** is a slender bone lies beneath the articular along the lower and inner border of posterior part of ramus. 5. **Supraangular** forms the dorsal part of about one third of the posterior part of the ramus. It bears a **coronoid process**. 6. **Splenial** is a thin bone lying on the inner surface of the middle of the ramus. 7. **Dentary** is the largest bone and forms about the complete distal half of the ramus. Anteriorly it is united with its counterpart of the other side forming the mandibular symphysis, while posteriorly it extends up to articular between angular and supraangular.

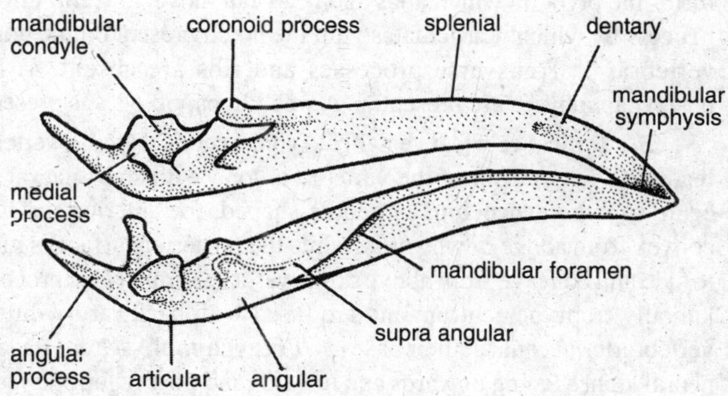

Fig. 29.3. Lower jaw of fowl (Lateral view).

10. Hyoid apparatus. 1. Hyoid apparatus lies in the floor of the mouth and supports the tongue. 2. It consists of a median arrow-shaped body formed of three bones : anterior **entoglossal (ceratohyal)** with pointed cartilaginous process in front, a median bony **basihyal**

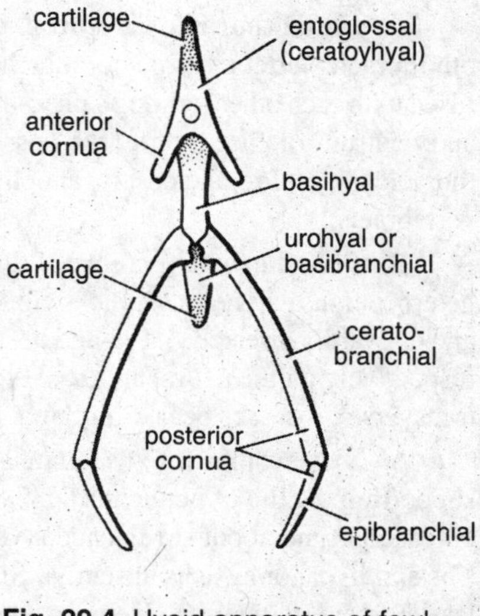

Fig. 29.4. Hyoid apparatus of fowl.

and a posterior **urohyal** or basibranchial with a pointed cartilage behind. 3. Entoglossal behind gives off a pair of processes **anterior cornua** which represent its free ends. 4. **Posterior cornua** arise from the junction of basihyal and urohyal which run backwards. Each cornu is made of two rod-like segments joined with cartilage: ceratobranchial and epibranchial.

VERTEBRAL COLUMN

Vertebrae lack epiphyses and centrum are **heterocoelous** (saddle-shaped). Cervical vertebrae are 14–16, thoracic are 7, synsacrum is fused vertebrae of thoracic, lumbar, sacral and few anterior of caudal region.

1. Atlas Vertebra. 1. Atlas is the first small ring-like vertebra of the cervical region. 2. Centrum, neural spine, transverse processes and prezygapophyses are absent. 3. Its thick ventral portion is the centrum. Its anterior side bears a cup-like cavity to provide articulation to the single occipital condyle of the skull. Centrum is notched above to receive the odontoid process of axis vertebra. 4. Atlas is formed of 3 fused pieces: a single and two **dorso-laterals ventral** which join mid-dorsally to form neural arch. 5. Posterior border of neural arch is produced into a pair of **postzygapophyses** for the articulation with axis vertebra.

2. Axis (Epistropheus) vertebra. 1. Axis is the second vertebra of the cervical region and forms the pivot on which atlas and head can move. 2. **Centrum** is produced in front into an **odontiod process** by which it articulates with the notch present on the dorsal side of the ventral piece of atlas vertebra. 3. Transverse processes and ribs are absent. 4. Neural spine is blunt. 5. Pre-and postzygapophyses are present. 6. A small hypapophysis is present below the centrum.

3. Typical cervical vertebra. Typical cervical (6–10) vertebrae have the following characteristic features : 1. Centrum of the vertebra is long with short neural arch and rudimentary spine. 2. The centrum is **heterocoelous** or saddle-shaped, the anterior surface is concave from side to side and convex from above downwards, while the posterior surface is more prominent and convex from side to side and concave from above downwards. 3. Transverse processes are short and irregular and arise laterally from the centrum and are fused with backwardly directed spine-like cervical ribs having vertebrarterial canal at their bases. 4. **Prezygapophyses** are flat articular facets on the anterior side of neural arches facing upwards and inwards and **postzygapophyses** are projecting backwards from the posterior side of the neural arch and facing downwards. 5. Behind 10th cervical vertebra, remaining cervicals have a ventral spine-like hypapophysis below the centrum and neural spine is distinct. Last cervical vertebra bear movably articulated two headed ribs.

4. Fused thoracic vertebrae. 1. Fused thoracic vertebrae are formed by the fusion of 2nd to 5th thoracic vertebrae. 2. Neural arches, neural spines, transverse processes and hypapophyses are fused with each other forming a plate-like ridges pierced by intervertebral gaps. 3. Transverse processes and centrum of fused vertebrae possess on either side the tubercular and capitular facets for the thoracic ribs. 4. Pre-and postzygapophyses are present on anterior and posterior sides of fused thoracic vertebrae.

5. Free thoracic vertebra. 1. First and VI vertebrae are free thoracic vertebrae. 2. **Centrum** is heterocoelous and neural spine is elongated and pointed. 3. Transverse processes are well developed and outwardly directed. 4. Pre-and postzygapophyses are present on the anterior and posterior surfaces respectively on the neural arch. 5. Hypapophysis is prominent beneath the centrum. 6. Centrum and transverse processes bear capitular and tubercular facets for the articulation with thoracic ribs.

6. Synsacrum. 1. Synsacrum is a compound bone formed by the fusion of about 16 vertebrae, supporting the ilia of pelvic girdle. 2. Various vertebrae of synsacrum are last thoracic, six lumbars, two sacrals and about seven caudal vertebrae. 3. Last thoracic vertebra bears a pair of free thoracic ribs and hypapophysis is absent. 4. Six vertebrae of lumbar region are fused together and also with last thoracic vertebra. 5. Neural spines of last thoracic and lumbar vertebrae are fused to form a

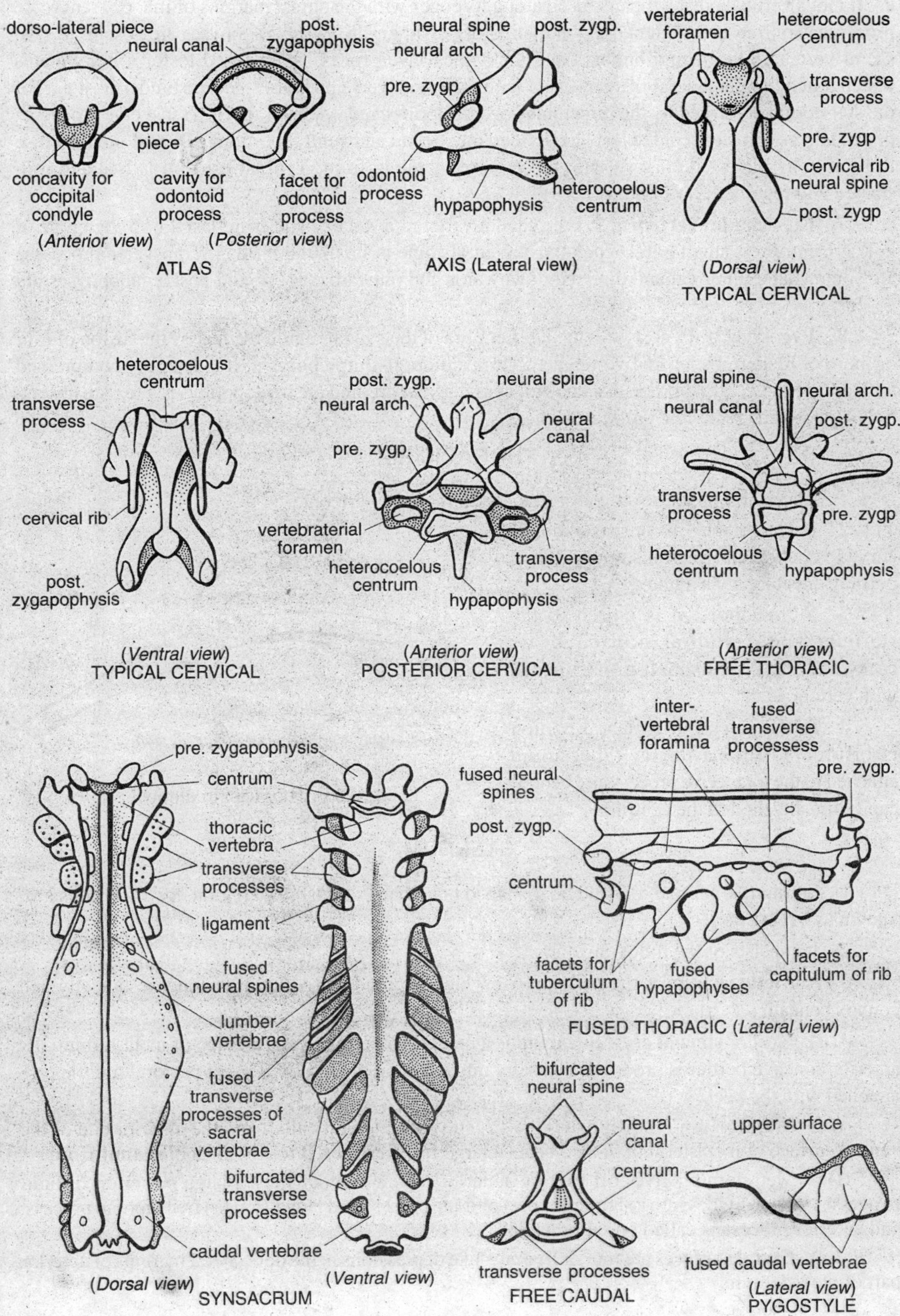

Fig. 29.5. Vertebrae of fowl.

continuous vertical crest which is fused along its edges with the dorsal margins of ilia. 6. Transverse processes are free and stout and touch the ilia of pelvic girdle. 7. Hypapophyses are absent. 8. Two sacral vertebrae lie behind lumbars and their transverse processes are fused to form bony plates which touch the ilia. Sacral ribs are fused with their transverse processes. 9. The hindermost part of the synsacrum is formed by the remaining seven anterior caudal vertebrae. 10. Transverse processes of caudal vertebrae except the last one divided into dorsal and ventral portions : **Dorsal portion** fuse to form bony plates continuous with those of vertebrae in front. **Ventral portion** is free and rod-shaped and regarded as rib.

7. Free caudal vertebra. 1. Free caudal vertebra is small . These are last 4 or 5 vertebrae of tail. 2. Centrum is small **heterocoelous**. 3. Neural spine is small and bifid. 4. Transverse processes are somewhat cylindrical and directed downwards and outwards. 5. Pre-and postzygapophyses are absent.

8. Pygostyle. 1. Pygostyle is the last vertebra of the caudal region formed by the fusion of four or more of hindermost caudal vertebrae. 2. It is a **plough-shape bone**. 3. It is laterally compressed with dorsal and ventral border. 4. Centrum, neural spine, pre-and postzygopophyses, etc., are absent. 5. It supports the rectrices (tail-feathers).

THORACIC RIB

1. Thoracic ribs are seven pairs attached to thoracic vertebrae.
2. Each thoracic rib consists of dorsal curved vertebral and ventral straight sternal portions meeting at an angle.
3. Sternal portion is united to the sternum and vertebral portion bears backwardly directed **uncinate process** united with vertebra.
4. Proximal end of vertebral portion is bifurcated into a dorsal tuberculum attached to the transverse process and ventral capitulum attached to the centrum.

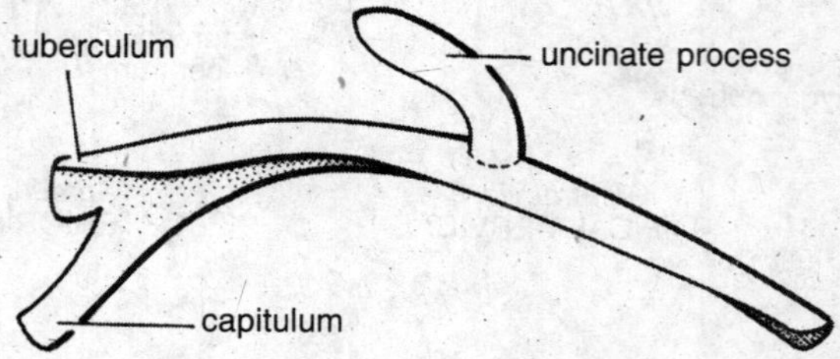

Fig. 29.6. Thoracic rib of fowl.

STERNUM

1. Sternum is a broad, flat and boat-shaped bone lying in the breast region, therefore, it is also known as **breast bone.**
2. It consists of the following parts :

(*i*) **Metasternum** is the body proper which is broad plate-like. It is concave dorsally and convex beneath.

(*ii*) **Keel** or **sternal crest** or **carina** is a median triangular, ventral ridge of metasternum. Its anterior concave border is produced ventrally into a cranial process. The great pectoral muscles are attached with the keel.

(*iii*) **Manubrium** is small plate-like structure arising vertically from the anterior end of the ventral surface of metasternum. It also bears a large foramen, **notch** for foramen of **manubrium**.

(*iv*) Metasternum gives off from its anterior region on both sides a large posterior **xiphoid processes** which divides into a shorter (upper) oblique process and larger (lower) free xiphoid process, and anterior process is called costal process.

(*v*) **Costal surfaces** are four or five cup-like depressions in the dors lateral margins of anterior part of metasternum.

(*vi*) **Coracoid grooves** are deep saddle-shaped grooves lying at the base of manubrium and communicate with each other through a foramen of manubrium. These grooves provide articular surface for the coracoid bones.

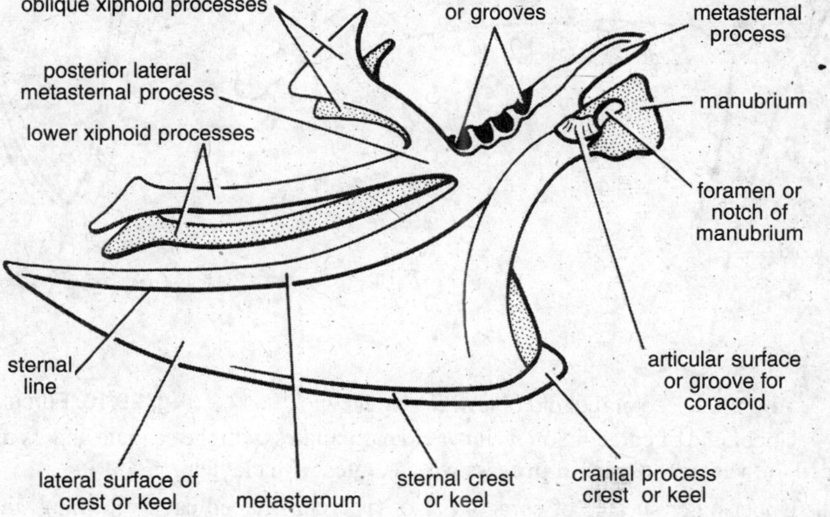

Fig. 29.7. Sternum of fowl. (Lateral view).

APPENDICULAR SKELETON

GIRDLES

1. Pectoral Girdle

1. Pectoral girdle (Fig. 29.8 and 29.9) is a very stout bony structure connected with the sternum on either side to support the wings.

2. Pectoral girdle on either side consists of a thin scapula, large coracoid and a slender clavicle.

3. **Scapula** is a long, flattened, sabre-shaped bone lying above the thoracic ribs parallel to the vertebral column.

4. Anterior expanded end of scapula is firmly united with coracoid by a ligament.

5. On its anterior outer side it bears shallow depression forming a part of glenoid cavity and its inner surface is produced into an **acromian process** to provide articular surface to the clavicle and to form a part of **foramen triosseum**.

6. **Coracoid** is a stout, straight bone directed downwards and its broad flat end articulates with coracoid groove on antero-lateral edge of sternum at the base of manubrium.

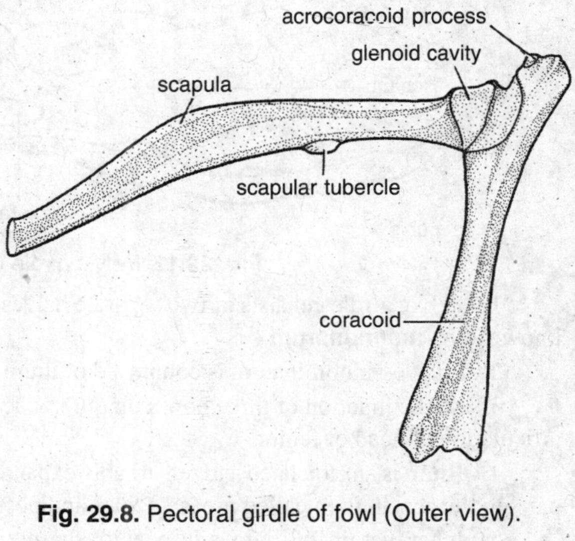

Fig. 29.8. Pectoral girdle of fowl (Outer view).

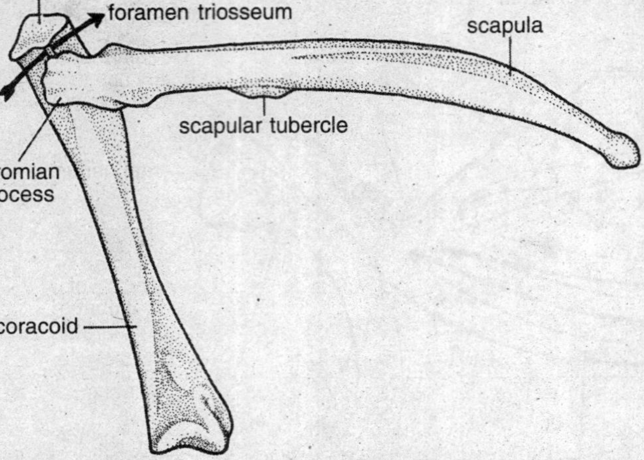

Fig. 29.9. Pectoral girdle of fowl (Inner view).

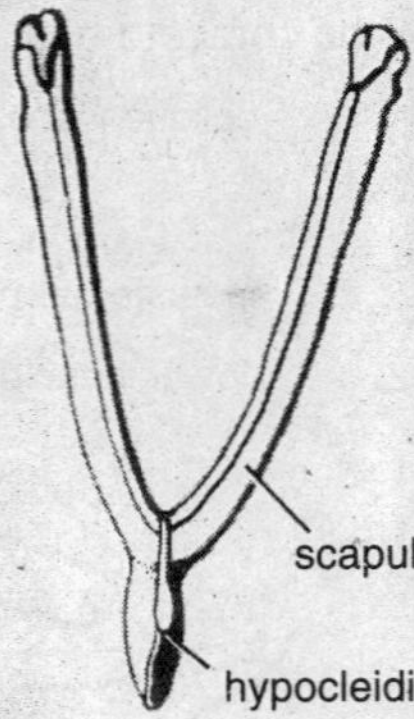

Fig. 29.10. Furcula of fowl.

7. Upper end of coracoid on its inner side articulates with the scapula and its upper end is also produced into an **acrocoracoid process** which unites with clavicle.

8. Upper outer surface of coracoid also has a cup-shaped cavity forming the larger part of glenoid cavity.

9. Clavicles are a pair of slender curved bones connected by their expanded upper ends with coracoid and scapula and enclose a smooth circular **foramen of triosseum** through which passes the tendon of pectoralis minor muscle. Ventrally the two clavicles are fused with the interclavicle to form a laterally compressed disc. The fork-shaped bone, thus, formed is known as "**furcula** or **merry thought**."

2. Pelvic Girdle

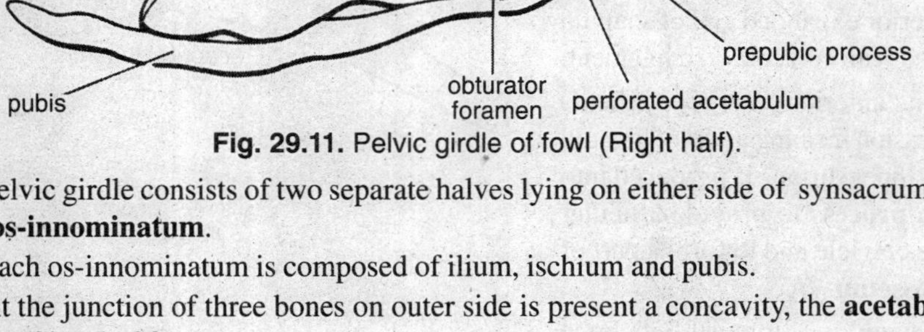

Fig. 29.11. Pelvic girdle of fowl (Right half).

1. Pelvic girdle consists of two separate halves lying on either side of synsacrum. Each half is known as **os-innominatum**.

2. Each os-innominatum is composed of ilium, ischium and pubis.

3. At the junction of three bones on outer side is present a concavity, the **acetabulum** for the articulation of head of femur.

4. **Ilium** is an elongated and remarkably expanded bone extending on both anterior and posterior parts to the acetabulum called preacetabular and postacetabular parts.

5 Inner margin of ilium is fused with synsacrum. Outer surface of its anterior part is concave and posterior part is convex. Ilium forms dorsal part of acetabulum.

6. On the outer surface above the acetabulum is a small projection called the **antitrochanter** which articulates with great trochanter of femur.

7. **Ischium** is a broad vertical flattened bone projecting backwards behind the acetabulum and forms about one third of acetabulum.

8. Ischium is fused posteriorly with ilium but separated anteriorly from it by an oval **ilio-ischiatic foramen**.

9. **Pubis** is a long slender curved bone directed backwards parallel to the ventral margin of ischium. It forms ventral part of acetabulum.

10. Just in front and outside acetabulum the pubis gives off a short and blunt process, the **prepubic process**.

11. Behind acetabulum pubis and ischium are separated by a slit-like opening, the **obturator foramen** and behind it both join with each other for a short distance.

FORELIMB BONES

1. Humerus. 1. Humerus (Fig. 29.12) is the bone of upper arm of forelimb. 2. It is an elongated stout bone expanded at both ends. 3. **Proximal end** is expanded into smooth convex surface, the **head** which fits into the glenoid cavity. 4. Head is bordered by two tubercles, smaller **preaxial** and larger **postaxial tuberosities.** (*i*)From preaxial tuberosity extends in front a short **deltoid ridge**. (*ii*) Postaxial tuberosity has a pneumatic foramen. 5. **Distal end** possesses a convex trochlear surface (condyles) for the articulation with radius and ulna.

2. Radius and ulna. 1. Radius and ulna (Fig. 29.12) are the two separate bones of forearm. 2. **Radius** is a short, slender, and straight bone. 3. Proximally it has a cup-shaped articular surface for the preaxial trochlea of humerus. 4. Distally it bears a knob that articulates with carpus (radiale). 5. **Ulna** is stouter and larger than the radius and also slightly curved bone. 6. Proximally it bears a small olecranon process and a large articular facet for the inner condyle of humerus. 7. Distally it articulates with ulnare carpal and radius (radiale).

3. Carpometacarpus. 1.Carpometacarpus (Fig. 29.12) is the bone of manus (palm). 2. It is a compound bone formed by the fusion of three metacarpals with the distal row of carpals. 3. **First** metacarpal is very short and a stumpy projection. 4.**Second** metacarpal is a strong and straight bone. 5. **Third** metacarpal is represented by thinner, slightly curved rod on the postaxial side and fused at both ends with second metacarpal. 6.**Phalanges.** Fingers are three clawless. First digit or pollex has one phalanx, second digit (index) has two phalanges and third digit has only a single phalanx.

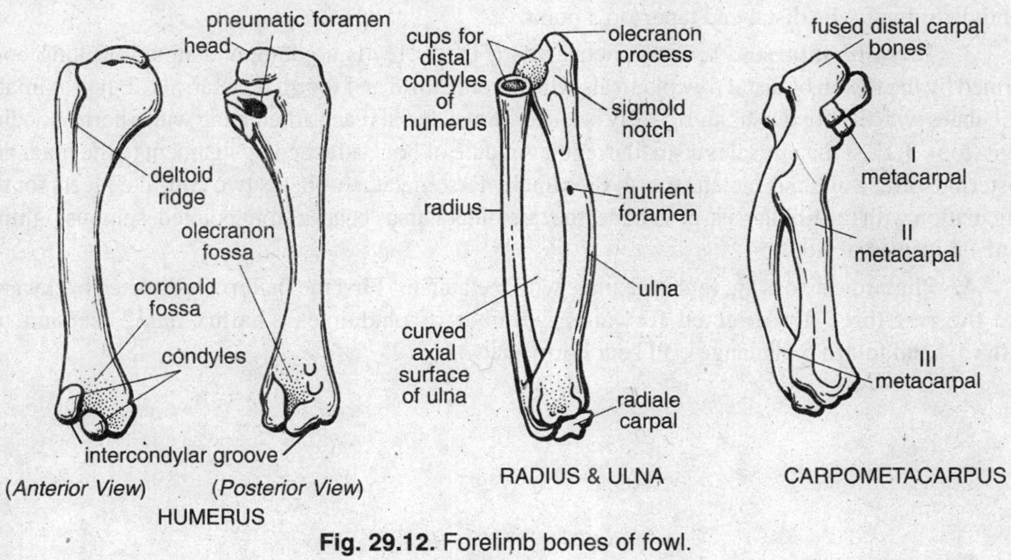

Fig. 29.12. Forelimb bones of fowl.

HINDLIMB BONES

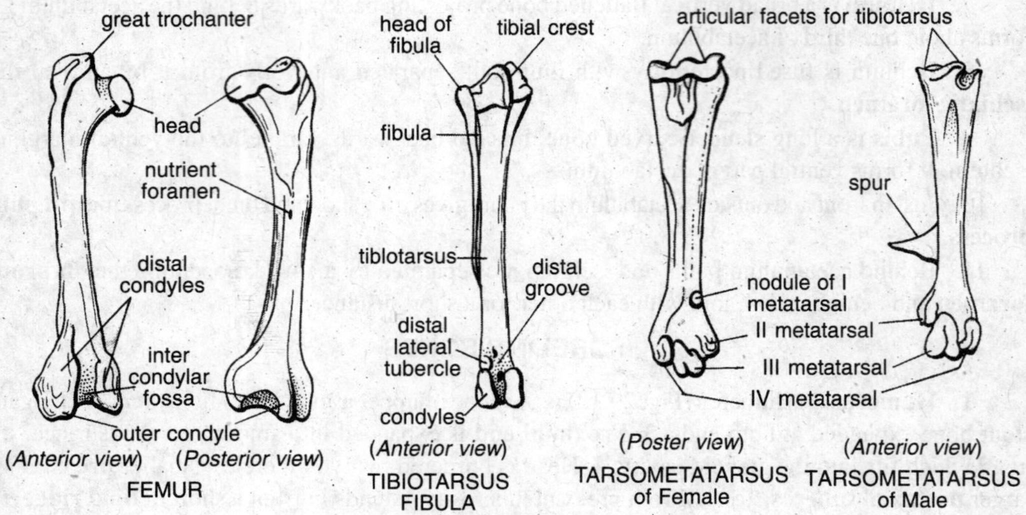

Fig. 29.13. Hindlimb bones of fowl.

1. Femur. 1. Femur (Fig. 29.13) is the bone of thigh of hindlimb. 2. It is short and powerful bone enlarged at both ends. 3. Proximally it is produced on the inner side into a rounded **head** which fits into the acetabulum. 4. On the outer side of head is a distinct irregular process, the **great trochanter**. 5. Between great trochanter and head is the articular surface for the antitrochanter of ilium. 6. Distally it has a pulley-like structure having in front a deep groove bounded laterally by two condyles for articulation with tibiotarsus. 7. Outer condyle is deeply grooved that receives upper end of fibula. 8. **Patella** is a sesamoid bone found in the tendon in front of tibiotarsus-femur joint.

2. Tibiotarsus and fibula. 1. **Tibiotarsus** and **fibula** (Fig. 29.13) are the bones of shank of hindlimb. 2. **Tibiotarsus** is a long, stout and nearly straight bone formed by the fusion of tibia with proximal row of tarsals (astragalus and calcaneum). 3. Proximal end of tibiotarsus is expanded and has on its anterior face a **tibial crest** and two articular surfaces for the condyles of femur. 4. Distally it articulates with the tarso-metatarsus by a pulley-like articular surface. 5. **Fibula** is a small slender bone closely applied to the outer surface of the tibiotarsus. 7. Proximally it articulates with the outer condyle of femur. Its distal end tapers to a point.

3. Tarso-metatarsus. 1. Tarso-metatarsus (Fig. 29.13) is a compound stout and long bone formed by the fusion of distal row of tarsals with second, third and fourth metatarsals. 2. It proximally articulates with the tibiotarsus and distally bears three metatarsals each articulating with a corresponding digit (toe). 3. First metatarsal is a small irregular nodule of bone attached by ligament to the inner and posterior surface of tarso-metatarsus. 4. Proximally tarsometatarsus bears two cup-like facets for the articulation with the tibiotarsus. 5. In male, the tarso-metatarsus bears a stout pointed, spinous fighting **spur** on its medial side

4. Phalanges. Toes are four, articulate with metatarsus. First toe (hallux) is directed backwards and the rest three are directed forwards. Number of phalanges : hallux has 2, second toe 3, third 4 and fourth 5 phalanges, all bear horny claw.

Chapter 30

SKELETON OF RABBIT

AXIAL SKELETON

1. Skull

1. Skull is completely ossified with distinct sutures.
2. Skull is **dicondylic**, *i.e.*, two occipital condyles bear by two exoccipitals.
3. Tropibasic, *i.e.*, inter-ortbital septum is present between two orbits. Cranium does not extend in front into orbital region.
4. **Cranial portion** is small, while the **facial portion** is elongated and deflected at angle of 60°.
5. **Occipital region** comprises four bones, *i.e.*, two lateral exoccipitals, one ventral basioccipital and one dorsal supraoccipital.
6. **Auditory region** is composed of periotic bone and tympanic bulla. Three bones prootic, epiotic and opisthotic are fused to form periotic bone. Tympanic bulla is a flask-shaped bone formed by the tympanic bone.
7. **Cribriform plate** is present between the cranium and olfactory chamber.
8. **Turbinals** or scroll bones are much folded to increase the olfactory or nasal chambers.
9. **Jugal** forms the outer boundary of the orbit.
10. **Zygomatic arch** and inter-parietal bone is present.
11. Each half of the lower jaw consists of a single bone, the **dentary**.
12. Suspensorium is **craniostylic**, *i.e.,* the lower jaw articulates with the skull by squamosal.
13. Teeth are **heterodont, *i.e.*,** differentiated into incisors, premolars and molars; **diphyodont**, *i.e.*, both milk and permanent teeth are present and **thecodont** embedded in the sockets.
14. Between incisors and premolars is present a gap called **diastema**.
15. Prefrontal, postfrontal, parasphenoid and quadratojugal are absent.
16. **Dental formula** is *i* 2/1, *c*%, *pm* 3/2, *m* 3/3 = 28.

DISARTICULATED BONES OF SKULL

1. Occipital segment. 1. Occipital segment is the posteriormost part of the skull (cranium) having a large **foramen magnum**. 2. It is composed of two lateral exoccipitals, one ventral basi-

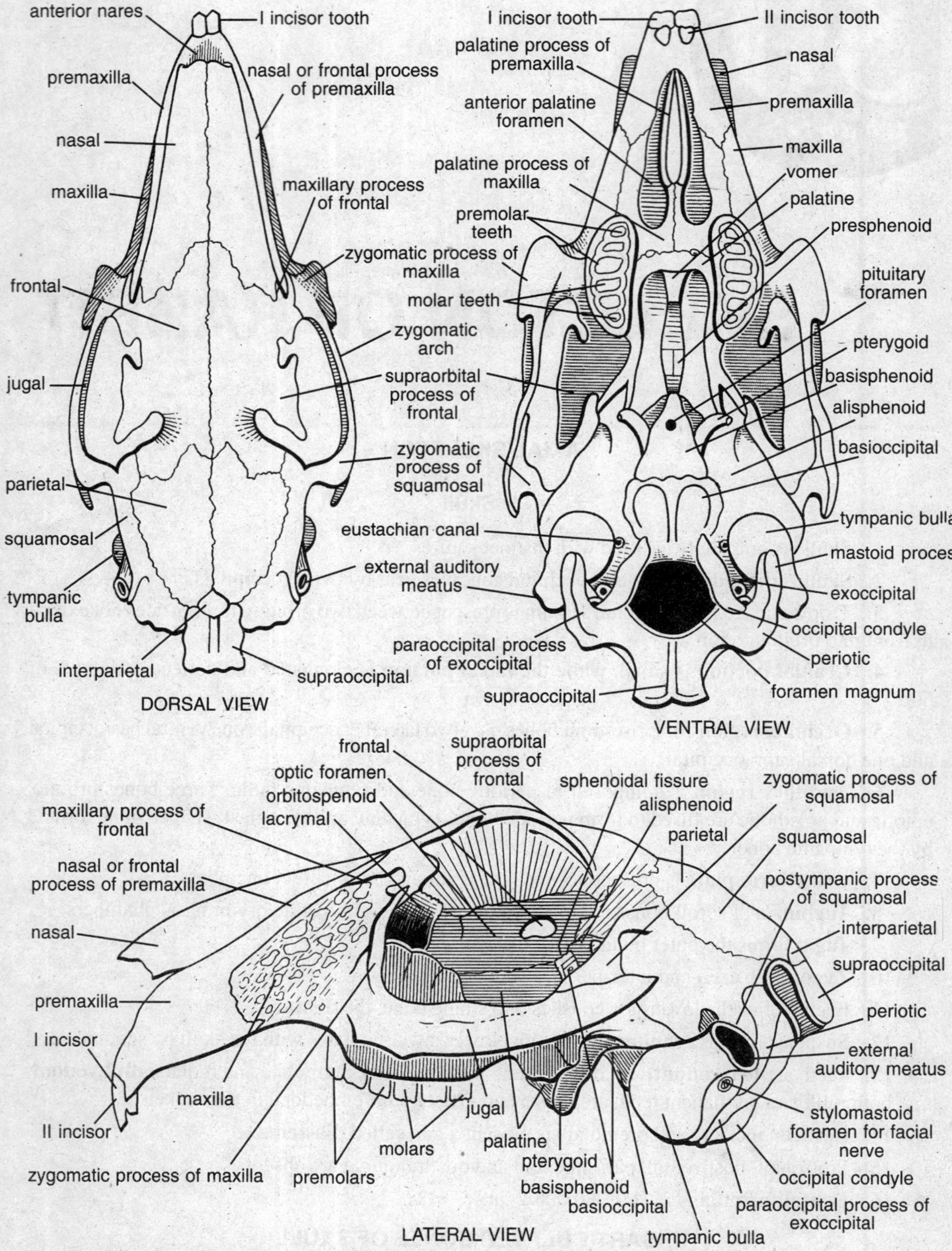

Fig. 30.1. Skull of rabbit in dorsal, ventral and lateral views.

occipital and one dorsal supraoccipital. 3. **Exoccipitals** bear oval occipital condyles and each exoccipital is produced on its outer side into a **paroccipital process** which is in close contact with tympanic bulla. 4. **Basioccipital** is a flat median bone forming the floor of foramen magnum. It articulates in front with basisphenoid and on sides with periotic. 5. **Supraoccipital** is a large median bone lying dorsal to the foramen magnum. 6. It articulates in front with inter-parietal and parietals and laterally with squamosals and periotics.

2. Basisphenoid and alisphenoid. 1. It is a compound bone of the **parietal region** of skull. 2. **Basisphenoid** is a median triangular bone lying just in front of basioccipital and united with its broad base by a thin plate of cartilage. 3. Its apex articulates in front with presphenoid and laterally with alisphenoids. 4. On its dorsal surface a depression **sella tursica** is present for lodging the pituitary body. 5. Near the middle surface it bears a small pituitary foramen. 6. **Alisphenoids** are paired wing-like bone attached obliquely to the sides of the basisphenoid. Each alisphenoid is differentiated into 3 parts : (*i*) **Postero-ventral part** in front of tympanic bulla is fenestrated. (*ii*) **Antero-ventral part** is plate-like bone united with palatine. (*iii*) Third plate-like part forms the floor of orbit.7. From the antero-ventral surface of each alisphenoid projects an **external pterygoid process** connected along the inner edge with the palatine. 8. At the anterior end between the alisphenoids and basisphenoid is a slit-like notch the **sphenoidal fissure** for the passage of 3rd, 4th and 6th nerves.

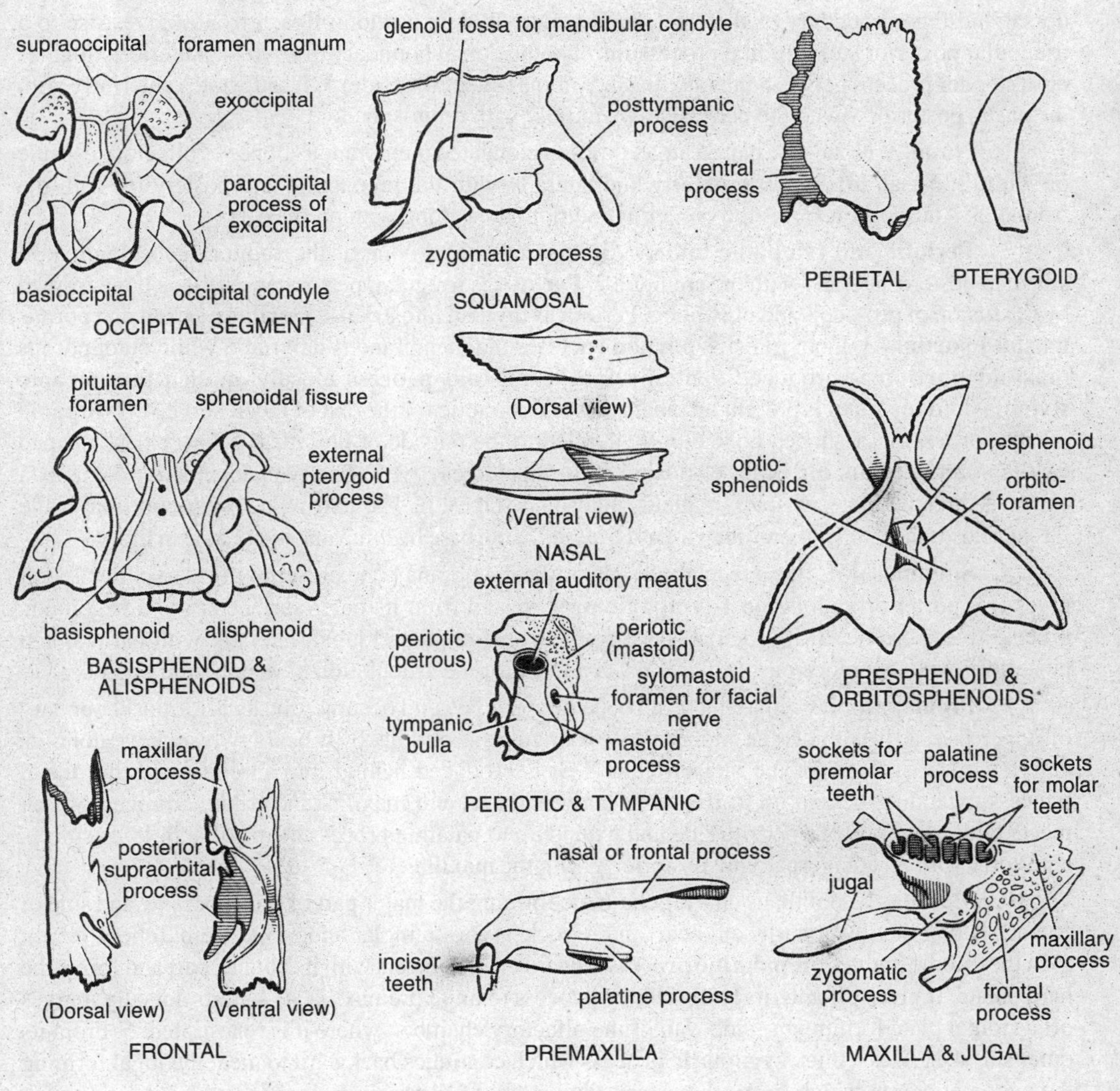

Fig. 30. 2. Disarticulated bones of skull of rabbit.

3. Parietal. 1. It is thin roughly rectangular membrane bone of parietal region of skull lying along the roof of the cranial cavity. 2. Parietal bones of both sides join with each other in mid-dorsal line by a prominent structure. 3. Posteriorly it gives rise to a thin plate-like **ventral process** that lies beneath squamosal. 4. It articulates anteriorly with frontal, laterally with squamosal and posteriorly with inter-parietal.

4. Presphenoid and orbitosphenoids. 1. It is a compound bone of frontal region of the skull 2. **Presphenoid** is a median, narrow and laterally compressed bone lying in front of basisphenoid and forms the floor of frontal region of cranium. 3. It forms the inner margin of optic foramen and articulates anteriorly and posteriorly with palatines. 4. **Orbitosphenoids** are paired wing-like thin bones intimately associated with the presphenoid on both sides and form the sides of frontal region of cranium. 5. Orbitosphenoids surround the large optic foramen for exit of optic nerve. 6. Orbitosphenoids articulate anteriorly with palatine, posteriorly with squamosals and alisphenoids and above with frontals.

5. Frontal. 1. Frontals are paired membranous bones forming the roof and sides of anterior part of cranial cavity. 2. Frontals of both sides are united mid-dorsally forming a suture and form the dorsal and inner boundary of the orbit. 3. The outer middle portion of each frontal gives rise to a triangular posterior supra-orbital process forming the dorsal boundary of orbit. 4. Anteriorly it gives off a slender process between the maxilla and premaxilla of that side. 5. It articulates anteriorly with the nasal, posteriorly with the parietal and ventrally with orbitosphenoid.

6. Nasal. 1. Nasals are paired large, thin and elongated membranous bones of olfactory capsule of skull. 2. Nasal articulates anteriorly and laterally with the premaxilla and posteriorly with the frontal. 3. Anteriorly it forms the border of anterior nares along with premaxilla.

7. Periotic and tympanic bulla. 1. It is a compound bone of the auditory capsule of skull attached postero-laterally with the cranium. 2. **Periotic** is irregular, perforated cartilage bone formed by the fusion of pro-, epi- and opisthotic. Periotic is divided into a dense **petrous portion** and porous **mastoid portion**. 3. Inner **petrous portion** encloses the membranous labyrinth, while outer porous **mastoid portion** is produced ventrally into a **mastoid process** closely applied to tympanic **stylomastoid foramen** is present on inner surface of periotic for the exit of facial nerve. 4. **Tympanic** is more or less flask-shaped bone closely applied to the outside of periotic. Its lower flask-shaped swollen part is **tympanic bulla** enclosing the tympanic cavity. 5. The tympanic cavity contains a chain of three auditory ossicles–malleus, incus and stapes. 6. The narrow neck-like portion of the flask encloses the **external auditory meatus**, closed at the base by tympanic membrane in life.

8. Squamosal. 1. It is a rectangular plate-like membrane bone present ventral to parietal. 2. It gives rise posteriorly along **post-tympanic process** and from its outer surface arises a zygomatic process over periotic. 3. It articulates anteriorly with frontal and jugal, laterally with parietal and posteriorly with supraoccipal. 4. It also bears a facet for the articulation of mandibular condyle.

9. Premaxilla. 1. Premaxilla is the bone of upper jaw and occupies the anteriormost boundary of upper jaw. 2. It is thick and more or less triangular in outline. 3. It bears two sockets along its anterior lower surface for the incisor teeth. 4. It is produced behind into an elongated thin **nasal process** extending behind up to frontal in between nasal and maxilla, and palatine process which meets with its fellow of opposite side and extend up to palatine process of maxilla. 5. It articulates anteriorly with its counterpart and posteriorly with the maxilla.

10. Maxilla. 1. Maxilla is an irregular bone forming the major part of the upper jaw and side of face. 2. Along the lower surface it bears many sockets for premolar and molar teeth. (cheek teeth). 3. In the middle it gives off **palatine process** inwards which meets with its counterpart and forms the hard palate. It gives off anteriorly **maxillary process** to unite premaxilla. 4. Antero-dorsally it gives off a frontal plate forming the side wall of the olfactory chamber, where it is fenestrated. 5. From its outer surface arises a thick **zygomatic process** which continues backward to meet the jugal forming the zygomatic arch which forms the anterior boundary of the orbit. 6. It articulates anteriorly with premaxilla, posteriorly with palatine and dorsally with nasal and frontal.

Lower Jaw (Mandible)

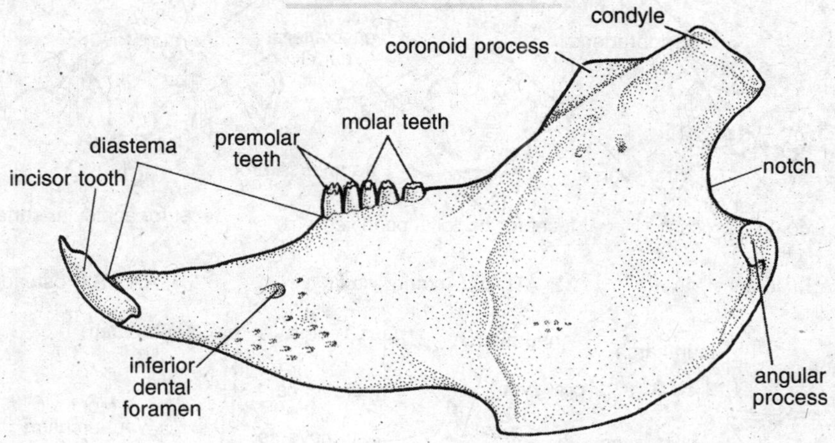

Fig. 30.3. Mandible of rabbit. (Lateral view).

1. Lower jaw comprises two rami, the **dentaries** (membrane bone) meeting in front at a mandibular symphysis.

2. Each dentary is more or less triangular and flattened vertical plate having along its upper edge sockets for teeth.

3. Posteriorly each dentary bears a thick elongated **condyle** upwards, a less prominent **coronoid process** anterior to condyle and a rounded **angular process** below.

4. A large mandibular foramen is present on the inner side just in front of the condyle and on outer anterior side is present a **mental** or **inferior dental foramen** for nerves and blood vessels.

5. The condyle articulates in the glenoid fossa of squamosal just below the base of its zygomatic process.

VERTEBRAL COLUMN

In mammals **epiphysis** is present at the ends of centrum of vertebrae and centrum is **acoclous** or **amphiplatyan** (flat). Cartilaginous **intervertebral discs** are present between adjacent vertebrae. There are 46 vertebrae in the vertebral column of rabbit which are differentiated into cervical (7), thoracic (12–13), lumbar (6–7), sacral (4) and caudal (16).

1. Atlas vertebra. 1. **Atlas** is the first ring-like cervical vertebra formed by neuroarch. Centrum and zygapophyses are absent. 2. Neural spine is in the form of a small ridge. 3. Transverse processes are broad horizontal plate-like. These are actually enlarged flat cervical ribs. 4. Each transverse process is having small openings the **vertebrarterial canal** at its base. 5. On the anterior side it bears two concavities for occipital condyles of skull and on the posterior side it carries two lateral and a mid-ventral facets for odontoid process of axis vertebra. 6. Neural canal is very long.

2. Axis vertebra. 1. **Axis** or **epistropheus** is the second cervical vertebra. 2. Centrum is **acoelous** or flat and at its anterior side is present a peg-like **odontoid process**, which fits into the posterior concavities of atlas. 3. Neural spine is ridge-like and laterally compressed. 4. Transverse processes of cervical ribs are small, posteriorly directed and basally perforated by vertebrarterial canals are entirely absent. 5. Postzygapophyses are present and prezygapophyses are absent.

3. Typical cervical vertebra. 1. The remaining five (3–7) cervical vertebrae are similar in structure and are, therefore, called typical cervical vertebrae. 2. **Centrum** is **acoelous** short and flat with broad sides. 3. Neural spine is small, pointed and backwardly directed. In 7th cervical neural spine is long and centrum at its posterior surface bears concave facets for thoracic ribs. 4. Small transverse processes are bifurcated and pierced basally by vertebrarterial canals and are formed by the fusion of small cervical ribs. 5. Pre- and postzygapophyses are present.

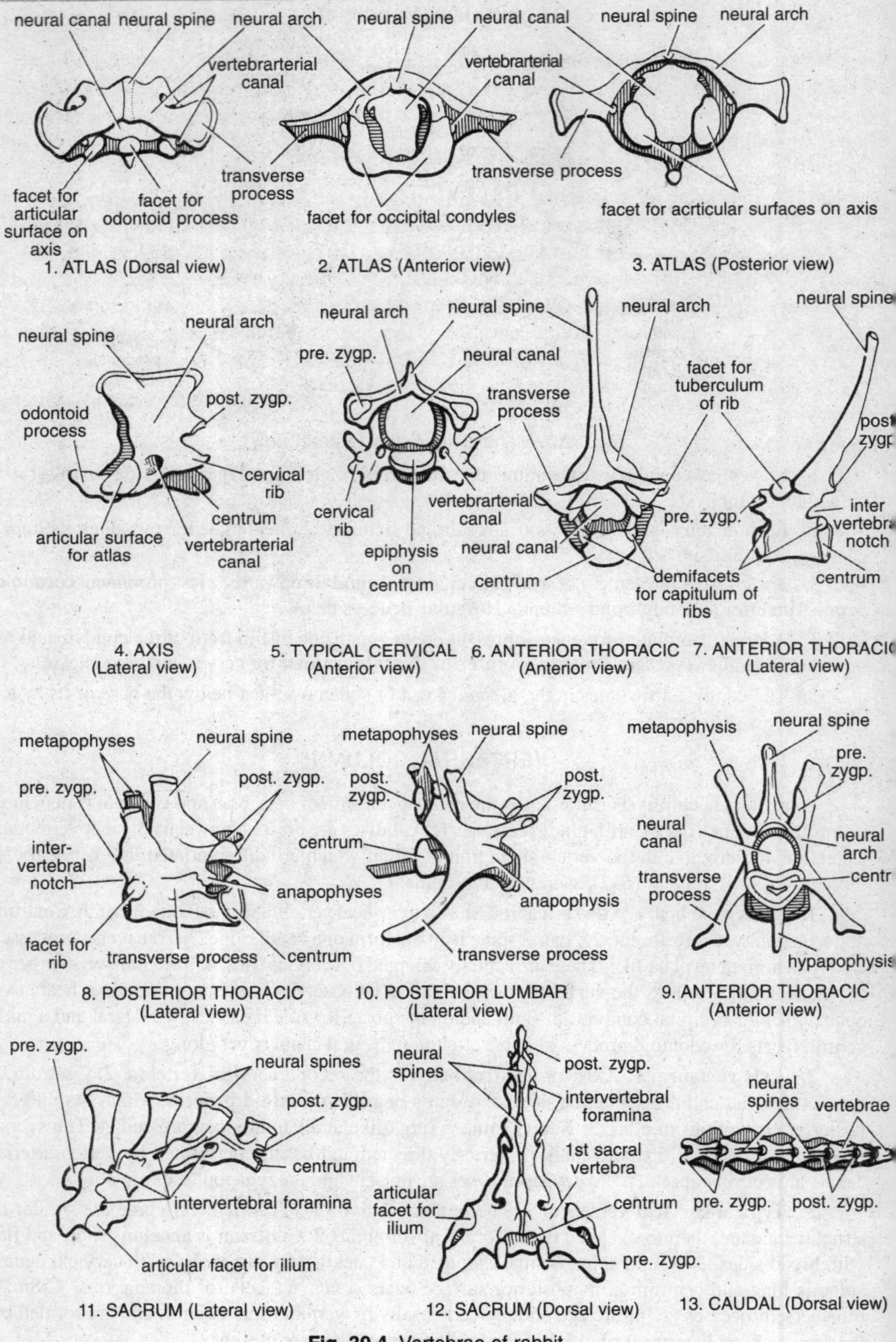

Fig. 30.4. Vertebrae of rabbit.

4. Anterior thoracic vertebra. 1. Centrum is short, thick and **acoelous** and bears a concave demifacet at each end for capitulum of a rib. 2. Transverse process are short, stout and horizontal carrying articular surfaces for the tuberculum of rib. 3. Neural spine is very long, slender and backwardly directed. 4. Pre- and postzygapophyses are present.

5. Posterior thoracic vertebra. 1. Centrum is large, stout and **acoelous** and bears at its anterior end facets for capitulum of a rib. 2. Neural spine is small, vertical and somewhat laterally compressed. 3. Pre-and postzygapophyses are present. 4. Transverse processes are reduced and without tubercular facets. 5. **Metapophyses** are present above the prezygapophyses and **anapophyses** are present below the postyzygapophyses.

6. Anterior lumbar vertebra. (First two). 1. It is comparatively large in size than those of thoracic. 2. **Centrum** is stout, well developed and **acoelous** and bears hypapophysis below. 3. Neural spine is crest-like and directed forward. 4. Transverse processes are long, flattened and stout and directed downwards and forwards. 5. Pre-and postzygapophyses are present. 6. At the anterior end is present a pair of large forwardly directed metapophyses arising above the prezygapophyses. 7. Below the postzygapophyses at the posterior end of neural arch are present less prominent backwardly directed **anapophyses**.

7. Posterior lumbar vertebra (3-7). 1. Posterior lumbar vertebra resembles the anterior vertebra in all the structural details except that the **hypapophysis** is absent (or represented by a small ridge).

8. Sacrum. 1. **Sacrum** is formed by the fusion of four sacral vertebrae. 2. Sacrum supports the pelvic girdle. 3. Each sacral vertebra is chracterised by the following features : (*i*) Neural spines are large and directed backwardly. (*ii*) Centrum is **acoelous** or flat. (*iii*) Pre- and postzygapophyses are present as tubercles (*iv*) Transverse processes are rudimentary. 5. First sacral vertebra is the largest and bears stout transverse processes attached with ilia bones of pelvic girdle. Its neural spine is upright. Metapophyses are present and anapophyses absent.

9. Caudal vertebra. There are nearly 16 caudal vertebrae in rabbit. They progresssively decrease in size backward. Their processes also become gradually shorter and finally the terminal vertebrae are merely rod-like centra alone.

Rib

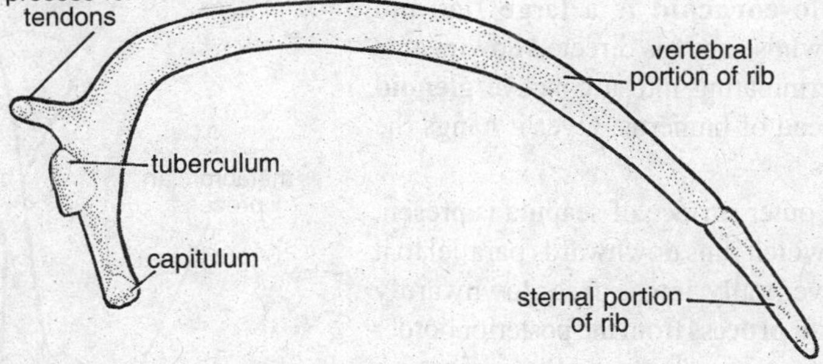

Fig. 30.5. Rib of rabbit.

1. Each rib is a curved rod-like bone and consists of a dorsal **vertebral portion** and a ventral **sternal portion**.

2. Vertebral portion is bony and has two processes, upper **tuberculum** and lower **capitulum**.

3. **Tuberculum** articulates with the tubercular facets of transverse process of vertebra, while **capitulum** articulates with demifacets on the centra of two adjacent vertebra.

4. Sternal portion is cartilaginous and articulates below with the sternum.

5. Uncinate process is absent in mammalian ribs.

6. Ribs are of two types : **True ribs** attach with sternum. **Floating ribs** (last 4–5) have no sternal parts and do not connect with sternum.

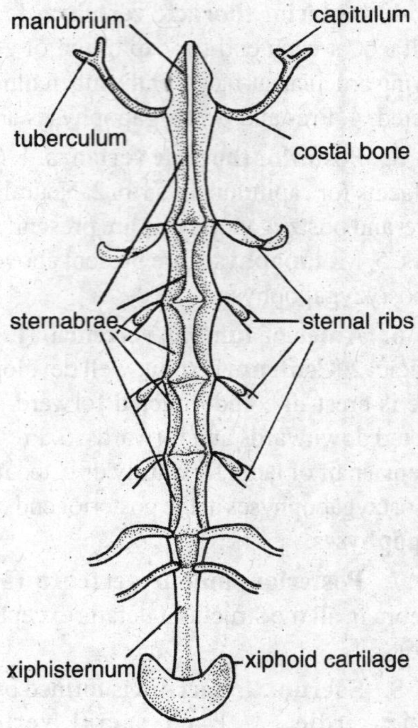

Fig. 30.6. Sternum of rabbit.

Sternum

1. **Sternum** is laterally compressed and is composed of six rod-like pieces, the **sternabrae** arranged in a straight line.

2. First piece is longest and called **manubrium.** It is ventrally produced into a **keel**.

3. Sixth sternabra is smallest and seventh or last piece is long and slender, called **xiphisternum** which terminates in an expanded plate of cartilage the **xiphisternal cartilage** or **xiphoid cartilage**.

APPENDICULAR SKELETON

1. Girdles

1. Pectoral Girdle

1. Two halves of the pectoral girdle are separate due to the presence of thoracic ribs.

2. Each half of pectoral girdle is composed of two bones : clavicle and scapulo-coracoid.

3. **Scapulo-coracoid** is a large flat and triangular bone whose apex is directed downwards and forwards terminating into a concave **glenoid cavity** for the head of humerus. Over it hangs the coracoid process.

4. On the outer surface of scapula is present a vertical **spine** which runs downwards parallel to it and continued ventrally into a free downwardly directed **acromian process** from the posterior border of which arises a long backwardly projecting **metacromian process**.

5. **Suprascapula** is a thin cartilaginous strip attached to the dorsal border of the scapula.

6. **Clavicle** is a slender curved bone whose one end articulates with manubrium of sternum and the other end with acromian process of scapula.

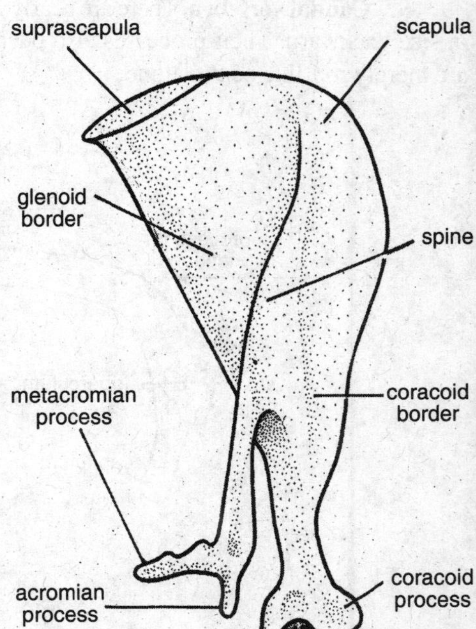

Fig. 30.7. Pectoral girdle of rabbit. (Right half in outer view).

2. Pelvic Girdle

1. Pelvic girdle (Fig. 30.8) is composed of two equal halves – os-innominata.

2. Two os-innominata united mid-ventrally with each other by a **pubic symphysis** and connected anteriorly with sacrum.

3. Each os-innominatum is made up of three pieces, **ilium, ischium** and **pubis** which are fused together to form a **hip bone**.

4. **Ilium** is a blade-like antero-dorsal bone running parallel to the vertebral column and bearing on its inner middle surface **articular facet** for sacrum.

5. **Ischium** is a postero-dorsal stout and straight bone. Its posterior thickened part is called ischial tuberosity.

6. At the point of union of ilium and ischium on outerside is present a concavity, **acetabulum** into which fits head of femur.

7. **Pubis** is the smallest slender bone, forming ventral median part of girdle.

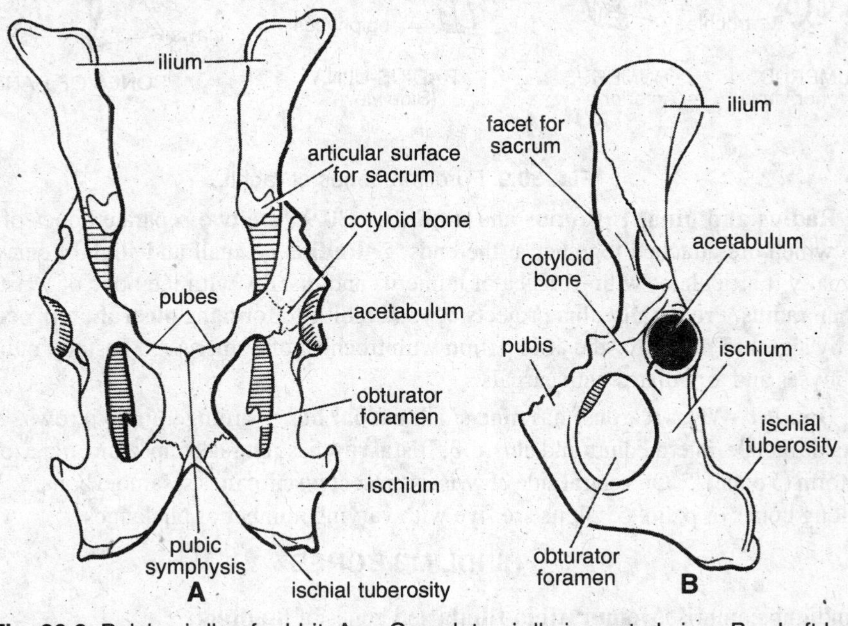

Fig. 30.8. Pelvic girdle of rabbit. A — Complete girdle in ventral view; B— Left half.

8. Between pubis and ischium is present a large space known as **obturator foramen**. Pubes of two sides join mid-ventrally at a pubic symphysis.

9. Pubis do not take part in the formation of acetabulum due to the presence of a small **cotyloid bone.**

LIMB BONES

Forelimb Bones

Forelimb comprises humerus, radius and ulna and bones of forefoot or hand.

1. Humerus (Fig. 30.9) is a stout rod-like bone of upper-arm of forelimb. 2. It is expanded at the proximal end, forming a rounded knob-like **head** that fits into the **glenoid cavity** of pectoral girdle. 3.**Greater** and **lesser tuberosities** are present at the outer and inner side of head respectively and between two tuberosities is present a **bicipital groove.** 4. A prominent **deltoid ridge** is present below the head on preaxial or anterior side. 5. Its distal end bears a pulley-like **trochlea** for articulation with ulna and olecranon fossa. 6.Above the trochlea are present two depressions or suprafossae which communicate with each other through a **supra-trochlear foramen**. 7. Anterior coronoid fossa is smaller and posterior olecranon fossa into which fits the olecranon process of ulna.

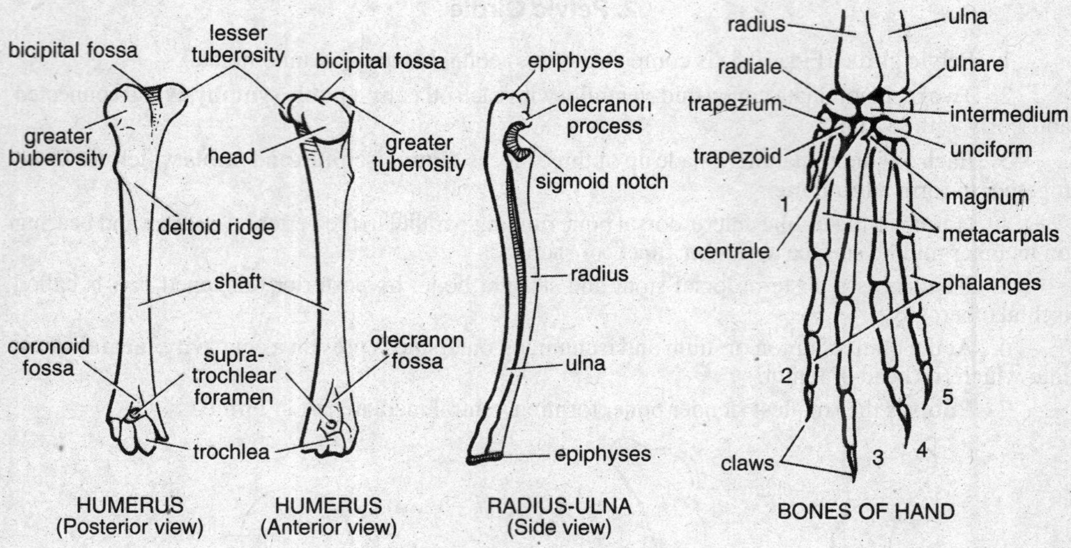

Fig. 30.9. Forelimb bones of rabbit.

2. Radius and ulna. 1. **Radius** and **ulna** (Fig. 30.9.) are two separate bones of forearm of forelimb which are attached together at the ends. 2. **Radius** is small and slightly curved forward. 3. Proximally it articulates with trochlea of humerus and distally with the bone of wrist. 4. **Ulna** is longer than radius. Proximally ulna projects beyond radius to form the **olecranon process** which is notched by **sigmoid notch** for the articulation with trochlea of humerus. 5. Distally radius and ulna bear epiphyses and articulates with carpals.

3. Hand. 1. **Wrist** (carpus) has nine small **carpal bones** arranged in two rows. 2. **Proximal row** contains radiale, intermedium and ulnare. 3. **Distal row** has centrale, trapezium, trapezoid, magnum and unciform (5 bone). 5. On ventral side of wrist is present **pisciform**, a sesamoid bone. 5. Metacarpals are five long bones of palm. 6. Digits are five with varying number of phalanges.

HINDLIMB BONES

Hindlimb comprises **femur**, **tibio-fibula** and bones of **hindfoot**.

1. Femur. 1. Femur (Fig. 30.10) is an elongated and stout bone of thigh of hindlimb. 2. It consists of a cylindrical shaft and enlarged proximal and distal ends. 3. Proximal end bears on inner side a rounded **head** which fits into the acetabulum of pelvic girdle. 4. Below head there are three projections, **greater trochanter** above the head, **third trochanter** lies below the great trochanter, while **lesser trochanter** is just below the head. Head and greater trochanter form the border of **digital fossa**. 5. Its distal end bears two large condyles separated by a **inter-condylar notch** or **groove** for articulation with **tibio-fibula**. 6. This groove on anterior (outer) side is called **patellar groove** in which slides **patella** (knee cap) with in ligaments.

2. Tibio-fibula. 1. **Tibio-fibula** (Fig. 30.10.) is the bone of shank of hindlimb made up of two bones **tibia** and **fibula** which are free proximally but fused distally. 2. **Tibia** is a stout and straight bone. 3. Proximal end of tibia is enlarged and bears two oval surfaces on its upper end for the articulation with two condyles of femur. 4. A prominent **cnemial crest** is also present near the proximal end anteriorly. 5. Distal end of tibia articulates with the tarsal bones. 6. **Fibula** is a small slender bone lying on posterior side of tibia and fused distally with tibia, but articulates proximally with tibia.

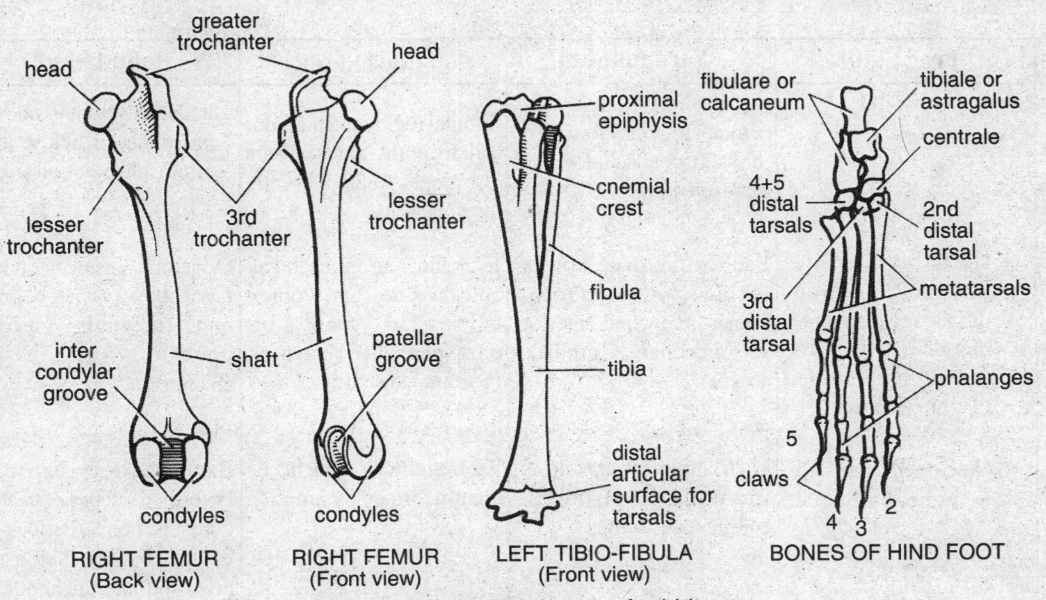

Fig. 30.10. Hindlimb bones of rabbit.

3. Foot. 1. **Ankle** (tarsus) contains 6 tarsal bones arranged in two rows 2. **Proximal row** has **astragalus** formed by the fusion of **tibiale** and **intermedium** lying on inner side, and **calcaneum** on the outer side and projected behind into a spur behind articulation with tibio-fibula. In front of astragalus lies a **centrale**. 3. **Distral row** has three bones–mesocuneiform, ectocuneiform and cuboid. 4. In the sole (pes) are present four **metatarsals**, first being absent. 5. **Hallux** (first toe) is absent.

COMPARISON OF SKULLS

Frog Skull	Varanus Skull	Fowl Skull	Rabbit Skull
1. Skull is triangular in shape and dorso-ventrally flat.	Skull is elongated and flattened.	Skull is elongated and triangular.	Skull is elongated, narrow anteriorly and broad posteriorly
2. Cranium is small.	Cranium is small.	Cranium is large due to large-sized brain.	Cranium is large.
3. Dicondylic	Monocondylic. Condyle formed by basioccipital.	Monocondylic.	Dicondylic. Condyle is formed by exoccipitals.
4. Platybasic-inter-orbital septum absent.	Tropibasic-inter-orbital septum is present,	Tropibasic.	Tropibasic-inter-orbital septum is present.
5. Sphenethmoid forms posterior wall of olfactory chambers.	—	—	Mesethmoid bone is present between two olfactory chambers.
6. Occipital segment formed of two bones-exoccipitals.	Occipital segment formed of 4 bones-supra,basi- and two exoccipitals.	Same	Same
7. Frontal and parital fused to form fronto-parietal (dorsal) and parasphenoid (ventral).	Parietal region has a basal basisphenoid and a dorsal parietal.	It includes 2 fused parietals, laterals, squamosals and alisphenoids and ventral basisphenoid.	Parietal region includes basisphenoid, ali-sphenoids and parietals and interparietal.

Frog Skull	Varanus Skull	Fowl Skull	Rabbit Skull
8. —	Frontal region is formed of dorsal frontals and ventral parasphenoid.	Frontal region includes dorsal frontals, sides ali-and orbitosphenoids and ventral presphenoid.	It includes frontals above, presphenoid below and orbitosphenoids on sides.
9. Olfactory capsule includes paired nasals above, septo-maxillaries and vomers on floor. (Turbinals absent)	Each olfactory capsule includes nasal above, vomer below and anterior septomaxillary. (Turbinals absent).	It includes a nasal above and on sides and vomer below in front of parasphenoid rostrum. (Turbinal absent).	Olfactory capsule includes nasal above, vomer below, and turbinals (scroll bones) within nasal chamber.
10. Auditory capsule is formed by prootic.	Each auditory capsule is formed by opisthotic, epiotic and prootic.	Each auditory capsule is mainly formed by prootic.	Each auditory capsule is formed of periotic and tympanic bones. Periotic is formed by the fusion of pro-, epi- and opisthotics.
11. Each eye orbit is bounded by fronto-parietal, parasphenoid, pterygoid and squamosal.	Each orbit is bounded in front by prefrontal, lacrymal, supra-orbital and post-orbital at postero-dorsal border.	Each orbit is bounded in front by frontal and posteriorly by post- orbital process of frontal and alisphenoid, and lacrymal.	Each orbit is laterally bounded by orbitosphenoid, dorsally by frontal and lacrymal.
12. **Upper jaw** includes premaxillae, maxillae, quadrato-jugals, quadrate cartilages, squamosals, pterygoids and palatines.	**Upper jaw** is formed by premaxillae, maxillae, jugals, septomaxillaries, palatines, pterygoids, transverse and epipterygoids.	**Upper jaw** is formed of premaxillae, maxillae, jugals, quadrato-jugals, quadrate, pterygoids and palatines.	**Upper jaw** includes premaxillae, maxillae, jugals, pterygoids and palatines.
13. **Lower jaw** includes mentomekelian, angulosplenial and dentary around Meckel's cartilage.	**Lower jaw** is formed of paired articular, angular, supra-angular, coronoid, splenial and dentary (6 bones).	**Lower jaw** is composed of five bones: articulars, angulars, supra-angulars, splenial and dentary.	**Lower jaw** is only formed of paired dentaries.
14. Jaw suspensorium is autostylic: lower jaw articulates with quadrate of upper jaw.	Same	Same	Craniostylic; mandible articulates with squamosal.

31

Chapter

SKELETON OF DOG

AXIAL SKELETON

1. Skull

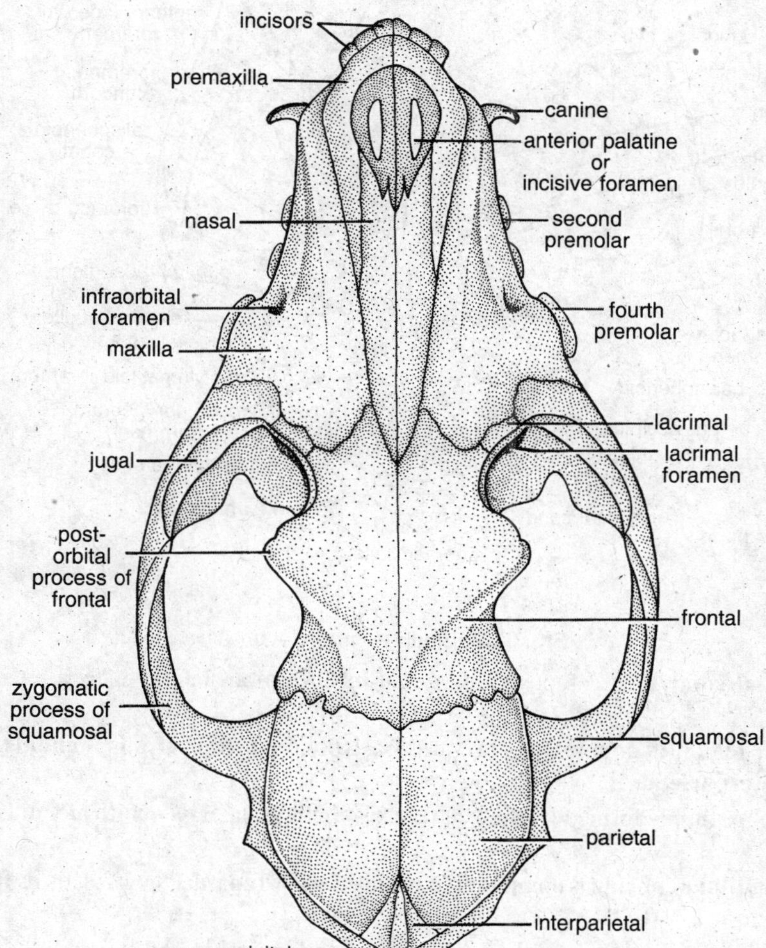

Fig. 31.1. Skull (Dorsal view).

1. Skull is **dicondylic**, *i.e.*, two occipital condyles.
2. The bones of skull are firmly united together having distinct sutures.

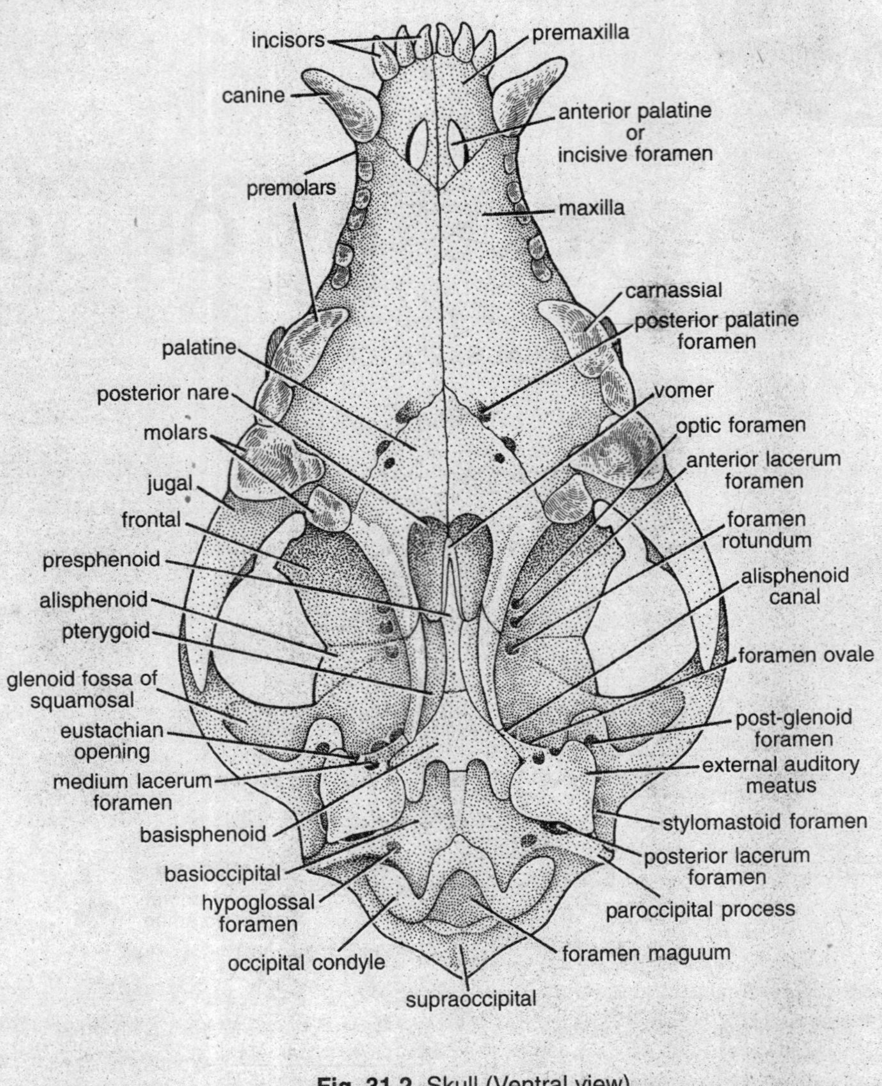

Fig. 31.2. Skull (Ventral view).

3. **Occipital condyle** is composed of ventral **basioccipital**, lateral **exoccipitals** and dorsal **supraoccipital**.

4. Parietal region is composed of ventral **basisphenoid**, **lateral alisphenoids** and dorsal **parietals** and **interparietal**.

5. **Frontal region** is formed by the ventral **presphenoid**, laterals **orbitospenoids** and dorsal **frontals**.

6. Each **auditory capsule** is composed of **periotic** and **tympanic**, the later forms flask-shaped **tympanic bulla**.

7. **Ear ossicles** form a chain of three bones, *i.e.*, **malleus**, **incus** and **stapes**.

8. The boundary of the orbit is formed by antero-dorsal, lacrymal, dorsal frontal and lower jugal forming zygomatic arch with squamosal.

9. Inter-orbital septum is present, *i.e.*, **tropibasic** type.

10. Palate is formed by the palatines, premaxillae, and maxilla.

11. Upper jaw is formed by the premaxillae, maxillae, jugals, pterygoids and palatines.
12. Zygomatic arch is formed by jugal and zygomatic process of squamosal.
13. Lower jaw consists of a single bone the **dentary** on each side.

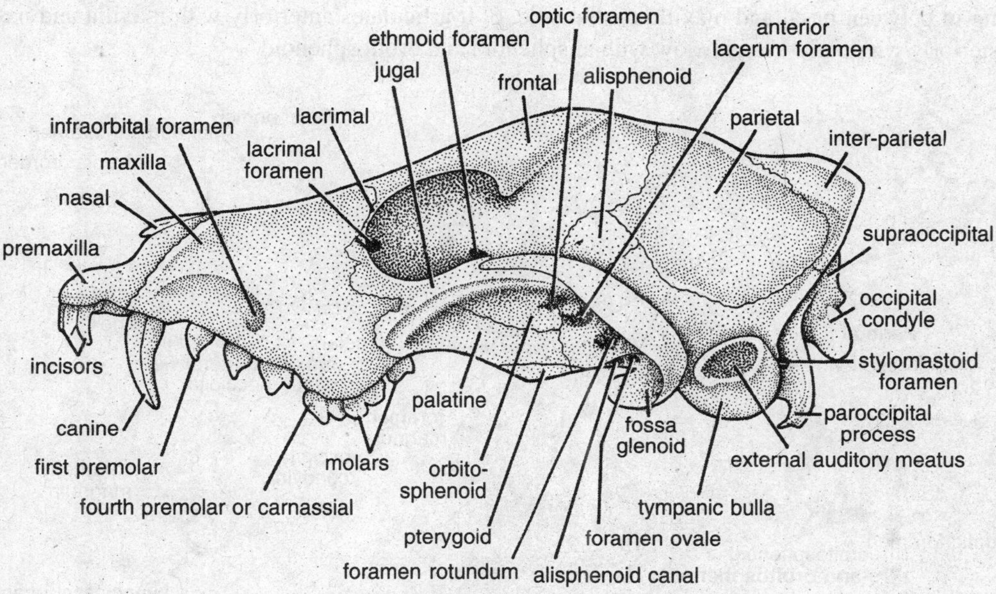

Fig. 31.3. Skull (Lateral view).

14. Suspensorium craniostylic type, *i.e.*, lower jaw articulates with squamosal without the intervention of quadrate.

15. Teeth are **heterodont,** *i.e*., incisors, canines, premolars and molars; **diphyodont** and **thecodont** type.

16. Dental formula is *i* 3/3, *c* 1/1, *pm* 4/4, *m* 2/3 = 42.

DISARTICULATED BONES OF SKULL

1. Occipital segment. 1. Occipital segment forms the hindermost part of the skull. 2. It is made up of four cartilaginous bones – basioccipital, exoccipitals and supraoccipital surrounding the foramen magnum. 3. **Basioccipital** is a flat median bone forming the base of foramen magnum. 4. Exoccipitals form the sides of the foramen magnum and ear bears oval occipal condyle gives off downwardly directed paroccipital process which is in close contact with tympanic bulla. 5. Supraoccipital is a large median bone forming the roof of foramen magnum. 6. Basioccipital articulates in front with basisphenoid and on sides with the exoccipitals and tympanic bone. 7. Supraoccipital articulates in front with inter-parietal, parietals and on sides with squamosal and periotics.

2. Parietal. 1. Parietal is the bone of parietal region of skull. 2. A pair of parietals meeting each other in mid-dorsal line forming the roof and lateral walls of cranial cavity. 3. Each parietal is a flattened and rectangular membrane bone. 4. It articulates in front with frontal, on sides with squamosal and supraoccipital and behind with inter-parietal.

3. Presphenoid and orbitosphenoids. 1. Presphenoid and orbitosphenoids are the bones of floor of frontal region of skull. 2. Presphenoid is a median, laterally compressed and narrow bone forming the floor of cranial cavity. 3. Presphenoid articulates anteriorly with mesethmoid and posteriorly with basisphenoid. 4. Orbitosphenoids are large paired wing-like bones forming the sides of frontal region, cranial cavity and orbits. 5. Each orbitosphenoid is perforated by a optic foramen. 6. Orbitosphenoids articulate posteriorly with aplisphenoids and squamosals, above with frontals and palatines in front.

4. Frontal. 1. Frontals are a pair of large membrane bones forming the roof and sides of cranial cavity in the frontal region. 2. Frontals unite with each other by sutures in mid-dorsal line and with parietals posteriorly. 3. Each frontal is an elongated irregular bone. 4. It projects on its outside into a prominent rounded postorbital or supraorbital process. 5. Anteriorly it gives off a long nasal process lying in between nasal and maxilla of that side. 6. It articulates anteriorly with maxilla and nasal, posteriorly with parietal and below with alisphenoid and orbitosphenoid.

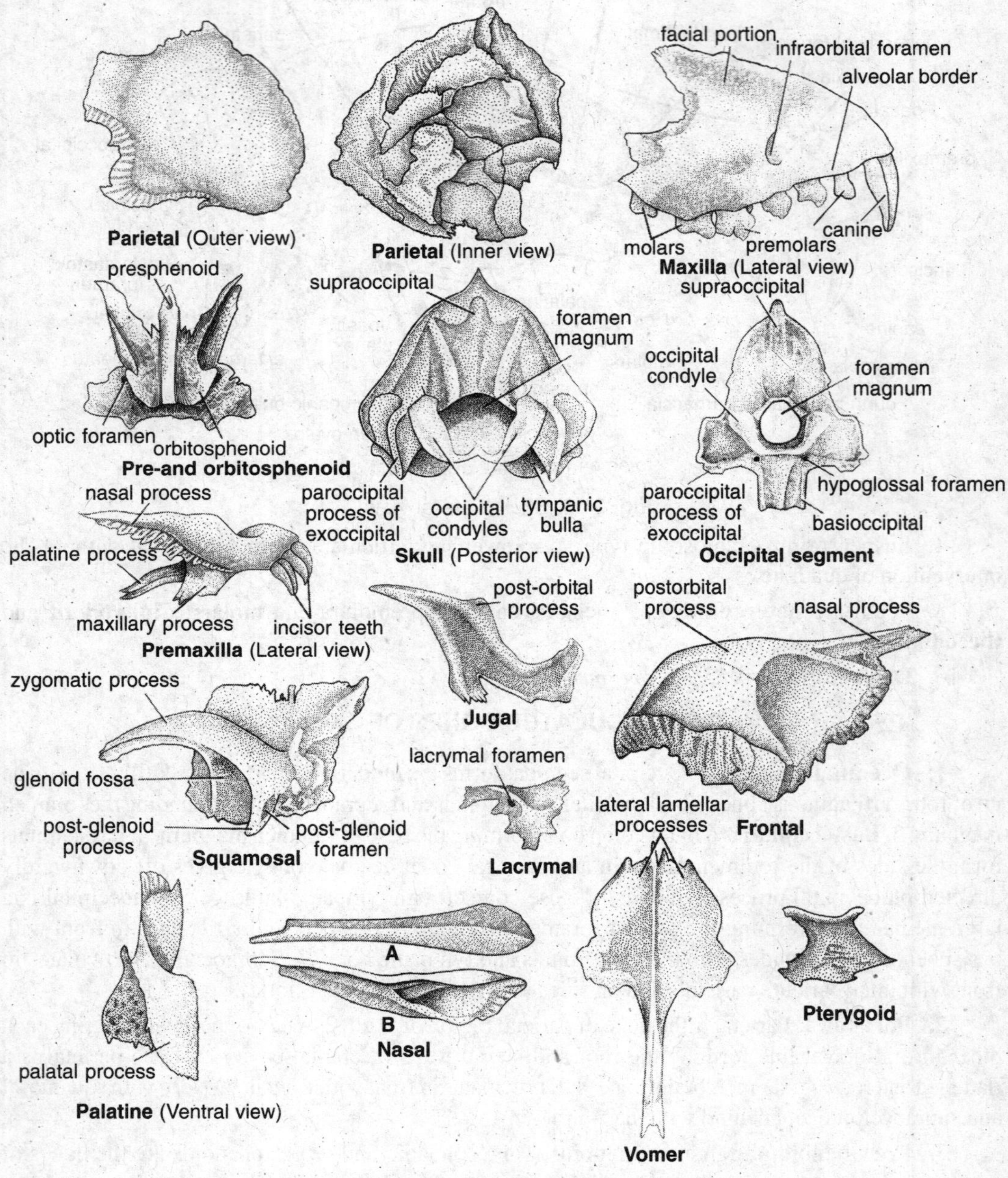

Fig. 31.4. Disarticulated bones of skull of dog.

5. Lacrymal. 1. Lacrymal is the bone of optic capsule of skull. 2. It is a small membrane bone present between the frontal and maxilla. 3. It is perforated by a lacrymal foramen opening in the orbit.

6. Nasal. 1. Nasals are the paired membrane bones of olfactory capsule of the skull. 2. Both the nasals unite with each other along their inner borders and form the roof of olfactory cavities. 3. It is a thin, narrow and flattened elongated bone. 4. It articulates in front with premaxilla, on sides with maxilla and behind with nasal process of frontal.

7. Vomer. 1. Vomer is the bone of olfactory capsule of skull. 2. It is a slender, vertical and unpaired membranous bone. 3. Anteriorly it forms trough-like structure and narrows posteriorly into a rod-like structure. 4. It lies between maxillo-turbinals, presphenoid and mesethmoid.

8. Pterygoid. 1. It is a thin narrow plate-like bone. 2. It lies attached to the base of skull at the junction of basisphenoid with alisphenoid. 3. It articulates anteriorly with palatine of its side and presphenoid.

9. Palatine. 1. It is a large irregular bone lying in front of pterygoid and forming the palate. 2. It consists of vertical and horizontal portions. 3. Vertical portion runs in front to meet the orbitosphenoid and frontal, while forming a plate-like structure on the inner side articulating with presphenoid and vomer. It forms the roof of nasal passage. 4. Horizontal portion is in the form of palatal process articulates with its counterpart of other side in mid-line forming the hinder part of palate and floor of nasal passage. 5. Palatal process also articulates in front with maxilla.

10. Maxilla. 1. Maxilla is a large irregular bone forming the major part of upper jaw. 2. It consists of vertical and horizontal portions and lies in front of palatine. 3. Vertical portion or facial portion bears deep sockets for canine, premolars, and molars on its lower margin. 4. Facial portion articulates in front with premaxilla, dorsally with nasal and frontal and posteriorly with lacrymal, jugal and palatine. 5. Horizontal or palatal portion meets its fellow of other side in middle forming the anterior part of bony palate.

11. Premaxilla. 1. Premaxilla forms the anterior part of upper jaw. 2. It is a small irregular bone and consists of vertical and horizontal portions. 3. Vertical portion bears three incisor teeth along its antero-ventral margin and is produced backwards into a long nasal process which runs between the nasal and maxilla and a small maxillary process. 4. Horizontal portion meets its fellow of other side in median line and is produced into palatine process.

12. Jugal. 1. Jugal is a straight and strong bone forming the anterior part of zygomatic arch. 2. It articulates anteriorly with maxilla and posteriorly with zygomatic process of squamosal. 3. It gives off a small post-orbital process at the point of union with zygomatic process of squamosal.

13. Squamosal. 1. Squamosal is a large quadrangular plate-like membrane bone. 2. On outerside it is produced into a curved zygomatic process which unite in front with jugal. 3. The lower surface of zygomatic process is hollow having a **glenoid fossa** into which fits the condyle of corresponding ramus of lower jaw. 4. The posterior margin of glenoid fossa is produced into a rounded post-glenoid process having a post-glenoid fossa. 5. It articulates posteriorly with supraoccipital, above with parietal, frontal and alisphenoid..

Lower Jaw (Mandible)

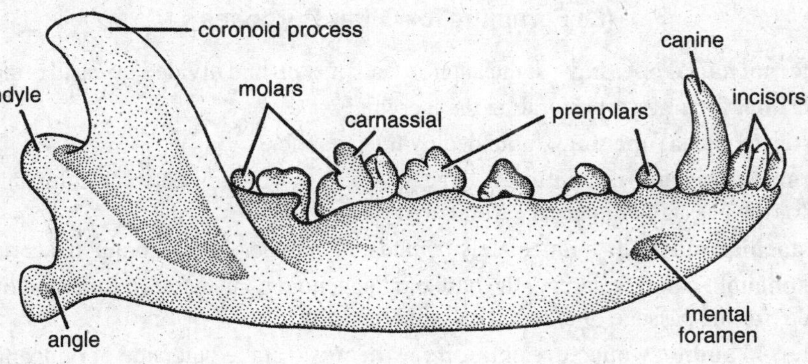

Fig. 31.5. Mandible (Lateral view).

1. The lower jaw consists of two rami, uniting anteriorly at a median symphysis.
2. Each ramus consists of a single bone the **dentary**.
3. Dentary is an elongated, stout and curved bone bearing incisors, canine, premolars and molars teeth in sockets along the upper or alveolar border.
4. Posteriorly it is produced into a condyle, a prominent dorsal coronoid process as well as an angular process.
5. It also carriers a pair of foramina, *viz.*, inferior dental foramen and mental foramen.
6. It articulates with the glenoid fossa of squamosal by the condyle.

Important Foramina of the Skull

The foramina may be divided into two groups : (A) foramina for the passage of cranial nerves and (B) foramina for other purposes.

(A) Foramina for Nerves

1. **Foramen magnum** – a median opening at the hind end of skull, through which brain is continued into spinal cord.
2. **Condylar foramen** – present in the exoccipitals for the passage of XII cranial nerve.
3. **Foramen lacerum posterius** – an irregular fissure between the tympanic bulla and exoccipital for the passage of IX, X and XI cranial nerve.
4. **Stylomastoid foramen –** a small opening on the external side of tympanic bulla and in front of pariotic process of VII cranial nerve.
5. **Foramen ovale** – at the hinder part of alisphenoid for mandibular branch of V cranial nerve.
6. **Foramen rotundum** – in front of foramen ovale for maxillary branch of V cranial nerve.
7. **Foramen lacerum anterius** – a large slit between the alisphenoid and orbitosphenoid for the passage of III, IV and VI cranial nerves and ophthalmic branch of V nerve.
8. **Posterior palatine foramen**– between the maxillary and palatine portions of the palate for the passage of V cranial nerve.
9. **Infra-orbital foramen** – in front of zygomatic process of maxilla for the passage of infra-orbital branch of maxillary nerve.
10. **Optic foramen** – in the orbitosphenoid for optic cranial nerve.
11. **Internal orbital foramen** – behind infra-orbital foramen on the inner wall of orbit for ophthalmic branch of V cranial nerve.
12. **Anterior palatine foramen** – in the anterior part of palate branch of V cranial nerve.
13. **Olfactory foramen –** in the cribriform plate for olfactory nerves.
14. **Mental foramen** – on the outer surface of lower jaw for dental nerve.

(B) Foramina for Other Purposes

1. **External narial opening** – at the anterior end of skull and divided by cartilaginous septum.
2. **Lacrimal foramen** - perforating the lacrymal bone.
3. **Posterior narial opening** - bounded by the palatines.
4. **Foramen lacerum medium** – a large space on lower side between tympanic bulla and alisphenoid for the passage of internal carotid artery.
5. **Eustachian opening** – immediately behind and on inner side of foramen lacerum medium.
6. **Alisphenoid canal** – piercing the base of alisphenoid bone between foramen rotundum and foramen ovale, for the passage of external carotid artery.
7. **External auditory aperture** or **auditory meatus** – at the outerside of tympanic bulla.
8. **Post-glenoid foramen** – between auditory meatus and glenoid fossa of squamosal. for the passage of a vein.

VERTEBRAL COLUMN

1. Atlas vertebra. 1. Atlas is the first vertebra of the cervical region. 2. It is ring-like in structure. 3. Centrum is reduced. 4. Neural spine is small and inconspicuous. 5. Transverse processes are large flattened and perforated at their bases by vertebrarterial canals. 6. A small backwardly directed hypapophysis is present. 7. Anteriorly it bears a pair of concave surfaces for the articulation with the occipital condyles. 8. On the posterior side it bears a pair of small articular facets for the articulation with second vertebra.

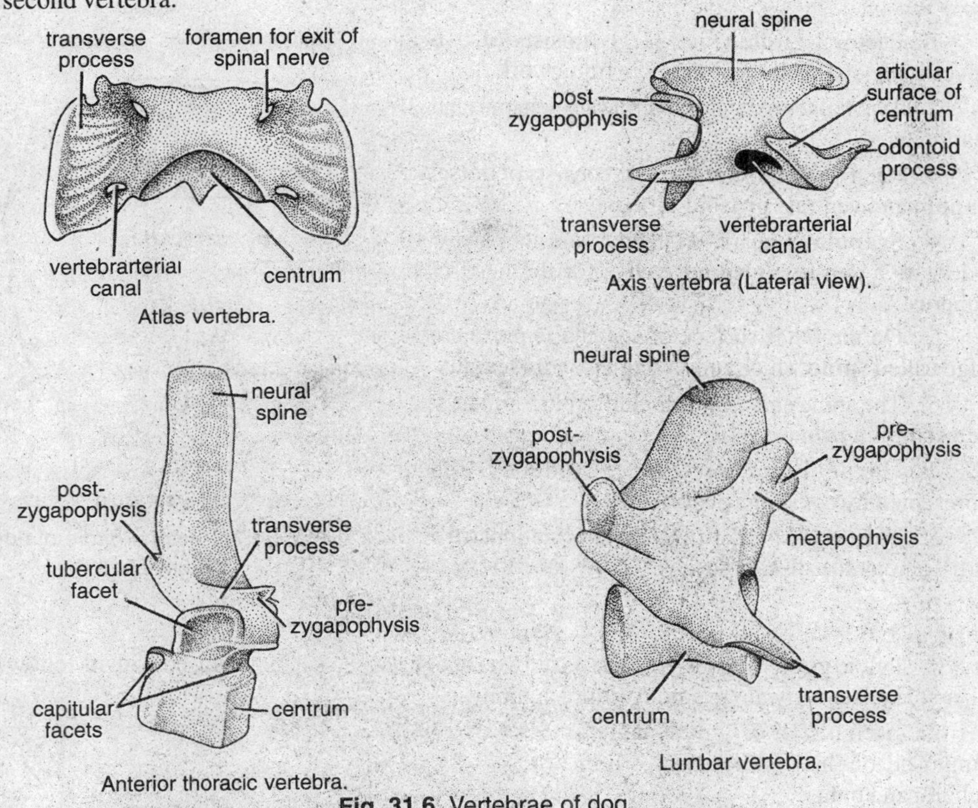

Fig. 31.6. Vertebrae of dog.

2. Axis vertebra. 1. Axis is the second vertebra of the cervical region. 2. Centrum is broad and **acoelous** (flattened) and produced in front into an anteriorly directed peg-like odontoid process which fits into the atlas. 3. Neural spine is elongated and crest-shaped. 4. Transverse processes are small directed backwards and perforated at their bases by vertebrarterial canals. 5. On each side of the odontoid process is an articular surface for atlas. 6. Only postzygapophyses are present.

3. Thoracic vertebrae. In the thoracic region, there are 12–13 vertebrae. About 5 vertebrae are considered anterior thoracic vertebrae and the remaining are posterior thoracic vertebrae.

3(a). Anterior thoracic vertebrae. 1. **Centrum** is short and bears capitular facets for the capitulum of thoracic rib at both the anterior and posterior ends. 2. Neural spine is very long, stout and backwardly directed. 3. Transverse processes are short and stout. 4. Pre-and postzygapophyses are present. 5. On either side below the transverse process are present tubercular facets for tuberculum of thoracic ribs.

3(b). Posterior thoracic vertebra. 1. **Centrum** is long, prominent and **acoelous**. 2. Neural spine is small and straight. 3. Transverse processes are small and irregular. 4. Presence of anapophyses and metapophyses. 5. Pre-and postzygapophyses are present. 6. Anterior end of centrum bears a pair of articular surfaces for capitulum of rib.

4. Lumbar vertebra. 1. Both the ends of the centrum are convex, although centrum is **acoelous**. 2. Neural spine is prominent running upwards and slightly forwards. 3. Transverse processes are small and strong running outwards, forwards and slightly downwards. 4. Pre-and postzygapophyses are present. 5. Metapophyses and anapophyses are also present.

APPENDICULAR SKELETON

GIRDLES

1. Pectoral Girdle

1. Pectoral girdle (Fig. 31.7) consists of two halves lying on the sides of thoracic framework.

2. Two halves of pectoral girdle are separate and unarticulated.

3. Each half of pectoral girdle consists of dorsal **scapula** and ventral **coracoid** process.

4. **Scapula** is nearly triangular and its narrow end bears a cup-like **glenoid cavity** for the head of humerus.

5. On the outer surface of scapula is present a ridge called **spine** which runs along its entire length.

6. The spine ends ventrally into a blunt process, the **acromian process**.

7. **Coracoid** is very much reduced and represented by **coracoid process**.

8. The dorsal border of scapula bears a narrow cartilaginous **suprascapula**.

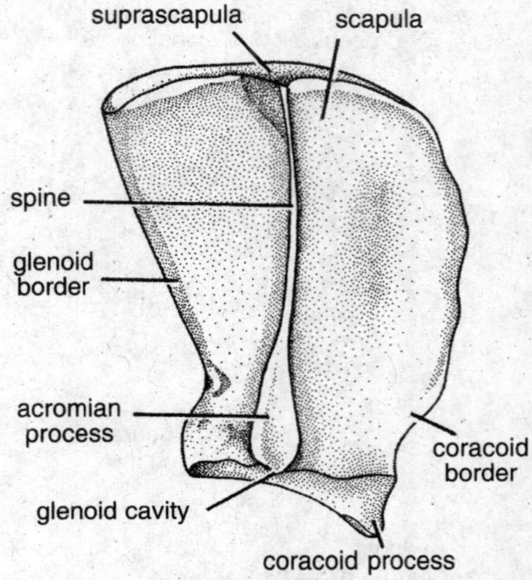

Fig. 31.7. Pectoral girdle of dog (Right half).

2. Pelvic Girdle

1. Pelvic girdle (Fig. 31.8) consists of two equal halves, each half is fused with its fellow of the other side posteriorly along a mid-ventral symphysis.

2. Each half or os-innominatum is composed of three bones, *viz.*, illium, ischium and pubis.

3. **Ilium** is a long bone expanded anteriorly and narrow posteriorly.

4. Ilium bears articular surface for sacrum on its inner side and also contributes in the formation of acetabulum.

5. **Ischium** is a stout flattened bone forming the hinder part of os-innominatum and contributes in the formation of acetabulum.

6. **Posteriorly ischium** bears a prominent **ischial tuberosity**.

7. **Pubis** is a smaller bone articulating anteriorly with ilium and posteriorly with ischium leaving a large gap the obturator foramen.

8. Pubis does not take part in the formation of acetabulum.

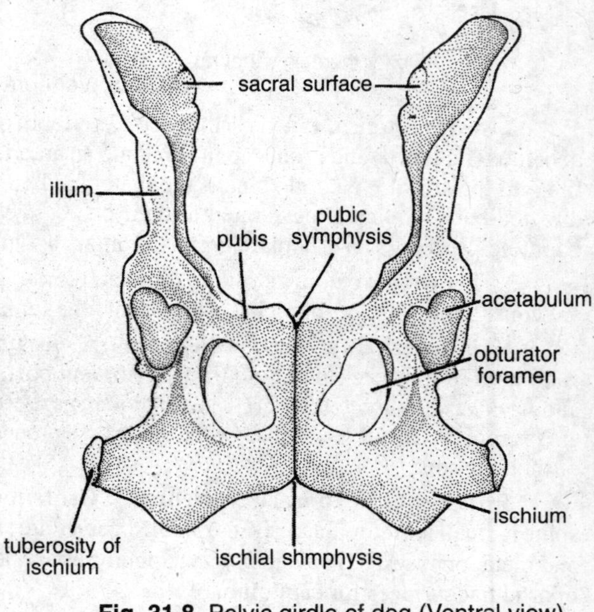

Fig. 31.8. Pelvic girdle of dog (Ventral view).

9. **Acetabulum** is a cup-shaped cavity which provides surface for the articulation to the head of femur.

LIMB BONES

Forelimb Bones

1. Humerus. 1.Humerus (Fig. 31.9) is the bone of upper arm of forelimb. 2. It is an elongated, stout and cylindrical bone. 3. Proximal end bears a distinct **head** and **greater** and **lesser tuberosities**. 4. Below the head on outerside of shaft is present a **deltoid ridge.** 5. The distal end bears a pulley-like **trochlea** and **supra-trochlear foramen** just above it. 6. Proximally it articulates with **glenoid cavity** of pectoral girdle by the head and distally with radius and ulna.

2. Radius and ulna. 1. Radius and ulna (Fig. 31.9) are bones of arm of forelimb. 2. Both the bones are movably articulated with each other instead of their fusion. 3. **Radius** is a straight and stout bone lying outside the ulna. 4. Proximally radius articulates with external surface of trochlea of humerous and distally with bones of carpus. 5. **Ulna** is also a straight and stout bone and longer than radius. 6. Proximal end of ulna is produced into an **olecranon process,** while its distal end is tapering. 7. Near the proximal end of ulna is also present a deep notch the **sigmoid notch.** 8. Ulna articulates proximally with humerus by receiving the outer condvle of trochlea in the sigmoid notch and distally with carpal bones.

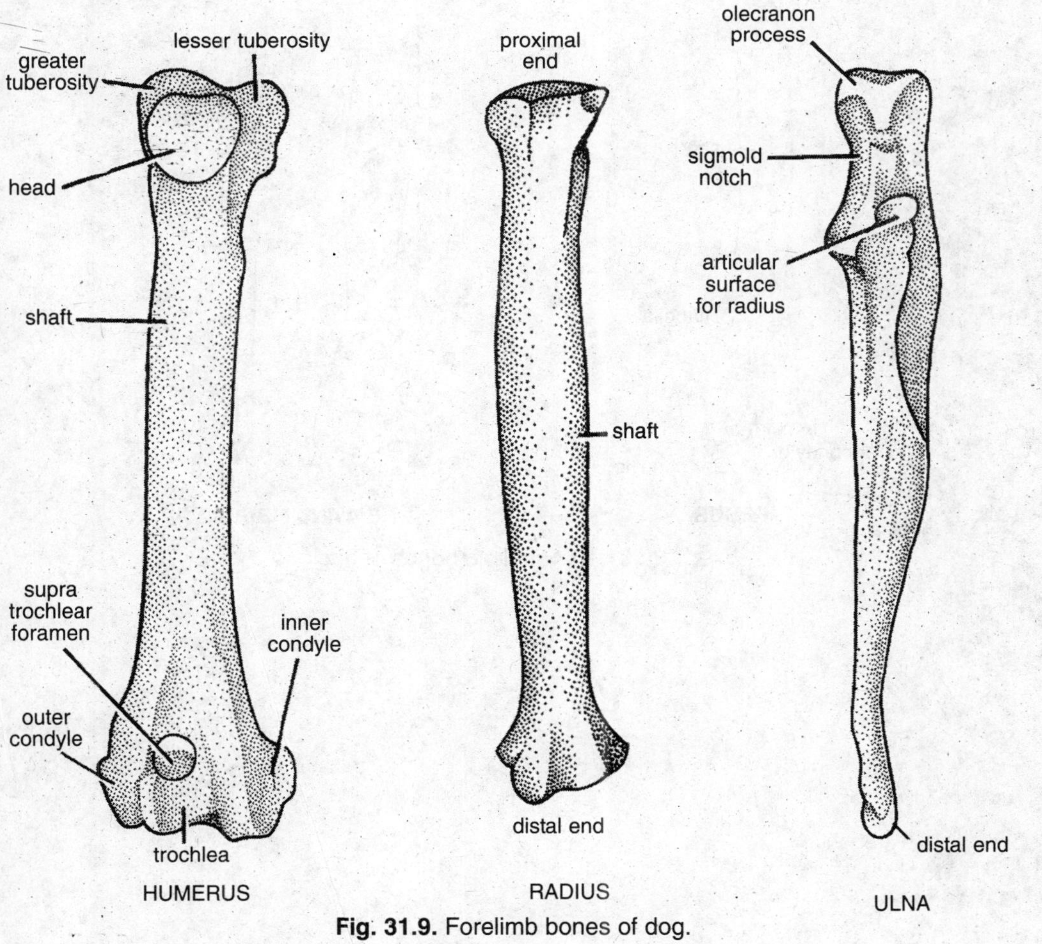

Fig. 31.9. Forelimb bones of dog.

Hindlimb Bones

1. Femur. 1. Femur (Fig. 31.10) is the bone of thigh of hindlimb. 2. It is a long and straight bone expanded at both the ends. 3. Its proximal end bears a rounded head and greater and lesser trochanters. 4. Distally it bears two prominent rounded condyles for articulation with tibia and fibula. 5. In between two condyles is present a large **patellar groove**. 6. Its distal end also carries an outgrowth the **flabella.** 7. Proximally it articulates with the pelvic girdle by gliding its head into the acetabulum.

2. Tibia and **fibula.** Tibia and fibula (Fig. 31.10) are bones of shank of hindlimbs. 2. **Tibia** is quite a thick and straight bone expanded at both ends. 3. A strong ridge, the **cnemial crest** is present on anterior surface of proximal end of tibia. 4. **Fibula** is a very slender bone having a small patella on proximal end. 5. Tibia articulates proximally with condyles of femur and distally with astragalus. 6. Fibula articulates proximally with head of tibia and distally with calcaneum.

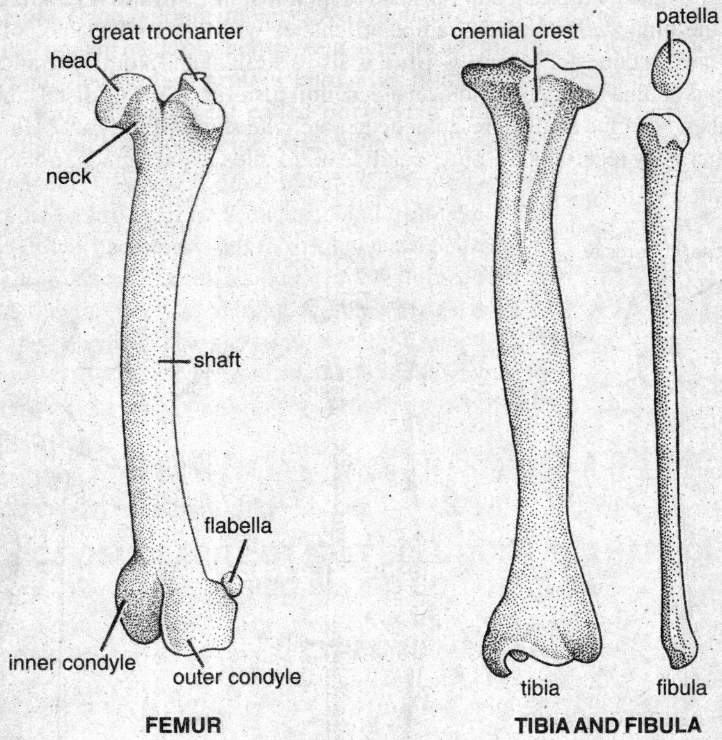

Fig. 31.10. Hindlimb bones of dog.

32

Chapter

EXPERIMENTAL BIOCHEMISTRY AND PHYSIOLOGY

LIST OF CHEMICALS AND GLASS APPARATUS

The following chemicals and glass apparatus are required for performing the colour reactions (tests of carbohydrates, proteins, fats and enzymes, etc.).

CHEMICALS

1. Absolute alconol.
2. Acetic acid (glacial).
3. Ammonium sulphate.
4. Benedict's solution.
5. Barfoed's reagent.
6. Barium chloride.
7. Bismuth subnitrate.
8. Chloroform.
9. Caustic soda.
10. Caustic soda (alcoholic).
11. Calcium chloride.
12. Copper sulphate 0.5 per cent.
13. Copper sulphate 1.0 per cent.
14. Chlorophenol red.
15. Ether.
16. Ethyl acetate.
17. Egg white 5 per cent.
18. Esbach's reagent.
19. Fehling's solution.
20. Fructose 0.2 per cent.
21. Fructose 0.5 per cent.
22. Ferric chloride 0.5.
23. Ferric chloride 1.0.
24. Glucose 1 per cent.
25. Glucose 2 per cent.
26. Glucose 0.2 per cent.
27. Glucose 0.5 per cent.
28. Glyoxalic acid reagent.
29. Glycerine.
30. Hydrochloric acid (conc).
31. Iodine solution.
32. Lactose 2 per cent.
33. Lead acetate.
34. Methylene blue.
35. Maltose 2 per cent.
36. Mercuric chloride (saturated).
37. Mercuric sulphate.
38. Mercuric nitrate.
39. Nylander's reagent.
40. Ninhydrin 1 per cent.
41. Olive oil.
42. Oleic acid.
43. Potassium ferricyanide.
44. Potassium tartarate.
45. Picric acid (saturated solution).
46. Phenyl hydrazine hydrochloride.
47. Potassium ferrocyanide solution.
48. Peptone.

49. Phenolphthalein.
50. Potassium dihydrogen phosphate.
51. Potassium hydrogen sulphate.
52. Sodium hydrogen sulphate.
53. Sodium chloride.
54. Sodium acetate.
55. Seliwanoff's reagent.
56. Silver nitrate.
57. Sulphosalicylic acid.
58. Sodium nitrate.
59. Sodium hypobromide.
60. Sodium hydroxide.
61. Sucrose 2 per cent.
62. Sulphuric acid (conc).
63. Tannic acid.
64. Thymol.
65. Thymol blue.
66. α–naphthol 1 per cent.

SOME IMPORTANT LABORATORY REAGENTS

1. Alpha naphthol. 1 gm of α-naphthol is dissolved in 95 per cent alcohol.

2. Barfoed's reagent. 13.3 gms of neutral copper acetate crystals are dissolved in 200 ml of distilled water. After filtering it add 1.8 ml of glacial acetic acid.

3. Benedict's reagent. Two solutions are prepared :

Solution A. 173 gms of sodium citrate and 100 gms of anhydrous sodium carbonate is dissolved in 800 ml of distilled water by heating.

Solution B. 17.3 gms of copper sulphate crystals dissolved in 100 ml of water.

Both the solutions are to be cooled. Solution B is added to solution A with constant stirring in 100 ml titration flask and diluted with distilled water up to the mark.

4. Esbach's reagent. 1 gm of picric acid and 2 gm of citric acid are dissolved in 100 ml titration flask in distilled water. Dilute with distilled water up to the mark.

5. Fehling's solution. There are two types of Fehling's solutions, *viz.*, A and B.

(*i*) **Fehling's solution A.** 7.93 gms of copper sulphate crystals are dissolved in 100 ml of distilled water.

(*ii*) **Fehling's solution B.** 250 gms of potassium hydroxide and 320 gms of sodium or potassium tartarate are dissolved in 1000 ml of distilled water.

6. Nylander's reagent. 2 gms of bismuth subnitrate and 4 gms of sodium or potassium tartarate in 100 cc of 10 per cent potassium hydroxide solution. It is filtered after cooling it.

7. Leishman's stain. It is prepared by dissolving 0.15 gm of Leishman's powder in 100 ml of methyl alcohol.

8. Wright's strain. Take 25 gm of Wright's stain powder and dissolve in 500 ml of methyl alcohol. The stain matures in four weeks.

9. Ammonium molyblate. In 300 ml of strong ammonia dissolve 75 gm of ammonium molybdate, crystals. This solution is added in 900 ml of conc. HNO_3 and later add 400 ml of water. Cool it and add more 1600 ml of water and then filter.

10. Barium chloride. Prepare 10% solution in water.

11. Basic lead acetate. Dissolve 180 gm of lead acetate in 700 ml water.

12. Chromic acid. It is for cleaning glass apparatus. Dissolve 60 gm potassium dichromate in 300 ml of water and then add 460 ml of conc. H_2SO_4. Cool it under running water and then use it for cleaning glass apparatus.

GLASS APPARATUS

1. Test-tubes.
2. Funnels.
3. Beakers.
4. Pipettes.
5. Measuring cylinders.
6. Glass stoppered bottles.
7. Spirit lamp.
8. Flasks.
9. Reagent bottles.

Besides these, test-tube holders, test-tube stands, funnel stands, cotton and filter papers are also required. These apparatuses should be cleaned before use for getting proper results.

Chapter 33

EXPERIMENTAL BIOCHEMISTRY AND PHYSIOLOGY

The experimental biochemistry includes colour and precipitation tests of carbohydrates, proteins, lipids (fats), enzymes, identification of amino acids, amines, peptides and carbohydrates by paper chromatography. Physiological experiments include estimation of haemoglobin, counting of R.B.Cs., W.B.Cs., differential count of W.B.Cs., study of coagulation of blood, bleeding time of blood, clotting time of blood, preparation of haemin crystals, effect of osmolarity of salt solutions and haemolytic agents on R.B.Cs., study of circulation of blood in the capillaries of foot of frog, recording of heart beat in situ, recording of muscular contraction, determination of rate of respiration, reflex action, wiping reflex, knee-jerk reflex, test of urine for urea, proteins, ketones and sugar, and demonstration of osmosis.

COLOUR AND PRECIPITATION REACTIONS ON FOOD STUFFS

A. CARBOHYDRATES

The carbohydrates are the compounds of the carbon, hydrogen and oxygen. Chemically the carbohydrates are polyhydroxy aldehydes or ketones. Carbohydrates are classified into **monosaccharides (monomers), oligosaccharides (oligomers)**, and **polysaccharides (polymers)**. Monosaccharides and oligosaccharides are soluble crystalline substances having a sweet taste and are collectively known as sugars. Polysaccharides are insoluble amorphous substances and are, thus, called non-sugars.

EXPERIMENT NO.1

Tests for Carbohydrates (Monosaccharides : Glucose and Fructose)

1. Fehling's test. Take 5.0 ml of Fehling's solution in a test-tube, add 1.0 ml of 0.2% sugar solution and boil. **Result** : The colour changes with the formation of yellow or brick-red precipitate. The precipitate is due to the formation of **cuprous oxide**. The precipitate indicates that reduction has taken place. It is reducing monosaccharide.

2. Benedict's test. Take 5.0 ml of Benedict's reagent in a test-tube, add 0.5 ml of sugar solution and mix it well. Heat to boil for two minutes then cool under tap-water. **Result** : Green, yellow or red precipitate is formed. The precipitate is due to the formation of cuprous oxide (Cu_2O). Test sample contains **reducing monosaccharides**.

3. Molisch's test. Take two test-tubes and mark them A and B. Take 5 ml solution of the given sample in test-tube A. Add 2 drops of Molisch's reagent in it and thoroughly shake the test-tube to mix both the solutions. Now take conc. H_2SO_4 in test-tube B and pour it gently in test-tube A along its side so that it does not mix the above solution. **Result** : A pink or purple ring is formed at the junction of two solutions that shows presence of **carbohydrate** in test sample.

4. Picric acid test. Take 3.0 ml of given solution of sugar in a test-tube, add 1.0 ml of saturated solution of picric acid and 1.0 ml of 1% or 4% NaOH and heat it. **Result** : Red colour develops. The red colour is due to the reduction of picric acid to picrylic acid. Sample contains **reducing monosaccharide.**

5. Take 3.0 ml of given solution of sugar in a test-tube, add 1.0 ml of 4% NaOH and boil for two minutes. **Result** : Solution becomes yellow and acquires light-brown colour on standing. The yellow colour is due to the formation of paramylin.

6. Tommer's test. In 2 ml of 0.5% of copper sulphate solution and 0.2 ml of 0.2 % sugar solution and mix both. Add 2 ml of 40 % NaOH solution. A clear blue solution is formed. Boil it later to get precipitate. **Result** : Yellow or red precipitate of CuO is formed showing **glucose**.

For **sucrose** perform Rapid Furfural and Scliwanof's test.

7. Fearson's test for sucrose. Take 4 ml of test solution in a test-tube and add 4 drops of 10% methylamine hydrochloride solution in it. Boil it for 30 seconds and then add 5 drops of 20% NaOH solution in it. **Result** : Yellow colour appears that turns red showing **sucrose** in test sample.

8. Rapid furfural test. Add 6 drops of α-naphthol in 1 ml of 2% sugar solution and then add 5 ml of conc. HCl in a test-tube. Boil it.

Result : Deep purple colour appears showing **fructose**.

9. Iodine test for starch. Take 2 ml of test solution or suspension in a test-tube, boil and cool at room temperature. Now add one or two drops of iodine solution in it.

Result : Blue colour appears that disappears on heating and again blue colour appears on cooling showing **starch** is solution.

Disaccharides

Lactose, maltose and sucrose are disaccharides. Lactose is found in milk, maltose is end product of starch hydrolysis and sucrose is found in cane sugar. For lactose and maltose perform Tommer's, Fehling's, Benedicts and Barfoed's test.

EXPERIMENT NO. 2

B. PROTEINS

Proteins are complex organic compounds. They are composed of carbon, hydrogen, oxygen, nitrogen and sulphur in their molecules. Proteins are the polymers of the amino acids. For protein test 5% egg white solution is prepared, *i.e.,* 5 ml of egg white in 95 ml of distilled water. Filter it through cotton.

Tests for Proteins : Colour Reactions

1. Biuret test. Take 2.0 to 3.0 ml of protein solution or egg albumin solution in a test-tube, add an equal volume of 10% NaOH, mix thoroughly and add 0.5% $CuSO_4$ solution drop-by-drop. **Result**: Mix it well until a purple-violet or pinkish-violet colour of protein develops.

2. Ring Biuret test. Take 3.0 ml of protein solution or egg albumin solution or any other solution of protein, add 1.0 ml of 40% NaOH and 1.0 ml of 1% $CuSO_4$ with the help of pipette. The two liquids should not mix. Rotate the test-tube gently. **Result** : A pink or violet ring is formed at the junction of two liquids showing presence of protein.

3. Take 3.0 ml of protein solution or egg-albumin solution in a test-tube, add 5 drops of lead acetate. A precipitate is formed. Add 40% NaOH drop-by-drop until the precipitate dissolves. Boil the solution of brown or black precipitate. **Result** : Protein appears.

4. Millon's reaction. Take 5.0 ml of dilute solution of protein or egg-albumin in a test-tube, add 3 to 4 drops of Millon's reagent. Mix and heat the mixture till it boils. **Result** : White precipitate of protein appears which gradually turns red upon heating. In case of other proteins the solution becomes red only.

5. Xanthoproteic test for tyrosine, phenylalanine and tryptophane. In a test-tube take 3 ml of 5% egg white solution and add 1ml of conc. HNO_3 and boil. Cool the tube and add excess of 40% NaOH or ammonia to make alkaline.

Result : Yellow colour appears that changes to orange. First white ppt. is formed that changes to yellow.

6. Sulphosalicyclic acid test. Take 3 ml of 5% egg white solution in a test-tube and add 20% sulphosalicyclic acid.

Result : White precipitate of **protein** appears.

7. Esbach's test. In a test-tube take 5% solution of egg white and add a little Esbach's reagent.

Result : Yellow precipitate of **protein** appears. (By this test albumen in urine can also be tested).

8. Tannic acid test. Take 3 ml of egg-white solution in a test-tube and add 5 drops of freshly prepared tannic acid solution.

Result: Brownish non-granular precipitate of **protein** occurs.

9. Acetic acid potassium ferrocyanide test. Take 3 ml of 5% egg-white solution in a test-tube and add 3 drops of glacial acetic acid and 3 drops of potassium ferrocyanide solution.

Result : White precipitate appears due to **proteoses**. It dissolves on boiling and reappears on cooling.

10. Ninhydrin test. Add 0.5 ml of ninhydrin solution in a test-tube containing 5 ml test solution. Boil it for 2 minutes and then cool.

Results : Blue colour of **protein** appears.

Note: Ninhydrin test gives positive results by all amino acids except proline and hydroxyproline.

11. Millon's test for tyrosine. In a test-tube containing 3 ml of 5% egg-white solution, add 2 ml of mercuric sulphate reagent and boil for one minute. A yellow precipitate generally appears. Cool the tube and add 2 drops of 1% sodium nitrate solution and heat again.

Result: Solution and precipitate becomes red showing **tyrosine**.

12. Aldehyde test for tryptophan. In a test-tube having 5% egg-white solution add one drop of 0.2 % of 40% formalin and then add 0.5 ml of mercuric sulphate reagent. Shake well and then add 2 ml of conc. H_2SO_4. Shake it.

Result : Violet (purple) colour develops showing **tryptophan** in test solution.

13. Arginine test for arginine. Take 3 ml of 5% egg-white solution in a test-tube and add 2 drops of 1% of naphthol solution and then add 1 ml of 40% NaOH solution and also add 2 drops of sodium hypobromide.

Result : Brick red colour appears showing **arginine** in test solution of protein.

14. Sulphur test for cystine and cysteine. In a test-tube containing 3 ml of 5% eggwhite solution, add 50 drops of lead acetate that causes precipitation. Now add 40% NaOH drop by drop till the precipitate dissolves. Boil it.

Result : Black or brown precipitate appears showing **cystine** and **cystein** amino acids in the protein solution.

C. LIPIDS (FATS)

The lipids are the organic compounds which are insoluble in water but soluble in ether, chloroform, benzene, hot alcohol and petroleum ether. They contain long chains of aliphatic hydrocarbons or benzene ring in their molecules. The lipids are non-polar and hydrophobic. Their caloric value is doubie than carbohydrates. Simple lipid is glycerol.

EXPERIMENT NO. 3

Tests of Lipids

1. Solubility test. In five test-tubes (A, B, C, D, E) take 3 ml of water, alcohol, ether, benzene and chloroform. To each test-tube add 2 or 3 drops of lipid. Shake well and note the solubility. **Result:** The lipid floats on water but it is miscible with ether, benzene and chloroform, while it sinks to the bottom in alcohol.

2. Formation of a transluscent greasy spot. Place a drop of olive oil on a piece of ordinary paper. A transparent greasy spot is produced at the point of contact with the oil.

3. Formation of acrolein. Take 0.5 gm of powdered potassium bisulphate ($KHSO_4$) or sodium bisulphate ($NaHSO_4$) in a clean and dry test-tube. Pour 5 ml of olive oil on the salt. Mix thoroughly with glass rod and heat it. **Result:** The irritating smell of **acrolein** is given off. In this reaction the glycerol of fat is dehydrated and acrolein or acrylic aldehyde is produced.

4. Take 3 ml of olive oil in a test-tube, add 2 drops of oleic acid. Mix them thoroughly. Rancid oil is obtained.

5. Emulsification of fat. Take 2 drops of rancid oil in a test-tube and add 3 ml of sodium hydroxide (NaOH). **Result:** The rancid oil dissolves in sodium hydroxide (alkali) forming the soap.

6. Salting out. Take 10 drops of olive oil in a test-tube, add 2ml of alcoholic caustic soda and boil. **Result:** A soft mass of soap separates which dissolves on cooling. This confirms the lipids (fats).

7. Colour test with Sudan III. In a test-tube containing 0.5ml of ether or chloroform, add 0.5 ml test sample drop by drop. When test sample is fully dissolved, add one drop of Sudan III reagent in the test tube.

Result : Red colour appears showing presence of fat.

D. ENZYMES

EXPERIMENT NO. 4

Test for Salivary Amylase

The enzyme **salivary amylase** is present in the saliva. It is also known as **ptyalin**. It changes starch into glucose and maltose. Salivary amylase acts steadily at body temperature but it acts optimally between 40 to 50°C. The optimum pH for activity of amylase is between 6 to 7 and that of salivary amylase of man is 6.6.

For the demonstration of amylase activity of salivary glands, saliva of man and salivary glands of cockroach are the best materials.

Preparation of enzyme extract. For the preparation of extract of salivary glands, 5 – 10 salivary glands are taken out from the cockroaches. The salivary glands are ground in a mortar with little thymol. The solution is diluted with 50 per cent glycerol to make 10 per cent solution of salivary gland extract. The extract is stored in a bottle with toluene.

Test for amylase activity on starch. Take 5 drops of extract in an **incubation tube A** and add 5 drops of 1% starch solution, mix it and then cover the mixture with a layer of toluene. In another incubation tube **B** take 5 drops of boiled extract, add 5 drops of starch solution. This will serve as the control experiment. Incubate both the tubes in a bath at 37°C for about one hour or keep both the tubes in a test tube stand for an hour or so.

Observation: 1. After an hour take a small part of the incubated mixture of **A tube** and add iodine solution. No change in colour of iodine solution will indicate the presence of **amylase enzyme**. **2.** Also take a small part of incubated mixture of B tube and add iodine. Solution turns purple colour. It shows starch is not digested due to boiling of extract which destroys the salivary amylase. **3.** Take another part of the A tube extract, add few drops of Benedict's solution and heat till boiling. A brick-red precipitate will confirm the presence of amylase enzyme in the extract.

Repeat this test with the control mixture of B tube. In the first test violet colour will confirm the presence of starch as such which has not been digested.

Starch always gives bluish-violet colour with iodine solution. In the experimental A solution, enzyme extract is mixed with iodine solution, amylase present in the enzyme extract hydrolyses the starch into maltose and glucose, therefore, there is no change in the colour of iodine.

In the control experiment B, amylase is destroyed by boiling the enzyme extract, starch is not hydrolysed into maltose and glucose, therefore, it gives bluish-violet colour with iodine solution.

In the second test with Benedict's solution a red precipitate of cuprous oxide appears. In this **glucose** is formed. Since glucose is a reducing sugar, it reduces copper sulphate into cuprous oxide.

EXPERIMENT NO. 5

Test for Pepsin Enzyme on Protein

Pepsin is the most important enzyme of the gastric juice of stomach. It hydrolyses proteins into **proteoses** and **peptones**. It is secreted by the gastric glands in the stomach in inactive form **pepsinogen**, which is converted into active **pepsin** by hydrochloric acid secreted in the stomach. Pepsin requires a rather definite pH for its optimum activity. The optimum pH for the activity ranges between 1.5 to 2.2.

Preparation of substrate. To test the presence of pepsin solution of casein protein is generally used as substrate. To prepare the substrate take 2.0 gm of vitamin free casein in a beaker, add 5.0 ml of NaOH of 0.1 normality, dilute it up to 50 ml with distilled water and heat it at 60°C. Stir the mixture well until the casein is completely dissolved. Transfer this mixture into another beaker containing a mixture of 5 ml of HCl of 1 normality and 4.5 ml of distilled water. In this way a 2% solution of casein having 1.5 pH is prepared.

Preparation of enzyme extract. For preparing the enzyme extract, dissect three or four frogs and take out the stomach portion, keep the stomach portion in a petridish, wash it thoroughly with distilled water and add a little thymol to check the bacterial action. Ground the stomach portion in a mortar and add 50% glycerine so as to prepare approximately 10% solution of stomach tissue. Centrifuge the extract and then filter. Store the filtrate in test-tube with toluene.

Test of enzyme activity. 1. Take 5 drops of extract in a test tube and add equal amount of substrate solution (acid solution of casein) and incubate it at 37°C for 8-12 hours. **2.** Perform control experiment by taking 5 drops of substrate with equal amount of distilled water and incubate it at 37°C. **3.** To confirm the presence of pepsin take 2 drops of incubated mixture, add 1 drop of 10% sodium acetate solution. Casein is precipitated by addition of sodium acetate but in the presence of pepsin casein is hydrolysed into **proteoses** and **peptones** which cannot be precipitated by sodium acetate. **4.**Repeat the confirmatory test in the control experimental tube and compare the results of the two tubes.

Test of pepsin on fibrin fibres of protein. The presence of pepsin can also be demonstrated by dissolving the **fibres of fibrin** in an acidic medium. Take a few fibres of fibrin, two drops of enzyme extract and two drops of buffer of pH 2.0 on a microslide and cover it by the cover slip. Keep the microslide in the incubation bath at 37°C for 8–12 hours. After incubation for 8–12 hours observe the microslide under the microscope; dissolved fibrin will indicate the presence of pepsin.

EXPERIMENT NO. 6

Test for Trypsin Enzyme on Protein

Trypsin is a proteolytic enzyme produced in the pancreas of animals. In pancreas it is secreted in the inactive condition called **trypsinogen**, but in intestine it is converted into the active **trypsin** by the alkaline action of bile secreted by liver. The optimum pH of trypsin is about 8.0 and it is more effective than pepsin. Trypsin hydrolyses the remaining proteins which are not hydrolysed by the gastric juice.

Preparation of substrate solution. Prepare 20% alkaline solution of casein. For this take 1 gm vitamin free casein and dissolve it in 50 ml of buffer of 7.2 pH.

Preparation of enzyme extract. Take a part of pancreas and prepare the enzyme extract as described in case of pepsin.

Test of enzyme activity. 1. Take 5 drops of extract in tube A, add 5 drops of substrate solution and to this also add 5 drops of buffer of 7.8 pH. **2.** Perform control experiment by taking 5 drops of substrate and drops of distilled water in a test-tube B. Incubate both the tubes in the incubation bath at 37°C for 4-6 hours. **3.**To confirm the presence of trypsin add few drops of 1% acetic acid in the experimental test-tube A. Absence of any precipitate will confirm the presence of trypsin. Acetic acid precipitates the casein. Since the casein has been hydrolysed, therefore, it does not form any precipitate and shows the presence of trypsin. **4.** In the control experimental tube B there is no extract, therefore, it does not contain trypsin and casein is not hydrolysed, and addition of acetic acid gives the precipitate.

EXPERIMENT NO. 7

Urine Test

To show presence of albumin, sugar and *ketone bodies in urine.*

1. For albumin (proteins). Take 20 ml of urine in a test-tube and boil it in such a manner that only upper layer is heated. If turbidity appears, it shows presence of albumin. Add 3–5 drops of acetic acid in the test tube.

Result : Turbidity shows presence of albumin.

2. For ketone bodies. Take urine in a test-tube, saturate it with ammonium sulphate and then add few drops of ammonia solution and few crystals of sodium nitroprusside. Shake the tube thoroughly to mix the contents.

Result : Formation of reddish purple colour shows presence of ketone bodies.

3. For sugars. Perform tests, for carbohydrates-Fehling's test. Benedict's test, etc.

EXPERIMENTS ON PHYSIOLOGY

1. ESTIMATION OF HAEMOGLOBIN BY HAEMOMETER

EXPERIMENT NO. 1

Object. *To estimate the amount of haemoglobin in (i) frog's blood, (ii) human blood and (iii) blood of any other vertebrate animal.*

Requirements. (*i*) Haemometer (Haldane's haemoglobinometer); (*ii*) decinormal (N/10) HCl (1.2 ml, of conc. HCl dissolved in 100 ml of distilled water); (*iii*) distilled water ; (*iv*) living frog or any other living vertebrate.

Apparatus. The haemometer (Fig. 33.1) consists of two sealed lateral comparison tubes containing a suspension of acid haematin. These are held in a black frame against a white ground glass. Besides, a graduated test-tube of the same diameter is also provided which can fit in the haemometer in between the two side tubes for comparison. A micropipette of 20 cmm is also provided. The other things provided are a small glass rod stirrer, a small bottle brush and a dropper and also a small bottle to contain the decinormal acid solution.

Principle. The method to estimate the haemoglobin content of blood is based on the principle of making an acid haematin solution of blood under experimentation in the graduated tube and then comparing it with the sealed comparison tubes containing the standard acid haematin.

Procedure. The graduated tube is first cleaned with distilled water and then with methylated spirit or 90% alcohol. It is thoroughly dried up before being used. Now with the help of a dropper the N/10 HCl solution is filled in the graduated tube up to 2 gms mark. The micropipette is now filled up by sucking fresh blood of the vertebrate under experimentation up to the mark of 20 cmm. The small amount of blood adhering to the outerside of micropipette should be wiped off by sterilised cotton. The blood of micropipette is now added to the N/10 HCl solution in the graduated tube. The pipette should be introduced carefully into the tube and its lower mouth should pass right up to the bottom into HCl solution. When blood has been expelled the pipette is rinsed twice or thrice by distilled water. Every time the contents of micropipette should be expelled into the graduated tube. The acid haematin solution is now thoroughly stirred with the help of a glass rod and then allowed to stand at least for 10 minutes. Afterwards the acid haematin solution is gradually diluted by adding distilled water in a dropwise manner. With the addition of each drop of distilled water the solution should be stirred and its colour matched with that of the standard sealed tubes. This should be continued till the colour of the acid haematin solution just fades away as compared to that of the standard comparison tubes. The reading before the colour just fades is taken as the correct and final reading.

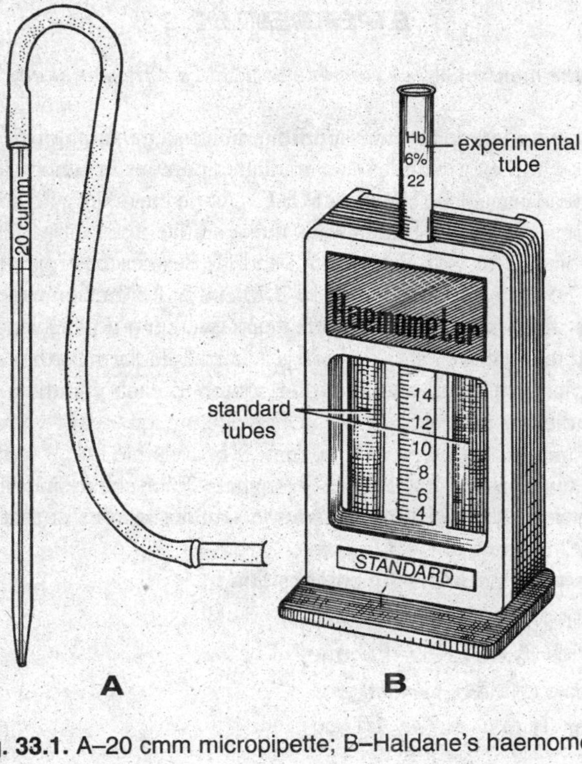

Fig. 33.1. A–20 cmm micropipette; B–Haldane's haemometer.

Observations. The readings are tabulated in the following manner :

Reading	*Frog's blood*	*Human blood*
I	7.8	12.45
II	7.10	12.60
III	7.9	12.50
Final =	7.9	12.50

Results : The blood of frog contains 7.9 mgm of haemoglobin in100 ml of blood.

The blood of man contains 12.50 mgm of haemoglobin in 100 ml of blood.

Precautions. The following precautions are to be taken.

1. In case human blood is being tested, give a light prick to the finger tip of the person providing blood.
2. The finger and the prick (needle) should be disinfected first with methylated spirit or alcohol (90% or absolute).
3. Avoid uncleaned tubes and pipettes, etc.
4. Avoid incorrect filling up of micropipette.
5. Avoid inclusion of blood sticking at the outer surface of the mouth of the pipette.
6. Readings should be taken always of the upper meriscus.
7. Experiment should be performed quickly without waste of time so that fresh blood is not allowed to coagulate before transfer to decinormal HCl.

2. ENUMERATION OF RED BLOOD CORPUSCLES BY HAEMOCYTOMETER

EXPERIMENT NO. 2

Object. *To determine the number of red blood cells (R.B.Cs.) present in one cubic mm volume of blood.*

Requirements. Haemocytometer, Hayem's solution, microscope and blood.

Apparatus. The haemocytometer includes two graduated pipettes in which dilution of blood is done. One pipette with a red bead is used for counting R.B.Cs, while the other with a white bead for counting W.B.Cs. Each pipette bears at least 3 graduations, those on the pipette for R.B.Cs are 0.5, 1 and 101. The graduations on the pipette for W.B.Cs. are 0.5, 1 and 11. Besides these pipettes the haemocytometer includes a glass slide, 7.5 cm or 3 inches long by 3.75 cm or 1.5 inches wide. It bears two counting chambers with a cover slip. One side of the slide bears two central platforms bordered by an edge-shaped groove. Beyond this groove on either side is a lateral platform further demarcated by a lateral groove. Each central platform bears one counting chamber, each counting chamber is formed by several straight perpendicular and horizontal lines enclosing squares of various measurements; the biggest square is 3×3 mm with an area of 9 sq. mm. It is divisible into 9 squares of 1×1 mm each, the central square is again divisible into 25 smaller squares. Each square measures $1/5 \times 1/5 = 1/25$ sq mm. Each of these squares is further divisible into 16 smaller squares so that there are 400 smallest squares in all. Each with an area of 1/400 sq mm.

Hayem's solution has the following composition :

(*i*) Mercuric chloride ($HgCl_2$) = 0.5 gm
(*ii*) Sodium chloride (NaCl) = 1.0 gm
(*iii*) Sodium sulphate (Na_2SO_4) = 5.0 gm
(*iv*) Distilled water (H_2O) = 200 cc

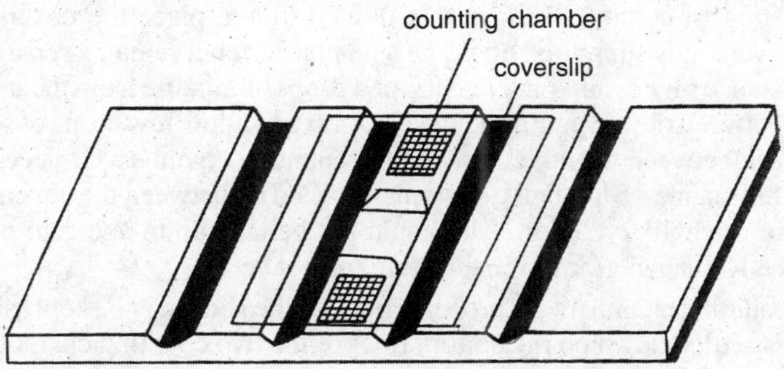

Fig. 33.2. Glass slide with counting chambers.

The mercuric chloride acts as corrosive sublimate and fixes the R.B.Cs, present in the blood. The other ingredients act isotonically so that the R.B.Cs. may not burst due to haemolysis. The Hayem's solution also serves to dilute the blood.

Procedure. Blood is obtained either directly from the heart of a frog or from some body part externally in case of man. Human blood is generally taken out from a finger (avoid thumb and first finger) which is thoroughly washed and cleaned by spirit or alcohol (absolute or 90%). It is now pricked by an ordinary injection needle which is also sterilised with spirit and dried before use. The finger is pricked quickly and effectively. It should not be pressed hard to let out blood so as to avoid other body juices also oozing out. For this reason also the first one or two drops of oozing blood should be avoided and wiped off.

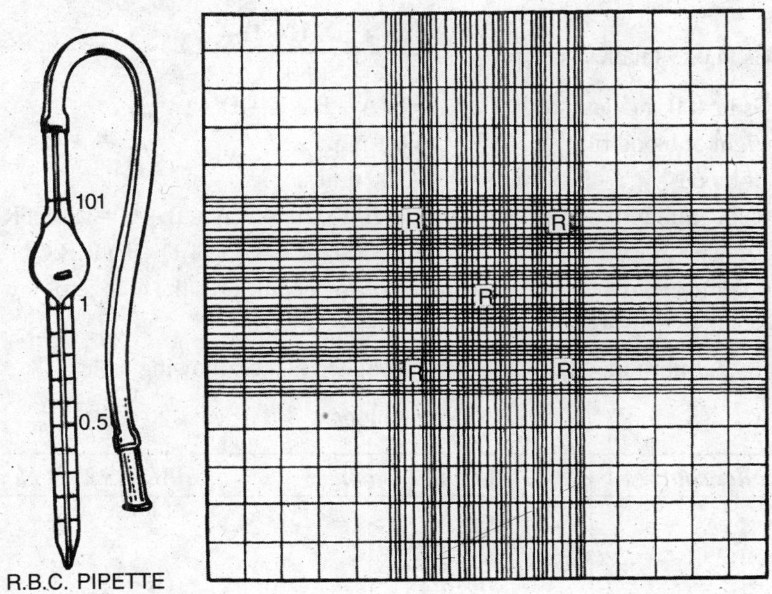

Fig. 33.3. R.B.C. pipette and small squares in the counting chamber. R denotes small squares for red cell count.

Now take the pipette meant for R.B.Cs which is already rinsed with alcohol or spirit or ether and thoroughly dried. Suck the blood in the pipette up to 0.5 mark taking care that air bubbles are not included. The excess of blood, if any, may be run out by touching the mouth of pipette to the palm. The blood which is sticking to the outer side of pipette should also be carefully cleaned. The pipette should now be transferred to the container of Hayem's solution which is carefully sucked up to 101 mark. The pipette is now held horizontally between the fore-finger and thumb or palm surfaces of the hand and rotated several times so that blood thoroughly mixes with Hayem's fluid. The red bead in the pipette also helps in mixing. In this way dilution of blood becomes 200 times.

Before starting to count R.B.Cs in this diluted blood, place the coverslip on the counting chambers. The coverslip is supported upon the side platform but remains separated from the central platforms by a distance of 0.1 mm. First reject 3 or 4 drops of mixture from the pipette. Now apply the tip of the pipette between the coverslip and the platform and allow few drops of blood mixture to flow in the narrow space between the coverslip and the counting chambers. If necessary both chambers may be filled in this manner. Blood mixture remains filled up between the coverslip and the counting chambers because of capillary action. Air should not be taken into and also pouring excess blood mixture so that the H-shaped groove remains free from it.

When the counting chambers are properly flooded the slide may be kept aside for a few minutes so that the R.B.Cs settled down on the bottom floor of the two counting chambers. Now transfer the slide gently and carefully under the microscope without disturbing the settled R.B.Cs and start counting them.

Counting. It is not necessary to count the R.B.Cs in all the 400 smallest squares or even in the 25 smaller squares, count them only in five smaller squares, *i.e.*, in the 1st, 5th, 13th, 21st and 25th. The R.B.Cs lying on the lower and right sides of a square are to be added in the total, while those lying on the upper and left sides are to be rejected.

Calculations. The calculations may be done by the following method :

$$\text{Number of R.B.Cs per cubic mm} = \frac{\text{Number of cell counted} \times \text{dilution} \times 4000}{\text{Number of small squares counted}}$$

Suppose total R.B.Cs for 5 smaller squares are equal to = A + B + C + D + E. Therefore, total R.B.Cs in 80 smallest squares = A + B + C + D + E.

$$\therefore \text{ R.B.Cs in one smallest square} = \frac{A+B+C+D+E}{80}$$

$$\therefore \text{ R.B.Cs in 400 smallest squares} = 5\,(A+B+C+D+E)$$

But the height of blood film = 0.1 mm

Dilution of blood = 200 times

So that 1 cubic mm of blood will contain $= 200 \times 10 \times 5\,(A + B + C + D + E)$ R.B.Cs

$= 10{,}000\,(A + B + C + D + E)$ R.B.Cs

Thus, in order to avoid the detailed calculations to get quick results four zeros are added to the total R.B.Cs of 5 smaller squares (80 smallest squares).

Tabulation. The readings are tabulated in the following table :

Frog's blood

Smaller squares	*R.B.Cs in I chamber*	*R.B.C's in II chamber*
I^{st} A	25	15
IInd B	30	20
IIIrd C	15	18
IVth D	20	25
V^{th} E	22	20

Human blood

Smaller squares	*R.B.Cs in I chamber*	*R.B.C's in II chamber*
I^{st} A	83	105
IInd B	87	97
IIIrd C	74	79
IVth D	96	86
V^{th} E	98	83

	Frog		Human	
Total R.B.Cs =	112	98	438	450
Average =	$\frac{112+98}{2}=105$		$\frac{438+450}{2}=444$	

Results : The result is given as follows :

(*i*) 1 cmm of frog's blood contains 1050000 R.B.Cs.

(*ii*) 1 cmm of human blood contains 4440000 R.B.Cs.

3. ENUMERATION OF WHITE BLOOD CORPUSCLES (W.B.Cs) BY HAEMOCYTOMETER

EXPERIMENT NO. 3

Object. *To determine the number of white blood cells (W.B.Cs) per cubic mm in the human blood.*

Requirements. Haemocytometer, glacial acetic acid, solution of gentian violet in water, distilled water. The glacial acetic acid haemolyses the red cells, while the gentian violet slightly stains the nuclei of white blood corpuscles (W.B.Cs), the W.B.C's may, therefore, be easily recognised.

Procedure. The procedure of counting of W.B.Cs is the same as that of the R.B.Cs. The W.B.Cs are counted in the four corners of 1 square millimetre in the central ruled area on both the sides of the counting chambers of the haemocytometer. The W.B.Cs are recognised by the refractile appearance and by the slight colour given to them by the stain contained in the diluting fluid. The cells touching the boundary lines are not counted.

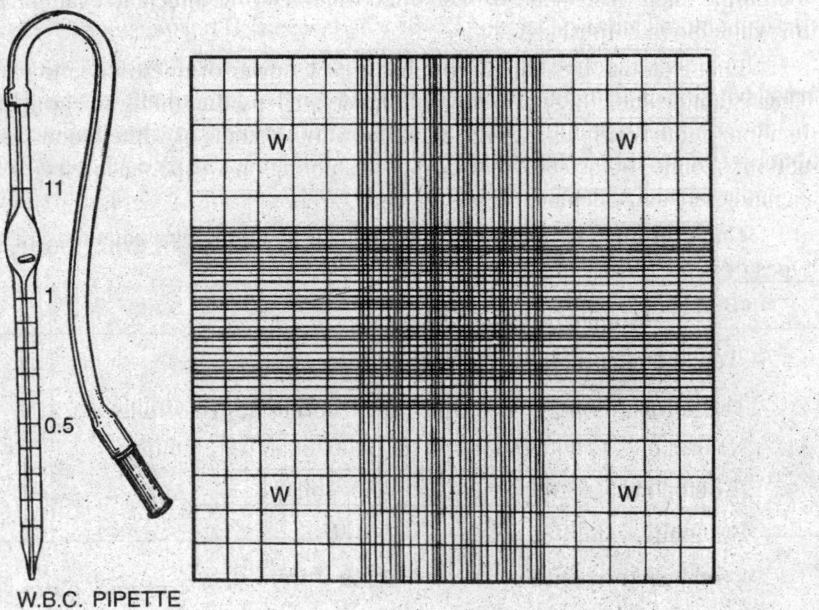

Fig. 33.4. W.B.C. pipette. Big squares in the counting chamber. W denotes big squares for white cell count.

Calculations. The calculations may be done by the following method :

$$\text{Number of W.B.Cs per cubic mm} = \frac{\text{Number of cells counted} \times \text{Dilution} \times 10}{\text{Number of 1 sq mm counted}}$$

Since the dilution is 20 times and the cubic capacity of the area counted is 1/10 cubic millimetre, the total volume is 1/200 cubic millimetre. Say for instance, the number of W.B.Cs in 4 outer squares is (10 + 12 + 13 + 10) = 45. In otherwords number of W.B.Cs in 1/200 cubic millimetre = 45.

Therefore, the number of W.B.Cs in 1 cubic millimetre = 45 × 200 × ½ = 4500 W.B.Cs.) Normal healthy man has 4000 to 6000 W.B.Cs per cubic millimetre of blood.

4. DIFFERENTIAL COUNT OF WHITE BLOOD CORPUSCLES

EXPERIMENT NO. 4

Object. *To determine the relative proportions (differential count) of white blood corpuscles in the human blood.*

Requirements. Sterilised needle, clean slides; Leishman's stain or Wright's stain and microscope.

Procedure. Prick the finger with a sterilised needle, wipe away the first drop of blood. Take the second drop on the end of a clean slide, place the slide on a smooth surface holding it steady with the left hand. Hold a second slide or drawing slide at 45° just in front of the drop of the blood, draw this slide in such a way that the blood spreads along its edge. Push the second or drawing slide forward 45° on the first slide without exciting any pressure, a film of blood will be made on the first slide. Dry the blood film in air, but do not heat it.

Stain the blood film with Leishman's stain or Wright's stain. Wait and after one minute carefully add distilled water to the stain, the water should be twice the volume of the stain used. Mix the water and stain well by sucking the mixture in and out of a pipette. After a time a greenish metallic scum forms on the surface of the mixture. Allow the diluted stain to act from 7 to 10 minutes. Drain off the stain and wash film for 10 seconds with distilled water. The film should be rose pink in colour, if it is too purple, then wash again with distilled water. Dry the slide and examine it under the microscope for white blood corpuscles.

First examine the stained film under low power of microscope to get an idea of leucocytes. Then examine under high power or oil immersion lens after thinly smearing the stained film with oil. In a longitudinal strip of the film count the various kinds of white blood corpuscles from top to the bottom. Count 100 or 200 white corpuscles noting on a paper each type of white blood corpuscles encountered, and calculate the percentage.

Calculation. The following formula can be adopted to calculate the percentage of different types of white blood corpuscles :

$$\frac{\text{Number of type cells}}{\text{Total number of white blood corpuscles}} \times 100$$

The normal proportion of white blood corpuscles is as follows :

Neutrophils.... 50–70 per cent. or	3000–6000 per cu mm	
Eosinophils 1.4	" " 150–300	" "
Basophils ... 0.5–1	" " 0–100	" "
Lymphocytes... 20–0	" " 1500–2700	" "
Monocytes...2–8	" " 300–600	" "

Precautions. It is absolutely essential for differential count of white blood corpuscles that the film of blood is well spread with no clogging of corpuscles.

5. COAGULATION OF BLOOD

EXPERIMENT NO. 5

Object. *To observe the coagulation of blood.*

Requirements. Test-tube and cotton wool.

Principle and procedure. Bleeding takes place if the skin is cut, but after a time blood coagulates to close the cut, thus, preventing much loss of blood. Take blood of a mammal in a large test-tube and close the tube with cotton wool. It will be seen that blood forms a soft jelly and a faint yellow liquid appears. In a few hours the liquid or **serum** increases in amount and the solid jelly or **clot** becomes smaller. The serum consists of blood plasma without its fibrin and almost no blood corpuscles. The clot is scarlet on the surface and bluish-black within. The clot is made of **fibrin**, an insoluble protein derived from the soluble fibrinogen of the blood, and entangled in the fibrin are almost all the blood corpuscles. The fibrin is a mass of fine interlacing fibres.

The blood contains an inactive enzyme **prothrombin** which is kept in check by **heparin** (antiprothrombin) found in blood. Hence, the blood flows freely in blood vessels and does not coagulate. If an injury occurs, blood begins to escape from it, but the injured tissue cells and blood platelets on contact with air release a substance called **thromboplastin**. Thromboplastin soon neutralises heparin, and it combines with calcium ions in the blood to act on prothrombin changing it into **thrombin** Thrombin is an active enzyme, it acts on the fibrinogen of blood converting it into **fibrin**. The fibrin forms a network of fibres which entangle the corpuscles to form a clot at the site of the injury.

6. BLEEDING TIME OF BLOOD

EXPERIMENT NO. 6

Object. *To determine the bleeding time of blood.*

Requirements. Sterilised needle, stop-watch and blotting or filter paper.

Procedure. Prick the base of the finger nail with a sterilised needle and let a drop of blood appear. Every 30 seconds dab the blood drop with a piece of blotting or filter paper without touching the skin otherwise the blood may be wiped off. The time from the appearance of the drop of blood to the time when bleeding stops and the filter paper is no longer stained is taken as the bleeding time. Normal bleeding time is 2 to 5 minutes.

7. CLOTTING TIME OF BLOOD

EXPERIMENT NO. 7

Object. *To determine the clotting time of blood.*

Requirements. Sterilised needle, small test-tubes, stop-watch, water bath, syringe and glass slides.

Principle and procedure. It is the time taken for blood to clot *in vitro* at a standard temperature, or it is the length of time from the moment blood is collected till the appearance of clotting.

Following three methods are used for counting the clotting time of blood.

A. Note the time, then take 1 ml venous blood in a small test-tube, put the tube in a water bath at 37°C. Every 30 seconds shake the tube and tilt it slowly. Note the time when blood stops flowing in the tube and its surface line no longer changes on tilting. The time taken is the clotting time, it is normally 5 to 10 minutes.

B. Take 5 ml venous blood in a syringe and start a stop watch. Put 1 ml of blood in each of two small equal-sized test-tubes, put the tubes in a water bath at 37°C. After some time tilt the tubes every half minute, when the tube can be tilted through 90° or even inverted without spilling the blood then the blood has coagulated, note the time.

C. Put a drop of blood on a clean slide and note the time. Every half minute pull a clean needle slowly through the drop of blood. When a fine thread of fibrin can be pulled up by the point of the needle, then coagulation has started, note the time.

8. PREPARATION OF HAEMIN CRYSTALS

EXPERIMENT NO. 8

Object. *To prepare haemin crystals.*

Requirements. *Sterilised* needle, glass slide, cover glass, acetic acid (glacial) and spirit lamp.

Principle. Haemoglobin imparts a red colour to blood, it is a conjugated protein made of a non-protein pigment **haematin** or haeme, and a colourless protein known as **globin**. The haematin is made of a pigment called **porphyrin** which is combined with iron. A hydrochloride of haematin forms **haemin.**

Procedure. Take some dried blood or a few drops of fresh blood, add a little glacial acetic acid and heat gently, on cooling dark brown haemin crystals (also known as Teichmann's crystals) are formed. Examine haemin crystals under a microscope, they appears, as shining rhombic plates, prisms and star-shaped clusters. These crystals contain haematin and iron. Haemin crystals test is used in medicolegal work to differentiate fresh or dried blood marks from any other red-colouring material.

Fig. 33.5. Haemin crystals.

Precaution. In the case of old dried blood it is necessary to boil with glacial acetic acid, and at boiling add a crystal of sodium chloride.

9. EFFECT OF OSMOLARITY OF SALT SOLUTIONS AND HAEMOLYTIC AGENTS ON RED BLOOD CORPUSCLES

EXPERIMENT NO. 9

Object. *To study the effect of osmolarity of salt solutions and haemolytic agents on red blood corpuscles.*

Requirements. Test-tubes, sodium chloride solutions of different percentages, ether or benzene or chloroform and potassium hydroxide.

Principle and procedure. Erythrocytes are membranous envelopes filled with haemoglobin. The envelope or plasma membrane is semipermeable, hence, substances in solution can pass in and out of the corpuscles. If normal red blood corpuscles are placed in normal sodium chloride solution which has the same concentration as the contents of the corpuscles, then no exchange takes place because the osmotic pressure of corpuscles and of the saline solution is the same. The salt solution is said to be **isotonic** or of equal concentration with the corpuscles.

If normal red blood corpuscles are put in a dilute solution of sodium chloride, they absorb water from the salt solution, and swell up, become spherical, and finally rupture. The dilute salt solution is **hypotonic** because its osmotic pressure is lower than that of the interior of the corpuscles.

If normal red corpuscles are placed in strong salt solution in which the concentration of sodium chloride is higher than in the red corpuscles, then the corpuscles shrink and become wrinkled. The strong salt solution is **hypertonic** because its osmotic pressure is greater than that of the interior of the corpuscles and the fluid passes from the corpuscles into the salt solution.

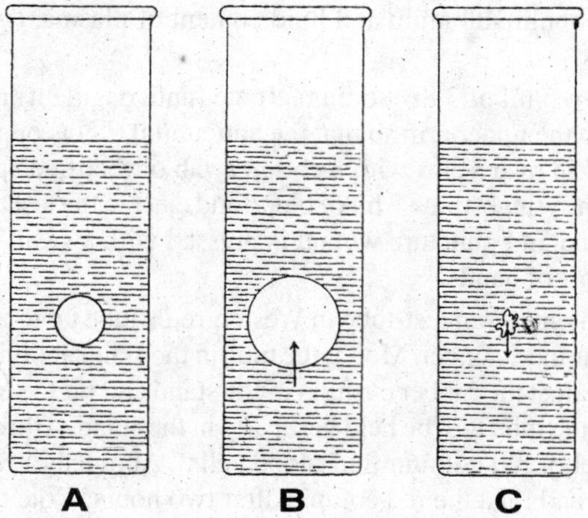

Fig. 33. 6. Osmolarity of red corpuscles in different saline solutions. A-isotonic; B-hypotonic; C- hypertonic.

A. Put 1 ml of venous blood in each of three small test-tubes : 1. To the first tube add 1 ml of 0.85% sodium chloride solution (isotonic). 2. To the second tube add 1 ml of 0.5% sodium chloride solution (hypotonic). 3. To the third tube add 1 ml of 1.5% sodium chloride solution (hypertonic). Shake the tubes and note the results.

In a **hypotonic solution** the red corpuscles rupture, in rupturing the haemoglobin passes from the red corpuscles into the plasma, the plasma becomes red in colour. This is called **haemolysis** or **laking of blood.** The substances inducing haemolysis are called **haemolytic agents**. Some common haemolytic agents are hypotonic solutions, fat solvents, snake venoms, and some toxins.

B. Take a drop of blood in a test-tube, add a fat solvent (such as ether, benzene, chloroform or bile salts), shake the tube, the blood will be haemolysed because the fat component of the plasma membrane of red corpuscles is dissolved and haemoglobin is set free.

C. Take a drop of blood in a test-tube, add a few drops of 0.2% potassium hydroxide, the red corpuscles are dissolved.

10. ERYTHROCYTE SEDIMENTATION RATE OF BLOOD (E.S.R.)

EXPERIMENT NO. 10.

Object. *To determine erythrocyte sedimentation rate (E.S.R.) of blood.*

Requirements. Westergren's apparatus with stand, a test-tube containing 3.8% sodium citrate solution as anticoagulant, a dry test-tube, rectified spirit, sterilised 2 ml syringe with needle, cotton and torniquet.

Description of apparatus. Westergren's apparatus consists of Westergren's tube which is graduated and open at both the ends and Westergren's stand. The stand can accomodate six Westergren's tubes and it has a rubber cushion at the base on which the lower ends of Westergren's tubes rest. There is a screw cap or a screw clip at the upper end of stand which fits upon the tubes (Fig. 33.7).

Principle. "Erythrocyte Sedimentation Rate (E.S.R.) is the rate at which the red cells settle". It is, therefore, governed by the factors which control rouleaux formation. When blood is mixed with a suitable anticoagulant and allowed to stand in a narrow vertical tube, the red cells settle down as a result of rouleaux formation. The red cells are negatively charged and move randomly in plasma. When they strike each other, they tend to stick to each other to form **rouleaux**, *i.e.*, they cling together and resemble a pile of coins. When a rouleaux is formed, the density of its mass increases and it sinks down quickly. In fact, rouleaux formation is affected by the following factors :

(*i*) Fibrinogen, globulin, albumin and lipid content of plasma, (*ii*) Shape and size of red cells and number of red cells.

Procedure. Take 0.5 ml of 3.8% sodium citrate (anticoagulant) in a syringe and eject in a test tube. Apply torniquet to the upper arm so that the anticubital (veins opposite to elbow joint) become prominent. Clean the skin over veins with a cotton swab of rectified spirit. Now select a prominent vein and puncture it with syringe, draw 2 ml of blood and eject in the test tube containing anticoagulant. The torniquet is removed and puncture wound is pressed with a swab of spirit. Rotate the test-tube gently between the palms of hands.

Now suck the contents from test-tube in Westergren's tube up to zero mark and close the upper end of the tube with your index finger. Mount the tube in the stand applying pressure by means of your hand to its lower end against the rubber cushion of the stand and fix the upper end of the tube by screw cap or screw clip. Now the tube will be held vertically in the stand. Allow the tube to stand for an hour and read the upper level of red column (red blood cells). Above the red column, there will be clear plasma. Again observe and read the red column after two hours. Note the readings for calculation.

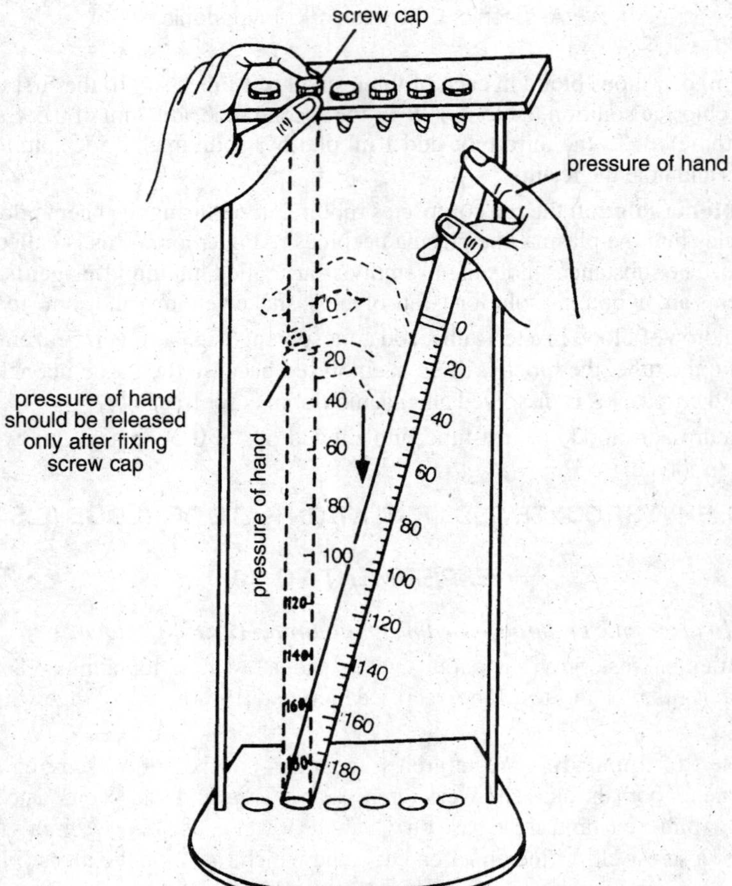

Fig. 33.7. Westergren's apparatus for the determination of E.S.R.

Result :

Suppose the reading at the end of 1st hour be X mm

Suppose the reading at the end of 2nd hour be Y mm.

Then, the mean average of E.S.R. will be $\frac{X+Y}{2}$ = Z mm per hour

Inference. The E.S.R.* is either normal, above normal or below normal.

Discussion. The normal range of E.S.R. determined by this method is as under :

At the end of 1st hour 3 to 5 mm in Males; 4 to 7mm in Females.

At the end of 2nd hour 7 to 15 mm in Males; 12 to 17 mm in Females.

If in a person, the E.S.R. value changes in two consecutive tests carried out at definite intervals, it suggests the increased activity of the diseased process. Thus, indication of change in E.S.R. is more significant than the actual rate observed for a person as this change suggests the presence of a disease, though it does not suggests the nature of disease.

However, an increase in E.S.R indicates quick rouleaux formation which is caused by a peculiar pattern of plasma molecules. The following physiological and pathological conditions cause a variation from normal E.S.R.

A decreased E.S.R. is very uncommon but increased E.S.R. suggests :

(*i*) **Pathological conditions** like rheumatic fever, tuberculosis, malignancy and similar other chronic conditions which are associated with tissue break down, and anaemia.

(*ii*) **Physiological conditions** like pregnancy, menstruration and parturition.

Precautions. While observing E.S.R. by the method mentioned above, following precautions should be taken :

(*i*) The tube should be kept perfectly vertical otherwise acceleration in E.S.R. will take place.

(*ii*) The blood sample taken should be used within two hours.

(*iii*) Air bubbles in the blood column of Westergren's tube should be avoided.

(*iv*) Blood showing any sign of coagulation should not be used.

(*v*) The test should be performed at a room temperature of 22°C to 27°C as temperature effects E.S.R.

(*vi*) The exposure of experiment to direct sunlight should be avoided.

Note: E.S.R. can also be determined by **Wintrob's method** which is based on the same principle as described above.

11. BLOOD PRESSURE AND ITS MEASUREMENT

Blood pressure. Arterial blood pressure is the force of pressure which the blood is exerting against the walls of the blood vessels in which it is contained. The blood pressure in the arteries varies during the **cardiac cycle**. During ventricular systole, when the left ventricle is forcing blood into the aorta the pressure rises to a peak, referred to as **systolic pressure**. During diastole the pressure falls, the lowest value it reaches is referred to as **diastolic pressure**.

The systolic blood pressure is produced by the heart muscles which drive the contents of the ventricle into the already stretched arteries. During diastole the arteries are kept partly distended because the peripheral resistance of the arterioles prevents all the blood running off into the tissues. Thus, the blood pressure depends partly on the force and volume of blood pumped by the heart and partly on the contraction of the muscles in the walls of the arterioles. This contraction is maintained

* Students while working in the class may note E.S.R. after every 15 minutes and plot a graph against time. This will help them in reporting E.S.R. at the end of an hour or so.

by vasoconstrictor nerves which are controlled by the vasomotor centre in the medulla oblongata of the brain.

The vasomotor centre adjusts the peripheral resistance to maintain the blood pressure relatively constant. It, however, changes slightly in physiological variations of exertion as during exercise, with mental changes of anxiety and emotion, in sleep and when eating. For this reason, it is always suitable to measure blood pressure when a person is relaxed and resting. The measurement of blood pressure is very significant in correctly estimating the state of blood circulation and, thus, the working of heart.

Measurement of blood pressure. Blood pressure is measured by a blood pressure instrument called **sphygmomanometer** and a stethescope is also needed.

Description of sphygmomanometer. Sphygmomanometer (Fig. 33.8) consists of a long glass **manometer** capillary tube. The upper end of this tube is open, while its lower end is slightly bend. The manometer is mounted in the middle of a **dark scale** having markings on its both sides. The lower bend end of manometer is connected to the ventral side of a large **bulb** filled with mercury. The upper open end of the manometer is exposed to atmospheric pressure but is guarded by a valve to check the spilling of mercury from the open end. The upper end of the bulb is connected with a **nozel** and the nozel is connected with a rubber tube having narrow lumen. The free end of rubber tube is connected to a flexible rubber bag enclosed in a **cuff**. The cuff is further connected to a **rubber pumping bulb** with a rubber tube. The neck of the bulb is provided with an **adjusting screw**; the tightening or untightening of this screw ensures the flow of air into the cuff (inflation) or out of the cuff (deflation).

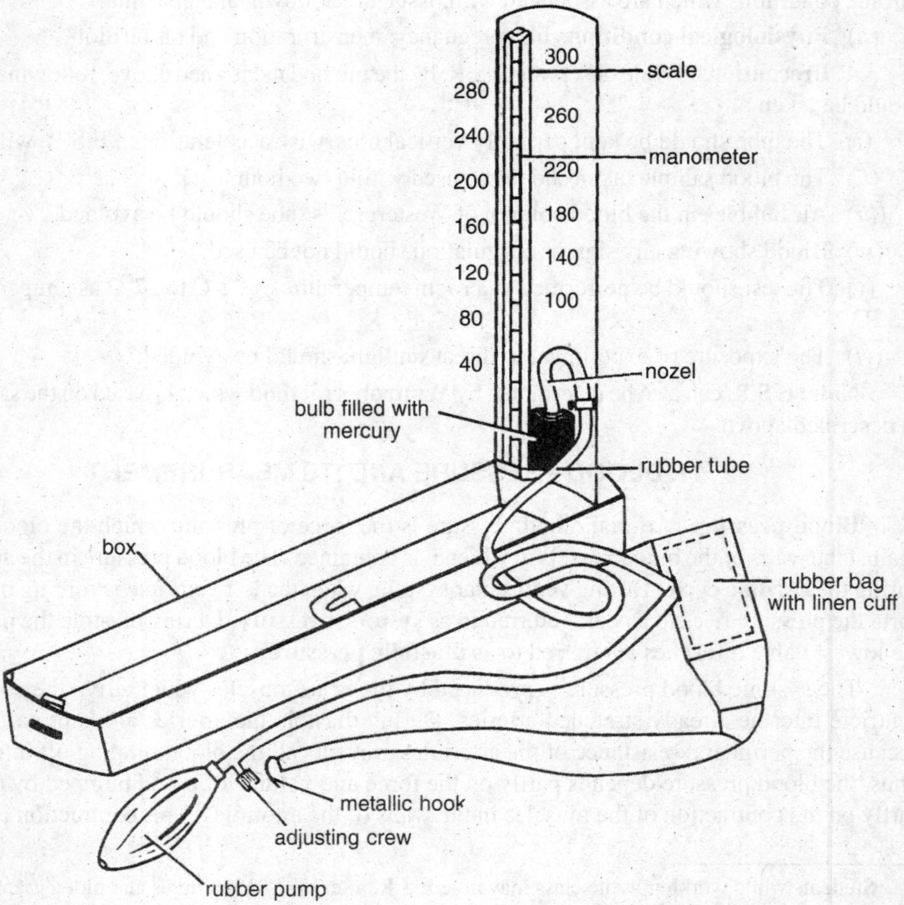

Fig. 33.8. Sphygmomanometer.

Method of using sphygmomanometer. Open the adjusting screw at the proximal end of the pumping bulb so that air is removed from the cuff and tubes. Then wrap the cuff around the upper arm just above the elbow joint and fix it with the help of attached hook in a manner that neither it is too tight nor too loose. While wrapping the cuff care should be taken that the tubes attached to it must face toward the lower end. Then tighten the adjusting screw and pump the air by pumping rubber bulb rapidly so that mercury is raised up to 110–120 mm Hg mark in the manometer. Observe the sound of arterial blood in brachial artery with the help of a stethescope. Then again pump the air slowly and keep on observing the sound of arterial blood in the brachial artery with stethescope to disappear. As and when sound is not felt, note the mercury level in the manometer and mark the reading. This is the **diastolic pressure**. Then start loosing the adjusting screw slowly and keep on observing carefully the return of normal arterial sounds through the stethescope. As and when it is felt, close the adjusting screw and mark the reading of mercury in manometer. Pump a little more air and again note the proper arterial sound to reappear. This would be the **systolic pressure**. Now, loose the screw, remove all air and open the cuff and keep the apparatus in a safe place.

The difference in pressure between systole and diastole is called the **pulse pressure** which is normally 30–50 mm Hg. The lower limit of systolic pressure in a normal adult is estimated about 105 mm Hg and the upper limit about 150 mm Hg. In women the blood pressure is from 5 to 10 mm Hg less than in men.

Normal Blood Pressure Range (In mm Hg)

		Diastolic	**Systolic**
In infants	...	50	70–90
In childhood	....	60	80–100
During adolescent stage	...	60	90–110
In young adult	...	60–70	110–125
As age advances it is increased	...	80–90	130–150

12. EXAMINATION OF HUMAN BLOOD GROUP

EXPERIMENT NO.11

Object. *To examine the blood group of human beings.*

Principle. Blood groups of human beings was discovered by **Karl Landsteiner**. Blood groups are classified on the basis of presence or absence of A or B agglutinogens or A and B antigens in red blood cells. These are of 4 types :

1. Blood group A. Persons of A blood group have agglutinogens A or antigen A in R.B.Cs and antibodies or agglutinin B in plasma.

2. Blood group B. Persons of B blood group have antigen B in R.B.Cs. and antibodies A in plasma.

3. Blood group AB. Persons with AB blood group have antigen A and B in R.B.Cs. and have no antibodies in plasma.

4. Blood group O. Persons with O blood group do not have any antigen in R.B.Cs, but contain antibodies A and B in plasma.

Landsteiner and **Jansky** have also demonstrated that serum antibodies of the same person do not agglutinate the antigens present in RBCs. This is called **Jansky-Landsteiner law.**

Requirements. Cavity slides, blood sample vials, antiserum A and antiserum B and pathological binocular microscope.

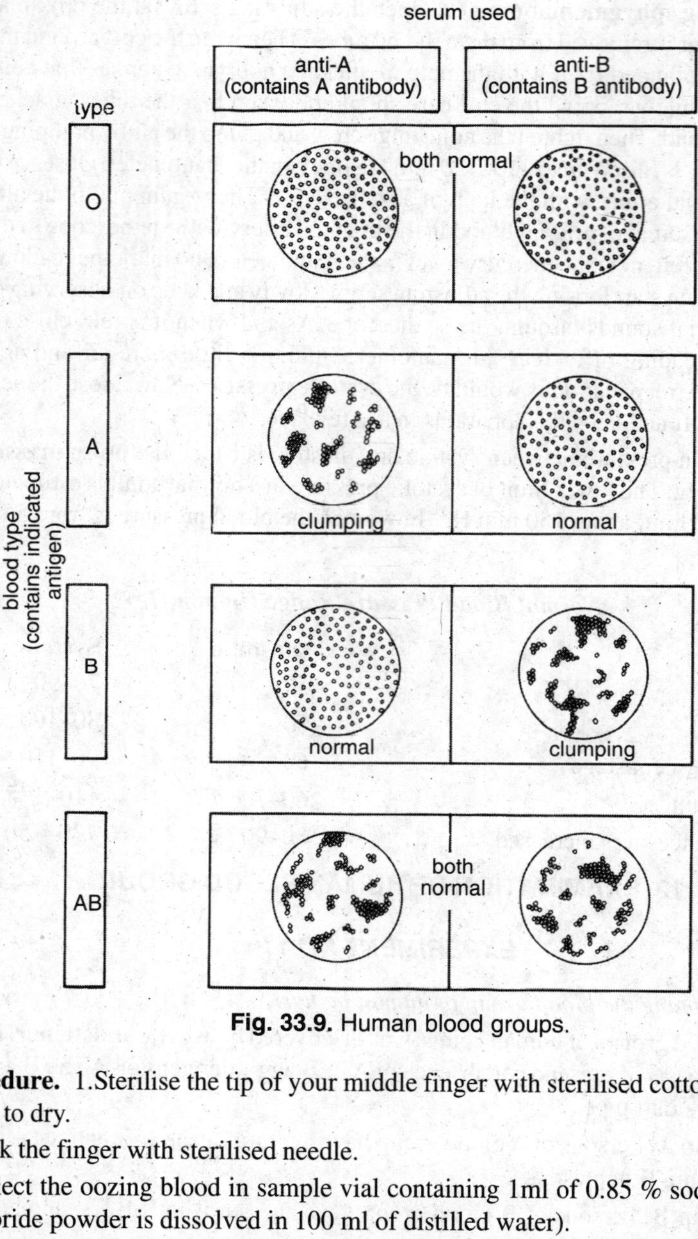

Fig. 33.9. Human blood groups.

Procedure. 1.Sterilise the tip of your middle finger with sterilised cotton wet in 90% alcohol and allow it to dry.

2.Prick the finger with sterilised needle.

3.Collect the oozing blood in sample vial containing 1ml of 0.85 % sodium chloride (0.85 g sodium chloride powder is dissolved in 100 ml of distilled water).

4.Take two cavity slides having 4 cavities in each. Wash and clean them and allow them to dry.

5. Label the cavities of the slides 1 and 2 as I, II, III and IV in slide I and V, VI, VII and VIII in slide 2.

6.Add a drop of blood of different samples of blood in each cavities as given below :

Cavities I – V-	First	sample	of	blood
" II–VI –	Second	"	"	"
" III–VII –	Third	"	"	"
" IV–VIII –	Fourth	"	"	"

7. In slide No. 1 add a drop of antiserum-A in cavities 1 – IV. In slide No.2 add a drop of antiserum-B in cavities V – VIII and then mix the blood sample and antisera with a sterilised needle gradually in each cavity slides for ten minutes.

Observations. 1.I – V cavities – No clumping, hence, blood group is O. First sample of blood.

2. II-VI cavities – Clumping in II, hence, blood group A. Second sample of blood. No clumping in VI.

3.III-VII cavities – Clumping in III and VII. Blood group B. Third sample.

4.IV-VIII cavities – Clumping in IV and VIII. Blood group AB. Fourth sample.

EXPERIMENT NO.12

Object. *Determination of Rh (Rhesus) factor in human blood.*

Principle. Rh^+ and Rh^- agglutinogen in RBCs of man are called *Rhesus* factor derived the name from *Rhesus macaque* monkeys. If red cells of Rhesus monkey are injected into the blood of rabbit, it agglutinise 36% of blood of rabbit and 86% blood of man. Thus, blood of rabbit and man are Rh^+. 14% RBCs of man remain unaffected. Blood that do not clamp are called Rh^-. When a person of Rh^- blood group is frequently given transfusion of Rh^+ blood, the former develops Rh^- antibodies. Because Rh^- antibodies of blood group become compatible to antibodies of blood group A and B and they cause haemolysis of R.B.Cs. Above blood groups are hereditary.

Rh factor has great medical significance during pregnancy. With Rh^+ father and Rh^- mother foetus may be Rh^+. Some R.B.Cs. (with Rh^+) may pass into the blood of mother due to which mother develops anti Rh^- antibodies. After the birth of first child mother becomes sensitised. In second pregnancy of sensitised mother some of Rh^- red blood cells would reach through placenta into the foetus and haemolyse the RBCs of foetus. In this way RBCs of foetus are destroyed resulting into anaemia in foetus causing its death. If any Rh^- woman is given transfusion of Rh^+ blood, then woman die. The above disease is called **erythroblast foetalis**.

Determination of Rh^+ and Rh^- Blood Groups

Presence of D-antigen on RBCs of man, blood group may be Rh^+ and absence of D-antigen on RBCs blood group is Rh^-. Presence of D-antigen on R.B.Cs. is agglutinated by antibodies. Rh^+ and Rh^- blood groups could be determined by following methods :

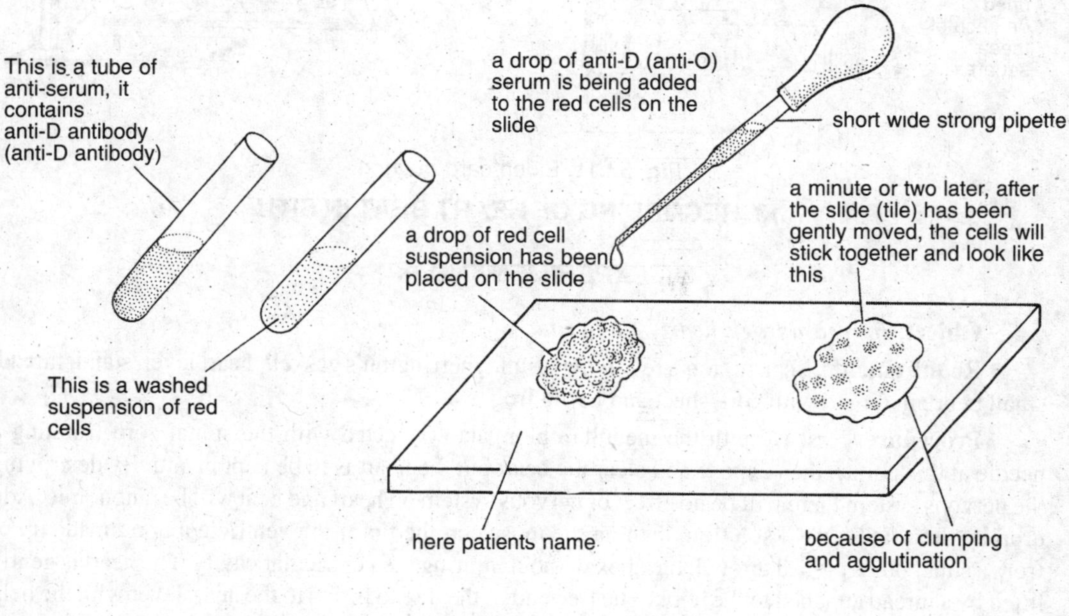

Fig. 33.10. Test tube method.

1. Slide Examination Method

Procedure. Take a clean glass cavity slide. Add a drop of anti-D serum in cavity and then a drop of blood. Leave it for 2 minutes and then observe.

Observation. If agglutination occurs it shows that D-antigen in serum has resulted clumping and thus, blood is **Rh^+.** If no clumping occurs blood is **Rh^-**.

2. Test Tube Method

Procedure. Take a clean test tube and pour in it 20 ml 0.9% saline. Add erythrocytes and make a suspension.

For **erythrocytes** take blood and keep it in centrifugation tubes and centrifuge at 3000 rpm. Erythrocytes settle at bottom and serum is separated. Pour serum in another tube and take erythrocytes. Again centrifuge the mixture of suspension having RBCs at 1000 rpm.

Observation. If R.B.Cs shows clumping then blood is **Rh^+** otherwise **Rh^-**.

3. Eldon Card Method

These are small whitish cards marked with a square area. Take 3 cards.

Prodecure. On one card smear with **anti-A serum**, on another card smear with **anti-B serum** and on third card smear with **anti-D serum**. Allow the smear to dry. Before testing the blood, wet the cards with distilled water and then add drop of blood in all the three square areas.

Observation.

Card one – If clumping occurs in anti – A serum, then blood group is **A.**

Card two – If clumping occurs in anti-B serum, then blood group is **B**.

Card three – If clumping occurs in anti-D serum then blood group is **Rh^+**.

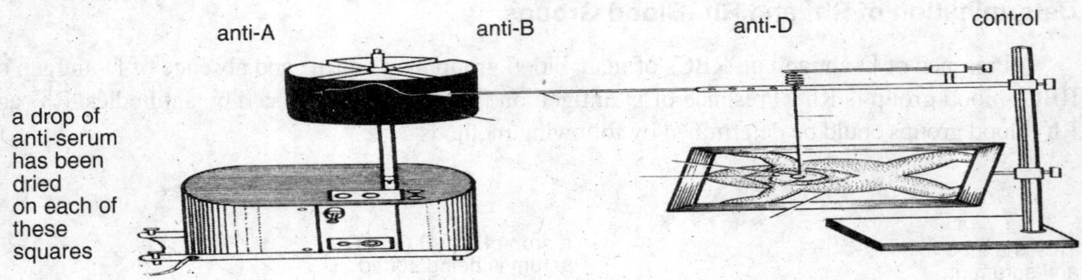

Fig. 33.11. Eldon card method

13. RECORDING OF HEART BEAT IN SITU

EXPERIMENT NO. 13

Object. *To record frog's heart beat in situ.*

Requirements. Sherrington's revolving drum, Sherrington's gaskell, heart lever, stand, thread, small bent pin, board with cork sheet and living frog.

Procedure. First we pith the medulla oblongata connected with the spinal cord (rotating a needle at this point), then expose and clear the heart but the heart is to be kept in situ. By destroying the nervous system there will be no effect of nervous system on heart and beat will be autonomic. Add drop by drop 0.6% NaCl solution, then on the muscle at the tip of the ventricle a hook made up of from a small pin is passed through the apex without injuring the ventricular cavity. The heart is gently lifted by a thread attached to the hook. The free end of the thread is tied to the heart lever with the help of the screw. The pointer of the lever is adjusted in a horizontal manner and pointer on the long arm of the lever is adjusted just to touch the smooth surface of the drum in order to reduce the friction of different speeds of the drum. The normal cardiographic records are registered on the smooth surface of drum.

Observations. The graphical records will show the following contractions :

(*a*) Auricular contraction. Denoted by downward facing on the smooth drum.

(*b*) Auriculo-ventricular pause. In the form of a small straight or slightly inclined line.

(*c*) Ventricular contraction. Just like an auricular contraction but the amplitude and height of contraction downward record is much longer than that of the auricular contraction.

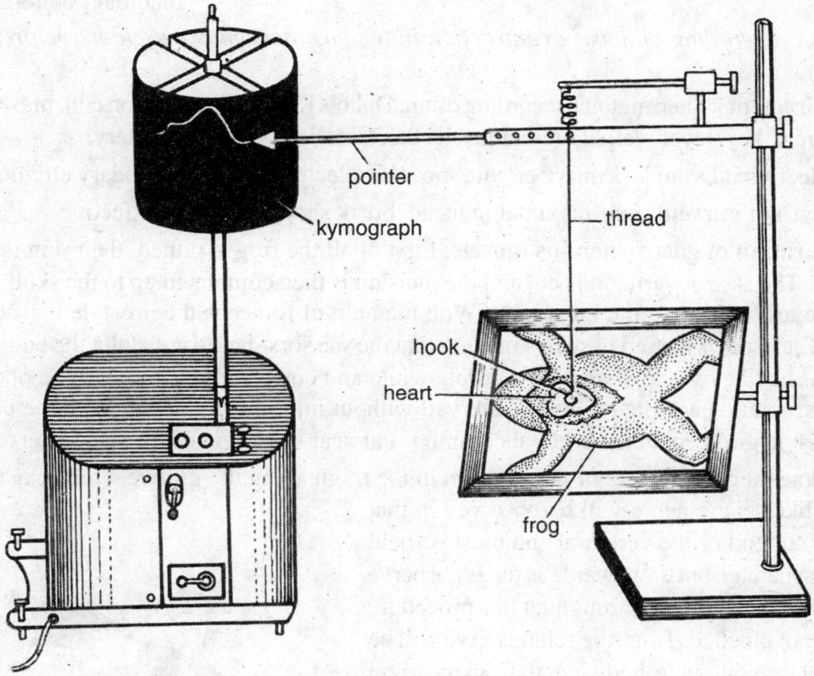

Fig. 33.12. Diagram showing the recording of heart beat of frog.

Interpretation. The contraction due to sinus venosus is not at all observed only because firstly sinus contraction is very weak and secondly auricular contraction is much more stronger as compared to sinus contraction.

Inference. Under normal conditions the contraction of sinus venosus is generally followed by auricular contraction and ventricular contraction. The downward tracing is the contraction and upward tracing is the relaxation of the heart, while small short tracings are the pause taken up by heart during series of each changes in each cardiac cycle.

Precautions. The following precautions are to be taken :

1. Rotations of kymograph should not be fast otherwise contractions of the auricles will not be recorded.

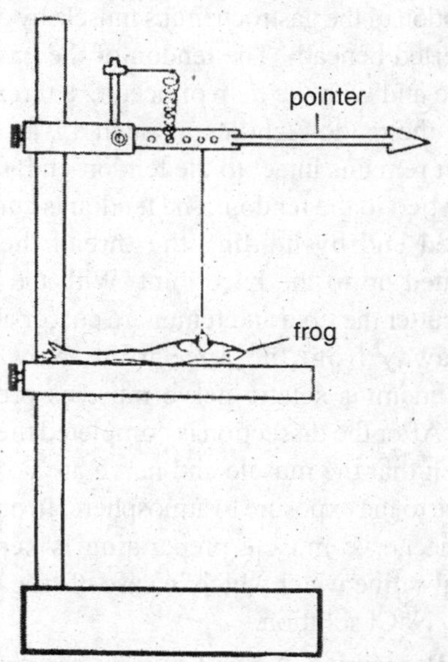

Fig. 33.13. Heart beat record of frog.

2. The paper should be blackened well.
3. Edge of the pointer should be accurate.
4. Current flow should be uniform.
5. Heart should be kept wet in 0.6% saline solution.

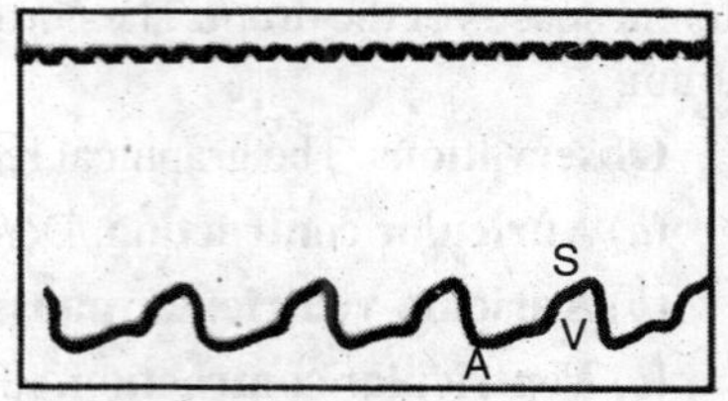

Fig. 33.14. Normal heart beat recording of frog.

14. RECORDING OF MUSCULAR CONTRACTION

EXPERIMENT NO. 14

Object. *Recording of muscle contraction in the gastrocnemius muscle-nerve preparation of frog.*

Requirements. Sherrington's recording drum, Dubois Rejmonds induction coil, muscle chamber, make and break key, lever weight, wire, muscle, 0.65% saline water and battery.

Connections. Drum is primary circuit, movable electrodes form secondary circuit.

Strength of current. Sub-maximal induced break shock should be effective.

Preparation of gastrocnemius muscle. First of all the frog is pithed, then skin is lifted from the urostyle. The skin covering it is cut and the incision is then completed up to the skull. The skin is cut from the urostyle up to the knee-joint. With the help of foreceps the urostyle is lifted up and a blunt pair of scissors is passed in such a manner that the scissors should cut along the body surface of the urostyle. Just below the dissection a purely white and cord-like structure will be observed. The course of the sciatic nerve can now be observed without injuring any nerve. With the help of bone cutter the pelvic girdle and the vertebral column is cut near the origin of the sciatic nerve.

On removal of the layers of the muscles in the thigh along the groove which exists there, the white cord-like sciatic nerve will be observed in that region. The cut end of the vertebral end must be held in hand and separated out downwards as far as the nerve gets bifurcated in the thigh. Throughout this procedure the nerve is not pinched. Thus, the sciatic nerve will be dissected out throughout its body length from its origin up to the knee. From the knee joint the skin is cut up to the tendon of the gastrocnemius muscle by a knife blade is inserted beneath. The tendon of the gastrocnemius muscle and with the help of needle, a thread is passed below the tendo-Achilles, in such a way that it does not but remains intact to the tendon-end, the thread is firmly tied to the tendon. The tendon is cut away from the tied end by holding the thread the muscle is separated up to the knee joint. With the help of the bone cutter the tibia and femur are cut for about half an inch away from the knee joint. Now the frog's gastrocnemius sciatic nerve muscles preparation is ready. After the dissection is completed the precaution is taken that the muscle and nerve are not allowed to dry due to the exposure to atmosphere. In order to avoid this the nerve muscle preparation is kept moist by normal saline water which in case of frog is generally 0.65% NaCl solution.

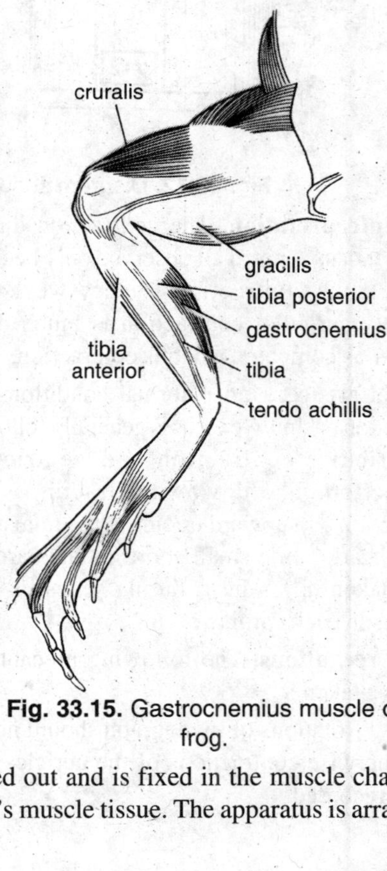

Fig. 33.15. Gastrocnemius muscle of frog.

Procedure. A nerve-muscle preparation is dissected out and is fixed in the muscle chamber containing 0.65% NaCl solution as it is isotonic with frog's muscle tissue. The apparatus is arranged for a single induction shock and the secondary induction shock will be effective. The drum is included in the primary circuit. A small weight say about 1 gm is attached to the pointer arm which keeps the pointer in a horizontal position. The strength and intensity of the current is adjusted in such a way that the muscle will be in a position to produce a **simple twich** about $1^1/_3$ inch amplitude. The primary as well as secondary make and break key is closed and the lever point is brought just to touch the writing surface in order to have the optimum pressure on the smoked drum. When the contact in the primary circuit is just broken due to revolving base electrodes on the Sherrington's drum the muscle will be stimulated via the sciatic nerve and it will record a **single twich.** As soon as the writing point has returned to the base line, the short circulating key and secondary make and break key is closed immediately. The drum is dropped with the help of stop key and the writing point is withdrawn immediately from the drum so that the tracing are not in a position to overlap one another. It will be observed that the twitching is in the form of a curve only because the drum is rotating with certain speed.

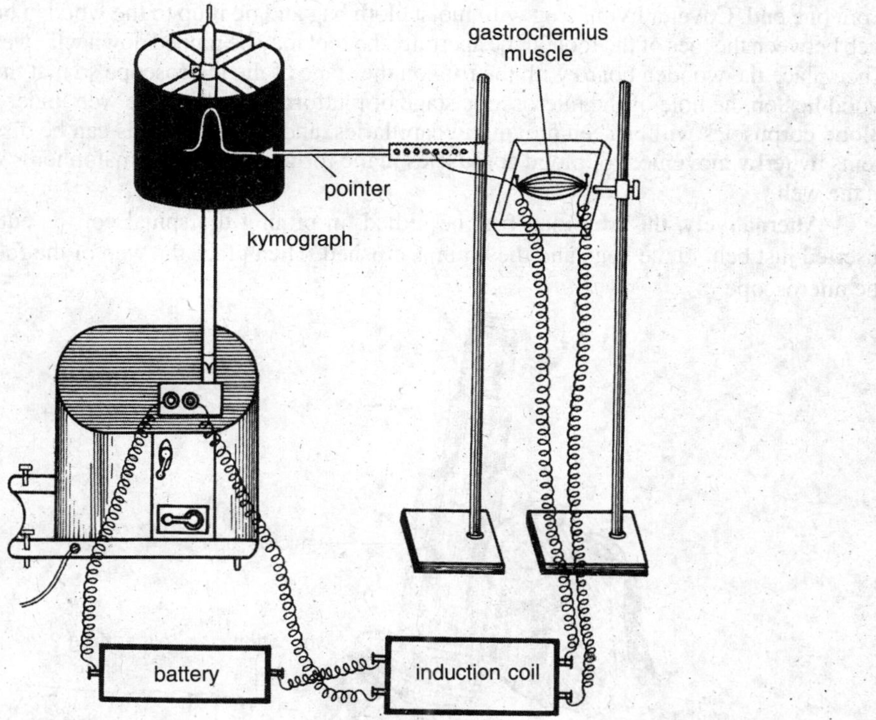

Fig. 33.16. Diagram showing the contraction of gastrocnemius muscle

In this experiment the tension due to weight on the muscle is merely constant but the muscle fibres are free to shorten in length. Hence, the contraction is said to be **isotonic** (change in length of muscle fibres not in tension).

The **latent period** in the phase during which no apparent changes takes place in the muscle. The latent period extends from the point at which the contracting muscle begins to raise the lever.

The contraction period extends from the point where the writing arm of the '*l*' shaped lever begins to rise to the point highest above the base line. The contraction period starts at the end of the latent period and is, thus, in active phase in the contraction.

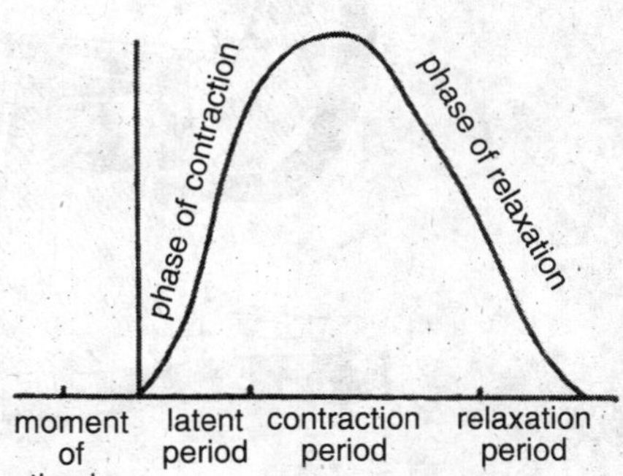

Fig. 33.17. Simple muscle twitch curve.

The **relaxation phase** or period is a passive phase in the muscular contraction and extends from the highest point of the simple muscle curve up to the point at which the simple muscle curve twitch retouches the base line.

Inference. An isolated twitch consists of three phases, *viz.,* **latent phase**, **contraction phase** and **relaxation phase**. Secondarily, a simple muscle twitch represents the response to a single stimulus and a stimulus causes a sudden change in the external as well as in the internal development.

15. CIRCULATION OF BLOOD

EXPERIMENT NO. 15

Object. *To study the circulation of blood in the capillaries of the web of foot of frog.*

Requirements. Soft wooden board, moist cloth, microscope and living frog.

Procedure. Take a soft wooden board and make an aperture of about 2 cm, about 4 cm away from one end. Cover a living frog with moist cloth bag and tie it up to the wooden board placing the web between the toes of the foot on the aperture, the foot may be pinned down with web well stretched. Then place the wooden board with the frog on the stage of the microscope so that the aperture in the wood lies on the hole of the microscope stage or platform. Examine the web under the microscope. Blood corpuscles will be seen moving in capillaries, and arterial vessels can be distinguished from veins by jerky movement of blood corpuscles in the former. Large chromatophores will also be seen in the web.

Alternatively, the frog may first be pithed, in pithing the spinal cord is cut by a sharp pin inserted just behind the head and the brain is crushed. Then place the web of the foot in position on the microscope.

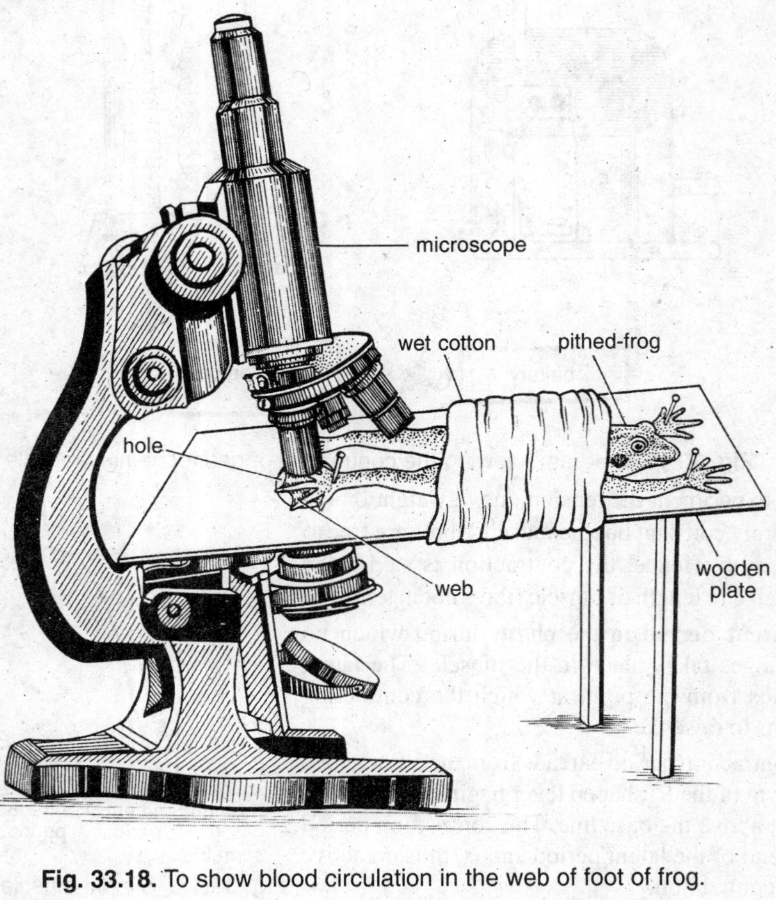

Fig. 33.18. To show blood circulation in the web of foot of frog.

Observations. Blood corpuscles are seen moving in capillaries. The arterial capillaries can be distinguished from veins by the jerky movement of blood corpuscles in the former.

16. RESPIRATION

EXPERIMENT NO. 16

Object. *To determine the oxygen consumption of a rat.*

Requrements. Haldane's respiration apparatus, moist soda lime, pumice stone socked in sulphuric acid, and a rat.

Procedure. Set up an apparatus as shown in the figure. This apparatus is known as **Haldane's respiration apparatus** and it is used to determine the rate of oxygen consumption in a small animal. Jars 1 and 4 contain moist soda lime for removing carbon dioxide, jars 2, 3 and 5 have pumice stone soaked in sulphuric acid for removing moisture. In the chamber rat is placed. M is a gas meter. J is a manometer having an inverted bell jar standing in a water trough,

It prevents an excess of negative pressure and indicates the pressure actually employed. P is an aspirator for drawing air through the entire apparatus.

Draw a current of air through the apparatus by means of the aspirator for a specified period of time. The air entering the rat chamber has been freed of carbon dioxide and moisture by jars 1 and 2 respectively, thus, the chamber receives only oxygen. This oxygen is used for respiration needs of the rat. The rat gives out carbon dioxide and moisture into the outgoing air. Jar 3 removes the moisture, jar 4 absorbs the carbon dioxide, jar 5 is to remove any moisture gained from soda lime of jar 4. The quantity of oxygen consumed by the rat is determined indirectly. The oxygen consumed is equal to the carbon dioxide and moisture given out. Thus, weigh jars 3, 4 and 5, and the chamber including the rat both before and after the experiment. The increase in weight during a specified period of time gives the quantity of oxygen consumed by the rat during that period. To find the quantity of carbon dioxide given out by the rat weigh jar 4 before and after the experiment.

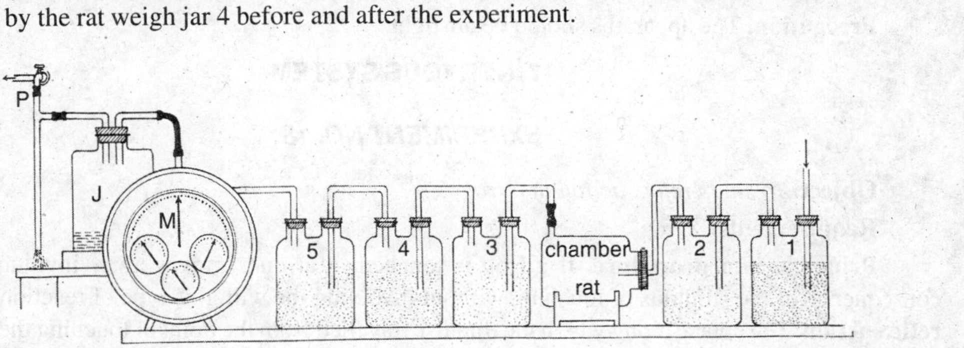

Fig. 33.19. Haldane's apparatus to determine oxygen consumption of a rat. 1 and 4 – soda lime, 2 , 3 and 5 – pumice stone and sulphuric acid; rat chamber; J-manometer; M-meter; P-aspirator.

EXPERIMENT NO. 17

Object. *To determine the rate of oxygen consumption of a rat.*

Requirements. One wide-mouth bottle, measured U-tube, cork, a small beaker, potassium hydroxide solution and a living rat.

Procedure. Take a volume of potassium hydroxide solution in a small beaker and keep it at the base of the wide-mouth bottle. Take a U-tube, fix it to the bottle and fill it with a known volume of water. Note the volume of water in the U-tube. Now take a living rat and put it in the bottle. Close the bottle with the cork and seal the apparatus with vaseline or grease to make it airtight.

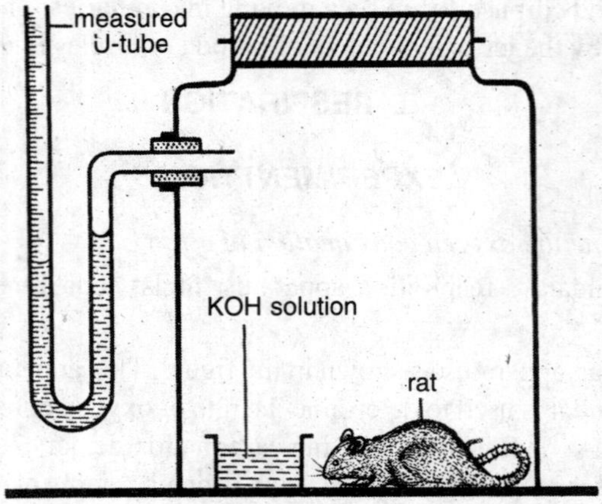

Fig. 33.20. Apparatus for the demonstration of rate of respiration.

Observations. Note the level of water in the U-tube after an hour. It is found to be lowered. This is due to the fact that the rat respires and consumes the oxygen from the air of the bottle. The carbon dioxide released is absorbed by the potassium hydroxide solution. This reduces the pressure in the bottle and, hence, the level of water in the U-tube falls.

Calculations. The fall in the level of water is measured in a definite period of time that is the rate of oxygen consumption per unit time suppose :

(*i*) Level of water in the beginning = 25 ml

(*ii*) Level of water after one hour = 20 ml

(*iii*) Fall of water level in 1 hour = 5 ml

(*iv*) Hence, the oxygen consumption by rat in 1 hour = 5 ml

Precaution. The apparatus should be air-tight.

17. NERVOUS SYSTEM

EXPERIMENT NO. 18

Object. *To study reflex action in a frog.*

Requirements. A frog.

Principle and procedure. If a frog is touched, it at once moves away involuntarily as a consequence of the stimulus of touching without involving the will, this type of reaction is called a **reflex action**. The course along which the impulse travelled from the point of touching the skin to the muscles which caused the animal to move away is known as **reflex arc**. The reflex arc is the structural basis of the reflex action. The impulse of touching the skin goes to the dendrites of a sensory neuron (afferent) and then into its cell body situated in the dorsal root ganglion of the nerve concerned, and then along its axon into the gray matter of the spinal cord where its terminal axon endings form a synapsis with the dendrites of an association or internuncial neuron which lies within the gray matter of the nerve cord. The internuncial neuron forms a synapsis with the dendrites of a motor neuron (efferent) whose cell body lies in the ventral part of the gray matter of the nerve cord, then from the cell body the axon of the motor neuron conveys the impulse through the ventral root of the spinal nerve concerned to the muscles which contract and the frog moves away. This is a very **simple reflex arc**, it involves a sensory neuron, an association neuron, and a motor neuron; thus, in a simple reflex arc only three neurons are involved and their synapses are in the nerve cord. But in reflex arcs of mammals usually there is a chain of several association neurons lying in the central nervous systems between a sensory and a motor neuron.

EXPERIMENT NO. 19

Object. *To study the wiping reflex in a frog.*

Requirements. A frog, stout pins, wire, hydrochloric acid and filter paper.

Procedure. First pith a frog by destroying the brain or by passing a stout pin just behind the head cutting between the brain and spinal cord. In pithing the brain has been severed and destroyed but the spinal cord is intact. Hang the frog with a wire from a stand and fix the other end of the wire to the lower jaw. Prick a pin in one toe, the frog will withdraw its foot. A more violent stimulation of one foot will make both legs move as in jumping. Place a filter paper soaked in hydrochloric acid on the thigh of the frog, the leg will jerk to move the acidified paper, if it fails then the other leg will wipe the paper away by reflex action.

EXPERIMENT NO 20

Object. *To study the knee-jerk reflex in man.*

Requirements. Man and a small hammer or book.

Procedure. Make a man sit on the edge of a chair or table with one knee crossed over the other, and with the legs suspended and relaxed. Tap firmly on the ligment just below the knee cap of the upper leg with the side of the hand or with the edge of a book or with the small hammer. The leg will straighten with a jerk. This is due to sudden contraction of the **quadriceps femoris muscle** by reflex action whose reflex arc involves the spinal cord. The knee-jerk response will be much more pronounced if this experiment is tried with the fists of the man clenched tightly, because the degree of response depends on the condition of the body.

Explanation. The knee-jerk is an automatic, unlearned and involuntary response to the stimulus which involves contraction of quadriceps femoris muscle stretching the leg forward.

18. OSMOSIS

EXPERIMENT NO. 21

Object. *To demonstrate the process of osmosis.*

Requirements. One big beaker or flask, one thistle funnel, strong cane sugar solution, pure water, semipermeable membrane (cellophane), one piece of gummed paper, wax, etc.

Principle. "Osmosis is a process in which water or any other solvent flows through a semipermeable membrane from a lower concentrated solute solution to one of higher concentrated solute solution." It is simply a case of diffusion, but the movement of solvent usually causes an obvious volume change, whereas the movement of solute has a negligible effect.

Procedure. Take a thistle funnel and the piece of a semipermeable membrane is tied tightly across the bottom of the thistle funnel and the joint is waxed with melted wax to make it water-tight. Now fill up the bowl of thistle funnel with strong solution of sugar and marked the upper level of the sugar solution on the outside of the thistle funnel by a piece of gummed paper stuck at that level. Now this sugar solution contained in the thistle funnel is stood in a beaker containing pure water with the help of some stand. This entire apparatus is put aside for some hours.

Observations. After some hours it is seen that water from the beaker has entered the thistle funnel with the result the solution in the thistle funnel has increased in its volume and rises in the thistle funnel, well beyond the original mark. This level of sugar solution in thistle funnel keeps rising till the osmotic equilibrium is attained. On attaining the osmotic equilibrium the same number of water molecules pass across the membrane in both directions.

Inference. The rise in volume of sugar solution in the thistle funnel is due to passing of water molecules from the beaker containing pure water of low concentration. The sugar molecules do not pass into the beaker because of semipermeable membrane. Thus, this experiment demonstrates the process of osmosis.

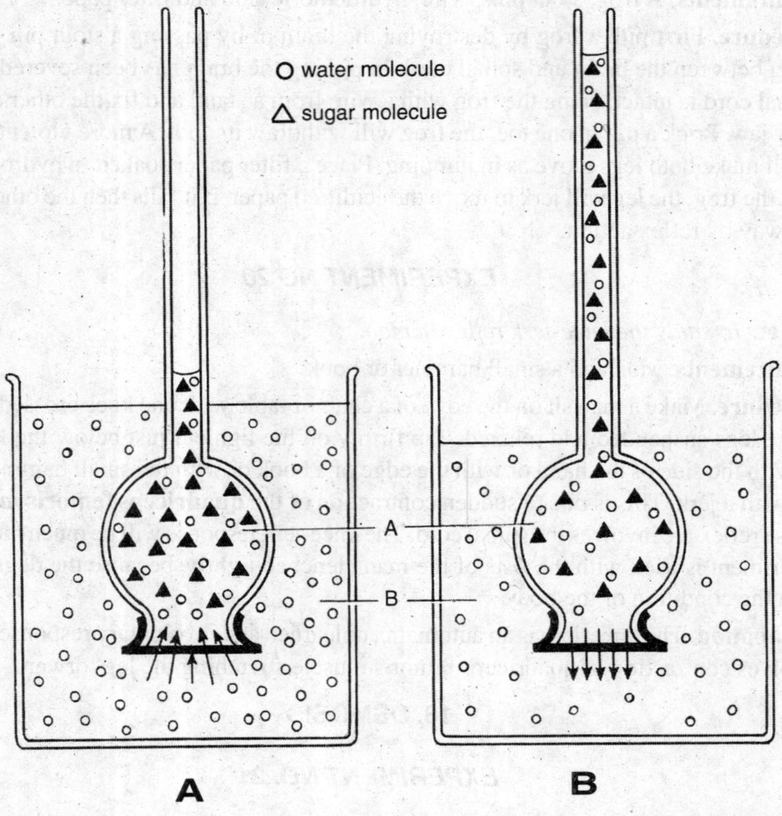

Fig. 33.21. Diagram illustrating the principle of osmosis. A–Beaker and thistle funnel which at the bottom is tied with semipermeable membrane and is filled with strong sugar solution. B–Beaker containing pure water. Arrows show direction of movement of water molecules, which can pass freely across the membrane, while the sugar molecules cannot. The right hand side diagram also expresses the state of osmotic equilibrium, at this state the same number of water molecules pass across the membrane in both directions.

Precautions. The membrane to be taken for the experiment should be semipermeable, *i.e.*, the water passes freely through it in either direction but not the solute such as sugar. If the membrane allows the solute to pass as well as water, then osmosis will not occur.

34

Chapter

CHROMATOGRAPHY

There are many methods which have been used to obtain and identify the substances with high state of purity. Besides chemical methods, many physical methods, like fractional precipitations, distillation and crystallisation have been effectively used for separation and purification of the chemical compounds. Sometimes it becomes necessary to follow a chemical method of separation by a physical method of purification and identification. These methods were applicable quite satisfactorily in many cases but great amount of difficulty was realised with compounds which had the composition of the individual components of very similar physical and chemical properties. Likewise in the fields of biological sciences and chemistry many substances are obtained where the classical methods for their identification and purification do not work satisfactorily, *e.g.*, protein is a complex mixture of several amino acids, which are very much alike to each other.

Chromatography plays a very important and significant role in solving all such problems. The complicated separations are achieved very rapidly and efficiently by chromatography. This method is so fruitful and common that by this time chromatography has been used in almost every type of compounds and fields. Biology, medicine, drawing and painting and even intelligence departments have used this method to their greatest advantage.

The chromatography was started with the work of botanist **M. Tswett** in the year 1896. His method consisted of pouring a solution of the pigments in the light petroleum through a column of powdered calcium carbonate in a glass tube. The pigments were absorbed on the top of the column which moved through the column with the solution. The pigments were separated by selective adsorption at different levels of the column and formed into different zones. The **calcium carbonate column** was carefully washed out of the tube and then divided into sections containing different pigments. Thus, he was led to the method of separating the pigments after their extractions from leaves.

Liesegang (1943) introduced the idea of using large sheets of adsorbent paper and is supposed to carry out **two way chromatography** for the first time. The present wide interest in chromatography must be attributed to the team of **Consden, Gordon, Martin** and **Synge** in U.K. (1944). Originally they were interested in the analysis of mixtures of **amino acids** by counter current extraction processes but experienced great difficulties through parts of two solvents which emulsify and was difficult to separate. They tried to use paper with the idea that pores in the paper might serve as basis of partition. Thus, they were successful in bringing about the separation of all the amino acids in a very short time.

Though the phenomena of ion-exchange was recognised much earlier but its use in the chromatography began with the work of **Adams** and **Holmes** (1935) who observed that certain synthetic resins were capable of exchanging ions. In 1941 **Martin** and **Synge** developed **partition column**

chromatography and in 1944 they replaced **silica gel support** by the strips of paper and developed widespread use of paper chromatography. In 1952 **James** and **Martin** gave an account for the **gas liquid chromatography** in which mobile phase was gas rather than liquid and they got Nobel Prize in recognition of their work. Afterwards there was rapid development in **adsorption chromatography** in which adsorbent in the form of thin layer on glass plate was used. In 1959 **Forth** and **Flodin** developed unique method of gel filtration in which gels of cross-linked dextrans were used, which separated the biological materials mostly by virtue of differences in their sizes. The wide applicability of chromatographic methods in the biological sciences is innumerable and unique.

What is Chromatography?

Chromatography is a procedure through which various substances in a mixture are separated from each other and identified. Separation involves the use of two phases, one of which is ***stationary*** *and the other* ***mobile****. Separation depends upon the differential movement of the test substances between two phases.* In a simple language chromatography can be defined as an analytical technique for separating compounds on the basis of differences in affinity for a stationary and mobile phase.

Chromatographic Methods

Chromatographic methods are classified into two categories, *viz.*,

1. Partition chromatography,
2. Adsorption chromatography.

1. Partition chromatography :

1. Partition column chromatography.
2. Paper chromatography.
3. Thin layer chromatography.
4. Gel filtration chromatography.

} Liquid as mobile and solid as stationary phase.

5. Gas liquid chromatography or vapour phase chromatography. } Gas as mobile and liquid as stationary phase.

2. Adsorption chromatography :

1. Adsorption column chromatography. } Liquid as mobile and solid as stationary phase.
2. Thin layer chromatography.
3. Ion exchange chromatography

} Electrolytic solution as mobile and some ionic polymer as stationary phase.

Techniques

Following techniques are employed in chromatography :

1. Partition column chromatography. In this technique, the column is packed with a porous solid of high surface area, *e.g.*, **silica gel** and **cellulose**, coated with water (stationary phase). The components of a mixture are separated by passing an organic solvent (mobile phase) through the column.

2. Paper chromatography. In this technique, the dissolved substances are applied as a small spot on a cellulose-bound filter paper which is then kept in a container dipped in an organic solvent. The mixtures are partitioned between (stationary phase) and organic solvent (mobile phase).

3. Thin layer chromatography. In this technique, the adsorbent (stationary phase) is spread over a glass plate in the form of a thin film of even thickness. The solvent (mobile phase) moves up the plate by capillary action and, thus, effects separation.

4. Gel-filtration chromatography. Sephadex contains the aqueous (stationary phase) which is distinguished from the mobile phase by its immobilisation. The component of the mixture are separated according to their size, by virtue of differential distribution between easily displaceable water present in the interbed space.

5. Gas liquid chromatography. In this technique, a column is packed with a porous inert solid coated with a thin layer of an involatile liquid as the stationary phase. Components of the mixture are separated by being partitioned between this phase and a gaseous (mobile phase).

6. Adsorption column chromatography. In this technique, the separation of a mixture is determined by the differential adsorption of the components on an active solid (alumina, silica gel, stationary phase) as an organic solvent (mobile phase) containing them which passes over and affects separation.

7. Ion exchange chromatography. In this technique, ionised compounds are separated in aqueous solutions (mobile phase) by virtue of their differences in affinity for ionised compounds which are an integral part of the insoluble solid phase (stationary phase).

Here only **paper chromatography** will be discussed in detail because only paper chromatography is prescribed in the syllabus of under-graduate and post-graduate classes of most of the Indian Universities.

PAPER CHROMATOGRAPHY

Amongst all the chromatographic techniques known so far, paper chromatography is considered to be the simplest and most widely used. Since the outstanding work by the team of **Consden**, **Gordon**, **Martin** and **Synge** in England in 1944, who later on were awarded Nobel Prize in recognition of their work, numerous workers have used this method so much so that at present no field, biology, biochemistry, chemistry and many other fields exists in which paper chromatography has not solved the most complicated problems.

What is Paper Chromatography?

The paper chromatography is the method in which analysis of an unknown substance is mainly done by the flow of solvents on specially designed filter paper. One of the two solvents is immiscible or partly miscible in the other solvent. The solvent rises up by the capillary action and by the adsorption on the filter paper, the separation is effected by the differential migration of the mixture of substances. This occurs due to the **partition coefficients**.

Types of Paper Chromatography

Paper chromatography is of the following two types :

1. Paper partition chromatography. The standard method of analysis in which paper is used as an inter-support with one solvent as **mobile phase** and the other solvent as **immobile phase** is called paper partition chromatography.

2. Paper adsorption chromatography. In certain cases the paper is impregnated with an adsorbent like alumina or silica. The paper so modified, is used as the adsorbent and the single solvent is allowed to flow over the unknown components. This process is known as paper adsorption chromatography.

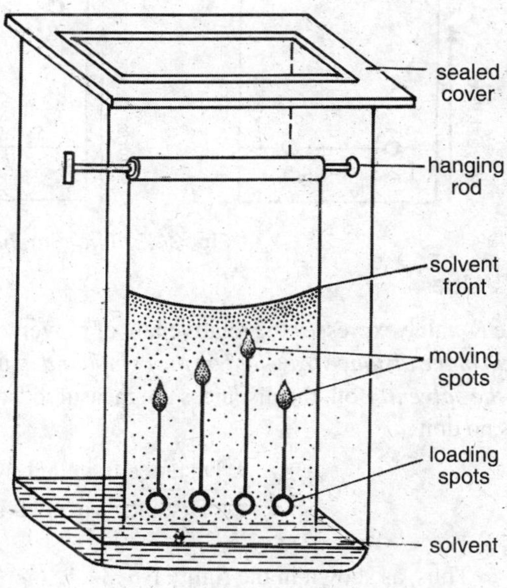

Fig. 34.1. Chromatography chamber.

Techniques of Paper Chromatography

The method of paper chromatography may be **one dimensional** or **two dimensional** depending on the type of complexity involved in the analysis.

In **one dimensional chromatography**, a drop of the solution containing the mixture of substances is placed near one end of the filter paper. The spot is then allowed to dry. The portion of the paper with the spot is then placed in a suitable solvent. The solvent passes the spot and carries the substance along with while moving up the filter paper. Different substances move with different rates, which depend on the size, nature of substances, solvent, diffusion coefficient and many other factors. The finished paper is known as **chromatogram** . The advancing rate is determined by their R_f values. The R_f value depends on the factors as enumerated above and differes from one substance to another.

But there are many substances, which cannot be separated under the above mentioned set conditions. This happens because the R_f values of many substances are very close and nearly the same. This problem can be solved by using the method of **two dimensional paper chromatography**.

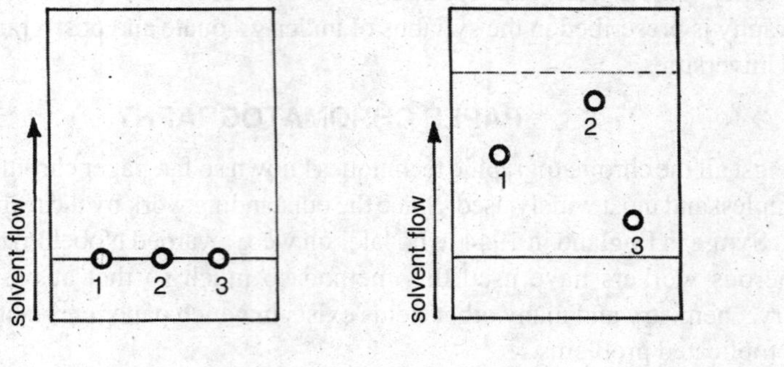

Fig. 34.2. One dimensional chromatogram.

In two dimensional chromatography, first the one dimensional chromatogram is obtained, the paper is turned at right angles to the first. The chromatography is done by using different solvents. By doing so overlapping R_f values are separated to a reasonable extent.

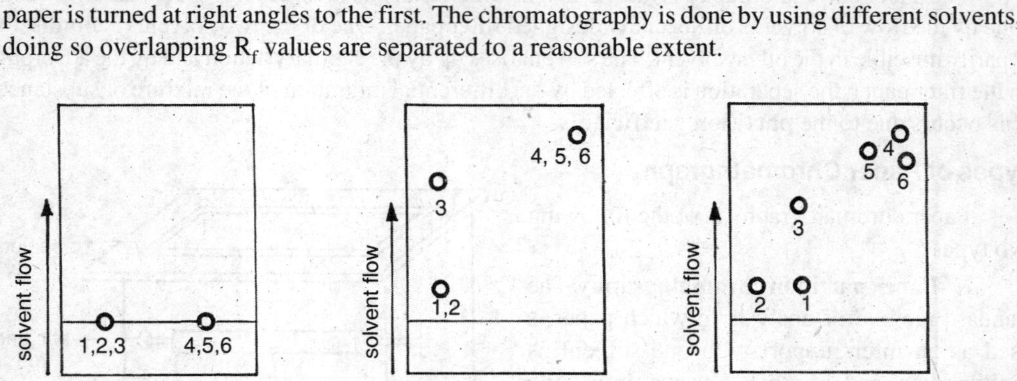

Fig. 34.3. Two dimensional chromatogram.

Rf Value

The R_f value expresses the relative rate of movement of solute and solvent. R_f value is defined as *the ratio of the distance travelled by the compound at its maximum concentration to the distance travelled by the solvent.* Both the distances are measured from the point of application of the sample. R_f value has no unit.

$$R_f = \frac{\text{Distance travelled by the substance from origin}}{\text{Distance travelled by the solvent from origin}}$$

Thus, as shown in the figure No 34.4, $R_f = \frac{A}{B}$

In some cases, it has been seen, that the solvent front runs off the paper. In such cases it is more convenient to express the movement of any substance just by comparing with the movement of some standard substances. This is represented by R_s as shown in the figure No. 34.5. R_s is defined as

$$R_s = \frac{\text{Distance moved by the substance}}{\text{Distance moved by the standard substance}} = \frac{C}{D}$$

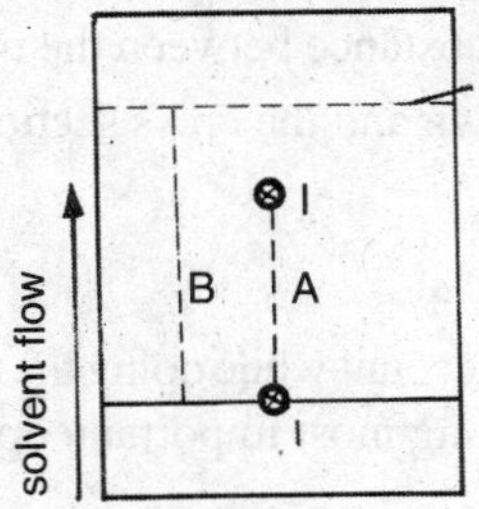

Fig. 34.4. Diagram showing the distance travelled by the substance from origin and the distance travelled by the solvent from origin.

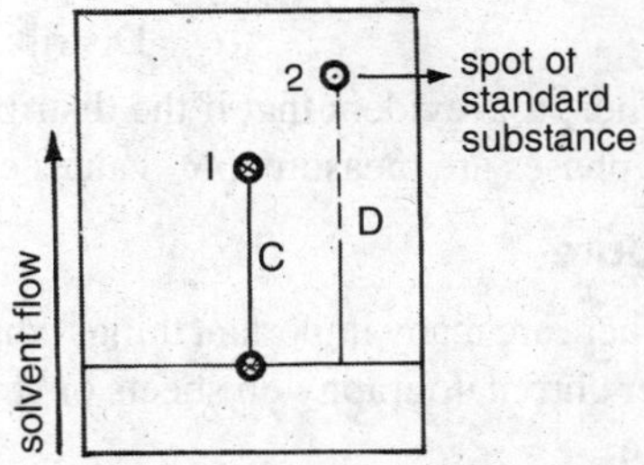

Fig. 34.5. Diagram showing the movement of any substance and its comparison with the movement of some standard substance.

R_f Variation

The R_f values of different compounds are entirely different. R_f values for different compounds are given in the literature and identification is made by **comparison method**. Thus, it can be said that R_f is quoted as the identifying characteristic. The values vary with the solvents used, *e.g.*, two solvents will give two values. Thus, R_f value is always quoted with reference of the solvent used. Some of the well known factors which affect R_f values are the temperature and the quality of the paper used. Other important factors on which R_f depends are following :

1. The pH of solution.
2. The distance through which solvent has moved on paper.
3. The type and quality of water used.
4. The direction of fibres of paper.
5. The method of development.
6. The distance through which spot travels.
7. The concentrations of the separated substances.
8. The method of drying, locating the colourless substances, and many other operations involved in the method.
9. The impurities present on the surface of the paper.
10. Irreversible adsorption on paper.
11. Chemical reactions between substances being partitioned and paper and ion exchance with acid groups of cellulose.

With the combined effects of 9, 10 and 11, it can be safely presumed that amongst all, the partition phenomena is most important. And if the distribution coefficient is known for the system, R_f value can be very easily calculated. They are related by the following expression :

$$= \quad R_f = \frac{A_i}{(Ai + \alpha A_s)}$$

or
$$\alpha = \frac{A_i}{A_s}\left(\frac{1}{R_f} - 1\right)$$

where A_i = cross sectional area of the mobile phase.

A_s = cross sectional area of the immobile phase.

α = Distribution coefficient of the substance between the two solverts.

Thus, it is evident that if the distribution coefficient is known and the cross sectional areas of the two phases are measured, R_f values can be calculated.

Procedure

There are many important things which should be taken into account while doing the experiments on paper chromatography on sheets or strips of paper. The following most important operations are involved :

1. Choice of filter paper.
2. Preparation of the paper.
3. Preparation of the sample.
4. Application of the sample to the paper.
5. Choice of solvents.
6. Development of the chromatograms.
7. Drying the chromatograph after development.
8. Development of spots.
9. Quantitative estimation.

1. Choice of filter paper. The basic material of the paper used in paper chromatography is α - cellulose of high purity which is 98–99 per cent. Besides, there are other substances which are also present; like β-cellulose ether soluble matter, ammonia, nitrogen like compounds and mineral matters. **Whatman filter paper** has been extensively used in chromatography. The chemical composition of Whatman paper is as follows :

(*i*)	α-cellulose	98–99 per cent
(*ii*)	β-cellulose	0.3–1.0 per cent
(*iii*)	Pentosans	0.4–0.8 per cent
(*iv*)	Ash	0.07–0.01 per cent
(*v*)	Ether soluble matter	0.015–0.1 per cent.

There are various grades and types of paper, the most suitable is the one which fulfils the following requirements :

1. The paper having maximum degree and clarity of the separation.
2. The paper having negligible diffuseness of the spot.
3. The paper having tne excellent rate of movement of the solvent front.

The choice of the filter paper mainly depends on the thickness flow rate, purity and net strength.

2. Preparation of paper. After the selection of the quality of the paper which has to be used, the shape and size of the paper is to be decided. This is particularly important in descending technique. The shape of the paper facilitates in giving better resolution of less polar components.

The shape and size of the paper depend on the type of the work involved. Though paper of shapes like rectangular and square have been used by many workers but the most common is **rectangular shaped paper**.

3. Sample preparation. In the preparation of samples for chromatographic work, the important question is to get and choose the proper solvent for making solution. The complexity arises in cases of qualitative or quantitative type of work involved therewith. For solids a weighed amount is dissolved in a volatile solvent and by careful means the minimum volume of the concentrated solution is applied on the paper. This is done simply to avoid diffusion through paper which ultimately results in larger zones. Extracts from plants and animal tissues are taken out with the help of some solvent which is then directly applied on the paper.

4. Application of sample to the paper. The amount of the sample to be applied is again an important factor. The factors which govern the amount of sample to be applied are : (*i*) the capacity of the solvent, (*ii*) optimum concentration required for the quantitative determination, (*iii*) the time required for development. With the above considerations in mind the final decision is made regarding the amount of sample to be applied.

The samples may be applied either as **spots** or **bands.** A line is drawn with the help of pencil at a suitable distance from one end of the paper. Numerous points are marked on the line at equal distances. A drop of each sample is then put on these points with the help of **capillary tubing**, **microsyringe** or **platinum loop**. It is also to be noted that the spots occupy small areas which should be at the most of 5 mm in diameter. The volume of the liquid may vary from 1–2μ or more as the case may be.

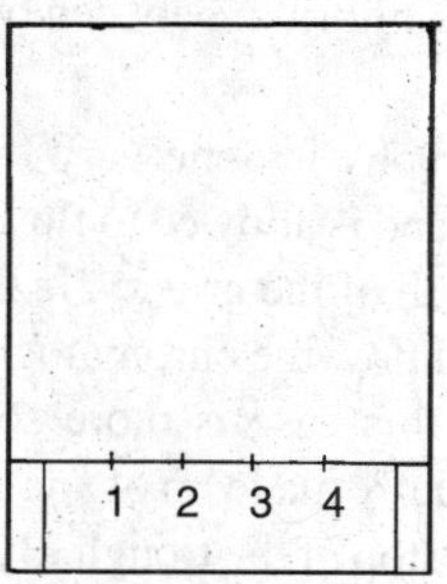

Fig. 34. 6. Diagram showing the marking of paper at equal distances.

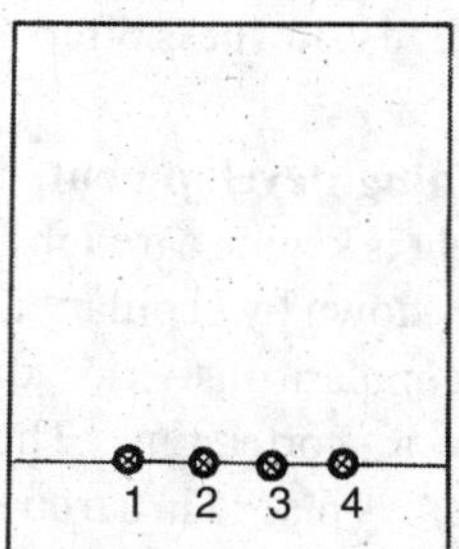

Fig 34.7. Diagram showing the application of samples at each spot.

5. Choice of solvents. There are numerous solvents which can be used in the paper chromatography. The final selection depends entirely on the nature of the substances to be separated. There are two ways to solve this problem. The first is the knowledge which one could get from literature and second is by trial and error method. The most commonly used is the **polar solvent**, which is adsorbed on the paper. A less polar solvent, which is called the **mobile phase**, advances ahead by capillary action and separates the applied substances. Water miscible solvents such as **furan propanol** and cellosolve have been extensively used. In most cases a mixture of **n-butanol acetic acid** and water has been used. But in place of n-butanol, some **tertiary alcohol** is preferred because of the tendency of the former to esterify with organic acids. The selection of the solvents should be such so that when the components are mixed together, two definite layers result. When using solvents it is essential that they are used in the specified ratios and are shaken well before use.

In general, one phase systems are extensively used in paper chromatography. Two-phase systems are, thus, avoided. This is simply because in that case we can get number of possible combinations and greater variation of compositions. Since the rate of movement of the solvent is governed also by its viscosity, surface tension and density, more than one phase may also be complicated.

As far as possible minimum number of solvents are used. This facilitates in making the chamber saturated with the solvent which prevents evaporation from the paper during the development. This is difficult if number of solvents are used.

6. Development of chromatogram. Numerous apparatus have been designed to carry on chromatography. The following things are to be taken into consideration :

(*i*) There is sufficient amount of solvent at the bottom of the reservoir.

(*ii*) There is proper arrangement for the free suspension of the paper.

(*iii*) Large temperature variations are to be avoided.

The reservoir is made from a variety of materials. The material should be insoluble in the solvent and for that purpose glass, polythene, stainless steel and many other materials have been successfully used. The paper may be held in position by clamps, etc.

The choice of the method of development depends upon the class of compounds to be investigated. The following are the types of developments :

A. Ascending development. This is the simplest of all the methods. It includes the dipping of the lower end of the paper into the solvent and allowing it to rise up the paper by capillary action. This method is preferred for quick analysis of a large number of substances. Proper care is to be taken to ensure that the lower end of the paper containing spots is so placed that it is above the solvent depth. The two ends of the paper are not allowed to touch with each other and are kept at a distance apart by clipping or by any other method.

As the solvent ascends, it does so against gravitational force, with the increasing height the rate becomes slower. This slowness offers more time for partition equilibrium between two phases to take place.

B. Descending development. This type of chromatography is done very frequently. In this method the solvent is kept in a trough at the top of the chamber and is allowed to flow down the paper. The liquid moves down by capillary action as well as by the pull of the gravity. In this case the flow is more rapid in comparison to ascending method. Due to rapid flow the chromatogram is completed comparatively in a shorter time. The apparatus required in this case is more sophisticated. The developing solvent is placed in a trough at the top, which is usually made up of an inert material. The paper is then suspended in the solvent and the lid is placed at the top of the trough. By this method two dimensional chromatography cannot be done. This is because the paper must be cut to fit into the trough. Compounds with low R_f values are not completely separated by ascending method, but this can be done by following descending technique.

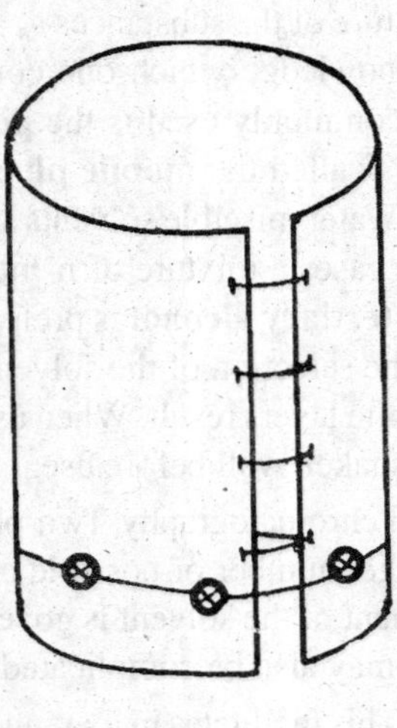

Fig. 34.8. Diagram showing the curling of paper. The two ends of the paper are curled into cylindrical form with a distance left in between them.

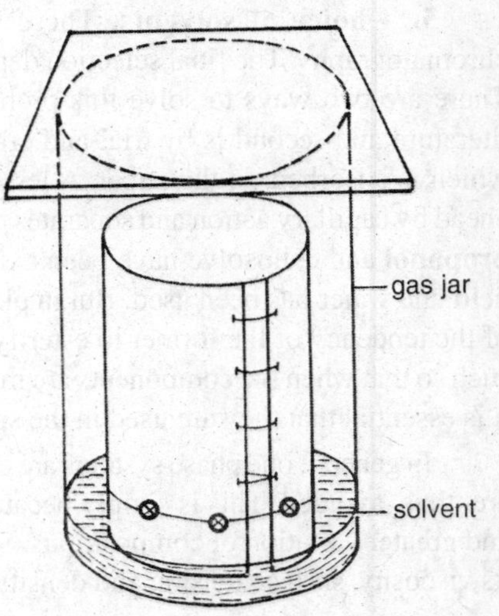

Fig. 34.9. Diagram showing the ascending technique.

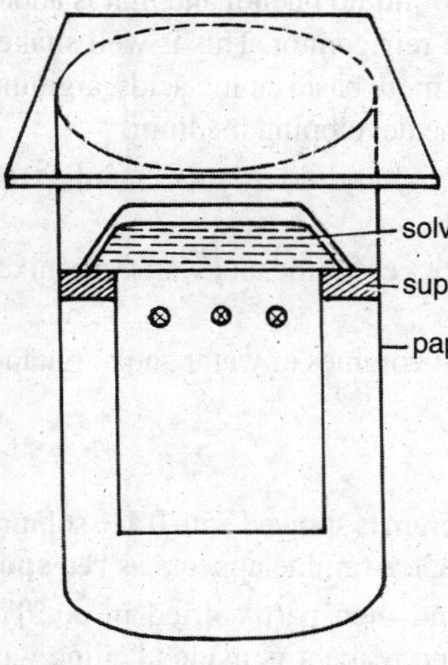

Fig.34.10. Diagram showing the descending technique.

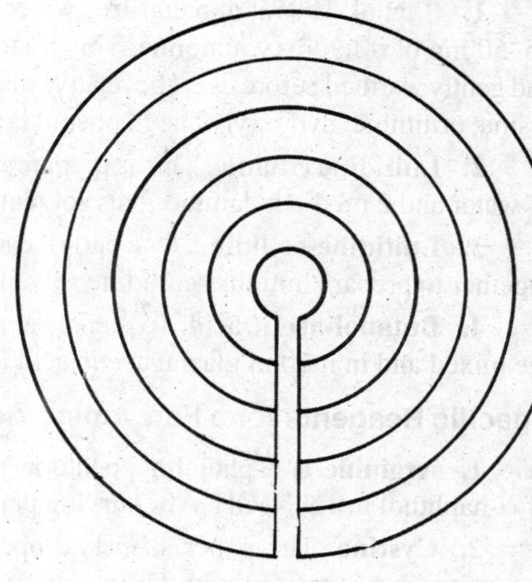

Fig. 34.11. Diagram showing radial or disc technique.

C. Radial or disc development. Radial or disc development (Fig. 34.11) is very rarely used and that too in very special cases. In this case, a circular piece of paper is taken which has a **wick cut** parallel to the radius, from edge to centre. The sample is deposited at the centre of paper and at the upper end of the wick. This paper is laid on the edge of a circular disc with the wick dipping into the solvent at the bottom of the dish. The liquid ascends the wick and flow radially through the paper. While moving it carries compounds with it. As the solvent moves ahead, the radial becomes slower and takes more time for this solvent to reach the last circle.

7. Drying the chromatogram. After the development by any of the above methods, the paper is removed from the reservoir and the position of the solvent front is marked with the pencil at the two edges of the paper. The chromatogram is then dried by keeping it in an oven or over a hot plate for few minutes. This is done so that the solvents are completely dried from the paper. This drying is best done by means of a fan or hair dryer.

8. Location of spots. After drying of the developed chromatogram it is very essential to locate the separated substances. If the substances are coloured, no difficulty arises but the problem comes with the colourless substances. There are several methods which have been used to locate the spots, all of which can be grouped into the following methods : (i) physical methods and (ii) chemical methods.

(*i*) **Physical methods.** The physical methods for locating the substances are more applicable in comparison to chemical methods in the sense that the substances on the paper are not converted into other compounds. Thus, they can be removed and studied.

(*ii*) **Chemical methods.** When physical methods are not applicable , the chemical methods are used. The colourless compounds are converted into coloured compounds by reaction with some chemical reagents. The chemicals so used are known technically as **locating reagents**. The locating reagents can be gas, liquid or solid, but mostly liquid and solid are used for this purpose.

EXPERIMENTAL PAPER CHROMATOGRAPHY

EXPERIMENT NO. 1

Paper Chromatography for Identification of Amino acids, Amines and Peptides

Solvents. Most useful solvents for the separation of amino acids and peptides are as follows :

1. Phenol. 100 ml of metal-free water is mixed with 400 ml of liquid phenol and in it is added 25–50 mg of 8-hydroxyquinoline. This is stored in a dark bottle in refrigerator. This is well shaked and gently warmed before use. The relative distance (R_f) travelled by more basic amino acids, arginine, lysine, ornithine, hydroxylysine in phenol is influenced by pH of the developing medium.

2. Luitidine ethanol. This is prepared by mixing 55 ml of 2, 6 -luitidine, 25 ml ethanol, 20 ml of water and 2 ml diethylamine. This solvent is completely miscible.

3. Luitidine-collidine. One part of each 2, 6-luitidine, 2, 4, 6, - collidine and water are mixed together to prepare luitidine-collidine solvent.

4. Butanol-acetic acid. To prepare a mixture of 500 ml, equal volumes of water and n- butanol are mixed and in it 60ml glacial acetic acid is added.

Specific Reagents for a Few Amino Acids

1. Arginine. α-naphthol-hypochlorite reagent. The chromatogram is sprayed with 0.1% solution of α-naphthol in 1N NaOH. After drying paper is sprayed with NaCl. Arginine appears as **red spot**.

2. Cystine. The paper strip is dipped into 1% Na_2SO_3 and then partly dried in air. The chromatogram is treated with Folin's-phospho-18 tungstic acid and reagent is made alkaline with $NaHCO_3$. Cystine gives deep blue colour.

3. Citrulline. This is near glutamine on two dimensional chromatograms. This is best detected by spraying with 1% solution of p-dimethylaminobenzaldeyde in 1N HCl. This gives yellow spot.

4. Glycine. Glycine, histidine and tryptophan give a colour when sprayed with Zimmerman's o-ophthaldialdehyde reagent.

5. Ornithine. Ornithine and glutamic acid give three and four spots on butanol-acid chromatograms after acid hydrolysis.

Separation of Free Amino Acids and Peptides in Tissues

For detection of free amino acids and peptides in plant and animal tissues, other constituents are to be removed as far as possible. Precipitation with trichlor acetic acid, tungstic acid, ferric hydroxide, zinc sulphate, barium hydroxide at pH 7.2–7.6, perchloric acid, acetone and ethanol is done.

Grind the tissues with enough absolute in a blender. Final concentration of ethanol should be 80% by volume. The insoluble material is removed by filtration with 80% ethanol.

PAPER CHROMATOGRAPHY FOR AMINO ACIDS

EXPERIMENT NO. 2

Experiment. *Detection of amino acid by two dimensional paper chromatography.*

Requirements. Chromatography chamber, water imbibed seeds or animal tissue, spray agent, capillary tubes, staples and stapler, spraying apparatus (atomizer), pistol mortom, conical flasks of 100 ml, pencil, Whatman No.1 paper, glass rod, spirit lamp or Bunsen's flame, ethyl alcohol, phenol, butanol, glacial acetic acid, ninhydrin and water.

Procedure. Grind in a clean pistol mortom or in blender 10 gm of animal tissue or germinating seeds in 50 ml of absolute ethyl alcohol. Filter and evaporate the filtrate to dryness. Again dissolve the evaporated solid filtrate in double distilled water to make up the volume of extract to nearly 2 ml. Glycine, aspartic acid, phenylalanine, tyrosine and tryptophane like known amino acids may be taken for beginners.

Take glass-walled chromatography chamber and fill one chamber with phenol and other with n-butanol-acetic water solvent (3:1:1). Keep the level of solvent half an inch above the bottom of the chamber.

Cut Whatman No.1 filter paper into strips according to the size of chromatography chamber. Draw a small circle on cut strips by pencil about one inch from the bottom and two inches from the left hand edge. Apply a small amount of extract on the above circle by glass rod.

Hang the chromatogram with extract dot at the base and allow the extract to dry. Dip the paper edge in solvent and keep the pencil mark above the solvent. Close the chamber and make it air tight. Allow the chromatogram to develop in 16 to 18 hours.

Now remove the chromatogram and allow it to dry, marking the position of the solvent. Turn the paper 90° and place the paper in this direction in another chamber containing solvent or n-butanol acetic acid water. Allow the chromatogram to develop in 10–12 hours in chromatography chamber. Remove the paper and spray uniformly with 0.1% ninhydrin (0.1 gm ninhydrin mixed in 100 ml saturated n-butanol water). Warm the paper at 90°C for two to five minutes and circle the spots with pencil.

Now calculate R_f value and compare it with standard values. Different amino acids give different colours.

PAPER CHROMATOGRAPHY FOR CARBOHYDRATES

For separation of various sugars one or two-dimensional paper chromatography may be done. Polysaccharides may be hydrolysed and hydrolysates may be treated like other simple sugar mixtures. The quantitative evaluation of sugar chromatograms can be done by a visual or photo-electric comparison of optical density of coloured spots.

Filter paper. Generally Whatman No.1 filter paper is used but Schleicher and Schnell No. 589 and white ribbon may also be used.

Solvents. For chromatographic development of sugars, ascorbic acid and related substances, the following solvents should be specially treated to retard the decomposition of ascorbic acid by heavy metal ions present in the paper or solvents.

1. 60 ml of n-butanol is shaken with 40 ml of water and an excess of oxalic acid. The mixture is allowed to settle in a separatory funnel. Use layer.

2. Dissolve 50 gm of phenol crystals in 10 ml of charcoal-filtered water and add excess of oxalic acid. The mixture is allowed to settle in a separatory funnel. Use lower layer. For more refinement filter paper is washed to solvent development with 2% oxalic acid.

3. Solvents n-propanol : ethyl acetate : water (7:1:2) gives excellent result for sugar chromatography.

4. For two-dimensional separation of many simple sugars, phosphorylated sugars, organic acids, the following pairs of solvents are recommended :

(*i*) **First direction.** Liquid phenol water (100 : 20). The mixture is neutralised to pH 5–6.

(*ii*) **Second direction.** (A) Butanol water, (B) Propionic acid water. Mix equal volumes of A and B and warm slightly to form a single phase solvent and now use it.

5. The following pairs may be used for two dimensional ascending chromatography :

(*i*) **First direction.** n=butanol : pyridine: water (3 : 2 : 1.5).

(*ii*) **Second direction.** Phenol saturated with ammonical water and small amount of 8-quinolinol.

6. For separation of sucrose, glucose, fructose and sorbitol, the following solvents can be used:

(*i*) **First direction.** Benzene : n-butanol : pyridine : water (1:5:3)

(*ii*) **Second direction.** Amyl alcohol : pyridine : water (1:1:2).

R_f values of some sugars (After Block)

Table 34.1. R_f Values of Sugar Derivatives 20°C.

Compounds	Solvent			
	1	2	3	4
D-Glucose	0.39	0.18	0.43	0.64
D-Mannose	0.45	0.20	0.46	0.69
D-Fructose	0.51	0.23	0.48	0.68
D-Arabinose	0.54	0.21	0.56	0.70
D-Ribose	0.59	0.31	–	0.76
Lactose	0.38	0.09	0.33	0.46
Sacrose	0.39	0.14	0.39	0.62
L-Ascorbic Acid	0.24	0.38	0.59	0.84
Dehydroascorbic Acid	0.28	0.48	–	0.88
D-Galacturonic Acid	0.13	0.14	0.40	0.31
Raffinose	0.27	0.05		0.45

1. Water-saturated phenol 1% NH_3 (HCN atmosphere)
2. n-Butanol : acetic acid : water (4 : 1 : 5 V/V) descending.
3. n-Butanol : acetic acid : water (4 : 1 : 5 V/V) horizontal.
4. i-Propanol: pyridine : water : acetic acid : (8 : 8 : 4 V/V) ascending.

Table. 34.2 R_f values of some Amino Acids.

Compounds	Solvents							
	1	2	3	4	5	6	7	8
α-Alanine	0.34	0.10	0.39	–	0.09	0.05	0.57	0.60
Arginine	0.34	0.02	0.19	–	0.05	0.01	0.22	0.89
Cystine	0.16	0.02	0.17	–	0.01	0.00	0.24	–
Glycine	0.21	0.07	0.33	–	0.05	0.03	0.45	0.41
Diiodotyrosine	–	–	–	0.70	–	–	–	0.61
Histamine	–	–	–	0.22	–	–	–	0.95
Norvaline	–	0.23	–	0.65	0.31	0.19	–	0.80
Valine	0.50	0.18	0.56	–	0.22	0.15	0.71	0.78
Serine	0.20	0.08	0.31	–	0.05	0.02	0.40	0.36
Aspartic acid	0.18	0.01	0.33	–	0.12	0.00	0.30	0.19
Glutamic acid	0.18	0.02	0.37	–	0.01	0.01	0.36	0.31
Lysine	0.25	0.02	0.18	–	0.02	0.01	0.20	0.81
Methionine	0.55	0.27	0.51	–.	0.05	0.21	0.76	0.81

1. Acetone : urea : water = 60 : 045 : 40 (V/W/V)
2. Tert. amyl alcohol saturated with water.
3. 1 Butanol : acetate acid : water = 250 : 60 : 250 : (V/V/V)
4. 1 Butanol : acetic acid : water = 40 : 10 : 60
5. 1 Butanol saturated with 2N NH_4OH.
6. 1 Butanol : Benzyl alcohol = 1 : 1. saturated with water in $(C_2H_5)_2$ NH atmosphere.
7. Pyridine : isoamyl alcohol : water : diethylamine = 10 : 10 : 7 : 0.3.
8. Phenol : water = 100 : 20 (V/V) use in presence of 1 : 1 acetic acid.

EXPERIMENT FOR SUGAR CHROMATOGRAPHY

EXPERIMENT NO 3

Experiment. *Separation of sugars by paper chromatography*

Requirements. Unknown 1% sugar solutions (preferably from some plant extract), 1% solutions of glucose, fructose, rhamnose, sucrose, arabinose, ribose and galactose, solvent 1 (t-butanol : glacial acetic acid : water (3 : 1 = 300 ml) solvent II (phenol : water, 4 : 1 = 300 ml), spray agent p-anisidine, 2 chromatography jars (10 by 18 inches) with ground glass lids, grease, oven maintained at 80°C, Whatman paper No.1, pencil, rubber scissors, micropipette and atomizer or other spray device.

Procedure. 1. Chromatography jars should be airtight by using grease over the mouth rim. Mark the jars A and B. Keep solvent I in jar A and solvent II in jar B *i.e.*, about one-half inch of solvents in both jars. Cover the jars and allow the internal atmosphere to equilibrate for several hours before use.

2. Cut two strips of Whatman No.1 paper, according to the size of chromatography jars. Avoid any contact with fingers or other substances with the paper.

3. Lay two cut strips of paper on a clean surface with long axis parallel to the table edge. Draw a pencil line two inches above the edge of one of the long sides. Make a series of pencil dots at one inch intervals along the line and label these dots as fructose, glucose, sucrose, arabinose, ribose and rhamnose. The two paper strips should be in duplicate which will be kept into separate solvents.

4. Arrange the sheets to overlap a support such as a glass tube or meter stick in such a way that the pencil spots are raised above the surface of the table.

5. Apply intermittently 3 micromilliliter of sugar solutions on the appropriate spot and allow them to dry.

6. Now staple the dry sheets separately in such a way as to form a cylinder and this chromatogram may be allowed to develop in solvent I and II separately for 10–12 hours. Cover the jars and leave the chromatogram until the solvent front approaches the upper edge of the paper.

7. After development mark the solvent and hang the papers to dry. When dry, spray the papers with p-anisidine reagent with atomizer and keep the chromatograms in 80°C maintained oven for 3 minutes. Circle the resulting spots with pencil and note the colour. Calculate R_f (Table 34.1) values of each sugar and compare with standard values. Make a table of yours results.

Result of Ascending Paper Chromatography of Unknown Sugars.

Compound	R_f value in solvent I-colour	R_f value in solvent II-colour
1. Glucose		
2. Fructose		
3. Rhamnose		
4. Arabinose.		
5. Sucrose, etc.		